COMMERCIAL BIOTECHNOLOGY

An International Analysis

Pergamon Titles of Related Interest

Halme MODELLING AND CONTROL OF BIOTECHNICAL PROCESSES

King BIORESOURCES FOR DEVELOPMENT

Messel THE BIOLOGICAL MANIPULATION OF LIFE

Moo-Young ADVANCES IN BIOTECHNOLOGY (3 Volumes)

Moo-Young COMPREHENSIVE BIOTECHNOLOGY AND BIOENGINEERING (3 Volumes)

Moo-Young & Cunningham BIOTECHNOLOGY AND WASTE TREATMENT

Morgan & Whelan RECOMBINANT DNA AND GENETIC EXPERIMENTATION

Rothman et al BIOTECHNOLOGY

Zadrazil & Sponar DNA: RECOMBINATION, INTERACTIONS AND REPAIR

Related Journals*

BIOTECHNOLOGY ADVANCES

BULLETIN OF MATHEMATICAL BIOLOGY

CELLULAR AND MOLECULAR BIOLOGY

CURRENT AWARENESS IN BIOLOGICAL SCIENCES

***Free specimen copies available upon request.**

COMMERCIAL BIOTECHNOLOGY

An International Analysis

Office of Technology Assessment
Congress of the United States
Washington, DC

PERGAMON PRESS
New York Oxford Toronto Sydney Paris Frankfurt

Pergamon Press Offices:

U.S.A.	Pergamon Press Inc., Maxwell House, Fairview Park, Elmsford, New York 10523, U.S.A.
U.K.	Pergamon Press Ltd., Headington Hill Hall, Oxford OX3 0BW, England
CANADA	Pergamon Press Canada Ltd., Suite 104, 150 Consumers Road, Willowdale, Ontario M2J 1P9, Canada
AUSTRALIA	Pergamon Press (Aust.) Pty. Ltd., P.O. Box 544, Potts Point, NSW 2011, Australia
FRANCE	Pergamon Press SARL, 24 rue des Ecoles, 75240 Paris, Cedex 05, France
FEDERAL REPUBLIC OF GERMANY	Pergamon Press GmbH, Hammerweg 6, D-6242 Kronberg-Taunus, Federal Republic of Germany

Library of Congress Cataloging in Publication Data
Main entry under title:

Commercial biotechnology.

Includes index.
1. Biotechnology industries--United States.
2. Biotechnology industries. 3. Competition, International. I. United States. Congress. Office of Technology Assessment. II. Title.
HD9999.B443U62 1984 338.4′76606′0973 84-1105
ISBN 0-08-031620-4

Published in cooperation with the Office of Technology Assessment, Congress of the United States, 1984

Printed in the United States of America

Contents

Office of Technology Assessment

Congressional Board of the 98th Congress

Foreword

This report assesses the competitive position of the United States with respect to Japan and four European countries believed to be the major competitors in the commercial development of "new biotechnology." This assessment continues a series of OTA studies on the competitiveness of U.S. industries. It was requested by the House Committee on Science and Technology and the Senate Committee on Commerce, Science, and Transportation. Additionally, a letter of support for this study was received from the Senate Committee on Labor and Human Resources.

New biotechnology, as defined in this report, focuses on the industrial use of recombinant DNA, cell fusion, and novel bioprocessing techniques. These techniques will find applications across many industrial sectors including pharmaceuticals, plant and animal agriculture, specialty chemicals and food additives, environmental applications, commodity chemicals and energy production, and bioelectronics. Over 100 new firms have been started in the United States in the last several years to capitalize on the commercial potential of biotechnology. Additionally, throughout the world, many established companies in a diversity of industrial sectors have invested in this technology.

A well-developed life science base, the availability of financing for high-risk ventures, and an entrepreneurial spirit have led the United States to the forefront in the commercialization of biotechnology. But although the United States is currently the world leader in both basic science and commercial development of biotechnology, continuation of the initial preeminence of American companies in the commercialization of biotechnology is not assured. Japan is likely to be the leading competitor of the United States, followed by the Federal Republic of Germany, the United Kingdom, Switzerland, and France. In the next decade, competitive advantage in areas related to biotechnology may depend as much on developments in bioprocess engineering as on innovations in genetics, immunology, and other areas of basic science. Thus, the United States may compete very favorably with Japan and the European countries if it can direct more attention to research problems associated with the scaling-up of bioprocesses for production.

Issues and options developed for Congress include Federal funding for the basic life sciences and for generic applied research, especially in the areas of bioprocessing engineering and applied microbiology, including the training of personnel in these areas. The United States may also need to be concerned with the continued availability of finances for new biotechnology firms until they are self-supporting. Additionally, there are changes in laws and policies that could improve the U.S. competitive position. These changes include clarification and modification of particular aspects of intellectual property law; health, safety, and environmental regulation; antitrust law; and export control laws.

OTA was assisted in the preparation of this study by an advisory panel of individuals representing a wide range of backgrounds, including science, economics, financial analysis, law, labor, and new and established firms commercializing biotechnology. Additionally, over 250 reviewers from universities, the private sector, and government agencies, both domestic and foreign, provided helpful comments on draft reports.

OTA expresses sincere appreciation to each of these individuals. As with all OTA reports, however, the content is the responsibility of the Office and does not necessarily constitute the consensus or endorsement of the advisory panel or the Technology Assessment Board.

John H. Gibbons

JOHN H. GIBBONS
Director

Commercial Biotechnology Advisory Panel

Michael Hooker, *Chairman*
Bennington College

Howard Bremer
Wisconsin Alumni Research Federation

Robert Fildes
Cetus Corp.

Julian Gresser
Massachusetts Institute of Technology

Ralph Hardy
E. I. du Pont de Nemours & Co., Inc.

Zsolt Harsanyi
E. F. Hutton & Co., Inc.

Peter Hutt
Covington & Burling

David Jackson
Genex Corp.

William Maxon
Upjohn Co.

Laura Meagher
North Carolina Biotechnology Center

Robert R. Miller
University of Houston

Dorothy Nelkin
Cornell University

Norman Oblon
Oblon, Fisher, Spivak, McClelland, & Maier

David Padwa
Agrigenetics Corp.

David Parkinson
Falk Clinic

Phillip A. Sharp
Massachusetts Institute of Technology

William J. Whelan
University of Miami

John Zysman
University of California, Berkeley

OTA Commercial Biotechnology Project Staff

H. David Banta* and Roger Herdman,** *Assistant Director, OTA*
Health and Life Sciences Division

Gretchen Schabtach Kolsrud, *Program Manager, Biological Applications Program*
Project Director through August 1982

Nanette Newell, *Project Director* from September 1982
OTA Congressional Science Fellow through August 1982

Thomas Bugbee, *Research Assistant* from February 1983

Susan Clymer, *Research Analyst*

Geoffrey M. Karny, *Legal Analyst*

Kerry B. Kemp, *Health and Life Sciences Division Editor*

Francis A. Packer III, *Research Analyst*

Kay Smith, *Analyst*

James A. Thomas, *Research Assistant* from January 1983

Louise A. Williams, *Senior Analyst*

Raymond Zilinskas, *Analyst* through October 1982

Special Contributor

Robert M. Cook-Deegan, *OTA Congressional Science Fellow*

Administrative Assistants

Susan Clymer through April 1982
Ted Wagner from May through August 1982
Fatimah Taylor from September 1982 through August 1983
Elma Rubright from September 1983

OTA Publishing Staff

John C. Holmes, *Publishing Officer*

John Bergling Kathie S. Boss Debra M. Datcher Joe Henson

Glenda Lawing Linda A. Leahy Cheryl J. Manning

*Until August 1983.
**From Dec. 26, 1983.

Major Contractors

Antonelli, Terry, & Wands
Washington, D.C.

L. W. Borgman
Borgman & Co.
Princeton, N.J.

Michael Borrus
University of California
Berkeley, California

Dike, Bronstein, Roberts, Cushman, & Pfund
Boston, Mass.

Foreman & Dyess
Washington, D.C.

Elmer Gaden
University of Virginia
Charlottesville, Va.

Genex Corp.
Gaithersburg, Md.

Sheila Jasanoff
Cornell University
Ithaca, N.Y.

Kaye, Scholer, Fierman, Hays, & Handler
Washington, D.C.

Management Analysis Center
Cambridge, Mass.

James Millstein
New York, N.Y.

Norine Noonan
Georgetown University
Washington, D.C.

William O'Neill
Poly-Planning Services
Los Altos, Calif.

Amelia Porges
Washington, D.C.

Gary Saxonhouse
University of Michigan
Ann Arbor, Mich.

Schwartz, Jeffery, Schwaab, Mack, Blumenthal, & Koch, P.C.
Alexandria, Va.

Southern Research Institute
Birmingham, Ala.

Philip Lee
University of California
San Francisco, California

Madeline Vaquin
Paris, France

Virginia Walbot
Stanford University
Stanford, Calif.

Steven Zimmer
Brooklyn, N.Y.

Chapter 1
Summary

Contents

Figures

Chapter 1
Summary

Introduction

In the past 10 years, dramatic new developments in the ability to select and manipulate genetic material have sparked unprecedented interest in the industrial uses of living organisms. Following the first successful directed insertion of foreign DNA in a host micro-organism in 1973, scientific researchers in the United States and other countries began to recognize the potential for directing the cellular machinery to develop new and improved products and processes in a wide diversity of industrial sectors. Potential industrial applications of those novel genetic techniques include the production of new drugs, food, and chemicals, the degradation of toxic wastes, and the improvement of agricultural products. Thus, these new techniques could have a major economic impact on industries throughout the world.

Beginning around 1976, many small entrepreneurial firms were formed in the United States specifically to build on the growing body of fundamental knowledge in molecular biology and to exploit it to a profitable end. Furthermore, large established American, Japanese, and European companies in a spectrum of industrial sectors expanded their research and development (R&D) programs to include the new genetic techniques. In the United States, private sector investments to commercialize these new techniques exceeded $1 billion in 1983.

This report assesses the competitive position of the United States with respect to Japan and four European countries—the Federal Republic of Germany, the United Kingdom, Switzerland, and France—believed to be the major competitors in the commercial development of "new biotechnology," as defined below. **Although the United States is currently the world leader in both basic science and commercial development of new biotechnology, continuation of the initial preeminence of American companies in the commercialization of new biotechnology is not assured.** Japan and other countries have identified new biotechnology as a promising area for economic growth and have therefore invested quite heavily in R&D in this field. Congressional policy options for improving U.S. competitiveness in new biotechnology are identified in this report.

Definitions

Biotechnology, broadly defined, includes any technique that uses living organisms (or parts of organisms) to make or modify products, to improve plants or animals, or to develop micro-organisms for specific uses. Biological processes and organisms have been used with great success throughout history and have become increasingly sophisticated over the years. Since the dawn of civilization, people have deliberately selected organisms that improved agriculture, animal husbandry, baking, and brewing. More recently, a better understanding of genetics has led to more effective applications of traditional genetics in such areas as antibiotic and chemical production.

This report focuses on the industrial use of recombinant DNA (rDNA), cell fusion, and novel bioprocessing techniques. To differentiate between biotechnology using these novel techniques and the more traditional forms of biotechnology, this report uses the terms "new biotechnology" and "old biotechnology," respectively. Thus, for example, traditional wine production is old biotechnology, but the use of yeast modified with rDNA techniques to produce wine with a higher alcohol content is new biotechnology. Where no specific distinction is made, the term biotechnology alone henceforth refers to new biotechnology.

Biotechnology is the most recent phase in a historical continuum of the use of biological organisms for practical purposes. Furthermore, developments arising from existing technologies are providing a base from which other technologies will emerge, and new technologies can make even the most potentially useful current technology obsolete in a short time. Of necessity, this assessment describes the development of biotechnology at a particular point in time, but it is important to emphasize that dynamic and progressive change has characterized biotechnology for the last decade. Figure 1 shows some prominent events that illustrate the rapid progress made in the development of biotechnology over the last decade. This pace is likely to continue into the 21st century.

The technologies

The novel techniques used in biotechnology are extremely powerful because they allow a large amount of control over biological systems. *Recombinant DNA technology,* one of the new techniques, allows direct manipulation of the genetic material of individual cells. The ability to direct which genes are used by cells permits more control over the production of biological molecules than ever before. Recombinant DNA technology can be used in a wide range of industrial sectors to develop micro-organisms that produce new products, existing products more efficiently, or large quantities of otherwise scarce products. This technology can also be used to develop organisms that themselves are useful, such as micro-organisms that degrade toxic wastes or new strains of agriculturally important plants.

Cell fusion, the artificial joining of cells, combines the desirable characteristics of different types of cells into one cell. This technique has been used recently to incorporate in one cell the traits for immortality and rapid proliferation from certain cancer cells and the ability to produce useful antibodies from specialized cells of the immune system. The cell line resulting from such

Figure 1.—Major Events in the Commercialization of Biotechnology

1973	First gene cloned.
1974	First expression of a gene cloned from a different species in bacteria. Recombinant DNA (rDNA) experiments first discussed in a public forum (Gordon Conference).
1975	U.S. guidelines for rDNA research outlined (Asilomar Conference). First hybridoma created.
1976	First firm to exploit rDNA technology founded in the United States (Genentech). Genetic Manipulation Advisory Group (U.K.) started in the United Kingdom.
1980	*Diamond* v. *Chakrabarty*—U.S. Supreme Court rules that micro-organisms can be patented under existing law. Cohen/Boyer patent issued on the technique for the construction of rDNA. United Kingdom targets biotechnology (Spinks' report). Federal Republic of Germany targets biotechnology (Leistungsplan). Initial public offering by Genentech sets Wall Street record for fastest price per share increase ($35 to $89 in 20 minutes).
1981	First monoclonal antibody diagnostic kits approved for use in the United States. First automated gene synthesizer marketed. Japan targets biotechnology (Ministry of International Trade and Technology declares 1981 "The Year of Biotechnology"). France targets biotechnology (Pelissolo report). Hoescht/Massachusetts General Hospital agreement. Initial public offering by Cetus sets Wall Street record for the largest amount of money raised in an initial public offering ($115 million). Industrial Biotechnology Association founded. DuPont commits $120 million for life sciences R&D. Over 80 NBFs had been formed by the end of the year.
1982	First rDNA animal vaccine (for colibacillosis) approved for use in Europe. First rDNA pharmaceutical product (human insulin) approved for use in the United States and the United Kingdom. First R&D limited partnership formed for the funding of clinical trials.
1983	First plant gene expressed in a plant of a different species. $500 million raised in U.S. public markets by NBFs.

SOURCE: Office of Technology Assessment.

a fusion, known as a hybridoma, produces large quantities of *monoclonal antibodies (MAbs),* so called because they are produced by the progeny, or clones, of a single hybridoma cell. MAbs can potentially be used for many purposes, including the diagnosis and treatment of disease and the purification of proteins.

The commercial success of specific industrial applications of rDNA and cell fusion techniques will hinge on advances in bioprocess engineering. *Bioprocess technology,* though not a novel genetic technique, allows the adaptation of biological methods of production to large-scale industrial use. Most industrial biological syntheses at present are carried out in single batches, and a small amount of product is recovered from large quantities of cellular components, nutrients, wastes, and water. Recent improvements in techniques for immobilizing cells or enzymes and in bioreactor design, for example, are helping to increase production and facilitate recovery of many substances. Additionally, new genetic techniques can aid in the design of more efficient bioreactors, sensors, and recovery systems. **In the next decade, competitive advantage in areas related to biotechnology may depend as much on developments in bioprocess engineering as on innovations in genetics, immunology, and other areas of basic science.**

The same technologies that yield commercial products will also provide new research tools. The new genetic technologies described above have ignited an explosion of fundamental knowledge. The widespread use of rDNA and cell fusion techniques in the investigation of a wide variety of biological phenomena in plants, animals, microorganisms, and viruses highlights the impact of these technologies on basic science research and the advances in fundamental knowledge that they make possible. This new knowledge, in turn, may reveal new commercial opportunities.

Industrial development

Biotechnology could potentially affect any current industrial biological process or any process in which a biological catalyst could replace a chemical one. As discussed in this report, **industrial applications of biotechnology will be found in several industrial sectors, including pharmaceuticals, animal and plant agriculture, specialty chemicals and food additives, environmental areas, commodity chemicals and energy production, and bioelectronics.**

The industrial sector in which the earliest applications of new biotechnology have occurred is the pharmaceutical sector. Reasons for the rapid diffusion of the new techniques into the pharmaceutical sector include the following:

- Recombinant DNA and MAb technologies were developed with public funds directed toward biomedical research. The first biotechnology products, such as rDNA-produced human insulin, interferon, and MAb diagnostic kits, are a direct result of the biomedical nature of the basic research that led to these new technologies.
- Pharmaceutical companies have had years of experience with biological production methods, and this experience has enabled them to take advantage of the new technologies.
- Pharmaceutical products are high value-added and can be priced to recover costs incurred during R&D, so the pharmaceutical sector is a good place to begin the costly process of developing a new technology.

Because of the rapid diffusion of the new genetic techniques into pharmaceutical R&D programs, the pharmaceutical sector is currently most active in commercializing biotechnology. For this reason, it serves as a model for the industrial development of biotechnology in much of this report. It is important to recognize, however, that the development of biotechnology in other industrial sectors will differ from its development in the pharmaceutical sector. Regulatory and trade barriers and a marketing and distribution system unique to the pharmaceutical sector limit its usefulness as a model. Furthermore, the techniques may not diffuse as rapidly into other industrial sectors, such as the chemical industry, because of difficulties companies may have in recovering investments in R&D and physical plants required to convert to biological methods of production.

Findings

Industrial applications of biotechnology

The earliest industrial applications of biotechnology (i.e., during the next 5 to 10 years) are likely to occur in pharmaceuticals, animal agriculture, and specialty chemicals. Applications of biotechnology to **pharmaceuticals** being pursued at present are in the production of proteins such as insulin, interferon, and human serum albumin; antibiotics; MAb diagnostics; and vaccines for viral, bacterial, and parasitic diseases. As more is learned about hormone growth factors, immune regulators, and neurological peptides, their importance in the treatment of disease may increase dramatically. Eventually, the production of such regulatory proteins may turn out to be the largest application of biotechnology in the pharmaceutical industry. U.S. companies pursuing biotechnological applications in pharmaceuticals include many of the established pharmaceutical companies* and a large number of small, entrepreneurial new biotechnology firms (NBFs).** Additionally, many established companies in other sectors are using biotechnology as a way to diversify into pharmaceuticals.

In **animal agriculture,** biotechnology is being used to develop products similar to those being developed in the pharmaceutical industry. However, since animal producers cannot afford to purchase expensive products made with new technology, biotechnologically produced products may initially be limited to products for "high value" animals such as pets and breeding stock. The most important products are likely to be vaccines and growth promotants.

Unlike the production of pharmaceuticals, the production of animal health products using traditional technologies is not dominated by a few large companies. Additionally, the animal agriculture industry differs from the pharmaceutical industry in that the regulatory requirements for animal health products, especially for vaccines and diagnostics, are significantly less stringent than for human health products; markets for animal products are smaller and more accessible; and the distribution and delivery systems are different. Because of these features, many NBFs are finding animal agriculture an attractive field for the application of biotechnology.

*Established companies pursuing applications of biotechnology are generally process-oriented, multiproduct companies in traditional industrial sectors such as pharmaceuticals, energy, chemicals, and food processing.

**NBFs, as defined in this report, are entrepreneurial ventures started specifically to pursue applications of biotechnology.

The potential applications of biotechnology are probably more varied for **specialty chemicals** (i.e., chemicals costing more than $1/lb) and **food additives*** than for any other industrial sector at the present time. Possible applications include improvements in existing bioprocesses, such as in the production of amino acids. Other products, such as vitamins and steroid compounds, are currently made in multistep production processes involving chemical syntheses. Biotechnology could provide one or more enzymatic conversion process to increase the specificity of currently used chemical conversions. Generally, complex products, such as enzymes and some polysaccharides, can only be made economically using bioprocesses. The production of specialty chemicals represents one of the largest opportunities for the application of biotechnology because of the diversity of potential applications. Several companies in the United States are pursuing biological production of specialty chemicals, but most specialty chemicals currently produced biologically are made almost exclusively in Japan and Europe, and these countries intend to pursue new applications for specialty chemical production.

Applications of rDNA technology to **plant agriculture** are proceeding faster than anyone anticipated 3 to 4 years ago. Some important traits of plants, including stress-, herbicide-, and pest-resistances, appear to be rather simple genetically, and it may be possible to transfer these traits to important crop species in the next few years. Other traits, such as increased growth rate, increased photosynthetic ability, and the stimula-

*Food additives are considered together with specialty chemicals because many (though not all) food additives are also specialty chemicals, e.g., amino acids and vitamins.

tion of nitrogen fixation, are genetically complex, and it is likely to be several years before plants with these characteristics developed with rDNA technology will be ready for field testing. Micro-organisms that interact with plants offer possibilities for genetic manipulation that may be more near-term. For instance, it may be possible to manipulate micro-organisms to produce pesticides or inhibit frost formation. Companies pursuing these applications include many NBFs and established companies in agricultural chemicals and seed production.

Environmental applications of biotechnology include mineral leaching and metal concentration, pollution control and toxic waste degradation, and enhanced oil recovery. These applications may take longer to reach the market, because little is known of the genetics of the most potentially useful micro-organisms. Additionally, regulation is expected to be a major factor influencing development of this area because these applications use micro-organisms that are deliberately released into the environment. The nature and extent of this regulation remains uncertain, and this uncertainty may deter some firms from entering the field, thus slowing development.

Commodity chemicals, which are now produced from petroleum feedstocks, could be produced biologically from biomass feedstocks such as cornstarch and lignocellulose. Commodity chemical production from cornstarch will probably occur before production from lignocellulose because of the high energy inputs necessary for the solubilization of lignocellulose. Although the technology exists now for the cost-effective biological production of some commodity chemicals such as ethanol, the complex infrastructure of the commodity chemical industry will prevent the replacement of a large amount of commodity chemical production using biotechnology for at least 20 years. This distant time horizon is due more to the integrated structure of the chemical industry, its reliance on petroleum feedstocks, and its low profit margins than to technical problems in the application of the biotechnology.

In the area of **bioelectronics,** biotechnology could be used to develop improved biosensors or new conducting devices called biochips. Sensors that use enzymes for detecting specific substances are available now. However, their use is limited by the narrow range of substances they detect and by their temperature instability. Biotechnology could be instrumental in the development of more versatile sensors that use enzymes or MAbs. Better sensors would be especially useful in the control of industrial bioprocesses. Biotechnology may also make it possible to construct devices that use proteins as a framework for molecules that act as semiconductors. The anticipated advantages of these biochips are their small size, reliability, and the potential for self assembly. The production of biochips, however, is one of the most distant applications of biotechnology.

The U.S. competitive position

A well-developed life science base, the availability of financing for high-risk ventures, and an entrepreneurial spirit have led the United States to the forefront in the commercialization of biotechnology. For the most part, the laws and policies of this country have made it possible for industrialists and scientists to capitalize rapidly on the results of basic research in biotechnology conducted in the university system and government laboratories. The relative freedom of U.S. industry to pursue a variety of courses in the development of products has also given the United States a comparative advantage. The flexibility of the U.S. industrial system and the plurality of approaches taken by entrepreneurial NBFs and established companies in the development of products have facilitated the rapid development of biotechnology in the United States.

Japan is likely to be the leading competitor of the United States for two reasons. First, Japanese companies in a broad range of industrial sectors have extensive experience in bioprocess technology. Japan does not have superior bioprocess technology, but it does have relatively more industrial experience using old biotechnology, more established bioprocessing plants, and more bioprocess engineers than the United States. Second, the Japanese Government has targeted biotechnology as a key technology of the future, is funding its commercial development, and is coordinating interactions among representatives from industry, universities, and government. **The United States may compete very favorably with Japan if it can direct more attention to research problems associated with the scaling-up of bioprocesses for production.**

The European countries are not moving as rapidly toward commercialization of biotechnology as either the United States or Japan, in part because the large established pharmaceutical and chemical companies in Europe have hesitated to invest in biotechnology and in part because of cultural and legal traditions that tend not to promote venture capital formation and, consequently, risk-taking ventures. Nevertheless, several of the large pharmaceutical and chemical houses in the United Kingdom, the Federal Republic of Germany, Switzerland, and France will surely be competitors in selected product areas in the future because of their prominent position in world sales of biologically derived products. Additionally, the increased interest shown recently by the British Government in biotechnology may speed its development in the United Kingdom.

The United States could have difficulty maintaining its competitive position in the future if several issues are not addressed. If U.S. Government funding for basic life science research continues its decline, the science base, which is the source of innovation in biotechnology as well as in other fields, may be eroded. U.S. Government funding of generic applied research,* especially in the areas of bioprocess engineering and applied microbiology, is currently insufficient to support rapid commercialization. U.S. Government funding for personnel training in these areas may also be insufficient. Additionally, clarification and modification of certain aspects of U.S. health, safety, and environmental regulation and intellectual property law may be necessary for the maintenance of a strong U.S. competitive position in biotechnology.

*Generic applied research, which is nonproprietary and bridges the gap between basic research and applied research, is aimed at the solution of generic problems that are associated with the use of a technology by industry.

Analysis of international competitiveness in biotechnology

Often international competitiveness is defined as the relative ability of firms based in one country to develop, produce, and market equivalent goods or services at lower costs than firms in other countries. Competitiveness is a matter of relative prices, and these usually reflect relative costs of developing, producing, and distributing goods and services. In the case of biotechnology, two factors preclude a traditional analysis of international competitiveness. First, standard analyses of competitiveness examine the marketing of products, but as of the end of 1983, only a few products of new biotechnology had reached the marketplace—notably human insulin, some MAb diagnostic kits, and some animal vaccines. Most of these products are substitutes for already existing products, and the markets are well defined and relatively limited. Furthermore, even the markets for some new animal vaccines are quite small when compared to potential markets for applications of biotechnology in the production of some chemicals or new crop plants. Thus, the biotechnology products that have reached the market to date may be inaccurate indicators of the potential commercial success in world markets of the much larger number of biotechnology products and processes still in R&D stages. Which of the biotechnology products and processes in development are likely to be marketed and when cannot be accurately predicted. Second, even with many more products on the market, a traditional competitive analysis might not be appropriate because an economic analysis of competitiveness usually addresses a specific industrial sector. The set of techniques that constitute biotechnology, however, are potentially applicable to many industrial sectors.

Since the technologies are still emerging and most biotechnology products and processes are in early development, most of this report focuses on *potential* rather than actual products and processes. In the case of biotechnology, knowledge about market size, distribution systems, customers, production processes, and learning curve economies is lacking. Thus, traditional parameters of competitiveness are difficult or impossible to estimate. Instead of examining the classical measures of competitiveness, this analysis of international competitiveness in biotechnology examines the aggregate industrial activity in biotechnology in both domestic and foreign firms and 10 factors that *might* be influential in determining the competitive position of the United States and other countries with respect to the commercialization of biotechnology.

In investigating competitiveness in biotechnology, this report analyzes the commercialization efforts of five countries in addition to the United States: Japan, the Federal Republic of Germany, the United Kingdom, Switzerland, and France. Although companies from many countries will have biotechnology products in world markets, these five countries were selected because of their research capabilities in biology and their existing capabilities in old biotechnology and because, as a whole, their companies are most likely to reach world markets first with biotechnology-produced products. Japan leads the world both in the microbial production of amino acids and in large-scale plant cell culture, and it has a strong position in new antibiotic markets. Japan is also the world leader in traditional bioprocess engineering. Furthermore, the Ministry of International Trade and Industry (MITI) in Japan has designated biotechnology for industrial development. The European pharmaceutical houses, notably in the United Kingdom, France, the Federal Republic of Germany, and Switzerland, lead the world in pharmaceutical sales. Like Japan, three of these European countries, the Federal Republic of Germany, the United Kingdom, and France, have national plans for the promotion of biotechnology. The Federal Republic of Germany and the United Kingdom have good basic biology research and especially good bioprocess engineering research.

The first step in the analysis of international competitiveness in biotechnology was to consider the aggregate level of industrial activity and the number and kinds of firms commercializing biotechnology in the competitor countries. OTA's industrial analysis, presented in *Chapter 4: Firms Commercializing Biotechnology,* was approached from three perspectives:

- the number and kinds of companies commercializing biotechnology,
- the markets targeted by industrial biotechnology R&D, and
- the interrelationships among companies applying biotechnology and the overall organization of the commercial effort.

The analysis began with the United States and comparisons were then made with other countries.

The second step in providing an overall picture of competitiveness in biotechnology involved the evaluation of the following 10 factors identified as potentially important in determining the future position of the United States and other countries in the commercialization of biotechnology:

- financing and tax incentives for firms;
- government funding of basic and applied research;
- personnel availability and training;
- health, safety, and environmental regulation;
- intellectual property law;
- university/industry relationships;
- antitrust law;
- international technology transfer, investment, and trade;
- government targeting policies in biotechnology; and
- public perception.

The relative importance of each of the factors was first evaluated to determine their importance to competitiveness today (see fig. 2) and which ones could be important as the technology matures and more products reach the marketplace. Then, each of the factors was analyzed for each of the six competitor countries: the United States, Japan, the Federal Republic of Germany, the United

Figure 2.—The Relative Importance of Factors Affecting the Commercialization of Biotechnology

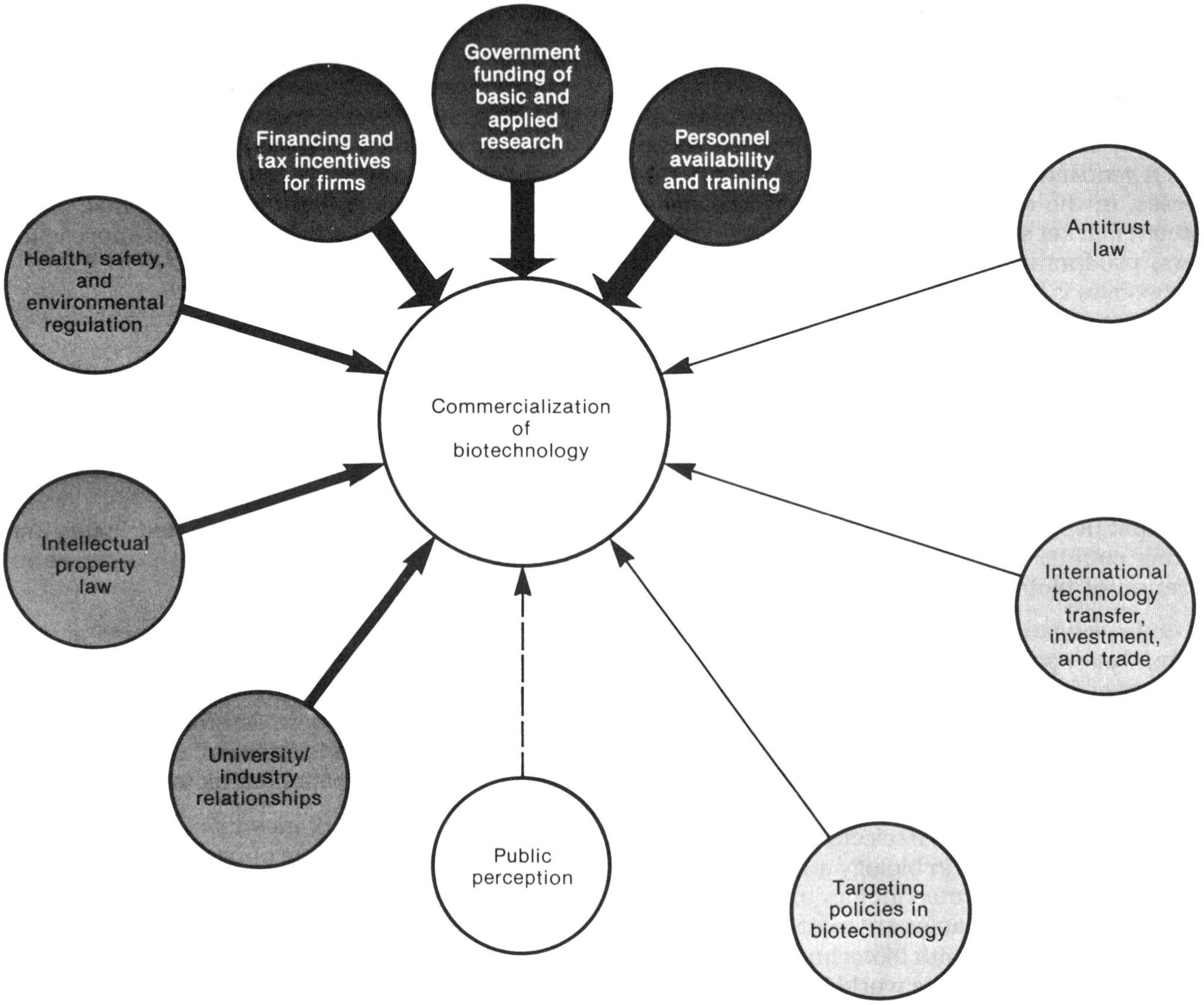

SOURCE: Office of Technology Assessment.

Kingdom, Switzerland, and France. Since the importance to competitiveness of any given factor is not necessarily the same for every industrial sector in which applications are being pursued—for instance, a country's intellectual property laws may protect pharmaceuticals better than plants—the importance of each factor was evaluated for different industrial sectors.

Additional considerations taken into account in the analysis are historical patterns of industrial commercialization, the lack or abundance of particular natural resources, and the tendency toward risk taking in each country. These other considerations were used as modifiers of the results of the analysis.

OTA's principal findings with respect to the types and activities of firms commercializing biotechnology, the factors potentially important to international competitiveness in biotechnology, and the other considerations just mentioned are presented below.

The importance of established and new firms in the commercialization of biotechnology

U.S. and foreign efforts to develop and commercialize biotechnology differ substantially in character and structure. In the United States, two distinct sets of firms are pursuing commercial applications of biotechnology—NBFs and established companies. Because NBFs were founded specifically to exploit perceived research advantages, they are providing the United States with a commercial edge in the current research-intensive phase of biotechnology's development. Through their R&D efforts, NBFs are contributing to innovation, expansion of the U.S. research base, technology diffusion, and encouragement of technical advances through the increased domestic competition they create. All of these contributions provide the United States with a competitive advantage.

Although NBFs have assumed much of the risk for biotechnology's early development in the United States, established U.S. companies are making substantial contributions to the U.S. commercialization effort. Through equity investments and licensing and contract research agreements with NBFs, established U.S. companies are providing many NBFs with the necessary financial resources to remain solvent. Through joint development agreements with NBFs, many established companies will also provide the necessary production and marketing resources to bring many NBF products to world markets. These resources could help to sustain the rapid pace of technical advance spurred by NBFs. Recently, more and more established U.S. companies have been investing in their own research and production facilities, so the role of established companies in the U.S. biotechnology effort is expanding.

U.S. efforts to commercialize biotechnology are currently the strongest in the world. The strength of U.S. efforts is in part derived from the unique complementarity and competition that exists between NBFs and established U.S. companies in developing biotechnology for wider commercial application. At present, most NBFs are still specializing in research-oriented phases of development, precisely the commercial stage where they excel. The established companies, on the other hand, have assumed a major share of the responsibility for production and marketing of, and, when necessary, obtaining regulatory approval for, many of the earliest biotechnology products— the commercial stages where their resources are strongest. **Since established companies control the later stages of commercialization for many new products being developed through production and marketing agreements with NBFs, they will also have considerable control over the pace at which these new products reach the market. Whether the dynamism arising from the competition and complementarity between NBFs and established companies will continue giving the United States a comparative advantage in the context of product introduction remains unclear.** Some established companies, for example, might have disincentives to market new products because the new products might compete with products they already have on the market.

In Japan, the Federal Republic of Germany, the United Kingdom, Switzerland, and France, biotechnology is being commercialized almost exclusively by established companies. **The Japanese consider biotechnology to be the last major technological revolution of this century, and the commercialization of biotechnology is accelerating over a broad range of industries, many of which have extensive bioprocessing experience.** The general chemical and petroleum companies especially are leaning strongly toward biotechnology, and some of them are making rapid advances in R&D through their efforts to make biotechnology a key technology for the future. In Europe, large pharmaceutical and chemical companies, many of which already have significant strength in biologically produced product markets, are the major developers of biotechnology. Their inherent financial, production, and marketing strengths will be important factors as the technology continues to emerge internationally.

The commercial objectives of biotechnology R&D vary across national boundaries. In the United States, commercial research projects appear primarily focused on pharmaceutical and plant and animal agriculture, and American competitive vigor in these application areas is correspondingly strong. Much of the investment in animal agriculture has been made by NBFs whereas much of the investment in plant agriculture has been made by major U.S. agrichemical companies.

In Japan, a competitive drive has been launched to enter international pharmaceutical markets. Furthermore, Japanese companies are world leaders in large-scale plant tissue culture, and MITI has identified secondary compound synthesis from plants as a major area for commercialization. Unlike the United States, Japanese companies appear to be dedicating a great deal of biotechnology R&D to specialty chemical production, an area where they are already internationally prominent.

To the extent that large companies in Europe began their commercialization efforts later than U.S. companies and may also lack the dynamism and flexibility to compete with the combined efforts of NBFs and established companies in the United States, European companies could initially be at a competitive disadvantage. The United Kingdom's major pharmaceutical companies are among the leading producers of biologically produced products, however, and their expertise in bioprocessing is impressive. Furthermore, the United Kingdom possesses some of the strongest basic research in interdisciplinary plant sciences. Whether or not the basic research will be commercialized successfully is difficult to predict.

U.S. competitive strength in biotechnology will be tested when large-scale production begins and bioprocessing problems are addressed. Pharmaceutical markets will be the first proving ground for U.S. competitive strength. The Japanese have extensive experience in bioprocess technology, and dozens of strong "old biotechnology" companies from several industrial sectors in Japan are using new biotechnology as a lever to enter profitable and expanding pharmaceutical markets. In addition to competing against Japanese companies, U.S. pharmaceutical and chemical companies will be competing against pharmaceutical and chemical companies of Western Europe, all of whom expect to recover their biotechnology investments through extensive international market penetration. There seem to be fewer European companies than Japanese companies strong in biotechnology now, but the competitive strength of European multinationals such as Hoechst (F.R.G.), Rhone Poulenc and Elf Aquitaine (France), ICI, Glaxo, and Wellcome (U.K.), and Hoffmann-La Roche (Switzerland) in the long run should not be underestimated.

Factors potentially important to international competitiveness in biotechnology

MOST IMPORTANT FACTORS

The three factors most important to the commercial development of biotechnology are financing and tax incentives for firms, government funding of basic and applied research, and personnel availability and training.

Financing and Tax Incentives for Firms.—The availability of venture capital to start new firms and tax incentives provided by the U.S. Government to encourage capital formation and stimulate R&D in the private sector are very important to development of biotechnology in the United States. Since 1976, private venture capital in the United States has funded the startup of more than 100 NBFs. Many of these firms have already obtained second- and third-round financing, while others, still seeking additional funds, are relying heavily on the currently strong stock market, R&D limited partnerships, and private placements to fund research, production scale-up, clinical trials, and early product development. Between March and July of 1983, 23 NBFs raised about $450 million. R&D limited partnerships in biotechnology are expected to total $500 million in 1983 and $1.5 billion by 1984. Corporate equity investment in NBFs, although now diminishing, has also been an important source of financing for the new firms. From 1977 to August 1983, corporate venture capital supplied over $350 million to NBFs in equity investments alone.

Current price/earnings ratios* for NBFs appear high, because most NBFs still have negative earn-

*A price/earnings ratio ($\frac{\text{market price per share}}{\text{company earnings per share}}$) reflects the stock market's anticipation of the company's future performance based on the earnings per share.

ings records. Continued reliance on the stock market and R&D limited partnerships to raise funds will place increased pressure on the new firms to begin showing profits. If NBFs do not begin showing profits within the time frame expected by investors, additional financing from public offerings and R&D limited partnerships may be difficult to obtain.

The future performance of NBFs now extensively using the stock market and R&D limited partnerships for financing may influence the availability of financing for other firms seeking capital in the future. If some of these companies do not begin to manufacture soon in order to generate product revenues, investors may lose confidence in many of the firms' ability to commercialize biotechnology.

In the United States, venture capital is generally more difficult to obtain for later rounds of financing than for initial rounds, in part because venture capitalists are more eager to invest in the earlier rounds to maximize their investment returns. The difficulty in getting subsequent financing for production scale-up may prove to be an insurmountable problem for some NBFs; the ability to self-finance may still be 5 to 10 years away.

Of all the six competitor countries, the United States has the most favorable tax environment for capital formation and financing small firms. Tax incentives, more than government funding, are used in the United States to stimulate business and encourage R&D expenditures. Thus, R&D limited partnerships, low capital gains tax rates, R&D tax credits (due to expire in 1985), and subchapter S provisions all benefit small firms.

In Japan and the European competitor countries, venture capital has played a very minor role in the commercialization of biotechnology, because these countries do not have tax provisions that promote the formation of venture capital and investment in high-risk ventures. As a consequence, few NBFs exist outside the United States. Instead, established foreign companies have initiated efforts to commercialize biotechnology because they generally can finance R&D activities through retained earnings. Established companies also have access to financing from bank loans. Additionally, the governments of Japan, the United Kingdom, the Federal Republic of Germany, and France have provided the private sector with public funds for biotechnology.

After the United States, Japan has the most financing available for companies using biotechnology. **The Japanese Government has made the commercialization of biotechnology a national priority and is financing cooperative interindustry biotechnology projects.** Most of the established companies commercializing biotechnology in Japan have at least one bank as a major shareholder that provides the company with low-interest loans for R&D. Wealthy individual investors in Japan, although few in number, have also provided some risk capital for new ventures.

Tax incentives relevant to established companies commercializing biotechnology are those which stimulate R&D investments and those which encourage capital formation. Corporate tax rates are also important. For purposes of international comparisons, the most reliable basis is the overall effective corporate tax rate. Unlike statutory rates, the effective rate takes into account different definitions of taxable income and treatments of depreciation. Available studies suggest that Switzerland, followed by Japan and the United Kingdom, have the lowest effective corporate tax rates. The effective rates in the United States, the Federal Republic of Germany, and France are higher and about equal.

Government Funding of Basic and Applied Research.—The objective of basic research is to gain a better understanding of the fundamental aspects of phenomena without goals toward the development of specific products or processes. Such research is critical to maintaining the science base on which a technology rests and to stimulating advances in a technology. Basic research is usually conducted by academic researchers who receive government funds. The objective of applied research is to gain the knowledge needed to supply a recognized and specific need, through a product or process. Such research is usually funded by industry. Generic applied science can be viewed as bridging a gap between basic science done mostly in universities and applied, proprietary science done in industry for the development of specific products. Such research is aimed at the solution of general problems that are associated with the use of a technology by industry. Generic applied research areas in biotechnology, for instance, include development of bioreactors, screening of micro-organisms for potential products, and better understanding of the genetics and biochemistry of industrially important micro-organisms. Support of basic science and of generic applied science is generally viewed as the responsibility of government, because it ultimately contributes to the public good and because it is high risk and too expensive for individual firms.

Controversy exists over the relative importance of national support of basic and applied science. Some argue that since the findings of basic research are readily accessible worldwide because they are published in journals with international distribution, strong government support for basic research is therefore not required for the maintenance of a leading position in the development of a technology. Others argue that the development of a technology within a country will progress faster if companies have access to local basic research scientists for consulting and contractual arrangements. Domestic technology transfer can help give industry a lead in innovation.

Of the competitor countries, the United States, both in absolute dollar amounts and in relative terms, has the largest commitment to basic research in biological sciences. Like the United States, the Federal Republic of Germany, the United Kingdom, and Switzerland have a strong basic science base. On the other hand, **the U.S. Government's commitment to generic applied research in biotechnology is relatively small.** The governments of Japan, the Federal Republic of Germany, and the United Kingdom fund a significant amount of generic applied science in biotechnology.

During the past few decades, the U.S. Government increased its commitment to basic biological sciences, although this commitment has decreased in the last few years. While the Government was increasing its commitment to basic science, there was a concomitant decrease in its commitment to generic applied fields such as bioprocess engineering and applied microbiology. The rationale for this policy has been that most applied science, regardless how general, is the responsibility of industry. This policy has contributed to a widening scientific gap between purely basic research funded by the U.S. Government and short-term, relatively product-specific applied research funded by private industry. **In fiscal year 1983, the Federal Government spent $511 million on basic biotechnology research* compared to $6.4 million on generic applied research in biotechnology. The relatively low level of U.S. Government funding for generic applied research in biotechnology may cause a bottleneck in this country's biotechnology commercialization efforts.**

The Japanese Government, in contrast, is devoting proportionately more public funding to the solution of generic applied science problems than to basic research. The pattern of funding in Japan may reflect a policy of placing a greater priority on generic applied research in lieu of basic research because the Japanese may rely on the United States and other countries to prove the early feasibility of new technologies for commercialization. This strategy worked well in the semiconductor industry, and Japan may very well attain a larger market share for biotechnology products than the United States because of its ability to rapidly apply results of basic research available from other countries.

Personnel Availability and Training.—Adequately trained scientific and technical personnel are vital to any country's industrial competitiveness in biotechnology. For the most part, countries with good science funding in a field also have a good supply of well-trained people in that field.

The commercial development of biotechnology will require several specific types of technical personnel. Especially important categories include specialists in rDNA and MAb technology such as molecular biologists and immunologists; specialists in scale-up and downstream processing such as microbiologists, biochemists, and bioprocess engineers; and specialists for all aspects of biotechnology such as enzymologists and cell culture

---

*From $20 million to $30 million of the $511 million may actually be generic applied research, because definitions of biotechnology differ among agencies.

specialists. Scale-up personnel will become more important as companies using biotechnology move into production.

The United States currently has a competitive edge in the supply of molecular biologists and immunologists able to meet corporate needs, in part because the U.S. Government has provided substantial funding since World War II for basic life sciences research in U.S. universities. The supply of Ph. D. plant molecular biologists and scale-up personnel in the United States, however, may be inadequate. Like the United States, the United Kingdom and Switzerland have funded life sciences well and have a sufficient supply of basic biological scientists. Unlike the United States, Japan, the United Kingdom, and the Federal Republic of Germany maintained a steady supply of both industrial and government funding for generic applied microbiology and bioprocess engineering in the past few decades and have adequate personnel in these fields. In Japan and the Federal Republic of Germany, slight shortages of molecular biologists and immunologists exist; Japanese companies are seeking to train personnel abroad. France appears to have shortages in all types of personnel.

The training of personnel is important to the continuing commercialization of biotechnology. The United States has, for the most part, good training programs for basic scientists. Specialists in plant molecular biology may be in short supply now, but training in this discipline can be readily achieved with interdisciplinary programs in biology departments in universities. On the other hand, the United States does not have more than a handful of training programs for personnel in the more applied aspects of biotechnology, nor does it have Government programs, such as training grants, to support training in these fields. The training of bioprocess engineers and industrial microbiologists will require greater interdisciplinary cooperation between engineering and biology departments within universities.

The United States promotes and funds the training of foreign nationals in laboratories in the United States, yet funds very little training of Americans abroad. Foreign countries have many significant research programs in biotechnology that U.S. researchers could be visiting were funding available.

FACTORS OF MODERATE IMPORTANCE

The three factors found to be of moderate importance to international competitiveness in biotechnology are health, safety, and environmental regulation; intellectual property law; and university/industry relationships.

Health, Safety, and Environmental Regulation.—The analysis of the effect of health, safety, and environmental regulation on competitiveness in biotechnology was made by determining how restrictive a country's laws would be with respect to marketing biotechnology products and whether there were any uncertainties about their application. The analysis focused on the drug laws for humans and animals and, to a lesser extent, on laws governing the production of chemicals and the deliberate release of novel organisms into the environment. **In all the competitor countries, there is some uncertainty as to the environmental regulation governing the deliberate release into the environment of genetically manipulated organisms.**

The only government controls directed specifically toward biotechnology are the rDNA guidelines adopted by the six competitor countries. They are essentially voluntary and directed primarily at research. Their containment and oversight provisions have been substantially relaxed since they were originally adopted, and this trend is expected to continue. The United States has the most liberal guidelines, whereas Japan has the most stringent.

Since companies generally approach domestic markets first, the countries with the least stringent regulation may have products on the market earlier. Japan has the most stringent health and safety regulation for pharmaceuticals and animal drugs, followed by the United States. Switzerland appears to be the most liberal. Thus, **the regulatory environment favors the European companies over those of Japan and the United States reaching their own domestic markets sooner for pharmaceuticals and animal drugs.** In the United States, the Food and

Drug Administration has taken the position that rDNA products whose active ingredients are identical to ones already approved or to natural substances will still need to go through the new product approval process. However, data requirements may be modified and abbreviated. This appears not to be the situation in the competitor countries, although there have not been definitive pronouncements by their regulatory agencies.

Regulation may also influence where companies locate their production facilities. A country with liberal regulation may attract production facilities and, as a consequence, gain access to technology. Alternatively, companies may set up facilities in the United States and Japan regardless of regulation because of market size and as a way to avoid certain nontariff trade barriers on imports. NBFs may not have the capital to establish foreign subsidiaries in order to avoid regulatory barriers. Thus, they may be at a competitive disadvantage with respect to larger firms for entering world markets.

Countries wishing to market their products abroad will have to abide by the regulations of the countries to which they are exporting. Thus, countries can control access to their domestic markets by the regulations they impose. This is a form of nontariff trade barrier. These barriers are considered further in the discussion of trade policy.

Intellectual Property Law.—The ability to secure property interests in or otherwise protect processes, products, and knowhow will encourage development of biotechnology, because it provides incentives for a private company to invest the time and money for R&D. Without the ability to prevent competitors from taking the results of this effort, many new and risky R&D projects would not be undertaken. Thus, a strong intellectual property law system will enhance a country's competitiveness in biotechnology.

The areas of intellectual property law most relevant to biotechnology are those dealing with patents, trade secrets, and plant breeders' rights. These areas work together as a system; an invention may be protected by one or more of them, and if one has disadvantages, a company can look to another. Thus, to the extent that a country's intellectual property law provides several alternative ways for companies to protect biotechnological inventions, it is more likely to be competitive in biotechnology.

The patent laws of the competitor countries provide fairly broad protection for biotechnological inventions, but the laws differ to some degree in the types of inventions that are protected, the effect of publication on patent rights, and the requirements regarding public disclosure of the invention, which is the quid pro quo for the grant of the patent. The United States provides the widest coverage. Patents are available for living organisms (including plants and possibly animals), their products, their components, and methods for making or using all of these. In addition, patents can be granted on therapeutic and diagnostic methods. In the United Kingdom, the Federal Republic of Germany, France, Switzerland, and Japan, patent coverage is almost as broad, but patents are not permitted on plants and animals nor on therapeutic and diagnostic methods. In addition, Switzerland does not permit patents on micro-organisms. In Japan, the relatively strict guidelines governing rDNA research also may bar patents on those genetically manipulated organisms viewed as hazardous.

With regard to the effect of publication on patent rights, the United States also has a slight advantage over the other countries analyzed here. The four European countries do not permit a patent to be granted to an inventor who has disclosed his or her invention in a publication before the patent application is filed, assuming the disclosure enables others to make it. This absolute novelty requirement is viewed as impeding the free exchange of scientific information and possibly providing a disincentive for scientists to seek patent rights. The United States, on the other hand, provides a 1-year grace period between the date that an inventor publishes an article and the date on which the patent application must be filed. Japan provides a 6-month grace period for certain activities, such as presenting scientific papers. The U.S. advantage is limited, however, because when U.S. inventors wish to secure patents in other countries, they must refrain from publication in order to protect their patent rights in those countries.

The patent law requirement that an invention be described in sufficient detail so that it could be replicated creates unique problems for biological inventions. Since a living organism generally cannot be described in writing with sufficient specificity to allow others to make and use it, granting of patents on such organisms and methods of using them generally is contingent on their deposit in a public depository. However, these deposits, in effect, turn over the factory for making a product to one's competitors, unlike patents in other technologies. The four European countries, and particularly the Federal Republic of Germany, place restrictions on access to such deposits that may be advantageous for their inventors.

Most aspects of biotechnology lend themselves to protection as trade secrets, and owners of such technology may rely on trade secrets when patent rights are uncertain or when they judge trade secrecy to be more advantageous. All of the competitor countries protect trade secrets relating to biotechnology, but the Federal Republic of Germany and, to a lesser extent, Switzerland, provide the greatest degree of protection. Japan appears to provide the least degree of protection.

All of the competitor countries recognize property rights in new varieties of plants, but the United States provides the greatest degree of protection. Protection in the United States is most favorable because the plant breeder has the greatest number of options among which to choose in securing property rights for a new variety of plant, including pursuing a patent under the traditional patent laws.

In the final analysis, **the U.S. intellectual property system appears to offer the best protection for biotechnology of any system in the world,** thus providing the United States with a competitive advantage with regard to this factor. This advantage results from the fact that the system provides the widest choice of options for protecting biotechnological inventions, the broadest scope of coverage, and some of the best procedural safeguards.

University/Industry Relationships.—A factor that has moderate overall importance is the relationship that exists between universities and industries. Interest in the commercial potential of biotechnology has dramatically increased university/industry interactions, especially in the United States. Established U.S. and foreign companies have invested substantial amounts of money in U.S. universities doing work in biotechnology in order to gain a "window on the technology." Many university/industry agreements in biotechnology focus on research directed toward applications of biotechnology in a specific industrial sector, whereas other university/industry agreements are directed at many applications of biotechnology. **The various agreements in the United States appear to be working well, and fears concerning conflict of interest and commingling of Government and industry funds have diminished.**

The increase of industry funding of university research in the United States in several disciplines came at a time when Federal funding of science was decreasing in constant dollars. Although the infusion of industry funds to the U.S. universities has been substantial, it accounts for only a small fraction (less than 10 percent) of the total funding of university research. In some university departments, however, such as electrical engineering, chemistry, and possibly now molecular biology, industrial funding of university research may exceed 10 percent. Even with the increase in industrial support, industrialists agree that private funding can never replace Federal funding of basic science research if past and current levels of basic research are to continue.

University/industry interactions are a very effective way of transferring technology from a research laboratory to industry. Such interactions promote communication between industrialists and academicians, a two-way interaction that benefits both sides. Industrial scientists learn the latest techniques and research results, while academicians gain increased familiarity with challenges of industrial R&D.

Neither Japan nor the European competitor countries identified in this assessment have as many or as well-funded university/industry relationships as the United States does, but varying degrees of cooperation do exist. In Japan, the ties between university applied research departments and industry have always been close. Additionally, the Japanese Government is implementing new policies to encourage closer ties between basic research scientists and industry. In the Federal Republic of Germany, the Federal Ministry of Science and Technology (BMFT, Bundesministerium für Forschung und Technologie) has a history of promoting close contact between academia and industry and is cosponsoring with industry many projects important to biotechnology. Switzerland encourages communication between individuals in academia and industry, and relationships are easy to maintain. The universities in both the United Kingdom and France have had very few ties with industry in biotechnology, but the governments of both countries have recently set up programs designed to encourage university/industry relationships.

Industrial funding for research in American universities is helping to promote the transfer of technology. However, the multimillion dollar arrangements that have characterized the initial relationships in biotechnology are most likely short term and will probably become less important as the firms develop in-house expertise and their research becomes more applied. **As in other fields, consulting and contractual research agreements are likely to predominate in university/industry relationships in biotechnology in the future.**

LEAST IMPORTANT FACTORS

The least important of the 10 factors analyzed were found to be antitrust law; international technology transfer, investment, and trade; government targeting policies in biotechnology; and public perception. Any of these factors, however, could become important as the technology develops and products reach the marketplace.

Antitrust Law.—Antitrust laws are based on the general economic assumption that competition among a country's industries will result in greater productivity, innovation, and general consumer benefits than will cooperation. Recently there has been much public debate about whether U.S. antitrust laws have, in fact, accomplished these goals in all cases and whether they place U.S. companies at a competitive disadvantage in the international marketplace when foreign companies face allegedly less restrictive antitrust laws.

The antitrust laws of the United States and the other major competitors in biotechnology are generally similar in that they prohibit restraint of trade and monopolization. However, the foreign laws generally provide for exemptions and vest much discretion with the enforcement authorities, especially in Japan. Thus, in practice, they are often less restrictive than in the United States. In addition, countries differ in the consequences to firms for failure to comply with antitrust laws. In the United States, the consequences of noncompliance can be more severe than in the competitor countries because private, in addition to Government, suits can be brought against alleged antitrust violators, and treble damages are assessed if a violation is found.

U.S. companies commercializing biotechnology face no major antitrust compliance problems, because the lack of concentration and the absence of measurable markets mean that most types of joint research arrangements would not be anticompetitive. Technology licensing agreements can raise antitrust concerns, but these generally are not unique to biotechnology. However, there is some degree of uncertainty about the scope and applicability of the antitrust laws to R&D joint ventures and licensing agreements. This uncertainty, plus the expense of litigation and the threat of treble damages, could deter some activities that might lead to innovation in biotechnology, thus limiting the ability of U.S. companies commercializing biotechnology to exploit their technology.* For these reasons, the current U.S. antitrust laws may have some modest adverse effect on biotechnology.

International Technology Transfer, Investment, and Trade.—Technology transfer across national boundaries can be promoted or inhibited by export control laws and by laws governing international joint ventures and technology licensing. Most export controls are directed at overseeing technology transfer for national security reasons, and the concept of national security is fairly narrowly interpreted in all of the competitor countries except the United States. Therefore export controls may not be very

*In addition, the rigid application of certain "per se rules" in the area of licensing may actually lead to anticompetitive results.

important for the international development of biotechnology. However, the export controls of the United States, which are the most restrictive of the competitor countries, include the control of pharmaceuticals and of many micro-organisms that potentially could be used in biotechnology product production. **These controls may have a slightly adverse affect on the competitiveness of U.S. companies commercializing biotechnology because they could cause delays that result in sales' being lost to foreign competitors.** U.S. export control laws may need clarification as biotechnology products proceed to the marketplace because there is some uncertainty as to what products or data will be restricted. In addition, the current U.S. export control law expired in October 1983. While it is virtually certain that a new law will be passed, the form that law will take is still unclear.

The U.S. Government has no laws governing international joint ventures and technology licensing among U.S. and foreign companies. As a consequence, technology can be transferred readily to other countries. The predominance of NBFs in the United States and their need for capital has led to the formation of many transnational joint ventures involving NBFs. Because of this, the United States appears to be transferring more technology outside of its national borders than are other countries at the present time. However, as biotechnology products reach the market, foreign firms will probably set up subsidiaries in the United States in order to have access to U.S. markets. If this happens, the United States could become a net importer of technology.

In contrast with the United States, France and Japan have Government programs for the review of potential transnational agreements, but it is uncertain whether such programs help or hinder the transfer of technology into those countries. As of now, laws governing the transfer of technology are not very important to the U.S. competitive position in biotechnology. However, if other countries establish themselves more favorably in world markets, the current outward flow of technology from the United States may hurt the U.S. competitive position.

Foreign exchange and investment control laws help prevent access to domestic markets and technology by foreign firms. The United States has the fewest controls, whereas Japan and France have the most control mechanisms. Japanese controls exist in the form of nontariff barriers such as ministerial review and screening of foreign investments and licensing agreements with respect to a number of criteria ranging from national security to competition with other Japanese business. Ministries also have the power to designate specific companies for special controls on foreign ownership. In France, the Government has the ability to object or order alteration of licensing agreements and foreign investments. Foreign direct investment in certain domestic industries is not encouraged. Thus, **U.S. markets are the most accessible to foreign firms and therefore the most vulnerable to foreign competition,** whereas Japanese and French markets are the least accessible and the most protected against foreign competition.

Trade policy was assessed by examining the competitor countries' abilities to protect domestic industries from imports and to control foreign investment in domestic industries. **Trade policy is not important for the commercialization of biotechnology today because of the small number of products that have reached the market and because trade in biotechnologically produced products is not likely to raise any unique trade issues.** However, trade policy will become increasingly important as more products reach the marketplace, especially in the area of pharmaceuticals, where significant nontariff barriers, such as conforming to country standards with appropriate testing data, quality control standards, and packaging requirements exist. Problems with nontariff barriers are now being negotiatied with Japan and other countries including the European Economic Community, and it apears as though some trade barriers may become less stringent.

Government Targeting Policies in Biotechnology.—The governments of four of the competitor countries—Japan, the Federal Republic of Germany, the United Kingdom, and France—have instituted comprehensive programs to help domestic companies develop certain areas of biotechnology. The targeting policies are intended to reduce economic risk and lessen corporate duplication in biotechnology R&D. A variety of policy measures are used within each country. In Japan and West Germany, the Governments carry out their policies mostly through projects that combine the resources of the Government and private companies to meet specific objectives set by the Government. The United Kingdom and France have adopted a different approach; they support startup of small firms, which are expected to commercialize the results of Government-funded basic and applied research.

At this early stage, any evaluation of the eventual success of foreign targeting programs is preliminary. History has shown that even the best thought-out targeting policies do not guarantee competitive success. Whether targeting policies of foreign governments in biotechnology are superior to the U.S. Government policy of funding basic research in the life sciences and encouraging R&D in all industries with tax credits remains to be seen. Though targeting policies are not of great importance when compared to other competitive factors, they could tip the balance of a competitive position in the future.

Public Perception.—Public perception of the risks and benefits of biotechnology is of greater importance in countries with representative, democratic forms of government than it is in countries with other forms of government, simply because of the greater attention paid to public opinion in democracies and the independence of the media. Therefore, **public perception could influence commercialization of biotechnology in all of the countries examined here.** As a factor influencing competitiveness, however, public perception is probably of greater importance in the United States than in the other competitor countries. Historically, the American public has been more involved than the public in Japan or the European countries with issues pertaining to genetic research and technology (e.g., issues regarding the safety of rDNA research).

In all countries, the importance of public perception as a factor influencing competitiveness will be greatly increased in the event of an accident or perceived negative consequence of biotechnology. Particularly in such a case, the level of scientific and technological literacy in the various competitor countries becomes important, as judgments must be made concerning complex issues. In the United States, survey data show that only a small fraction of the public is fully informed about genetics in general and therefore, probably, about biotechnology in particular. Survey data also suggest that there is public apprehension concerning applied genetics. Thus, an accident associated with biotechnology could arouse strong public reaction in the United States, a reaction that might be greater than in the competitor countries.

Given the lack of public knowledge in the United States, it is particularly important that the media play a responsible role with respect to biotechnology. The role of the media already extends beyond mere reporting of the facts, by virtue of the events and issues the media elect to cover.

At the current time, public perception is not an important factor in the commercialization of biotechnology. However, the volatility of a potential public response must be noted. Were there to be an accident due to commercial biotechnology, the public's reaction could be extremely important to the future of biotechnology.

Other influences on competitiveness in biotechnology

Three other considerations that should be noted in evaluating competitive positions in the commercialization of biotechnology are, for each country, historical patterns of industrial commercialization, the availability of natural resources, and cultural attitudes toward risk-taking.

Historically, industries in some countries have moved research results into commercialization rapidly, while industries in other countries have moved more slowly. This observation is especially important in this analysis of biotechnology. For instance, the United Kingdom has a good science base, trained personnel, and industries that could be using these new technologies; however, the United Kingdom may not be a major contender in the commercialization of biotechnology mainly because it does not have a history of rapid commercialization. On the other hand, both the United States and Japan historically commercialize scientific advances rapidly.

Another historical consideration is the quantity of sales of specific products in a country. For example, Japan's per capita consumption of pharmaceuticals is significantly higher than that of the other competitor countries; therefore, Japan may have more interest than other countries have in applying biotechnology to the production of pharmaceuticals. In other words, cultural differences will probably play a role in determining the markets each country will attempt to dominate.

The absence or presence of certain natural resources may also determine how quickly a country moves into the commercialization of biotechnology. For instance, Japan does not have domestic petroleum resources. Because biomass can potentially replace petroleum as a feedstock in the chemical industry, Japan may be more interested in applying biotechnology in the chemical industry than a country, such as the United Kingdom, which has domestic petroleum resources. The United States, a country that produces excesses of grain each year, may find commercialization of processes that can use grain as a feedstock particularly attractive. However, it is too early to predict the degree to which natural resources will determine the commercial applications of biotechnology a country may undertake.

The United States, as a general rule, is not averse to risk-taking in business. Risk-taking is a part of the American lifestyle. European countries are more risk averse. Since investment in biotechnology is considered risky, countries that are more risk averse are less likely to move rapidly to commercialize biotechnology.

Conclusion

The unique complementarities between established and new firms, the well-developed science base, the availability of finances, and an entrepreneurial spirit have been important in giving the United States its present competitive advantage in the commercialization of biotechnology. In order to maintain this advantage, increased funding of research and training of personnel in basic and generic applied sciences, especially bioprocess engineering and industrial microbiology, may be necessary. The United States may also need to be concerned with the continued availability of finances for NBFs until they are self-supporting. On most of the other factors influencing competitiveness, the United States rates very favorably, although there are changes in laws and policies that could potentially improve or help maintain the U.S. competitive position. These changes include clarification and modification of particular aspects of intellectual property law; health, safety, and environmental regulation; antitrust law; and export control law.

Japan will be the most serious competitor of the United States in the commercialization of biotechnology. Japan has a very strong bioprocess technology base on which to build, and the Japanese Government has specified biotechnology as a national priority. The demonstrated ability of the Japanese to commercialize rapidly developments in technology will surely manifest itself in biotechnology.

The Federal Republic of Germany, the United Kingdom, Switzerland, and France lag behind the United States and Japan in the commercialization of biotechnology. The European countries generally do not promote risk-taking, either industrially or in their government policies. Additionally, they have many fewer companies commercializing biotechnology. Thus, **the European countries are not expected to be as strong general competitors in biotechnology as the United States and Japan.** In markets for specific products, including some pharmaceuticals, specialty chemicals, and animal agriculture products, however, some European companies will undoubtedly be strong international competitors.

Issues and options

Congressional issues and options for improving the competitive position of the United States in biotechnology are presented at the end of most of the chapters in part IV. To improve the competitive position of the United States, legislation could be directed toward any of the 10 factors OTA identified as influencing competitiveness, although coordinated legislation directed toward all of the factors might be more effective in promoting U.S. biotechnology efforts. The chapters in part IV discuss only those options that are specific to the development of biotechnology. Some of the options presented in part IV are limited and straightforward, such as some options concerning health and safety regulation and R&D limited partnerships. Other options are much broader with potentially large political, ethical, and financial considerations. Some examples of the latter include establishing university/industry cooperative research centers, regulating the deliberate release of genetically manipulated organisms into the environment, and changing patterns of research funding. Thus, the adoption of some options may occur more rapidly than others.

Policy options in some areas are not specific to biotechnology but apply to high technology or industry in general. These options are to:

- improve U.S. science and engineering education and the retraining of industrial personnel,
- change U.S. antitrust law to promote more research collaboration among domestic firms,
- regulate imports into the United States to protect domestic industries,
- regulate the transfer of technology from the United States to other countries, and
- target specific industries or technologies for Federal assistance.

There are many arguments for and against these options that are beyond the scope of this report. Because of their broad applicability to industry in general, these options are not discussed in part IV. It is important to note, however, that legislation in any of these areas could affect the development of biotechnology and potentially have a large influence on the U.S. competitive position.

Chapter 2
Introduction

Contents

Chapter 2

Introduction

This report assesses the international competitive position of the United States with respect to the development and commercialization of industrial applications of new biotechnology. New biotechnology is defined as the use of novel technologies—recombinant DNA (rDNA) technology, monoclonal antibody (MAb) technology, and new techniques used in bioprocess engineering—to develop commercial products and processes that use living systems.

Despite its rather narrow focus on new biotechnology, this report can be viewed as an introduction to the entire subject of biotechnology, a field that will become increasingly important in industrial production during the next few decades. Developments associated with new biotechnology could spur a renaissance in traditional biotechnology. The lure of profitability in new biotechnology, for instance, will very likely attract students to bioprocess engineering, and an increase in the number of engineers will probably improve bioprocess technologies applicable to the traditional uses of biotechnology. Another reason biotechnology may increase in importance is the movement, albeit not very rapid, toward the use of renewable resources. Diverse micro-organisms able to convert biomass into useful chemicals, some of which are a source of energy, are known, and these micro-organisms have yet to be exploited to the fullest extent. Furthermore, the industries that use traditional biotechnology are showing interest in the novel techniques mentioned above, and many of these industries will probably be using these techniques, because of their broad applicability, in some aspect of their operations in the future.

Impact of biotechnology on the research community

A point to be mentioned that does not relate directly to this report is the impact of the novel technologies, especially rDNA technology, on the biological research community. Recombinant DNA technology has already allowed a greatly increased understanding of the basis of life, and thus, of the genetic basis of disease. Research over the next 10 years may yield an increased understanding of the mechanism of carcinogenesis, genetic susceptibility to disease, the functioning of the immune system, the basis of debilitating diseases such as diabetes and arthritis, and some knowledge of brain function. Additionally, gene transplantation technology may reach a stage where some genetic diseases could be cured. It may be that the main benefit of the new biological technologies will be the advances in fundamental knowledge that accrue. Thus, even if no commercial products were to result from them, these technologies would still have a substantial impact on the quality of life.

The multidisciplinary nature of biotechnology

Biotechnology is unusual among most technologies in that it spans an array of scientific disciplines. Individuals seeking to be well versed in applications of biotechnology must have interdisciplinary training. Bioprocess engineers, for example, need some knowledge of biochemistry and microbiology as well as knowledge of engineering design so that the most efficient combination of micro-organism and bioreactor can be determined. Similarly, plant molecular biologists need to know both plant physiology and molecular genetics. People working in microbial enhanced oil recovery need training in microbiology as applied to a specific geological environment.

The multidisciplinary nature of biotechnology has extensive implications for educational and industrial structures. To excel in biotechnology, universities will need to draw on the resources of several departments. Diversified companies may have an inherent advantage over other companies, because technologies perfected for the production of one product (e.g., a pharmaceutical product) can be modified and used for the production of another (e.g., a food additive).

Biotechnology in developing countries

One area where biotechnology could certainly have an impact, though not considered extensively in this report, is in developing countries. Plants that have been genetically manipulated for growth in tropical and desert climates could improve agricultural production. Vaccines that do not need refrigeration could have widespread influence on the health of the people and their livestock. Small local factories that convert biomass to ethanol could help solve the problem of costly petroleum imports for energy.

The applications of biotechnology to developing countries was discussed in a workshop held by the National Academy of Sciences and the U.S. Agency for International Development (1). The proceedings of this workshop include suggested priorities for research and time frames for development of various biotechnology products important to developing countries. Additionally, the United Nations Industrial Development Organization has proposed the construction of an international center for biotechnology (2). The proposed center would have 50 staff scientists, 26 postdoctoral fellows, and 100 visiting scientists; the annual budget would be $8.6 million; and the research would concentrate on problems specific to developing countries.

This report does not cover developing countries for two reasons. First, developing countries are not likely to compete with the United States for market shares in biotechnology in the near future. Second, all countries in a competitive position generally have equal access to markets in developing countries, allowing them equal access to international market shares. Some developing countries give preferential treatment to the first company to market a product in that country, but all countries have equal access for first introductions.

Local efforts to promote the development of biotechnology in the United States

Many State governments are actively promoting the establishment of local high-technology centers to stimulate the local economy, and many of these include centers for biotechnology. The oldest and best known of these is the North Carolina Biotechnology Center. This report does not analyze the development of these centers because they are analyzed in another OTA report, *Technology, Innovation, and Regional Economic Development,* due to be published in 1984. It is important to note, though, that it will take several years to recoup the costs of initiating one of these centers. Local biotechnology centers cannot be viewed as a short-term solution to economic problems.

Organization of the report

This report is organized into four parts. Part I introduces the scientific background of the new technologies and forms a basis for discussion of the commercialization of new biotechnology. The three chapters consider the construction of rDNA, the formation of MAbs, and the relevant engineering principles for the large-scale growth of micro-organisms and the use of immobilized enzymes to perform specific catalytic functions. Each emphasizes the industrial use of the technologies and identifies the problems yet to be solved.

Part II is an overview of the companies using biotechnology in the United States and its five major competitors in biotechnology: Japan, the Federal Republic of Germany, the United Kingdom, Switzerland, and France. The discussion considers the relative importance of and level of collaboration between established companies and new biotechnology firms in determining a competitive advantage. This part also includes a discussion of the firms producing the necessary reagents and equipment for the commercial use of biotechnology. Joint ventures among firms, both foreign and domestic, are analyzed.

How specific industrial sectors are applying biotechnology is the subject of the several chapters in Part III. The sectors discussed are pharmaceuticals, agriculture, specialty chemicals and food additives, environmental applications, commodity chemicals and energy, and bioelectronics. The order of the chapters corresponds to the approximate time frames for the development of products and processes in the various sectors—beginning with the sectors in which developments are likely to occur first. Priorities for future research to promote the development of biotechnology in each of the specific industrial sectors are outlined at the end of each chapter.

Part IV is an analysis of specific factors believed to influence a country's competitiveness in biotechnology. It considers only those factors that government policies could potentially affect. The first chapter of Part IV describes the framework used for the analysis. Subsequent chapters analyze specific factors, more or less in order of their importance: private sector financing and tax incentives, government funding of basic and applied research, personnel availability and training, health, safety, and environmental regulation, intellectual property law, university/industry relationships, antitrust law, international technology transfer and trade policy, targeting policies in biotechnology, and public perception. The analysis of the relative importance of each factor in determining a country's competitive position in biotechnology and where the United States stands is presented in *Chapter 1: Executive Summary*. Throughout Part IV, issues of congressional interest and a range of policy options are examined with respect to improving the U.S. competitive position in biotechnology.

This report is a follow-on study to OTA's April 1981 report entitled *Impacts of Applied Genetics: Micro-Organisms, Plants, and Animals* (3). Much useful information is contained in that report and is not repeated here. The reader is advised to read the earlier report for more information on the biological technologies and market forecasts.

Chapter 2 references

1. Board on Science and Technology for International Development, Office of International Affairs, National Research Council, *Priorities in Biotechnology Research for International Development: Proceedings of a Workshop* (Washington, D.C.: National Academy Press, 1982).
2. Newmark, P., "International Biotechnology: U.N. Center To Be Based in India," *Nature* 302:100, 1983.
3. U.S. Congress, Office of Technology Assessment, *Impacts of Applied Genetics: Micro-Organisms, Plants, and Animals*, OTA-HR-132, Washington, D.C., April 1981.

PART I
The Technologies

Chapter 3
The Technologies

Contents

Tables

Figures

Chapter 3
The Technologies

Introduction

This chapter reviews the scientific bases for the technologies discussed in this assessment. The most publicized and broadly applicable of these technologies is recombinant DNA (rDNA) technology, which includes gene cloning, and is explained first. The second technology discussed is monoclonal antibody (MAb), or hybridoma, technology. This technology, used to prepare complex molecules known as MAbs which can be used to recognize or bind a large variety of molecules, has an expanding number of applications. The last technology discussed, bioprocess technology, allows the scaling-up of a biological production process so that large quantities of a product can be made. Bioprocess technology is, in many respects, the most difficult and least understood of the technologies, so it receives a more intensive discussion in this chapter. Because of the lack in the United States of broadly applicable knowledge in bioprocess engineering, the section on bioprocess technology also ends with priorities for future research, giving a focus to where Federal research funds might best be spent.

Recombinant DNA technology

The development of rDNA technology—the joining of DNA from different organisms for a specific purpose—has allowed a greatly increased understanding of the genetic and molecular basis of life. This technology has also led to the founding of many industrial ventures that are addressing the production of numerous compounds ranging from pharmaceuticals to commodity chemicals. This section introduces some aspects of the scientific basis of rDNA technology, discusses methods that are used to construct rDNA, and notes several additional features of the commercial use of rDNA technology.

Structure and function of DNA

Throughout the spectrum of life, the traits characteristic of a given species are maintained and passed on to future generations, preserved simply and elegantly by the information system contained within DNA. DNA can be thought of as a library that contains the complete plan for an organism. If the plan were for a human, the library would contain 3,000 volumes of 1,000 pages each. Each page would represent one gene, or a unit of heredity, and be specified by 1,000 letters. As shown in figure 3, DNA, a double-stranded, helical molecule, is composed, in part, of four nucleotide bases—adenine (A), cytosine (C), guanine (G), and thymine (T)—which are the letters of the chemical language. A gene is an ordered sequence of these letters, and each gene contains the information for the composition of a particular protein and the necessary signals for the production of that protein.

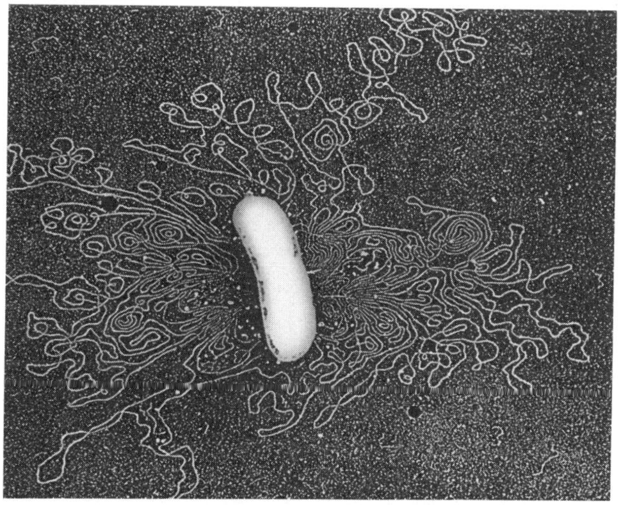

Photo credit: Science Photo Library and Porton/LH International

The DNA of the bacterium *Escherichia coli*

The mechanism by which DNA replicates is inherent in the structure of DNA itself. As can be seen from figure 3, the nucleotide bases are paired to form the rungs of the twisted DNA ladder. This pairing is absolutely specific: A always pairs with T and C always pairs with G. The pairing is accurate, but not very strong. Thus, in cell division, the DNA can "unzip" down the middle, leaving a series of unpaired bases on each chain. Each free chain can serve as a template for making a complementary chain, resulting in two identical DNA molecules, each a precise copy of the original molecule. Figure 4 illustrates the replication of DNA.

The DNA present in every cell of every living organism has the capacity to direct the functions of that cell. Gene expression, shown in figure 5, is the mechanism whereby the genetic directions in any particular cell are decoded and processed into the final functioning product, usually a protein. In the first step, called transcription, the DNA double helix is locally unzipped near the gene of interest, and an intermediate product, messenger RNA (mRNA), a single-stranded, linear sequence of nucleotide bases chemically very similar to DNA, is synthesized. The transcription process dictates the synthesis of mRNA that is complementary to the section of unzipped DNA in a manner that is somewhat similar to the replication of DNA. In the second step of gene expression, translation, the mRNA, after release from the DNA, becomes associated with the protein-synthesizing machinery of the cell, and the sequence of nucleotide bases in the mRNA is decoded and translated into a protein. The protein goes on to perform its particular function, and when the protein is no longer needed, the protein and the mRNA coding for that protein are degraded. This mechanism allows a cell to "fine tune" the quantity of its proteins while keeping its DNA in a very stable and intact form.

Proteins perform most of the necessary functions of a cell. By far the most diverse group of proteins is the enzymes, which are the proteins

Figure 3.—The Structure of DNA

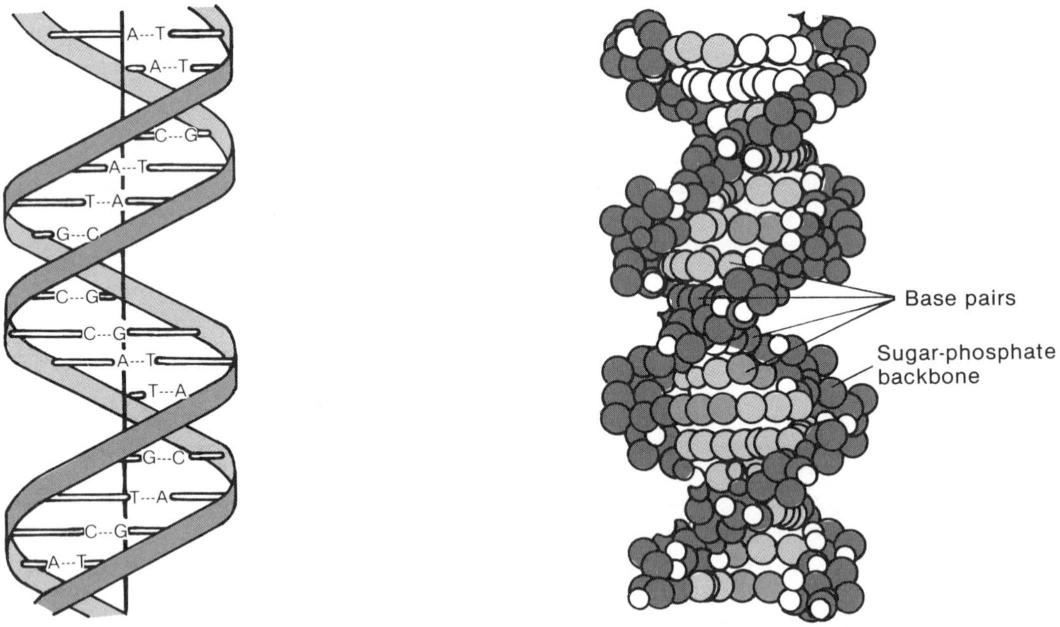

A schematic diagram of the DNA double helix. A three-dimensional representation of the DNA double helix.

The DNA molecule is a double helix composed of two chains. The sugar-phosphate backbones twist around the outside, with the paired bases on the inside serving to hold the chains together.

SOURCE: Office of Technology Assessment.

Figure 4.—The Replication of DNA

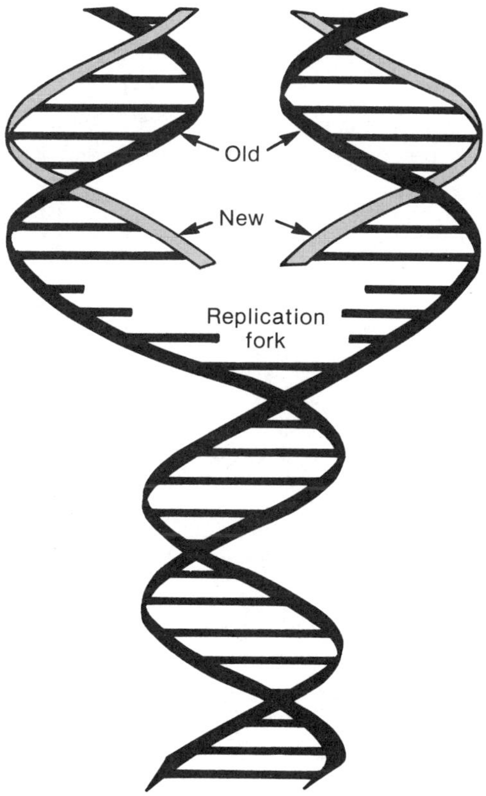

When DNA replicates, the original strands unwind and serve as templates for the building of new complementary strands. The daughter molecules are exact copies of the parent, with each having one of the parent strands

Figure 5.—Mechanism of Gene Expression

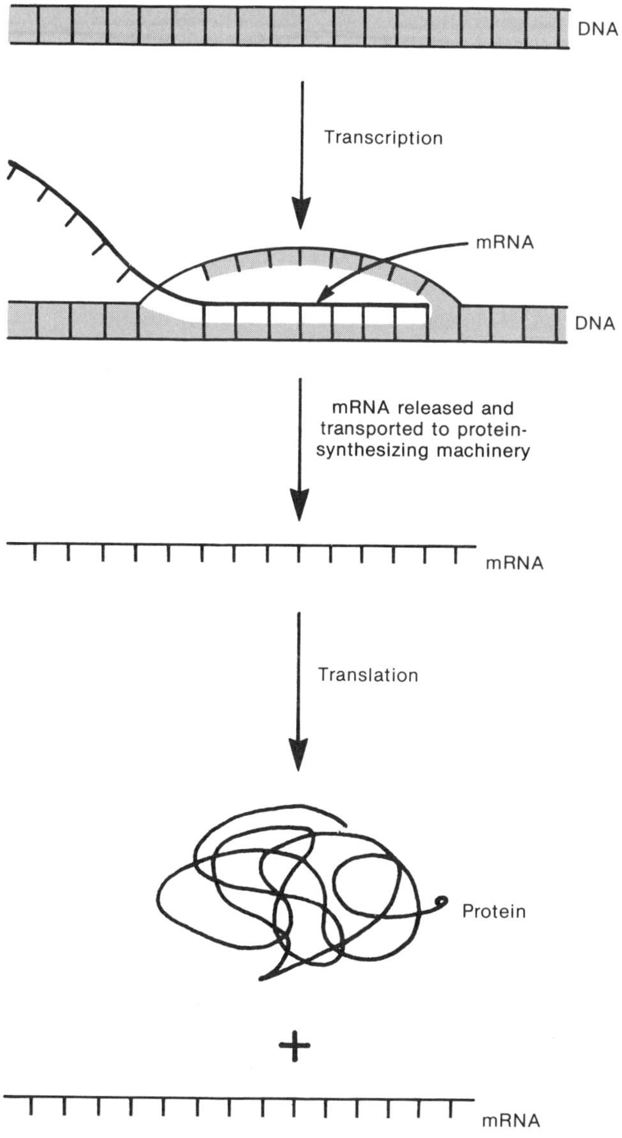

SOURCE: Office of Technology Assessment.

that catalyze biological reactions. Another group is the structural proteins, which are found, for instance, in cell membranes. Other proteins have regulatory functions; these include some hormones. Still others have highly specialized functions (hemoglobin, for example, carries oxygen from the lungs to the rest of the tissues).

The code by which genetic information is translated into proteins is the same for all organisms. Thus, because all organisms contain DNA and all organisms interpret that DNA in the same manner, all organisms, in essence, are related. It is this concept that forms the basis for the industrial use of DNA. In nearly every instance, a production process using rDNA technology depends on the expression of DNA from one species in another species. Only a universal genetic code would allow DNA to be used in this manner.

Despite the existence of a universal genetic code, regulatory signals indicating starts and stops of genes are known to vary among species. Thus, a gene removed from one organism and placed in another will code for the same protein as it did in its native system, but its synthesis needs to be induced by the proper host signal. One of the great challenges of rDNA technology is to construct DNA molecules with signals that optimally control the expression of the gene in the new host.

Preparing recombinant DNA

The amount of DNA present in each cell of a human (or most higher animals) is approximately 3 billion base pairs (2), and an average gene is about 1,000 base pairs, or about one millionth of the DNA. It is extremely difficult to study one gene in a million. Therefore, powerful tools have been developed to isolate genes of interest, place them in a foreign, simpler system, and replicate them many times to give a large amount of a single gene. The isolation of genes from higher organisms and their recombination in simple cells has already yielded a wealth of information, including insight into how genes determine the differences between different types of cells, how gene expression is regulated, and how genes may have evolved. For industrial uses, however, not only must the gene be cloned (reproduced), but that gene must also be expressed (the protein based on the gene be produced).

The basic technique of preparing rDNA is shown in figure 6. Preparations of restriction enzymes (enzymes that are made in certain bacteria and cut DNA at specific sites) are used to cut donor DNA (usually from a higher organism) into fragments, one of which contains the gene of interest. The resulting DNA fragments are then inserted into a DNA "vector," which is most often a plasmid.* Each plasmid vector will contain a different donor DNA fragment. These rDNA plasmids are introduced into host cells in a process called "transformation." Once inside the host cells, the rDNA plasmids replicate many times, thus providing many copies of each donor DNA fragment. Of the many bacteria transformed by plasmids containing donor DNA, only a few will contain the DNA fragment of interest. The desired

*A plasmid is a circular, double-stranded piece of DNA which replicates in cells apart from the chromosome.

Figure 6.—Recombinant DNA: The Technique of Recombining Genes From One Species With Those of Another

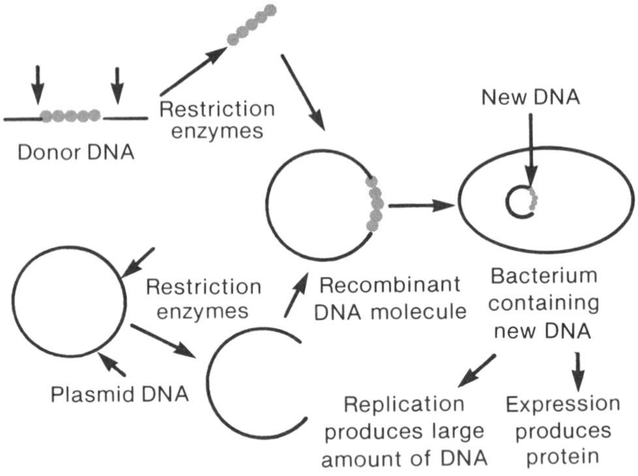

Restriction enzymes recognize certain sites along the DNA and can chemically cut the DNA at those sites. This makes it possible to remove selected genes from donor DNA molecules and insert them into plasmid DNA molecules to form the recombinant DNA. This recombinant DNA can then be cloned in its bacterial host and large amounts of a desired protein can be produced.

SOURCE: Office of Technology Assessment.

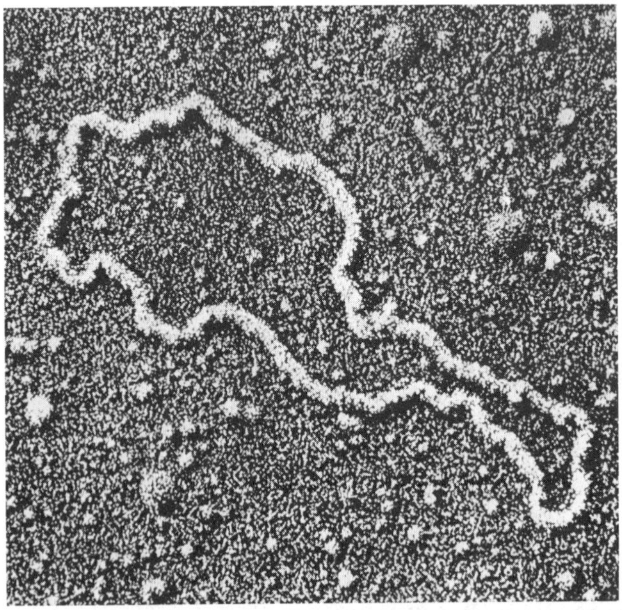

Photo credit: Science Photo Service and Porton/LH International

Bacterial plasmid

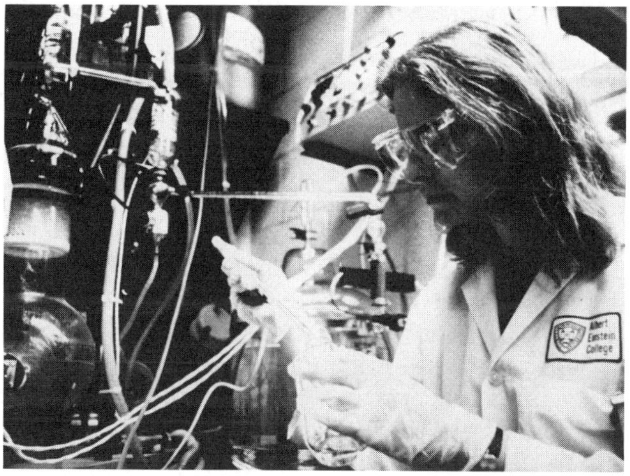

Photo credit: Science Photo Service and Porton/LH International

Molecular biologist in laboratory

gene is located among the vast number of bacteria containing plasmids with a suitable probe.* Vectors other than plasmids can be used for cloning DNA. One method uses the DNA of viruses, and another uses cosmids, artificially constructed hybrids of plasmids and viruses. Another method uses transposable elements, fragments of DNA that can insert themselves into the host cell's chromosomes.

All rDNA methods require the following:

- a suitable vector that is taken up by the host, is capable of autonomous replication, and, during the process, replicates the segment of donor DNA faithfully;
- an adequate selection system for distinguishing among cells that have, or have not, received rDNA; and
- an appropriate probe for detecting the particular DNA sequence in question.

The most difficult part of the cloning process is isolating an appropriate probe. The genes that were first cloned were those that, in certain cells, produced large quantities of relatively pure mRNA. Since the mRNA was complementary to the gene of interest, the mRNA could be used as a probe. This method severely limited the number of genes that could be cloned, however, because most genes do not produce large amounts of mRNA. More recently, a different technique has been used that allows a much greater diversity of genes to be cloned. If the amino acid sequence of a protein is known, then, working backwards through the gene expression scheme, the nucleotide base sequence can be determined. Because of the advent of automated DNA synthesizers, a portion of DNA can be synthesized that is complementary to the gene. This piece of DNA can then be used as a probe. Thus, if enough of a particular protein can be isolated and sequenced, its corresponding gene can be cloned.

At present, rDNA is grown principally in simple micro-organisms such as bacteria and yeast. Yeasts, in addition to bacteria, are being used as hosts for rDNA cloning because they more closely resemble cells of higher organisms. Yeasts perform functions similar to those of higher eukaryotic cells. These functions include adding sugar groups to some proteins. For the function of many proteins, these sugar groups are essential. Recently, scientists have learned how to introduce novel genetic material into higher plants and animals. The special techniques that pertain to cloning DNA in plants are discussed in *Chapter 6: Agriculture*.

Recombinant DNA technology in industrial processes

The commercial use of rDNA technology has several features in addition to those just discussed. In order to produce a product or improve a process, the cloned gene must be expressed to give a functional product. Since the signals that regulate gene expression vary from species to species, achieving the expression of a gene in a foreign cell may be difficult. The commercial development of biotechnology is highly dependent on the ability to achieve gene expression, for it is proteins (or their metabolites) that either are the marketable products themselves or establish the cellular environment necessary for performing such practical tasks as degrading toxic wastes or increasing the efficiency of photosynthesis. To a large extent, the problem of gene expression has been addressed through the manipulation of the adjacent vector DNA so that it contains the host's regulatory sequences. The cloned gene can

*A probe is a sequence of DNA that has the same sequence as the desired gene and has been prepared in such a way that it can be identified after it base pairs with that gene.

then be "switched on" by using host-regulated controls (1). Moreover, it is possible to alter specifically the regulatory sequences so that the gene is expressed at higher levels or so that its expression is more readily controllable in an industrial situation (3).

The purification of a protein from an industrial bioprocess is greatly simplified if the protein is secreted from the cell into the growth medium. If the protein is secreted, it does not have to be purified away from all the other cellular components. It is possible to attach additional regulatory signals to the vector DNA that direct the cell to secrete the protein and, thus, simplify its purification. The successful development of methods to enhance gene expression and product function and secretion will undoubtedly enhance the commercial applicability of rDNA technology.

The computer-aided design of proteins is another technology that will expand the use of rDNA molecules industrially. In the past, enzymes were modified by mutagenizing the host cell and then selecting or screening for mutants that contained an altered enzyme. Now, through use of the techniques of X-ray crystallography, protein sequencing, and computer modeling, the amino acid sequence and three-dimensional structure of the protein can be determined and amino acid changes that should bring about altered enzyme properties can be selected. The DNA sequence of the cloned gene for an enzyme can then be modified to incorporate the amino acid changes. Specific gene modification is made possible because of technical advances resulting in rapid and inexpensive synthesis of small DNA segments that can be used to change specific base pairs in a DNA sequence. Near-term protein modification experiments could result in enzymes with increased temperature and pH stability. Longer term experiments could define the structure of active sites of enzymes to be used for specific catalytic functions.

Monoclonal antibody technology

The production of antibodies in higher animals is one aspect of a complex series of events called the immune response. Specialized cells called B lymphocytes, present in the spleen, lymph nodes, and blood, recognize substances foreign to the body, or antigens, and respond by producing antibodies that specifically recognize and bind to those antigens. Any given B lymphocyte can recognize only one antigen. Thus, when a B lymphocyte meets and recognizes an antigen for the first time, the B lymphocyte is stimulated and becomes committed to producing a single type of antibody for the duration of its life. The end result of this aspect of an immune response is the antigen's removal from the body.

Antibodies bind to antigens and carry out their functions by virtue of the antibody's unique structure. All antibodies are comprised of four protein chains in a precise orientation, as shown in figure 7. One end of the antibody (the constant, or effector region) is nearly identical among antibodies. This effector region is associated with functions such as the secretion of antibodies from the B lymphocyte and "signaling" to the immune system after the antibody binds with the target antigen. The other end of the antibody, the variable region, contains the site that recognizes and binds to a particular antigen, and the structure of this end varies greatly from antibody to antibody to accommodate a wide range of antigens.

Apart from their natural functions in the protection of organisms via the immune response, antibodies have long been important tools for researchers and clinicians, who use an antibody's specificity to identify particular molecules or cells and to separate them from mixtures. Antibodies also have a major role in diagnosis of a wide variety of diseases. Antibodies that recognize known antigens are used to detect the presence and level of drugs, bacterial and viral products, hormones, and even other antibodies in sensitive assays of blood samples.

Figure 7.—Structure of an Antibody Molecule

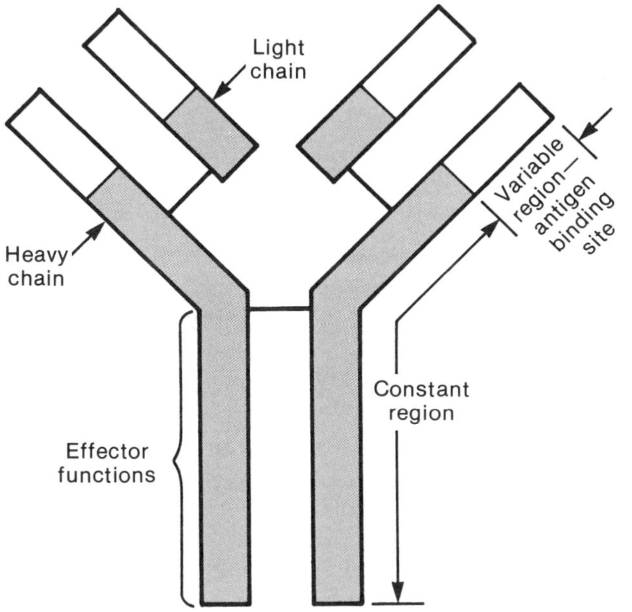

SOURCE: Office of Technology Assessment.

The conventional method of producing antibodies for diagnostic, therapeutic, and investigational purposes is to inject an antigen into a laboratory animal and, after evoking an immune response, to collect antiserum (blood serum containing antibodies) from the animal. Although this method has been and continues to be widely used, there are several problems associated with conventional antibody technology. These include:

- minor contamination of the injected antigen with other molecules, so that the antiserum collected from the animal contains a mixture of antibodies against both the target antigen and the contaminating molecules;
- heterogeneous populations of antibodies with concomitant differences in activity, affinity for the antigen, and biological functions, especially when a number of different animals are used to prepare the antiserum; and
- the limited supply of quality antisera for any given purpose (10,28,32).

Since these difficulties are almost unavoidable in standard antibody preparations, the standardization of immunoassays and the accumulation of large amounts of reference antisera have been difficult. Such problems, although time-consuming and expensive, have not prevented the effective use of antibodies as diagnostic, therapeutic, and investigational tools for both research scientists and clinicians, but the search for new methods for continual production of large amounts of pure antibodies has continued.

By what Cesar Milstein calls a "lucky circumstance," he and Georges Köhler began experimenting with the well-established technique of cell fusion in myeloma (antibody-producing tumor) cells adapted for cell culture. Milstein and Köhler fused myeloma cells with antibody-producing spleen B lymphocytes from mice that had been immunized with sheep red blood cells (SRBCs), and they found that some of the resulting hybrid cells, called hybridomas, secreted large amounts of homogeneous (monoclonal) antibodies directed against SRBCs (20,21). The myeloma parent cell conferred on the hybridoma the ability to grow permanently in cell culture and thus to support almost unlimited antibody production, while the B lymphocyte parent contributed the genes coding for the specific antibody against an SRBC antigen.

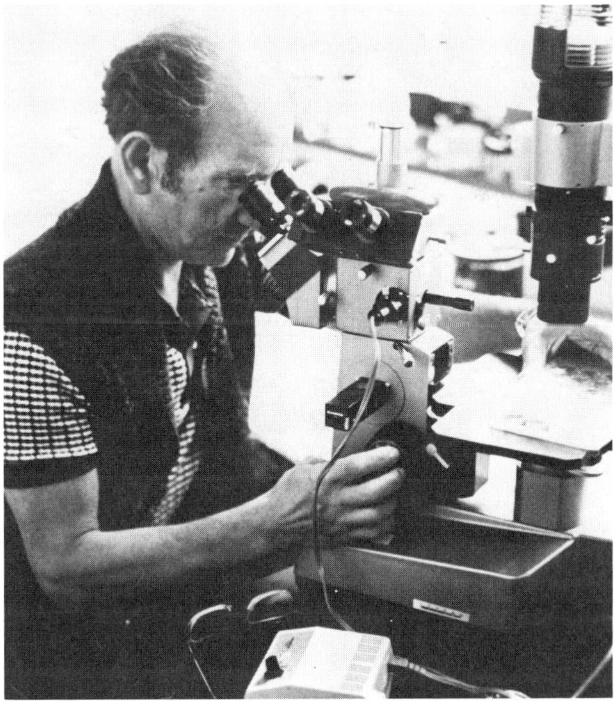

Photo credit: Science Photo Service and Porton/LH International

Dr. Cesar Milstein, discoverer of monoclonal antibodies

By using the method of hybridoma, or MAb, technology, it is now possible to "immortalize" individual antibody-producing cells by fusion with tissue culture-adapted myeloma tumor cells in the laboratory (4,5,8,13,22,25).

Preparing monoclonal antibodies

The method used to prepare MAbs is summarized in figure 8. The purified antigen of choice is injected into a mouse, and a few weeks later, the spleen of the mouse is removed. The B lymphocytes (antibody-producing cells) are isolated from the spleen and fused with myeloma cells. The resulting cells are placed in a cell culture medium that allows only the hybridomas to grow. The many hybridomas that result are cloned, and each clone is tested for the production of the antibody desired. A particular hybridoma clone either may be established in an in vitro culture system or may be injected into mice, where the hybridoma grows in abdominal cavity fluid (ascites) from which the antibodies are readily collected.

This method allows the preparation of large quantities of highly specific MAbs against almost any available antigen. The antibodies produced by MAb technology are homogeneous, and their production is predictable and repeatable, as compared to polyclonal antibodies produced with conventional immunological methods.

Preparation of Monoclonal Antibodies

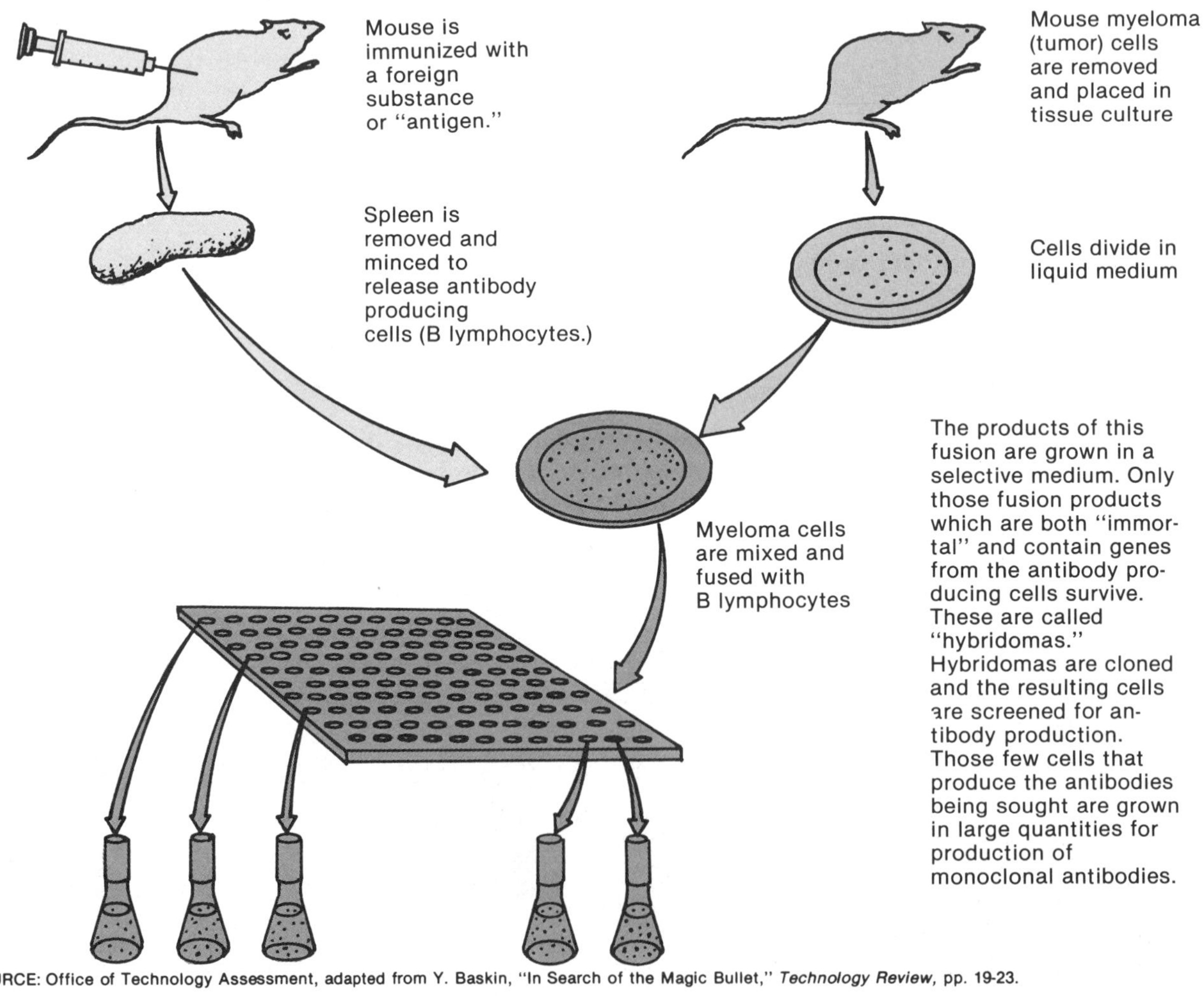

SOURCE: Office of Technology Assessment, adapted from Y. Baskin, "In Search of the Magic Bullet," *Technology Review*, pp. 19-23.

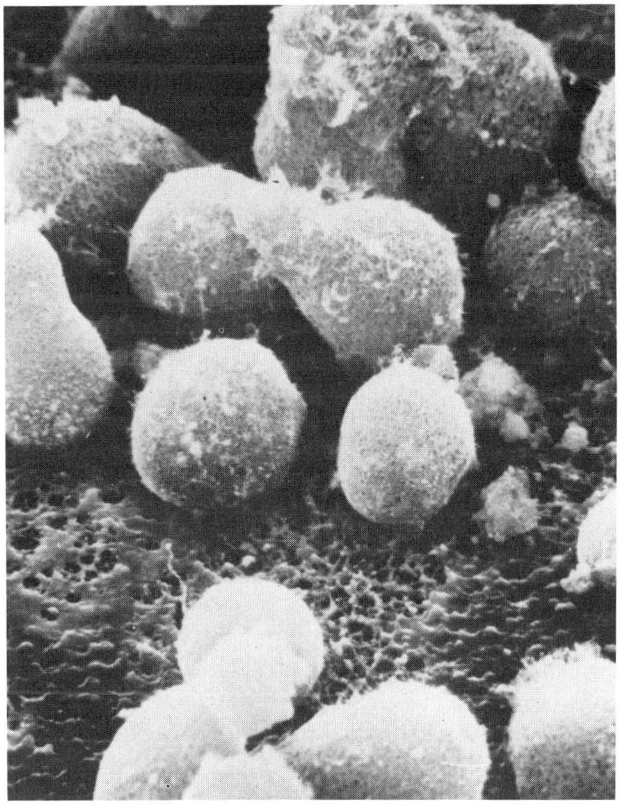

Photo credit: Science Photo Service and Porton/LH International

Scanning electron micrograph of human hybridoma cells

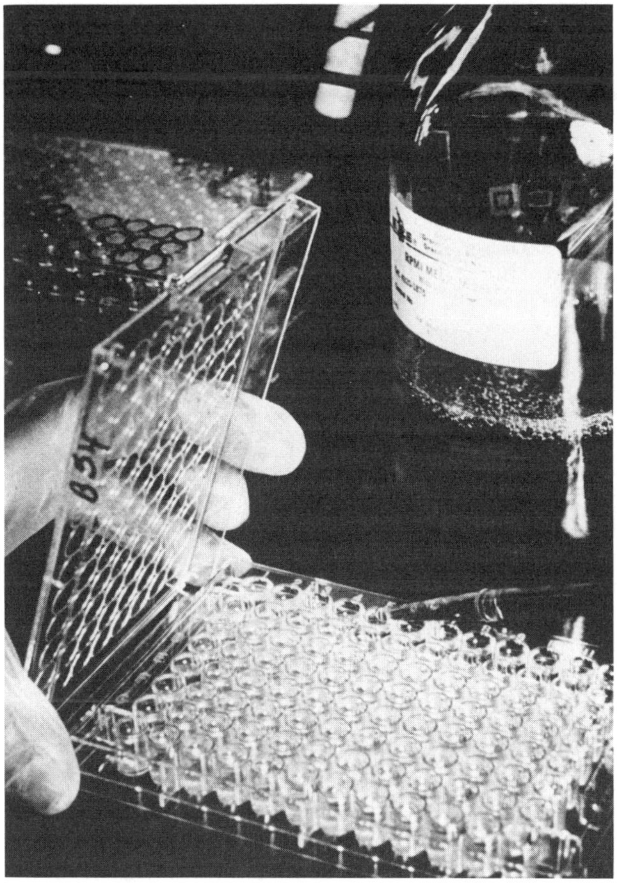

Photo credit: Science Photo Library and Porton/LH International

One step in the isolation of hybridomas

Despite the great promise of MAbs, there are several persistent technical problems to be considered:

- obtaining MAbs against certain weak antigens (antigens that do not produce a large immune response) remains difficult (11,24);
- homogeneous antibodies cannot perform some functions such as forming a precipitate with other antigen-antibody complexes, a necessary function for some diagnostic assays;
- low frequency of fusion is a continuing problem in the preparation of hybridomas, as is the stability of the hybridomas and antibodies (14); and
- some MAbs are sensitive to small changes in pH, temperature, freezing and thawing, and can be inactivated during purification.

Many of these problems are being alleviated or solved as research with MAbs progresses.

Another problem being addressed is the development of hybridomas for specific species. Some suitable myeloma cell lines exist for mice, rats, and humans (12,20,27), but a wider variety of human cell lines and cell lines for other species are needed if wider applications of MAb technology are to be made. Hybridomas are often made with cells from two different species, but these fusions regularly result in the preferential loss of the spleen B lymphocyte chromosomes, resulting in an absence of antibody production (24). For therapeutic applications, it is desirable to treat people with human antibodies to avoid allergic reactions and other problems of antibody cross-reactivity. Thus, MAbs from a human myeloma/human spleen cell fusion are needed. Several investigators have reported the development of human myelomas that are suitable for

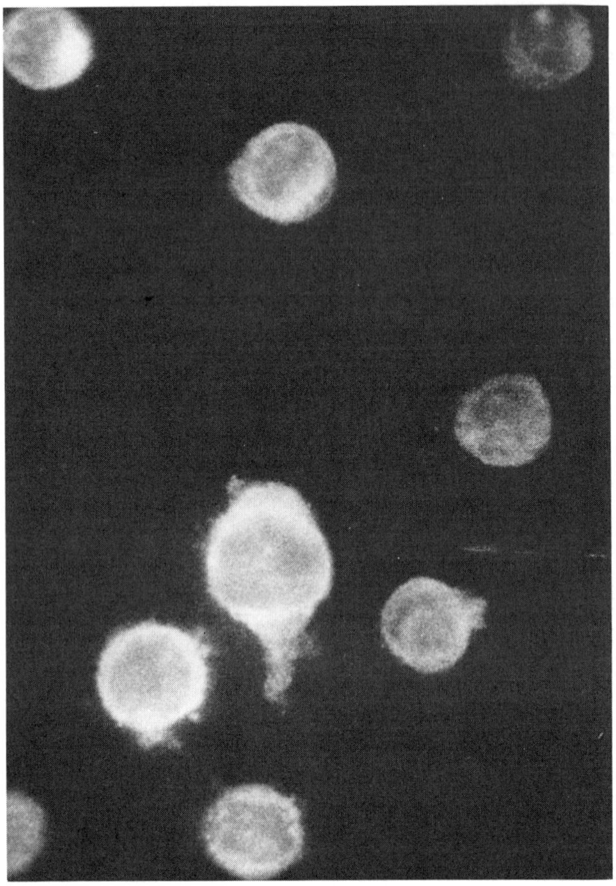

Photo credit: Science Photo Library and Porton/LH International

In vitro identification of specific cells using fluorescently labeled monoclonal antibodies

hybridoma preparation (9,17,23,27). Successful fusions apparently result from using these cell lines (6).

Monoclonal antibodies and recombinant DNA technology

The combination of MAb technology and rDNA technology offers intriguing possibilities for further technological exploration. Recombinant DNA techniques could be used to produce portions of antibody molecules in bacteria to circumvent some of the problems (e.g., hybridoma instability) associated with MAb production in mice or tissue culture. Additionally, these MAbs would be free of impurities, such as viruses, found in animal cells and possibly could be produced in large amounts with a concurrent savings in cost.

The first cloning and expression of a complete antibody molecule in a bacterial system was announced recently by the U.S. firm Genentech and the City of Hope Medical Center and Research Institute (19). The protein chains were expressed separately in bacteria and reconstituted by the researchers. The pharmaceutical applications of bacterially produced antibody genes will be limited. Antibody molecules must be modified by the cell to function in most diagnostic and therapeutic applications. Bacteria do not perform the modifications necessary for proper function. However, it may be possible to clone the antibody genes in a cellular system such as yeast where the proper modifications can be made.

The production of MAbs in rDNA systems may prove useful for making reagents used in industrial applications where only the antigen-binding function may be necessary. With genes cloned for the antigen-binding regions of the antibody, portions of MAbs may be produced more economically in bacterial rDNA systems than in a large-scale mouse ascites or cell culture protocols.

Large-scale production of monoclonal antibodies

Although MAbs can be produced by several methods, manufacturers primarily use mouse ascites to produce the modest amounts of MAbs needed to service current diagnostic and research markets. As applications for MAbs to human therapy are developed, the need for larger quantities of MAbs (free from mouse-derived contaminants that might cause allergic reactions) may encourage a switch to the use of large-scale cell culture to produce MAbs. If MAbs are to be used in industrial applications (e.g., in the purification of proteins), production methods will be needed to produce even larger quantities of antibodies. In these cases, efficient cell culture or microbial bioprocess techniques will probably be necessary to provide enough antibodies to fill these needs.

Improved, more controllable cell culture systems will be needed for the production of MAbs in the future. A crucial need for large-scale cell culture is either the isolation of hybridoma cell lines that attach to surfaces or the use of techniques for immobilizing cells on a solid matrix.

Immobilized cells could be grown in large quantities in culture; the MAbs secreted from the cells could then be routinely collected from the medium. Immobilized cell methods may prove valuable for large-scale MAb production. Such methods are already used industrially, for example, in growing cells that produce polio virus for subsequent vaccine production (26,31).

Damon Biotech Corp. (U.S.) has recently introduced the technique of microencapsulation to MAb technology (7). This method uses a porous carbohydrate capsule to surround the hybridoma cells and to retain the antibodies while allowing the circulation of nutrients and metabolic wastes. After several days in culture, the encapsulated colonies are harvested and washed to remove the growth medium, the capsules are opened, and the antibodies are separated from the cells. According to Damon Biotech, 40 to 50 percent (by weight) of the harvested medium is made up of MAbs. The company claims the microencapsulation method for producing MAbs is significantly less expensive than the ascites method, provides a high concentration of antibodies, and does not require the maintenance of animals (18).

Industrial uses for monoclonal antibodies

Because of their unique properties of homogeneity, specificity, and affinity, MAbs can be used effectively in downstream purification systems for molecules, especially proteins. A MAb purification system relies on the binding of a target molecule to a MAb immobilized on a solid support such as a bead. The beads are packed in a column, and a mixture containing the target molecule is passed through the column. The MAb binds the molecule while the impurities wash through the column. Then the binding is reversed, and the target molecule is released and collected from the column.

Before MAb-based purification systems can be used in large-scale, several practical and technical factors must be optimized. These include cost, purification of the antibody itself, and elution of the desired product after purification by the antibody. Elution requires the use of an antibody of somewhat lower affinity than one would use for diagnostic or therapeutic applications so that the binding can be reversed easily.

Various important proteins, including alphafetoprotein and leukocyte interferon, are now purified using MAbs (29,30). MAb purification systems may be used in the future to purify a vast number of compounds, particularly substances present in small amounts.

A simple extension of the procedure just described involves using MAbs to bind unique surface proteins and, with them, the cells to which they are attached. This permits separation of cells with surface proteins of interest and is carried out by passing the cells over a suitable matrix to which the antibodies have been bound. In another procedure, fluorescence-activated cell sorting, cells are mixed with fluorescently labeled MAbs, and the mixture is passed through a special instrument called a flow cytometer, which responds to the fluorescent marker and sorts the cells into labeled populations at rates of 50,000 cells per minute (15,16). So far, fluorescence-activated cell sorting has been used mostly for research purposes, but as the method is improved, it may be employed in a range of clinical applications.

Conclusion

Many fields of biological research are being affected by MAb technology. Researchers now use MAbs to study problems in endocrinology, biochemistry, cell biology, physiology, parasitology, and many other fields, because the products of MAb technology are easily standardized and reproduced. Furthermore, many diagnostic, therapeutic, and industrial uses for MAbs are becoming apparent, and, as outlined in subsequent chapters of this report, several U.S. and foreign firms are developing these applications. Industrial purification applications of MAbs and the widespread advantages of MAb technology in preparing pure and easily standardized antibodies offer substantial benefits in industrial, research, and clinical laboratories. Recombinant DNA and MAb technologies can complement each other, because rDNA technology can lead to the production of new compounds, and MAbs can aid in their identification and purification.

Bioprocess technology*

Bioprocesses are systems in which complete living cells or their components (e.g., enzymes, chloroplasts, etc.) are used to effect desired physical or chemical changes.** Since the dawn of civilization, bioprocesses have been used to produce alcoholic beverages and fermented foods. Until the 19th century, alcoholic fermentation and baker's yeast production were carried out in the home or as local cottage industries. As industrialization occurred, both these bioprocesses moved into factories.

Although other minor products made with bioprocesses were added over the years, bioprocesses did not become significant in the overall spectrum of chemical technology in the United States until the introduction of commercial acetone and butanol production during and after World War I. Somewhat later, large-scale microbial production of citric acid was introduced, and by the beginning of World War II, the U.S. bioprocess industry was thriving, with solvent alcohols and related low molecular weight compounds comprising the bulk of bioprocess production. The rapid growth of the petrochemical industry during World War II caused the displacement of microbial production of industrial solvents, however, and by 1950, microbial production of such solvents (including nonbeverage alcohol) had virtually disappeared in the United States.

This contraction of bioprocess manufacturing might have been the death-knell for old biotechnology had it not been for the introduction of, and the proliferation of markets for, antibiotics during the 1940's. The unique qualifications of biological processes for the synthesis of complex molecules such as antibiotics rapidly became apparent. Microbial production of a number of vitamins and enzymes was initiated at about this time, although only on a small scale. Thus, in the decade from 1940 to 1950, there occurred a complete transformation of industrial bioprocesses. Production of high-volume, low-value-added industrial chemicals (e.g., acetone, butanol) by anaerobic processes employing primarily yeasts and bacteria was largely replaced by more modest-scale production of high-value-added products (e.g., pharmaceuticals, vitamins, enzymes) made by highly aerobic processes in a variety of less familiar bacteria (e.g., the actinomycetes) and some fungi (see table 1). These aerobic processes are generally quite vulnerable to contamination by other micro-organisms and require much closer control of process conditions. Such aerobic processes continue to be used in industry today.

The advent of new biotechnology has sparked renewed interest in the industrial use of bioprocesses. The discussion that follows examines the dependence of new biotechnology, including rDNA and MAb technology, upon bioprocess technologies. Two aspects of the interrelationship between new genetic technologies and bioprocess technologies are emphasized:

- the engineering problems unique to genetically modified organisms, and
- the ways in which genetically modified organisms or parts of organisms may be used to enhance the efficiency and usefulness of bioprocesses.

In order to be viable in any specific industrial context, bioprocesses must offer advantages over

Table 1.—Volume and Value of Biotechnology Products

Category	Examples
High volume, low value . . .	Methane, ethanol, animal feed, waste treatment
High volume, intermediate value	Amino and organic acids, food products, polymers
Low volume, high value . . .	Pharmaceuticals, enzymes, vitamins

SOURCE: Office of Technology Assessment, adapted from A. T. Bull, G. Holt, and M. D. Lilly, *Biotechnology: International Trends and Perspectives* (Paris: Organisation for Economic Co-Operation and Development, 1982).

*This section is based largely on a contract report prepared for the Office of Technology Assessment by Elmer Gaden, University of Virginia. The information in that report was extensively reviewed and added to by James Bailey, California Institute of Technology; Harvey Blanch, University of California, Berkeley; and Charles Cooney, Massachusetts Institute of Technology.

**The term bioprocess is used here in preference to the more familiar term "fermentation" because it more correctly identifies the broad range of techniques discussed. A fermentation process, though often used to denote any bioprocess, strictly speaking refers only to an anaerobic bioprocess.

competing methods of production. In most cases, bioprocesses will be used industrially because they are the only practical way in which a desired product can be formed. Biological processes may be desirable:

- in the formation of complex molecular structures such as antibiotics and proteins where there is no practical alternative,
- in the exclusive production of one specific form of an isomeric compound,
- because micro-organisms may efficiently execute many sequential reactions, and
- because bioconversions may give high yields.

Examples of the categories of current uses of bioprocesses are the following:

- production of cell matter ("biomass" itself) (e.g., baker's yeast, single-cell protein);
- production of cell components (e.g., enzymes, nucleic acids);
- production of metabolites (chemical products of metabolic activity), including both primary metabolites (e.g., ethanol, lactic acid) and secondary metabolites (e.g., antibiotics);
- catalysis of specific, single-substrate conversions (e.g., glucose to fructose, penicillin to 6-aminopenicillanic acid); and
- catalysis of multiple-substrate conversions (e.g., biological waste treatment).

Bioprocesses may offer the following advantages over conventional chemical processes:

- milder reaction conditions (temperature, pressure, and pH);
- use of renewable (biomass) resources as raw materials for organic chemical manufacture, providing both the carbon skeletons and the energy required for synthesis;
- less hazardous operation and reduced environmental impact;
- greater specificity of catalytic reaction;
- less expensive or more readily available raw materials;
- less complex manufacturing facilities, requiring smaller capital investments;
- improved process efficiencies (e.g., higher yields, reduced energy consumption); and
- the use of rDNA technology to develop new processes.

Some of the conceivable disadvantages of bioprocesses, on the other hand, are the following:

- the generation of complex product mixtures requiring extensive separation and purification, especially when using complex substrates as raw materials (e.g., lignocellulose);
- problems arising from the relatively dilute aqueous environments in which bioprocesses function (e.g., the problem of low reactant concentrations and, hence, low reaction rates;* the need to provide and handle large volumes of process water and to dispose of equivalent volumes of high biological oxygen demand wastes; complex and frequently energy intensive recovery methods for removing small amounts of products from large volumes of water);
- the susceptibility of most bioprocess systems to contamination by foreign organisms, and, in some cases, the need to contain the primary organism so as not to contaminate the surroundings;
- an inherent variability of biological processes due to such factors as genetic instability and raw material variability; and
- for rDNA systems, the need to contain the organisms and sterilize the waste streams, an energy-intensive process.

Solutions to some of these problems through the use of biotechnology may make bioprocesses more competitive with conventional chemical syntheses. Genetic intervention may be used in some instances to modify micro-organisms so that they produce larger amounts of a product, grow in more concentrated media, have enzymes with increased specific activity, or grow at higher temperatures to help prevent contamination. Recombinant DNA technology may lead to the development of completely new products or modification of important existing ones. In the past, some potentially useful bioprocesses have

*It is often said that biochemical catalysis is many times more effective than conventional chemical catalysis. This contention is based on the very high specific activities observed for individual enzymes in vitro. Such rates are seldom encountered under large-scale conditions. In general, bioprocesses are extremely slow in comparison with conventional chemical processes.

not been economical. Now, however, a combination of improved engineering design and procedures and rDNA technology may yield bioprocesses that are more efficient than they have been in the past and therefore more competitive.

*Bioprocess essentials**

The steps in bioprocessing are presented schematically in figure 9. The substrate and nutrients are prepared in a sterile medium and are put into the process system with some form of biocatalyst—free or immobilized cells or enzymes. Under controlled conditions, the substrate is converted to the product and, when the desired degree of conversion has been achieved, byproducts and wastes are separated.

Water is the dominant component of the medium for virtually all current bioprocesses. Even when micro-organisms are grown on solid materials, an unusual processing mode, the substrate must be dampened in order to permit microbial growth and enzyme action. Products must usually be purified from dilute, aqueous solutions.

*The bioprocesses discussed here exclude uncontrolled environmental applications.

Figure 9.— Steps in Bioprocessing

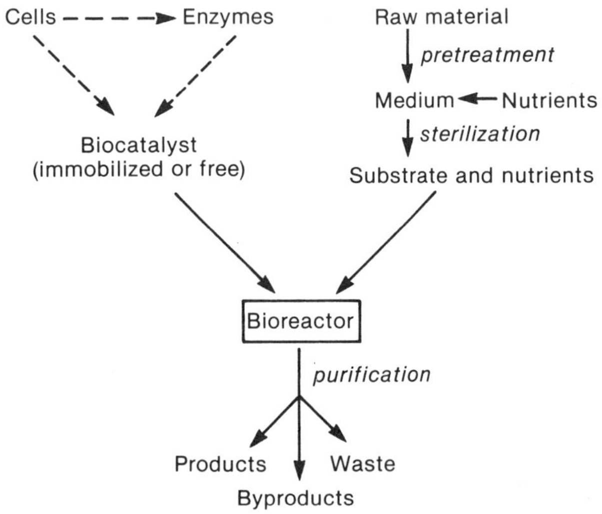

SOURCE: Office of Technology Assessment.

Bioprocesses require a closely controlled environment, and this necessity markedly influences their design. Biocatalysts generally exhibit great sensitivity to changes in temperature, pH, and even concentrations of certain nutrients or metal ions. The success of a bioprocess depends on the extent to which these factors are controlled in the medium where interaction between biocatalyst and substrate takes place.

SUPPLY OF NUTRIENTS

In addition to establishing a suitable environment, the medium must provide for the nutritional needs of living cells. A primary requirement is a source of carbon. In addition to supplying the energy needed for metabolism and protein synthesis, carbon sources contribute structural elements required for the formation of complex compounds. Often, the carbon source may itself be the substrate for the catalyzed reaction, as in the fermentation of sugar to ethanol. Sugars, starches, and triglycerides, and, to a lesser extent, petroleum fractions, serve as carbon sources.

Other important nutrients required by living cells are nitrogen, phosphorus, and sometimes oxygen. Nitrogen and phosphorus are incorporated into structural and functional molecules of a cell and may also become part of product molecules. Most of the micro-organisms currently used by industry are highly aerobic and require an adequate supply of oxygen, but others are strictly anaerobic and must be protected from oxygen. A number of other nutrients, such as vitamins and metal ions, though required only in very small amounts, are nevertheless essential. Some of these nutrients, especially metals, may appear in the product.

In order to make the substrate and nutrients accessible to the biocatalyst, the medium must be thoroughly mixed. Most bacteria and some yeasts used in bioprocesses commonly grow as individual cells or as aggregates of a few cells suspended in the medium, whereas fungi and actinomycetes grow in long strands. As they grow, all these types of cells increase the viscosity of the fluid in which they are growing in a batch process, making the fluid more difficult to mix, and thus more difficult for nutrients to reach them.

Since most of the micro-organisms currently used by industry perform their conversions aerobically, they demand a constant supply of oxygen. Oxygen's low solubility in water represents a significant stumbling block to efficient bioprocessing. Since oxygen is depleted during conversion, the medium must be constantly aerated; the more viscous the medium, however, the more difficult it becomes to supply oxygen. Approaches to maintaining an adequate oxygen supply include:

- increasing reactor pressure to increase oxygen solubility,
- the use of oxygen-rich gas for aeration, and
- changes in process design and operation.

PURE CULTURES AND STERILIZATION

Most of the products of bioprocesses are formed through the action of a single biocatalyst, either a micro-organism or an enzyme.* If foreign organisms contaminate the process system, they may disrupt its operation in a variety of ways. They can directly inhibit or interfere with the biocatalyst, whether it is a single enzyme or a complete cell, and they may even destroy the biocatalyst completely. Alternatively, contaminating organisms may leave the catalyst unaffected, but modify or destroy the product. Foreign organisms can also generate undesirable substances that are difficult to separate from the primary product. In the manufacture of pharmaceutical products, the risk of toxic impurities is of particular concern.

To avoid or minimize contamination, most current bioprocess technologies employ pure culture techniques. The medium and its container are sterilized, and a pure culture consisting of a population of a particular species is introduced. In order to avoid subsequent contamination, all materials entering the system, including the large amounts of air required for aerobic processes, are sterilized. The apparatus must be designed and operated so that opportunities for invasion by unwanted organisms are minimized.

*A significant exception to this generalization is the broad group of biological waste treatment processes. These processes use mixed and varied populations of micro-organisms developed naturally and adapted to the waste stream being treated.

Processing modes

Bioprocesses may, in principle, use any of the operating modes employed by conventional chemical technology. These modes range from batch processing to continuous steady-state processing.

In batch processing, the reaction vessel is filled with the medium containing the substrate and nutrients, the medium is sterilized, the biocatalyst is added, and conversion takes place over a period ranging from a few hours to several days. During this period, nutrients, substrates, agents for pH control, and air are supplied to, and product gases are removed from, the reaction vessel. When conversion is complete, the reaction vessel is emptied, and the purification process begins. Turnover time between batches can account for a significant portion of total processing time.

In continuous steady-state processing, which lies at the other end of the operational spectrum from batch processing, raw materials are supplied to, and spent medium and product are withdrawn from, the reaction vessel continuously and at volumetrically equal rates. Potential advantages offered by continuous processing over batch processing include significantly higher productivity, greater ease of product recovery due to the lack of contaminating biocatalyst, and lower cost due to reuse of biocatalyst.

The simplest approach to the implementation of a continuous processing system is to modify a batch reactor so that fresh substrate and nutrients can continually be added while a product stream is removed. This simple arrangement has one serious drawback: the biocatalyst leaves the reactor continuously with the outlet stream and must be separated from the product. Several techniques, all of which involve fixing the biocatalyst in some manner, have been developed to avoid the biocatalyst's escape with the reaction mixture and allow its repeated use. The development of techniques for the immobilization of biocatalysts has greatly expanded the possibilities for continuous bioprocesses. Although still not widely employed for large-scale bioprocesses, the biocatalyst immobilization techniques now available offer a diversity of new opportunities for more effective bioprocessing (see *Box A.—Continuous Bioprocessing*).

Box A.—Continuous Bioprocessing

Biocatalyst Immobilization Techniques

In recent years, several techniques for immobilizing both enzymes and whole cells have been devised (see fig. BXA-1 and table BXA-1). The oldest technique involves simple *adsorption*. Enzymes or whole cells adhere loosely (without chemical bonding) to the surface of a support material such as alumina, charcoal, clay, or cellulose. Eventually such adsorbed agents will wash away, but some last a surprisingly long time.

For isolated enzymes, *chemical bonding* between the enzyme molecule and the surface of the support material achieves a much firmer hold and results in a stable biocatalyst preparation capable of extended service. Binding supports include cellulose, glass, and various synthetic polymers. In some cases, fixation of the enzyme seems to interfere somewhat with its catalytic activity. Other reports claim increased activity after immobilization.

Entrapment in a polymer matrix represents a third method of immobilization, applicable to both enzymes and cells. Starch, silica gel, collagen, and a variety of synthetic polymers readily form gels containing microscopic cages through which water can pass, but which trap enzyme molecules or microbial cells. This technique reduces the rate of reaction, because substrate, nutrient, and product molecules must all diffuse through the solid matrix structure. A compensating advantage, however, is that living cells can be firmly retained without damage. At least one Japanese pharmaceutical and fine chemical manufacturer (Tanabe Seiyaku) has successfully employed this method for the commercial production of a number of specialty chemical products.

Finally, in the technique called *microencapsulation,* enzymes or cells are enclosed in a spherical polymer membrane. The resulting capsules range in diameter from 5 to 300 micrometers and look much like enlarged cells. The membrane is semipermeable; the relatively small substrate and product molecules pass readily through the encapsulating membrane, but the larger enzyme molecules and the still larger structure of a complete cell cannot escape.

Continuous Bioprocessing Modes

Both packed- and fluidized-bed reactors, similar to those widely employed in conventional chemical technology, can be employed in continuous bioprocesses using immobilized catalysts (see fig. BXA-2). Continuous packed-bed bioreactors have, in fact, long been used for making vinegar. Diluted wine or fermented cider percolates through a bed containing wood shavings or some similar material. The microorganisms that convert ethanol to acetic acid form a slimy film on the surface of the solid support and so remain in the reactor. The treatment of sewage and other wastes is similar. The waste stream trickles through a bed of stone, ceramic, or plastic on which a microbial film has grown. Waste particles are adsorbed and converted to nonpolluting compounds. More sophisticated packed-bed reactors are beginning to be employed to perform a variety of conversions.

The development of better methods of immobilization has allowed the introduction of fluidized-bed reactors. Biocatalyst particles are suspended in an upward flowing stream of medium, with the flow rate adjusted to maintain suspension while preventing the loss of excessive amounts of catalyst from the reactor. Though a variety of designs have been proposed and several have been carried to the pilot scale, no industrial application of this approach to bioprocesses has been realized as yet.

Technical Limitations on Continuous Bioprocessing

The technical factors that limit the adoption of continuous as opposed to batch processing fall into the following categories:

- *Kinetic.* In principle, it is seldom possible to establish optimal conditions for economic secondary metabolite production (e.g., antibiotics) in a single-stage process. Separate growth and product formation stages require different process conditions.

Figure BXA-1.—Techniques for Immobilizing Enzymes and Whole Cells

Enzymes can be immobilized by adsorption or chemical bonding (a), by entrapment in a polymer matrix (b), or by microencapsulation (c).

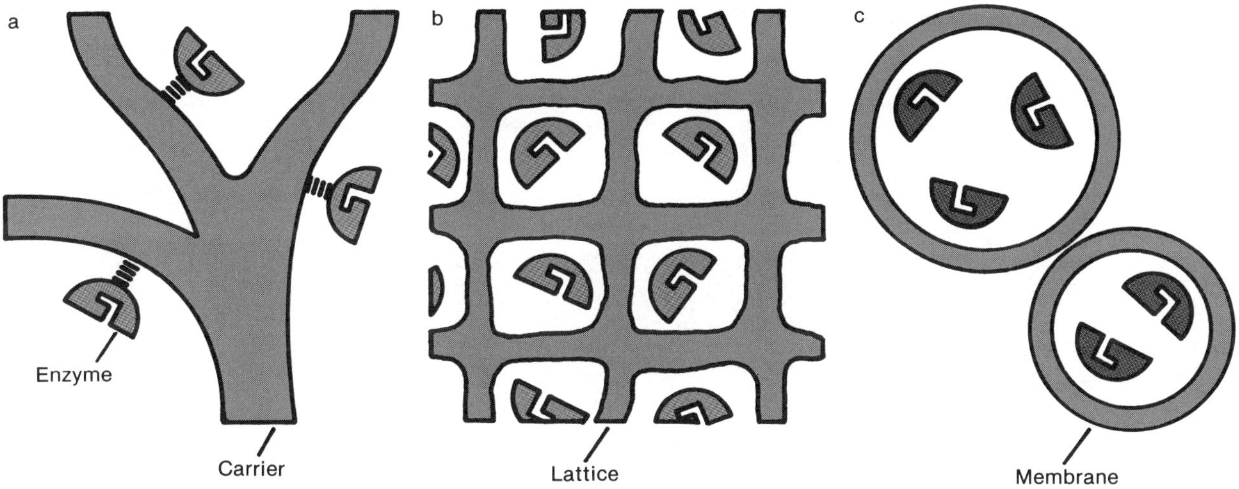

SOURCE: Adapted from E. Gaden, "Production Methods in Industrial Microbiology," *Scientific American*, September 1981, p. 182.

Table BXA-1.—Characteristics of Immobilization Methods for Enzymes and Cells

	Immobilization method		
Characteristic	Physical adsorption	Chemical bonding	Entrapment; encapsulation
Preparation	Easy	Difficult	Moderate
Activity	Low	High	High
Specificity	Unchanged	Changeable	Unchanged
Binding force (retention)	Weak	Strong	Strong
Regeneration	Possible	Impossible	Impossible
Cost	Low	High	Low

SOURCE: Gaden, personal communication, 1983.

- *Biological.* Biocatalyst stability may be difficult to maintain for long periods of continuous operation. The phenomenon of "culture degeneration," reported in many instances, deserves careful study. The results of such studies will surely be case-specific and may simply reflect inadequacies in the knowledge of nutrient requirements necessary to sustain long-term productivity. As the use of rDNA organisms grows, this matter will require close attention because of concerns over the stability of these types of organisms.
- *Operational.* The primary technical factors acting to limit continuous bioprocessing in the past have been difficulties in maintaining sterile conditions and in handling biocatalytic suspensions, especially those of filamentous fungi or large cell clumps. The perplexing contamination problem has focused improvement efforts on the deficiences of equipment (mainly pumps) for moving liquids and slurries and on valves and transfer lines. Many specific difficulties have already been overcome in connection with batch operations, and improved equipment design and more rigorous operating procedures may result in successful continuous processes.

Figure BXA-2.—Packed-Bed and Fluidized-Bed Reactors

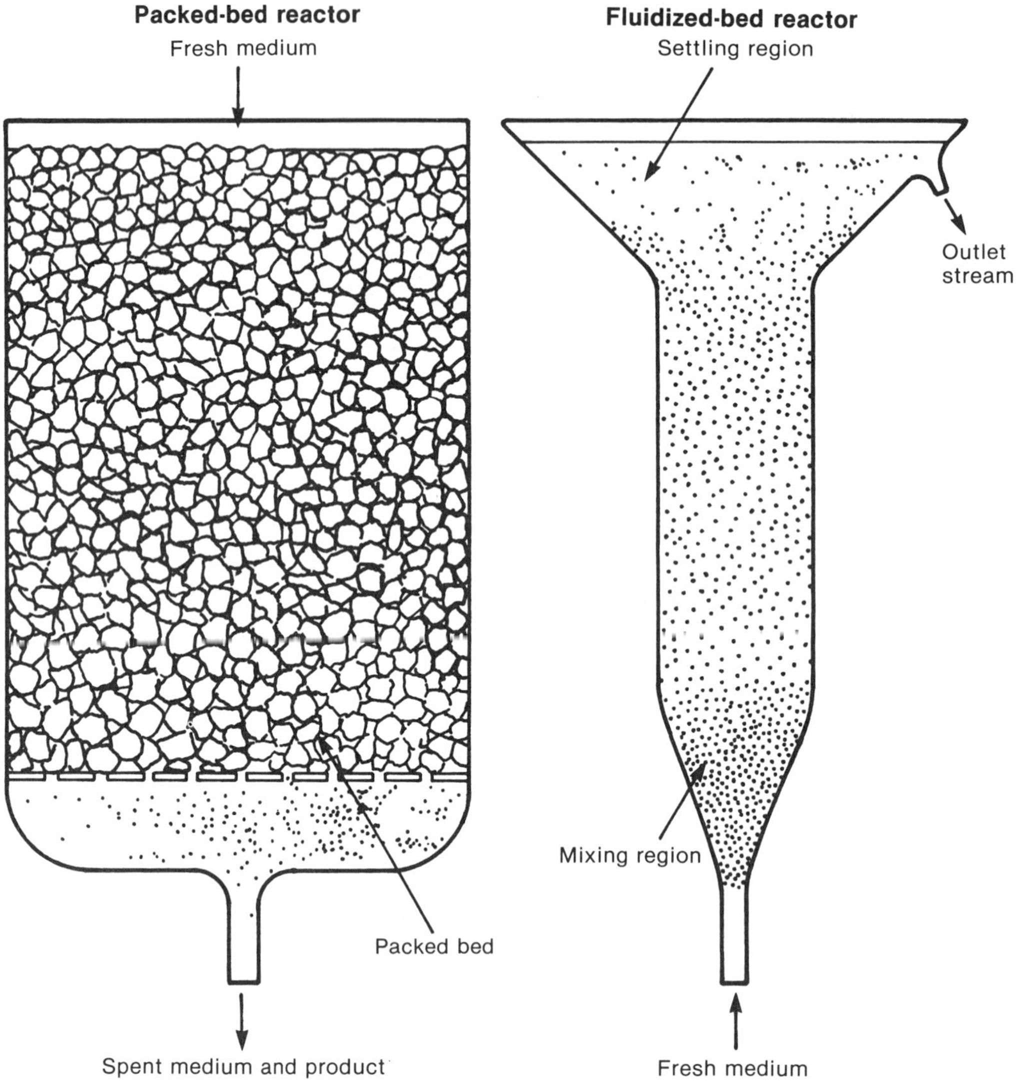

SOURCE: Adapted from E. Gaden, "Production Methods in Industrial Microbiology," *Scientific American*, September 1981, p. 196.

Batch operation currently dominates specialty chemical and pharmaceutical bioprocesses and is likely to continue to do so in the near future. In addition to technical limitations on continuous processing (see box A), other considerations have led manufacturers to choose the batch mode. Batch processing is often used, for example, because it offers the operational flexibility needed when a large number of products are manufactured, each at fairly low production levels; each process unit, more or less standard in design, can easily and rapidly be switched from one product line to another. Furthermore, a switch from batch to continuous processing is expensive, and, if a company has unused batch equipment, it may find that a switch to continuous processing is not economically warranted in the near term.

Increased use of genetically manipulated biocatalysts could affect the design and operation of bioconversion units. Harvey Blanch points out (33):

> . . . one of the difficulties which arises from the insertion of foreign DNA into the organism is re-

version. This can be minimized by placing the cell in an environment in which cellular replication is minimized, while cellular activity, such as the production of enzymes and products, is maintained at high levels.

Achieving the dual objective of minimal growth and maximum conversion activity requires restrictive nutrient supplies and high cell densities. Immobilized biocatalysts could be used to achieve these objectives.

Bioprocesses, unlike petroleum refining or petrochemical operations which completely convert raw materials to products or consume them as process fuels, regularly produce large amounts of waste, mainly cell matter and residual nutrients. Bioprocesses also require large volumes of clean water, discharge equivalent amounts of dilute, high biological oxygen demand wastes, and produce products in low concentrations. One solution to problems associated with bioprocessing might be the use of cleaner, more defined media, which produce fewer byproducts. Another solution might be the use of more concentrated media. The latter option is normally considered in bioprocess development, but the micro-organisms now in use are limited in their tolerance for high nutrient concentrations. Genetic manipulation may provide micro-organisms that are less sensitive to increased product concentration.

Raw materials

Current bioprocess technology uses an extremely limited range of raw materials. Just a few agricultural commodities—starch, molasses, and vegetable oil—are employed as raw materials in many of the existing industrial bioprocesses. Industry chooses these feedstocks for several reasons. There are established markets for these materials and, for primary products like starch, reasonably defined quality standards and assay procedures. Several competing suppliers guarantee uniform quality and fairly stable prices. Bioprocess applications constitute only a relatively small fraction of the market for agricultural commodities. The need for raw materials for bioprocesses, however, could become a major factor in commodity grain markets if bioprocesses find a place in large-scale fuel or chemical production.

Less important raw materials are some byproducts of agricultural and food processing, such as "corn steep liquor" and "distillers solubles." Petroleum hydrocarbons are little used because of their high cost. The potential for relatively pure cellulose (e.g., delignified wood) remains unrealized.* For various carbohydrate wastes—agricultural, food, industrial, or municipal—in spite of frequent claims of their availability and low cost, no economical bioprocess applications have yet been found.

Biocatalysts

The substances that actually cause chemical change in bioprocesses are the enzymes produced by a living cell. For simple enzymatic conversions, isolated enzymes can be used as biocatalysts. When biological transformation of the substrate involves several sequential and interrelated chemical reactions, each catalyzed by a separate enzyme, however, whole cells (most commonly, but not exclusively, micro-organisms—bacteria, yeast, or fungi) are used as biocatalysts. Bioprocesses used for the synthesis of complex molecular structures (e.g., antibiotics or proteins such as insulin), for example, require entire systems of enzymes. Such systems do not yet function in concert outside a living cell. Indeed, when the desired product is the cell itself (e.g., baker's yeast or single-cell protein), all the enzymes comprising the cell's growth machinery are components of the catalytic system.

An inspection of the immense spectrum of organisms whose biochemical capabilities have been reasonably well defined reveals that bioprocesses employ only a small, select group of biocatalysts. If one eliminates those organisms considered "natural populations" in food fermentation or biological waste treatment, the range of biocatalysts employed in bioprocesses is even more limited. Some animal cells and tissues are employed for vaccine production and related activities, but the catalytic capabilities of plant cells, except for some algae, have not yet been employed commercially. It is possible that biotechnology will provide a means whereby important catalytic activities from poorly understood

*See *Chapter 9: Commodity Chemicals and Energy Production*.

organisms can be transferred to cells whose large-scale growth is well understood.

Wider availability of thermotolerant biocatalysts is important for all industries using bioprocesses. Recent research on the development of thermotolerant biocatalytic agents has advanced the potential efficiency of bioprocesses. The advantages of thermotolerance include:

- reduced susceptibility to contamination;
- easier removal of metabolic heat;
- more complete and rapid conversions when volatile inhibitors are present (but oxygen solubility is reduced); and
- easier recovery of volatile products (e.g., ethanol).

Biocatalysts that can withstand high pressure may also be useful industrially. For instance, higher pressures will increase the solubility of oxygen.

Finally, research investigating the relationship between the structure and the function of enzymes is proceeding. Ultimately, the aim is to be able to design, with the help of computers, an enzyme to perform any specific catalytic activity under given conditions. Although this procedure will not be done routinely for many years, it will soon be possible, using rDNA technology, to modify the structure of an enzyme to improve its function in a given condition, such as at a particular pH or temperature. Thus, biotechnology could greatly affect the efficiency of bioprocesses.

Bioprocess monitoring and associated instrumentation

Despite the need for close control of process variables during a bioprocess operation, the techniques available for making measurements on-line are extremely limited. Existing equipment can readily monitor only temperature, pH, dissolved oxygen concentration, and evolution of gases. Although many other sensors have been developed to measure other variables (e.g., glucose levels), all are sensitive to steam sterilization. Thus, their usefulness in monitoring most bioprocesses is limited. Many critical variables are able to be monitored only by withdrawing samples from the reaction vessel and analyzing them off-line, and, even then, it is difficult to determine key characteristics accurately. When measuring cell mass (an indicator of growth), for example, most process operators simply note such crude indicators as packed cell volume, turbidity, or, at best, dry weight.

It is possible to measure the compositions and flow rates of gaseous streams entering and leaving the reactor and to use the values obtained from such measurements to help estimate key process conditions indirectly. Such procedures have been greatly facilitated by the use of computers. The real potential of computer control, however, will not be realized until a greater range of reliable on-line sensors becomes available.*

A number of European, Japanese, and American groups have developed improved sensors for bioprocess control, but, so far, most devices require removal of samples for off-line analysis because the sensors cannot withstand sterilization. Continuous sampling combined with various types of rapid instrumental analyzers offers a reasonable compromise, but, with this approach, there is a time lag between the actual sample time and the time at which the assay information becomes available.

Sophisticated instrumentation will have increasing use in bioprocess monitoring. High performance liquid chromatography, for example, is used to identify particular compounds in a mix of compounds and is one of the fastest growing instrumentation fields. Flow cytometry has potential use in measuring process variables such as cell size (an indicator for adjusting nutrient flows) and cell viability. Other instrumentation will surely be used as bioprocess monitoring becomes more widely investigated.

Computer-coupled bioprocesses can greatly improve monitoring and controlling the growth conditions during a bioprocess run. Computers can be used to analyze the data from sensors and other monitoring instrumentation and respond to these data by adjusting process variables, such as nutrient flow. Additionally, computer interfaces can be used:

- to schedule efficiently the use of equipment;
- to alarm operators when necessary;

*For a discussion of biosensors, see *Chapter 10: Bioelectronics*.

Photo credit: Porton/LH International

100-liter pilot plant bioreactor with computer controls

- to log, store, and analyze data; and
- to inventory raw material depletion and product synthesis.

These functions optimize the methodology and organization of bioprocessing within a plant. Companies are only now starting to use computer-controlled bioprocesses because of cost, lack of good sensors, and interfacing problems. Yet advances in this field are sure to occur soon because of increased interest in bioprocesses by electronics experts, as evidenced by the recent joint venture between Genentech and Hewlett-Packard.*

The automation of bioprocessing will be of critical importance in the future as companies compete for shares in biotechnology product markets.

*See *Chapter 4: Firms Commercializing Biotechnology*.

As automation reduces the labor intensity of laboratory tasks, the pace of competition will quicken, and countries with sophisticated software to direct the automation will possess an advantage in the commercialization of biotechnology.

Separation and purification of products

Separation and purification techniques used in bioprocesses are the aspect of bioprocess engineering most in need of attention, especially for the production of novel products such as proteins. Research is needed to find highly selective recovery techniques that leave as little residual product as possible in the medium and thus lessen the labor intensity associated with downstream processing. An example of the effort expended in downstream processing is provided by the new plant Eli Lilly built to produce human insulin (Humulin®). The plant employs 220 people, 90 percent of whom are involved in recovery processes.

Some of the possibilities for improving recovery techniques now under consideration include the following:

- *Ultrafiltration.* Membranes and other filtration systems, such as porous metals, offer many advantages, and considerable experience in other areas of chemical technology is already available. Some U.S. companies, such as Millipore, Amicon, and Nucleopore are making advances in this area.
- *Continuous chromatography and high performance liquid chromatography.* If these approaches, already available on the laboratory scale, could be scaled-up, it would be possible, in principle, to collect a crude product from the medium and then, by selective elution, recover product, reusable nutrients, and inhibitory substances separately. One American manufacturer (Waters, a Millipore subsidiary) claims to have developed a pilot-scale chromatographic unit.
- *Electrophoresis.* Electrophoretic methods, especially continuous flow, can separate proteins, peptides, and nucleic acids on the basis of their electrical charge. The advantage of this separation method over some others is that it can run continuously and can effectively separate molecules in large sample volumes. The potential of continuous-flow electrophoresis for producing commercial quantities of high purity substances such as pharmaceuticals was demonstrated on a recent space shuttle mission. The electrophoresis experiment, cosponsored by McDonnell Douglas and the National Aeronautics and Space Administration, demonstrated that under weightless conditions an electrophoresis system, identical to one tested on Earth, separated about 700 times more material in a given period of time and also achieved four times the purity while processing 250 times more material.
- *Monoclonal antibodies.* Immobilized MAbs are being used as purification agents for protein products (see "Monoclonal Antibody Technology" section above). This technique best suits large molecular weight and high-value-added products such as proteins.

Genetic modifications of micro-organisms used in bioprocessing could also aid in recovery processes. Two changes in particular would greatly improve the yield and ease recovery of proteins. First, micro-organisms could be developed that have minimal intracellular protein-degrading enzymes. The presence of these enzymes will decrease the yield of protein product. Second, a protein is much more easily purified if it is secreted from the cell into the surrounding medium. The genetic incorporation of protein secretion mechanisms will lower production costs dramatically.

Although purification and separation protocols have been developed for existing bioprocesses, new bioprocesses will present new challenges. For example, rDNA technology has led to a new set of bioprocesses that synthesize protein products, and substantial work is needed to improve recovery strategies for large-scale protein purifications. In addition, one of the factors that restricts the use of bioprocesses for producing commodity chemicals is the expense of recovering these low-value-added chemicals from dilute aqueous solutions.

Culture of higher eukaryotic cells

The organisms used most extensively in large-scale bioprocesses are prokaryotes (e.g., bacteria) or simple eukaryotes (e.g., yeast). These are hardy organisms which grow to high cell densities and consequently give high product yields.

Certain products can be obtained in some situations only from the large-scale cultivation of higher eukaryotic cells. As noted in table 2, for instance, many proteins that are potentially useful (e.g., in medicine) have not been isolated in large enough quantities to study adequately. If eukaryotic cell culture made these proteins available in larger quantities, their amino acid sequence could be determined, their genes cloned, and even more of the proteins could be produced. Furthermore, some proteins probably need "post-translational modifications" (changes in protein structure after the protein is made from mRNA) that only higher eukaryotes can perform. These modified proteins may only be made in eukaryotic cells. Also, in many cases, the production of secondary metabolites in plant cells is a function of several enzymatic functions, most of which are not known. Therefore, the growth of plant cells in culture might be the easiest way to produce useful plant compounds. Finally, many individuals think that the growth of hybridomas would be easier and more economic in culture if the culture technology were better developed (see "Monoclonal Antibody Technology" section above). As biotechnology becomes more integrated into the industrial structure, the development of more efficient and economic bioprocess technologies for higher eukaryotic cells will increase in importance.

Photo credit: Science Photo Library and Porton/LH International

Laboratory tissue culture production

The technologies developed for the growth of micro-organisms have limited applicability to the growth of higher eukaryotic cells because of differences between microbial and mammalian cells (see table 3). Mammalian cells are larger, more fragile, and more complex than microbial cells.*

*Most cell culture research has been done with mammalian cells, so the work reported here focuses on those cells. Problems with plant cell culture are similar to those of mammalian cell culture.

Table 2.—Situations Potentially Requiring Large-Scale Eukaryotic Cell Culture

Cell culture system	Reason for large-scale eukaryotic cell culture
Cells producing useful proteins	Not large enough quantity to determine amino acid sequence; therefore cannot use rDNA technology
Cells producing modified proteins	Modification systems present only in higher cells
Plant cells producing useful secondary metabolites	Enzymatic pathways for metabolic production not well understood; therefore cannot use rDNA technology
Hybridomas	Mouse acites system has limited capacity and applicability

SOURCE: Office of Technology Assessment.

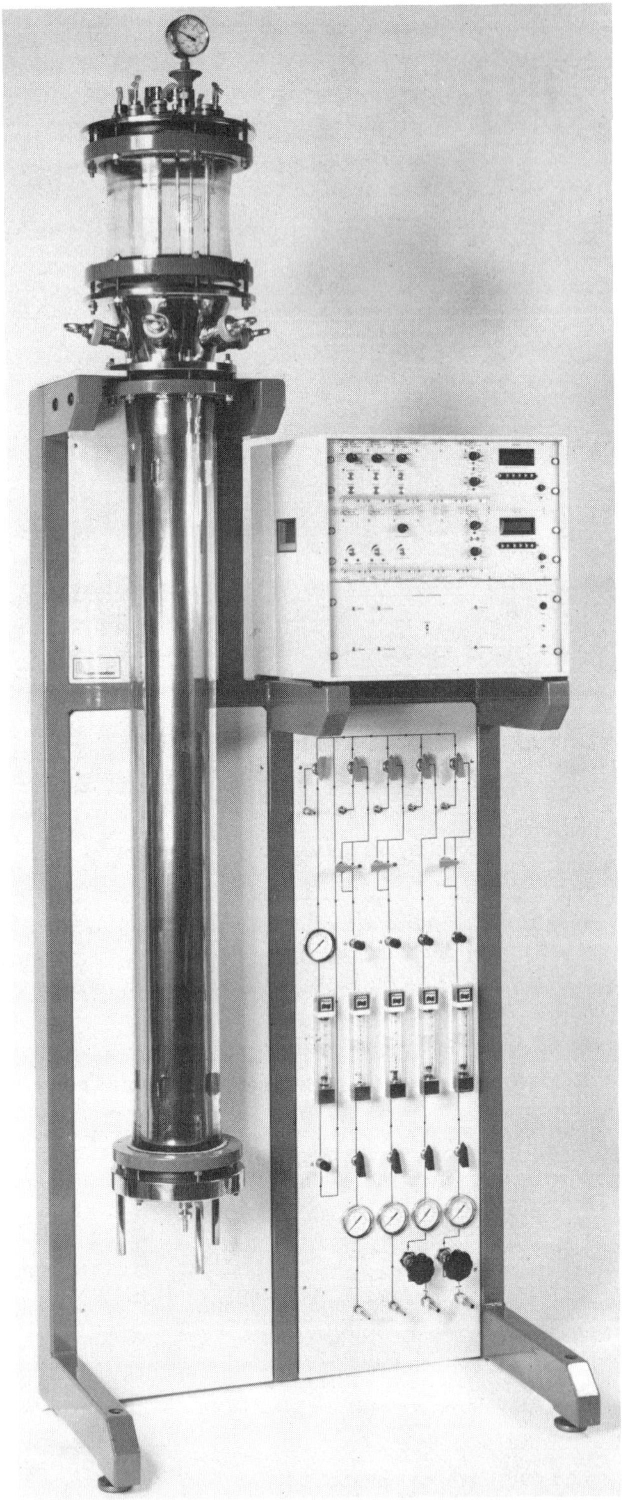

Photo credit: Porton/LH International

Bioreactor specially designed for the growth of plant or animal cells

Furthermore, mammalian cells have very complex nutritional requirements, which have not been completely defined. They require serum from blood for growth, and the essential composition of serum is not well characterized. In contrast to microbial cells, mammalian cells are not normally exposed to the environment, but are constantly surrounded by a circulatory system that supplies nutrients and removes wastes. When these cells are grown in culture, the medium is initially clean and nutritionally balanced; as the cells take up the nutrients and excrete waste products, however, the medium becomes much less like the cells' normal environment. This problem, along with the problem of fragility, requires modified reactor design (39).

Some mammalian cells grow in suspension like microbial cells, but most higher cells must attach to a solid surface. A major problem with large-scale cell growth of mammalian cells has been the availability of large, accessible surfaces for cell growth. The attachment of cells to microcarriers, or very small beads, has begun to solve many of the problems associated with large-scale mammalian cell culture. The beads provide a large amount of surface area and can be placed in a column where either a continuous-flow or fluidized-bed bioreactor can be used for cell growth and product formation (see box A). Either of these bioreactors is gentler than a stirred tank reactor. Additionally, because of the continuous nature of these bioreactors, fresh nutrients are added and wastes are removed continuously.

The instrumentation requirements for mammalian cell growth are different than those for microbial growth. The lower rates of metabolism and lower density of mammalian cells require more sensitive sensor systems than for microbial cell growth. Additionally, because the nutritional requirements are so much more complex, different strategies are needed to monitor and control cell growth. These problems are just beginning to be addressed (40).

Priorities for future research

Priorities for future generic research in bioprocess engineering that would be applicable

Table 3.—Comparison of Microbial and Mammalian Cells

Characteristic	Microbial cells	Mammalian cells
Size (diameter)	1-10 microns	10-100 microns
Metabolic regulation	Internal	Internal and hormonal
Nutritional spectrum	Wide range of substrates	Very fastidious nature
Doubling time	Typically 0.5-2.0 hours	Typically 12-60 hours
Environment	Wide range of tolerance	Narrow range of tolerance
Other characteristics		Limited life span of normal cells Lack of protective cell wall

SOURCE: Office of Technology Assessment, adapted from R. J. Fleischaker, Jr., "An Experimental Study in the Use of Instrumentation To Analyze Metabolism and Product Fermentation in Cell Culture," thesis, Massachusetts Institute of Technology, Cambridge, Mass., June 1982.

to all industries using biotechnology include research in the following areas:

- continued work on the practical use of and design of bioreactors for immobilized cell and enzyme systems;
- development of a wider range of sterilizable sensors for process monitoring and control;
- improved product recovery techniques, especially for proteins;
- general reactor design and practical approaches to better oxygen transfer;
- inhibition of intracellular protein-degrading enzymes;
- improving the genetic stability of rDNA organisms;
- protein secretion mechanisms;
- improved methods for heat dissipation during bioprocessing; and
- biochemical and physiological mechanisms for temperature and pressure tolerance.

The large-scale culture of eukaryotic cells is beginning to receive some research attention. Because of the complex nutritional requirements of eukaryotic cells, the cost of the medium is high. If industry is going to adopt eukaryotic cell culture technology, the development of economic artificial media is important. Also important is the development of new bioreactor design and instrumentation for the control of cell growth.

Chapter 3 references

Recombinant DNA technology references

1. Guarente, L., Roberts, T., Ptashne, M., et al., "A Technique for Expressing Eukaryotic Genes in Bacteria," *Science* 209:1428-30, 1980.
2. Ris, H., and Kubai, D. F., "Chromosome Structure," *Ann. Rev. Genetics* 4:263-94, 1970.
3. Shortle, D., DiMaio, D., and Nathans, D., "Directed Mutagenesis," *Ann. Rev. Biochem.* 15:265-94, 1981.

*Monoclonal antibody technology references**

*4. Baskin, Y., "In Search of the Magic Bullet," *Technology Review,* October 1982, pp. 19-23.

*5. *Business Week,* "Biotechnology's New Trust in Antibodies," May 18, 1981, pp. 147-156.

6. Cavagnaro, J., and Osband, M., "A Status Review—Human-Derived Monoclonal Antibodies," *Genetic Engineering News,* May/June 1983, pp. 6-7.

7. *Chemical and Engineering News,* "Method May Boost Monoclonal Antibody Output," Jan. 11, 1982, p. 22.

*8. Chisholm, R., "On the Trail of the Magic Bullet," *High Technology,* January 1983, pp. 57-63.

9. Croce, C. M., Linnenbach, A., and Hall, W., et al., "Production of Human Hybridomas Secreting Antibodies to Measles Virus," *Nature* 288:488, 1980.

10. Diamond, B. A., Yelton, D. E., and Scharff, M. D., "Monoclonal Antibodies: A New Technique for Producing Serological Reagents," *New Eng. J. Med.* 304:1344, 1981.

11. Fox, P. L., Berenstein, E. H., and Berganian, R. P.,

*References of general interest are marked with asterisks.

"Enhancing the Frequency of Antigen-Specific Hybridomas," *Eur. J. Immunol.* 11:431, 1981.
12. Galfre, G., Milstein, C., and Wright B., "Rat x Rat Hybrid Myelomas and a Monoclonal Anti-Fd Portion of Mouse IgG," *Nature* 277:131, 1979.
*13. Gatz, R. L., Young, B. A., Facklam, T. J., and Scantland, D. A., "Monoclonal Antibodies: Emerging Product Concepts for Agriculture and Food," *Bio/Technology,* June 1983, pp. 337-341.
14. Gefter, M. L., Margulies, D. H., and Scharff, M. D., "A Simple Method for Polyethylene Glycol-Promoted Hybridization of Mouse Myeloma Cells," *Somat. Cell Genet.* 3:231, 1977.
15. Herzenberg, L. A., and Herzenberg, L. A., "Analysis and Separation Using the Fluorescence Activated Cell Sorter (FACS)," *Handbook of Experimental Immunology,* vol. 2, D. M. Weis (ed.) (London: Blackwell Scientific Publications, 1978).
16. Hoffman, R. A., and Hansen, W. P., "Immunofluorescent Analysis of Blood Cells by Flow Cytometry," *Int. J. Immunopharmac.* 3:249, 1981.
17. Karpas, A., Fischer, P., and Swirsky, D., "Human Plasmacytoma With an Unusual Karyotype Growing *in Vitro* and Producing Light-Chain Immunoglobulin," *Lancet* 1:931, 1982.
18. Kindel, S., "The Birth of Bioindustry," *Forbes,* July 18, 1983, pp. 130-132.
19. Klausner, A., "Genentech Makes Monoclonal Precursors From *E. coli,*" *Bio/Technology,* July 1983, pp. 396-397.
20. Kohler, G., and Milstein, C., "Continuous Cultures of Fused Cells Secreting Antibody of Predefined Specificity," *Nature* 256:495, 1975.
21. Kohler, G., and Milstein, C., "Derivation of Specific Antibody-Producing Tissue Culture and Tumor Lines by Cell Fusion," *Eur. J. Immunol.* 6:511, 1976.
*22. Langone, J., "Monoclonals: The Super Antibodies," *Discover,* June 1983, pp. 68-72.
23. Lewin, R., "An Experiment That Had To Succeed," *Science* 212:767, 1981.
24. Melchers, F., Potter, M., and Warner, N. L. (eds.), "Lymphocyte Hybridomas," *Curr. Top. Microbiol. Immunol.,* vol. 81 (New York: Springer-Verlag, 1978).
*25. Milstein, C., "Monoclonal Antibodies," *Scientific American* 243:66-74, 1980.
26. Montagnon, B. J., et al., "Large-Scale Culture of Vero Cells in Microcarrier Culture for Virus Vaccine Production: Preliminary Results for Killed Polio Virus Vaccine," *Inter. Devel. Biol.* 47:55, 1980.
27. Olsson, L., and Kaplan, H., "Human-Human Hybridomas Producing Monoclonal Antibodies of Predetermined Antigenic Specificity," *Proc. Natl. Acad. Sci. (U.S.A.)* 77:5429-5431, 1980.
28. Scharff, M. D., Roberts, S., and Thammana, P., "A Better Cell Line for Making Hybridomas Secreting Specific Antibodies," *Nature* 276:269, 1978.
29. Staehelin, T., "The Use of Monoclonal Antibodies in Interferon Production," presented at Robert First Conference on Monoclonal Antibodies, Rye, N.Y., September 1981.
30. Tsung, Y., Milkunsky, A., and Alpert, E., "Derivation and Characterization of a Monoclonal Hybridoma Antibody Specific for Human Alpha-Fetoprotein," *J. Immunol. Meth.* 39:363-368, 1980.
31. Van Wezel, A. L., et al., "New Approach to the Production of Inactivated Polio Vaccine on Serially Cultured Kidney Cells From Captive-Bred Monkeys," *Inter. Devel. Biol.* 46:151, 1980.
*32. Yelton, D. E., and Scharff, M. D., "Monoclonal Antibodies," *Amer. Scientist* 68:510, 1980.

Bioprocess technology references**

33. Blanch, H., "Engineering Challenges in Genetic Engineering," *The Victor Cape Lecture on Biotechnology,* McGill University, Montreal, Canada, May 14, 1982.
34. Blanch, H., "Biotechnology: Doing Well in the Lab But Struggling in the Plant," *Ind. Res. Dev.,* August 1983, pp. 76-79.
35. Bull, A. T., Holt, G., and Lilly, M. D., *Biotechnology: International Trends and Perspectives* (Paris: Organisation for Economic Co-Operation and Development, 1982).
36. Chibata, I., and Tosa, T., "Immobilized Microbial Cells and Their Applications," *Trends in Biological Sciences,* April 1980, pp. 88-90.
37. Cooney, C. L., "Bioreactors: Design and Operation," *Science* 219:728-733, 1983.
38. Demain, A. L., "Industrial Microbiology," *Science* 214:987-995, 1981.
39. Feder, J., and Tolbert, W. R., "The Large-Scale Cultivation of Mammalian Cells," *Scientific American* 248:36-43, 1983.
40. Fleischaker, R. J., Jr., "An Experimental Study in the Use of Instrumentation To Analyze Metabolism and Product Formation in Cell Culture," thesis, Massachusetts Institute of Technology, Cambridge, Mass., June 1982.
41. Gaden, E. L., "Production Methods in Industrial Microbiology," *Scientific American,* September 1981, pp. 181-196.

**Most of these references are of general interest for those wishing to pursue the subject of bioprocess technology further.

42. Gaden, E. L., "Bioprocess Technology: An Analysis and Assessment," contract paper prepared for the Office of Technology Assessment, U.S. Congress, July 1982.
43. Gaden, E. L., University of Virginia, personal communication, 1982.
44. Hochhauser, S. J., "Bringing Biotechnology to Market," *High Technology,* February 1983, pp. 55-60.
45. Johnson, I. S., "Human Insulin From Recombinant DNA Technology," *Science* 219:632-637, 1983.
46. Kindel, S., "Enzymes—The Bioindustrial Revolution," *Technology,* November/December 1981, pp. 62-74.
47. Klibanov, A. M., "Immobilized Enzymes and Cells as Practical Catalysts," *Science* 219:722-727, 1983.
48. National Academy of Engineering, "Genetic Engineering and the Engineer" (Washington, D.C.: National Academy Press, 1982).
49. Wang, D. I. C., Cooney, C. L., Demain, A. L., et al., *Fermentation and Enzyme Technology* (New York: John Wiley & Sons, 1979).

PART II
Firms Commercializing Biotechnology

Chapter 4

Firms Commercializing Biotechnology

Contents

Tables

Figures

Chapter 4

Firms Commercializing Biotechnology

Introduction

Biotechnology has the technical breadth and depth to change the industrial community of the 21st century because of its potential to produce substantially unlimited quantities of:

- products never before available,
- products that are currently in short supply,
- products that cost substantially less than products made by existing methods of production,
- products that are safer than those now available, and
- products made with raw materials that may be more plentiful and less expensive than those now used.

By virtue of its wide-reaching potential applications, biotechnology lies close to the center of many of the world's major problems—malnutrition, disease, energy availability and cost, and pollution. It is because of biotechnology's promise that the developed countries of the world have commenced a competitive battle to commercialize its applications.

Nowhere in the world are efforts to commercialize biotechnology stronger than in the United States.* Large established U.S. companies in industries ranging from pharmaceuticals to petroleum have followed the lead in developing biotechnology that was set by entrepreneurial new biotechnology firms (NBFs) in the United States whose dedication to biotechnology is unmatched anywhere. Major competitive challenges to the United States in current product markets, as well as in new biotechnology markets yet to be defined, will be mounted by established companies in the Federal Republic of Germany, United Kingdom, Switzerland, and France—but the most formidable challenge will come from established companies in Japan. The Japanese consider biotechnology to be the last major technological revolution of this century (58). More immediate than its promise of helping to alleviate some world problems, biotechnology offers Japan an important opportunity to revitalize its structurally depressed basic industries whose production processes are reliant on imported petroleum.

This chapter provides an overview of U.S. and foreign private sector research and development (R&D) and commercialization efforts in biotechnology to help answer the broader question being addressed by this report: Will the United States be able to translate its present technological lead into worldwide commercial success by securing competitive shares of biotechnology-related product markets? The first section of the chapter provides an overview of the types of companies that are commercializing biotechnology in the United States and the five foreign countries expected to be the major competitors in the area of biotechnology. This section briefly examines the four fields where biotechnology is being applied most vigorously—pharmaceuticals, animal health, plant agriculture, and specialty chemicals. The second section analyzes and compares the strength of the U.S. support base with that of the competitor countries, using three important product areas for comparison: biochemical reagents, instrumentation, and software. The third section analyzes the respective roles of the firms applying biotechnology in the United States—NBFs and established companies—in the domestic and international development of biotechnology. It also describes collaborative ventures between NBFs in the United States and established U.S. and foreign companies that are seeking to commercialize biotechnology. The chapter concludes by summarizing major findings with respect to the role of NBFs and established companies in the U.S. commercialization effort.

*For a summary of activities in biotechnology in countries other than the United States, see *Appendix B: Country Summaries.*

Overview of U.S. and foreign companies commercializing biotechnology

U.S. and foreign efforts to develop and commercialize biotechnology differ substantially in character and structure. The manner in which the United States and other countries organize their development efforts is important for two reasons: it can influence their respective commercial capabilities; and it will ultimately shape the character of international competition.

In the United States, two distinct sets of firms are pursuing commercial applications of biotechnology—NBFs and established companies. NBFs, as defined by this report, are entrepreneurial ventures started specifically to commercialize innovations in biotechnology. For the most part, they have been founded since 1976—the same year the U.S. firm Genentech was founded to exploit the recombinant DNA (rDNA) technology patented in the United States by Cohen and Boyer.* Typically, NBFs are structurally organized specifically to apply biotechnology to commercial product development. The established companies pursuing applications of biotechnology are generally process-oriented, multiproduct companies in traditional industrial sectors such as pharmaceuticals, energy, chemicals, and food processing. These companies have undertaken in-house biotechnology R&D in an effort to determine how and where best to apply biotechnology to existing or new products and processes. Table 4 provides a list of NBFs and established companies currently applying biotechnology in the United States and the targeted commercial areas of their research. Figure 10 illustrates the percentage of U.S. firms pursuing biotechnology R&D in specific application areas.

Sixty-two percent (135) of the 219 U.S. companies for whom commercial application areas are known* are pursuing applications of biotechnology in the area of pharmaceuticals; 28 percent are pursuing applications in animal agriculture, and 24 percent in plant agriculture.** In the area of specialty chemicals and food additives, commodity chemicals and energy, the environment, and electronics, respectively, relatively fewer U.S. firms are pursuing commercial applications of biotechnology. In some of these sectors, conventional technologies are working well or existing investments in capital equipment are very substantial. In others, much uncertainty still surrounds the potential of biotechnology or the research needed to develop applications of biotechnology is long term.

In Japan, the Federal Republic of Germany, Switzerland, France, and the United Kingdom,*** biotechnology is being commercialized almost exclusively by established companies. Most European nations and Japan, unlike the United States, tend, for different reasons, to emphasize the importance of large companies instead of small ones. Thus, the development of biotechnology in these countries is biased considerably toward the large pharmaceutical and chemical companies.

It should not be assumed that the small number of NBFs in the European countries or the lack of

*Two U.S. firms, Cetus and Agrigenetics, though established before 1976, are considered to be NBFs. Cetus was founded to capitalize on classical genetic techniques for product development, but showed early interest in biotechnology and began aggressively pursuing product development with the new techniques. Agrigenetics was formed in 1975 to link new genetic research with the seed business. Thus, the behavior and research focus of both Cetus and Agrigenetics place them in the new firm category despite their early founding dates.

*This figure does not include the companies listed that are specializing in bioprocessing, because the bioprocessing R&D may not be associated with specific products. See *Appendix D: Firms Commercializing Biotechnology in the United States* for an explanation of how the list was obtained.

**These percentages add up to more than 100 percent because many of the firms are engaged in more than one area of commercial application.

***In the United Kingdom, some NBFs, not including Celltech and Agricultural Genetics, are beginning to form on the periphery of universities. Plant Science, Ltd., for example, is linked to the University of Sheffield; Imperial Biotechnology, Ltd., is linked to the Imperial College in London; IQ (Bio) was formed by some Cambridge University biochemists; Boscot, Ltd., a joint venture between two Scottish institutions, was established by the University of Edinburgh and Heriot-Watt University, and Cambridge Life Sciences pursues biosensors based on work at Southampton University. As an indication of the increased number of NBFs forming in Britain, Biotechnology Investments, Ltd., the venture fund managed by N.M. Rothschild (the bank) now has for the first time since the fund was established more proposals from British firms than from companies in the United States (56).

Table 4.—Companies Commercializing Biotechnology in the United States and Their Product Markets[a,b]

Company (date founded)	Commercial application of R&D[c]	Ph.D.s[d]
Abbott Laboratories	Ph	
Actagen (1982)	Ph	5
Advanced Biotechnology Associates, Inc. (1981)	Ph	
Advanced Genetic Sciences, Inc. (1979)	PA	27
Advanced Genetics Research Institute (1981)	AA	8
Advanced Mineral Technologies, Inc. (1982)	Env	
Agrigenetics Corp. (1975)	PA,SCF	46
Allied Chemical Corp.	PA	
Alpha Therapeutic Corp.	Ph	
Ambico, Inc. (1974)	AA	
American Cyanamid Co.	Ph,PA,AA	
American Diagnostics Corp. (1979)	Ph	
American Qualex (1981)	Ph,AA	
Amgen (1980)	Ph,PA,AA,SCF	45
Angenics (1980)	Ph	5
Animal Vaccine Research Corp. (1982)	AA	
Antibodies, Inc. (1960)	Ph,AA	
Applied DNA Systems, Inc. (1982)	Ph,SCF,CCE,Env	
Applied Genetics, Inc. (1981)	AA	
ARCO Plant Cell Research Institute	PA	18
Atlantic Antibodies (1973)	AA	2
Axonics	Ph	
Baxter-Travenol Laboratories, Inc.	Ph	
Becton Dickinson & Co.	Ph	
Bethesda Research Laboratories, Inc. (1976)	Ph,AA	
Biocell Technology Corp. (1980)	Ph	
Biochem Technology, Inc. (1977)	Bioprocessing	
Bio-con, Inc. (1971)	AA	
BioGenex Laboratories (1981)	Ph	
Biogen, Inc. (1980)	Ph,AA,CCE,Env	79
Biological Energy Corp. (1981)	CCE,SCF	3
Bio Response, Inc. (1972)	Mass cell culture	6
Biotech Research Laboratories, Inc. (1973)	Ph,CCE	11
Biotechnica International, Inc. (1981)	PA,CCE,SCF,Env, AA,Ph	12
Bio-Technology General Corp. (1980)	PA,AA,Ph	5
Brain Research (1968)	Ph	
Bristol-Myers Co.	Ph	
BTC Diagnostic, Inc. (1980)	Ph	3
Calgene, Inc. (1980)	PA	21
California Biotechnology, Inc. (1982)	Ph,AA	21
Cambridge Bioscience Corp. (1982)	Ph,AA	
Campbell Institute for Research & Technology	PA	
Celanese Corp.	CCE	
Cellorgan International, Inc. (1972)	Ph	
Celtek, Inc. (1980)	Ph	5
Centaur Genetics Corp. (1981)	Ph,PA,AA	4
Centocor (1979)	Ph	14
Cetus Corp. (1971)	Ph,AA,CCE	45
Madison (1981)	PA	25
Palo Alto (1980)	Ph	2
Immune (1980)	Ph	
Chiron Corp. (1981)	Ph,AA	26[e]
Ciba-Geigy	Ph	
Clonal Research (1970)	Ph	3
Codon (1980)	CCE	15
Collaborative Genetics, Inc. (1979)	Ph,SCF,CCE	12
Collagen, Inc. (1977)	Ph	
Cooper Diagnostics, Inc.	Ph	
Cooper-Lipotech, Inc. (1981)	Ph	
Corning Glass Works	SCF	

Table 4.—Companies Commercializing Biotechnology in the United States and Their Product Markets[a,b] (Continued)

Company (date founded)	Commercial application of R&D[c]	Ph.D.s[d]
Crop Genetics International (1981)	PA	
Cutter Laboratories, Inc.	Ph	
Cytogen Corp. (1981)	Ph	7
Cytox Corp. (1975)	Env	
Damon Biotech, Inc. (1981)	Ph	10
Dairyland Foods Corp.	SCF	
Dart and Kraft, Inc.	SCF	
Davy McKee Corp.	Bioprocessing	
DeKalb Pfizer Genetics (1982)	AA	
Diagnon Corp. (1981)	Ph	
Diagnostic Technology, Inc. (1980)	Ph	
Diamond Laboratories	AA	
Diamond Shamrock Corp.	AA,CCE	
DNA Plant Technology (1981)	PA	10
DNAX Corp.	Ph	
Dow Chemical Co.	Ph,PA,CCE,SCF, AA,Env	
Ean-tech, Inc. (1982)	El,Env,Ph	3
Eastman Kodak Co.	Ph,Env	
Ecogen (1983)	PA	
E. I. du Pont de Nemours & Co., Inc.	Ph,PA,CCE,SCF	
Electro Nucleonics Laboratories, Inc.	Ph	
Eli Lilly & Co.	Ph,PA	
EnBio, Inc. (1975)	Bioprocessing	
Endorphin, Inc. (1982)	Ph	
Engenics, Inc. (1981)	Bioprocessing	25
Enzo Biochem, Inc. (1976)	Ph,AA,CCE,SCF,PA	
Enzyme Bio-systems, Ltd.	SCF	
Enzyme Center, Inc.	SCF	
Enzyme Technology Corp.	SCF	
Ethyl Corp.	CCE,SCF,Env	
Exxon Research & Engineering Co.	CCE,Env,SCF	
Fermentec Corp. (1978)	Bioprocessing	
FMC Corp.	Ph	
Frito-Lay, Inc.	PA	
Fungal Genetics, Inc. (1982)	Ph,SCF	
Genencor (1982)	SCF,CCE	
Genentech, Inc. (1976)	Ph,AA,CCE,El	75
General Electric Co.	El,Env,Ph,SCF	
General Foods Corp.	PA	
General Genetics (1982)	Ph	
General Molecular Applications (1981)	Ph	
Genetic Diagnostics Corp. (1981)	Ph	3
Genetic Replication Technologies, Inc. (1980)	Ph,AA	
Genetic Systems Corp. (1980)	Ph	14
Genetics Institute (1980)	Ph,PA,SCF,Env	24
Genetics International, Inc. (1980)	AA,Ph,SCF,CCE, Env,El	17
Genex Corp. (1977)	Ph,AA,SCF,Env	48
Gentronix Laboratories, Inc. (1972)	El	
Genzyme (1981)	SCF	6
W. R. Grace & Co.	AA,SCF,Env,PA,Ph	
Hana Biologics, Inc. (1978)	Ph	
Hem Research (1966)	Ph,AA	
Hoffmann-La Roche, Inc.	Ph	
Hybridoma Sciences, Inc. (1981)	Ph	
Hybritech, Inc. (1978)	Ph	13
Hytech Biomedical, Inc. (1981)	El,Ph	10
IBM Corp.	El	
IGI Biotechnology, Inc. (1975)	Ph	
Immulok, Inc. (1980)	Ph	

Table 4.—Companies Commercializing Biotechnology in the United States and Their Product Markets[a,b] (Continued)

Company (date founded)	Commercial application of R&D[c]	Ph.D.s[d]
Immunetech, Inc. (1981)	Ph	
Immunex Corp. (1981)	Ph	18
Immuno Modulators Laboratories, Inc. (1982)	Ph	
Immunogen (1981)	Ph	
Immunotech Corp. (1980)	Ph	
Imreg, Inc.	Ph	
Indiana BioLab (1972)	PA,AA,SCF,CCE	
Integrated Genetics, Inc. (1981)	Ph	17
Interferon Sciences, Inc. (1980)	Ph	7
International Genetic Engineering, Inc. (Ingene) (1980)	Ph,PA,CCE	16
International Genetic Sciences Partnership (1981)	PA,AA	
International Minerals & Chemical Corp.	AA,PA,Env,CCE	
International Plant Research Institute (IPRI) (1978)	PA	35
Kallestad Laboratories, Inc.	Ph	
Kennecott Copper Corp.	Env	
Lederle Laboratories	Ph,AA	
The Liposome Co., Inc. (1981)	Ph,AA	
Liposome Technology, Inc. (1981)	Ph,AA	
Litton Bionetics	Ph	
3M Co.	Ph	
Mallinckrodt, Inc.	Ph	
Martin Marietta	SCF,PA	
Meloy Laboratories, Inc. (1975)	Ph	
Merck & Company, Inc.	Ph,AA	
Microlife Genetics (1981)	SCF,Env	
Miles Laboratories, Inc.	Ph,SCF,CCE,AA	
Miller Brewing Co.	PA	
Molecular Biosystems, Inc. (1980)	Ph	7
Molecular Diagnostics (1981)	Ph	
Molecular Genetics, Inc. (1979)	Ph,PA,AA	20
Monoclonal Antibodies, Inc. (1979)	Ph,AA	7
Monsanto Co.	PA,AA	
Multivac, Inc.	Ph,PA,AA,SCF	
Nabisco, Inc.	PA	
National Distillers & Chemical Co.	CCE	
NPI (1973)	PA,CCE,SCF	25
Neogen Corp. (1981)	PA,AA	
New England Biolabs	Ph	
New England Monoclonal Resources (1982)	Ph	
New England Nuclear Corp.	Ph	
Norden Laboratories	AA	
Novo Laboratories, Inc.	Ph,SCF	
Nuclear & Genetic Technology, Inc. (1980)	Ph	
Ocean Genetics (1981)	SCF	
Oncogen (1982)	Ph	
Oncogene Science Inc. (1983)	Ph	
Organon, Inc.	Ph	
Ortho Pharmaceutical Corp.	Ph	
Petrogen, Inc. (1980)	Env	7
Pfizer, Inc.	Ph,PA,CCE,AA, SCF,Env	
Phillips Petroleum Co.	Env,SCF,CCE	
Phytogen (1980)	PA	5
Phyto-Tech Lab	PA	
Pioneer Hybrid International Corp.	PA	
Plant Genetics, Inc. (1981)	PA	11
Polybac Corp.	Ph,SCF,Env	
PPG Industries	SCF	
Purification Engineering, Inc.	Bioprocessing	
Quidel Home (1982)	Ph	

Table 4.—Companies Commercializing Biotechnology in the United States and Their Product Markets[a,b] (Continued)

Company (date founded)	Commercial application of R&D[c]	Ph.D.s[d]
Replicon (1982)	Ph,SCF	
Repligen Corp. (1981)	Ph,AA,CCE,SCF	
Ribi Immunochem Research, Inc. (1981)	AA,Ph	3
Rohm & Haas	PA	
Salk Institute Biotechnology/ Industrial Associates, Inc. (1981)	Ph,AA,CCE	9
Sandoz, Inc.	Ph,PA,AA	
Schering-Plough Corp.	Ph,AA	
SDS Biotech Corp. (1983)	AA	
G. D. Searle & Co.	Ph,SCF	
Serono Laboratories, Inc.	Ph	
SmithKline Beckman	Ph,AA	
E. R. Squibb & Sons, Inc.	Ph	
A. E. Staley Manufacturing Co.	AA,PA,SCF	
Standard Oil of California	Env	
Standard Oil of Indiana	Ph,PA	
Standard Oil of Ohio	PA	
Stauffer Chemical Co.	PA	
Summa Medical Corp.	Ph	
Sungene Technololgies Corp. (1981)	PA	4
Sybron Biochemical	Env	13
Synbiotex Corp. (1982)	Ph,AA	
Syncor International	Ph	
Synergen (1981)	AA,SCF,CCE,Env	21
Syngene Products and Research, Inc.	AA	8
Syntex Corp.	Ph,AA	
Syntro Corp. (1982)	AA,CCE	5
Syva Co. (1966)	Ph	
Techniclone International Corp. (1982)	Ph	6
Unigene Laboratories, Inc. (1980)	Ph,AA	12
Universal Foods Corp.	SCF,PA	
University Genetics Co. (1980)		
Genetic Clinics	Ph	
U.O.P., Inc.	SCF,CCE	
The Upjohn Co.	Ph,AA,PA	
Viral Genetics (1981)	Ph	
Wellcome Research Laboratories	Ph	
Worne Biotechnology, Inc. (1982)	PA,CCE,Ph,AA, Env,SCF	10
Xenogen, Inc. (1981)	Ph,PA	
Xoma Corp. (1981)	Ph	
Zoecon Corp. (1968)	PA,AA	
Zymed Laboratories	SCF,CCE	5
Zymos Corp. (1982)	Ph,SCF	

[a]Does not include support firms.
[b]See *Appendix D: Index of Firms in the United States Commercializing Biotechnology* for a description of how the data were collected.
[c]Ph: Pharmaceuticals, PA: Plant Agriculture, AA: Animal Agriculture, SCF: Specialty Chemicals and Food, CCE: Commodity Chemicals and Energy, Env: Environmental (Microbial Enhanced Oil Recovery, Microbial Mining, Pollution Control, and Toxic Waste Treatement), El: Electronics.
[d]As of March 1983.
[e]M.D.s and Ph.D.s.

SOURCE: Office of Technology Assessment.

Figure 10.—Percentage of Firms in the United States Pursuing Applications of Biotechnology in Specific Industrial Sectors*

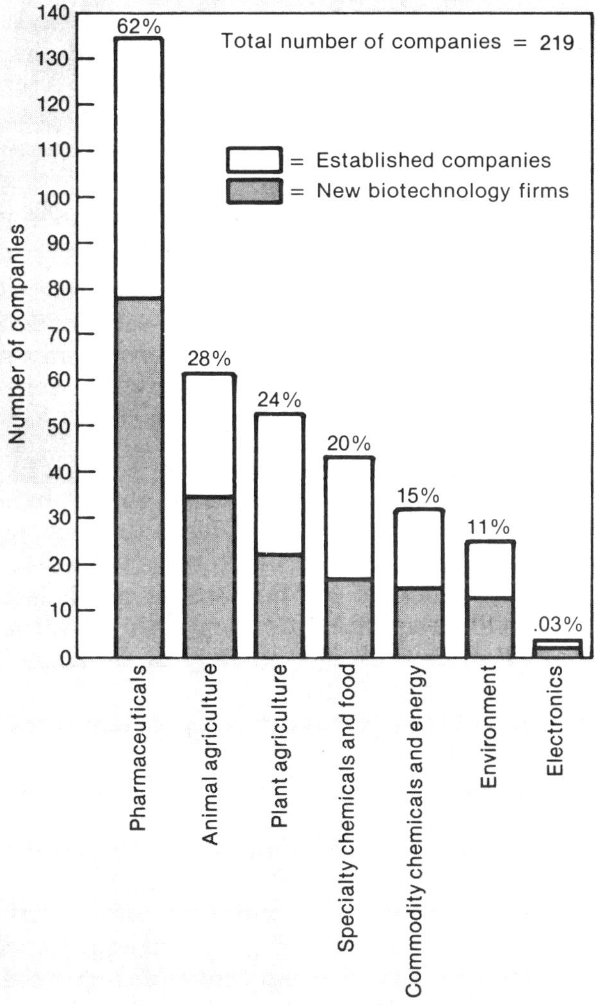

*The total percentage of firms exceeds 100 percent because some companies are applying biotechnology in more than one industrial sector.

SOURCE: Office of Technology Assessment.

NBFs in Japan will retard those countries' development of biotechnology. Varying strategies, organizational differences, and cultural factors all contribute to the competitive strengths of foreign countries' established companies. It is important to note, however, that the complementary efforts of NBFs and established companies in the United States have been a major factor in providing the United States with an early competitive advantage in the commercialization of biotechnology.

Although there are few NBFs outside the United States at present, some European countries are beginning to sense that small firms can make important contributions to innovation, particularly in high-technology fields such as biotechnology. Thus, in contrast to the West German Government, which believes that the development of biotechnology in West Germany is the province of the large chemical companies for which the country is noted and that NBFs are "not in line with the German mentality" (5), the British and French Governments have aided in the establishment of small firms such as Celltech, (U.K.), Agricultural Genetics (U.K.), and Transgene (France's leading biotechnology venture company).

Efforts in support of small company formation are also being undertaken by organizations elsewhere in Europe. The Organisation for Economic Co-Operation and Development, for example, in an effort to spur technological innovations, has made several proposals designed to support small firm development (65). These proposals encompass the promotion of new sources of venture capital, assistance to new startups in developing high quality feasibility studies, and diverse measures to encourage high-technology startups.

Venture capitalization is almost exclusively an American phenomenon (5,69). Many would agree that the formation of venture capital and entrepreneurial drive necessary to start small high-technology firms and vigorously commercialize inventions has been inhibited in much of Europe by a historical labor attitude that gives priority to job security and a predictable business environment rather than to aggressive risk-taking. In Japan, individualism and the creation of small, entrepreneurial and independent high-technology firms appears to be discouraged by cultural traits emphasizing group identity and acceptance. Large, very successful firms typical of Japan provide workers with a group identity and a sense of security, and it is these firms that are commercializing biotechnology in that country.

The biotechnology-related activities of U.S. and foreign companies in the pharmaceutical and animal and plant agriculture sectors are introduced below. Also discussed are foreign companies' biotechnology-related activities in specialty chemicals. Discussion of U.S. private sector activities in specialty chemicals, commodity chemicals, and the environmental and electronics fields is reserved for the chapters in part III. It is important to recognize that there is no "biotechnology industry." Biotechnology is a set of technologies* that can potentially benefit or be applied to several industries.

The industrial sector in which the earliest applications of biotechnology have occurred is the pharmaceutical sector. Because of the rapid diffusion of the new genetic techniques into pharmaceutical R&D programs, the pharmaceutical sector is currently the most active in commercializing biotechnology. For this reason, the pharmaceutical sector serves as a model for the development of biotechnology in this chapter and in much of this report. It is important to recognize however, that the development of biotechnology in other industrial sectors will differ from its development in the pharmaceutical sector. Regulatory and trade barriers and a marketing and distribution system unique to the pharmaceutical sector limit the applicability of the model to other industrial sectors.

*Pharmaceutical industry***

The pharmaceutical industry is one of the most successful high-technology sectors of the world economy (80). Because research is the foundation of competitive strength for modern pharmaceutical companies (55), and because pharmaceuticals are the first products to which biotechnology has been applied, the first and perhaps most intense proving ground for U.S. competitive strength in biotechnology will be in the area of pharmaceuticals.

U.S. COMPANIES

The first applications of biotechnology have emerged in the area of pharmaceuticals for several reasons. First, rDNA and MAb technologies were developed with public funds directed toward biomedical research. The first biotechnology products—MAb in vitro diagnostic kits, rDNA-produced human insulin, and interferon—are a direct result of the biomedical nature of the basic research that led to these new technologies. Second, pharmaceutical companies have had years of experience with biological production methods, and this experience has enabled them to take advantage of the new technologies. Finally, since some pharmaceutical products, such as large polypeptides and antibiotics, can only be produced by biological methods, there are no competing production methods that might inhibit the application of biotechnology to their production.

Pharmaceuticals are profitable products because they are low volume, high-value-added products.* This and other financial considerations such as the following have led many U.S. companies to apply biotechnology to the pharmaceutical field.

- The time required to develop some pharmaceutical applications of biotechnology, in particular MAb or DNA probe in vitro diagnostic products for humans, is much less than that required to develop other industrial applications (except possibly some animal health applications).
- Many of the pharmaceutical products being developed with biotechnology are replacements for or improvements in pharmaceutical products already on the market, and they can quickly generate income to finance the development of additional products.
- The pharmaceutical industry offers high rates of return on both sales and equity and is thus an attractive and profitable industrial sector into which firms might diversify.
- Many of the biotechnology pharmaceutical markets may be relatively small. Small firms with limited production and financial resources are able to compete more equally with large firms in small product markets rather than in large markets, because economies of scale and costs of marketing in small product markets are small.

*See *Chapter 3: The Technologies*.

**Applications of biotechnology to the area of pharmaceuticals are discussed further in *Chapter 5: Pharmaceuticals*.

*Value added is the value that a company adds to goods and services that it purchases from other companies. It is the difference between the sales revenues and the cost of resources that it has purchased from other companies. For a "high-value-added" product, therefore, the difference between the resources expended to produce the product and the sales revenues generated by the product is greater than average.

U.S. pharmaceutical companies are quite active internationally. Table 5 illustrates the distribution of sales by the top 20 U.S. and foreign pharmaceutical companies in 1981. Sales by the U.S. companies listed represented almost 60 percent of the total pharmaceutical sales for the top 20 pharmaceutical companies in the world. On the average, almost 42 percent of the sales by these U.S. companies were foreign sales. According to the Institute for Alternative Futures, foreign sales accounted for roughly 43 percent of total U.S. prescription drug sales in 1980 (45), and U.S. pharmaceutical subsidiary sales in foreign countries exceeded $10 billion in 1980.* Given established U.S. pharmaceutical companies' strong export performance in the past, the U.S. posture in world pharmaceuticals markets will be a subject of great interest as biotechnology develops.

Up until about 1976, the average participant in the U.S. pharmaceutical industry could be described as a research-based, integrated, multinational company that spent (and still does) approximately 11.5 percent of its annual pharmaceutical sales on R&D (67). Since about 1976, the profile of the participants has changed considerably. Approximately 70 new U.S. companies have entered the pharmaceutical field just to apply biotechnology. Many of these NBFs are wagering their existence on the success of commercial pursuits of biotechnology in nascent pharmaceutical product markets. In total, about 135 U.S. companies—78 NBFs and 57 established companies—are known to be pursuing pharmaceutical product and process development using biotechnology.*

Since the early 1960's, the U.S. share of world pharmaceutical research, innovation, production, sales, and exports has declined, as has the number of U.S. companies actively participating in the various ethical drug markets compared to the

*This figure is from a survey of Pharmaceutical Manufacturers Association member companies that had not been published as this report went to press.

*The high level of U.S. firms' interest in pharmaceutical applications of biotechnology is in part a reflection of the large number of old and new firms producing MAbs. Many companies included in table 4 are using hybridoma technology to produce MAbs for the markets traditionally addressed by the pharmaceutical industry. In some cases, OTA did not have sufficient information to determine the specific application for MAbs. For example, some companies indicated that they were engaged in the production of MAbs, but would not specify their intended use (i.e., research, separation and purification, diagnostic or therapeutic products for humans, animals, or plants). Because a majority of firms producing MAbs are manufacturing MAbs for pharmaceutical use, OTA placed firms for whom data were incomplete in the pharmaceutical sector, even though hybridoma technology is also essential to fundamental molecular research on plants, animals, and bacterial systems.

Table 5.—Distribution of Sales by the Top 20 U.S. and Foreign Pharmaceutical Companies, 1981

Company	Home country	Percent of sales in home country	Percent of sales in other countries	1981 total pharmaceutical sales (millions of dollars)	Share of pharmaceutical sales
American Home Products	U.S.	66%	44%	$2,303	
Merck	U.S.	53	47	2,266	
Bristol-Myers	U.S.	71	29	2,190	
Warner Lambert	U.S.	55	45	2,045	
SmithKline Beckman	U.S.	59	41	1,782	
Pfizer	U.S.	43	57	1,777	
Eli Lilly	U.S.	62	38	1,664	58%
Johnson & Johnson	U.S.	56	44	1,308	
Upjohn	U.S.	62	38	1,242	
Abbott	U.S.	65	35	1,182	
Schering-Plough	U.S.	51	49	924	
Hoechst	F.R.G.	28	72	2,555	
Bayer	F.R.G.	24	76	2,400	19%
Boehringer-Ingleheim	F.R.G.	37	63	1,197	
Ciba-Geigy	Switz.	2	98	1,891	
Sandoz	Switz.	5	95	1,515	16%
Hoffmann-La Roche	Switz.	3	97	1,629	
Takeda	Japan	94	6	1,195	4%
Rhone-Poulenc	France	41	59	1,008	3%

SOURCE: Adapted from Arthur D. Little, estimates based on publicly available company data.

number of foreign firms (80). At least one study has suggested that substantially fewer U.S.-originated new chemical entities will appear on the market in the mid to late 1980's than are appearing today because of a decline in self-originated investigational new chemical entities since the mid-1970's (83). Table 6 indicates the number of new pharmaceutical products introduced by the United States, four European countries, and Japan in the period 1961-80 and each year since. As the figures in that table show, the United States and France were the leaders in 1961-80, with 23.6 and 18.1 percent of new product introductions, respectively. They were followed by West Germany, Japan, Switzerland, and the United Kingdom. The world leader for the years 1981-83 is Japan, with an average of 27 percent of new product introductions. Although the United States had an average of only 16 percent of new product introductions for the years 1981-83, the drive by NBFs and established U.S. companies to apply biotechnology to the development and production of pharmaceuticals could help reverse the downward trend in U.S. innovation and thereby contribute to the competitive strength of U.S. companies in world pharmaceutical markets.

FOREIGN COMPANIES

Established European and Japanese companies, following the lead of NBFs and established companies in the United States, are now vigorously pursuing pharmaceutical applications of biotechnology.* On average, European companies' biotechnology R&D budgets lag somewhat behind the budgets of established U.S. companies and some U.S NBFs as well (see table 7). As biotechnology processes gain wider acceptance in the pharmaceutical industry, however, European manufacturers—e.g., the West German companies Bayer AG and Hoechst, the Swiss companies Hoffmann-La Roche, Ciba Geigy, Sandoz, and the French company Rhone Poulenc—are expected to challenge U.S. companies, if for no other reasons than their prevailing strength in bioprocessing, their strength in international pharmaceutical markets (see table 5)* and their intentions to maintain this strength,

*Japanese companies are though to have begun making a serious commitment to biotechnology as early as late 1981 (70). West German companies were among the last European companies to begin commercializing biotechnology and did not intensify their R&D efforts in biotechnology until late 1982. Other European countries have paralleled the Japanese in their date of entry into biotechnology.

*Although no British pharmaceutical companies appear in table 5, British companies such as Beecham, Wellcome, Glaxo, and ICI are important international manufacturers of biologically produced products and are applying biotechnology to product development. Additionally, Beecham and Glaxo are among the world's largest producers of biologically made products (48).

Table 7.—Biotechnology R&D Budgets for Leading U.S. and Foreign Companies, 1982[a]

Company[b]	Biotechnology R&D budget (millions of dollars)
Hoechst (F.R.G.)	$4.2[c]
Schering A.G. (F.R.G.)	4.2
Hoffmann-La Roche (Switz.)	59
Schering-Plough (U.S.)	60
Eli Lilly (U.S.)	60
Monsanto (U.S.)	62
DuPont (U.S.)	120
Genentech (U.S.)*	32
Cetus (U.S.)*	26
Genex (U.S.)*	8.3
Biogen (U.S.)*	8.7
Hybritech (U.S.)*	5
Sumitomo (Japan)	6+
Ajinomoto (Japan)	6+
Suntory (Japan)	6+
Takeda (Japan)	6+
Elf-Aquitaine (France)	4+

[a]Biotechnology R&D figures for British companies not available.
[b]Companies with asterisks are NBFs.
[c]1983 figure.
SOURCE: Office of Technology Assessment.

Table 6.—Introduction of New Pharmaceutical Products by Country of Origin Between 1961 and 1983

Country	Number of new products introduced by year[a]			
	1961-80	1981	1982	1983 (est.)
Japan	155 (10.3%)	15 (23.1%)	9 (23.1%)	17 (35.4%)
West Germany	201 (13.4%)	8 (12.3%)	1 (2.6%)	7 (14.6%)
United States	353 (23.6%)	9 (13.9%)	9 (23.1%)	6 (12.5%)
France	271 (18.1%)	3 (4.6%)	5 (12.8%)	5 (10.4%)
United Kingdom	— (—)	3 (4.6%)	— (—)	3 (6.2%)
Switzerland	109 (7.3%)	6 (9.2%)	4 (10.2%)	— (—)

[a]Numbers in parentheses indicate share of total number of new pharmaceutical products introduced for the years indicated.
SOURCE: Nomura Research Institute, "Trends of Biotechnology in Japan," Tokyo, July 1983.

and their increasing shares of worldwide pharmaceutical R&D expenditures as compared to U.S. companies. (Pharmaceutical R&D expenditures by country for the years 1964, 1973, and 1978 are shown in table 8).

The average European company's involvement in biotechnology is largely characterized by research contracts with universities and research institutes rather than by investments in new in-house biotechnology facilities.* Some of the large pharmaceutical companies of Switzerland have, however, begun to make substantial investments in biotechnology facilities. Hoffmann-La Roche, for example, spent $59 million on biotechnology R&D in 1981 (26) and ranks eleventh in worldwide pharmaceutical sales (28). Ciba-Geigy, which commands 3.1 percent of the global drug market, is building a $19.5 million biotechnology center in Switzerland and a $7 million agricultural biotechnology laboratory in North Carolina (11,12).

West German chemical and pharmaceutical companies have been among the last foreign companies to move into biotechnology. Many of the companies have signed contracts with universities instead of investing in facilities to support their research (10). Some West German companies, including Schering AG and Boehringer Ingleheim, however, are making significant contributions to the German biotechnology effort. Schering AG, for example, in a joint agreement with the State of Berlin is establishing a $10.7 million institute of "genetic engineering," which is regarded as an important step for biotechnology research in Germany (29).

In terms of total sales, pharmaceutical companies in the United Kingdom are not among the world's top 20, and historically, the United Kingdom has been slow to commercialize the results of much of its basic research. It is important to note, however, that some British pharmaceutical companies (e.g., Glaxo and Beecham) possess substantial bioprocessing knowledge, a capability that may provide them with a competitive advantage as biotechnology develops. Furthermore, some British pharmaceutical companies have made in-house investments in biotechnology. ICI and Wellcome appear to be among the most strongly committed of the British pharmaceutical companies commercializing biotechnology. ICI, for example, has the world's largest continuous bioprocessing plant and is considered an international leader in bioprocessing technology. This company recently developed a new biodegradable thermoplastic polyester, Biopol®, formed by a genetically manipulated micro-organism. Although Biopol® is not a pharmaceutical, it does give some indication of ICI's innovative capacity in the biotechnology field.

The pharmaceutical and chemical companies of France appear less aggressive than British companies in developing biotechnology expertise. Three major French companies have R&D programs in biotechnology—Elf Aquitaine (67-percent Government-owned), Rhone Poulenc (100-percent Government-owned), and Roussel Eclaf (40-percent Government-owned and a Hoechst subsidiary). Of these three, Elf Aquitaine has committed the most to biotechnology. It owns Sanofi, a phar-

*Many of the established U.S. companies have made substantial investments in new in-house facilities. See section below on "Established Companies."

Table 8.—Pharmaceutical R&D Expenditures by Country: 1964, 1973, and 1978

	1964		1973		1978	
	Level (millions of dollars)	World share (percentage)	Level (millions of dollars)	World share (percentage)	Level (millions of dollars)	World share (percentage)
United States	$282	60%	$640	34%	$1,159	28%
Federal Republic of Germany	40	8	310	16	750	18
Switzerland	38	8	244	13	700[a]	17
Japan	27	6	236	13	641	15
France	28	6	166	9	328	8
United Kingdom	29	6	105	6	332	8

[a] Estimated.

Note: Data are in current dollars and represent expenditures for both human and veterinary research.

SOURCE: National Academy of Sciences, *The Competitive Status of the U.S. Pharmaceutical Industry*, Washington, D.C., 1983.

maceutical company that is applying biotechnology to human and animal health in areas including diagnostics, neuropeptides, serums, vaccines, and antibiotics, and has established Elf-Bioindustries and Elf-Bioresearch to develop biotechnology in the foodstuffs and agriculture sectors. To support some of its new biotechnology R&D, Elf is currently building a $10 million "genetic engineering" plant (5). Rhone Poulenc is the world's second largest producer of animal health products (84) and is considered to be the second most committed of the three French companies actively commercializing biotechnology (50). To support its biotechnology effort, in 1980, Rhone Poulenc established a small specialty biotechnology subsidiary named Genetica.

Despite the efforts of companies such as Elf and Rhone Poulenc, the initial hesitation France expressed in the early stages of biotechnology development has put French companies at a distinct disadvantage internationally, particularly vis-a-vis U.S. companies. The French Government has a formal policy designed to promote biotechnology, but it is not clear that whatever impetus this policy provides will be great enough to compensate for France's slow entry into biotechnology. Historically, the French Government's plans to promote national champions (e.g., the Plan Calcul, the Concord) have failed. As the pace of biotechnology commercialization quickens, a strong private sector effort may be necessary in order to launch France into a more competitive position.

Overall, Europe is considered to be farther behind the United States in the application of biotechnology to product-related research areas than in fundamental research (23). Strong commercialization efforts by the major chemical companies of West Germany or by the pharmaceutical companies of Switzerland or the United Kingdom, however, could significantly improve West Germany's, Switzerland's, or the United Kingdom's current competitive positions in the commercialization of pharmaceutical applications of biotechnology.

Some would argue that large companies have an inertia that is difficult or impossible to change, making rapid changes in research policy and direction impracticable (5). To the extent that large companies pursuing pharmaceutical applications of biotechnology in Europe lack the dynamism and flexibility to compete with the combined efforts of NBFs and established companies in the United States, Europe could initially be at a competitive disadvantage. If the timing of market entry for therapeutic and diagnostic products becomes the most important factor in competition for market share and market acceptance, however, the marketing strength of the European multinationals could help balance competition in pharmaceuticals between the United States and Europe.

The potential competitive challenge that will be mounted by Japan in the area of pharmaceuticals is more difficult to estimate than the challenge from the European countries for two reasons: 1) Japanese pharmaceutical companies such as Takeda, Sumitomo Chemical, Mitsubishi Chemicals traditionally have not had a significant presence in world pharmaceutical markets (55); and 2) present Japanese commercialization efforts, most being proprietary, are difficult to assess either quantitatively or qualitatively. One set of factors characterizing Japanese efforts to apply biotechnology to pharmaceutical development suggests a rather formidable challenge facing U.S. companies in future biotechnology-related pharmaceutical markets, while a different set of factors suggests less of a future challenge. Each set of factors is discussed in turn below.

Factors that suggest that Japan will have international competitive advantages in the application of biotechnology to pharmaceutical development include the following:

- The application of biotechnology to pharmaceuticals in Japan has stimulated the involvement in pharmaceuticals of many Japanese companies from a broad variety of bioprocess-based industries. Table 9 shows the diversification of Japanese chemical, food processing, and textile and pulp processing companies into pharmaceuticals.

A 1982 Keidanren* survey of 132 Japanese companies using biotechnology found that 83 percent

*Keidanren, the Japan Federation of Economic Organizations, is a national organization composed of about 700 of the largest

Table 9.—Diversification of Japanese Chemical, Food Processing, Textile, and Pulp Processing Companies Into Pharmaceuticals

Company	Pharmaceutical field of entry
Chemical companies:	
Sunstar	Antibiotics, interferon
Hitachi Chemical	Antibiotics, vaccines
Hokko Chemical Industry	Antibiotics
Mitsubishi Chemical Industries	Physiologically active agents, anticancer drugs, diagnostic reagents, monoclonal antibodies
Denki Kagaku Kogyo	Physiologically active agents
Sumitomo Chemical	Monoclonal antibodies, interferon, growth hormone
Daicel	Anticancer drugs
Mitsubishi Petrochemical Industries	Diagnostic reagents
Chisso	Diagnostic reagents
Mitsui Toatsu Chemical	Urokinase
Food processing companies:	
Ajinomoto	Antibiotics
Suntory	Antibiotics, interferon, anticancer drugs, drugs for treatment of high blood pressure
Meiji Seika Kaisha	Antibiotics, interferon
Sanraku-Ocean	Antibiotics
Kikkoman Shoyu	Physiologically active agents, antibiotics, immune suppressors
Takara Shuzo	Physiologically active agents
Meiji Milk Products	Physiologically active agents, interferon
Yakult Honsha	Physiologically active agents, anticancer drugs, diagnostic reagents for liver cancer
Kyowa Hakko Kogyo	Physiologically active agents, interferon
Kirin-Seagrams	Interferon
Kirin Brewery	Anticancer drugs
Sapporo Breweries	Anticancer drugs
Toyo Jozo	Immune suppressors
Morinaga & Co.	Diagnostic reagents for liver cancer, drugs for treatment of high blood pressure
Snow Brand Milk Co.	Interferon
Textile and pulp companies:	
Asahi Chemical Industry	Interferon
Toray Industries	Interferon
Teiji Limited	Interferon
Kirin-Seagrams	Interferon

SOURCE: Office of Technology Assessment.

Japanese companies. It enjoys the regular and active participation of the top business leaders working closely with a large professional staff to forge agreements on behalf of business as a whole. It often surveys its members on issues of economic importance.

of the 60 companies that responded were pursuing applications in the area of pharmaceuticals (70), compared to only 62 percent of U.S. companies (see table 4). Intensified competition is expected to push technical advances in the area of pharmaceuticals along in Japan at a rate that is comparable to or greater than the rate in the United States. Among the companies using biotechnology in Japan, it is already a widely accepted view that Japan can catch up with the United States within 5 years. This point is very well illustrated by the *Nikkei Sangyo Shimbun* (Japan Industrial Daily) survey undertaken in June 1981. According to the survey, 48 percent of the 128 responding firms thought Japan could catch up to the United States in the commercial development of biotechnology in 5 years, and 24 percent estimated that catching up would take only 2 to 3 years (57).

- The Government of Japan, which has targeted the pharmaceutical industry for international expansion, has improved the environment for pharmaceutical innovation, and thus, for the application of biotechnology.

The Japanese Government through targeting of the pharmaceutical industry, changes in patent laws to prevent imitation, and pricing policies in the Government-administered national health insurance system has begun an effort to coordinate trade, pricing, and health care policies to promote pharmaceutical innovation and overseas expansion (74). These Government efforts are expected to facilitate the application of biotechnology in the Japanese pharmaceutical industry.

- Joint pharmaceutical research projects and collaborative arrangements among companies, sometimes in conjunction with Government research institutions, promote biotechnology transfer throughout Japanese industry and accelerate the pace of technical advances. Table 10 provides a list of some Japanese joint ventures in pharmaceuticals derived from the Keidanren survey of 1982.

As early as 1979, the Japanese Ministry of Health set up a study group between Green Cross and Toray Industries to speed the development of interferon, because the Ministry had concluded

Table 10.—Japanese Joint Ventures in Pharmaceutical Applications of Biotechnology

Companies	Product area
Otsuka Pharmaceutical/Hayashibara/Mochida Pharmaceutical	Production of alpha, beta, and gamma interferon
Yamanouchi Pharmaceutical/Ajinomoto	Large-scale production of thrombolytic agent
Yoshitomo Pharmaceutical/Takeda Chemical	Large-scale production of thrombolytic agent
Ajinomoto/Morishita Pharmaceutical	R&D on pharmaceuticals
Yoshitomi Pharmaceutical Industries, Ltd./ Yuki Gosei Kogyo Co., Ltd.	Developing rDNA products for circulatory system
Takara Shuzo/Taiho Pharmaceutical	Development of heart drugs using rDNA
Toray Industries/Kyowa Hakko Kogyo/Gan Kenkyu Kai (Cancer Research Association)	Development of beta and gamma interferon by rDNA
Asahi Chemical Industry/Dainippon Pharmaceutical/Tokyo University	R&D on alpha and gamma interferon
Toray Industries/Daiichi Seiyaku Co., Ltd.	Using rDNA to produce gamma interferon
Ajinomoto/Takeda Chemical Industries, Ltd.	Development of interleukin-2
Asahi Chemical Industries Co., Ltd./ Dainippon Pharmaceutical Co.	Development of tissue necrotic factor

SOURCE: Office of Technology Assessment.

that the separate approach being taken was costly both in terms of funds expended and time taken (73). Many other examples of technical collaboration in biotechnology in Japan can be cited, and many more Japanese companies have intentions to cooperate with one another in research or development and/or in commercialization in the future. In 1981, a scientist from the Fermentation Research Institute of Japan's Ministry of International Trade and Technology acknowledged that almost half of the companies who work or intend to work in "genetic engineering" will cooperate or have already cooperated in some biotechnology activities (79). Joint ventures such as those listed in table 10 might provide Japanese companies with commercial advantages for two reasons: 1) each firm participating in the venture brings different resources and expertise to the project, thereby making the group effort more efficient; and 2) the intention of some of the joint ventures is to secure patents in fields not yet pre-empted by foreign competition (e.g., new host-vector systems and sophisticated sensors for bioprocessing) or to undertake joint clinical testing (70).

- Japan's share of world pharmaceutical R&D expenditures has been increasing steadily since 1964 (see table 8) as has its share of the worldwide total of newly introduced pharmaceuticals (see table 6).

In 1981, Japanese companies ranked first in terms of the largest number of major new drugs introduced into world markets, being responsible for 15 (23 percent) of the 65 newly introduced pharmaceuticals (see table 6). In 1982, Japanese companies again accounted for roughly 23 percent of the new pharmaceuctical products introduced. They also accounted for over 16 percent of all U.S. patents issued for pharmaceutical and medicinal products and for 38 percent of all U.S. pharmaceutical and medicinal patents granted to foreign firms (14).

- Japanese companies applying biotechnology to pharmaceutical development (in contrast to U.S. companies) appear to be dedicating relatively more research effort to the later stages of commercialization (i.e., bioprocessing) and cancer treatment. Seventy-five percent of all Japanese medical and drug companies are engaged in MAb research, and a large proportion of the MAb R&D is targeted toward developing a "magic bullet" for cancer treatment, monitoring bioprocesses, and recovering proteins (70).

Factors that suggest that the Japanese may not have significant advantages in future biotechnology-related pharmaceutical markets include the following:

- Barriers to entering foreign pharmaceutical markets are high, and Japanese companies at present have neither distribution channels in place nor a sufficient sales force to permit aggressive marketing of pharmaceutical products in Western markets.

Japanese companies' lack of distribution channels in Western pharmaceutical markets is one factor that has limited Japanese companies' ability to penetrate these markets. It is expected that the mode by which Japanese companies will penetrate these markets in the future will be through joint ventures with U.S. or European companies that allow Japanese companies to take advantage of existing distribution channels.* Although Japanese companies tend to seek opportunities to penetrate foreign markets directly through manufacturing subsidiaries rather than through licensing contracts, only two Japanese companies have established equity joint ventures with U.S. firms** and only three have established U.S. subsidiaries.*** However, the international expansion of Japan's pharmaceutical industry has only just begun.

- Almost half of the Japanese companies now using biotechnology expect to "catch up" technologically to the United States in 5 years. These companies therefore intend to set their own R&D and commercialization targets beyond the 5-year catch-up period at considerable commercial risk.

The intention of Japanese companies to catch up to U.S. companies and to set their own R&D targets is a unique phenomenon. In the past, even in high- technology fields such as computers and electronics, the R&D and commercialization targets have been demonstrated in advance by U.S. and Western European companies, so Japanese companies have not had to worry about the marketability of their R&D and commercialization efforts. By selecting the best technology available and refining it, Japanese companies have been able to minimize the time required to catch up with the front runners and sometimes surpass them at the product marketing stage (70). Given the lack of established commercial targets in biotechnology and considering the barriers to entering foreign pharmaceutical markets mentioned above, it cannot be assumed that the Japanese will be major competitors in biotechnology-related pharmaceutical markets.

- Japan's traditional bioprocess-based industries, including pharmaceuticals, rely largely on conventional microbiology, genetics, and bioprocess feedstocks. These traditional approaches in bioprocessing could be challenged by new biotechnology (41).

Japan is considered to be behind the United States in fundamental biology. This weakness in fundamental biology could reduce the potential competitive threat of Japanese companies applying biotechnology to pharmaceutical development.

- Biotechnology R&D investments by Japanese companies are still low in comparison to the investments by U.S. companies.

Although Japan's aggregate investment in pharmaceutical R&D has increased steadily since 1964, investments by individual Japanese companies in biotechnology R&D are still low compared to investments by NBFs and established companies in the United States (see table 7). According to the *Nikkei Sangyo Shimbun* survey (June 1981) and the Keidanren survey (1982), only 5 Japanese companies spent more than $6 million per year on biotechnology R&D. The average R&D expenditure of 49 of the 60 Japanese companies that responded to the Keidanren survey was under $1 million. Although it is difficult to translate R&D investment into commercial success, on a quantitative basis, Japan falls far behind the United States in terms of industrial expenditures on biotechnology research.

*Animal agriculture industry**

U.S. COMPANIES

The animal agriculture industry encompasses companies engaged in the manufacture of products, the prevention and control of animal diseases, animal husbandry, growth promotion, and genetic improvement of animal breeds. The companies that dominate the production of most animal health products are established U.S. and

*In support of this expectation is a study by the Japanese Productivity Center in 1982 of the potential for Japanese drug firms in the United States. The study estimated that the establishment of a U.S. subsidiary by a Japanese company would require an investment of about $80 million over a 4-year period. The study recommended that Japanese companies form joint ventures with U.S. companies rather than establish a Japanese company or purchase a U.S. company (75).

**Takeda with Abbott (U.S.) and Fujisawa with SmithKline (U.S.).

***The three U.S. subsidiaries are Daiichi Pharmaceutical Corp., Otuska Pharmaceutical, and Alpha Therapeutics (subsidiary of Green Cross).

*Applications of biotechnology to animal agriculture are discussed further in *Chapter 6: Agriculture*.

foreign pharmaceutical and chemical companies.* Most of these companies have global marketing and distribution networks and undertake animal drug production as a diversification of their principal activities. In recent years, the advent of biotechnology, the rising industrialization of animal agriculture, and changing dietary habits in foreign countries have increased the demands for improvements in old products and for completely new products. NBFs may have a major role to play in expanding animal health markets.

Sixty-one companies in the United States are known to be pursuing animal health related applications of biotechnology, as shown in table 4. Thirty-four (56 percent) of these companies are NBFs. Of special note is the role new firms appear to be playing in three major segments of the industry—diagnostic products, growth promotants, and vaccines. Possible explanations for why some NBFs might be interested in these three animal health markets include the following:

- Recombinant DNA methods used to make human vaccines are suited to making safe and effective animal vaccines against both viral and bacterial infections, just as the MAb or DNA probe technology used to produce human products is suited to making passive vaccines or diagnostic products for animals.**
- The markets for many animal health products (e.g., vaccines or diagnostic products) are relatively small and therefore allow NBFs to compete equally with larger companies without suffering from scale disadvantages.
- The commercial introduction of veterinary vaccines can generally be achieved more quickly than can that of human therapeutic products. The regulatory process allowing veterinary vaccines to enter the market typically can be completed in about 1 year (17). Thus, the lower costs of commercialization for veterinary vaccines in comparison to human pharmaceuticals and the potential for short-term product revenues may reduce NBFs' financial need to collaborate extensively with established companies.*
- Some veterinary vaccine research (e.g., on feline leukemia vaccines) could serve as a model for developing human vaccines for similar viruses that could launch some NBFs into the more profitable human pharmaceutical markets.

The fact that 34 of the 61 U.S. companies pursuing applications of biotechnology in animal agriculture are NBFs suggests the evolution of an expanding animal health market in which NBFs such as Molecular Genetics, Inc. (MGI), Amgen, Chiron, Bio-Technology General and Cetus perceive opportunities. In contrast to human pharmaceutical products, animal vaccines and diagnostic products are in many cases being developed by NBFs independently of established U.S. or foreign companies.

In the development of animal growth promotants, however, established U.S. companies are more involved. The market for animal growth promotants is the second fastest growing market in the animal health field, and because it may be the most significant commercial development area (26), it is also one of the most competitive. Global sales for growth promotants are expected to reach $515 million by 1985 (84). Several established U.S. companies, including American Cyanamid, Eli Lilly, Monsanto, and Norden (a subsidiary of SmithKline Beckman), have displayed interest in the field by sponsoring research contracts with NBFs such as MGI, Biotechnica International, Genentech, and Genex. Other established U.S. companies have shown interest by conducting initial evaluations of growth promotants developed by NBFs, as Eli Lilly did in the case of a product developed by the NBF BioTechnology General.

In an effort to expand their own technical capabilities and reach new product markets, some es-

*Major U.S. producers of animal health products include Syntex, Pfizer, Eli Lilly, Upjohn, SmithKline Beckman, American Cyanamid, Merck, Johnson & Johnson, Tech America, and Schering-Plough. Major foreign producers include Burroughs-Wellcome (U.K.), Rhone-Merieux (France), Hoechst AG (F.R.G.), Bayer AG (F.R.G.), Connaught (Canada), Beecham (U.K.), Solvay (Belgium), Boehringer Ingleheim (F.R.G.), Intervet (Netherlands), and Elf Aquitaine (France).

**The NBFs Chiron and Cetus both became involved in the veterinary products business as extensions of their research in the field of human health care (17,20). The NBF Monoclonal Antibodies, Inc., as a spinoff from research on detection kits for human pregnancy and ovulation, is developing an ovulation detection kit for large animals which will be useful in animal breeding management.

*Collaborative ventures between NBFs and established U.S. and foreign companies are discussed further below.

tablished pharmaceutical and chemical companies have contracted with NBFs for animal health projects including the development of animal growth promotants and vaccines for foot-and-mouth disease, rabies and colibacillosis (a diarrheal disease that kills millions of newborn pigs and calves each year). Norden, for example, funded research by the NBF Cetus to develop a vaccine to prevent colibacillosis in hogs. This vaccine received the U.S. Food and Drug Administration's (FDA's) approval in 1982. As other examples, American Cyanamid and Merck have both contracted with NBFs for projects involving bovine growth hormone and a vaccine for foot-and-mouth disease. Many of the products under joint development are already undergoing testing.

Several NBFs are in a strong competitive position vis-a-vis established U.S. and foreign companies in animal-related biotechnology. Most of the established U.S. companies have made relatively small investments in this area—equal to or less than investments in animal health by most of the leading NBFs (54). As established U.S. companies in the animal health field increase their biotechnology investments, the U.S. competitive position in domestic as well as foreign animal health markets should strengthen.

FOREIGN COMPANIES

Established U.S. and European companies control world animal health product markets, but collectively, European companies' efforts to produce new or replacement animal vaccines or growth promotants using biotechnology do not appear to be as strong as the collective efforts underway in the United States. European companies appear on the basis of reported research projects almost exclusively dedicated to the development of products for the world's two largest animal vaccine markets, rabies and foot-and-mouth disease. U.S. companies dominate the world market for animal growth promotants, and few European animal health companies have indicated an interest in entering the growth promotants product market. Furthermore, few European companies have established R&D joint ventures with the leading U.S. NBFs engaged in growth promotant R&D.

Japanese companies have exhibited relatively little commercial interest in the area of animal health, probably because meat does not constitute as large a portion of the Japanese diet as it does of the diets in Western European countries and the United States. Recently, however, the Japanese chemical company Showa Denko and the U.S. company Diamond Shamrock set up a biotechnology joint venture, SDS Biotech Corp., in Ohio exclusively for animal health research (13).

*Plant agriculture industry**

U.S. COMPANIES

The plant agriculture industry encompasses companies engaged in R&D activities to modify specific plant characteristics (e.g., tolerance to stress, nutritional content, yield, and growth rate) or to modify traits of micro-organisms that could be important to plant agriculture (e.g., nitrogen fixation, disease suppression, and insecticide production). The importance of plants as a food source and renewable resource and the potential of biotechnology to alter plant characteristics has attracted a diverse set of firms to the plant agriculture industry. Fifty-two U.S. firms listed in table 4, 30 established companies and 22 NBFs, are applying biotechnology to plants. Table 11 provides some examples of the diverse application areas that NBFs are pursuing.

Established U.S. companies from industries ranging from oil and chemicals to food and pharmaceuticals appear to be dominating the U.S. investment in biotechnology R&D in plant agriculture (25). U.S. chemical companies that have made considerable in-house investments in plant-related biotechnology research include American Cyanamid, Dow, Allied, DuPont, and Monsanto. These companies already produce chemical pesticides and herbicides and already have research using plant cell and molecular biology techniques directed toward increasing the resistance of crop plants to these chemicals (15). American Cyanamid, which has the expertise to synthesize herbicides, and the NBF MGI, which has the expertise to develop novel corn strains tolerant to new herbicides, have a joint program to develop herbicide-resistant corn. New corn strains developed for herbicide resistance might make it possible

*Applications of biotechnology to plant agriculture are discussed further in *Chapter 6: Agriculture*.

Table 11.—Applications of Biotechnology to Plant Agriculture for Seven New Biotechnology Firms

Advanced Genetic Sciences, Inc.:
- —Development of plant varieties with increased resistance to disease, stress, herbicides and pests, and tolerance to extreme weather conditions
- —Development of antagonistic bacteria that do not contain ice nucleation properties to optimize frost protection

Biotechnica International:
- —Improvement of nitrogen-fixing capability of *Rhizobium*
- —Introduction of nitrogen-fixing capability in plants that rely on fertilizer
- —Herbicide resistance in selected plants
- —Improved protein content in alfalfa

Bio-Technology General Corp:
- —Development of a biofertilizer, *Azospirillum*
- —Development of several strains of trichoderm, a micro-organism that controls soil-borne fungi that cause damage to many plants.

International Genetic Engineering (Ingene):
- —Modification and production of bacteria that are lethal to four specific weeds and three groups of insects
- —Production in micro-organisms of plant growth regulators—hormones that affect many biological functions including flowering, fruit ripening, and water loss.
- —Modification of organisms that are responsible for ice nucleation in an effort to interfere with the organisms' ability to adhere to plants

Cetus Madison:
- —Development of soybean and corn hybrids to increase vigor
- —Development of microbial inocculant for corn, soybean, cotton, wheat, and rice to protect plants against fungal and insect diseases and to increase plant growth through nitrogen fixation and other biological processes
- —Exploration of ways to add genes to make plants unpalatable to insect pests and to make plants resistant to diseases

Ecogen, Inc.:
- —Development of microbial and viral pesticides

Molecular Genetics, Inc.:
- —Development of herbicide-resistant corn and nutritionally enhanced field corn

SOURCE: Company prospectuses and annual reports.

to develop markets for broad-spectrum herbicides that might not otherwise be used. Some U.S. chemical companies may be investing in plant-related biotechnology to compensate for possible reductions in future sales due to the development of plants that do not require chemicals (e.g., plants that fix nitrogen, plants that produce pesticides) or due to competition from microbial insecticides or nonchemical treatments (30). Some pharmaceutical companies may invest in plant-related biotechnology, for example, to seek new sources of therapeutically active substances or to develop a commercial process for producing secondary products from plants (e.g., morphine and codeine).

One route by which some established U.S. companies have entered the plant agriculture field is through the acquisition of seed companies. Seed companies provide both an in-place marketing system and high-quality, commercially successful gene pools, often representing as much as 10 to 20 years of R&D. Through their ownership of seed companies and investments made both in-house and through sponsored research with NBFs, some established companies are assuming active roles in the modern research impetus for seed improvement. By assuming stronger roles in basic plant science research, U.S. companies like ARCO, Shell, Allied, Monsanto, and DuPont hope to play a leading role in the development of future agriculture markets.

FOREIGN COMPANIES

The commercialization of plant-related biotechnology is occurring more slowly in the European competitor countries than in the United States. For example, most West German plant tissue culture research is going on in universities (6). Some of the large European pharmaceutical companies are reportedly interested in plant tissue culture, but only Boehringer Mannheim (F.R.G.) and the Society for Biotechnology Research (GBF, Gesellschaft für Biotechnologische Forshung) have made their interests public. Boehringer Mannheim is also engaged in research to produce digitalis using immobilized plant cells (10). Although excellent basic research is conducted in centers such as the Max Planck Institute for Plant Research in Cologne,* few commercial pursuits are known.

Great Britain possesses some of the strongest basic research in interdisciplinary plant sciences and recently a new firm launched by the British Technology Group, Agricultural Genetics, was established to exploit discoveries made at the Agricultural Research Council. Whether or not the basic research will be commercialized successfully is difficult to predict.

*Bayer signed a 3-year agreement with the Max Planck Institute for research in plant cultivation with special attention to rDNA to improve plant resistance to phytotoxins.

The Japanese are very interested in the development of amino acids and high-value compounds by selecting and engineering plant cells to produce secondary metabolites in vat culture. MITI has identified secondary compound synthesis as a major area for commercialization, and this area of plant-related biotechnology research will receive approximately $150 million from MITI during the next 10 years (15). With their experience in large-scale bioprocessing, the Japanese are well ahead of the United States in this aspect of plant biotechnology. Japanese companies have already reported repeated success in growing plant cells in 15,000 liter batches (68). The upper limit in the United States is only 300 liters (68).

Although biotechnology is not expected to provide foreign countries with an ability to reduce U.S. dominance in world grain markets, it may provide foreign countries with opportunities to seize specific agricultural markets. In both France and Italy, for example, there are major commercial activities in plant tissue culture techniques for eliminating viruses and propagating fruit and nut trees (15).

*Specialty chemicals industry**

The specialty chemicals industry promises to be a particularly competitive industry as biotechnology develops, because large chemical companies from both Japan and the Federal Republic of Germany as well as the United States are hoping to switch from the stagnant commodity chemicals industry into the more profitable specialty chemicals industry.

The general chemical and petrochemical firms of Japan are leaning strongly to biotechnology, and some of them are making rapid advances in R&D through their efforts to make biotechnology a key technology for the future. Japanese companies are expected to be especially strong competitors in future specialty chemical markets for reasons including the following:

- Japanese bioprocess-based companies are known to possess highly developed enzyme technology, a prerequisite for efficient biological production.
- Japanese chemical companies view specialty chemicals as a profitable area in which to diversify. Showa Denko, a leading chemical company in Japan, is expecting to become a major world producer of the amino acid tryptophan, first by using a new low-cost semisynthetic production method, and second by rDNA production.
- Two Japanese companies, Kyowa Hakko and Ajinomoto, are currently the world's major producers of amino acids. Both companies have operating production plants in the United States, and both have strong biotechnology R&D programs in Japan. Ajinomoto, for example, has succeeded in improving the production of the amino acid threonine by rDNA technology using *E. coli*, and Showa Denko has cut in half the production cost for tryptophan through a semisynthetic production process.

The commercialization of biotechnology will require many small, incremental improvements in bioprocess technology, superb quality control, and mass production to progress along the learning curve. As biotechnology development reaches large-scale production stages, well-developed bioprocessing skills will be necessary to compete in world product markets. Nowhere is the art of bioprocessing better refined than in Japan. Certainly Japan's expertise in this area will provide competitive strengths in many future biotechnology product markets.

Two West German companies that have experienced declining profits for the last 10 years because of poor chemical sales are Hoechst and Bayer, the world's largest chemical exporters and the world's two largest pharmaceutical companies (see table 5). These two companies spend more on R&D than any other pharmaceutical companies. Both these companies have targeted specialty chemicals as an area where biotechnology might increase corporate sales and profits (10). Bayer has a longstanding collaboration with its two U.S. subsidiaries, Miles and Cutter, and these two subsidiaries help keep Bayer informed of biotechnology developments in the United States. Much

*Applications of biotechnology to specialty chemicals are discussed further in *Chapter 7: Specialty Chemicals and Food Additives*.

of Bayer's specialty chemical research is taking place in the United States through these two subsidiaries. Bayer has opted for specialty chemicals as its main R&D focus; Miles is important in the enzyme and organic acid field using bioprocessing, and Cutter is expanding its R&D activity in purifying enzymes and proteins on a large scale (10). Two other German companies, Schering AG and BASF AG, are also actively applying biotechnology to the production of specialty chemicals. Schering's main research focus is on the genetic manipulation of micro-organisms to produce amino acids such as lysine (10), and BASF is building a $24 million "Biotechnicum," a combination of research laboratory and pilot plant with a product focus on optically active intermediate chemicals and vitamins. Schering has also signed two research agreements with Genex, one of which involves the development of a genetically manipulated microbe to produce an amino acid.

U.S. and foreign support firms

Companies engaged in biotechnology research have increased and expanded the demands placed on the infrastructure that has traditionally supplied biochemical reagents, instrumentation, and software for biological research and production. As "scaled-up" production of biotechnology products comes on line, the demand for these supplies as well as for new production instrumentation is likely to increase further.

The United States, with an assortment of companies supplying biochemical reagents, instrumentation, and software, has the strongest biotechnology support sector in the world. The U.S. biotechnology support sector is characterized by a large number of small specialty firms that compete in small specialty product markets such as biochemical reagents used in rDNA research (e.g., BioSearch, Vega, P-L Biochemicals (a subsidiary of the Swedish company Pharmacia), Bethesda Research Laboratories,* Collaborative Research, New England BioLabs, Applied Biosystems, Creative Biomolecules, and Intelligenetics) and several medium-sized to large firms that produce analytical and preparative instrumentation as well as bioprocess equipment** for larger, more diverse product markets (e.g., Beckman, Perkin Elmer, Varian, Hewlett Packard, Waters, New Brunswick).

In most support areas, European and Japanese support sectors are underdeveloped compared to that of the United States, although both are expanding quickly. Two factors might account for weak support sectors in Japan and Europe as compared to that of the United States:

- The United States is a recognized leader in basic biomedical research, and over the years, public funds, notably from the National Institutes of Health, have created a large well-defined market for specialty products used in biological research (1).
- Because so many large and small U.S. companies are currently applying biotechnology, the specialty research product needs are greater in the United States than in any other country, and opportunities exist for many small manufacturers. In fact, the U.S. market for custom oligonucleotides (DNA fragments) and biochemical reagents for synthesis of DNA is equal to that of the rest of the world (51).

In Europe and Japan, there are few biotechnology support firms supplying biochemicals. Thus, European and Japanese companies developing biotechnology generally have to manufacture oligonucleotides and other biochemical reagents in-house. Consequently, the expense for biochemicals in European countries and Japan is often greater than in the United States, where many support firms have achieved significant economies of scale (51). The alternative to in-house production of support materials in Europe and

*Bethesda Research Laboratories was recently purchased by Dexter Corp.'s GIBCO division. The new name for the merged company will be Life Technologies, Inc.

**See *Chapter 3: The Technologies* for a discussion of bioprocess equipment.

Japan is reliance on a foreign supplier. Such reliance could impede technical advances (21) and retard commercialization in the short run. Although there are Japanese and European instrumentation manufacturers, U.S. instrumentation is considered superior to both Japanese and European instrumentation and dominates the European market (51). The Japanese instrumentation market is supplied by Japanese manufacturers, which have not made significant inroads in foreign markets (52).

Important product areas

For purposes of analysis, OTA examined three product areas thought to have significant short-term implications for research developments and technical developments in the biotechnology field:

- biochemical reagents used specifically in rDNA research (e.g., oligonucleotides and restriction enzymes);
- instrumentation used in product R&D (e.g., DNA and peptide synthesizers) and separation and purification instruments such as high-performance liquid chromatography (HPLC); and
- software designed to drive the microprocessors that automate instruments as well as software designed to analyze DNA and protein sequence data in data banks.

The United States is a world leader in all three product areas. If adequate supplies of the above products and services can sustain the present rate of growth of biotechnical advancement, the United States could possess a short-term advantage in bringing biotechnology products to international markets.

BIOCHEMICAL REAGENTS

The availability of quality biochemical reagents such as oligonucleotides (DNA fragments) and restriction enzymes (enzymes used to cut DNA) is crucial to sustaining the rapid development of the new biotechnology field and making it viable on a large scale. Between 1980 and 1990, sales of biochemicals for DNA and peptide synthesis in the United States are expected to increase at an annual rate of 20 percent (81). As more research is undertaken in plant agriculture, sales are expected to rise further. The total synthetic DNA market for 1983 to 1984 is estimated at $3 million to $4 million, and demand is expected to increase 25 to 30 percent a year (36).

Until rather recently, most oligonucleotides were made in-house in the United States; however, as demand for these materials has increased, small specialty support firms have been started to exploit these small markets. One source believes that the evolution of small support firms in the United States is gradually shifting many skilled biochemists in U.S. companies commercializing biotechnology from routine laboratory duties to basic research and that the net result has been an increase in the progress of biotechnology research in the United States (51).

Small U.S. support firms are estimated to supply about 25 percent of the total reagents used in biotechnology research in the United States at present (51). Some expect this figure to increase to about 50 percent as small firms achieve economies of scale, and their prices become lower than those of in-house manufacture. Others believe an estimate of 50 percent might be somewhat high, because some of the major users of reagents, in order to control availability, quality, and cost, are opting for in-house manufacture rather than purchase (40). In-house manufacture may in fact limit the growth of the reagent market. The Canadian firm Bio Logicals no longer manufactures oligonucleotides at all, because the market is smaller than was originally estimated, and the business is becoming one of low profit margin (4).

The unavailability of specific DNA sequences will clearly slow any research development on those sequences. Research at the U.S. firm Genentech was slowed, for example, when the company had to wait weeks for a reagent that is only available from Sweden (43). In the United States, the existence of many small custom reagent suppliers makes delays of this kind rare. In Europe, however, delays of 1 to 2 months occur more often. Nonetheless, there is little competition in Europe among firms making custom synthesized fragments, because European researchers are willing to wait a couple of months for special reagents (51). DNA probes (small pieces of DNA that recognize specific genes) are not even manufactured there (21).

The biochemical supply situation is somewhat different in the United Kingdom, a nation strong in basic research but weak in commercial applications (51,69). As early as 1980, a well-known British Government biotechnology report, the Spinks' report, recognized that the United Kingdom had a shortage of suppliers of suitable equipment and reagents for biochemical laboratories (2). The number of new small British suppliers of biochemical reagents and restriction enzymes is increasing, but British firms using these products as well as instrumentation still purchase much of them overseas.* British firms' reliance on foreign biochemical suppliers could be reduced as an increasing number of small supply companies are beginning to form in the United Kingdom.

The demand for support materials in Japan has increased significantly since MITI designated biotechnology a priority area for the 1980's. In anticipating the increased demand for research supplies, the Science and Technology Agency (STA) sponsored an industrial research team** whose objective is "DNA extraction, analysis, and synthesis technology development" (70).

Until recently, oligonucleotides were produced in Japan only on an experimental basis and foreign products were used for domestic consumption. Now, three Japanese companies and their affiliated trading firms produce and market synthetic DNA in Japan,*** and two of them are members of the MITI research team. Only two Japanese companies, Nippon Zeon Co. and Takara Shuzo, produce restriction enzymes for the estimated $4.5 million Japanese market (35). Nippon Zeon Co., a subsidiary of Kongo Pharmaceutical Co., is manufacturing 35 kinds of restriction enzymes and 87 different synthetic DNA fragments mostly for research institutes in Japan (37). Takara Shuzo, in addition to supplying enzymes to the Japanese market, is exporting them to the United Kingdom. Because of the increasing rate at which biotechnology research is being carried out in Japan, and because of the underdeveloped support industry there, the current supply of oligonucleotides and restriction enzymes for biotechnology research in Japan is inadequate. In fact, Japanese distributors are still looking for U.S. suppliers (40).

The biotechnology support structure in Japan is expected to develop differently from that of the United States, because most companies commercializing biotechnology in Japan will continue to manufacture or import their own specialty biochemical supplies. In order to meet their own needs, Japanese companies have integrated vertically and are increasing their efforts to develop products such as reverse transcriptase and other enzymes that will reduce the cost and speed up the rate of biotechnology R&D. This pattern of vertical integration and in-house manufacture is not likely to change in the short term. The Japanese supply structure could retard research and create an early commercial disadvantage for Japanese companies in the short run.

INSTRUMENTATION

The instrumentation field includes all the instrumentation used in biotechnology from the analysis and synthesis of DNA molecules to the monitoring and control of large-scale separation and purification of commercially important biological compounds. Of particular importance to the pace of biotechnical development is the newly designed or recently modified instrumentation that is meeting the special needs of biotechnology research and production. Two of the most important instrument areas are DNA and peptide synthesizers and bioprocessing separation and purification instruments such as HPLCs.

Automated DNA and Peptide Synthesizers.—Automated DNA and peptide synthesizers significantly reduce the number of personnel and the amount of time required for synthesis. Such synthesizers will have significant impacts on the timing of research outputs and technical developments in biotechnology in the United States (61). An increased availability of specifically synthesized gene fragments arising from automated

*The British firm Amersham recently launched new product lines to meet the growing need for restriction enzymes in the United Kingdom, but rather than manufacturing the enzyme itself, Amersham will be supplied with 22 restriction enzymes by the Japanese firm Takara Shuzo Co. (9). Typically, Japanese companies do not pursue small foreign markets; in this case, however, Amersham's distribution network provided easy access to the European enzyme market.

**Ajinomoto, Wakinaga Yukuhin, Yamasu Shoyu, Yuki Gosei Yakuhin Kogyo, Toyo Soda Manufacturing Co., Ltd.

***Nippon Zeon Co.-Mitsui Trading Co., Yamasa Shoyu-Sumitomo Shoji, Yoshitomi-Yuki Gosei.

synthesis may give researchers more flexibility in the manipulation of genetic information. Automated synthesizers can, among other things, expand the availability and variety of linkers and adapters* for cloning DNA, provide probes for finding messenger RNA and DNA gene sequences, or manufacture whole genes themselves.

The United States leads the world in synthesizer technology. The support companies that manufacture DNA and/or peptide synthesizers in the United States include Vega Biotechnologies, BioSearch, Beckman Instruments, Sys-Tec, Applied BioSystems, P-L Biochemicals, Syncor, Genetic Design, and Sequemat. Generally, these companies have very good communication with the U.S. companies and laboratories they supply. BioSearch customers, for example, keep BioSearch continually informed of their needs so that automation can be designed based on these needs. Communication networks between European instrument suppliers and their European customers are not so well developed.** U.S. companies might, therefore, gain some competitive lead time in biotechnology, because they will be among the first to benefit from automation developments in the United States.

There are no Japanese companies actually manufacturing DNA or peptide synthesizers for commercial use (21,81), but some U.S. manufacturers of DNA and peptide synthesizers have established distribution agreements in Japan.*** The reasons given most often for the dearth of Japanese manufacturers are the high risks of bringing synthesizers to market and the small size of the Japanese synthesizer market. A 1982 market survey by American Commercial Co. (Vega Biotechnoloy's Japanese trading company) found the Japanese market at that time to be approximately 150 machines (81). Without automation to synthesize the genes or fragments necessary for research, the Japanese may find it difficult in the short run to keep pace with American research advances. Additionally, if future markets develop for total gene synthesis, Japanese research could be slowed because Japanese companies have not developed their own automation.

The only two DNA synthesizer manufacturers in Europe are Celltech and Cruachan Chemicals Co., Ltd. (U.K.). However, companies in France, the Federal Republic of Germany, Switzerland, and the United Kingdom have introduced peptide synthesizers to the market or plan to soon. Sempa (France) is not aggressively marketing its machine in the United States. The relatively small size of the European market discourages many potential large European manufacturers from entering the market. The inherent risks of introducing a new product might also discourage small European companies from entering the market as well.

Over the next 5 years, the U.S. market for automated DNA synthesizers is expected to grow to between approximately 500 (81) and 1,000 units (21). Since March 1983, Applied BioSystems (U.S.) has shipped 30 synthesizers, and in just over a year, BioSearch (U.S.) has shipped about 50 (37). Some observers expect that the largest biotechnology support markets in the near term will be those for synthesized whole genes and purification systems (21). Though some firms doubt that a market for whole genes is developing, other firms, including Creative BioMolecules (U.S.), have already begun to market whole genes. Creative BioMolecules' synthetic gene for human pancreatic growth hormone releasing factor.

New developments in continuous-flow peptide synthesizers have led to an upsurge in interest in this different type of instrument technology. The U.S. market for peptide synthesizers 5 years from now is expected to be 500 units—the same size market that is forecast for DNA synthesizers (81).

In a situation of rapidly changing technology, the United States is at a clear advantage in the short run because of the supply of automated instrumentation, an automated synthesis instrument supply standpoint, because many small U.S. companies are willing to address these small, high-risk markets. In Europe, few small or large firms are willing to do the same.

---

*Short nucleotide sequences that encode restriction enzyme sites.
**See the Spinks' report recommendations.
***A U.S. synthesizer manufacturer contacted by OTA was not aware of any Japanese companies that manufacture synthesizes (40).

Bioprocessing Separation and Purification Instrumentation.—Technical advances in separation and purification as well as monitoring will affect both laboratory research and commercial production and ultimately the U.S. competitive position in biotechnology (61).* The use of rDNA technology to produce low-volume, high-value-added products as well as high-volume products has greatly increased the need to develop more economic bioprocesses. As large-scale production draws closer, the ability to isolate and purify large quantities of desired products will be a determinant in how fast companies can reach international product markets. Those countries that possess the most advanced technology to separate and purify commercially important compounds might gain some commercial advantages in the early stages of production. Without more economic production, financial and commercial success in biotechnology may be difficult to achieve.

In the United States, Europe, and Japan, there is intense competition in R&D to develop improved large-scale separation and purification methods for biological compounds as well as methods for monitoring and controlling a bioprocess itself.** There is widespread effort to apply HPLC, continuous-flow electrophoresis, and flow cytometry to bioprocesses to decrease the manufacturing costs of compounds such as proteins. Increasingly, R&D efforts are being undertaken to scale-up analytical instruments, particularly HPLCs, for use in larger volume production processes. The United States is a recognized leader in analytical instrumentation used in biological research and thus stands at the forefront of many of the technical innovations being made in the bioprocess field. As automation and the use of sophisticated instrumentation to monitor and control the production process begins to transform bioprocessing from an art to a science, thereby making production more economic, U.S. companies will be in a strong competitive position.

*The reader is directed to *Chapter 10: Bioelectronics* for a discussion of sensor technology.

**See the discussion of bioprocess technology in *Chapter 3: The Technologies*.

HPLC is one of the most commonly used separative techniques and also one of the fastest growing instrumentation fields in the world (76). The growing sales are due in part to its expanded use in both analytical and preparative areas. HPLCs are considered standard analytical tools in the laboratory to accurately isolate and purify organic molecules, drugs, and some peptide hormones. More recently, HPLCs have been scaled-up successfully to monitor bioprocesses and purify large quantities of proteins such as leukocyte interferon.

Half of the $300 million worldwide HPLC market belongs to U.S. producers, and the European HPLC market is dominated by three U.S. companies, Varian, Beckman Instruments, and Waters. Japanese and European companies have tried with little success to penetrate segments of the U.S. instrument market. Pharmacia, a Swedish company, is the only exception. Large American companies such as Hewlett Packard, Perkin Elmer, and Beckman are so firmly entrenched by virtue of their service and applications networks that foreign firms (e.g., Shimadzu, a Japanese company) are having a difficult time making inroads. An absence of major foreign companies in the U.S. market and the dominance of American companies abroad highlights the prominent U.S. position in instrumentation markets.

Although U.S. companies dominate world HPLC markets, the Swedish company Pharmacia is a major competitor in separation and purification technology, especially chromatography (52). In fact, it is the only company in the world doing large-scale industrial chromotography. Waters and Beckman are thought to be catching up (52). According to John McTaggart of Tag Marketing, U.S. companies are catching up to Pharmacia in procedures for reducing the bulk of material at initial stages of isolation and purification (52). The gap is narrowing, because U.S. companies strong in hardware support (i.e., advanced solid matrix, membrane, and hollow fiber design) such as Millipore, Amicon, and Nuclepore are making advances in product recovery through ultrafiltration. The United States is considered the technological leader in hollow fiber and membrane technology.

SOFTWARE

The United States holds a commanding position in software designed for molecular biology and bioprocessing. With a superior capability to analyze and manipulate sequence data or to purify large quantities of valuable products, for example, the United States might gain some commercial lead by hastening research in some product development areas.

Automation will be necessary to develop more efficient bioprocesses and to lower the costs of biological production. U.S. instrumentation and software manufacturers such as Perkin-Elmer and Fisher Scientific are designing a wide range of software for use in biological research and production processes. The United States is the recognized leader in software design in general and in sophisticated computer applications to biological research specifically. Because of the dominant role U.S. companies play in instrumentation markets, and because of the increasing importance microprocessors and automation are having in biological research and production, the United States is expected to gain some short-term advantages in the commercialization of biotechnology.

Software controls all processes automated by microprocessors. Current software applications in biotechnology are wide ranging and include the manipulation of DNA sequence data contained in data banks, the automatic ordering of nucleotide bases to synthesize pieces of DNA, the modeling of protein structures, and the monitoring and control of large-scale bioprocessing. On the analytical level, purification of peptides and DNA fragments, for example, is expected to become more sophisticated through technical advances in automation (40). On a preparative level, the utility of HPLCs, for example, is being increased by interfacing HPLCs with other instruments (e.g., infrared and mass spectrometers) and computers.

The availability in the United States of software designed to analyze the data in the private and public DNA and protein data banks that have been created worldwide may give U.S. companies commercializing biotechnology some competitive advantages. Both public and private DNA sequence banks exist in the United States. The two largest private and public banks respectively are: the Nucleic Acid Sequence Database (1,200,000 nucleotide bases), operated by the National Biomedical Research Foundation, Georgetown University Medical Center; and the Genetic Sequence Data Bank (GENBANK) (1,800 DNA sequences totaling 2 million nucleotide bases) founded on data collected, organized, and annotated by the Los Alamos National Laboratory and developed through funding from the U.S. National Institutes of Health. The latter data bank will be a repository for all published nucleic acid sequences of more than 50 nucleotide base pairs in length. Georgetown also operates the world's largest protein sequence data base, which currently contains 2,100 sequences and about 360,000 amino acids.

The United States is not unique in its creation of such data bases; however, in terms of size, there are no foreign equivalents. The Europeans have their own nucleic acid data base, the Nucleotide Sequence Data Library (operated by the European Molecular Biology Laboratory, EMBL), and the Japanese will have their own equivalent soon. In addition to these foreign DNA data bases, small private foreign protein data banks exist for the exclusive use of the institutions with which they are affiliated.

A research advantage for the United States is expected to arise not only from the availability of data bases, but also from the software being designed by academic institutions, nonprofit research foundations, and private companies to analyze the data in the banks. Since GENBANK's development was made possible through public money, the data are available to the public, domestically as well as internationally. Additionally, subscribers to Georgetown's Nucleic Acid Database can use the accompanying programs to access both the GENBANK and EMBL's bank. With equal international accessibility to the data bases, competitive advantage will flow to the country that has the ability to perform sophisticated sequence manipulation through specially developed software. In fact, the utility of the data bases will be defined by the available software.

The U.S. company Intelligenetics is specializing in the application of data processing and artificial intelligence techniques to biological problems, and this company has created specific software packages to assist researchers with molecular genetics analysis. Some of the subscribers include SmithKline Beckman, DNAX, Hoffmann-La Roche, Biogen, and Pfizer.

Conclusion

The U.S. support sector provides competitive as well as commercial advantages to U.S. companies developing biotechnology through: 1) the timely and sufficient supply of biochemicals such as oligonucleotides and restriction enzymes for rDNA R&D, 2) new or modified instrumentation such as DNA and peptide synthesizers as well as large-scale purification instruments such as HPLCs, 3) the design of new software for research and production, and 4) a continuous exchange of information between suppliers and companies using biotechnology that results in the creation of new products and in constant improvements in existing instrumentation, equipment, and software used in biotechnology R&D.

The first advantage, timely and sufficient supply of biochemical reagents for rDNA R&D, can affect the rate at which some biotechnology research is carried out. An increasing number of small U.S. companies specializing in custom DNA synthesis has made available sufficient supplies of reagents in the United States that are priced lower than European or Japanese supplies. In Europe, although the number of companies supplying custom reagents has increased, supplies still are not adequate and delivery is slow, especially when reagents are imported (43).

The second and third advantages, new or modified instrumentation and new software design, may provide U.S. companies with a short-term advantage through more efficient research methods and production processes. DNA and peptide synthesizers, for example, are beginning to automate the long and tedious manual task of assembling DNA and peptides, thereby creating greater efficiency in the early stages of research. The scale-up of HPLCs for use in purification of commercially important compounds may also provide greater production efficiency. Software used to drive the microprocessors used in synthesizers or bioprocessing equipment, or to manipulate sequence data in data banks, or to direct computer modeling of proteins may also give U.S. companies short-term advantages in the earlier stages of commercialization. It should be noted, however, that these materials can be exported without difficulty, and that any U.S. advantage derived from their manufacture in the United States is short term.

The fourth advantage, information exchange between support firms and the companies developing biotechnology, promotes technology transfer within the United States and stimulates improvements in instrumentation and software design for biotechnology application. Not only do support companies constantly improve on the products that they themselves manufacture, but the companies that they are supplying in turn strengthen the U.S. support base by developing customized and automated instrumentation and equipment for in-house use, which they may then make available to other companies once their proprietary position has been secured. Examples of companies in the latter category include Genentech, Cetus, and Bio Logicals (Canada). Bio Logicals' DNA synthesizer grew out of in-house technology to produce oligonucleotides for itself. Cetus recently established a new subsidiary, Cetus Instrument Systems, to capitalize on the commercial value of novel instrumentation and computer systems developed for its own in-house R&D. Genentech and Hewlett Packard started a joint venture company, HP Genenchem, to develop for themselves and other companies automated instrumentation for use in biotechnology R&D. Genentech will provide the joint venture with instrumentation already developed and add early insights for research and commercial instrument opportunities (37). Possible areas of automation include DNA and protein sequencers and synthesizers and industrial-scale HPLC and flow cytometers for bioprocess monitoring and control.

In the current stage of biotechnology development, there is considerable interaction between suppliers and potential users, particularly in the area of sophisticated instrumentation. Ideas for new products are developed through in-depth conferences with customers and potential customers to determine or anticipate what kinds of R&D problems they might have. Also, in response to customers' needs, U.S. support firms are constantly upgrading and modifying instrumentation to maximize its utility. These interactions and tailoring of instrumentation and equipment to meet industrial needs will be critical to surmounting the numerous problems anticipated in the design, scale-up, control, and optimization of industrial biotechnological processes (22).

The U.S. biotechnology support sector currently provides a sufficient and *timely* supply of biochemicals, instrumentation, and software to U.S. firms using biotechnology. By virtue of its support strength, the United States holds research advantages over other countries—advantages that may or may not be translated into commercial products. For the United States to retain these advantages in the future, U.S. support firms must remain poised to meet the immediate and expanding supply needs of the U.S. firms commercializing biotechnology.

U.S. firms commercializing biotechnology and their role in competition

As noted at the beginning of this chapter, the commercial development of biotechnology in the United States is being advanced by two types of firms: NBFs and large established U.S. companies. It is important to keep in mind throughout this report the organizational nature of the U.S. biotechnology development and commercialization effort and the strength that the present NBF-established firm competition and complementarity lends to this effort. NBFs and established U.S. companies both have important roles to play in the present phase of biotechnology development. Not until the technology is more fully developed will the parameters of responsibility for each group of firms be more clearly defined.

New biotechnology firms

The development of biotechnology is still at an early stage, and competition at present, both in the United States and abroad, is largely in research and early product development (e.g., vector selection and gene expression). Development and commercialization have not yet progressed to a point where competition for market shares is of immediate concern. In the present research-intensive stage of biotechnology's development, NBFs are providing the United States with competitive advantages in biotechnology through contributions to innovation. In the early stages of a new technology, small firms in the United States tend to dominate an industry and contribute most to product innovation. As a group, it is the small companies that have most "quickly and successfully taken new technologies from the laboratory and adapted them for large-scale production" (78). Small firms move much more aggressively to market than do established companies that have built-in disincentives to advance the state-of-the-art quickly because of existing investment in established product lines and production processes.* As a technology matures, many established companies, as later entrants, begin to play a larger role in innovation, as well as production and marketing.

That small firms contribute significantly to technological innovation is widely accepted, although there is disagreement over the amount of their contribution. Some U.S. studies suggest that small businesses play a more important role in technological innovation than do large firms. A recent study prepared for the Small Business Administration by Gellman Research Associates, Inc., for example, holds that: 1) small firms produce 2.5 times as many innovations as large firms, relative to the number of people employed; and 2) small firms bring their innovations to market much more rapidly than do large firms (32). Another study undertaken by Human Services Research for the National Science Foundation found that small firms (i.e., firms with fewer than 1,000 employees) pro-

*For example, a pharmaceutical firm with a vested interest in symptomatic treatment of colds may have little incentive to develop a vaccine against the cold-causing viruses, since it would diminish the company's sales of decongestants.

duced 24 times as many major innovations per R&D dollar as did large firms and 4 times as many as did medium-sized firms (44). Finally, an Office of Management and Budget study concluded that small firms (i.e., firms with fewer than 1,000 employees) had a ratio of innovations to employment in R&D 4 times as great as that of larger firms (19). In combination, the results of these studies suggest that small firms appear to be more efficient than large companies in the way they use the R&D funds available to them (32).

THE EMERGENCE AND FINANCING* OF NBFs

Since 1976, more than 100 NBFs have been formed in the United States. The founders of many NBFs recognized early that most developments in biotechnology would flow from basic research carried out in academic institutions. For this reason, they formed their companies around a nucleus of talented university scientists, frequently using nonproprietary technology. Several NBFs (e.g., Genentech, Centocor, Genetic Systems) got started by placing R&D contracts with academic researchers for the commercial development of a laboratory discovery.

The character and record of the chief scientists in a new firm is important for several reasons: the amount of venture capital made available to the firm might be determined by the scientist's reputation in the scientific community; the scientist may have some influence over the flow of other well-respected scientists and skilled technicians to the company; and his or her reputation might attract the endorsement of established companies which provides valuable reinforcement to the NBF (e.g., Genentech's early relationships with the U.S. company Eli Lilly and the Swiss company Hoffmann-La Roche).

NBFs must be able to attract and retain qualified personnel if they wish to attract venture capital,** develop marketable products, and maintain their domestic competitive position. Competition in the United States for skilled personnel is intense.***

*The financing of NBFs is discussed in detail in *Chapter 12: Financing and Tax Incentives for Firms.*

**Because most NBFs are unable to meet many of the standard investor requirements for such things as earnings, sales, rate of growth, etc., sometimes potential investors use the number of Ph. D.s per firm as a measure of future earning power.

***See *Chapter 14: Personnel Availability and Training* for a more detailed discussion of personnel needs and availability in the United States.

According to the First Annual Technical Staffing Survey conducted by Scherago Associates in New York, the average biotechnology firm* in the United States more than doubled its staff of scientists between 1980 and 1982 from 3.1 to 7.3 (72). Scherago expects the number of Ph.D.s to almost double again by 1984. The results from the OTA/NAS survey of firms' personnel needs** substantiate the Scherago survey findings, but they also show that the average number of scientists per firm might be growing at a faster rate than originally estimated. The average number of Ph. D.s for the NBFs listed in table 4 as of March 1983 was already 15.7.***

NBFs, by virtue of their size, incentive plans, and innovative and academic-like environment have been able to attract many talented scientists. It is expected that NBFs will continue forming, in part because new firms will continue to be able to attract good scientists.

The formation of the loosely organized and highly competitive structure within which biotechnology is developing in the United States has been shaped largely, but not exclusively, by the availability of venture capital and the willingness of scientists to pursue commercial gain through small, newly formed entrepreneurial companies. The emergence and growth of venture-capital-backed NBFs in the United States began around 1976. As shown in figure 11, not until late 1982, when venture capitalists had satisfied much of their portfolio requirements for biotechnology stock (42) and over 100 new companies had been formed, did startup activity begin to taper off.†

Many of the first NBFs (e.g., Genentech, Genex, Cetus) financed their own proprietary research by providing large established U.S. and foreign companies with research services for initial product development or by entering into licensing agreements with such companies that would re-

*Scherago defines a biotechnology firm as a gene manipulation company.

**See *Appendix E: OTA/NAS Survey of Personnel Needs of Firms in the United States.*

***This average is based on the firms in table 4 for whom Ph. D. figures are given.

†The pace of new biotechnology startups may also have been slowed because many of the top university scientists who wanted to join new firms probably had already done so. A year or two ago a survey done by an investment company looking for an unaffiliated molecular biologist reportedly approached 20 researchers before it found one without a commercial tie (16).

Figure 11.—Emergence of New Biotechnology Firms, 1977-83

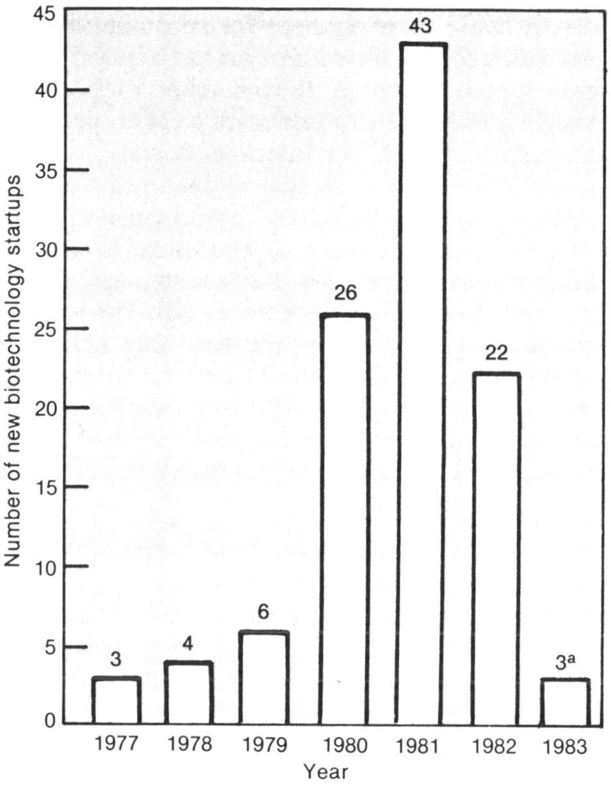

[a]As of November 1983.

SOURCE: Office of Technology Assessment.

sult in future product royalty income. Product development contracts between NBFs and established companies generally provide for periodic cash payments from the established company to the NBF during the stages of research and early product development and for additional payments to the NBF (royalties income) following product sales. Following early product development by the NBF, the established company is generally responsible for obtaining the necessary regulatory approvals, manufacturing, and marketing of the product.

In the last couple of years, more and more NBFs have begun shifting away from developing products for larger companies for reasons including the following:

- NBFs have decided to concentrate more on proprietary research,
- profit margins from licensing technology to established companies are low and may not provide sufficiently substantial earnings (26), and
- most NBFs do not want to be dependent on another company for financial survival.

Instead of relying on contract revenues many NBFs are now obtaining financing through R&D limited partnerships, public stock offerings, or private placements. By retaining the rights to produce and market some of the products they develop (rather than developing products for established companies), some NBFs are seeking to become fully integrated producers and marketers. Genentech, for example, is hoping to manufacture and market four new products (human growth hormone, tissue plasminogen activator, and two types of interferon), and a large portion of Genentech's capital expenditures since 1981 has gone into a production plant for these products (24). Similarly, the NBF Amgen is building a $10 million pilot plant in Chicago for preclinical and clinical studies, and the NBF Genex has just purchased a manufacturing plant in Kentucky to produce phenylalanine and aspartic acid (the two amino acids used to produce the sugar substitute aspartame).

COMMERCIAL PURSUITS OF NBFS

Most NBFs are applying biotechnology to the development of pharmaceutical products or products for use in animal and plant agriculture. For several reasons, the most popular area of commercial pursuit among NBFs at present is the development of MAbs for research and in vitro diagnosis of human and animal diseases.*

- MAb in vitro diagnostic products require much shorter development times than do many rDNA-produced pharmaceutical products, because the technological development of MAb products is less complex. Furthermore, FDA's premarketing approval process is less costly for in vitro products than for products intended for internal use.

*Pharmaceutical applications of MAbs are discussed in *Chapter 5: Pharmaceuticals*. The applications of MAbs in the diagnosis, prevention, and control of animal diseases are discussed in *Chapter 6: Agriculture*.

- Relatively short development times and modest capital requirements for MAb in vitro diagnostic products afford NBFs opportunities to generate short-term cash flow from these products with which to fund the more time-consuming and costly R&D on pharmaceutical products intended for internal use.*
- Entering the MAb in vitro diagnostic products market is relatively easy for NBFs, because the diagnostic market is highly fragmented and the individual diagnostic markets relatively small. Thus, NBFs are likely to encounter few scale disadvantages in competition with large established companies.
- The markets for in vitro MAb diagnostic products are growing, thus providing expanding opportunities for entry by NBFs. The clinical immunodiagnostic market has grown at an annual rate of approximately 20 percent for the past few years, and this rate of growth is expected to continue or increase in the future (63). The 1982 market was valued at $5 million to $6 million (77). Table 12 provides 1982 and 1990 estimates for the size of various MAb markets in the United States.

Oppenheimer & Co. expects the clinical immunodiagnostics market to be the most important source of revenue to NBFs in 1983 (63). Many of the in vitro MAb diagnostic products now being developed or sold are replacement products that offer improved (more accurate) detection, shorter test times, and lower production costs (63)—and as might be expected, competition for market shares and scientific and financial resources is intense. Since 1980, more than 12 new U.S. companies (e.g., Xoma, Quidel, Techniclone, New England Monoclonal Resources) have formed specifically to exploit hybridoma technology, and most of them either already have MAb diagnostic kits on the market or are seeking FDA's approval. In 1982 alone, FDA approved some 30 in vitro MAb diagnostic kits (26).

To increase their chances for commercial success, NBFs solely dependent on MAb-based diagnostic products must find market niches. Although, a focused strategy such as MAb production could bring NBFs financial success with a smaller investment of dollars and scientific expertise in a shorter time frame than a more diverse strategy typical of some of the more heralded, multipurpose companies, such a strategy could also limit their growth potential (26). The worldwide diagnostic market represents only $2 billion out of the $80 billion annual human drug market (24). Until NBFs are capable of entering the larger drug markets, however, diagnostic products may prove crucial in supporting the high costs of pharmaceutical development.

Some NBFs are developing MAb therapeutic and in vivo diagnostic products, although the number of NBFs developing these products is less than the number developing in vitro MAb diagnostic products.* In addition to MAb therapeutics to treat cancer, MAb therapeutic products are being developed to treat bacterial infections that are sometimes difficult to treat with antibiotics and viral infections for which no antibiotics exist. As will be discussed in the section below entitled "Collaborative Ventures Between NBFs and Established U.S. Companies," the regulatory environment for pharmaceuticals imposes heavy long-term financial burdens, which many NBFs may be unable to bear alone. Since many of the new firms aspire toward short-term earnings and independent production and marketing, it is not surprising that in vitro MAb diagnostic products are the area of application most widely chosen by NBFs.

Many small markets exist for NBFs in animal agriculture, and for replacement as well as new products, the barriers to market entry are low. Furthermore, the costs of obtaining regulatory approval for most animal health products are lower than those for human pharmaceuticals. However, in order to market some animal health products, including vaccines, a large and highly

---

*Cetus Corp. (U.S.), for instance, is developing diagnostic products for detecting blood-borne pathogens such as hepatitis B virus with funding from Green Cross of Japan and for detecting cytomegalovirus. Cetus is also developing readily marketable biotechnology products for animal agriculture until its more profitable products, particularly anticancer drugs, are developed. Likewise Hybritech (U.S.) and Genetic Systems (U.S.) are producing MAb diagnostic products to support other longer range R&D activities such as MAb therapeutics.

*An even smaller number are developing MAbs for use in separation and purification.

Table 12.—Estimates of U.S. Monoclonal Antibody Markets, 1982 and 1990 (1981 dollars in millions)

Application	1982 market size	1990 market size
Diagnostics:		
In vitro diagnostic kits	$5 to $6	$300 to $500 ($40)[a,b]
Immunohistochemical kits (examination of biopsies, smears, etc.)	Nil	$25
In vivo diagnostics (primarily imaging)	Nil	Small to $100[b,d]
Therapeutics (includes radiolabeled and toxin-labeled reagents)	Nil	$500 to $1,000[b,c]
Other:		
Research	Small	$10
Purification	Small	$10

[a]High number indicates market for total kit, number in parentheses indicates value of antibody alone for kit (includes patent licensing fees).
[b]Variation depending on industry source, although the range has been corroborated by at least two sources.
[c]This number could be much higher or lower depending on regulatory process.
[d]Based on current pricing (1981 dollars) for diagnostic tests of the same type.

SOURCE: Office of Technology Assessment.

specialized sales force may be necessary. Some NBFs do not expect to hire their own marketing force. Genentech, for example, does not expect to market its own animal vaccines. Some NBFs hope to use existing distribution networks for animal health products instead of developing their own specialized marketing force.

NBFs pursuing plant agriculture applications of biotechnology seem to have found sponsors for longer term research in areas such as enhanced protein content and nitrogen fixation, but a number of new firms are conducting proprietary research in areas such as the regeneration of inbred crop lines from tissue culture. NBFs pursuing plant biotechnology are already using cell culture technologies rather successfully to introduce new plants to the market. One firm, Ecogen, has been formed to focus exclusively on microbial and viral pesticides and other novel pest control methods. As the more frontier techniques such as gene transfer are developed, they can be incorporated into ongoing product lines (15).

FUTURE PROSPECTS OF NBFs

Almost 2 years ago, skeptics forecast a "shake-out" among the NBFs (18,31,60,66). Even though the commercialization of biotechnology now may be more time-consuming, more expensive, and less profitable than was initially hoped, such a shake-out has not yet occurred. A shake-out will occur, however, when new markets develop and present trends in financing, established firm involvement, and technical capability change.

NBFs were formed to exploit research advantages in biotechnology, and many NBFs still possess such advantages. Given their research advantages, and assuming good management and adequate financing, many NBFs may continue to compete successfully with both larger companies and other NBFs as long as competition in biotechnology remains focused in research. Eventually, however, perhaps within 2 or 3 years, most NBFs will have to manufacture and market their own products in order to finance future growth and achieve some level of commercial success. A change from a research-oriented strategy to a more production-oriented strategy will mark a new stage in development for the average NBF, because in the past (and to some extent even now) NBFs out of need for capital have sold their processes to established companies.

NBFs that are wholly dependent on biotechnology for revenues cannot spread the risk of product development over a broad range of products made by traditional methods (unlike the established companies that have several product lines to generate revenues). Many NBFs will fail if markets for the biotechnology products now being commercialized do not develop. Furthermore, many NBFs will fail if capital for production scale-up, clinical trials (if necessary), and marketing is not available when markets develop.

The commercialization of biotechnology in the United States and other countries at present is characterized by a large number of companies, many small, some medium, and many large, applying biotechnology to a very narrow range of products.* Most of the products are rDNA-pro-

*Examples of such products are interferon, interleukin-2, human growth hormone, tissue plasminogen activator, and MAb-based diagduced pharmaceuticals and MAb-based diagnostic products. Because of the large number of companies and small range of biotechnology products, most of the initial product markets are likely to be very crowded, costly to enter, and highly competitive. The sharp decline in the formation of NBFs in 1983 might be explained in part by the currently high levels of competition. How many producers the initial biotechnology product markets might ultimately accommodate is uncertain. Thus, the factors likely to affect the future commercial success of the NBFs most immediately are the timing of market introduction, product performance, and product quality. Price, and hence production costs, will be of greater importance later.

The major determinant to the commercial future of NBFs, assuming they are able to maintain a research advantage, will be their ability to obtain financing and their ability to enter the newly developing product markets. NBFs must manufacture and market their own products not only to generate sufficient revenues to fuel growth but also to be in control of the timing of their own product introduction. It remains unclear whether NBFs will have the financial resources and marketing strength to enter some of the new markets. Large established pharmaceutical companies, for example, normally employ some 500 people just to market their drugs (24), while Genentech, one of the largest NBFs, has a total of about 500 employees.

Some of the most difficult markets for NBFs to enter will be those for human therapeutics, in part because of the regulatory costs associated with product approval and in part because of the market competition posed by established U.S. pharmaceutical companies, which could control some of the early channels of distribution. Entering the markets for in vitro diagnostic products, as mentioned earlier, is relatively easy and does not require large capital investments, but because these markets are currently very crowded, survival may be difficult.

The specialty chemicals market appears relatively easy to enter, both because little competition exists at present and also because the regulatory environment does not impose high costs on product development. Research is near term for many of the products, 3 to 5 years, and an NBF would experience few production scale disadvantages in competition with larger companies.

The safety regulations applicable to animal health products are significantly less stringent than those applicable to pharmaceutical products intended for internal human use, and many market niches exist for small firm entry. Additionally, relatively little competition from established companies exists at present. However, the need for an extensive sales force to market some of the products might pose a considerable barrier to some NBFs wishing to enter animal health markets.

The availability of venture capital and financing for NBFs has been sufficient thus far to fuel the growth of many NBFs. The public market, particularly for new issues, and R&D limited partnerships continue to provide capital to NBFs for use in further research, pilot plant construction, clinical trials, and product development. From August 1982 to May 1983 alone, NBFs raised $200 million through R&D limited partnerships (6). One analyst estimates that R&D limited partnerships will raise a total of $500 million in 1983 (7). The public stock market has also been receptive to NBF issues. Between March and July 1983, 23 NBFs raised about $450 million (39). As long as the public market and R&D limited partnerships make financing available to NBFs, they can continue developing independent strategies, thereby reducing their reliance on established companies.

Paralleling the emerging desire by some NBFs to become integrated producers and marketers is an apparent reduction from 1982 to 1983 in the number of research contracts sponsored by established U.S. companies* and an increase in the amount of capital established U.S. companies

---

nostic products for detection of venereal diseases and pregnancy. Tables 18 and 23 in *Chapter 5: Pharmaceuticals* provide a list of firms engaged in cloning projects for interferon and human tissue plasminogen activator, respectively, and exhibit a rather high level of competition for the two products. Additionally, at least eight NBFs are cloning interleukin-2 (Chiron, Genex, Biogen, Cetus, Genetics Institute, Immunex, Interferon Sciences, and Quidel).

It is impossible to quantify the number and value of all established company sponsored research contracts because not all of are devoting to in-house biotechnology programs. Although the pattern is beginning to change, research contracts sponsored by established companies still provide a large portion of the NBFs' revenues. If the decline in number of research contracts sponsored by established companies continues, which is likely, NBFs must begin finding other sources of revenue. Increases in the amount of capital established U.S. companies are devoting to in-house biotechnology programs portend greater competition in R&D from the larger companies. Equipped with greater financial and marketing resources, more regulatory and, in some cases, production expertise, many U.S. established companies will be formidable competitors in the long run as biotechnology product markets develop. Not all NBFs will survive the competition of the established companies; provided they have adequate financing, however, some NBFs will be able to commercialize their early research advantages before the established companies commercialize theirs.

As biotechnology continues to emerge, and further technical advances are made, new generations of NBFs undoubtedly will evolve to develop the technologies. Within the next several years, a second generation of NBFs is likely to emerge as the result of developments such as the following:

- intensified competition that forces some firms out and creates new opportunities for more entrants,
- a major technological advance in some area of biotechnology such as computer-assisted protein design, which encourages the entry of more new companies,
- the diffusion of advances in bioprocessing, which enables small firms to assume responsibility over their own production, and
- the development of the technologies to the point where scientists from present companies or young scientists from universities will start their own companies.

public. However, on the basis of those that have been reported, most observers would probably agree that the number of new outside research contracts sponsored by established companies in 1983 has dropped significantly from 1982 levels.

*See *Chapter 12: Financing and Tax Incentives for Firms* for further discussion of the sources of NBF revenues.

ROLE OF NBFS IN U.S. COMPETITIVENESS IN BIOTECHNOLOGY

The development of biotechnology is still at an early stage, and competition at present is predominantly in the areas of research and early product development. This early stage of biotechnology development is precisely where NBFs are playing the largest role in competition. Later, however, as the technology develops further and enters a large-scale, capital-intensive production stage, the science may become less important vis-a-vis production expertise, and the dominant role NBFs currently play in the U.S. biotechnology effort may diminish.

The launching of embryonic high-technology industries by entrepreneurial firms is a phenomenon unique to the United States. Historically, small new firms in the United States have had a major role in shaping the competitive position of the United States in emerging technologies.* As discussed further below, NBFs have thus far assumed a similar role in biotechnology:

- by contributing to the expansion of the U.S. basic and applied research base for future biotechnology development,
- by transferring the technology to several industries through joint agreements with other companies,
- by decreasing investment risk by advancing learning curves for later entrants, such as established companies or other NBFs,
- by developing markets, and
- by increasing the level of domestic competition in the United States and thereby accelerating the pace of technology advance.

The formation in the United States of over 110 NBFs that have various links to the network of university biology, chemistry, and engineering departments has extended the basic research base beyond the universities and has expanded the applied research base beyond just a few companies. While the basic and applied research base is being broadened for future biotechnology development, joint agreements and licensing arrangements between NBFs and large established U.S.

*See *Appendix C: A Comparison of the U.S. Semiconductor Industry and Biotechnology.*

companies are effectively diffusing biotechnology across many industrial sectors.

With the help of venture capitalists, NBFs started much earlier to evaluate the commercial potential of biotechnology than did large established U.S. or foreign companies. As early as 1976, NBFs were willing to risk their very existence on the undemonstrated potential of biotechnology. A survey conducted by OTA indicated that most established U.S. companies did not begin in-house biotechnology R&D until 1981 or later.* This finding suggests that the early burden of risk was carried by NBFs. Although many established U.S. companies have now made substantial commitments to biotechnology through investments in plant and equipment for in-house biotechnology R&D programs, others are still hesitant to make such investments and many NBFs continue to function as a litmus test for the new technologies. In Europe and Japan, most companies did not make major investments in biotechnology until after 1981. Thus, it might be suggested that the early R&D activity of NBFs has given the United States a competitive lead in the early stages of biotechnology's commercialization.

The NBF initiative to commercialize biotechnology not only has spurred the development of new product markets but also is expected to expand existing markets through the introduction of products with increased effectiveness and decreased cost. For example, diagnostic kits using MAbs and DNA probes are being developed to detect venereal diseases (e.g. chlamydia and herpes) that are difficult and time-consuming to detect by existing methods. Vaccines are being developed for diseases that now have no reliable prevention (e.g., hepatitis and herpes in humans and colibacillosis in calves and pigs).

The NBFs' entry into the traditional markets served by established companies, where NBFs have taken the risks of developing new products or potentially reducing the production costs of existing ones, has prompted many established U.S. companies to explore potential applications of the new technologies. The market uncertainty created by the new firms and the perceived competition they represent to the established companies is healthy in a competitive context, because it increases the aggregate level of industrial R&D in the United States. The perceived competitive threat that NBFs pose to established companies could become even greater as NBFs such as Biogen, Genentech, and Genex begin to shift away from developing products for large corporate clients and begin to turn toward independent production and marketing of their own products.

Because of their technological expertise and early role as contract research companies, the NBFs have helped established U.S. companies evaluate the feasibility and suitability of using the new technologies in their existing lines of business. They have also helped the established companies evaluate new avenues for diversification. Frequently, the established U.S. companies maintain multiple research contracts with the NBFs to evaluate several applications simultaneously or to evaluate the same application from different perspectives. In this way, the established companies can "ride along" the NBF learning curves while minimizing expenses and risk. In a competitive context, this relationship between NBFs and established U.S. companies is important because it may help to position both types of U.S. firms in international product markets.

From the standpoint of U.S. competitiveness, the innovative lead taken by NBFs in the United States might seem to be a handicap because of the potentially adverse consequences from the transfer of technology from the United States to foreign countries. But the United States, at first through the new firms and now with the combined effort of the established companies, has the ability to maintain its lead by continuing to innovate and develop at a pace equal to or faster than its competitor countries. While competition remains mostly in research, the ability of the United States to remain competitive and in the forefront of biotechnology development rests heavily on NBFs. As biotechnology reaches production stages, the bioprocessing, regulatory, and marketing experience of the established companies will be crucial to a strong U.S. position.

*The survey questionnaire is reproduced in *Appendix E: OTA/NAS Survey of Personnel Needs of Firms in the United States.*

Established U.S. companies

The proliferation of many NBFs and the developments in biotechnology that have been made thus far have prompted many established U.S. companies to re-evaluate the competitive and technological environments in which they have been operating. To some extent, U.S. corporate investment in biotechnology has been both an aggressive and defensive response to the potential market threat represented by NBFs such as Biogen, Genex, Cetus, and Genentech. Although a few pharmaceutical and chemical companies such as Monsanto, DuPont, and Eli Lilly have had biotechnology research efforts underway since about 1978, most of the established U.S. companies now commercializing biotechnology did not begin to do so until about 1981.*

INVESTMENTS IN BIOTECHNOLOGY BY ESTABLISHED U.S. COMPANIES

The motivations underlying established U.S. companies' decisions to invest in biotechnology and the forms that each investment takes vary from company to company. When biotechnology first began to receive commercial attention, many established U.S. companies, particularly those without a major in-house biotechnology program, elected to gain in-house expertise by obtaining technology through research contracts with NBFs or universities,** R&D contracts with NBFs,*** or equity investments in NBFs. For some established U.S. companies, contracts with or equity positions in NBFs are still a major route by which to expand their knowledge of biotechnology.† However, several of the established U.S. companies that initially entered the field through R&D joint ventures are now increasing their commitment to biotechnology through internal expansion.

Since 1978, equity investments in NBFs, often accompanied by research contracts, have been a popular way for established U.S. companies to gain expertise in biotechnology. Table 13 lists many established U.S. companies that have made equity investments in NBFs and the NBF in which they have taken the equity position.* Although only individual corporate strategies can specifically explain why established U.S companies have taken positions in NBFs, some of the investments may have been viewed by the established companies as:

- a defensive strategy against market share losses to unknown technologies,
- an avenue for diversification and greater return on investment, and
- a means of gaining a "window on the new technology."

Figure 12 provides the aggregate equity investment figures for 1977 to 1983 based on table 13. Review of table 13 and figure 12 shows that:

- equity investments in NBFs range from $0.5 million to $20 million;
- some established companies have made multiple investments in the same NBF;
- a number of established companies have made investments in more than one NBF;
- equity investments, in some cases, have led to the formation of another firm (e.g., Genentech and Corning Glass formed Genencor, and Diamond Shamrock and Salk Institute/Biotechnology Industrial Associates formed Animal Vaccine Research Corp.); and
- equity investments have tapered off since 1982.

The years 1978 and 1979 appear to have marked the beginning of general U.S. corporate

*This statement is based on the responses to a survey conducted by OTA and the National Academy of Sciences. The survey questionnaire is reproduced in *Appendix E: OTA/NAS Survey of Personnel Needs of Firms in the United States.*

***Major* university contracts in biotechnology appear to have been declining over time. University/industry relationships in biotechnology are discussed in *Chapter 17: University/Industry Relationships*.

***For a more detailed discussion of R&D joint ventures, see the section below entitled "Collaborative Ventures Between NBFs and Established U.S. Companies."

†In 1982, Monsanto, for example, committed approximately $40 million to outside contracts in biotechnology; however, the overall number of newly formed research and licensing agreements is waning as more and more established companies commit large amounts to in-house staff and facilities.

*A much smaller number of foreign established companies have taken equity position in American NBFs. They are not included in table 13. The notable foreign investors are Sandoz (in Genetics Institute), Novo (in Zymos), a group of Japanese and Swedish investors (in Genentech), C. Itoh (in Integrated Genetics), and Bayer in Molecular Diagnostics).

**The percentage of NBFs purchased by the established companies listed in table 13 range from 1.6 to 100 percent, with 10 to 30 percent being the most common.

Table 13.—Equity Investments in New Biotechnology Firms by Established U.S. Companies, 1977-83[a]

Date	U.S. established company	New biotechnology firm	Equity (millions of dollars)
1980	Abbott Laboratories	Amgen	$5
1981	Allied Corp.	Calgene	2.5
1983		Genetics Institute	10
1981	American Cyanamid	Molecular Genetics, Inc.[b]	5.5
		Cytogen	6.75
1981	ARCO	International Genetic Engineering, Inc. (INGENE)	0.75
1982	Baxter-Travenol	Genetics Institute	5
1982	Beatrice Foods	International Genetic Engineering, Inc. (INGENE)	3.0
1980	Bendix	Engenics	1.75
1982	Bendix/Genex	Proteins Association	16.5[e]
1983	BioRad	International Plant Research Institute (IPRI)	1
1981	Campbell Soup	DNA Plant Technologies	10
1981	Continental Grain	Calgene	1
1981	Cooper Labs/Liposome Tech. Corp.	Cooper-Lipotech	2.7
1982	Corning/Genentech	Genencor	20
1983	Cutter Laboratories	Genetic Systems	9.5
1982	DeKalb	Bethesda Research Laboratories	0.6
1980	Dennison Manufacturing Corp.	Biological Technology Corp.	2
1983	Diamond Shamrock/Salk Institute Biotechnology Industrial Associates	Animal Vaccine Research Corp.	N.A.[d]
1981	Dow	Collaborative Research	5
1981	Dow	International Genetic Engineering, Inc. (Ingene)	N.A.
1981	Ethyl	Biotech Research Labs	0.95
1981	Fluor	Genentech	9
1981	FMC/Centocor	Immunorex	4.9
1980	General Foods	Engenics	0.5
1982	Getty Scientific Corp.	Synergen	4
1982	Gillette	Repligen	N.A.
1983	Hewlett-Packard Co./Genentech	HP Genenchem	N.A.
1978	INCO, Inc.	Biogen[e]	0.35
1979	INCO, Inc.	Biogen	1.25
1980	INCO, Inc.	Biogen	4.61
1981	INCO, Inc.	Biogen	2.5
1981	INCO, Inc.	Immunogen	1
1981	INCO, Inc.	Plant Genetics	N.A.
1981	INCO, Inc.	Liposome Co.	N.A.
1977	Innoven[f]	Genex; Genentech	2
1981	Johnson & Johnson	Quadroma	0.7
1982	Johnson & Johnson	Enzo Biochem	14
1983	Johnson & Johnson	Immulok[g]	18
1982	Kellogg	Agrigenetics	10
1979	Koppers	Genex	3
1980	Koppers	Genex	12
1981	Koppers	Engenics	1.25
1981	Koppers	DNA Plant Technologies	1.7
1982	Eli Lilly	International Plant Research Institute (IPRI)	5
1979	Lubrizol	Genentech	10
1980	Lubrizol	Genentech	15
1982	Lubrizol	Sungene	4
1980	McLaren Power & Paper Co.	Engenics	1.25
1982	Martin Marietta	Molecular Genetics, Inc.	9.7
1982	Martin Marietta	NPI	5
1982	Martin Marietta	Chiron	5
1983	Martin Marietta	Chiron	2
1980	Mead Co.	Engenics	1.25
1980	Monsanto	Biogen	20
1980	Monsanto	Collagen	5.5

Table 13.—Equity Investments in New Biotechnology Firms by Established U.S. Companies, 1977-83[a] (Continued)

Date	U.S. established company	New biotechnology firm	Equity (millions of dollars)
1978	National Distillers	Cetus	5
1980	National Patent Development Corp.	Interferon Sciences	0.6
1980	Nuclear Medical Systems	Genetic Replication Technologies	0.95
1981	Phillips Petroleum	Salk Institute Biotechnology/Industrial Associates	10
1981	Rohm & Haas	Advanced Genetic Sciences	12
1979	Schering-Plough	Biogen	8
1980	Schering-Plough	Biogen	4
1982	Schering-Plough	DNAX[g]	29
1978	Standard Oil of California	Cetus	12.9
1978	Standard Oil of Indiana	Cetus	14
1982	Syntex[h]/Genetic Systems	Oncogen	9.5
1982	Syntex/Syva	Genetic Systems	9.5
1980	Tosco	Amgen	3.5

[a]As of May 1983.
[b]American Cynamid sold 375,000 shares of MGI to Moorman Manufacturing in 1983.
[c]Investment over a 6-year period.
[d]N.A. = Information not available.
[e]Biogen is only 80 percent U.S.-owned.
[f]Monsanto & Emerson Electric.
[g]Acquisition.
[h]Incorporated in Panama.

SOURCE: Office of Technology Assessment.

Figure 12.—Aggregate Equity Investments in New Biotechnology Firms by Established U.S. Companies, 1977-83[a]

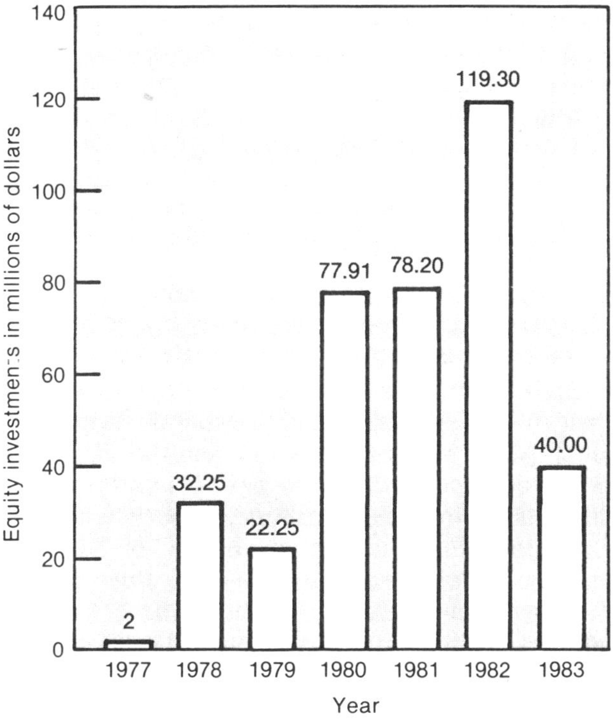

[a]As of May 1983.

SOURCE: Office of Technology Assessment.

interest in biotechnology, with equity investments made by a number of oil and mining companies in the NBFs Biogen, Cetus, Genex, and Genentech. By 1980, commercial applications of biotechnology were advancing in industrial areas where some established companies had no prior R&D commitment, and from 1979 to 1980, there was a dramatic increase in the number and size of equity investments. Equity investments in NBFs have been made by U.S. companies from a variety of industrial sectors: Monsanto (chemicals), for example, invested $20 million in Biogen and $5.5 million in Collagen; Lubrizol (chemicals) made a second equity investment in Genentech totaling $15 million, Fluor (engineering) invested $9 million; and Koppers (mining) expanded its equity position in Genex by investing $12 million.

In 1981, the amount of equity capital invested in NBFs barely exceeded the amount invested the previous year, but in 1982, equity investments soared to a record high of $119 million, an increase of 52 percent over 1981, and the highest level of equity investments in biotechnology ever made. In 1983, the level of equity investments in NBFs dropped significantly. A growing commitment among established U.S. companies to in-house R&D programs in conjunction with previously made equity investments may have contributed to the sharp decline.

In 1982, established U.S. companies not only increased their equity investments in NBFs but they also dramatically increased their in-house investments in biotechnology R&D programs. Capital investments for in-house R&D programs generally reflect the highest level of commitment to biotechnology, as new facilities and employees are often needed to start the new effort. Several U.S. pharmaceutical companies are spending large amounts on new facilities: G.D. Searle, for example, is building a $15 million pilot plant to make proteins from rDNA organisms; DuPont is building an $85 million life sciences complex; Eli Lilly is building a $50 million Biomedical Research Center with emphasis on rDNA technology and immunology and a $9 million pilot plant and lab for rDNA products; Bristol Myers is building a new $10 million in an alpha interferon production plant in Ireland.* Companies from other sectors have also made substantial investments in biotechnology. See table 7 for a list of the 1982 biotechnology R&D budgets for some of the established U.S. and foreign companies most actively supporting biotechnology.

The product areas in which established U.S. companies have directed their biotechnology R&D efforts are as diverse as the industrial sectors they represent. Established companies, however, appear to be playing a dominant role in the development of biotechnology in the areas of plants (25) and commodity chemicals—two rather long-term and costly research areas (see table 4).

ROLE OF ESTABLISHED COMPANIES IN U.S. COMPETITIVENESS IN BIOTECHNOLOGY

Many established U.S. companies manufacture several product lines and are therefore concurrently evaluating different biotechnology application areas. DuPont, for example, is evaluating applications of biotechnology to food production, health care, and renewable resources. Broad strategies such as DuPont's will have a positive effect on the development of biotechnology in the United States by diffusing applications throughout many industrial sectors.

Unlike the many NBFs that have taken a relatively short-term approach to biotechnology in order to generate income for longer term research, many established U.S. companies have several product lines and are taking a longer term approach to biotechnology research; some established companies are not expecting commercial development for 10 to 20 years (27). The long-range research orientation of established U.S. companies will be very important to the long-term competitive position of the United States.

Established U.S. companies will play a major role in the first biotechnology product markets. Because many NBFs have licensed technology to established U.S. companies hoping to finance future growth from the royalties received from the future sale of the products, the established companies will be responsible for the production and marketing of many early biotechnology products. For example, two NBFs, Petroferm and Interferon Sciences, have already solicited the production expertise of Pfizer and Anheuser Busch, respectively. Pfizer's chemical division is the foremost producer of biopolymers and xanthan gums and will produce Petroferm's new bacterial oil emulsifier. Anheuser Busch, through beer production, has accumulated years of experience using yeast and will produce interferon using Interferon Science's genetically manipulated yeast.

The most important element in competition for pharmaceutical market acceptance and market share might be the timing of product entry. Although some NBFs have recently begun funding their own clinical trials and product development, most NBFs still have rather limited financial resources. Most NBFs also have limited production, marketing, and regulatory experience. Such limitations may hinder the ability of NBFs to become major participants in early pharmaceutical product markets. Although the U.S. competitive position in pharmaceutical markets has been declining since the mid-1970's, established U.S. companies appear strategically positioned to compete effectively in international biotechnology product markets as such markets develop.

*Schering-Plough is expected to spend more than $40 million on interferon R&D alone in 1983.

Established U.S. companies also have a competitive role to play in research, because continuous technical advances will be necessary to maintain the present competitive strength of the United States. As the established U.S. multinational companies, along with the other later entrants, expand their in-house research and production facilities they will undoubtedly make substantial contributions to the U.S. commercialization of biotechnology.

Collaborative ventures between NBFs and established U.S. companies

As suggested previously, the development of biotechnology in the United States is unique from the standpoint of the dynamics of the interrelationships between NBFs and the large established U.S. companies. NBFs and established U.S. companies not only compete with one another, but they also, through joint ventures of many kinds, complement one another's skills. In addition to delaying a "shake-out" among NBFs, joint ventures between NBFs and established companies have allowed NBFs to concentrate on the research-intensive stages of product development, the area in which they have an advantage in relation to most established U.S. companies.

A joint venture is a form of association between separate business entities that falls short of a formal merger, but that unites certain agreed upon resources of each entity for a limited purpose.* Joint ventures between NBFs and established companies are attractive for at least three reasons:

- they assist NBFs and established companies in overcoming resource limitations which may prevent them from developing or marketing a product themselves;
- they offer established companies and NBFs less costly methods by which to develop expertise in areas in which they lack in-house capability; and
- they provide established companies with an opportunity to achieve economies of scale in R&D for complex technological problems that might not otherwise be obtainable.

Considerable expenditures in time and money are required to research, develop, and market biotechnologically produced products. The NBFs, started exclusively to exploit innovations in biotechnology, have initially concentrated their activities on research. As a rule, therefore, NBFs have limited financial resources with which to fund production scale-up activities beyond the laboratory or pilot plant stage, not to mention the financing required for regulatory approval and marketing should their research activities in biotechnology yield pharmaceuticals and to a lesser extent, animal drugs and biologics, food additives, chemicals, or micro-organisms for deliberate release into the environment. Established companies have an advantage over NBFs in that they have relatively more financial strength, regulatory experience, and product distribution channels that are already in place, although many established companies are at a disadvantage compared to NBFs with respect to the possession of technical expertise in biotechnology. R&D joint ventures and contracts between NBFs and established companies, therefore, reflect a mutual search for complementary skills and resources.

Examples of the collaborative agreements that are taking place between NBFs and established U.S. and foreign companies are shown in table 14.* R&D contracts accompanied by product licensing agreements form the basis for most joint ventures between NBFs and established U.S. companies in the area of pharmaceuticals. Furthermore, equity investments in NBFs by established companies are often accompanied by R&D contracts. Equity joint ventures wherein equity capital is provided by both partners (e.g., Genencor) for R&D or marketing are less common. Since research contracts and product licensing agreements characterize most joint ventures, three points should be kept in mind throughout this section:

- Licensing agreements and future royalties provide NBFs with financing to do their proprietary research.

Chapter 18: Antitrust Law explores some of the legal considerations surrounding R&D joint ventures, and *Chapter 12: Financing and Tax Incentives for Firms* highlights joint ventures from a financial perspective.

*The large proportion of pharmaceutical joint agreements presented in table 14 reflects the commercial emphasis by companies on pharmaceutical development.

Table 14.—Some Collaborative Ventures Between New Biotechnology Firms and Established U.S. and Foreign Companies[a]

New biotechnology firm*—Established company*

Biogen N.V. (Netherlands Antilles)[b]:

—*Meiji Seika Kaisha, Ltd.* (Japan) has license and development agreement with Biogen N.V. for the scale-up of a still unnamed agricultural chemical which Meiji could bring to market by 1984-85.

—*International Minerals Corp.* has exclusive marketing rights to Biogen's rDNA-produced swine and bovine growth hormones. Biogen will receive royalties.

—*Shionogi & Co., Ltd.* (Japan) will conduct clinical trials and pursue the commercial development in Japan of Biogen's gamma interferon for human therapeutic use.

—*Merck* is developing Biogen's hepatitis B vaccine.

—*Shionogi* (Japan) has a license from Biogen to develop and market Biogen's human serum albumin in Japan and Taiwan.

—*Shionogi* (Japan) has a license and development agreement with Biogen to develop interleukin-2. Shionogi will conduct Japanese clinical trials.

—*INCO* has a contract with Biogen to do studies of the feasibility of bioextraction of nonferrous metals from low-grade ores and other sources of minerals.

—*Fujisawa Pharmaceutical Co.* (Japan) has an agreement to develop and produce Biogen's tissue plasminogen activator in Japan, Taiwan, and South Korea.

—*Monsanto* will fund Biogen's developments of a technique to produce and purify tissue plasminogen activator.

—*KabiVitrum* (Sweden) is collaborating with Biogen in the development of commercial products based on Factor VIII. Biogen intends to market the products in the United States and Canada, and KabiVitrum will have the right to market such products in certain other countries.

—*Green Cross* (Japan) has a license from Biogen to manufacture hepatitis B vaccine. Green Cross has exclusive license to market in Japan.

—*Suntory, Ltd.* (Japan) has an agreement with Biogen under which Biogen will develop rDNA micro-organisms to produce tumor necrosis factor, to scale-up production, and to support clinical trials, and Suntory will have exclusive marketing rights in Japan and Taiwan.

—*Teijin, Ltd.* (Japan) has a license to develop and market Biogen's Factor VIII in Japan, South Korea, Taiwan, Australia, and New Zealand.

Calgene:

—*Allied Chemical Corp.* has a contract with Calgene under which Calgene will do research in nutrient efficiency in plants.

Cambridge Bioscience:

—*Virbac,* a French animal health care company, has a contract with Cambridge Bioscience under which Cambridge Bioscience will develop feline leukemia virus vaccine.

Centocor:

—*FMC Corp.* has 50/50 joint venture to develop human-derived monoclonal antibodies (MAbs).

—*Toray/Fujizoki* (Japan) have signed an agreement to manufacture and market Centocor's hepatitis diagnostic in Japan.

New biotechnology firm*—Established company*

Cetus:

—*Roussel Uclaf* (France) has a contract with Cetus under which Cetus produces vitamin B12. Cetus is receiving royalties.

—*TechAmerica* has a contract with Cetus under which Cetus will develop a rDNA antigen to be used as a vaccine against calf bovine diarrhea. TechAmerica will perform clinical research, manufacture, and market.

—*Norden Labs, Inc.* has a contract with Cetus under which Norden will produce and market rDNA colibacillosis vaccine. Cetus receives royalties.

—*Cooper* will market a MAb from Cetus Immune that is used in tissue typing for organ transplants.

—*Shell Oil Co.* gave a research contract to Cetus under which Cetus will develop human beta-l (fibroblast) interferon.

Chiron:

Merck possesses option for exclusive worldwide license for the use, manufacture, and sale of Chiron's hepatitis B vaccine.

Collaborative Genetics:

—*Akzo N.V.* (Netherlands) gave Collaborative Genetics a research contract to develop genetically manipulated micro-organisms to produce bovine growth hormone.

—*Green Cross* (Japan) has licensed from Collaborative and Warner-Lambert the process by which urokinase is microbially produced.

—*Dow* has given a research contract to Collaborative under which Collaborative will produce rennin via genetically manipulated micro-organisms.

Cytogen:

—*American Cyanamid* has an agreement with Cytogen to develop a MAb that will deliver a chemotherapeutic agent to cancer cells.

Damon Biotech:

—*Hoffmann-La Roche* (Switz.) has contracted Damon to apply its microencapsulation system to the production of MAbs. Hoffmann-La Roche will retain the marketing rights to the interferon produced by this process.

Enzo Biochem:

—*Meiji Seika Kaisha* (Japan) obtained worldwide marketing rights to products based on Enzo's hybridoma technology, including a newly developed pregnancy test.

Genentech:

—*Monsanto* is testing Genentech's bovine and porcine growth hormones. Commercialization and production will be joint effort.

—Genentech has a joint development contract with *Hoffmann-La Roche* for the production of leukocyte and fibroblast interferons. Hoffmann-La Roche will conduct testing to determine its effectiveness. Genentech will supply part of Roche's requirements and receive royalties on sales.

—*KabiVitrum* (Sweden) has worldwide (except in the United States) marketing rights for Genentech's human growth hormone.

—*Fluor* will develop commercial production operations for Genentech to scale-up new biotechnology products.

Table 14.—Some Collaborative Ventures Between New Biotechnology Firms and Established U.S. and Foreign Companies[a] (Continued)

New biotechnology firm*—Established company*

—*Eli Lilly* has been granted exclusive worldwide rights to manufacture and market Genentech's human insulin.

—*Corning* and Genentech have a joint venture (Genencor) to manufacture and market rDNA-produced enzymes for food processing and chemical industries. Corning provides expertise in immobilization of enzymes.

Genetics Institute:

—*Sandoz* (Switz.) is funding research by Genetics Institute to clone monokines and lymphokines in bacteria, i.e., interleukin-2.

Genetic Systems Corp.:

—*Cutter Labs* and Genetic Systems have a $2.5 million joint venture to develop human MAbs for the diagnosis and treatment of *Pseudomonas* infections. For other MAb products, Genetic Systems will do R&D and market the diagnostic products, and Cutter will market therapeutic products.

—*Syva* has a research, development, and marketing agreement with Genetic Systems which will finance some of Genetic Systems' R&D activities related to diagnostic tests for sexually transmitted diseases such as herpes, gonorrhea, and chlamydia. Genetic Systems receives 5 percent royalties on sales.

—*Daiichi Pure Chemicals Co., Ltd.* (Japan) (a subsidiary of Daiichi Seiyaku Co.) entered into an agreement with Genetic Systems to collaborate on the R&D of a diagnostic test kit for blood disorders in the human immune system. Daiichi will receive the exclusive manufacturing and marketing rights in Japan, Taiwan, Mainland China, and Southeast Asia, for the products for treating blood disorders. Genetic Systems will receive royalties.

—A separate marketing agreement with Daiichi grants the exclusive right to purchase and sell, for research products only, in Japan and other Asian countries, certain MAbs developed by Genetic Systems.

—A joint venture between *Syva Co.* (a subsidiary of Syntex Corp.) and Genetic Systems to develop MAbs for the diagnosis and treatment of human cancer.

—*New England Nuclear* (E. I. du Pont de Nemours & Co.) has the rights to market Genetic Systems' MAbs for the identification of different types of human blood cells to the research market throughout the world, with the exception of Japan, Taiwan, People's Republic of China, and Southeast Asia, which are covered by Daiichi Pure Chemicals Co., Ltd.

Genex:

—*Yamanouchi Pharmaceutical Co.* (Japan) will manufacture and sell a biological product developed by Genex which dissolves fibrin. Yamanouchi will market the product for 15 years, paying Genex a licensing fee of 8 percent of sales for development and scale-up. Genex will retain the patent rights.

—*Bristol-Myers Co.* has a contract with Genex under which Genex will develop genetically modified micro-organisms that will produce leukocyte (alpha) and fibroblast (beta) interferons. Bristol-Myers owns all rights. Genex receives royalties.

—A Japanese company (proprietary) has a contract with Genex under which Genex will develop a genetically modified micro-organism to produce L-tryptophan. All discoveries will be the sole property of the Japanese customer.

—*Vineland Laboratories* and Genex have a joint development project to produce a vaccine against coccidiosis.

—*Koppers* has a contract with Genex under which Genex will develop genetically modified micro-organisms to do biocatalytic transformations of aromatic chemicals from coal distillate derivatives. All micro-organisms and research findings are the sole property of Koppers. Genex will receive royalties.

—*Schering AG (F.R.G.)* has a contract with Genex under which Genex will develop a microbe that will produce a blood plasma protein. Schering AG will receive worldwide exclusive license.

—*Green Cross* (Japan) has a contract with Genex under which Genex will develop a microbial strain that produces human serum albumin (HSA). Green Cross will receive an exclusive license to sell, for at least 15 years, all microbially produced HSA under the contract in Japan, Southeast Asia, India, China, Australia, New Zealand, North America, and South America. Genex receives royalties.

—*KabiVitrum* (Sweden) has a contract with Genex for HSA similar to that of Green Cross except Kabi's rights are limited to Africa, Europe, and the Middle East.

—*Yoshitomi Pharmaceutical Industries* (Japan) has a contract with Genex under which Genex will develop genetically modified micro-organisms to produce interleukin-2.

—*Mitsui Toatsu Chemicals Inc.* (Japan) contracted Genex to develop a microbial strain that produces human urokinase. Genex will retain the patent and Mitsui Toatsu will receive an exclusive license with the right to make, use, and sell the product for the royalty period, about 15 years.

—*Mitsubishi Chemical Industries, Ltd.* (Japan) will develop and market Genex's HSA.

—*Pharmacia* has a contract with Genex under which Genex will develop a nonpathogenic strain of bacteria that would produce a protein with potential therapeutic applications.

Hana Biologics, Inc.:

—*Recordati S.p.A.* (Italy) has an agreement with Hana under which Hana will develop and distribute biomedical research and MAb diagnostic products.

—*Fujizoki Pharmaceutical Co.* (Japan) has a joint venture with Hana under which Hana will develop new immunodiagnostic tests. Also, Fujizoki has a distribution agreement with Hana under which Fujizoki will market Hana products in Japan.

Hybritech:

—*Teijin, Ltd.* (Japan) has an agreement with Hybritech under which Hybritech will develop human MAbs for treatment of lung, breast, colorectal, prostate, and certain leukemia-lymphoma type cancers. The goal of the

Table 14.—Some Collaborative Ventures Between New Biotechnology Firms and Established U.S. and Foreign Companies[a] (Continued)

***New biotechnology firm**—Established company*

joint venture is to combine Hybritech's MAb manufacturing technique and Teijin's unique technique of binding a cytotoxic substance to an antibody for cancer therapy.

—*Travenol Laboratories, Inc.* will provide $1 million for research and $1.9 million for stepwise benchmark payments to Hybritech to develop MAbs for treating major bacterial infections. Hybritech will receive royalties on Travenol's worldwide sales.

Immunex:
—*Diamond Shamrock* has a license to commercialize Immunex's lymphokines for use in animals.

Integrated Genetics, Inc.:
—*Connaught Laboratories, Ltd.* (Canada) has an R&D agreement with Integrated Genetics to produce hepatitis B surface antigen in yeast or mammalian cells.

Interferon Sciences:
—*Bristol-Myers* has a licensing and supply agreement with Interferon Sciences under which Bristol-Myers will commercially develop interferon for the treatment of herpes zoster.
—*Green Cross* (Japan) has a $2.5 million R&D and supply agreement with Interferon Sciences under which Interferon Sciences will supply Green Cross with gamma and alpha interferon.
—*Collaborative Research* is synthesizing interferon in yeast. Collaborative provides Interferon Sciences with the alpha-interferon producing clones. Interferon Sciences is involved in the product end and plans to optimize the bioprocess.

Interferon Sciences, Inc./Collaborative Genetics:
—Both companies have a license agreement under which *Green Cross* shares results of a study evaluating application of rDNA technology to the production of interferon by yeast or other micro-organisms.

Molecular Genetics, Inc.:
—*American Cyanamid* has an R&D contract and licensing agreement with Molecular Genetics under which Molecular Genetics will develop bovine growth hormone. Cyanamid is conducting scale-up and testing.
—*American Cyanamid* has sponsored an R&D contract and formed a licensing agreement with Molecular Genetics to select herbicide-resistant corn in tissue culture.

***New biotechnology firm**—Established company*

—*American Cyanamid* sponsored an R&D contract and formed a licensing agreement with Molecular Genetics under which American Cyanamid will conduct human testing, secure regulatory approvals, and manufacture and market any products developed from Molecular's human herpes simplex vaccine research. *Lederle* has begun preclinical testing.
—*Philips-Roxane* (subsidiary of Boehringer-Ingleheim (F.R.G.)) sponsored research and has exclusive license to manufacture and market bovine papilloma virus vaccine developed by Molecular Genetics. Philips-Roxane is responsible for obtaining government approval.

Monoclonal Antibodies:
—*Ortho Pharmaceuticals* has an agreement with Monoclonal Antibodies under which Monoclonal Antibodies will develop and manufacture an innovate diagnostic product that will be marketed by Ortho.

Petrogen, Inc.:
—*Magna Corp.* has a 10-year joint venture with Petrogen under which Magna will field test micro-organisms developed by Petrogen for use in shallow, low-pressure stripper wells.

ARCO Plant Cell Research Institute:
—*H. J. Heinz* and ARCO Plant Cell Research Institute have a joint venture to develop a tomato with high solids content.

Schering-Plough:
—*Yamanouchi* (Japan) will manufacture alpha interferon using Schering-Plough's technology.

University Genetics:
—*Kureha Chemical Industry* (Japan) has a license to develop bovine interferon based on University Genetics' technology.

Worne Biotechnology:
—*Omni Biotech* (Canada) and Worne are in a joint project to extract usable petroleum from Canadian oil sands using micro-organisms.

Zymos, Inc.:
—*Cooper Laboratories* funded research and has the rights to alpha-1 antitrypsin developed by Zymos for possible treatment in emphysema.

[a]Major public contracts, agreements, and ventures.
[b]Biogen is only about 80-percent U.S. owned.
SOURCE: Office of Technology Assessment.

- NBFs in many cases are still reliant on established companies for working capital, whether it be through research contract revenue or equity investments.
- Licensing agreements diffuse technology to different industrial sectors and promote the development of biotechnology in the United States.

Typically, an NBF will enter into an R&D contract, joint venture, or licensing agreement with an established U.S. company to secure funds for proprietary R&D, or, in the case of some pharmaceutical products, to obtain a partner to do clinical evaluations, obtain regulatory approvals, and undertake marketing. Furthermore, the revenues make the new firm attractive to investors if and when the firm wants to use the public market as a source of financing. Typically, the research objective of the NBF in many R&D joint ventures is to develop a micro-organism and the related bioprocessing, extraction, and purification processes needed to produce the desired product in quantities sufficient to proceed with testing. The established company then organizes and implements clinical trials (if necessary) and takes responsibility for the production and marketing of the product. Joint venture partners are usually sought by NBFs to share the risk in new technological areas that appear to have significant commercial applications but that require large investments and have long development times. Joint venture partners are usually sought by established companies because they can provide a "window on the new technology" in addition to oftentimes providing products. Corporate equity investments in NBFs, in addition to providing "windows on the new technologies," can also provide the corporate investor with the possibility of a large return on its investment when (and if) the NBF goes public, or, if the NBF is already publicly held, with potential profit if the stock increases in value.

NBFs in general retain the rights to any patents resulting from the contract research performed, and should the product be marketed, the NBF obtains income through the royalties, which over a range of products may enhance the NBF's financial position so as to enable it to later enter future markets independently. The established company often obtains an exclusive license to the technology developed through the contract and also gains access to that specific product market. If the contract has been preceded by an equity investment, the established company might serve as a marketing partner to the NBF in diverse product areas.

R&D contracts also enable the established company to minimize the risks and costs associated with biotechnology R&D. Should the research not produce desirable results, the contract can be canceled and someone else has paid for the infrastructure. By sponsoring several companies at one time, as Schering-Plough, Koppers, and Martin Marietta have done, the sponsor can spread the risk of not finding the most relevant technology—in essence, portfolio diversification. Additionally, the research effort can be either short or long term depending on the desire of the contracting firm. By minimizing the front end costs and the risk, contracts serve as a kind of feasibility study (49). Successful contracts with NBFs or universities can lend credibility to the commercial potential of the new technology and can help obtain the corporate support necessary to fund future projects in the same field.

Established companies suffer no disadvantages in joint ventures with NBFs except a loss of risk capital should the research be unsuccessful. In fact, as the only buyers of the technology and the major group with the financial resources to commercialize it, established companies exert a great deal of control over the rate at which biotechnology is being developed in the United States.

NBFs do suffer disadvantages as a consequence of their own resource deficiencies, which necessitate their reliance on established companies. These financial reliances of NBFs on established companies will play a crucial role in the future viability of the entire NBF sector for three reasons:

- The low profit margins from licensing technology do not generally provide NBFs with adequate financing for growth and expansion.
- Contract relationships, and thus revenues, are very likely to be transitory. There is a strong economic incentive for established companies to exercise a high degree of "control" over their own product development efforts and to bring their own work in-house.
- The commercial success of many NBF products is reliant on the amount and timing of resources that licensees and partners (established companies) devote to clinical testing (when necessary), obtaining regulatory approval, and marketing.
- Some of the contracts with established companies are tightly written, making it difficult for some NBFs to pursue interesting research findings which might occur in the course of the contracted work.

NBFs with a heavy reliance on contract revenue could face uncertain futures unless their own proprietary research yields marketable products in the near term. Most NBFs are not assured that operating revenues from established companies will be sufficient to fund projected product development. The reliance on established firms for manufacturing and royalty incomes could also jeoprdize the future earning power of many small firms. Those NBFs that have licensed to established companies the right to manufacture and market their products do not control the timing of market entry for these products. If royalties are expected to be the major source of an NBF's operating revenue, then the NBF's correct choice of a marketing partner is crucial for financial success. It might not be wise, for example, for an NBF to choose a marketing partner whose own products stand to be displaced by the new product.

The NBF Genentech, for example, licensed Eli Lilly to produce the new human insulin product Humulin® On the one hand, because Lilly controls the insulin market in the United States, an effective distribution network is already in place and Humulin® sales could be substantial. On the other hand, Humulin® is a competitor of Eli Lilly's animal-derived insulins, and Eli Lilly holds about 85 percent of the U.S. insulin market. In other words, the pace of market development for Humulin® is controlled by the very company whose monopoly position Humulin® sales otherwise might challenge. For example, Eli Lilly could be threatened by the introduction of the new product, and delay the marketing of Humulin®, or if the costs of producing Humulin® are not competitive with Eli Lilly's existing insulin product, then Eli Lilly could also delay the market introduction of Humulin®. Other arrangements of this kind between NBFs and established companies could slow the market entry of new products and reduce the flow of royalties to NBFs.*

An obvious disadvantage common to all NBFs is the sale of technology to ensure survival. By transferring technology to established companies, some NBFs could be canceling the comparative advantage they currently possess in domestic markets. If the competitive pressures arising from the technology transfer to established companies grow too strong, many NBFs will not survive. Additionally, since the most important factor in market acceptance and market share competition may be the timing of market introduction of competitive therapeutic and diagnostic products, the correct choice of partners could be crucial to the U.S. competitive strength.

*See *Chapter 5: Pharmaceuticals* and *Appendix C: A Comparison of the U.S. Semiconductor Industry and Biotechnology* for a more general discussion of the Eli Lilly-Genentech joint agreement.

Collaborative ventures between NBFs and established foreign companies

The observations made concerning NBFs' reliance on established U.S. companies apply equally to R&D arrangements between NBFs and established foreign firms. But the same situation has greater implications for U.S. competitiveness when viewed in the context of international technology transfer.*

Joint ventures between NBFs and established foreign companies are motivated in part by a foreign need for American technology and in part by NBFs' desire to retain U.S. marketing rights—rights often ceded in joint ventures with established U.S. companies. Most observers would agree that the United States is currently the leader in developing commercial applications of biotechnology. Reflecting the strong technological position of some U.S. companies is the increasing number of established foreign companies that are seeking R&D contracts with NBFs. Between 1981 and 1982, for example, the NBF Biogen experienced a 948-percent increase ($520,000 to $5.5 million) in R&D fees from Japanese companies (3), while Genentech experienced a 504-percent increase ($2.6 million to $15.7 million) (33). NBFs often seek joint marketing agreements with established foreign companies for access to foreign markets. On the basis of publicly available R&D joint venture agreements, it appears that the United States is a net exporter of technology.

Foreign companies' joint ventures with NBFs generally take the form of licensing agreements for R&D, and few foreign companies seem to be taking equity positions in the NBFs. From the NBFs' point of view, the same advantages (e.g., the

*There are enormous difficulties in assessing the degree of technology inflow and outflow because of the many ways technology can be transferred; however, most observers would probably agree that the current net flow of biotechnology is outward from the United States.

revenues) and disadvantages (e.g., reliance on royalty income instead of product sales and a loss of technological advantage) are associated with licensing agreements with foreign companies as are associated with licensing agreements with U.S. companies. From the standpoint of the U.S. competitive position in biotechnology, however, the advantages and disadvantages of such agreements are not at all the same. In the case of domestic-domestic licensing agreements, technology is diffused within the United States and U.S. biotechnology development is promoted. In the case of domestic-foreign agreements, technology is transferred out of the United States and thus contributes to the foreign development of technology.

Agreements in the pharmaceutical industry between established U.S. and foreign companies are more difficult to evaluate than agreements between NBFs and established foreign firms. Licensing in the pharmaceutical industry is standard practice to overcome the complexities of clinical testing, registration, and marketing in foreign countries. It is common for licensors to barter, so that they can obtain privileges to market in their territories some products developed by the licensee. The established U.S. companies applying biotechnology are in a position to be able to barter without a loss to their competitive position. The NBFs, if in need of financing or in pursuit of foreign markets, are not in such an advantageous position. The only bargaining chip they have is their proprietary research.

NBFs that because of their initial inability to finance development and clinical trials license some of their proprietary research to foreign companies may be ceding an indirect advantage to foreign companies. However, the licensing strategy and future royalty income may also provide some NBFs with the needed working capital to commercialize other research advantages. At this time, it remains unclear both how technology export will affect the commercial success of the NBFs and how it is likely to influence the U.S. competitive position in biotechnology.

Findings

U.S. efforts to commercialize biotechnology are currently the strongest in the world in part because of the unique dynamism and complementarity that exists between NBFs and established U.S. companies in developing biotechnology for wider commercial application and in part because of a strong U.S. support sector that supplies reagents, instrumentation, and software to the companies applying biotechnology. At present, most NBFs are still specializing in research-oriented phases of product and process development, precisely the commercial stage where they excel. The established companies, on the other hand, have assumed a major share of the responsibility for producing and marketing, and, when necessary, obtaining regulatory approval for, many of the earliest biotechnology products, the commercial stages where their resources are strongest.

Whether the dynamism arising from the competition and complementarity betwen NBFs and established companies will continue giving the United States a comparative advantage in the context of product introduction remains unclear. Since the established U.S. companies, through production and marketing agreements with NBFs, control the later stages of commercialization for many new products being developed, they will have considerable control over the pace at which these new products reach the market. Some established companies may have disincentives to market the new products that might compete with products they are already producing.

Biotechnology is still in an early stage of commercial development, and competition remains largely in research and early product development. In the current research-intensive phase of development, the new entrepreneurial firms founded specifically to exploit innovations and research advantages are providing the United States with a competitive edge in the commercial development of biotechnology. Through their R&D efforts, NBFs are contributing to biotechnology's commercial development in the United States through innovation, technology diffusion, product market development, and encouragement of technical advances because of the increased domestic competition they generate.

The financial constraints faced by the NBFs in the United States have led NBFs into R&D joint ventures and licensing agreements that are diffusing NBF-generated innovations to established U.S. and foreign companies. The collaborative ventures between NBFs and established U.S. companies, by broadening the U.S. technology base for future biotechnology development, in the short run have promoted competitive vigor among U.S. companies commercializing biotechnology. Increasing domestic competition arising from established company R&D, however, stands to threaten the survival of many NBFs and, consequently, the source of much of the current innovation in biotechnology. Since the established U.S. companies now have some control over the later aspects of product development, they can control the rate at which some of the early products are introduced to the marketplace. It is not clear what this situation may do to the U.S. competitive position.

Although NBFs have assumed much of the risk associated with biotechnology's early development, established U.S. companies are making substantial contributions to the U.S. commercialization effort. Through equity investments and licensing and contract agreements with NBFs, established U.S. companies are providing many NBFs with the necessary financial resources to remain solvent. Through joint development agreements with NBFs, many established companies will also provide the necessary production and marketing resources to bring many NBF products to world markets. These resources, in turn, are helping to sustain the rapid pace of technical advance spurred by NBFs. Recently, more and more established U.S. companies have been increasing their in-house investments in biotechnology research and production facilities, so the role of established U.S. companies in the U.S. biotechnology commercialization effort is expanding.

U.S. competitive strength in biotechnology will be tested when large-scale production begins and bioprocessing problems are addressed. The Japanese have extensive experience in bioprocess technology, and dozens of strong "old biotechnology" companies from a variety of industrial sectors in Japan are hoping to use new biotechnology as alever to enter profitable and expanding pharmaceutical markets. Japanese companies, which already dominate biologically produced amino acid markets, are also major competitors in new antibiotic markets; in the future, they could dominate other specialty chemical and pharmaceutical markets as well.

Pharmaceutical markets will be the first proving ground for U.S. competitive strength. International competition will be intense, and the American drug and chemical companies, as well as some NBFs, will be competing against not only the Japanese companies but also the major pharmaceutical and chemical companies of Western Europe, all of whom expect to recover their biotechnology investments through extensive international market penetration. Although there seem to be fewer European companies than Japanese companies commercializing biotechnology, the potential of European pharmaceutical companies such as Hoechst (F.R.G.), Rhone Poulenc and Elf Aquitaine (France), ICI, Wellcome, and Glaxo (U.K.), and Hoffmann-La Roche (Switzerland) is impressive. Thus, to remain competitive internationally and to compete effectively in the future, it is crucial for U.S. companies to rely on rapid innovation made possible by NBFs, rapid product development made possible by established companies, and the accumulated and combined experience of both groups of firms.

Chapter 4 references

1. Abelson, H., "Biotechnology: An Overview," *Science* 219(4585):611, 1983.
2. Advisory Board for the Research Councils, and the Royal Society, *"Biotechnology"* (Spinks' Report), Report of a Joint Working Party of the Advisory Council for Applied Research and Development (London: H. M. Stationery Office, March 1980).
3. Biogen N.V., *Biogen Annual Report,* Curacao, Netherlands Antilles, 1982.
4. Bio Logicals, *Bio Logicals Form 10-K,* Ottawa, Ontario, Dec. 31, 1982.
5. *Bio/Technology,* "New Trends in Financing Biotechnology," September 1983, p. 556.
6. *Bio/Technology,* "Harvesting Profitable Products From Plant Tissue Culture," October 1983, p. 654.
7. *Bio/Technology,* "Predicted Bankruptcies Fail To Materialize," October 1983, p. 646.
8. *Biotechnology News,* "Genentech Makes Monoclonals via rDNA, Forms New RDLP," 3(10):7, May 15, 1983.
9. *Biotechnology Newswatch,* "A Top Radiochemical Maker Moves Into rDNA Products," Sept. 6, 1982, p. 2.
10. *Biotechnology Newswatch,* "Biotechnology R&D in West Germany," Jan. 3, 1983, p. 6.
11. *Biotechnology Newswatch,* "Biotechnology in Switzerland," Apr. 18, 1983, p. 6.
12. *Biotechnology Newswatch,* "No Letup in End-of-Summer Biobusiness Undertakings," Sept. 19, 1983, p. 3.
13. *Biotechnology Newswatch,* "Biobusiness," Nov. 21, 1983, p. 3.
14. *Biotechnology Patent Digest,* Feb. 14, 1983, cited in National Academy of Sciences, *The Competitive Status of the U.S. Pharmaceutical Industry,* Washington, D.C., 1983.
15. Bollinger, H., Vice President of NPI (formerly Native Plants, Inc.), Salt Lake City, Utah, personal communication, June 1983.
16. Bouton, K., "Academic Research and Big Business: A Delicate Balance," *New York Times Magazine Section, New York Times,* Sept. 11, 1983, p. 63.
17. Cetus Corp., Cetus Annual Report, Berkeley, Calif., 1983.
18. *Chemical Week,* "Biotechnology—Seeking the Right Corporate Combinations," Sept. 30, 1981, p. 40.
19. Chilton, K., and Hatfield, D., "Big Government and Small Business: The Changing Relationship," *Chemical Times and Trends,* October 1981, p. 13.
20. Chiron Corp, Chiron Prospectus, Emeryville, Calif., 1983.
21. Cook, R., BioSearch, San Rafael, Calif., personal communication, February 1983.
22. Cooney, C. L., "Bioreactors: Design and Operation," *Science* 219:728-733, 1983.
23. Davies, J., Biogen S.A., Geneva, Switzerland, personal communication, July 1982.
24. *The Economist,* "Biotechnology is Moving From Lab Bench to Marketplace," May 21, 1983, p. 101.
25. *The Economist,* "Tasting Better," Nov. 5, 1983, p. 79.
26. E. F. Hutton, "Biotechnology Overview, Focus on Heavily Traded Biotechnology Stocks," Washington, D.C., July 7, 1983.
27. Ellis, P., "Biotechnology: Industry Emergence, Development, and Change," Thesis, Alfred P. Sloan School of Management, Massachusetts Institute of Technology, Cambridge, Mass., May 14, 1982.
28. Erikson, T., Arthur D. Little, Inc., "Worldwide Pharmaceutical Industry Overview: The Context for Shifting R&D," presented at Seminar on Research and Development in the Global Pharmaceutical Economy, Institute for Alternative Futures, Alexandria, Va., Apr. 11, 1983.
29. *Europa Chemie* (in German), "Genetic Engineering Institute Established in Berlin," Oct. 8, 1982.
30. Flinn, J., Battelle Memorial Institute, Columbus, Ohio, personal communication, July 1983.
31. Fox, J., "Biotechnology: A High-Stakes Industry in Flux," *Chemical and Engineering News,* Mar. 29, 1982, p. 10.
32. Gellman Research Associates, Inc., "The Relationship Between Industrial Concentration, Firm Size, and Technological Innovation," Jenkintown, Pa., May 11, 1982.
33. Genentech, Genentech Annual Report, South San Francisco, Calif., 1982.
34. *Genetic Engineering News,* "Japanese Produce New Gene-Splicing Mechanism," January/February 1983, p. 18.
35. *Genetic Engineering News,* "Japanese Firms and Government Making Major Efforts in Enzymes," March/April 1983, p. 26.
36. *Genetic Engineering News,* "Quality and Consistency Improved, the Automated Gene Synthesizer Market is on the Rebound," September/October 1983, p. 1.
37. *Genetic Engineering News,* "Hewlett-Packard,

Genentech Sign Joint Venture Agreement," September/October 1983, p. 6.

38. *Genetic Engineering News,* "Nippon Zeon Invests in Biotech Operations," September/October 1983, p. 23.
39. *Genetic Technology News,* "Public Invests Nearly Half Billion in Genetic Engineering," 3(8):5, August 1983.
40. Groginski, C., Vega BioTechnologies, Inc., Tucson, Ariz., personal communication, May 1983.
41. Hartley, B., "The Biology Business," *Nature* 283:122, 1980.
42. Harsanyi, Z., Vice President, E. F. Hutton, N.Y., N.Y., personal communication, June 1983.
43. Hooper, G., Vice President of Marketing, Genentech, South San Franciso, Calif., personal communication, February 1983.
44. Human Services Research, Inc., "Small Business Guide to Federal R&D," prepared for the Office of Small Business R&D, National Science Foundation, Washington, D.C., 1980.
45. Institute for Alternative Futures, announcement of seminar, "Research and Development in the Global Pharmaceutical Economy," Foresight Seminars on Pharmaceutical R&D, Washington, D.C., Apr. 11, 1983.
46. *International Herald Tribune,* "Sweden's Fortia Tools a Biotech Revolution," Dec. 7, 1982, pp. 8-9.
47. Jasanoff, S., "Public and Private Sector Activities in Biotechnology: The Federal Republic of Germany," contract report prepared for the Office of Technology Assessment, U.S. Congress, January 1983.
48. King, S., Remora Associates, Palo Alto, Calif., "Companies in Order of Sales From Fermentation Products," submitted at AAAS/EPA meeting Dec. 21, 1982.
49. Luccesi, P., Vice President of Corporate Research, Exxon Research and Engineering Corp., Clinton, N.J., personal communication, August 1981.
50. *L'usine Nouvelle* (in French), "Elf Reinforces Its Presence," Feb. 3, 1983, p. 42.
51. Macki, H., Cruachem, Inc., Bend, Oreg., personal communication, February 1983.
52. McTaggart, J., Tag Marketing, Erie, Pa., personal communication, February 1983.
53. *Marketplace Sweden,* "Swedish Pharmaceuticals: Cresting on Product Development," 14(4):11-12, October/December 1982.
54. Muscoplat, C., Vice President, Molecular Genetics, Inc., Minnetonka, Minn., August 1983.
55. National Academy of Sciences, *The Competitive Status of the U.S. Pharmaceutical Industry,* Washington D.C., 1983.
56. *New Scientist,* "Biotech Fund Finds Hope in Britain," 99(1375):748, Sept. 15, 1983.
57. Nikkei Sangyo Shimbun Biotechnology Survey, June 1981, cited in G. Saxonhouse "Biotechnology in Japan," contract report prepared for the Office of Technology Assessment, U.S. Congress, June 1983.
58. *Nikkei Sangyo Shimbun* (Japan Industrial Daily), "Advanced Technology Battleground—Biotechnology," Aug. 18, 1982, pp. 2-3.
59. Nomura Research Institute, "Trends of Biotechnology in Japan," Tokyo, July 1983.
60. Nossiter, D., "Designer Genes: Biotechnology Is Big in Promise and Pitfalls for Investors," *Barrons,* Feb. 22, 1982, p. 24.
61. O'Brien, T. C., "NBS and Industrial Biotechnology: Instrumentation and Associated Measurement Needs," NBSIR, March 1983.
62. O'Neill, W., "Pharmaceutical Applications of Biotechnology," contract report prepared for the Office of Technology Assessment, U.S. Congress, March 1983.
63. Oppenheimer & Co., Inc., "Biotechnology," Report No. 83-276, New York, Mar. 21, 1983.
64. Organisation for Economic Co-Operation and Development, Directorate for Science, Technology, and Industry, "Impact of Multinational Enterprises on National Scientific and Technical Capacities," Paris, 1977.
65. Organisation for Economic Co-Operation and Development, "Innovation in Small and Medium Sized Firms," Paris, 1982.
66. Patterson, W., "The Rush To Put Biotechnology To Work," *Industry Week,* Sept. 7, 1981, p. 67.
67. Pharmaceutical Manufacturers Association, *Annual Survey Report, U.S. Pharmaceutical Industry, 1979-1980,* Washington, D.C., 1980.
68. Roberts, L., "Gardening in Test Tubes," *Mosaic,* May/June 1982, p. 16.
69. Rothwell, R., and Zegveld, W., *Innovation and the Small and Medium Firm* (Hingham, Mass.: Kluwer-Nijhoff Publishing, 1981).
70. Saxonhouse, G., "Biotechnology in Japan," contract report prepared for the Office of Technology Assessment, U.S. Congress, June 1983.
71. Scherago Associates, Inc., New York, N.Y., HPLC Brand/Company Preference Survey conducted among subscribers to *Science,* Oct. 13, 1982.
72. Scherago Associates, Inc., New York, N.Y., First Annual Technical Staffing Survey, 1982.
73. *Scrip 408,* "Japanese Ministry Coordinating Interferon Research," Aug. 1, 1979, p. 9.
74. *Scrip 625,* Sept. 14, 1981, cited in National Academy of Sciences, *The Competitive Status of*

the U.S. Pharmaceutical Industry, Washington, D.C., 1983.
75. *Scrip 686,* "U.S. Potential for Japanese Drug Firms," Apr. 21, 1982, p. 9.
76. Stinson, S. C., "Analytical Instrument Sales Set To Bounce Back After Recession," *Chem. & Eng. News* Feb. 7, 1983, p. 16.
77. Taylor, R., Hybritech Inc., San Diego, Calif., personal communication, September 1983.
78. Tilton, J., *International Diffusion of Technology: The Case of Semiconductors* (Washington, D.C.: The Brookings Institution, 1971).
79. Tomizuka, N., "Industrial Aspects and Government Policy Concerning Genetic Engineering in Japan," *Genetic Engineering: Commercial Opportunities in Australia,* Proceedings of a symposium held in Sydney, Nov. 18-20, 1981.
80. U.S. Department of Commerce, International Trade Administration, "An Assessment of U.S. Competitiveness in High Technology Industries," Washington, D.C., February 1983.
81. Ward, G., Vega Biochemicals, Tucson, Ariz., personal communication, February 1983.
82. Waldholz, M., "Ballyhoo Has Faded But Interferon Still Has Boosters at High Levels," *Wall Street Journal,* Sept. 30, 1983, p. 1.
83. Wardell, W., and Sheck, L., "Is Pharmaceutical Innovation Declining?: Interpreting Measures of Pharmaceutical Innovation and Regulatory Impact in the USA, 1950-1980," Center for the Study of Drug Development, Department of Pharmacology, University of Rochester Medical Center, Rochester, N.Y., presented at the Arne Ryde Symposium on Pharmaceutical Economics, Helsingborg, Sweden, Sept. 27-28, 1982.
84. Zimmer, S. J., "The Impacts of Applied Genetics in Animal Agriculture," contract report prepared for the Office of Technology Assessment, U.S. Congress, August 1982.

PART III
Applications of Biotechnology in Specific Industrial Sectors

Chapter 5
Pharmaceuticals

Contents

Tables

Figures

Chapter 5

Pharmaceuticals

Introduction

In the United States, many industrial biotechnology developments rest on the broad base of knowledge generated by university research in the biological sciences. Such research has been funded largely by the National Institutes of Health (NIH) and other public health-oriented sponsors. As a consequence, the first areas of application of new biotechnology in the United States have been in the pharmaceutical field. As research using the new genetic techniques has progressed, the pharmaceutical industry has been the leader in industrial applications.

Perhaps the most important application of biotechnology is to facilitate further biomedical research. Among the most intriguing areas of research using biotechnology are those pertaining to the nervous system, the immune system, the endocrine system, and cancer. As research in these areas yields insight into mechanisms of disease and healthy body function, basic questions about the organization and function of the brain, the nature of behavior, and the regulation of body functions may be answered. The illumination of these phenomena, in turn, may generate new possibilities for pharmaceutical products.

Pharmaceutical production may be improved with biotechnology in many ways. In some instances, production of pharmaceutical products by chemical synthesis or tissue extraction methods may be replaced by production from cloned genes. In other instances, applications of recombinant DNA (rDNA) technology may supplant traditional bioprocess methods for the production of antibiotics and other pharmaceutical compounds. Perhaps most importantly, new biotechnology provides a means of producing for the first time large amounts of compounds that are otherwise scarce. Thus, biotechnology may give rise to the development of entirely new pharmaceutical products.

Whatever the intended impact of a new pharmaceutical product, profit expectations usually govern the selection of projects for development. In considering the use of biotechnology to produce substances by new means, manufacturers must make multifaceted decisions that include the following considerations:

- the possibility of making products superior to those already marketed for a given purpose (i.e., more effective, convenient, safe, or economical);
- the technical feasibility of applying new methods (e.g., in rDNA applicatons, the feasibility of cloning DNA that directs synthesis of desired substances);
- the cost of the conventional method (e.g., chemical synthesis, tissue extraction, or traditional bioprocessing) and the potential to reduce costs with rDNA technology or other new methods;
- the nature of the market (i.e., whether it is of high enough value or volume to justify the substantial start up costs of new production methodology and regulatory approval);
- the possible loss of production of other substances with the change in methods (e.g., substances that were coproduced in the old method), as well as the potential for developing new, useful byproducts; and
- the possibility that the new methods employed will serve as useful models for preparing other compounds (whereby the new technology may justify high startup costs and the loss of formerly coproduced products).

Although biosynthesis may eventually reduce production costs of widely used compounds by several orders of magnitude (from millions of dollars per kilogram for chemical synthesis to several thousand dollars per kilogram for biosynthesis), chemical synthesis often suffices for production of low molecular weight compounds for testing. In many cases, substantial research and development (R&D) costs and high product attrition rate in pharmaceutical development may not justify initial exploration of some compounds with biotechnology.

This chapter introduces the scientific and commercial bases of a number of pharmaceutical developments that exemplify biotechnology's promise in the pharmaceutical industry. Some examples include human insulin (hI), the first rDNA-manufactured product of biotechnology to reach the marketplace, interferon (Ifn), human growth hormone (hGH), and human serum albumin (HSA) rDNA projects. Other examples discussed are monoclonal antibodies (MAbs) and DNA hybridization probes, which are already being marketed for in vitro diagnostic use. Discussions include market profiles for each of these compounds, many of which will compete with products made by other methods.

Several important points are raised in this chapter that are discussed throughout this report. The first is that government regulation and licensing of pharmaceuticals play a major part in the development of these new products. With the rapid progress taking place in biotechnology, technical barriers may in some instances become secondary to regulatory barriers. Regulatory considerations that have shaped the use of biotechnology in the pharmaceutical industry are noted in this chapter.*

A second point is that in assessing the potential for biotechnology's use throughout the pharmaceutical industry, it is important to examine the receptivity of established companies to the adoption of new production methods. Traditionally, funding for most of the applied research and development of new pharmaceutical products in the United States has been provided by large pharmaceutical manufacturers. Since these manufacturers generally command the markets for products made by conventional means, they may have vested interests in established products that will impede the development and marketing of new products. This situation might perpetuate the problem of decreasing innovation in the pharmaceutical industry and contribute to the underdevelopment of biotechnology applications to pharmaceuticals.

*For a further discussion of regulatory factors that affect the use of biotechnology in the pharmaceutical and other industrial sectors, see *Chapter 15: Health, Safety, and Environmental Regulation*.

Regulatory proteins

The use of biotechnology to manufacture pharmaceutical products can be viewed in several ways. First, biotechnology may be used as a substitute for conventional methods of production, which include chemical synthesis and extraction from tissue. The successful cloning projects and microbial production of the proteins hI, Ifns, and hGH in rDNA systems, outlined below, are valuable as paradigms for biotechnology's role in developing competitive pharmaceutical substitutes. Second, biotechnology may be used to produce unprecedented amounts of scarce biological compounds, of which certain regulatory proteins provide the leading examples. Finally, the use of biotechnological methods yields basic knowledge on which future research can be based.

Human insulin

The first therapeutic agent produced by means of rDNA technology to achieve regulatory approval and market introduction is hI, marketed under the name Humulin®.* Although Humulin® may be the debutant of rDNA produced drugs, the extent to which rDNA-produced hI will be substituted in the marketplace for animal insulin is uncertain. Insulin derived from animals has long been the largest volume peptide hormone used in medicine. Human insulin differs only slightly from that of pigs and cows, and its incremental benefits have yet to be demonstrated (82).

*Humulin® has been approved in both the United States and the United Kingdom.

A profile of insulin markets and sales by Eli Lilly & Co. (U.S.)—the dominant producer and marketer of insulin, and licensee from Genentech Corp. (U.S.) of the new rDNA product—in the United States and Europe is shown in table 15. By 1985, as indicated in that table, both U.S. and European markets for insulin are expected to double. Eli Lilly is expected to retain a sizable portion of the U.S. market, but its greatest potential lies in penetrating foreign markets with Humulin®.

The development and commercialization of Humulin® establishes several important precedents of general significance to the introduction of biotechnology to industry:

- *Liaison between industry and academic scientists.* The original bacterial production of polypeptide chains of insulin at the new biotechnology firm (NBF)* Genentech made use of nucleic acid sequences synthesized by collaborators at City of Hope Medical Center, an academic laboratory that had capabilities not otherwise available to Genentech at the time (31).
- *Collaboration between NBFs and established companies.* Early in the development of Humulin®, Genentech entered a collaborative arrangement with Eli Lilly. Under the agreement, Genentech performed the rDNA work and received financial support for the work from Lilly. Lilly, in addition to providing this financial support, was responsible for manufacturing, marketing, and obtaining regulatory approval for the hI product that resulted from Genentech's work. This arrangement capitalized on Lilly's decades of experience in large-scale bioprocessing and the purification of insulin. Most significantly, Lilly was thoroughly familiar with insulin and the procedures of regulatory agencies, marketing, and distribution. Lilly was able to satisfy the Food and Drug Administration's (FDA's) requirements for approval of Humulin® in record time—4 years after the first bacterial preparation of hI. Under their arrangement, Genentech receives royalties from Lilly on the sale of Humulin®. Lilly, in turn, has access to improvement inventions by Genentech. Proinsulin, for example, produced from genes cloned by Genentech (disclosed in March 1980), may provide a more efficient route for the production of hI or may have clinical value of its own (see below). This pattern of collaboration between NBFs and established pharmaceutical firms is common.*
- *International joint ventures.* Though Eli Lilly has had little competition in the U.S. insulin market until now, the company has been only a minor factor in insulin markets outside of the United States. Recently, however, Lilly has licensed Swedish and Japanese firms to facilitate penetration of overseas markets (121). The leading insulin supplier abroad is the Danish firm Novo Industri A/S (142). Novo countered Lilly's rDNA hI effort by commercializing an enzymatic process devised in the early 1970's to transform insulin from swine into a form identical to hI.** Novo's symisynthetic hI product was approved for marketing in the United Kingdom shortly before Lilly's Humulin® attained approval there. To compete with Lilly in the United States for insulin markets, Novo formed a joint venture with an established American pharmaceutical company, E. R. Squibb (116). Novo also con-

*NBFs, as defined in *Chapter 4: Firms Commercializing Biotechnology,* are firms that have been started up specifically to capitalize on new biotechnology. Most NBFs are U.S. firms.

Table 15.—U.S. and European Markets for Insulin: Eli Lilly's Estimated Sales (millions of dollars)

	1981	1985 estimate
U.S. market:		
Lilly's sales	$133	$205[a]
Total market	$170	$345
European market:		
Lilly's sales	$ 12	$100[a]
Total market	$140	$285

NOTE: In 1981, approximately three-quarters of a ton of pure insulin for about 1.5 million diabetics was sold in the United States. The number of American diabetics is expected to increase to 2.1 million people between 1981 and 1986 (*Scrip*, 10/4/82).

[a]Includes sales of Humulin®.

SOURCE: Office of Technology Assessment, based on estimates from D. L. Smith, *Eli Lilly and Company: A Basic Study* (New York: Smith Barney Harris Upham & Co., Inc., September 1982).

*For a further discussion of collaboration between NBFs and established firms, see *Chapter 4: Firms Commercializing Biotechnology.*

**Hoechst (F.R.G.) and Nordisk (Denmark) have subsequently introduced semisynthetic hI products, and Shionogi (Japan) has developed a significant process improvement involving an immobilized bacterial enzyme (94).

tracted with Biogen S.A. (Switzerland)* to develop an alternative rDNA process for the production of hI (11).

- *Refinement of process technology.* The race to supply international insulin markets has spawned further biotechnological innovation in the pharmaceutical industry. The A and B protein chains of insulin can join in several ways, only one of which is correct. Combining the two chains by nonbiological chemistry is generally regarded as the "hard way" to make insulin. In the body, a connecting peptide in proinsulin (the precursor of insulin) positions the chains appropriately for joining to make the biologically active form of insulin. The connecting peptide is deleted when proinsulin is converted to insulin within pancreatic cells. Work to design bioprocesses using immobilized enzymes** to transform rDNA-produced proinsulin into insulin and to separate the products is currently underway. Lilly has reported the production of human proinsulin in bacteria through rDNA technology and the efficient conversion of proinsulin to hI (27). The NBF Cetus (U.S.) also has an improved proinsulin process, and Hoechst (F.R.G.) is reported to be developing one (10).
- *Clarification of related problems.* The injection of insulin has saved the lives of many diabetics, but the delivery of insulin by injection is thought to cause complications.*** Initial hopes for rDNA-produced hI centered on avoiding allergic reactions to impurities in insulin preparations, but these hopes have not been realized. Although results with patients switching from animal insulin to hI are encouraging, substantial allergic responses sometimes occur in patients taking hI for the first time (79). These problems probably arise because insulin is administered by subcutaneous injection. Thus, improvements in the mode of delivering insulin to patients may be at least as important to commercial implementation as technical advances in rDNA production of hI. (See *Box B.—Recent Work on Drug Delivery Systems.)*

Some diabetic complications may not be caused simply by insulin deficiency. Human proinsulin, for example, may have therapeutic value. Animal proinsulin, which differs significantly from its human counterpart, is considered a contaminant in preparations of animal insulin. However, some scientists hypothesize that administration of human proinsulin may be beneficial to diabetic patients. Human proinsulin's availability through rDNA technology is allowing Eli Lilly to evaluate this hypothesis (27).

Interferons

Ifns, a class of immune regulators or lymphokines, are proteins that regulate the response of cells to viral infections and cancer proliferation. These extraordinarily potent substances are the subject of the most widely publicized, well-funded applications of rDNA technology to date, but details of their functions remain unknown. Until recently, the study of Ifns was limited by the extremely small amounts of Ifn that could be obtained from cultured cells. Now, however, rDNA technology allows production of large quantities of Ifn-like proteins for testing as pharmaceutical products. Despite certain structural differences from native Ifns,* rDNA-produced Ifns appear to have identical effects on cultured cells.

The cloning and production of Ifns illustrate several aspects of the commercialization of biotechnology:

- the use of rDNA technology to produce a scarce product in quantities sufficient for research on the product's effects;
- a massive, competitive scale-up campaign by

*Biogen N.V., the parent company of the Biogen group, is registered in the Netherlands Antilles. Biogen S.A., one of Biogen N.V.'s four principal operating subsidiaries, is a Swiss corporation that conducts R&D under contract with Biogen N.V.

**Immobilized enzymes are enzymes bound to solid supports so that they can exert their catalytic effects on dissolved substances without becoming inextricably mixed up with the reactants and products. For further discussion, see *Chapter 3: The Technologies.*

***In spite of daily injection of insulin, long-term complications continue to plague many diabetics. After 20 to 30 years of disease patients often develop blindness, need for leg amputations, kidney failure, stroke, heart disease, and/or nerve damage. About 10 percent of all hospital days (21 million per year) are consequences of diabetes, and the disease accounts for 19 million physician visits per year (49).

*Ifns produced by rDNA in bacteria lack carbohydrate (sugar) groups found on native Ifns. It is not known to what extent the absence of these groups affects protein function.

Box B.—Recent Work on Drug Delivery Systems

Oral ingestion is the most convenient route of administration for drugs, but many drugs, especially proteins, are destroyed by the digestive system before they can be absorbed from the stomach and gut. This problem could be especially serious for products of biotechnology, since many of them are proteins. Additionally, the effect of administration of drugs in a specified quantity one or more times a day gives the patient an initial high dose which tapers off with time. Pills and injections, therefore, can cause blood levels so high that undesirable side effects can occur and levels so low that the drug has no effect. Thus, for many years, researchers have sought to develop devices that deliver drugs continuously. Most of these delivery systems also aim to bypass the intestinal tract. Some possible drug delivery systems are discussed below:

- Several devices that deliver drugs through the skin have been developed. These transdermal devices work well for drugs with low molecular weights; research is proceeding on devices that will transfer higher molecular weight drugs, such as proteins, through the skin. Some methods being investigated use electricity or magnetism to speed the delivery (72).
- Continuous delivery of insulin is the optimal way to treat diabetes. External pumps that infuse insulin at a constant, low rate (with rate adjustments according to patients' eating and exercise schedules), combined with patient self-monitoring of blood glucose level, are gaining favor among diabetologists and a limited sector of the diabetic population. Pumps are fairly expensive and require sophisticated user attention. There are an estimated 5,000 to 7,000 users of external insulin infusion pumps (111).
- Implantable pumps are probably the best way to deliver proteins in most instances. Preprogramed implantable plants have been supplying insulin to diabetics for several years in the United States, France, and West Germany (49). More sophisticated devices are being developed by industry and by academic institutions with funding from NIH and the National Aeronautics and Space Administration (111). Animal trials with some models are underway. It would ultimately be most therapeutic if the device could respond automatically to a need in the patient and thus supply the drug in response to that need. For instance, diabetics need varying amounts of insulin depending on what they have eaten, whether they have exercised, and many other factors. Thus, if the pump could also have a sensor that detected the amount of blood sugar, the pump could supply appropriate amounts of insulin in response to need.
- A drug entrapped in an artificially constructed set of membrane layers (liposome) may be effective when given orally or injected into the bloodstream. The layers of membranes are slowly stripped off, causing a slow release of the drug. There is some evidence that insulin can be absorbed through the mucosal membranes of the upper and lower ends of the digestive tract (102,159), permitting consideration of liposome systems to control insulin release to areas of the mouth, nose, or rectum (3,21), but a reliable formulation has not yet been devised.

pharmaceutical manufacturers in advance of demonstrated uses of the product;

- the attempt to produce economically a functional glycoprotein (protein with attached sugar molecules) in an rDNA system;
- a pattern of international R&D investment that reflects the differing needs and medical practices of various nations; and
- the establishment of a U.S. national effort, via research grants and procurement contracts administered through the National Cancer Institute, the American Cancer Society (ACS), and other organizations, to support testing of Ifns toward a national goal (cure of cancer).*

Ifns are being considered for various pharmaceutical applications, but are not yet approved as

*In general, Ifn projects in the United States have received massive public funding. Studies in Sweden, and to a limited extent in the United States, stimulated appropriations of $5.4 million by the nonprofit ACS for extended clinical trials in the early 1980's. This was by far the greatest single commitment ever made by ACS, and it was followed by a boost in NIH funding for Ifn research from $7.7 million to $19.9 million for fiscal year 1980.

pharmaceutical products. There is some evidence that Ifns are effective in preventing certain viral infections, but more clinical trials are necessary to demonstrate their preventive abilities (81).* Most evidence that Ifns cure viral infections is anecdotal. In combination with other drugs, however, Ifns may prove useful in treatment of viral diseases (50,81,130,157). Extensive clinical trials to determine Ifns' effectiveness in the treatment of herpes and other viral infections are underway, some which are listed in table 16. The availability of Ifns made with rDNA technology has allowed many of these clinical trials to be undertaken.

Several clinical trials to evaluate Ifns' effectiveness in the treatment of cancer have taken place, but, at present, only limited conclusions can be drawn from the data. In some cases, Ifns inhibit tumor cell growth and may stimulate immune

*Assuming the safety criterion can be satisfied for the use of Ifn in a prophylactic mode, the immediate market may be for persons whose natural defenses are weakened by illness or medication, such as those undergoing cancer therapy with drugs or radiation. Other early markets could be for patients entering elective surgery or persons at high risk of viral exposure, such as teachers and certain medical personnel. Since Ifns apparently will be available from many sources, the dosage forms or delivery systems may be crucial for widespread acceptance and efficacy.

Table 16.—Some Ongoing Clinical Trials Using Alpha or Beta Interferons To Treat Human Viral Diseases

Disease	Interferons (source)	Sponsors	Remarks
Herpes genitalis	Alpha (rDNA, *E. coli*)	NIAID (U.S.)[a] and Schering-Plough (U.S.)[b]	Intramuscular injection for infection
	Alpha (blood buffy coat)	Enzo Biochem (U.S.)[c]	Topical ointment (Enzoferon®)
	Beta (cultured fibroblasts)	Inter-Yeda (Israel)[d]	Cream formulation (Frone®)
Herpes labialis	Beta (cultured fibroblasts)	Inter-Yeda[d]	Cream formulation (Frone®)
	Alpha (blood buffy coat)	Enzo Biochem[c]	Topical ointment (Enzoferon®)
Herpes keratitis and adenovirus conjunctivitis	Alpha (rDNA, *E. coli*)	Schering-Plough[b]	Topical ointment
Periocular herpes	Beta (cultured fibroblasts)	Inter-Yeda[d]	Cream formulation (Frone®)
Herpes zoster	Beta (cultured fibroblasts)	Bioferon (F.R.G.)	Approved for marketing in West Germany
	Alpha (rDNA, *E. coli*)	Hoffmann-La Roche (Switz.)[e]	100 immunosuppressed patients in trial
	Alpha (blood buffy coat)	NIAID[f]	Spread of shingles inhibited by injection
Herpes infections	Alpha (rDNA, *E. coli*)	Takeda Chem. (Japan)	Own mfr. after use of Hoffmann-La Roche's Ifn for Phase I
Genital warts	Beta (cultured fibroblasts)	Inter-Yeda[d]	Direct injection superior to topical application
Warts	Alpha (rDNA, *E. coli*)	Takeda Chem.	Many unreported tests underway
Laryngeal papillomas[g]	Lymphoblastoid and alpha	Wellcome (U.K.) & others	Injection following surgery
Cytomegalovirus	Alpha (rDNA, *E. coli*)	Hoffmann-La Roche	Injection for life-threatening infantile infections
Hepatitis B[h]	Alpha (rDNA, *E. coli*)	NIAID[a]	Alternated with Vidarabine® in 150-patient, 5-year, wide dose range trials
	Alpha (rDNA, *E. coli*)	Takeda Chem.	
Multiple sclerosis	Alpha (blood buffy coat)	National Multiple Sclerosis Society	Subcutaneous injection of Ifn from K. Cantell, Finnish Red Cross
Amyotrophic lateral sclerosis (Lou Gehrig's disease)[i]	Alpha (rDNA, *E. coli*)	Hoffmann-La Roche	Intravenous or intrathecal injection at two U.S. centers

[a]NIAID=National Institute of Allergy and Infectious Diseases.
[b]Schering-Plough's Ifn produced for clinical trials outside of the United States is synthesized microbially from genes cloned by Biogen S.A.
[c]Enzo Biochem obtained natural alpha-Ifn from New York Blood Center and sponsors trials at Sloan Kettering.
[d]Inter-Yeda is an Israeli firm conducting clinical trials primarily in Israel, Europe, and Canada.
[e]Genentech (U.S.) cloned and produces the Ifns being evaluated by Hoffmann-La Roche (Switzerland).
[f]Phase III studies at Stanford with Ifn obtained from K. Cantell, Finnish Red Cross, completed in 1982.
[g]Regrowth of these wart-like growths, apparently caused by virus, has been inhibited by Ifns in Danish studies.
[h]NIAID-sponsored trials indicate that Ifn alone is ineffective for the carrier state in males, but combinations with other drugs show promise.
[i]Viral origin suspected but not proved.

SOURCE: Office of Technology Assessment.

cells to destroy cancerous cells; their effects on inhibiting tumor metastasis are better established than their ability to cause regression of primary tumors (8). With some exceptions, the tumors that respond to Ifn treatment (certain lymphomas, benign human esophogeal papillomavirus tumors, and leukemia, in particular) are also the most responsive to established chemotherapeutic agents. Some subtypes of interferon (e.g., alpha-Ifn) occasionally induce tumor regression in patients who are resistant to radiation and multiple drug therapy (95).

Several problems have been noted in initial clinical trials designed to test Ifns' effectiveness in the treatment of cancer. For example, side effects (fever, fatigue, and influenza-like symptoms) caused by injections of Ifn made in cell cultures were thought to be toxic reactions to impurities of the culture medium, but pure rDNA-produced Ifns show similar side effects (95). Thus, despite extensive research, numerous questions remain concerning Ifns' anticancer potential. Some ongoing clinical trials for Ifns' anticancer properties are listed in table 17.

Perhaps the most enlightening results stemming from Ifn research will concern cellular function during immune responses. Such results may prove extremely valuble in medicine. Better understanding of immune mechanisms, for example, may provide insight into the etiology of the recently problematic acquired immune-deficiency syndrome (AIDS). Substantial supplies of Ifns to conduct such research can now be produced with rDNA technology.

Though most rDNA-made Ifns currently under evaluation are produced in the bacterium *E. coli*, yeast are being increasingly employed as production organisms. Yeast require less stringent culture conditions than do most bacteria, have long records of reliability and safety in large-scale bioprocessing, and are more adaptable to continuous culture production than are many bacteria. Furthermore, because yeast more closely resembles higher organisms than bacteria, yeast can add sugar molecules to protein when necessary. Thus, modified products made in yeast are more likely to be pharmaceutically useful than unmodified products made in bacteria. Several groups have recently reported progress with Ifn production from yeast, including secretion of the Ifn polypeptide into the culture medium from which it can more easily be purified (45). Academic workers funded by the British firm Celltech, Ltd., have reported yields of alpha-Ifn as high as 15 milligrams (3 billion units*) per liter of yeast culture (139). Numerous genetic techniques are being devised to increase production: 1) amplification of the number of Ifn genes, 2) enhancement of gene expression by placing it under control of regulatory elements which can be varied without hampering cell growth, 3) limitation of product degradation, 4) inducement of product secretion, and 5) stabilization of microbial strains. Additionally, the Swiss company Hoffmann-La Roche has reported a MAb system for alpha-Ifn purification that gives in excess of 1,000-fold purification with 95 percent recovery of biological activity (133).

Many U.S. and foreign companies using biotechnology are working toward large-scale Ifn production. Some of the companies with Ifn gene cloning projects are listed in table 18. The large number of companies involved in Ifn production reflects the large market potential so widely publicized in the late 1970's. Since clinical trials have not supported many of the claims made for Ifns, companies are beginning to draw back from Ifn R&D.

The international pattern of interest and investment in the use of rDNA technology to produce Ifn reflects to some extent international differences in medicine and, possibly, movements to reduce national dependence on pharmaceutical imports. Japan, for instance, has long been the largest market in the world for cancer drugs, today exceeding $375 million in annual sales (compared to $210 million in the United States) (127), and is actively investigating the production of anticancer pharmaceutical products using new biotechnology.**

*A single dose of Ifn ranges from 1 million to 100 million units.

**Protein agents are especially popular for cancer treatment in Japan. Immunotherapeutic concepts which are regarded as experimental hypotheses in the West provide the rationale for administration in Japan of hundreds of millions of dollars worth of agents, such as Krestin® (an orally administered fungal glycoprotein that accounted for Japanese sales in 1981 of $230 million) and urokinase (which is used in Japan for indications not even suggested in the United States). Sales of over $117 million were recorded in 1981 for a streptococcal "vaccine," called Picibanil®, which Japanese physicians regard as an immunostimulant (118).

Table 17.—Some Ongoing Clinical Trials of the Use of Interferons To Treat Cancer

Interferon supplier	Sponsor	Cancer	Phase	Institution
Natural lymphoblastoid (produced from cultured cells; contains mixture of interferon types):				
National Cancer Institute (NCI)	NCI	Broad range of advanced cancers	I	University of Wisconsin
NCI	NCI	Melanoma	II	Georgetown University
Wellcome Foundation	NCI	Ovary	II	Gynecological Oncology Group East Coast Oncology Group
Wellcome Foundation	NCI	Lymphoma, non-Hodgkin's	II	Southeast Oncology Group
NCI	NCI	Breast, metastatic	II	UCLA
NCI	NCI	Breast, recurrent	II	Duke University
Wellcome Foundation	NCI	Breast, recurrent	II	National Surgical Adjuvant Breast Project
NCI	NCI	Multiple myeloma	II	UCLA Duke University Memorial Sloan Kettering Cancer Center
NCI	NCI	Kidney (renal cell)	II	Duke University
Wellcome Foundation	NCI	Kidney (renal cell)	II	Southwest Oncology Group East Coast Oncology Group
Wellcome Foundation	NCI	Leukemia, childhood acute lymphocytic	I-II	Children's Cancer Study Group
NCI	NCI	Kaposi's sarcoma	II	NCI-Clinical Oncology Program
NCI	NCI	Colorectal	II	Memorial Sloan Kettering Cancer Center
rDNA-produced alpha-interferon:				
NCI	NCI	Broad range of advanced cancers	I	NCI-Frederick Cancer Research Facility
NCI	NCI	Lymphoma, non-Hodgkin's	II	NCI-Frederick Cancer Research Facility
NCI	NCI	Lymphoma, Burkitt's	II	NCI-Frederick Cancer Research Facility
NCI	NCI	Leukemia, chronic (CLL)		NCI-Frederick Cancer Research Facility
NCI	NCI	Mycosis fungoides		NCI-Frederick Cancer Research Facility
NCI	NCI	Leukemia, acute	I-II	University of Maryland
Schering-Plough (S-P)	NCI	Multiple myeloma	II	Wake Forest University
S-P	NCI	Bladder cancer	I-II	Northern California Oncology Group
S-P	S-P	Melanoma	II	Yale University University of Wisconsin University of Rochester M. S. Hershey Medical Center University of Missouri
S-P	S-P	Lymphoma, non-Hodgkin's	II	Roswell Park University of Maryland Harper Grace Hospital Yale University University of Chicago
S-P	S-P	Lymphoma, Hodgkin's	II	Yale University University of Chicago Wilford Hall Medical Center
S-P	S-P	Breast cancer	I	Bowman-Gray Hospital Harper Grace Hospital USC Cancer Center
S-P	S-P	Multiple myeloma	II	University of Texas (Galveston) Roswell Park Bowman-Gray Dartmouth-Hitchock
S-P	S-P	Leukemia, acute	II	UCLA
S-P	S-P	Kaposi's sarcoma	II	San Francisco General Hospital UCLA
S-P	S-P	Lung, small cell	II	USC Cancer Center Bowman-Gray
S-P	S-P	Head and neck cancer	II	University of Texas (Galveston)
S-P	S-P	Colorectal	II	Lombardi Cancer Center
Hoffmann-La Roche (HLR)	HLR	Broad range of advanced cancers	II	University of Arizona
HLR	HLR	Melanoma	II	University of Arizona Mayo Clinic

Table 17.—Some Ongoing Clinical Trials of the Use of Interferons To Treat Cancer (Continued)

Interferon supplier	Sponsor	Cancer	Phase	Institution
HLR	HLR	Ovary	II	Dana Farber Cancer Institute
HLR	HLR	Lymphoma, non-Hodgkin's	II	University of Arizona Minneapolis VAH Mayo Clinic
HLR	HLR	Multiple myeloma	II	M. D. Anderson Hospital
HLR	HLR	Kidney (renal cell)	II	University of Arizona
HLR	HLR	Leukemia, chronic	II	George Washington University
HLR	HLR	Kaposi's sarcoma	II	University of Arizona Memorial Sloan Kettering Cancer Center
HLR	HLR	Osteogenic sarcoma	II	Mayo Clinic
HLR	HLR	Breast cancer	II	Georgetown University USC Cancer Center
Cultured cell-produced gamma-interferon:				
Revlon	NCI	Broad range of advanced cancers	I	NCI-Frederick Cancer Research Facility

SOURCE: Office of Technology Assessment, adapted from R. K. Oldham, U.S. National Cancer Institute, "Update on Clinical Trials With Interferon and Monoclonal Antibodies," memorandum, May 4, 1983.

Human growth hormone

As suggested by the preceding discussion, rDNA technology is increasingly being used to produce large amounts of otherwise scarce biological compounds. In addition to supplying compounds for basic research, rDNA technology is likely to contribute to the discovery of many new pharmaceutical products. Some of the promising protein compounds actively being developed with rDNA technology—human growth regulators, neuroactive peptides, and lymphokines, for instance—are listed in table 19.

The development of hGH with rDNA methods is another model for biotechnology's use in the pharmaceutical industry. Human growth hormone is one of a family of at least three, closely related, large peptide hormones secreted by the pituitary gland. These peptide hormones are about four times larger than insulin (191 to 198 amino acids in length). All three hormones possess a wider variety of biological actions than do most other hormones. The primary function of hGH is apparently the control of postnatal growth in humans. Whereas insulin derived from slaughtered animals can be used for treating diabetics, only growth hormone derived from humans is satisfactory for reversing the deficiencies of hypopituitarism in children (65).

Although the established market for hGH is small and current supplies from tissue extracts are sufficient,* hGH was one of the first targets for the applications of rDNA technology. Workers at both Genentech and the University of California, San Francisco (UCSF) reported cloning and expression of hGH in 1979 (39). Genentech's work was supported by the Swedish firm KabiGen AB, while partial funding for the UCSF work was provided by Eli Lilly, which is believed to be the licensee for the product (39). Genentech has such high aspirations of proving sufficient utility for hGH in medical applications beyond those currently treated with cadaver hGH that it has announced its intent to make the development of hGH from rDNA one of the cornerstones of its integrated pharmaceutical enterprise (9). To this end, Genentech is raising capital through an R&D limited partnership specifically to support clinical testing of hGH and is investigating a variety of possible new clinical indications for hGH use. The NIH National Pituitary Agency has been enthusiastic about these investigations, which were not practical when the supply of hGH was limited by the availability of human cadaver pituitaries (104).

*Most pharmaceutical hGH is obtained from human pituitaries removed at autopsy. In the United States, isolation and distribution of hGH has been managed primarily by the National Pituitary Agency (under the auspices of NIH and with the cooperation of the College of Pathologists). Under programs of the National Institute of Arthritis, Diabetes, and Digestive and Kidney Diseases, hGH is provided, without charge, for approximately 1,600 children per year for treatment of hypopituitarism. Another several hundred patients are treated with commercial hGH imported from abroad, which is also obtained from tissue extracts (39).

Table 18.—Some U.S. and Foreign Companies Involved in Interferon Gene Cloning Projects

Alpha interferons:
Amgen (U.S.)
Biogen S.A. (Switzerland)/Schering-Plough (U.S.)[a]
Burroughs Wellcome (U.K.)
Cetus (U.S.)
Collaborative Research (U.S.)/Green Cross (Japan)[b]
Enzo Biochem (U.S.)
Genentech (U.S.)/Hoffmann-La Roche (Switzerland)[a]
Genex (U.S.)/Bristol-Myers (U.S.)
Life Sciences (U.S.)
Meloy Labs (U.S.)
New England Enzyme Center (U.S.)
Beta interferons:
Cetus (U.S.)/Shell Oil (U.S.)[c]
Collaborative Research (U.S.)
E. I. du Pont de Nemours (U.S.)
Genex/Bristol-Myers
Hem Research (U.S.)
Serono Labs (Italy)/ARES Applied Research Systems (Switzerland)
Toray Industries (Japan)[d]
Gamma interferons
Advanced Biotechnologies, Inc. (U.S.)
Amgen (U.S.)
Biogen/Shionogi (Japan)
Bristol-Myers
Cetus
Collaborative Research/Green Cross
Genentech[e]/Daiichi Seiyaku & Toray Industries (Japan)
Genetics Institute (U.S.)
Genex
Hoffmann-La Roche
ImmunoModulators Labs (U.S.)
Interferon Sciences (U.S.)
Revlon (U.S.)[f]
G. D. Searle (U.S.)
Suntory (Japan)[g]
Takeda (Japan)

[a]This alpha-Ifn lacks carbohydrate groups, but lack of glycosylation does not appear to influence activity.
[b]Attempting production in yeast.
[c]Clinical trials began early in 1983.
[d]Toray is scaling-up to a capacity of $3 = 10^{12}$ units per month and expects approval from Japan's Ministry of Health and Welfare soon for beta-Ifn as an anticancer drug (122).
[e]Genentech retained all manufacturing rights and only licensed its Japanese collaborators to sell in Japan and, perhaps, other Asian markets (32).
[f]Revlon's subsidiary, Meloy Laboratories was the first firm to supply both alpha-Ifn and gamma-Ifn to the National Cancer Institute.
[g]Using Genentech's published gamma-Ifn gene sequence (450 bases long), Suntory, a Japanese beverage company, took only 3 months to synthesize and clone the gamma-Ifn gene (119). Suntory has also succeeded in producing gamma-Ifn in yeast.

SOURCE: Office of Technology Assessment; and S. Panem, *The Interferon Crusade: Public Policy and Biomedical Dreams,* Brookings Institution, Washington, D.C., in press.

KabiVitrum AB, a firm owned by the Swedish Government, is the world's largest producer of hGH from frozen human pituitaries (113). KabiVitrum owns 50 percent of KabiGen AB, which has the sole rights to manufacture and market hGH made by the Genentech process anywhere in the world, except in the United States and Canada, where Genentech has sole rights (31). KabiGen researchers are among the long-term leaders in the study of other growth-promoting hormones, especially the polypeptides known as somatomedins (30,100).

Although it is premature to judge the likelihood of success, hGH is being evaluated for: 1) treating constitutionally delayed short stature; 2) improving healing of burns, wounds, and bone fractures; and 3) treating the deficiency of nitrogen assimilation known as cachexia (9). Approximately 3 percent of all children are thought to have constitutionally delayed short stature, and Genentech advisors speculate that as many as one-third of these might benefit from hGH treatment (136).*

Neuroactive peptides

Several important biosynthetic discoveries in recent years have involved identification of polypeptides in the body that act at the same cellular receptors that are affected by drugs. Some of the body's neuroactive peptides, for example, bind to the same receptors affected by opiate drugs and produce analgesic effects in the nervous system similar to those produced by these drugs. Two of the body's own "opiates," enkephalins and endorphins, appear to be structurally related to many other polypeptides that play various roles in the nervous and endocrine (hormonal) systems (41). Another neuroactive peptide that may affect neurological processes, including attention span, is melanocyte stimulating hormone (MSH). Some evidence suggests that MSH enhances the ability of test animals to pay attention to their environment, and MSH treatment has improved the health of some mentally retarded patients as well (53). Initial hopes raised by the treatment of schizophrenic patients with beta-endorphin have not withstood more rigorous testing. Results of testing some other peptides as antidepressants, after encouraging earlier studies, are also disappointing (53).

*Genentech, Lilly, Amgen, Monsanto, and other firms are also interested in applications of rDNA-produced GHs for food production purposes, and those investigations may prove complementary to the medically oriented studies (see *Chapter 6: Agriculture).*

Table 19.—Some Proteins With Possible Pharmaceutical Applications Being Developed With Recombinant DNA Technology

Class/substance	Size (number of amino acids)	Function	R&D status	Project sponsors	Applications
Human growth regulators:					
Growth hormone (GH)	191-198	Promotes growth	Cloned, expressed, 1979	Genentech (U.S.)/ Kabigen AB (Sweden) UCSF/Eli Lilly (U.S.)	Growth promotion; healing burns, fractures; cachexia
Somatostatin	14	Inhibits GH secretion	Cloned, expressed, 1977	UCSF/Genentech	Adjunct to insulin
Somotomedins	44-59	Mediates action of GH	Cloned, expressed, 1982	Chiron (U.S.)	Growth promotion, regulation
Growth hormone releasing factor (GRF)	44	Increases pituitary GH release	Isolated, sequenced, synthesized, 1982	Salk Institute (U.S.)	Growth promotion
Calcium regulators:					
Calmodulin	148	Mediated calcium's effects	Determined to be unprofitable[a]	None	Numerous applications in basic research; hypertension
Calcitonin	32	Inhibits bone resorption	rDNA production	Genentech, Amgen (U.S.)	Bone disease therapy
Parathyroid hormone (PTH)	84	Mobilizes calcium; prevents calcitonin excretion	Cloned, but no production	Massachusetts General Hospital	Osteoporosis therapy; calcium metabolism
Reproductive hormones:					
Luteinizing hormone (LH)	Beta chain; 115[b]	Females: induces ovulation Males: stimulates androgen secretion	Cloning in progress (glycoprotein)	Integrated Genetics (U.S.)/Serono Labs (Italy)	Antifertility
Follicle-stimulating hormone (FSH)	Beta chain; 115	Induces ovarian growth	Cloning in progress (glycoprotein)	Integrated Genetics/ Serono Labs	Reproductive services
Human chorionic gonadotrophin (HCG)	Beta chain; 147	Like LH; more potent	Cloning in progress (glycoprotein)	Integrated Genetics/ Serono Labs	Pregnancy testing
Relaxin	52	Dilation of birth canal; relaxation of uterus	Cloning in progress (non-glycoprotein)	Genentech	Soften bone connective tissue of reproductive tract; antiarthritic (?)
Neuroactive peptides:					
β-Endorphin	31	Analgesia	Cloned, expressed	Amgen, others	Analgesia
Enkephalins	5	Analgesia	Cloning in progress	Amgen, others	Analgesia
Pancreatic endorphin	N.A.[c]	Undetermined	Cloning in progress	Endorphin, Inc.	Analgesia, particularly in childbirth
Lymphokines and immunoactive peptides (other than interferons):					
Interleukin-2	133	Promotes T-cell growth, activity	Cloned, expressed	Ajinomoto Co. (Japan) Japanese Cancer Institute Immunox (U.S.) Cetus (U.S.) Chiron Genex (U.S.) Biogen (U.S.) Genetics Institute (U.S.) Interferon Sciences (U.S.) Quidel (U.S.)	Maintain T-cell cultures; immunotherapy
Thymosin (fraction 5)	10-150	Promotes maturation of bone marrow cells, T-cell differentiation	Purified, sequenced	George Washington University	Immunodeficiency diseases
Thymosin (alpha 1)	28	Promotes T-helper and T-amplifier functions	Purified, sequenced cloned, 1979	Hoffmann-La Roche (Switz.) Genentech	Systemic lupus erythmatosis; other immune disorders
Thymic hormone factor (THF)	9	Promotes T-helper and T-amplifier functions	N.A.	N.A.	Antiviral protection in immunosuppressed patients
Thymic factor (TFX)	40	Restores delayed-type hypersensitivity	N.A.	N.A.	Cancer treatment
Thymopoietins	49	Inhibits B-cell differentiation	N.A.	Ortho Pharms. (U.S.)	Reversing immunodeficiencies

Table 19.—Some Proteins With Possible Pharmaceutical Applications Being Developed With Recombinant DNA Technology (Continued)

Class/substance	Size (number of amino acids)	Function	R&D status	Project sponsors	Applications
Macrophage inhibitory factor (MIF)	N.A.	Inhibits macrophage migration	Cell fusion	Denki Kagaku (Japan)	Immunotherapy
Respiratory system regulators:					
Alpha-1-antitrypsin	45,000 molecular weight	Prevents destruction of alveolar walls by elastase	rDNA in yeast	Zymos Corp. (U.S.)/ Cooper Laboratories (U.S.)	Emphysema treatment

[a]Armor Pharmaceutical Co., the source of salmon calcitonin in the United States, does not believe that rDNA technology offers significant advantages over chemical synthesis for the production of salmon calcitonin at the present time. A New Drug Application is pending for human calcitonin, but this product is 20 times less than salmon calcitonin for the same effects. Hence, the economics of human calcitonin production are less advantageous than those of salmon calcitonin production.
[b]Most reproductive hormones thus studied are glycoproteins consisting of two polypeptide chains. All share a common (89 amino acids long) alpha chain. Biological activity is manifested in the beta chain, and most cloning efforts focus on producing the biologically active component.
[c]N.A. = Information not available.

SOURCE: Office of Technology Assessment.

Despite the setbacks noted above, many investigators are confident that neuroactive peptides are among the most promising potential advances in medicine; thus, a great deal of research is being done on synthetic analogs of neuroactive peptides (e.g., 26,41) to identify structures that may have research or pharmaceutical applications. Lilly and Burroughs-Wellcome (U.K.) are investigating the use of enkephalin analogs in clinical trials in the United States. Foreign companies with major research programs concerning neuroactive peptides include Abello (F.R.G.), Hoechst (F.R.G.), Hoffmann-La Roche (Switzerland), Organon (Netherlands), Reckitt & Colman (U.K.), Roussel Uclaf (France), Sandoz (Switzerland), and Takeda (Japan). In addition to screening neuroactive peptides compounds for analgesic and anesthetic activity, researchers are attempting to recognize those compounds that might suppress coughing or diarrhea or might counteract asthenia, cerebral vascular disorders, failing memory, mental depression, Parkinson's disease, and forms of dementia, including senility.

Much basic research remains to be done before substantial use is made of neuroactive peptides as pharmaceutical compounds in medicine (53). Studies of these substances and their chemical analogs are expected to result in the development of new drugs, some of which may be produced with biotechnology. Companies vigorously pursuing the production of neuroactive peptides with biotechnology include Amgen (U.S.), which has cloned and obtained expression of the genes for the neuroactive peptide beta-endorphin (126), and Endorphin, Inc. (U.S.), which is primarily concerned with compounds active in both the nervous and digestive systems.

Lymphokines

Lymphokines are proteins produced by lymphocytes (cells of the immune system) that convey information among lymphocytes. With the exception of Ifn, lymphokines are only beginning to be characterized, but these proteins appear to be crucial to immune reactions. Some lymphocytes, for example, produce lymphokines that engage other lymphocytes to boost the immune response to a foreign substance (antigen) and repel foreign invasion or disease. Other lymphocytes produce lymphokines that act in tandem with the antigen to stimulate the secretion of antibodies. Lymphokines may also help to ensure that only the antigen is attacked during an immune response, not the body's own tissues.

The importance of lymphokines in preventing disease and understanding cellular function (including aberrant cell function such as cancer growth) is fostering widespread research on these compounds (for review, see 47). Investigations of the complex interactions among lymphocytes have been hampered in the past by impure lymphokine preparations, which have led to ambiguous findings. Recent progress, including the establishment of lymphocyte cell lines that produce various classes of lymphokines (e.g., 37) and cloning of lymphokine-producing genes into rDNA systems for production in bacteria (24,137), has been made possible with the use of biotechnology. The availability of pure lymphokine samples from such systems may enable researchers to answer more questions concerning cell biology and immune function. Lymphokines may also be useful in the culture of certain cell lines. Eventually, these efforts may lead to the use of lymphokines in medicine to stimulate the patient's own immune system to combat disease.

Leading commercial efforts to produce lymphokines with biotechnology are centered in Japan, Switzerland, and the United States. In Tokyo, Dr. Tadatsugi Taniguchi of the Japanese Cancer Institute is collaborating with Ajinomoto Company to produce the lymphocyte growth factor, interleukin-2 (13). In Switzerland and the United States, numerous firms using biotechnology are engaged in lymphokine research, especially in the production of interleukin-2, but their efforts are largely proprietary at this time (24).

Other regulatory proteins

In addition to hormones and other regulatory proteins, a number of protein "growth factors" for a variety of somatic (body) cells have been isolated and are currently being characterized with the possibility that they may soon be candidates for production by rDNA technology as well (see table 20). Perhaps the most important use of growth factors will be in preparing culture media for growing higher eukaryotic cells, thereby facilitating further research with more complex cells.

Table 20.—Some Protein "Growth Factors" With Potential Pharmaceutical Applications

Factor	Function
CSF (colony stimulating factor)	Stimulate granulocyte differentiation
ECGS (endothelial cell growth supplement)	Required by vascular lining cells
EDGF (endothelial-derived growth factor)	Stimulates cell division in blood vessels
EGF (epidermal growth factor)	Stimulates growth of epidermal cells and many tumors
FGF (fibroblast growth factor)	Stimulates fibroblast cell growth
FN (fibronectin)	Stimulates adhesion and proliferation of fibroblast cells
MDGF (macrophage-derived growth factor)	Stimulates cell division near inflammation
NGF (nerve growth factor)	Stimulates nerve growth and repair
PDGF (platelet-derived growth factor)	Stimulates division of fibroblast-like cells
SGF (skeletal growth factor)	Stimulates bone cell growth
WAF (wound angiogenesis factor)	Stimulates wound healing
TAF (tumor angiogenesis factor)	Stimulates blood vessel proliferation in tumors

SOURCE: Office of Technology Assessment, 1983.

Blood products

Products derived from the fractionation of human blood represent the greatest volume of biological pharmaceutical products sold today and comprise a world market of $1 billion yearly. The three main plasma commodities are human serum albumin (HSA), gamma globulin (GG), and antihemophilic factor (AHF), which accounted for 41 percent, 25 percent, and 13 percent, respectively, of the global plasma component market in 1978. North America and Japan each consume 25 percent of the world's blood products (106).

The United States now enjoys a favorable trade balance with respect to blood products. Because blood donation is more widely practiced in the United States than elsewhere, the United States supplies blood components to many other countries. Japan obtains 50 percent of its HSA and 60 percent of its GG* from the United States. The plasma production of Europe is about 60 percent of that of the United States (105).

The blood products industry is characterized by large markets and strong incentives for biotechnological innovation on a nationwide basis. Currently, the industry is troubled by the disease AIDS. Although the etiology of AIDS is not yet understood, the strong possibility that it can be transmitted in blood products lowers the marketability of such products. Thus, the industry is seeking new methods for the production of blood products.**

Human serum albumin

HSA, a single polypeptide chain of 585 amino acids, is the protein used in the largest quantities in medicine. HSA is used primarily during surgery and to treat shock, burns, and other physical trauma. In 1979, worldwide HSA consumption exceeded 90,000 kg, with U.S. consumption accounting for 80 percent (72,500 kg) of this amount. Although the United States consumed large amounts of HSA relative to most other countries in the past, foreign HSA consumption is rising, as shown in table 21. Worldwide HSA consumption is expected to exceed 250,000 kg by 1984 (64,106,143) with the largest increases of HSA consumption taking place in foreign countries. The United States has experienced an overcapacity of HSA production from blood fractionation since 1975 (143) and is currently the world's leading exporter of HSA.

HSA's trememdous markets make it an attractive target for production with biotechnology. However, HSA's substantial molecular size (585 amino acids) and its relatively low cost of conventional production present formidable challenges to biotechnology. In November 1981, Genentech announced successful HSA production in bacteria and yeast through rDNA manipulation (63). This achievement is a landmark in several respects:

- HSA is the largest protein (585 amino acids) yet produced by rDNA technology.
- Planners and technologists aim to manufacture tons rather than grams of injectable products using rDNA systems.
- Competitive product costs are more than an order of magnitude lower per unit weight of product than those for previously considered rDNA pharmaceuticals (e.g., less than $1/

*GG is a fraction of serum that contains antibodies. Boosting a patient's antibody level generally is thought to help prevent infectious disease. This treatment is used especially for hepatitis prevention. The ability to produce specific antibodies (MAbs) may make GG a less desirable therapy and increase the effectiveness of antibody prophylaxis.

**These efforts are to be discussed in a forthcoming OTA report, *Blood Banking Policy and Technology*.

Table 21.—Human Serum Albumin Production and Consumption in the United States

	1971	1976	1979	Forecast 1984
Plasma processed in the United States (thousands of liters)	1,950	2,910	3,950	6,920
HSA production in the United States (millions of grams)	39	67	91	159
HSA consumption:				
Domestic (millions of units)	2.9	4.6	5.8	8.5
Foreign (millions of units)	0.3	0.7	1.5	4.2
Total (millions of units)	3.2	5.3	7.3	12.7
Domestic	94%	87%	80%	67%
Foreign	6%	13%	20%	33%
HSA revenues:				
Domestic (millions of dollars)	$58	$133.4	$168.2	$300
Foreign (millions of dollars)	4	20.3	43.5	148
Total (millions of dollars)	62	153.7	211.7	448

SOURCE: Office of Technology Assessment, based on data and estimates in M. M. LeConey, "Who Needs Plasma?" *Plasma Quarterly* 2:68-93, September 1980.

gram, compared to somewhat less then $50/ gram for insulin).

- The companies that successfully produce HSA with rDNA technology will amass knowledge of certain related processes, including purification of large amounts of product. This knowledge might allow them to dominate the production of other proteins made by similar processes.

Since cloning the HSA gene, Genentech has entered into an agreement with Mitsubishi Chemical Industries, Ltd. (Japan) to cooperate in continued R&D for manufacturing and commercialization. The partnership hopes to produce 10 metric tons (tonnes) of HSA per year by 1985 (121). Mitsubishi will probably ask Green Cross, which is the largest Japanese blood products company, to distribute the rDNA-produced product, thus avoiding discrimination against the present distributor of HSA. In 1981, HSA sales in Japan were $60 million (¥14.2 billion) (118), compared to about $200 million in the United States (64). The corporate arrangements between Genentech, Mitsubishi, and Green Cross may lead to the reduction of Japanese imports, the establishment of a blood product industry in Japan, and advances in Japanese technology for producing and purifying proteins.

Genex (U.S.) and Biogen S.A. (Switzerland) also have established arrangements with Japanese firms to conduct R&D on rDNA production of HSA (115). Genex made a contract in 1981 with Green Cross. In exchange for research funding, Genex agreed to grant Green Cross exclusive licenses to make, use, and sell all microbially produced HSA developed under the contract in the Far East, South America, and North America. Genex made a similar agreement with the Swedish firm KabiVitrum, with licensing pertaining to Europe, Africa, and the Middle East. Biogen S.A. negotiated a similar agreement in late 1981 to cooperate with Shionogi (Japan) in the development of rDNA techniques for HSA production.

Only one major American drug company, Upjohn Pharmaceuticals, shows evidence of developing a fully in-house large-scale biosynthetic HSA process. Upjohn is making HSA in both *E. coli* and yeast.

Antihemophilic factor

AHF, a class of proteins contained in the fraction of blood used to treat hemophilia (a set of hereditary disorders that prevent blood clotting), is used by approximately 14,000 hemophiliacs in the United States on a routine basis (143). Type A hemophilia, which affects about 5 people in every 100,000, is caused by a deficiency of factor VIII, and type B hemophilia (which is much rarer but equally severe) by a lack of factor IX.

AHF is separated during the fractionation of whole blood to obtain HSA. As shown in table 22, U.S. AHF production has multiplied faster than consumption in recent years, and AHF comprises sizable exports for U.S. firms and nonprofit organizations. With AHF selling for over $1 million per gram and AHF use growing at a rate of 14 percent per year, AHF is the blood fractionation industry's most profitable product (64).

Table 22.—Antihemophilic Factor Production and Consumption in the World

	1971	1976	1979	Forecast 1984
Plasma processed globally for AHF (thousands of liters)	365	1,600	2,750	5,320
AHF units processed (millions)	80	400	688	1,330
Domestic consumption:				
Millions of units	72	300	412	648
Average price (cents/unit)	15	10	10	14
Sales (millions of dollars)	10.8	30	41.2	91
Foreign consumption:				
Millions of units	8	100	275	682
Average price (cents/unit)	40	30	30	27
Sales (millions of dollars)	3.2	30	82.5	184
Total AHF sales (million of dollars)	14	60	123.9	275

SOURCE: Office of Technology Assessment, based on data and estimates in M. M. Le Coney, "Who Needs Plasma?" *Plasma Quarterly* 2:68-93, September 1980.

Efforts to produce AHF with biotechnology are underway. The gene for factor IX has recently been cloned and expressed in *E. coli* (18,61). The availability of factor IX produced by rDNA technology facilitates studies concerning the genetic basis of type B hemophilia (e.g., 35). However, quantities of factor IX necessary to treat the relatively uncommon type B hemophilia are adequately provided by whole blood fractionation, and the rDNA product is not now a competing alternative.

Significantly stronger medical and commercial reasons motivate efforts to clone factor VIII genes, since the majority of hemophiliacs are type A. At present, difficult problems surround factor VIII gene cloning. Not only is factor VIII present in low concentrations in plasma, making its isolation and purification difficult, but this molecule is an extremely large and labile glycoprotein (over 300,000 molecular weight, about 20 times the size of Ifn). Recent progress in factor VIII research includes development of MAbs to aid in AHF isolation (86,132) and localization of AHF-producing cells in the liver (134).

The rDNA production of factor VIII is an elusive goal, but the implications of success are substantial. Apart from providing more economic treatment for hemophiliacs, results of factor VIII cloning may lead to a better understanding of the most common type of hemophilia and prove useful for prenatal screening for the disease.

Biosynthetic AHF may lower costs of treatment for the expanding population of hemophiliacs throughout the world. Furthermore, if the production of HSA from rDNA technology proves competitive with fractionation, the need to produce AHF with rDNA may be paramount, since AHF is copurified with HSA from plasma.*

Research laboratories working towards AHF microbial biosynthesis include the following (12,128):

- Armour Pharmaceutical (U.S.)/Scripps Clinic and Research Foundation (U.S.),
- Baxter Travenol Laboratories (U.S.)/Genetics Institute (U.S.),
- Biogen S.A. (Switzerland)/Teijin (Japan),
- Speywood Laboratories (U.K.)/Katherine Dormandy Hemophilia Centre and the Royal Free Hospital of London (U.K.)/Genentech (U.S.), and
- Connaught Laboratories (Canada)/Canadian Government.

Thrombolytic and fibrinolytic enzymes

Thrombosis, the blockage of blood vessels, is the leading cause of death in industrialized nations. Blood clots in the vessels that supply the heart (coronary heart disease), brain (stroke), or lungs (pulmonary embolism) account for more than half of all deaths in the Western Hemisphere.

The search for substances that dissolve blood clots is a major undertaking of the pharmaceutical industry. At present, the most popular compounds are thrombolytic and fibrinolytic enzymes. These substances initiate the dissolution process by converting plasminogen, a plasma protein, into plasmin, which then attacks fibrin, the protein that comprises most of the blood clot.

The two most widely used thrombolytic enzymes are streptokinase and urokinase. Streptokinase is manufactured from colonies of *Streptomyces* bacteria, while urokinase is obtained either from cultured human kidney tissue or from human urine. Recent improvements in large-scale cell culture techniques and purification methods (including the use of MAbs for the purification of protein) now yield good quantities of thrombolytic enzymes (57). Despite the great usefulness of these enzymes, however, several problems diminish their clinical value. In prolonged therapy with streptokinase, chances of allergic reactions arise. In addition, streptokinase and urokinase appear to act nonspecifically throughout the body, thus raising risks of internal hemorrhaging in patients. To circumvent this risk, carefully placed catheters must be used to deliver the enzyme to its target. Finally, high costs of manufacturing and therapy also restrain more widespread use (streptokinase treatment costs $275, while urokinase costs about $3,000 per patient) (57). Because of

*The price of factor VIII controls the price of serum albumin (64). The worldwide growth rate for AHF, about 14 percent per year (64), is twice the growth rate of HSA. Thus, any major shift of HSA production to rDNA technology with a concomitant loss of AHF production may drive the price of AHF (produced from fractionation) to higher levels.

these problems, alternative thrombolytic enzymes and more economic production methods are being sought.

A group of fibrinolytic enzymes called tissue plasminogen activators (tPAs) may solve some of the problems associated with streptokinase and urokinase. Although tPAs are generally not well characterized and are only available in limited quantities at present, they appear to work specifically at blood clots over a prolonged time (59), reducing both the risks of hemorrhage and the doses necessary for thrombolysis, thus lowering costs of treatment.

Advances in culturing tPA-secreting cells and isolating tPA using MAbs indicate that manufacturing costs may be reduced in the future. Moreover, Genentech, in collaboration with investigators at the University of Lueven (Belgium), recently succeeded in cloning the gene that produces tPA (108), and a number of other companies are working to produce tPA from rDNA systems (see table 23). Cloned genes in bacteria or yeast may provide a means for economically producing large quantities of tPA. The biochemical effectiveness and commercial viability of rDNA-produced tPAs remain to be demonstrated. In particular, questions concerning the stability of the cloned genes in bacterial strains, scale-up costs, and importance of sugar residues found on native tPA remain to be answered.

At present, the extent to which thrombolytic enzymes are used by different countries varies substantially. German and Japanese physicians prescribe streptokinase and urokinase extensively, often in conjunction with cancer chemotherapy (on the premise that fibrin shields tumors from drugs and the body's immune defenses and hence must be removed). American medical practices, on the other hand, discourage the use of streptokinase and urokinase because of the problems mentioned earlier. Thus, the annual market for thrombolytic enzymes in the United States represents a modest $8 million, whereas the annual market for urokinase in Japan, where it is the seventh largest selling drug, represents $150 million (57).

The widespread sponsorship of tPA projects by Japanese companies, as shown in table 23, reflects these national differences in thrombolytic enzyme use. In addition to underwriting clinical testing and marketing costs of enzymes produced from cultured cells, Japanese companies such as Green Cross are active in sponsoring tPA production using rDNA techniques.

The development of tPA illustrates biotechnology's role in providing new pharmaceutical agents.

Table 23.—Thrombolytic and Fibrinolytic Enzymes: Companies Involved in Development and Marketing

Protein	Company	Project description
Streptokinase	Hoechst-Roussel (F.R.G.)	Production from bacteria
	KabiVitrum (Sweden)	Production from bacteria
Urokinase	Abbott Laboratories (U.S.)	Extraction from cultured kidney cells
	Genex (U.S.)/Mitsui Toatsu Chemicals, Inc. (Japan)	Production from rDNA
	Genentech (U.S.)/Grunenthal (F.R.G.)	Production from rDNA
Human tissue plasminogen activator	Genentech/University of Leuven (Belgium)/ Mitsubishi Chemical Industries, Inc. (Japan)/Kyowa Hakko Kogyo (Japan)	Production from rDNA
	Biogen S.A. (Switz.)/Fujisawa (Japan)	Production from rDNA
	Integrated Genetics (U.S.)/ Toyobo Pharmaceutical (Japan)	Production from rDNA
	Chiron (U.S.)	Production from rDNA
	Collaborative Resarch (U.S.)/ Green Cross (Japan)	Extraction from cultured kidney cells
Anticoagulant and fibrinolytic agents	Genentech/Yamanouchi Ltd. (Japan)	Development of microbial strains that produce a fibrinolytic agent
	Genex/Yamanouchi Ltd.	

SOURCE: Office of Technology Assessment.

Through the use of improved bioprocess systems, purification methods, and rDNA technology, large quantities of scarce materials are becoming available for study, possibly leading to substantial changes in medical practices in the United States.

Given successful economic development of tPA (i.e., at one-half the cost of urokinase production) and improved mode of action, industry experts estimate that U.S. markets for tPA could climb swiftly to $125 million per year (57).

Vaccines

The combined techniques of biotechnology find perhaps no greater promise for medicine than in the preparation of vaccines and other pharmaceutical products to combat infectious diseases. There are several approaches to disease control using biotechnology, including the use of rDNA and MAb technology, artificial vaccine synthesis, and protoplast fusion to prepare novel antibiotics.

Most vaccines used at present consist of the organisms that cause the particular disease that the vaccine is intended to prevent. These organisms (pathogens) are killed or otherwise treated ("attenuated") in an effort to make them nonvirulent, and the killed or attenuated mixture is then injected into the person to be vaccinated. Ideally, the recipient's immune system responds to the introduction of the vaccine by producing antibodies that bind to particular molecules (antigens) on the surface of the vaccine organism and identifying it for destruction by other components of the immune system. The antibodies produced by the recipient remain in circulation for a period of months to years, protecting the recipient against the live pathogen should it be encountered later. Thus, the recipient becomes "immune" to the disease. Immunity thus induced, since it uses the recipient's immune system for constant surveillance and defense against the disease, is known as "active immunity." The administration of foreign antibodies or immune products that themselves protect the recipient from the disease, on the other hand, provides what is known as "passive immunity." Passive immunization usually confers only short-term protection against a disease.

Killed and attenuated vaccines represent one of the highest achievements in medicine. Nevertheless, several problems with these vaccines persist. One substantial problem is that killed and attenuated vaccines contain the complete genetic material of the pathogen. If the pathogen is not killed or attenuated completely, the vaccine itself may be capable of causing the disease it is intended to prevent. Another problem with conventional vaccines is that, in many instances, they do not immunize the recipient against all of the various strains of the pathogen. Finally, many conventional vaccines are not stable enough for use where they may be most needed, as in areas without refrigeration.

Subunit vaccines—vaccines that contain only portions of the pathogens—may solve some of the problems associated with killed and attenuated vaccines. Subunit vaccines do not contain the pathogen's genetic material, and, thus, they cannot themselves cause infection. Furthermore, subunit vaccines may be more stable for storage and of greater purity than most conventional vaccines, although these qualities remain to be demonstrated in most cases. Two new methods are being developed to prepare subunit vaccines: rDNA technology to produce all or part of a surface protein molecule of the pathogen and chemical synthesis of short polypeptides that represent surface proteins. Both of these new approaches have the added advantage that subunit vaccine manufacture does not require large-scale culture of the infectious organism.

Viral disease vaccines

Because of the relatively simple, well-understood structure of viruses, the most preeminent biotechnology efforts for the development of new vaccines are focused on viral diseases (51,135). As shown in table 24, biotechnology is being used to develop vaccines for influenza types A and B, herpes, polio, hepatitis A and B, and a number

Table 24.—Some Current Viral Vaccine Biotechnology Projects

Viral disease	Company	Project description
Influenza virus......	Numerous investigators	Improved attenuated strains
	Numerous investigators	Modifications of viral genome through rDNA manipulations
	Scripps (U.S.)	Synthesis of short peptides corresponding to fragments of influenza virus surface proteins
	Scripps	Attachment of viral subunit to larger carrier to evoke broader immune response
Polio virus	Numerous investigators	Modifications of viral genome through rDNA manipulations
Hepatitis B virus....	Merck (U.S.)	Purification of viral particles from blood
	Institut Pasteur Production (France)	
	Chiron Corp (U.S.)/Merck/University of Washington, UCSF	Production of viral surface proteins from rDNA in yeast
	Takeda/Osaka and Hiroshima Universities (Japan)	
	Amgen (U.S.)	
	Biogen/Green Cross (Japan)/University of Edinburgh	
	Integrated Genetics (U.S.)/Connaught (Canada)	
Herpes viruses	Merck	Purification of surface glycoprotein from herpes simplex viruses
	Molecular Genetics (U.S.)/Lederle Labs (U.S.)	Production of viral proteins in bacteria
	Institut Merieux (France)/University of Chicago	Production of nonpathogenic viruses by the deletion of specific genes

SOURCE: Office of Technology Assessment.

of other human viral diseases. The two main methods used to prepare subunit vaccines for viral diseases are summarized in figure 13.

Hepatitis B subunit vaccines, in particular, illustrate the use of biotechnology in vaccine improvement. Using the rDNA approach, a number of groups have cloned genes that encode portions of the hepatitis B surface antigen (HBsAg) and have shown that isolated surface antigens behave similarly to the whole virus when used as a vaccine (25,74,131,146). Merck (U.S.), which supports work done at UCSF and Chiron Corp. (U.S.) and has built an in-house molecular genetics group of nearly 50 scientists since 1978, expects to market a hepatitis B vaccine made from rDNA in yeast by 1987 (44). Biogen S.A. (Switzerland) has successfully immunized chimpanzees against hepatitis B using its yeast-grown vaccine, and a license to Biogen's work with hepatitis vaccines has been acquired by Green Cross (Japan). It has been estimated that Biogen's hepatitis B vaccine will sell for only $10 to $30 per dose as compared with $100 per dose for Merck's vaccine made from virus particles extracted from blood of hepatitis B carriers (14,71). How well these rDNA-produced hepatitis B subunit vaccines will compete with vaccines made by traditional methods is not yet known, but the need for an effective and inexpensive hepatitis B vaccine is great.*

Using chemical synthesis, other researchers have prepared synthetic polypeptides which may be useful as subunit vaccines. These synthetic peptides are based on known amino acid sequences of virus surface proteins. The amino acid sequences and their molecular shapes are analyzed by computer, and peptide sequences that are likely to elicit immune responses are defined (for review, see 68)). Researchers have synthe-

*In the United States, there are 80,000 to 100,000 cases of hepatitis B and about 1,000 deaths each year. The incidence in some other parts of the world runs 10 times as high. Between 3 and 15 percent of healthy blood donors in Western Europe and the United States show serological evidence of past infection, and 0.1 percent are chronic carriers of the type B virus. In many African and Asian countries the majority of the adult population have been infected, and 5 to 10 percent of the population are clinically ill with hepatitis. A very strong association has recently been demonstrated between the carrier state of hepatitis and liver cancer. In areas of the world where hepatitis B is endemic, primary liver tumors account for 20 percent of cancer, in contrast to the 1 percent level of liver tumor incidence in the United States (150). A costly hepatitis B vaccine was brought to market by Merck in 1982 in the United States. Although not made with new biotechnology, this vaccine consists of natural subunits—particles of the virus coat protein which are isolated and purified from the blood of relatively rare suitable donors (34,44).

Figure 13.—Methods Used to Prepare Subunit Vaccines for Viral Diseases: Recombinant DNA Technology v. Chemical Synthesis

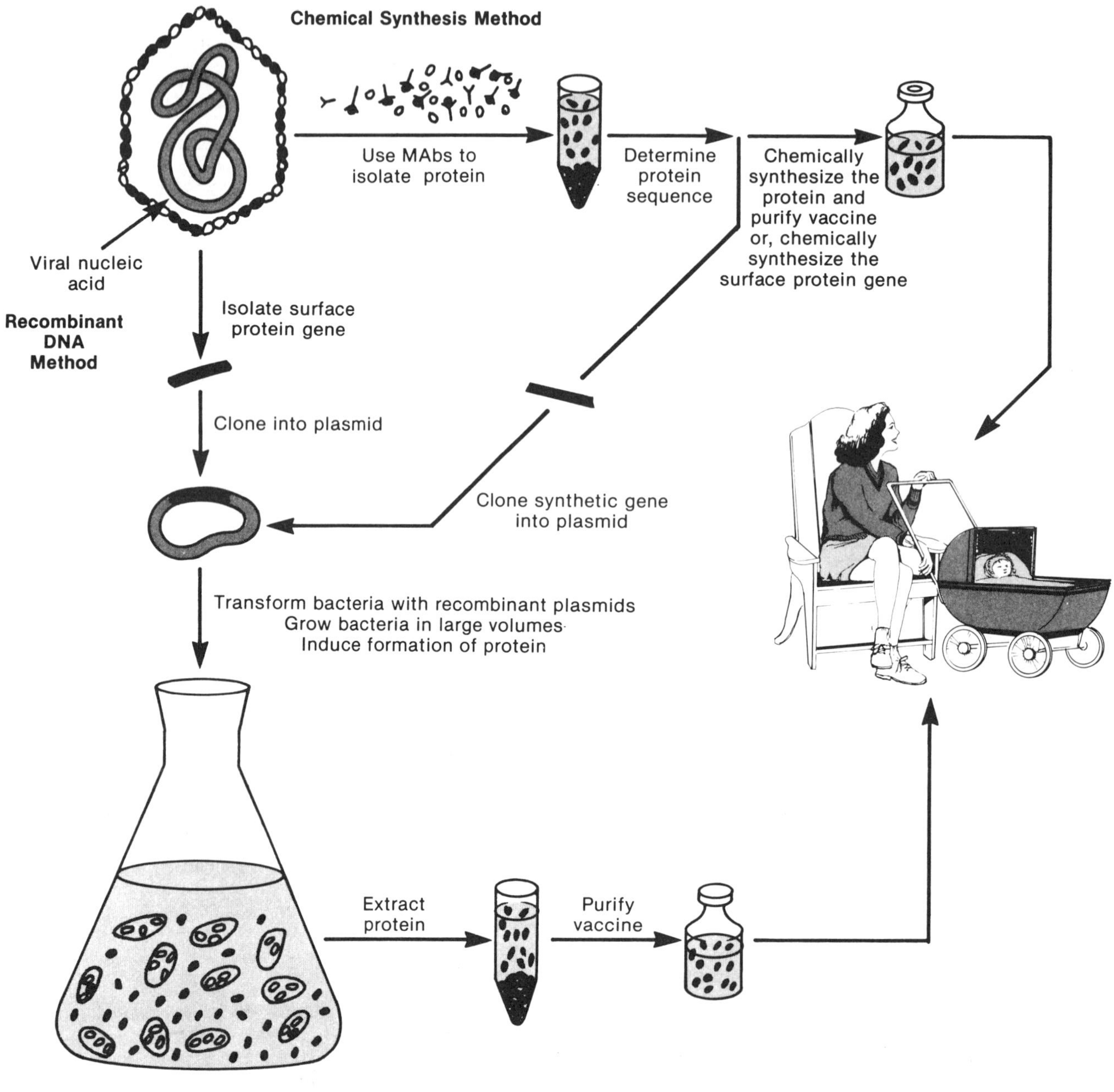

In the **chemical synthesis method,** proteins that comprise the viral surface are isolated, often with the use of monoclonal antibodies. The protein sequence is then determined. Based on the sequencing information, large amounts of the protein or portions of the protein are made chemically for use as the vaccine; alternatively, the sequencing information may allow chemical synthesis of the gene that encodes the protein (or a small portion of the protein). This synthetic gene is cloned via rDNA techniques.

In the **recombinant DNA method,** the gene that encodes the viral surface protein is isolated and cloned into an appropriate vector (such as plasmid), transformed into a host (such as a bacterium or yeast), and the host is grown in large quantities. Formation of the protein by the rDNA and isolation of the protein results in the subunit vaccine.

SOURCE: Office of Technology Assessment.

sized both linear and cyclic peptides that stimulate immunity similar to the complete virus for hepatitis B and influenza (23,46,66, cf. 68). Preliminary evidence indicates that a synthetic influenza subunit vaccine adequately protects animals against several strains of the live virus, but more tests must be done before synthetic subunit vaccines are ready for clinical evaluation.

If synthetic vaccines prove effective, they may be produced in rDNA systems by cloning the DNA corresponding to the synthetic polypeptide and producing the vaccine using microbial bioprocesses. Fairly small amounts of protein may be required, with a few kilograms sufficing for millions of vaccine doses. However, it remains to be seen whether economics favor development of microbial bioprocesses over chemical synthesis. On the other hand, multivalent vaccines (vaccines that protect against several diseases) may be created by combining a number of peptide sequences to elicit responses to several different antigens and thus broaden the range of synthetic subunit vaccines. Such multivalent vaccines may be more economically produced using biotechnology.

In order for both synthetic and rDNA-produced subunit vaccines to be more effective, better immunizing systems must be devised to promote active immunity. Live (attenuated) vaccines proliferate within the body, thus sustaining immune responses that are necessary for long-term protection. On the other hand, subunit vaccines are destroyed rapidly. Delivery systems are being formulated by coupling the subunit proteins with larger carrier proteins that evoke better immune responses (e.g., 2), and by encapsulating subunit vaccines in lipid packages that allow the vaccine to diffuse slowly throughout the body and prolong exposure (92).

A potential live virus vector system is being investigated using vaccinia virus, a virus not pathogenic to humans (131). DNA encoding HBsAg is joined to DNA sequences ("vaccinia virus promotors") which control transcription of the HBsAg DNA. This rDNA construct is inserted into vaccinia virus, and a "living" vaccine that synthesizes and secretes the HBsAg is produced. Rabbits receiving injections of this live vaccine rapidly produce antibodies against HBsAg, and the vaccine is currently being tested in chimpanzees. The investigators are doing further work on the use of this live virus vector system for other vaccines. Such live vaccines may prove useful after a single, easily administered dose of the vaccine where subunit vaccines fall short in achieving a sufficient immune response.

Bacterial disease vaccines

Unlike viruses, whose surfaces are relatively simple and offer protein targets to which vaccines can be directed, bacteria and other microbial pathogens have complex, dynamic surfaces which in many cases defy vaccine development. Most bacterial surfaces are composed mainly of lipids and polysaccharides, which are molecules derived from complex biosynthetic pathways determined by many genes. Hence, bacteria are not as amenable as viruses to genetic manipulation techniques used in subunit vaccine technology.

Biotechnology is being used in several ways to create novel vaccines against bacterial infections, but the results with bacterial vaccines at present are not as extensive as those with viral vaccines. It is necessary first to identify targets that might be suitable for vaccine development. On the surface of some bacteria, such as *Gonococci* and several pathogenic *E. coli* strains, for example, there are certain proteins which perform functions essential to the disease mechanisms. Though subunit vaccine technology has not been widely explored in bacteria, these proteins may provide targets for subunit vaccines comparable to those being made against viruses.

The genes responsible for a bacterium's virulence can be genetically manipulated to create viable, harmless mutants. These mutant bacteria, which outwardly resemble the pathogenic form, can be introduced into the body, where they elicit the production of antibodies against both mutant and pathogenic bacteria.* Such mutant bacteria might be used to colonize body spaces prone to infection and to provide long-lasting immunity (51).

*As discussed in *Chapter 6: Agriculture,* such bacterial vaccines are currently being introduced to the animal agriculture industry to treat colibacillosis, a common bacterial infection in newborn farm animals.

A similar method involves using mutation/selection procedures on pathogenic bacteria to select bacteria that die after a short period of time in the body. For instance, a mutant of the typhoid-causing bacterium, *Salmonella typhi,* type Ty-21a, accumulates toxic amounts of galactose during growth and causes its own death. This mutant can proliferate within the body for a short time, and its presence elicits an immune response that protects against the disease. The Swiss Serum and Vaccine Institute, in association with the French Institut Pasteur, has developed an oral typhoid vaccine of this type.

Other workers have taken this typhoid vaccine strain and incorporated a plasmid with a gene encoding a protein normally produced by *Shigella sonnei,* one of the bacteria which cause dysentery. In mice, this "hybrid" strain elicits immune responses that protect against both the dysentery and typhoid organisms. Thus, it may be possible to construct a multipurpose oral, attenuated typhoid-dysentery vaccine organism that will produce "protective" antigens for both dysentery and typhoid (51).

Parasitic disease vaccines

Diseases caused by parasites, including protozoa, pose major barriers to acceptable health standards for millions of people throughout the world (see table 25). Many of these organisms exhibit even more extraordinary degrees of complexity than bacteria, however, and lack of basic knowledge restrains new vaccine development in virtually all cases (51). As basic knowledge accrues, immunization against diseases caused by parasites may eventually be the greatest breakthrough in health care provided by biotechnology.*

Table 25.—Estimated Worldwide Populations Affected by Parasitic Diseases in 1971

Type of parasite	Diseased population (in millions)
Intestinal parasites:	
Ascariasis	650
Ancyclostomiasis	450
Amoebiasis	350
Trichuriasis	350
Periocular parasites:	
Trachoma	Greater than 400
Systemic parasites:	
Filariasis	250
Schistosomiasis	180
Malaria	100
Leishmaniasis	N.A.[a]
Trypanosomiasis	7

[a]N.A. = Information not available.

SOURCE: Office of Technology Assessment, based on data from World Health Organization, *Report for the Special Programme for Research and Training in Tropical Diseases,* Geneva, 1976.

Progress in developing malaria vaccines exemplify efforts to realize biotechnology's potential in combating parasitic diseases. Because of the lack of a vaccine, combined with parasitic resistance to the drugs used in malaria control (e.g., chloroquine), malaria remains the most prevalent infectious disease in the world.** Historically, the search for malaria vaccines has been hampered by difficulties in growing the malarial parasite *Plasmodium* (which is transmitted by female *Anopheles* mosquitoes) in the laboratory. Other difficulties stem from *Plasmodium's* complex lifecycle and the apparent ability of the parasite to evade the body's immune system. In addition, vaccines based on killed, injected whole *Plasmodia* presently require the use of powerful adjuvants (additional components of vaccines that boost immune responses) in test animals which are too strong for human use.

The complexity of *Plasmodium's* lifecycle hints at the difficulties in developing a vaccine that protects against all forms of malaria. As shown in figure 14, the sporozoites, injected into the blood

*The U.S. National Academy of Sciences and the Agency for International Development convened meetings in July and December 1982 on the applications of biotechnology most significant for the developing world. Recommendations were made with respect to research priorities on the basis of applicability of the new technologies and other considerations (88,145). The only human parasitic diseases that ranked among the top priorities for development at this time were leishmaniasis and malaria. Leishmaniasis is a family of diseases, caused by sandfly-transmitted protozoa, which is considered to have grossly underestimated public health importance in South America, Africa, and the Middle East. It was identified for special attention because there is evidence that immunity can be developed by people in sandfly-infested areas over a period of time. An understanding of this immunity may provide ways to prevent leishmaniasis.

**There are now an estimated 300 million malaria cases per year and a very high mortality rate for children (1 million deaths in Africa alone per year) (158). About 850 million people live in areas where malaria continues to be transmitted despite activities to control it. An additional 345 million people reside in areas with little or no active malaria control efforts. Over half of the health budget of India is spent on malaria control. Resistance to both drugs and insecticides and the number of new malaria cases are all increasing at alarming rates (155). No vaccine is currently available.

Figure 14.—The Lifecycle of *Plasmodium,* the Malarial Organism: Possibilities for Development of Vaccines for Malaria

The malarial infection begins when a person is bitten by an *Anopheles* mosquito that bears *Plasmodia.* Sporozoites (1) are injected into the bloodstream, where they may remain for only 30 minutes before they infect liver cells . Within the liver cells, each sporozoite divides into six to twenty-four merozoites, the next *Palsmodium* life-stage. Merozoites burst from the infected liver cell (2) destroying it, and enter the blood stream, where they infect red blood cells and proliferate. In subsequent waves of infection, merozoites burst from the red blood cells and spread to other red blood cells. Red blood cells infected with merozoites may produce new cell surface molecules which allow them to bind to blood vessel walls (3). Some of the merozoites go on to become gametophytes, the next life-stage (4). These gametophytes are picked up by another *Anopheles* mosquito in another bite; they reproduce within the mosquito and form sporozoites, which may be injected into another person to begin the cycle anew.

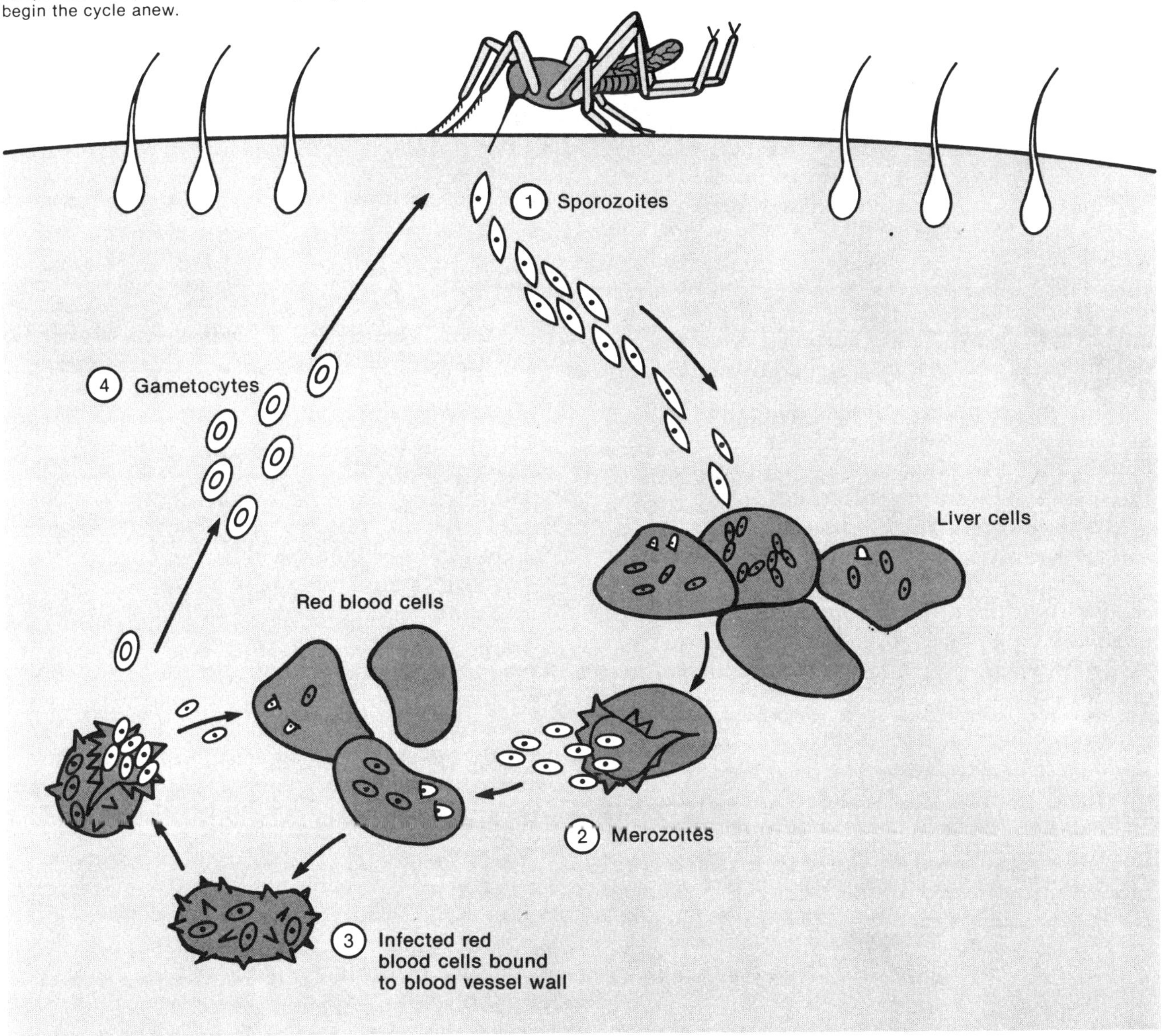

Vaccine possibilities:

1. *Anti-Sporozoite*—Vaccines against the sporozoite, whether antibodies that react with the sporozoite or peptides that mimic the sporozoite surface would probably be ineffective since they must kill every sporozoite to prevent infection.
2. *Anti-Merozoite*—Both passive (antibody) and active (subunit) vaccines against the merozoite might be effective in preventing malaria since the merozoite is often exposed to circulation and because the merozoite is most directly responsible for ongoing malaria infection.
3. *Anti-Malaria-infected red blood cell*—Because red blood cells infected with merozoites may be differentiated by new surface molecules, vaccines (particularly antibodies) against these surface molecules may help in reducing the spread of merozoites to other cells.
4. *Anti-Gametocyte*—Vaccines against the gametocyte would reduce the transmission of malaria since they would lower the number of gametocytes carried by mosquitoes, but such vaccines would not reduce the severity of the disease in its earlier stages.

SOURCE: Office of Technology Assessment.

stream during the mosquito bite, infect liver cells to initiate infection. Large numbers of merozoites, the next life-stage, proliferate within the liver cells and, bursting into the blood stream, successively infect large numbers of red blood cells. Some of the merozoites remain blood-borne; other merozoites develop into gametocytes, which are picked up by mosquitoes, reproduce to form new sporozoites, and begin the cycle anew. Additionally, *Plasmodium* has the ability to evade the immune system over time.

Since the pathology of malaria is caused largely by *Plasmodia* in the merozoite stage, the merozoite appears to be the best target for vaccines. Even one sporozoite reaching a liver cell is capable of causing malaria, so vaccines against this stage must kill every sporozoite to be effective. The gametocyte itself is not pathogenic; an antigametocyte vaccine, therefore, would serve only to reduce the transmission of the disease.

Many investigators (particularly in the United States, the United Kingdom, and Switzerland) are developing MAbs that may be useful in malaria research (153). Antisporozoite and antimerozoite MAbs that inhibit the in vitro multiplication of *Plasmodia* and antigametocyte MAbs that inactivate male gametes have been developed (153). Also, MAbs that destroy merozoite-infected red blood cells have been developed. Such MAbs may prove useful as vaccines that confer passive immunity (19,87,160).

The most promising use of such MAbs is in the isolation of surface antigens which might be used for the development of subunit malarial vaccines. Though quantities of surface antigens obtained by MAb precipitation are too small for use as vaccines, these purified antigens provide a starting point for developing other MAbs with an even greater affinity for *Plasmodium* for use as passive vaccines. They may also provide a starting point for using rDNA technology to isolate large amounts of antigen. Workers at New York University (NYU) recently reported the successful cloning and expression in *E. coli* of a surface antigen from the sporozoite stage of one species of *Plasmodium* using rDNA technology (28), and similar efforts to obtain quantities of antigen from other *Plasmodium* species and life stages using rDNA technology are underway (54). These rDNA-produced surface antigens may serve as protective malarial vaccines.

NYU's "antisporozoite vaccine" has been the subject of a widely publicized dispute between NYU; Genentech (U.S.) (the proposed manufacturer of the vaccine); and the World Health Organization (WHO) (which, with the U.S. Agency for International Development, sponsored NYU's basic research with the standard provision that all WHO-funded work must be "publicly accessible").* When it became clear that Genentech would not obtain an exclusive license to commercialize the vaccine, the company bowed out of negotiations. At present, no other arrangements to pursue large-scale rDNA production of the sporozoite antigen have been made.

As mentioned earlier, a vaccine effective against only the sporozoite stage of a single *Plasmodium* species may not prove to be fully protective against malaria. Ultimately, malaria vaccines may include a variety of stage-specific antigens that result in combined sporozoite and merozoite neutralization, accelerated removal of infected red blood cells, and prevention of gametocyte transmission to the mosquito (158). The delay of further development of NYU's potential milestone sporozoite vaccine imposed by the turmoil over commercialization, however, has raised concern that, in the future, profit motivations may delay the development of urgently needed pharmaceutical products made possible by biotechnology (75,90). Despite their promise, the development of effective malarial vaccines appears to be several years away.

For a variety of reasons, biotechnology holds less promise for vaccine solutions for other parasitic diseases than for malaria. For most of the parasites, there are formidable problems related to culture of the pathogenic organisms and establishment of meaningful models of the human disease in animals. For example, the parasite that causes schistosomiasis, a disease that ranks second only to malaria as a cause of morbidity and

*A similar situation arose with regard to the cloning of several more malarial surface antigens at Walter and Eliza Hall Institute of Medical Research in Australia. This research was also partially funded by WHO (110).

mortality from parasitic organisms, is difficult to culture in the laboratory. The ability of this parasite to alter its susceptibility to host immunological responses and the difficulty in obtaining sufficient quantities of an antigen have hampered efforts to develop a vaccine for schistosomiasis.

Much basic research on parasites is needed in order to develop effective antiparasite vaccines using rDNA technology. The techniques of biotechnology have accelerated the study of parasitic diseases, but urgently needed pharmaceutical applications in this area are still in early stages.

Antibiotics

For the past three decades, antimicrobial agents for the treatment of infectious diseases caused by bacteria have consistently led worldwide sales of prescription pharmaceuticals. Novel antibiotics, produced mainly by traditional microbial bioprocesses, continue to be developed and introduced each year (especially in Japan in recent years). Methods of biotechnology such as the following offer strong innovative possibilities for producing new antibiotics:

- *"Sexual" recombination.* A technique known as protoplast fusion, whereby the contents of two micro-organisms are fused to give one cell, enables researchers to induce rapid improvements in bacterial germplasms. Protoplast fusion allows the rejuvenation of strains of industrial microbes that have lost vigor as a result of mutation and selection procedures that have been performed to increase their antibiotic productivity. The fusion of micro-organisms is beginning to yield new (hybrid) antibiotics (22).*
- *Recombinant DNA technology.* Gene coding for enzymes and other metabolic proteins can be cloned into antibiotic-producing micro-organisms to add steps to existing biosynthetic pathways that improve products or manufacturing processes. Ongoing research includes: 1) the rDNA-mediated transfer of acyltransferase genes among species of bacteria to obtain solvent-extractable cephalosporins (149); 2) the combination of genes via rDNA technology and transformation to obtain direct, efficient synthesis of the antibiotic amikacin (149); and 3) Eli Lilly's utilization of rDNA technology to improve the production of the antibiotic tylosin (4).

The combination of new and traditional technology in the pharmaceutical industry holds tremendous potential for improvement of micro-organisms used in antibiotic production and the isolation of new antibiotic products. Japanese pharmaceutical companies, with their extensive bioprocessing resources, are placing great emphasis on new antibiotic research (114). This emphasis may be due to the fact that antibiotics comprise 25 percent of (1981) ethical drug sales in Japan (compared to about 8 percent in the United States) and that at least 28 percent of the antibiotic sales in Japan now arise from antibiotics produced in the United States (120,125).

*Through protoplast fusion and selection, researchers at Bristol-Myers (U.S.) developed an improved method of producing purer penicillins that has accounted for 8 percent per year improvement in penicillin productivity over the past 4 years. Other genetic approaches produced 60 to 70 percent improvements in yields of cephalosporins (a class of antibiotics) in the same period. Genetic research by Pfizer, Inc., at laboratories in the United Kingdom and United States, have gradually lowered costs of producing oxytetracycline, a long established antibiotic, to costs similar to bulk chemical production, to give prices of several dollars per kilogram (73).

Monoclonal antibodies

MAb technology currently leads other forms of biotechnology in commercial use, as measured by numbers of products on the market. Its lead is largely due to MAb in vitro diagnostic products. In vitro diagnostic products do not have to undergo the same rigorous safety testing required of pharmaceuticals used within the body (in vivo).* The increasing number of MAb-based products also stems from advances in knowledge about hybridoma technology and antibody functions. Further refinements of MAb technology will allow MAbs to be used in numerous applications in the pharmaceutical industry, including in vivo diagnosis, prophylaxis, and therapy.

Hybridomas (MAb-secreting cell lines) derived from human (rather than rodent) cells have only recently become available for use in the pharmaceutical industry. The use of human-cell-derived MAbs in in vivo pharmaceutical applications should give fewer adverse immune reactions than the use of mouse-derived MAbs. Though the preparation of human hybridomas is in its technical infancy, as described in *Chapter 3: The Technologies,* advances in producing MAbs from human cell lines will encourage MAb-based applications for new and replacement medicines.

Diagnostic products

IN VITRO DIAGNOSTIC PRODUCTS

The roster of MAb-based in vitro diagnostic products is growing rapidly. Table 26 provides a list of the products approved for use in the United States as of June 1983.** MAb technology is being used to make both novel diagnostic products and products to replace conventional, polyclonal diagnostic products. Although the competitive advantages of MAb products must ultimately be demonstrated in the marketplace, such products may prove superior to traditional methods used to identify infectious diseases, hormonal changes, or the presence of cancer.

Recently developed applications of MAbs for in vitro diagnosis include the following:

- *Diagnosis of human venereal diseases.* Conventional diagnosis of several common venereal diseases—gonorrhea, chlamydia, and herpes simplex virus—is hampered by time-consuming cell culture requirements. A speedy, sensitive MAb-based diagnostic kit for chlamydia has been produced by Genetic Systems Corp. (U.S.), in collaboration with Syva Co. (U.S.) and the University of Washington (93), and MAb-based diagnostic kits for all three types of infections may be used in the clinic in the near future (38,93).*
- *Diagnosis of hepatitis B and other viral infections.* MAb-based diagnosis of hepatitis B infection is reportedly 100 times more sensitive than conventional diagnosis based on polyclonal antibodies (6,151). The MAb diagnostic product, developed by Centocor (U.S.) with Massachusetts General Hospital, may benefit the blood banking industry, where unambiguous screening for hepatitis is crucial. MAbs are also proving satisfactory for diagnosing rotavirus and cytomegalovirus infections and for distinguishing between strains of influenza viruses that have until now been indistinguishable by conventional methods (80).
- *Diagnosis of bacterial infections.* The recuperation of hospitalized patients is often jeopardized by infections with bacteria such as *Pseudomonas aerouginosa,* and diagnosis may take several days before treatment is begun. Also, group B streptococcal infections are the most common serious infections of newborn infants in the United States. Prior to availability of MAbs, there was little application of immunoassays to the diagnosis of bacterial infections. Genetic Systems, in a joint venture with Cutter Laboratories (U.S.) and its parent company Bayer (F.R.G.), is developing diagnostic and therapeutic MAb products for *Pseudomonas* infections (124). Researchers at the University of Pennsylvania report that diagnosis times for streptococcal

*The regulation of pharmaceutical products in the United States and other countries is discussed in *Chapter 15: Health, Safety, and Environmental Regulation.*

**A longer list of approved MAb products for research and diagnostic use appears in *Monoclonal Antibodies in Clinical Medicine* (77).

*New infections of gonorrhea, chlamydia, and herpes simplex virus type 2 (HSV2) are estimated to exceed 15 million per year in the United States. Approximately 1 million new cases of gonorrhea are reported annually to the U.S. Centers for Disease Control. It is estimated that the true prevalence of gonorrhea in the United States is 3 million cases annually. Chlamydia infections are not reported and their prevalence can only be estimated. Clinically, the infection rate is estimated to be three to four times that of gonorrhea (approximately 10 million cases annually in the United States). Separately or in combination, chlamydia and gonorrhea are responsible for an estimated 200,000 to 300,000 cases of pelvic inflammatory disease per year resulting in infertility in 50,000 to 100,000 women. HSV2 infections are becoming increasingly common, with approximately 200,000 to 300,000 new cases occurring each year. These new cases accrue on an estimated base of 10 million individuals who are already infected (38).

Table 26.—In Vitro Monoclonal Antibody Diagnostic Products Approved in the United States[a]

Manufacturer	Analyte	Date approved
Hybritech, Inc.	IgE	5/29/81
Hybritech, Inc.	PAP	9/1/81
Hybritech, Inc.	HCG	10/13/81
Hybritech, Inc.	T Cell	7/26/81
Hybritech, Inc.	Ferritin	10/19/81
Abbott	PAP	1/19/82
Abbott	CEA	3/3/82
Abbott	CEA	3/29/82
Ortho III	OKT-11	4/6/82
Centocor	Anti-Rabies	4/16/82
Hybritech, Inc.	HCG	4/23/82
Hybritech, Inc.	HGH	6/8/82
Mallinckrodt, Inc.	Total T_4	6/9/82
Hybritech, Inc.	Prolactin	6/10/82
Clinical Assays	^{125}I-IgE	6/18/82
Biogenex Laboratories	β-HCG	7/13/32
Hybritech, Inc.	HCG (EIA)	7/22/82
New Horizons	Gonogen	8/4/82
Monoclonal Antibodies, Inc.	UCG	9/24/82
Hybritech, Inc.	TSH	10/8/82
Allergenetics (Div. of Axonics)	IgE-Fast® (Specific IgE)	11/10/82
Becton Dickinson & Co.	T_4	12/7/82
Syva Co.	Chlamydia	12/10/82
Miles Laboratories	Gentamicin	12/14/82
Allergenetics (Div. of Axonics)	Total IgE-FAST	1/13/83
Carter-Wallace, Inc.	β-HCG	1/20/83
Hybritech, Inc.	Tandem-E® Ferritin	2/24/83
Ortho	Rubella	3/15/83
PCL-RIA	HCG	4/5/83
Quidel Home	HCG[b]	4/14/83
Ventrex Labs, Inc.	Enzyme TSH	4/26/83
Quidel Home	HCG[b]	4/26/83
Diagnon	Ferritin	4/28/83
BTC Diagnostics	HCG	4/28/83
Immunlok	Chlamydia	4/29/83
Monoclonal Antibodies	HCG	5/25/83
Ventrex Labs., Inc.	IgE (total)	5/25/83
Organon Inc.	HCG	5/26/83
BioGenex Laboratories	RIA Gen β-HCG RIA Kit	5/26/83
Micromedia System, Inc.	Micromedia β-HCG RIA	6/1/83
Organon Inc.	Neo-Pregnosticon Duoclon Tube Kit	6/3/83

[a]As of 6/14/83.
[b]These kits are for home use.

SOURCE: U.S. Department of Health and Human Services, Food and Drug Administration, National Center for Devices and Radiological Health, 1983.

infections may be reduced to 2 hours using MAb-based products. Additionally, Becton Dickinson (U.S.) has introduced a MAb kit that detects the bacteria responsible for meningitis infection. The bacterial strains can be detected in 10 minutes, and the company's price for each test is $2 (17).

- *Pregnancy testing.* Products composed of polyclonal antibodies have long been used to detect high levels of human chorionic gonadotropin (hCG) in the blood as an indicator of pregnancy. Large amounts of antisera are required to circumvent the need for radioactive isotope labels, which often accompany immunoassays. MAb technology is an economic means of producing the high quantities of antibody required in pregnancy testing.*

- *Cancer detection.* The detection and quantitation of indicators related to malignant tumors is potentially one of the most important applications of immunoassay and MAbs. A great deal of work on tumor markers is underway, and a few MAb-based products have been approved for marketing. In some cases (e.g., prostatic acid prosphatase and CEA), MAbs are used used to detect blood-borne antigens shed by the tumor; in others, the MAb reagents are used to identify tumor cells by staining tissue specimens.

*Antibodies against human chorionic gonadotropin (hCG) were among the first products made by hybridoma technology, and several firms have made anti-hCG MAbs available commercially. To date, little advantage has been shown in comparison to polyclonal sera derived from animals. Nevertheless, Monoclonal Antibodies, Inc. (U.S.), received approval in September 1982 (based on "substantial equivalence" to previously marketed products) for ModEL, a urine hCG test for pregnancy, to be marketed to clinical labs and doctors' offices. Eventually products for consumer self-testing may be based on the same antibodies. Hybritech, Inc. (U.S.), also has an FDA-approved MAb hCG kit. It utilizes Hybritech's enzymatic amplification technology rather than radioisotopes, but an instrument is required to read the results.

IN VIVO DIAGNOSTIC PRODUCTS

Diagnosis of some diseases requires identification and localization of the disease within the body. Antibodies with detectable markers (e.g., radioactive chemicals) provide highly specific means for accomplishing these ends. Antibodies injected into the body, although used in diagnostic applications, are considered drugs; thus, they require extensive testing prior to approval for marketing.

MAb technology provides quantities of antibodies for testing, and MAbs are being evaluated in an increasing number of in vivo diagnostic appli-

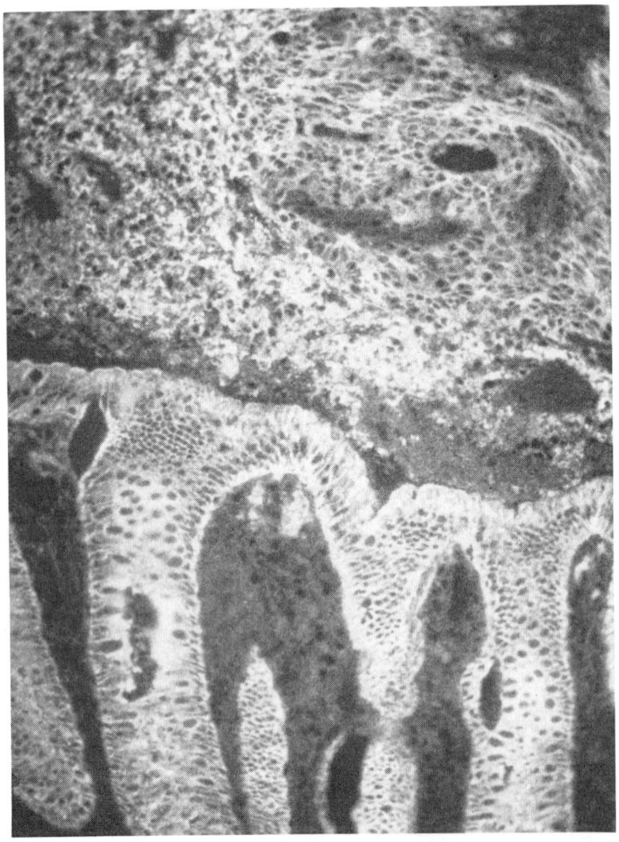

Photo credit: Science Photo Library and Porton/LH International

The spread of colonic cancer is visualized with the use of fluorescently labeled MAbs specific for cancer cells

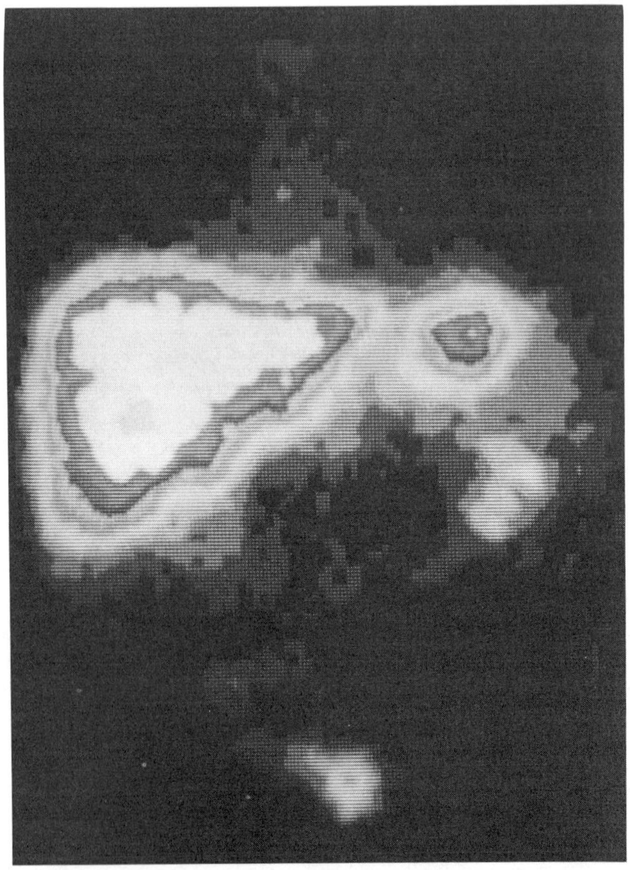

Photo credit: Science Photo Library and Porton/LH International

Liver scan after injection of MAbs shows metastases of colonic cancer cations. One application involves radioisotope-labeled MAbs that bind to cardiac myosin (a major heart muscle protein) to locate and characterize myocardial infarcts (the most common type of "heart attack") (55,56). Another application involves the use of radioisotope-labeled MAbs that bind to antigens on cancer cells, but results to date have not been encouraging. As yet, no antigen that occurs on cancer cells exclusively has been found. A few clinical trials of in vivo diagnosis using MAbs have been undertaken, but experts agree that clinically useful products will require 5 or more years of further development (48). Success in this work could provide useful information prior to and following surgery.

In certain types of malignancies, such as plasmacytomas whose surface immunoglobulins are homogeneous and particular to the tumor, MAbs can be made against these proteins and then used as diagnostic or therapeutic agents. The therapeutic approach has been used in clinical trials for some types of cancer with encouraging results (20,109).

Preventive and therapeutic products

Applications of MAbs to prevent or treat diseases are being pursued on two fronts: 1) administration of MAbs as passive vaccines to protect against specific diseases, and 2) coupling cytotoxic agents (e.g., diptheria toxin, ricin, or cobra venom) to MAbs that direct the agents to diseased cells (7).

Much of the technology being developed that uses MAbs as diagnostic reagents may lead to development of MAb-based (passive) vaccines. This is especially true in the case of the viral diseases hepatitis B, herpes, and cytomegalovirus. Until recently, no cell culture system for hepatitis B has been available; however, a human liver tumor has been adapted to cell culture, and these tumor cells secrete the HBsAg (23). The availability of this HBsAg may make MAb preparation possible, leading to MAbs that neutralize the virus and are effective as a passive vaccine. Infants born to women with hepatitis B apparently benefit from treatment with human serum that contains antibodies against hepatitis B (78), and such serum is used prophylactically in many parts of the world. MAb technology provides a means for producing large quantities of antibodies against hepatitis B.

Scientists at Genetic Systems have produced human MAbs against *Pseudomonas, Klebsiella,* and *E. coli,* all gram negative bacteria which account for serious problems in patients with depressed immune system function (83). Clinical trials of these MAbs as passive vaccines are underway.

Trials of MAb-directed cytotoxic agents against tumor cells indicate that while cytotoxic agents such as cobra venom factor can be made to direct their activity in a very specific fashion against their targets, problems with finding cancer-specific antigens noted earlier restrain such applications of MAbs (36,60,147,148,161). Other problems associated with the use of MAbs in either chemotoxic or direct anticancer therapy include the following:

- toxicity problems associated with rapid administration of antibodies,
- cancer defense mechanisms that apparently involve shielding of target antigens by tumor cells (109),
- the difficulty of getting the cytotoxic agents inside the tumor cells, and
- the difficulty of getting the agent to the majority of cells of a solid tumor.

MAbs will undoubtedly play a major role in the pharmaceutical industry in the future, both as products and reagents for pharmaceutical research. R&D in the use of MAbs as pharmaceuticals is proceeding rapidly in the United States, where several MAb-based biotechnology companies have emerged, in the United Kingdom, where MAb technology was invented, and in Japan.

DNA hybridization probes

DNA "hybridization" occurs when two single strands of DNA join to reform the double helix (see *Chapter 3: The Technologies).* The DNA strands must have exact, corresponding sequences of nucleotide bases for hybridization to occur; thus, a given strand can hybridize only with its complementary strand.

DNA hybridization is a powerful tool in molecular biology. Radioactive phosphorus is commonly incorporated into one of the DNA strands, the "probe," so that the hybridization process can be followed using the radioactive label. DNA hybridization is used to identify and isolate for further study particular DNA sequences (and cells that bear this DNA). DNA hybridization is also used to determine where certain DNA sequences are located on chromosomes. In addition, DNA probes are being tested as reagents in clinical medicine. Probe DNA obtained from a pathogenic organism such as a virus, for example, can be used to identify the presence of that virus within human cells, thus allowing specific diagnosis based on whether or not the radioactive DNA probe hybridizes with DNA in the cells.

Radioactive labeling of DNA hybridization probes raises problems of safety, handling, and disposal that in many cases limit the use of such probes to the research laboratory. Furthermore, since radioactive phosphorus loses its radioactive strength rapidly, only small batches of probes may be practically labeled with radioactivity at any given time.

Several methods to label DNA probes with nonradioactive substances are emerging. The most predominant new method, developed and patented by Dr. David C. Ward and his colleagues at Yale University's School of Medicine, is to couple chemically the molecule biotin to DNA. Biotin-labeled DNA probes hybridize with the target DNA and the hybrids are identified using compounds that recognize biotin (62) (see fig. 15). With such detection systems, only a few hours are required to identify DNA sequences using biotin-labeled probes, whereas 1 or more days are required when radioactive phosphorus labels are used. Additionally, biotin-labeled probes have the potential to be more sensitive than radioactive probes (70).

Nonradioactively labeled DNA is stable and safe to handle, so these probes can be prepared in large (manufacturer's level) quantities and stored for long periods of time. Almost any given short DNA fragment can now be chemically synthesized for use as a probe rather than isolating the fragment from a natural source. Another method of preparing DNA for use as probes is the isolation of DNA fragments made by restriction enzymes. Several companies (e.g., Applied Biosystems (U.S.), University Genetics (U.S.)) are working toward producing a large repertoire of DNA fragments for use as probes.

The ready availability of DNA probes and the convenience of nonradioactive labeling is likely to encourage widespread use of DNA hybridization probes in the near future. While many uses for DNA probes exist in basic research, developers such as Enzo Biochemicals (U.S.) and Cetus Corp. (U.S.) are striving to produce probes for clinical use, where much larger markets exist. Some promising clinical applications of DNA probes include the following:

- *Diagnosis of infectious diseases.* DNA probes that identify and differentiate among species of bacteria that cause diarrheal diseases have been made. Other DNA probes may prove useful in diagnosing human sexually transmitted diseases. DNA probes to detect infections of rotavirus, cytomegalovirus, hepatitis, herpes, and other viruses are being developed (29). In some clinical situations, DNA probes may be more useful than MAbs for diagnosis.
- *Prenatal diagnosis of congenital abnormalities* such as sickle cell anemia (97), beta-thalassemia (101), and duchenne muscular dystrophy.
- *Diagnosis of disease susceptibility.* Researchers in several laboratories are developing DNA probes that recognize DNA abnormalities leading to such conditions as atherosclerosis, the leading cause of death in the United States (5).

Figure 15.—DNA Probe Filter Assay

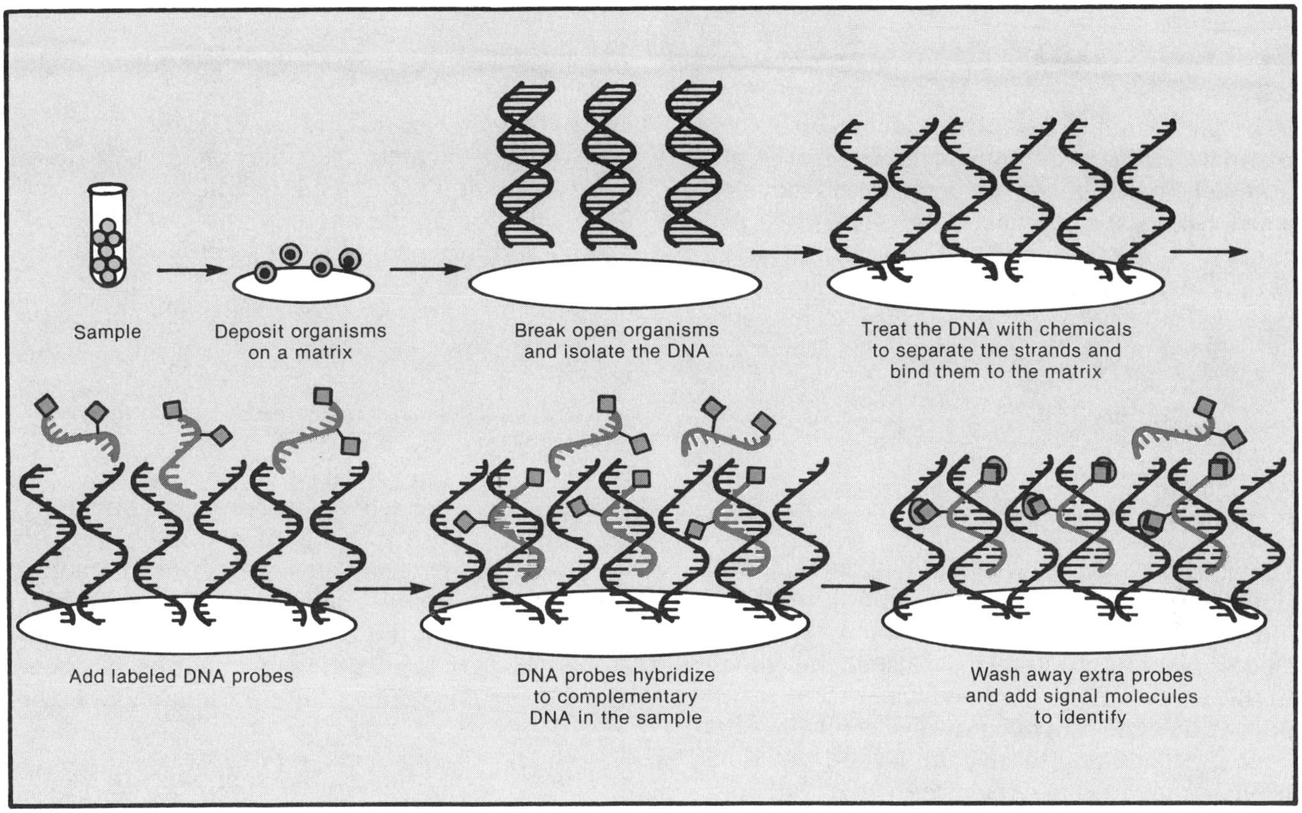

SOURCE: A. Klausner and T. Wilson, "Gene Detection Technology Opens Doors for Many Industries," *Biotechnology*, August 1983; Ron Carboni, N.Y., N.Y., artist.

The success of DNA probes for clinical use probably depends most on convenience and safe labeling of the DNA. Enzo Biochem (U.S.), capitalizing on Ward's biotin labeling technique, markets kits for labeling any given DNA sequence with biotin for use as a probe. Enzo has granted Ortho Diagnostics, a subsidiary of Johnson & Johnson (U.S.), exclusive worldwide marketing rights for its human diagnostic products. Cetus (U.S.), the exclusive licensee of a patent that involves diagnostic applications of DNA probes stemming from work at the University of Washington, is also emphasizing diagnostic applications of probes (91). Other NBFs that have announced their intentions to develop commercial diagnostic products based on DNA probe technology are Amgen (with backing by Abbott Laboratories) and Integrated Genetics (in collaboration with the University of California, San Diego).

The development of DNA hybridization probes represents a challenge to MAb technology for clinical diagnostic applications. MAb kits for diagnosing human venereal diseases are now on the market, but proponents of DNA hybridization probes claim that DNA hybridization offers an even more specific method of diagnosing infections (58). DNA hybridization can be performed with a minimum of tissue handling and may be used on some fixed tissues that are not amenable to MAb use. Ultimately, the relative strengths of DNA hybridization probes and other diagnostic products must be assessed on an individual basis.

Commercial aspects of biotechnology in the pharmaceutical industry

The path leading from the concept for a drug to a marketable product is arduous, costly, and extremely speculative. Discovery and development costs alone in the United States are estimated to range from $54 million to over $70 million per drug (43). Despite the generally low returns on the majority of potential drugs, however, high investments in pharmaceutical R&D continue. With an average of 11.5 percent of annual sales invested in R&D (99), the U.S. pharmaceutical industry ranks only below the information processing and semiconductor industries in terms of R&D as a percentage of annual sales (16).

During the past 40 years, the pharmaceutical industry has given increasing attention to R&D, and extensive government regulation of pharmaceutical products has evolved. Despite the increasing R&D commitments, however, recent trends indicate that the rate of innovative return to pharmaceutical companies throughout the world has declined (89). In short, fewer new drug introductions are emanating from larger research commitments by the public and industry (40).

Reasons most often cited for this decline in the United States center on the burdens imposed by Government legislation, including high costs of obtaining FDA approval, brevity and insufficiency of patent protection for new drugs, sponsorship of competition and product undercutting by State substitution laws and maximum allowable cost programs, and other regulatory factors that act as disincentives for renewed industrial R&D for new drugs. Other popular hypotheses for lower pharmaceutical innovation are decentralization of R&D resources by pharmaceutical companies to other industries such as specialty chemicals, cosmetics, and agricultural products, and increased displacement of R&D in industrial countries by R&D in less developed countries, emphasizing substitution rather than innovation.

Although biotechnology should not be viewed as a panacea for the problem of diminishing innovation in the pharmaceutical industry, it does offer promise in augmenting existing technologies in the pharmaceutical industry. In addition to allowing improvements in pharmaceuticals themselves, the adoption of biotechnology may provide ways for companies to streamline R&D costs for such things as biological screening, pharmacological testing, and clinical evaluation of new products. To a large degree, pharmaceutical development involves the correlation of function and molecular structure, and biotechnology may aid in making such correlations. Prior knowledge about the structure of drug receptor molecules, as gained partially from gene cloning and DNA sequencing research, for example, could supply investigators with information about which structures of new drugs might be effective in reacting with these receptors. This predictive ability may be increased by the use of computers to select appropriate drugs for development, as has been done in the development of synthetic subunit vaccines (67,68).

The costs of applying biotechnology to the development of new pharmaceutical entities cannot be readily determined at this time. In most instances, however, biotechnological methods of production are probably not yet cost-competitive with conventional methods. With biotechnology, as with other new technologies, there are costs associated with learning the technology that will diminish as facilities and skills are acquired. Achieving the limited goal of supplying MAbs successfully to manufacturers of in vitro diagnostic products, it has been estimated, will require a cumulative 3-year investment of $3.5 million to $4 million, and final immunodiagnostic product development may require 5 to 10 times this amount (138). The costs of commercial rDNA work are considerably higher. Although expenditures are rarely disclosed, indications of the cost of production for rDNA produced products can be gleaned from Schering-Plough's (U.S.) $6 million investment in a pilot-scale bioprocessing and purification facility (52), Genentech's drive to raise $32 million to sponsor clinical testing and development of its rDNA produced tPA (32), and Eli Lilly's $60 million investment in facilities to produce hI (129).

The international pharmaceutical market represents a major source of trade between nations, and foreign sales are comprising increasing percentages of total sales in the developed countries. From 1975 to 1981, for example, U.S. companies' control of their domestic market fell to 73 percent from 85 percent, and Japanese companies' share of their domestic market fell to 69 percent from 87 percent (120). Foreign sales account for 43 percent of total sales by U.S. ethical drug manufacturers. West German and Swiss companies are even more oriented toward foreign markets than their U.S. counterparts (40).

Many companies conducting biotechnology R&D are considering markets on a global scale, and for that reason, international market differences are likely to have strong effects on the pattern of biotechnology's introduction to the pharmaceutical industry. These differences are suggested by the fact that the most widely used pharmaceuticals in the U.S. market are neuroactive drugs, while those most widely used in foreign markets are anti-infective compounds. Thus, national preferences lead to differences in the choice of R&D ventures among leading companies, as exemplified by Japanese companies' interest in thrombolytic compounds. The potential of these new agents is more readily appreciated by Japanese drug firms than their U.S. counterparts, and thrombolytic agent R&D efforts by U.S. NBFs are underwritten largely by Japanese companies.

International differences of pharmaceutical use may also make the high costs associated with developing new methods such as biotechnology more acceptable in certain nations. In Japan, where blood products are imported because of cultural barriers to domestic collection, the Government may choose to subsidize the costs for domestic production of HSA by rDNA technology (which is likely to exceed the current price of HSA on the world market) rather than perpetuate the import trade. Such an action might enable firms involved with HSA biotechnology in Japan to move more rapidly along the manufacturing learning curve with generally applicable technology. Ultimately, this could reverse Japan's substantial pharmaceutical trade debt with the United States.

Biotechnology is likely to augment the international stature of the pharmaceutical industry through international corporate arrangements that combine research, production, and licensing in ways that best satisfy market needs in various countries. Because biotechnology offers possibilities of creating novel pharmaceutical compounds in large quantities and at reduced costs (e.g., Ifns, growth hormones, vaccines, and other biological response modifiers) and because many small new companies are involved in pharmaceutical R&D, the demands of "less glamorous" markets for products such as parasitic vaccines may have greater chances of being met than they have in previous years. Thus, biotechnology provides the pharmaceutical industry with a variety of new sources of R&D possibilities.

Priorities for future research

Funding from NIH has been and will continue to be instrumental in developing biotechnology for pharmaceutical use. The new biological techniques have dramatically increased the understanding of many disease mechanisms. Areas of research that would benefit pharmceutical innovation in biotechnology including the following:

- clarification of the functions and mechanisms of action of immune regulators such as Ifn and interleukin-2,

- investigation into the clinical use of neuroactive peptides and thrombolytic and fibrinolytic peptides,
- development of improved drug delivery systems,
- clarification of the mechanisms of acquired immunity leading to better vaccine development procedures,
- development of the ability to culture and an increased understanding of the lifecycle of the world's more debilitating protozoan parasites, and
- acquisition of a better understanding of the physiology and genetics of cancer.

Chapter 5 references

1. Aikawa, M., Yoshida, N., Nussenzweig, R. S., et al., "The Protective Antigen of Malarial Sporozoites (*Plasmodium berghei*) is a Differentation Antigen," *J. Immunol.* 126:2494-2495, 1981.
2. Arnon, R., Sela, M., Parant, M., et al., "Antiviral Response Elicited by a Completely Synthetic Antigen With Built-in Adjuvanicity," *Proc. Natl. Acad. Sci. (U.S.A.)* 77:6769-6772, 1980.
3. Arrieta-Molero, J. F., Kyrieckos, A., Madhur, K. S., et al., "Orally Administered Liposome-Entrapped Insulin in Diabetic Animals," *Hormone Research* 16:249-256, 1982.
4. Baltz, R. H., "Genetics and Biochemistry of Tylosin Production: A Model for Genetic Engineering in Antibiotic-Producing *Streptomyces*," *Genetic Engineering of Microorganisms for Chemicals,* A. Hollaender, R. D. DeMoss, S. Kaplan, et al. (eds.) (New York: Plenum Publishing, 1982).
5. Banks, P., "HDL Deficit, Gene Linked," *Clinical Chemistry News,* April 1983, pp. 1-5.
6. Ben-Porath, E., Isselbacher, K. J., and Wands, J. R., "Improved Detection of Hepatitis B Virus by Monoclonal Antibody," presented at *Second Annual Congress for Hybridoma Research,* Philadelphia, Feb. 7, 1983.
7. Bernstein, L. D., "Monoclonal Antibodies: Prospects for Cancer Treatment," *Immunological Approaches to Cancer Therapeutics* (New York: John Wiley & Sons, 1982).
8. Billiau, A., "Perspectives in Cancer Research: The Clinical Value of Interferons as Antitumor Agents," *Eur. J. Cancer & Clin. Oncology* 17:949-967, 1981.
9. *Bioengineering News,* "Genentech Partnership Closes—A Detailed Look," Jan. 7, 1983, p. 1.
10. *Biotechnology Newswatch,* "Lilly, Cetus Try New Routes To Scale Up rDNA Insulin," July 6, 1981, p. 1.
11. *Biotechnology Newswatch,* "As Lilly's Synthetic Insulin Gets GDA OK, Novo, Biogen Join To Clone Their Own," Nov. 15, 1982, p. 2.
12. *Biotechnology Newswatch,* "Genentech, Speywood Close Ranks in Move To Clone Factor VIIIc," Nov. 15, 1982, p. 3.
13. *Biotechnology Newswatch,* "Japanese Clone, Express Key T-Cell Growth Factor," Feb. 21, 1983, p. 1.
14. *Biotechnology Newswatch,* "Biogen's Hepatitis B Vaccine Headed for Clinical Trials in 1984," Aug. 1, 1983, p. 1.
15. Bloom, A. L., "Benefits of Cloning Genes for Clotting Factors," *Nature* 303:474-475, 1983.
16. *Business Week,* "R&D Scoreboard," July 5, 1982, pp. 55-74.
17. *Chemical Engineering News,* "Monoclonal Antibodies Test for Meningitis," Aug. 8, 1983, p. 4.
18. Choo, K. H., Gould, K. G., Rees, D. J. G., and Brownlee, G. G., "Molecular Cloning of the Gene for Human Anti-Haemophilic Factor IX," *Nature* 299:178-180, 1982.
19. Cochrane, A. H., Santoro, F., Nussenzweig, V., et al., "Monoclonal Antibodies Identify the Protective Antigens of Sporozoites of *Plasmodium Knowlesi*," *Prod. Natl. Acad Sci. (U.S.A.)* 79:5651-5655, 1982.
20. Cosimi, A. B., Colvin, R. G., Burton, R. C., et al., "MCA's for Kidney Transplants," *N. Eng. J. Med.* 305:308-14, 1981.
21. deLeede, L. G. J., deBoer, A. G., and Breimes, D. D., "Rectal Infusion of a Model Drug Antipyrine With an Osmotic Delivery System," *Biopharmaceutics & Drug Disposition* 2:131-136, 1981.
22. Demain, A. L., "Industrial Microbiology," *Science* 214:987-995, 1981.
23. Dreesman, G. R., Sanchez, Y., Ionescu-Matiu, I., et al., "Antibody to Hepatitis B Surface Antigen

After a Single Innoculation of Uncoupled Synthetic LHLBs^{Ag} Peptides," *Nature* 295:158-160, 1982.

24. Dybel, M. W., "Lymphokines: Activators of Lymphocyte Growth and Differentation," *Bio/Technology,* July 1983, pp. 412-413.
25. Edman, J. C., Hallewell, R. A., Valenzuela, P., et al., "Synthesis of Hepatitis B Surface and Core Antigens in *E. coli,*" *Nature* 291:503-506, 1981.
26. *Eighth American Peptide Symposium,* "Peptides: Structure and Function" (Rockford, Ill.: Pierce Chemical Co., 1983).
27. Eli Lilly Co., *Annual Report,* Indianapolis, Ind., 1981.
28. Ellis, J., Ozaki, L. S., Gwadz, R. W., et al., "Cloning and Expression in *E. coli* of the Malarial Sporozoite Surface Antigen Gene From *Plasmodium knowlesi,*" *Nature* 302:536-539, 1983.
29. Flores, J., Purcell, R. H., Perez, I., et al., "A Dot Hybridization Assay for Detection of Rotavirus," *Lancet,* Mar. 12, 1983, pp. 555-559.
30. Fryklund, L., Skottner, A., and Hall, K., "Chemistry and Biology of the Somatomedins" *Growth Factors,* K. W. Kastrup, and J. H. Nielsen (eds.) (New York: Pergamon Press, 1978).
31. Genentech, "The Genentech Story," South San Francisco, Calif., Apr. 28, 1980.
32. Genentech, *First Quarter Report,* South San Francisco, Calif., May 9, 1983.
33. Genex Corp., *Prospectus,* Rockville, Md., Sept. 29, 1982.
34. Gerety, R. J., and Tabor, E., "Newly Licensed Hepatitis B. Vaccine," *J. Am. Med. Assoc.* 249:745-6, 1983.
35. Giannelli, F., Choo, K. H., Rees, D. J. G., et al., "Gene Deletions in Patients With Hemophilia B and Anti-Factor IX Antibodies," *Nature* 303:181-182, 1983.
36. Gilliland, D., Gary, Z., Steplweski, R. J., et al., "Antibody-Directed Cytotoxic Agents: Use of Monoclonal Antibody To Direct the Action of Toxin A Chains to Colorectal Carcinoma Cells," *Proc. Natl. Acad. Sci. (U.S.A.)* 77-4539-4543, 1980.
37. Gillis, S., "Interleukin 2: Biology and Biochemistry," *J. Clin. Immunol.* 3:1-13, 1983
38. Goldstein, L., "Infectious Diseases," presented at *Robert First Conference on Biotechnology in Health Care,* Rye, N.Y., Oct. 11, 1982.
39. Gonzalez, E. R., "Teams Vie in Synthetic Production of Human Growth Hormone," *J. Am. Med. Assoc.* 242:701-2, 1979.
40. Grabowski, H. G., and Vernon, J. M., "The Pharmaceutical Industry," *Government and Technical Progress, A Cross-Industry Analysis,* R. R. Nelson (ed.) (New York: Pergamon Press, 1983).
41. Guillemin, R., "Peptides in the Brain: The New Endocrinology of the Neuron," *Science* 202:390-402, 1978.
42. Hansen, R. W., "The Pharmaceutical Development Process: Estimates of Current Development Costs and Times and Effects of Regulatory Changes," *Issues of Pharmaceutical Economics,* R. I. Chien (ed.) (Cambridge, Mass.: Lexington Books, 1979).
43. Hayes, A. H., Jr., Commissioner, Food and Drug Administration, statement in *Patent Term Extension and Pharmaceutical Innovation,* hearings before the Subcommittee on Investigations and Oversight, House Committee on Science and Technology, U.S. Congress, Feb. 4, 1982 (Washington, D.C.: U.S. Government Printing Office, 1982).
44. Hilleman, M. R., Senior Vice President, Virus and Cell Biology, Merck Research Laboratories, Rahway, N.J., personal communication to W. P. O'Neill, Feb. 10, 1983.
45. Hitzeman, R. A., Leung, D. W., Jeanne-Perry, L., et al., "Secretion of Human Interferons by Yeast," *Science* 219:620-25, 1983.
46. Hopp, T. P., and Woods, K. R., "Prediction of Antigenic Determinants From Amino Acid Sequences," *Proc. Natl. Acad. Sci. (U.S.A.)* 78:3824-8, 1981.
47. Howard, M., and Paul, N. E., "Regulation of B-Cell Growth and Differentiation by Soluble Factors," *Ann Rev. Immunol.* 1:307-33, 1983.
48. *International Radioimaging Symposium,* "Radioimmunoimaging and Radioimmunotherapy" (New York: Elsevier, 1983).
49. Ismach, J. M., "Type 1 Diabetes: Ripe for an Early Cure?" *Medical World News* 23:47-63, 1982.
50. Jacobs, L., O'Malley, J., Freeman, A., et al., "Intrathecal Interferon in Multiple Sclerosis," *Arch. Neurology* 39:609-615, 1982.
51. Jordan, W. S., U.S. National Institute of Allergy and Infectious Diseases, "Accelerated Development of New Vaccines," memorandum, November 1982.
52. Karanikas, M., "Schering-Plough Corp. Prepares To Scale Up Production of Leukocyte Interferon," *Genetic Engineering News* 2(4):3, July-August, 1982.
53. Kastin, A. M., "Potential Role of Peptides in Disorders of the Central Nervous System," presented at *Robert First Conference on Biotechnology in Health Care,* Rye, N.Y., Oct. 11, 1982.
54. Kemp, D. J., Coppel, R. L., Cowman, A. F., et al., "Expression of *Plasmodium Falciparum* Blood-Stage Antigens in *Escherichia coli:* Detection With Antibodies From Immune Humans," *Proc. Natl. Acad. Sci. (U.S.A.)* 80:3787-3791, 1983.

55. Khaw, B. A., Fallon, J. T., Beller, G. A., et al., "Specificity of Localization of Myosin-Specific Antibody Fragments in Experimental Myocardial Infarction," *Circulation* 60:1527-1531, 1979.
56. Khaw, B. A., Fallon, J. T., Haber, E., "Myocardial Infarct—Imaging With Indium-III-Diethylene Triamine Pentacetic Acid—Anticanine Cardiac Myosin Antibodies," *Science* 209:295-297, 1980.
57. Klausner, A., "Activating the Body's Blood Clot Dissolvers: Biotech's New Role," *Bio/Technology,* June 1983, pp. 331-336.
58. Klausner, A., and Wilson, T., "Gene Detection Technology Opens Doors for Many Industries," *Bio/Technology,* August 1983, pp. 471-478.
59. Korninger, C., Matsuo, O., Suy, R., et al., "Thrombolysis With Extrinsic (Tissue-type) Plasminogen Activator in Dogs With Femoral Vein Thrombosis," *J. Clin. Invest.* 69:573-580, 1982.
60. Krolick, K. A., Villemex C., Isakson, P., et al., "Selective Killing of Normal or Neoplastic B Cells by Antibodies Coupled to the A Chain of Ricin," *Proc. Natl. Acad. Sci. (U.S.A.)* 77:5419, 1980.
61. Kurachi, K., and Davie, E. W., "Isolation and Characterization of a cDNA Coding for Human Factor IX," *Proc. Natl. Acad. Sci. (U.S.A.)* 791:6 461-6464, 1982.
62. Langer, P. R., Waldrop, A. A., and Ward, D. C., "Enzymatic Synthesis of Biotin-Labeled Polynucleotides: Novel Nucleic Acid Affinity Probes," *Proc. Natl. Acad. Sci. (U.S.A.)* 78:6633-7, 1981.
63. Law, R. M., Adelman, J., Block, S. C., et al., "The Sequence of Human Serum Albumin cDNA and Its Expression in *E. coli*," *Nucleic Acids Research* 9:6103-6127, 1981.
64. LeConey, M. M., "Who Needs Plasma?" *Plasma Quarterly* 2:68-93, September 1980.
65. Lee, J., and Laycock, J., *Essential Endocrinology* (New York: Oxford University Press, 1978).
66. Lerner, R. A., Green, N., Alexander, H., et al., "Chemical Synthesis of Peptides Predicted From the Nucleotide Sequence of the Hepatitis B Virus Genome Elicit Antibodies Reactive With Native Envelope Protein of Dane Particles," *Proc. Natl. Acad. Sci. (U.S.A.)* 78:3403-3407, 1981.
67. Lerner, R. A., "Tapping the Immunological Repertoire To Produce Antibodies of Predetermined Specificity," *Nature* 299:592-596, 1982.
68. Lerner, R. A., "Synthetic Vaccines," *Scientific American* 248:66-74, 1983.
69. Levin, S., and Hahn, T., "Evaluation of the Human Interferon System in Viral Disease," *Clin. Exp. Immun.* 46:475-483, 1981.
70. Lewin, R., "Genetic Probes Become Ever Sharper," *Science* 221:1167, 1983.
71. Linnebank, G., "Biogen's Inside Track," *Nature* 304:297, 1983.
72. Lyons, R. D., "Drugs: New Method Proves Effective," *New York Times,* July 19, 1983, p. C1.
73. Mabry, D. S., Proctor, A. R., Wernau, W. C., et al., Fermentation/Recovery R&D, Pfizer, Inc., Groton, Conn., personal communication to W. P. O'Neill, Oct. 13, 1982.
74. MacKay, P., Pasek, M., Magazin, M., et al., "Production of Immunologically Active Surface Antigens of Hepatitis B Virus by *E. coli*," *Proc. Natl. Acad. Sci. (U.S.A.)* 78:4510-14, 1981.
75. Marshall, E., "NYU's Malaria Vaccine: Orphan at Birth?" *Science* 219:466-67, 1983.
76. Martinis, J., et al., "Monoclonal Antibodies With Dual Antigen Specificity," presented at the *30th Annual Colloquium of the Protides of the Biological Fluids,* Brussels, Belgium, May 1982.
77. McMichael, A. J., and Fabre, J. W., *Monoclonal Antibodies in Clinical Medicine* (New York: Academic Press, 1982).
78. *Medical World News,* "Shots Stop Tots' Hepatitis B," May 11, 1981, p. 83.
79. *Medical World News,* "Cloned Human Hormones May Cause Antibodies But Don't Seem Allergenic," Aug. 16, 1982, p. 65.
80. Melchers, F., Potter, M., and Warner, N. L. (eds.) "Lymphocyte Hybridomas," *Current Topics in Microbiology and Immunology,* vol. 81 (New York: Springer-Verlag, 1978).
81. Merigan, T. C., "Present Appraisal of and Future Hopes for the Clinical Utilization of Human Interferons," *Interferon,* vol. 3 (New York: Academic Press, 1982).
82. Miller, H. I., "Designer Genes for Producing Drugs: Will They Wash?" *DNA* 1:101-2, 1982.
83. Miller, J. A., "Antibodies for Sale," *Science News* 123:296-302, 1983.
84. Miller, R. A., Maloney, D. G., Warnke, R., et al., "Treatment of B-cell Lymphoma With Monoclonal Anti-Idiotype Antibody," *N. Eng. J. Med.* 306: 517-522, 1982.
85. Modell, W. M., "Clinical Pharmacology: Reflections in My Rearview Mirror," *Clin. Pharm. & Ther.,* May 1978, p. 497-504.
86. Muller, H. P., Van Tilburg, N. H., Derks, J., et al., "A Monoclonal Antibody to VIIIc Produced by a Mouse Hybridoma," *Blood* 58:1000, 1982.
87. Nardin, E. H., "Circumsporozoite Proteins of Human Malaria Parasites *Plasmodium Falciparum* and *Plasmodium Vivax*," *J. Exp. Med.* 156:20-30, 1982.
88. National Academy of Sciences, *Priorities in Biotechnology Research for International Devel-*

opment: Proceedings of a Workshop, Washington, D.C., July 26-30, 1982.

89. National Academy of Sciences, *The Competitive Status of the U.S. Pharmaceutical Industry* (Washington, D.C.: National Academy Press, 1983).
90. Newmark, P., "What Chance of a Malaria Vaccine?" *Nature* 302:473, 1983.
91. Noel, K., Director, Diagnostic Business, Cetus Corp., Berkeley, Calif., personal communication, 1983.
92. North, J. R., Epstein, M. A., and Morgan, A. S., "Glycoprotein Induces Potent Virus-Neutralizing Antibodies When Incorporated in Liposomes," *Proc. Natl. Acad. Sci. (U.S.A.)* 79:7504-7508, 1982.
93. Nowinski, R. C., Tam, M. R., Golstein, L. C., et al., "Monoclonal Antibodies for Diagnosis of Infectious Diseases in Humans," *Science* 219:637-643, 1983.
94. Oka, T., Muneyuki, R., and K. Morihara, "Enzymatic Semisynthesis of Human Insulin: A Proposed Procedure Using Immobilized Enzymes," presented at the *Eighth American Peptide Symposium,* Tucson, Ariz., May 23, 1983.
95. Oldham, R. K., U.S. National Cancer Institute, "Update on Clinical Trials With Interferon and Monoclonal Antibodies," memorandum, May 4, 1983.
96. O'Neill, W. P., "Implications of Molecular Genetics for Medicine," Part A in *Impacts of Applied Genetics: Micro-Organisms, Plants, and Animals, Vol. II—Appendixes* (Washington, D.C.: U.S. Congress, Office of Technology Assessment, July 1981).
97. Orkin, S. H., Little, P. F. R., Kazazian, H. H., and Boehm, C. D., "Improved Detection of Sickle Mutation by DNA Analysis," *N. Eng. J. Med.* 307:32-36, 1982.
98. Panem, S., *The Interferon Crusade: Public Policy and Biomedical Dreams* (Washington, D.C.: Brookings Institution, in press).
99. Pharmaceutical Manufacturers Association, *Annual Survey Report, U.S. Pharmaceutical Industry, 1979-1980,* Washington, D.C., 1980.
100. Phillips, L. S., and Vassilopoulou-Sellin, R., "Somatomedins," *N. Eng. J. Med.* 302:371-380 and 438-446, 1980.
101. Pirastu, M., Kan, Y., Cao, A., et al., "Prenatal Diagnosis of Beta-Thalassemia," *N. Eng. J. Med.* 309: 284-287, 1983.
102. Pontrioli, A. E., Alberetto, M., Secchi, A., et al., "Insulin Given Intranasally Induces Hypoglycemia in Normal and Diabetic Subjects," *Br. Med. J.* 284:303-306, 1982.
103. Post, L. E., and Roizman, B., "A Generalized Technique for Detection of Specific Genes," *Cell* 25:227-232, 1981.
104. Raiti, S., Director, National Pituitary Agency, NIH, personal communication to W. P. O'Neill, 1980.
105. Randolph, H. G., "Plasma, Its Derivatives and Market," *Plasma Quarterly* 1:74-93, 1979.
106. Reasor, J., "International Demand for Therapeutic Plasma Fractions," *Plasma Quarterly* 3:8, 1980.
107. Reilly, R. W., "Speak Out! What's Going On in the Plasma Industry?" *Plasma Quarterly* 3:36-62, 1981.
108. Rijken, D. C., and Collen, D., "Purification and Characterization of the Plasminogen Activator Secreted by Human Melanoma Cells in Culture," *J. Biol. Chem.* 256:7035-7041, 1981.
109. Ritz, J., Sallan, S. E., Bast, R. C., et al., "Autologous Bone-Marrow Transplantation in CALLA-Positive Acute Lymphoblastic Leukaemia After *In Vitro* Treatment With J5 Monoclonal Antibody," *Lancet* 2:60-63, 1982.
110. Sarma, V., "Conflicting Interests at Work," *Nature* 304:7, 1983.
111. Saudek, C. D., "Technology in the Treatment of Metabolic Diseases," presented at *Robert First Conference on Biotechnology in Health Care,* Rye, N.Y., Oct. 11, 1982.
112. Schaumann, L., "Pharmaceutical Industry Dynamics and Outlook to 1985," Stanford Research Institute, Menlo Park, Calif., 1976.
113. *Scrip 555,* "KabiVitrum Receives Swedish Approval for Recombinant DNA Production," Jan. 12, 1981, p. 6.
114. *Scrip 662,* "Nine New Antibiotics for Japan," Jan. 27, 1982, p. 12.
115. *Scrip 664,* "Shionogi and Biogen Cooperate," Feb. 3, 1982, p. 5.
116. *Scrip 669,* "Squibb & Novo Form U.S. Company," Feb. 22, 1982, p. 16.
117. *Scrip 676,* "Healthy U.S. Drug Trade Balance," Mar. 17, 1982, p. 7.
118. *Scrip 684,* "Leading Japanese Companies and Products," Apr. 14, 1982, p. 11.
119. *Scrip 689,* "Gamma-Interferon Gene Synthesized," May 3, 1982, p. 5.
120. *Scrip 690,* "Japanese Drug Production Up 5.7% in '81," May 5, 1982, p. 15.
121. *Scrip 698,* "Genentech-Mitsubishi Chem. Joint Venture," June 2, 1982, p. 11.
122. *Scrip 704,* "Toray Upgrading Beta-Interferon Production," June 23, 1982, p. 8.
123. *Scrip 704,* "Lilly's New Enkephalin Analgesic," June 23, 1982, p. 16.
124. *Scrip 713,* "Cutter/Genetic Systems Joint Venture," July 26, 1982, p. 7.
125. *Scrip 715,* "Leading Japanese Antibiotics Surveyed," Aug. 2, 1982, p. 13.

126. *Scrip 739,* "AMGen Makes Significant Progress," Oct. 25, 1982, p. 6.
127. *Scrip 740,* "Japanese Anticancer R&D," Oct. 27, 1982, p. 10.
128. *Scrip 748,* "Biogen and Teijin in Factor VIII Deal—Other Developers," Nov. 24, 1982, p. 7.
129. Smith, D. L., *Eli Lilly and Company: A Basic Study* (New York: Smith Barney Harris Upham & Co., Inc., September 1982).
130. Smith, C. I., Kitchen, L. W., Scullard, G. H., et al., "Vidarabine Monophosphate and Human Leukocyte Interferon in Chronic Hepatitis B Infection," *J. Am. Med. Assn.* 247:2261-2265, 1982.
131. Smith, G. L., Mackett, M., and Moss, B., "Infectious Vaccinia Virus Recombinants That Express Hepatitis B Surface Antigen," *Nature* 302:490-495, 1983.
132. Sola, B., Avner, P., Sultan, Y., et al., "Monoclonal Antibodies Against Human Factor VIII Molecule Neutralize Antihemophilic Factor and Ristocetin Cofactor Activities," *Proc. Natl. Acad. Sci. (U.S.A.)* 79:183-187 1982.
133. Staehelin, T., "The Use of Monoclonal Antibodies in Interferon Production," presented at *Robert First Conference on Monoclonal Antibodies,* Rye, N.Y., September 1981.
134. Stel, H. V., van der Kwast, Th. H., and Veerman, E. C. I., "Detection of Factor VIII/Coagulant Antigen in Human Liver Tissue," *Nature* 303:530-532, 1983.
135. Sutcliffe, J. G., Shinnick, T. M., Green, N., et al., "Antibodies That React With Predetermined Sites on Proteins," *Science* 219:660-6, 1983.
136. Swift, R., Director of Clinical Affairs, Genentech Corp., South San Francisco, Calif., personal communication to W. P. O'Neill, Apr. 5, 1983.
137. Taniguchi, T., "Use of Recombinant DNA Technology for the Study of Lymphokines," *Taisha* 19:1695-704, 1982.
138. Treble, M. J., "Scale-Up of Hybridoma Business Ventures: Investment Requirements and Perspectives," *Genetic Engineering News,* July/August 1982, p. 5.
139. Tuite, M. F., Dobson, M. J., Roberts, N. A., et al., "Regulated High Efficiency Expression of Human Interferon-Alpha in *Saccharomyces cerevisiae,*" *EMBO J.* 1:603-8, 1982.
140. U.S. Congress, Office of Technology Assessment, *Impacts of Applied Genetics: Micro-Organisms, Plants, and Animals,* OTA-HR-132, Washington, D.C., April 1981.
141. U.S. Department of Commerce, *1982 Industrial Outlook,* Washington, D.C., 1982.
142. U.S. Department of Health, Education, and Welfare, "Appendix to a Study of Insulin Supply and Demand," DHEW Publication No. (NIH) 78-1589, Washington, D.C., 1978.
143. U.S. Department of Health, Education, and Welfare, "Study To Evaluate the Supply Demand Relationships for AHF and PTC Through 1980," DHEW Publication No. (NIH) 77-1274, Washington, D.C., 1980.
144. U.S. Department of Health and Human Services, Food and Drug Administration, National Center for Devices and Radiological Health, 1983.
145. U.S. Department of Health and Human Services, National Institutes of Health, "Targeted Opportunities for Biomedical Research in Developing Countries," memorandum for the Collaborative Workshop on Expanded Biomedical Research Opportunities in Developing Countries, Washington, D.C., Dec. 13-17, 1982.
146. Valenzuela, P., Medina, A., Rutter, W. J., et al., "Synthesis and Assembly of Hepatitis B Surface Antigen Particles in Yeast," *Nature* 298:347-350, 1982.
147. Vitetta, E. S., Krolick, K. A., Miyama-Inaba, M., et al., "Immunotoxins: A New Approach to Cancer Therapy," *Science* 219:644-649, 1983.
148. Vogel, C. W., and Muller-Eberhard, H. J., "Induction of Immune Cytolysis: Tumor-Cell Killing by Complement Is Initiated by Covalent Complex of Monoclonal Antibody and Stable C3/C5 Convertase," *Proc. Natl. Acad. Sci. (U.S.A.)* 78:7707-7711, 1981.
149. Vournakis, J. N., and Elander, R. P., "Genetic Manipulation of Antibiotic Producing Microorganisms," *Science* 219:703-709, 1983.
150. Vyas, G. N., Cohen, S. N., and Schmid, R., *Viral Hepatitis: A Contemporary Assessment of Etiology, Epidemiology, Pathogenesis and Prevention* (Philadelphia: Franklin Institute Press, 1978).
151. Wands, J. R., Carlson, R. I., Schoemaker, H., et al., "Immunodiagnosis of Hepatitis B With High Affinity IgM Monoclonal Antibodies," *Proc. Natl. Acad. Sci. (U.S.A.)* 78:1214-1218, 1981.
152. Wardell, W. M., et al., "The Rate of Development of New Drugs in the United States," *Clinical Pharmacology and Therapeutics,* May 1978, cited in H. G. Grabowski and J. M. Vernon, "The Pharmaceutical Industry," *Government and Technical Progress, A Cross-Industry Analysis,* R. R. Nelson (ed.) (New York: Pergamon Press, 1983).
153. Wernsdorfer, W. H., "Prospects for the Development of Malaria Vaccines," *Bulletin of the World Health Organization* 59:335-341, 1981.
154. Wetzel, R., Heyneker, H. L., Goeddel, D. V., et al., "Production of Biologically Active N Alpha-Desace-

tylthy-Myosin Alpha 1 in *Escherichia coli* Through Expression of a Chemically Synthesized Gene," *Biochemistry* 19:6096-104, 1980.

155. World Bank, *Poverty and Human Development* (New York: Oxford University Press, 1980).
156. World Health Organization, *Report for the Special Programme for Research and Training in Tropical Diseases,* Geneva, 1976.
157. World Health Organization, "Interferon Therapy," WHO Technical Report 676, Geneva, 1982.
158. Wyler, D. J., "Malaria—Resurgence, Resistance, and Research," *N. Eng. J. Med.* 308:934-940, 1983.
159. Yamasaki, Y., Shiehiri, M., Kawamori, E., et al., "The Effect of Rectal Administration of Insulin on the Short-Term Treatment of Alloxan-Diabetic Dogs," *Can. Physiol. Pharmacol.* 59:1-6, 1981.
160. Yoshida, N., Nussenzweig, R. S., Potocnjak, P., et al., "Hybridoma Produces Protective Antibodies Directed Against the Sporozoite Stage of Malaria Parasite," *Science* 207:71-73, 1980.
161. Youle, R. J., and Neville, D. M., "Anti-Phy 1.2 Monoclonal Antibody Linked to Ricin is a Potent Cell-Type-Specific Toxin," *Proc. Natl. Acad. Sci. (U.S.A.)* 77:5483, 1980.

Chapter 6
Agriculture

Contents

Tables

Figure

Chapter 6
Agriculture

Introduction

As the world population grows, agriculture will need to provide more and more food. Biotechnology may yield methods and products that improve agriculture in many ways. In animal agriculture, biotechnology offers promise in the following areas:

- *diagnosis, prevention, and control of animal diseases* with the use of monoclonal antibody (MAb) technology to diagnose, monitor, and better understand disease and the use of recombinant DNA (rDNA) to expand the pharmacopoeia of vaccines and other animal health products;
- *animal nutrition and growth promotion* through the use of growth hormones and feed additives to improve animal feed usage and animal health; and
- *genetic improvement of animal breeds* by using MAb and rDNA technology to better understand the bases of animal productivity and disease resistance and by the direct transfer of "beneficial" genes from one animal breed to another.

Though the potential for using biotechnology to improve animal agriculture is exciting, the commercial feasibility and actual impacts of using biotechnology in this area at present remain largely speculative. In some cases, existing animal health products may be replaced by improved, biotechnologically made materials. In other cases, entirely new products may become available to solve formerly intractable problems. In almost all instances, efficacy, safety, and practicality must be demonstrated for each new product. Only a few products for practical use in animal agriculture have been produced to date, so the success of biotechnologically produced compounds compared with conventionally made products remains to be demonstrated. For many animal agriculture products made with new biotechnology, the speed and scale of adoption by producers will be determined by the ease with which the products can be integrated into existing production systems (20).

In addition to the potential applications of biotechnology in animal agriculture, there are several potential applications in the area of crop improvement. The potential applications generally fall into two categories:

- *improvement of specific plant characteristics,* for example, through the introduction or manipulation of genes that confer resistance to disease and environmental factors, that increase the amount and quality of primary and secondary products from plants, that enhance plant growth rate, or that increase photosynthetic efficiency; and
- *genetic manipulation of micro-organisms,* for example, to enhance the process of nitrogen fixation, to produce insecticides, or to suppress disease or promote growth in plants.

The genetic manipulation and modification of plants presents some special challenges, but research is proceeding rapidly. There is a great deal of research interest at present in the use of biotechnology to improve plant resistances to disease and environmental factors. If plants are made more resistant to disease and certain environmental factors, greater crop yields or a reduction in the cost of crop production may result. Furthermore, unlike most plant traits, some of these specific crop improvements may be accomplished with only one or a few gene modifications. It is likely that there will be considerable research progress in this area in the next 5 to 10 years.

The applications and commercial aspects of biotechnology in animal and plant agriculture, respectively, are discussed in more detail in the next two sections of this chapter. A separate section at the end of the chapter indicates priorities for future research in each of these areas.

Animal agriculture

The commercial use of biotechnology in animal agriculture is affected by several often-contradictory forces. Favorable forces include the extensive use of animals as test models in basic research and, as is discussed in *Chapter 15: Health, Safety, and Environmental Regulation,* less stringent regulatory approval processes for animal health products than for pharmaceutical products intended for human use. Because animals are used during the development of pharmaceutical and biologic products for humans, veterinary medicine stands to benefit from biotechnology research and development (R&D) such as that described in *Chapter 5: Pharmaceuticals.* Biotechnologically made products for use in animal agriculture, such as MAb diagnostic products, growth hormones (GHs), and vaccines, are becoming available on a limited basis.

Among the factors that inhibit commercial applications of biotechnology in animal agriculture is the fact that the low value-added nature of individual farm animals limits veterinary costs per animal, veterinary medicine sales, and funding for veterinary R&D. In addition, some biotechnologically made products do not suit current animal husbandry practices. Commercialization of at least one rDNA-made vaccine, the vaccine for foot-and-mouth disease (FMD), for example, awaits successful applied research results to achieve protection against several strains of the disease, fewer dosage requirements, lower costs, and other improvements that make the vaccine amenable to animal husbandry practices in the developing nations where FMD exists (20).

Biotechnological developments in the areas of animal disease control, animal nutrition and health, and genetic improvement of animal breeds are discussed further below. Distinctions between the use of biotechnology to expand fundamental knowledge and to develop specific products for commercial use are noted.

Diagnosis, prevention, and control of animal diseases

Losses due to animal diseases exceed hundreds of millions of dollars yearly in the United States,* giving strong impetus to efforts to improve animal health. A combination of the techniques of biotechnology is being used to better understand viral, bacterial, protozoan, and parasitic infections that affect animal productivity throughout the world. MAbs, for example, are being used as research tools to gain a better understanding of the molecular biology of animal diseases. MAbs may also be used for diagnosis of diseases, for monitoring the efficacy of drugs, and for providing short-term passive immunity against animal diseases. In addition, recombinant DNA technology and polypeptide synthesis may be used to develop vaccines for long-term immunization against certain animal diseases.

MONOCLONAL ANTIBODY DIAGNOSTIC PRODUCTS

The diagnosis of animal diseases can be accomplished by the identification in the laboratory of specific antigens displayed by the infectious agent. As discussed in *Chapter 3: The Technologies,* MAbs that recognize specific antigens can be prepared readily. MAbs for several animal diseases are now being made, and in vitro MAb diagnostic products for a number of animal diseases may be used in the near future. MAb-based diagnostic tests are currently being developed for bluetongue, equine infectious anemia, and bovine leukosis virus. Furthermore, diagnostic MAbs are be-

*"Animal losses" are described by a number of parameters, including dollar value of animals lost, losses in productivity due to morbidity, and value of potential progeny lost due to sickness or death of breeders. In this report, the dollar value of animals actually lost to disease (as a primary cause of death) is used for the sake of comparison in describing animal diseases. These estimates are based on data collected for U.S. Department of Agriculture's Animal and Plant Health Inspection Service, Veterinary Services, Hyattsville, Md., and by Doane Agricultural Services, Inc., St. Louis, Mo.

ing sought for canine parvovirus, canine rotavirus (a potentially fatal viral diarrhea in puppies), feline leukemia virus, and canine heartworm disease. For MAb diagnostic products to be effective diagnostic tools and hence commercially viable, they must recognize the large variety of disease strains likely to be encountered (20).

The acceptance of MAbs for field use by veterinarians and animal owners remains to be demonstrated. Whether MAb products will have a large role in the diagnosis of specific animal diseases is unclear. Since livestock producers and poultry growers attempt to spend as little money as possible per animal raised, the markets for individual MAb diagnostic tests initially may be limited. Applications of MAb diagnostic as well as therapeutic products initially may be restricted to high-profit animals, animal products for export, and companion animals such as dogs, cats, and horses. Although individual diagnostic kits are not costly, the farmer's narrow margin of return on other animals may prevent the routine use of diagnostic products.

In the future, diagnostic MAbs could substantially assist large-scale disease control programs in both developed and less developed countries (16). Such reagents might be used to detect disease in order to select an appropriate vaccine and monitor the level of disease during the course of a control program.

Apart from potentially being used as diagnostic reagents by animal producers, MAbs can be used as purification tools to isolate compounds (antigens) that may prove to be effective animal vaccines. They can also be used to provide "passive immunity" to certain animal diseases. The applications of biotechnology to the development of animal vaccines is described further below.

ANIMAL VACCINES

Prevention of a number of animal diseases is being sought with rDNA subunit vaccines in efforts similar to human vaccine programs described in *Chapter 5: Pharmaceuticals.* Subunit vaccines may solve some of the problems associated with conventional vaccines. One problem, for example, is that "attenuated" and killed whole vaccines contain the genetic material of the pathogen and therefore have the potential to cause the infection they are supposed to prevent. Subunit vaccines do not contain the pathogen's genetic material and therefore cannot cause infection. Subunit vaccines may also be more stable, more easily stored, and of greater purity than conventional vaccines, but these qualities remain to be demonstrated. Despite their potential advantages, subunit vaccines raise new technical problems, as mentioned above, and these must be overcome if the vaccines are to prove useful in the field (20).

Viral Animal Diseases.—The development of improved vaccines may allow the prevention of several problematic animal diseases caused by viruses (34). Most subunit vaccine research for animals to date has been focused on viral diseases, particularly FMD and rabies, but some of the findings can be generalized to other viral diseases. Table 27 shows some viral diseases in animals against which subunit vaccines may prove effective and economic.

The development of subunit vaccines for FMD is currently receiving much attention from researchers (2). Although the disease is nonexistent in the United States, FMD affects livestock productivity and exportability throughout South America, Africa, and the Far East. The world market for FMD vaccine is larger than that of any other vaccine, either animal or human. In 1981, 800 million doses of inactivated FMD virus vaccine worth $250 million were used (36). Vaccines for all types of FMD commonly encountered exist at present, but these vaccines vary in effectiveness against different FMD field strains. Evolution of new field strains is a continuing problem, because a vaccine may lose its effectiveness against such strains. The impetus for developing a subunit vaccine for FMD is the hope that such a vaccine will offer enhanced protection with greater safety than conventional vaccines. The degree of protection offered, however, will only become clear over the next few years as research and field evaluations progress (9).

Three research groups have cloned the gene that codes for the major FMD viral surface protein (5,14,15). The new biotechnology firm (NBF)*

*NBFs, as defined in *Chapter 4: Firms Commercializing Biotechnology,* are firms that have been started up specifically to capitalize on new biotechnology.

Table 27.—Viral Animal Diseases and Potential Vaccine Production

Disease	Potential for new biotechnology	Company	Current vaccine status	Potential for new vaccine
Viral diseases:				
Foot-and-mouth disease	+	Genentech (U.S.)/USDA (U.S.) Pirbright (U.K.) Biotech Gen (Israel) MGI (U.S.)[a]	Medium	Relacement
Rabies	+	Wistar Transgene (France) Genentech (U.S.) Inst. Pasteur (France)	Variable	Replacement
Parvovirus:				
Swine	+	MGI	Poor	Replacement
Canine	+	TechAmerica (U.S.)	Medium	Replacement
Bovine leukosis virus	+	MGI	N.A.[b]	Replacement
Bovine papilloma virus	+	MGI	N.A.	Export animals
Rift Valley fever	+	MGI/U.S. Department of Defense	Good	Replacement
Marek's disease (fowl)	+	BRL (U.S.)[c]	Medium	Replacement
Infectious bovine rhinotracheitis	+	MGI	Medium	Replacement
Pseudorabies	+	MGI	Medium	Replacement
African swine fever	+	Spanish Government	None	New product
Rota viruses	−	Vido Institute University of Saskatchewan	None	New product
Bluetongue	+	Bio-Tech Gen. (Israel) USDA	Poor to medium	Export animals
Hog cholera	−	N.A.	Good	Replacement
Newcastle disease	+	USDA	Poor in some areas	Replacement
Bacterial diseases:				
Tuberculosis	N.A.	N.A.	None	New product
Neonatal diarrhea	+	Cetus (U.S.)/Norden (U.S.) Intervet (Netherlands/ Akza (U.S.) MGI	Poor	Replacement
Bacterial respiratory disease	N.A.	N.A.	Poor	Replacement
Anaplasmosis	N.A.	N.A.	None	New Product
Parasitic diseases:				
Babesiosis	+	IMC (U.S.)[d]	None	Replacement
Trypanosomiasis	+	American Cyanamid (U.S.) Genex (U.S.) Hoffmann-La Roche (Switz.)	None	New product
Coccidiosis	+	Eli Lilly (U.S.)	Good	Replacement
Helminithic diseases	+	Merck (U.S.)	Fair	Replacement

[a] MGI = Molecular Genetics, Inc.
[b] N.A. = Information not available.
[c] BRL = Bethesda Research Laboratories.
[d] IMC = International Minerals & Chemicals Corp.

SOURCE: Board of Science and Technology for International Development, et al., "Priorities in Biotechnology Research for International Development—Proceedings of a Workshop" (Washington, D.C.: National Academy Press); and the Office of Technology Assessment.

Genentech Corp. (U.S.), in collaboration with the U.S. Department of Agriculture (USDA), cloned the DNA that encodes the protein of one strain of FMD into bacteria, made the protein product in large enough quantities for field trials, and tested it at USDA's Plum Island Animal Disease Facility (14). The FMD subunit vaccine protected animals against infection by the particular strain against which the vaccine was made (although the field trial was not extensive), but it was less effective than the whole inactivated vaccine. The two other research groups working on a subunit FMD virus vaccine are a Swiss-West German team (University of Heidelberg, Federal Research Institute for Animal Virus Diseases at Tubingen, Max Planck Institute for Biochemistry, and Biogen S.A.) and a British team (Animal Virus Research Institute and Wellcome Research Laboratories) (9).

Cloning of the genes that code for the surface proteins of viruses of fowl plague, influenza, vesicular stomatitis, herpes simplex, and rabies also has been achieved, and the cloned genes may lead to the development of effective subunit vaccines for these animal diseases (2). Cloning projects for virus proteins that cause gastroenteritis, infectious bovine rhinotracheitis, Rift Valley fever, and paramyxovirus currently are underway (2). Different challenges are associated with each project. Rabies projects, for example, have encountered problems with the consistent expression of the surface protein from rDNA plasmids (34). Influenza virus projects, among others, face problems in that the natural viruses spontaneously change their surface proteins to evade the immune system, making the choice of optimal genes for cloning difficult.

Another method being used to prepare new subunit vaccines for animals, aside from the use of rDNA technology, is chemical synthesis of peptides. Synthetic peptides corresponding to part of one viral surface protein of FMD protect test animals against live FMD virus (3), and efforts are underway to prepare synthetic rabies vaccines (28). As noted in *Chapter 5: Pharmaceuticals,* most synthetic vaccines are prepared with the use of MAbs as purification tools. Chemically synthesized peptides may prove useful in rapid screening programs to determine which peptides act as the best vaccines; subsequently, the DNA corresponding to these fragments may be cloned for large-scale production in microbial systems.

Whether produced from rDNA or chemical synthesis, subunit vaccines for viral animal diseases must satisfy several requirements to be effective. In most instances, subunit vaccines must contain antigens from a sufficient number of different strains of virus to offer comprehensive protection against field challenge. The new vaccines must induce a protective immune response to the same or greater degree than conventional vaccines if they are to compete for market shares. Proper dosage and timing of vaccination must be determined. Ideally, the vaccines should be administered in a single injection to be amenable to most husbandry practices throughout the world where animals are dispersed over wide tracts of land. Also, long shelf storage life and stability when stored at room temperature are desirable features of the new vaccines for use in all the countries affected by the particular diseases.*

In addition to subunit vaccines that provide active immunity, MAbs may be used to provide passive immunity against a variety of viral animal diseases. Several MAb-based products currently are being developed. For instance, antirabies MAbs that protect mice from active rabies virus have been made (19). The use of these products, however, is likely to be limited to herds (e.g., dairy animals) where the passive vaccines can be readily and repeatedly administered.

Bacterial Animal Diseases.—The potential for biotechnology in fighting bacterial diseases in animals is less clear than its potential in fighting viral diseases, but several promising advances are currently being made. In developing new methods to prevent these diseases, an understanding of the natural and pathogenic roles bacteria play in domestic animals is important. Numerous types of bacteria are normal inhabitants of both human and animal gut. In general, disease may result when animals, especially those predisposed to infection (e.g., young, weak, or stressed animals), either succumb to pathogenic bacteria or suffer from overgrowths of their own native bacteria. Bacterial infections often occur simultaneously with other infections, including viral invasions. Because of the complexity of most of the currently important animal diseases in which bacteria are involved, the effectiveness of bacterial vaccines produced by biotechnology is difficult to predict.

Bacterial vaccines against colibacillosis (scours), a widespread disease that causes diarrhea, dehydration, and death in calves and piglets, are be-

*At present, the rabies subunit vaccine is most promising in meeting the criteria for becoming a competitive vaccine. There appear to be only slight variations in surface protein sequences between rabies virus strains, and these surface proteins elicit large immune responses. The RNA encoding several viral surface proteins has been cloned and expressed in *E. coli* (34). Questions that remain concerning the efficacy of this vaccine include: 1) the need for glycosylation of the rDNA-product for proper functioning (see *Chapter 5: Pharmaceuticals);* and 2) proper delivery systems, primarily to wild animal reservoirs such as skunks and foxes, where rabies proliferates, and to dispersed animal herds such as those in South America, where the death of cattle infected by the bites of rabid vampire bats results in an estimated yearly loss of more than $29 million (34).

ing made with biotechnology. Recombinant DNA technology is used to change bacterial plasmids found in pathogenic strains of enteric bacteria from a virulent to a harmless state. This approach is used by both Intervet (Netherlands) and Cetus Corp. (U.S.) to prepare vaccines against colibacillosis. These vaccines have been successfully tested in pregnant cows, which transferred immunity against colibacillosis to their offspring, and the products are now available commercially.*

Using another approach to fight colibacillosis, the NBF Molecular Genetics, Inc. (U.S.) uses hybridomas to produce MAbs against the attachment antigens of the bacteria responsible for the disease. Incorporating these MAbs in milk fed to young calves within 36 hours of birth protects the animals through the period for which they are most susceptible (36). This product is approved for use in the United States and Canada.

The development of biotechnological solutions to bacterial animal diseases, as well as viral infections, will require much basic research. Pasteurellosis (a lower respiratory tract infection in cows, sheep, and pigs) and swine dysentery (which causes annual losses of $75 million in the United States) are among the major animal bacterial infections about which more knowledge is needed before applications of biotechnology are possible. The potential for biotechnological production of new bacterial vaccines and the development of successful delivery systems is largely unexplored.

Protozoan and Parasitic Infections of Animals.—Coccidiostats (compounds that prevent coccidiosis in poultry) and anthelmintics (substances that fight helminthic parasites such as roundworms, tapeworms, lung worms, and liver flukes) constitute large, rapidly expanding animal health product markets. In 1985, the global market for coccidiostats is expected to be $500 million (compared to $300 million in 1981), and the global market for anthelmintics may exceed $900 million (compared to $450 million in 1981) (35). At present, coccidiostats and anthelmintics are synthesized by either chemical synthesis or microbial bioprocess methods. These agents, many of which have been discovered serendipitously, are commonly administered in animal feed (10).

The widespread use of coccidiostats, anthelmintics, and antibacterials in animal feed raises concerns about the nurturing of drug resistance among populations of micro-organisms. These risks are outlined in a 1979 OTA report entitled *Drugs in Livestock Feed* (30). As described in that report, the genes in bacteria that encode resistance to most drugs are located on plasmids. Resistance to drugs may be shuttled via these plasmids into pathogenic micro-organisms such as *Salmonella*. Widespread use of antibacterials selects for bacteria, including *Salmonella,* that contain resistance genes, perpetuating drug resistance among bacteria. Thus, wide use of antibacterials in animal feed eventually may compromise the effectiveness of the same drugs in treatment of human diseases. Drug resistance among the protozoa and parasites is even less well understood than is resistance among bacteria. Such resistance is difficult to quantify but may be increasing (13,30).

Fundamental knowledge may be gained by using rDNA technology to explore the structure and function of genes that confer resistance to drugs. MAb technology and other conventional methods may be used to isolate, purify, and better understand antigens found on parasitic cells, perhaps resulting in vaccines effective against these parasites. The increased use of vaccines would decrease the use of feed additives and presumably lessen the problems of drug resistance.

Strong needs, large market potentials, and safety considerations characterize the further development of compounds effective against protozoa and parasites that afflict animal populations. Because of the complexity of most parasitic infections, however, biotechnological solutions may not be forthcoming immediately. In addition, the recent introduction of potent new antiparasitic feed additive compounds, such as the avermectins (which are microbially produced) (8), may lower incentives to explore new antiparasitic possibilities with biotechnology in the near term.

*These bacterial vaccines were made by replacing a "virulence gene" (a gene which encodes a protein that regulates cellular water loss and is responsible for the diarrhea) located on a plasmid with a harmless gene and "infecting" animals with bacteria containing these harmless plasmids. The bacteria continue to produce surface antigens, but they do not produce the virulence protein. The surface antigens stimulate an immune response that prevents adherence of natural virulent bacteria (18).

One serious worldwide rickettsial disease that requires urgent attention is anaplasmosis. Anaplasmosis, which is caused by blood-borne micro-organisms transmitted to cattle by ticks, causes severe anemia and subsequent death in afflicted animals. In the United States, annual losses due to anaplasmosis are estimated to exceed tens of millions of dollars. At present, an unsatisfactory attenuated vaccine exists, and attempts to culture the micro-organism and prepare better vaccines have been only marginally effective (36).

Animal nutrition and growth promotion

Practices and products that promote animal nutrition and growth have the potential to produce direct, substantial returns on investments. Animal scientists seek better animal nutrition and feed-use efficiency in several ways, including the study of gut bacteria that participate in animal digestion, feed additives that enhance absorption of nutrients, and substances such as GHs that may directly stimulate growth and animal productivity.

Synthetic steroids and natural hormones are used widely to promote animal growth, as indicated in table 28. Furthermore, as noted above, health- and growth-promoting compounds from industrially grown micro-organisms constitute a large share of feed additives (30). Some of these compounds act by enhancing the growth of beneficial micro-organisms in the gut, others by reducing the prevalence of harmful micro-organisms and parasites throughout the gastrointestinal tract; still other compounds directly provide animal nutrition. In cases where microbial metabolic pathways and products are known, biotechnology may augment the production of compounds used as feed additives by increasing the production of specific microbial metabolites.* At present, however, applications of biotechnology to the production of metabolites largely remain unexploited (10).

GHs produced by rDNA technology, in contrast, currently are undergoing trials in humans and animals in efforts to demonstrate safety and effectiveness in stimulating growth. Several U.S. NBFs, including Genentech Corp. (in collaboration with Monsanto Corp.), Molecular Genetics, Inc. (for American Cyanamid), Bio-Technology General, Amgen, and Genex Corp., are producing GHs for various animal species. In addition to yielding potential commercial products, rDNA GH projects are stimulating widespread research into the nature of growth, development, and animal produc-

*The production of compounds used as feed additives is discussed in *Chapter 7: Specialty Chemicals and Food Additives.*

Table 28.—World and U.S. Sales of Growth Promotants (millions of dollars)

	Sales								
	1979		1980		1981E[a]		1985E[a]		Compound annual growth
Products	World	U.S.	World	U.S.	World	U.S.	World	U.S.	1981-85E[a]
Hormones:									
Synovex (Syntex)	$ 14	$ 8	$ 16	$ 8	$ 19	$ 8	$ 23	$ 6	9%
MGA (Upjohn)	12	11	12	10	12	9	12	0	No change
Ralgro (IMC)	16	15	24	22	32	29	55	45	15%
Compudose (Eli Lilly)	—	—	—	—	4	—	100	50	N.A.[b]
Other:									
Rumerain (Eli Lilly)	$ 60	$ 55	$ 65	$ 55	$ 75	$ 60	$200	$125	28%
Feed	60	55	65	55	75	60	125	100	14%
Bolus	—	—	—	—	—	—	75	25	N.A.
Avoparcin ("Avotan") (American Cyanamid)	15	—	20	—	25	—	50	10	19%
Other	33	—	38	—	43	—	75	—	15%
Total	$150	$ 29	$175	$ 95	$210	$106	$515	$246	25%

[a]E = estimated.
[b]N.A. = Information not available.

SOURCE: S. J. Zimmer and R. B. Emmitt, "Industry Report: Animal Health Products Market" (New York: F. Eberstadt & Co., Inc., 1981). Modified by the Office of Technology Assessment.

tivity. The results of experiments pertaining to GHs' mode of action to date have yielded results that suggest caution. Previous observations that injections of bovine pituitary gland extracts enhance lactation in cows led to the finding that purified GHs increase milk yield by 10 to 17 percent, without a concurrent change in feed intake (24). Other experiments with sheep and pigs have shown rapid growth following GH treatment (36). However, other evidence indicates that GHs stimulate growth and feed-use efficiency at the expense of body-fat deposition (24). Thus, critics argue, GHs may impair long-term animal health and productivity (24).

Substantial hurdles must be overcome before rDNA-produced GHs become commercially available. In addition to requiring regulatory approval, the commercial success of GHs requires an adequate drug delivery system that introduces GH slowly to animals. Oral administration of GHs, although most convenient and marketable, is an inadequate system of delivery because polypeptides such as GHs are degraded by digestive enzymes. The hormones must be made available to the body's circulation, where they can reach endocrine organs. Slow-releasing ear implants may be used as alternatives to frequent injections (injections are not amenable to most husbandry practices except those for dairy cattle), but, at present, dose requirements are too high for such implants to be practical (21). Eli Lilly (U.S.) is developing a long-lasting bolus to be used in the rumen. Presumably, enough GH is released directly through gastrointestinal tract walls to avoid the problem of enzymatic degradation. With the development of convenient delivery systems, better field trials to investigate the efficacy of GH may result.

Genetic improvement of animal breeds

Throughout the history of animal agriculture, breeders have sought to improve animal productivity by selecting animals with desirable traits for breeding. Recent increases in the understanding of animal reproductive biology and the genetic basis of traits have fostered new animal breeding technologies (31). As a result of increased knowledge due to biotechnology, the identification of genes and gene products that influence traits of productivity, vigor, and resistance to certain diseases may be possible.

In the future, animal breeding programs may be augmented by biotechnology to achieve desired changes with unprecedented speed and selectivity. Biotechnology may be used in ongoing breeding programs first to identify animals with desirable genes (e.g., genes that make the animals resistant to certain infections), and second, to transfer these genes directly into the germ line (cells that contain the genes that will be passed on to future generations) of other animals. Possible applications of biotechnology include the use of MAbs to identify and isolate gene products correlated with certain traits, the use of rDNA technology to produce large quantities of desired gene products for further study, gene transfer (micro-injection of isolated DNA into embryo cells), and implantation of the embryo cells to which genes have been transferred into surrogate mothers.

The technology of gene transfer is in its infancy. To date, it has been used only in laboratory animals. In most instances, the gene(s) to be studied is fused within a plasmid to a gene with a known "housekeeper" function required for growth. The plasmid is injected into a host cell that is deficient in the housekeeper function. Only host cells whose chromosomes incorporate the foreign DNA have the restored housekeeping activity and survive. These cells then are screened for activity of the desired gene. The GH gene has been the subject of many recent gene transfer experiments.

Thus far, gene transfer experiments in animals have increased fundamental understanding in several areas. Scientists have made great gains in preparing receptive host cells, transferring genes from one animal cell to another, and recognizing the successful recombination of foreign DNA in host chromosomes (1,6,32,33). Fundamental understanding of mechanisms of gene control in mammals has also burgeoned in recent years. Several investigations have revealed that the host tissues surrounding the cells that contain implanted genes affect expression of the foreign genes (as surrounding tissue may regulate gene expression in normal cells) and that this "tissue-specific gene regulation" continues through successive generations (7,12,23,25,27). Finally, gene transfer experiments have allowed the study of the expression of single genes that, with other genes, comprise traits that might be too complex for study by other methods.

Gene transfer studies may reveal much about the function of single gene products. For instance, the transfer of genes implicated in immune responses and resistance/susceptibility to disease are being studied (some of these genes encode immunological cell-surface proteins called the HLA antigens) (11). The ability to transfer such genes into foreign cells to distinguish the production and function of their products may lead to valuable knowledge about animal diseases.

In the future, gene transfer may prove to be the sole means of overcoming certain animal diseases that defy preventive vaccine technology and/or veterinary treatment. An example of such a disease is trypanosomiasis ("nagana" in cattle and "sleeping sickness" in humans). Trypanosomiasis is caused by parasites borne in the saliva of certain insects and impedes livestock productivity throughout Africa (where the disease is transmitted by tsetse flies). Strains of cattle and sheep with resistance to trypanosomiasis (trypanotolerance) exist, and their resistance may be traceable to several distinct genes (26,29). Gene transfer may prove useful in better identifying these genes and selecting animals for breeding programs designed to encourage trypanotolerance in affected areas. In the future, transfer of these genes into cattle germ lines may rapidly foster widespread trypanotolerance where most other programs to control trypanosomiasis have failed. The application of knowledge gained from gene transfer experiments to animal agriculture will not be immediate, but such knowledge eventually may lead to considerable agricultural advances.

Commercial aspects of biotechnology in animal agriculture

Although field trials of several biotechnology products for animals are underway and a few products (e.g., vaccines for colibacillosis) have been approved for use, it is not yet clear to what extent biotechnologically made products will be adopted for use in animal agriculture. Most of the nascent products will require more convenient, cost-effective delivery systems, greater demonstration of effectiveness, and appropriate publicity before they are used widely.

If these challenges are successfully met, biotechnology may affect animal agriculture in numerous ways. Some novel products, such as rDNA and synthetic vaccines, in addition to lowering the costs of animal health care, may create new markets. Other products, such as diagnostic MAbs, may replace conventional diagnostic tests. At present, the animal health product markets are skewed against biologics such as vaccines in favor of pharmaceuticals such as antibiotics, mostly because biologics have not demonstrated high levels of efficacy. Until recently, commercial investment in vaccine research has been relatively small, but the wide-ranging applications of biotechnology to animal agriculture is prompting increasing amounts of investment in vaccine research. Applications of biotechnology to products for highly valued animals, such as companion animals and breeding stock, may help support substantial R&D and licensing costs associated with the first new animal drugs and biologics made using biotechnology.

Existing global markets for animal health products are shown in table 29, which differentiates major markets for nutritional products, antibacterials, and other compounds from the market for vaccines. As shown in that table, the markets for vaccines, anthelmintics, and growth promotants are expanding the most rapidly.

The companies that dominate the production of most animal health products are primarily major chemical and pharmaceutical manufacturers.* Most of these companies possess global marketing and distribution networks and undertake animal drug production as a diversification of their principal activities. As shown in table 30, animal health

*Major U.S. producers include Syntex, Pfizer, Eli Lilly, Upjohn, SmithKline Beckman, American Cyanamid, Merck, American Home Products, Johnson & Johnson, Tech America, and Schering-Plough. Major foreign producers include Burroughs-Wellcome (U.K.), Rhone-Merieux (France), Hoechst AG (F.R.G.), Bayer AG (F.R.G.), Connaught (Canada), Beecham (U.K.), Solvay (Belgium), Boehringer Ingelheim (F.R.G.), Intervet (Netherlands), and Elf Aquitaine (France).

Table 29.—Global Animal Health Product Markets

	Estimated sales, 1981 (millions of dollars)	Estimated annual growth, 1981-85
Nutritional products	$2,500	10-15%
Medicinal products:		
Biologics/Vaccines	1,000	20-25%
Antibacterials	2,000	10-15%
Anthelmintics	450	25-30%
Ectoparasiticides	400	10-15%
Coccidiostats	300	15-20%
Growth promotants	200	24-30%
Other	650	15-20%
Subtotal	5,000	15-20%
Total	$7,500	15-20%

SOURCE: S. J. Zimmer and R. B. Emmitt, "Industry Report: Animal Health Products Market" (New York: F. Eberstadt & Co., Inc., 1981).

Table 30.—U.S. Producers of Animal Health Products

	Estimated animal health sales, 1981 (millions of dollars)	Percent of corporate sales	Percent of corporate operating income	Estimated animal health sales annual growth, 1981-85
Pfizer	$ 440	13%	13%	10%
Eli Lilly	365	13%	15%	20%
American Cyanamid	265	7%	7%	11%
Merck	200	7%	7%	27%
SmithKline	155	7%	5%	17%
Upjohn	134	7%	7%	11%
Syntex	83	12%	N.A.[a]	11%

[a] N.A. = Information not available.

SOURCE: S. J. Zimmer and R. B. Emmitt, "Industry Report: Animal Health Products Market" (New York: F. Eberstadt & Co., Inc., 1981).

product sales by the U.S. companies that produce such products constitute a fairly low percentage (an average of 11 percent) of the companies' total sales. Investments in animal-related biotechnology R&D in those companies probably average about the same or less than the investments by the leading NBFs that are applying biotechnology to animal agriculture (22).

As noted in *Chapter 4: Firms Commercializing Biotechnology,* most major pharmaceutical and veterinary medicine companies are investing in biotechnology R&D, but there is some question as to their motivation for producing new products for large, established animal health care markets. Such markets include those for antibiotics, anthelmintics, and coccidiostats. Established companies with existing lines of conventionally made, widely marketed animal health products may have strong interests in maintaining these products. In many cases, therefore, their primary interests do not lie in R&D to produce new animal health products. As described earlier, applications of biotechnology to the production of animal products involve a substantial investment in basic research. In some cases, healthy sales of conventionally made products may dissuade a company from pursuing basic research that could lead to the development of a competing product. In other cases, corporate developers may choose to pursue human pharmaceutical innovations of new biotechnology, rather than applications of biotechnology to animal health. Because of these considerations, innovation and new product development in animal agriculture might be slowed.

Innovation in smaller product market areas, such as animal vaccines and diagnostic products, however, is widespread. New or replacement animal vaccines are among the most promising applications of biotechnology, as are MAb-based diagnostic products. Much of the innovation in developing these products is attributable to NBFs in the United States.

At present, the extent to which biotechnology will be used for the development of animal vaccines is uncertain. Most individual vaccine markets are relatively small—as shown in table 31, most U.S. vaccine markets for animals range from $5 million to $10 million per year—and sales of a single vaccine line would probably be insufficient to sustain a company. Therefore, most companies must market a broad range of vaccines to be competitive. The practice of maintaining a diverse selection of these products may facilitate the development of vaccines for diseases that alone might be unprofitable, such as diseases endemic to developing countries (16). Ultimately, the application of biotechnology to animal vaccine development depends on technical feasibility and the ability of vaccine developers (currently, mostly NBFs) to obtain funding for further work.

In addition to improving vaccines for a broad range of animal diseases, biotechnology may shift the sites of vaccine production from several large foreign producers (e.g., Rhone-Merieux (France) and Burroughs-Wellcome (U.K.))* to smaller U.S. producers. Currently, as shown in table 32, three foreign manufacturers control approximately 25 percent of the U.S. veterinary vaccine market (which accounts for one-fourth of the world's yearly $1 billion veterinary vaccine market). With the successful development of subunit veterinary vaccines by U.S. NBFs, competition may result in a redistribution of worldwide vaccine production. Collaborative arrangements between NBFs and local producers for the development of safe subunit vaccines effective against local strains of animal diseases may increase in the future.

Conclusion

The use of biotechnology to improve animal feed, nutrition, and health promise to improve production of food from animals. MAb-based diagnostic products exemplify this promise. Other products, such as new vaccines, may face technical problems of dosage, formulation, and delivery before they are suitable to animal husbandry practices. Until these problems successfully are resolved, the impact of biotechnology on improving animal productivity will not be realized. Applications of biotechnology such as gene transfer experiments and investigations into the nature of growth using rDNA-produced GHs currently serve to increase basic knowledge about animal biology.

*Rhone-Merieux and Wellcome command the international markets for rabies and FMD vaccines. Together these two vaccines comprise 30 to 35 percent of global animal vaccine sales. Other leading FMD vaccine producers are Bayer (F.R.G.), Pfizer (U.S.), Hoechst (F.R.G), and Rosenbusch (Argentina). State agencies serve about one-half of the rabies market, and Rhone-Merieux, Wellcome, and Connaught (Canada) dominate the rest.

Table 31.—Sales of Major U.S. Animal Vaccine Products, 1981 (millions of dollars)

	Sales
Cattle products:	
Clostridium	$ 16.0
Infectious bovine rhinotracheitis and bovine leukosis virus	13.0
Leptospirosis and combinations	6.0
Vibriosis and combinations	3.0
Swine products:	
Atrophic rhinitis (Bordetella)	$ 6.0
Pseudorabies	5.0
Erysipelas	3.5
Pet products:	
3-way feline virus disease	$ 4.5
Rabies	12.0
Canine parvovirus and combinations	9.0
Poultry products:	
Marek's disease	$ 12.0
Newcastle disease and combinations	9.0

SOURCE: S. J. Zimmer, "The Impacts of Applied Genetics in Animal Agriculture," contract report prepared for the Office of Technology Assessment, August 1982.

Table 32.—Major Producers of Animal Vaccines Sold in the United States

Company	1981 sales (millions of dollars)	Market share
Norden (SmithKline) (U.S.)	$40	27%
Philips-Roxane (Boehringer Ingelheim) (F.R.G.)	18	12
Fort Dodge (American Home Products) (U.S.)	14.5	10
Beecham (U.K.)	11	7
Jensen Salsbery (Wellcome) (U.K.)	9	6
Dellen (TechAmerica) (U.S.)	8.6	6
Pitman-Moore (Johnson & Johnson) (U.S.)	1.8	1
Syntex Agribusiness (U.S.)	1.6	1

SOURCE: S. J. Zimmer, "The Impacts of Applied Genetics in Animal Agriculture," contract report prepared for the Office of Technology Assessment, August 1982.

Plant agriculture

There are hundreds of forms of crop improvement whose purpose generally falls into one of three categories. The first is to increase crop yields by increasing resistances to pests or environmental conditions such as drought or soil salinity or by developing more productive plants. The second is to improve crop quality by enhancing such features as nutritional value, flavor, or processability. The third is to reduce agricultural production costs by reducing a crop's dependence on chemicals or by making harvesting easier (55, 56).

During the last century, plant breeders have been efficiently and successfully addressing all of these goals. The use of new biotechnology in crop improvement, as in other areas, is not a new beginning, but an extension of previously evolved skills. New biotechnology alone will not produce better crop plants, but combined with knowledge from other plant science and microbiological disciplines, biotechnology will develop techniques that could be very powerful in improving agriculturally important crops. Thus, the greatest advances in crop improvement are likely to be made using an interdisciplinary approach.

The genetic manipulation and modification of plants presents some special challenges. Most molecular genetics to date has been done with simple unicellular organisms and, to a lesser extent, with laboratory animals. The application of molecular genetics to plants is relatively more recent and consequently at an earlier stage of technical development. Furthermore, there are fewer studies of the physiology and biochemistry of plants than there are studies of these aspects of animals. The recent application of the new techniques of molecular biology to plants has produced astoundingly rapid results, however, and these techniques are sure to have an impact on crop improvement in the next several years.

Of the several hundred domesticated plants in the world today, only about 30 are of great economic significance. Of these, eight domesticated grains, including rice, corn, and wheat, produce most of the calories and protein consumed by humans and agriculturally important animals. The legumes, which include soybeans, represent the second most common source of food for human and animal consumption. There are two philosophies, which are not incompatible, with regard to improving crop plants. One is that there should be diversification of crop plants and attention given to the domestication and breeding of new major crop plants. Another philosophy is that plant breeding, tissue culture, and biotechnology efforts should be devoted to the most successful crop plants. The genetic diversity of some of the world's current crop plants is not great. Consequently, even if the major crop plants are the focus of research efforts, some genetic material from exotic sources may be required to effect the desired improvements. In any case, the techniques discussed here are equally applicable to the improvement of both common and exotic species.

Research on plants has shown that the genetic organization of plants exhibits striking similarity to that of animals. The universal genetic code is used, and most genes contain intervening sequences and are surrounded by very similar regulatory sequences. Unlike animals, however, plants have a characteristic called totipotency that, for many species, indicates the potential for regeneration of a single cell into a complete plant. Because plants have this totipotent characteristic, certain genetic manipulations can be done in cell culture, and, after selection of cells with the appropriate qualities, the cells can be regenerated into parental plants (for breeding programs). Adjusting the laboratory variables to achieve regeneration from single cells is evolving from an art to a science and has yet to be accomplished consistently for the principle cereal grains (monocots*), but regeneration research is proceeding at a rapid rate. It is likely that many important crop plants will be able to be regenerated from single cells in the next few years.

There are several potential applications of new biotechnology for plants that may help in the im-

*Early in the evolutionary history of flowering plants, two main types of plants, monocots and dicots, diverged. Cereal grains (corn, wheat, rye, barley, rice, etc.) are monocots, whereas legumes (soybeans, etc.) are dicots.

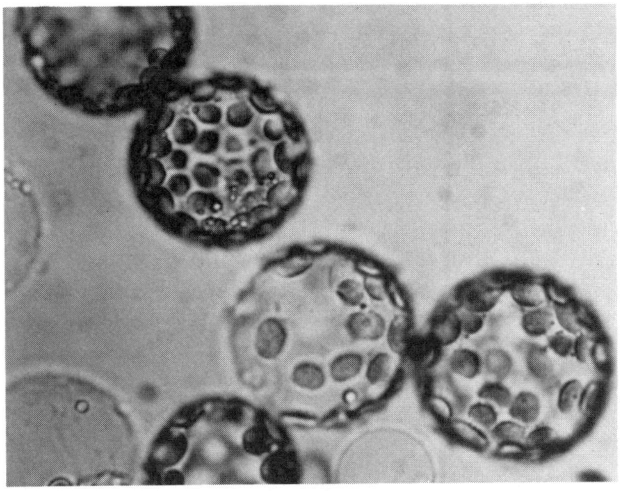

Photo Credit: Dean Engler, Agrigenetics Corp.

Freshly isolated plant protoplasts

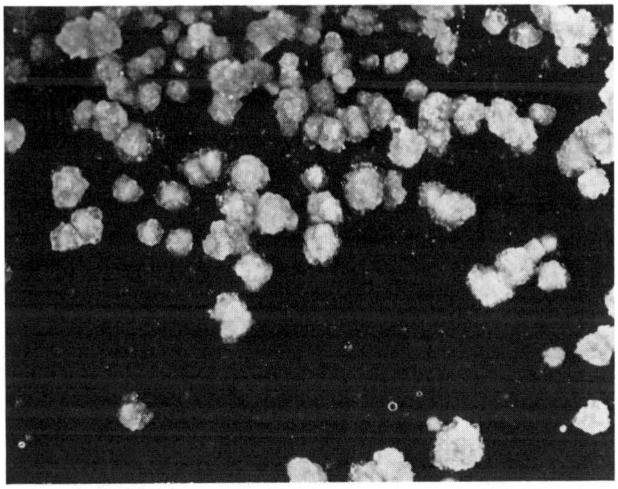

Photo Credit: Dean Engler, Agrigenetics Corp.

Masses of plant cells (calli) resulting from the divisions of individual protoplasts

Photo Credit: Dean Engler, Agrigenetics Corp.

Plant shoots arising from protoplast-derived calli

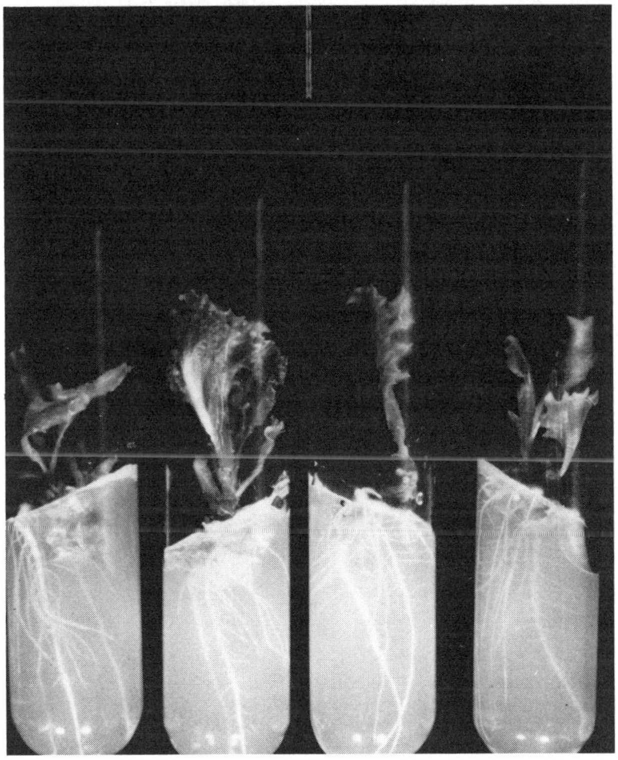

Photo Credit: Dean Engler, Agrigenetics Corp.

Complete plants, each regenerated from a single protoplast

provement of crop species, as shown in figure 16. New technologies for testing for the presence or absence of traits, for example, will save years of plant breeding time. Many applications to plant agriculture will be in the regulation of endogenous genes, and other improvements will be made using techniques such as the following, which transfer DNA from one species to another:

- The fusion of cells from two different plant species can be used to overcome species hybridization barriers. In order to be useful, the resulting cell fusion product must be regenerated to form a whole plant. To date, regenerated plants have only resulted from fusions between closely related genera (62). The regenerated plants are selected to express beneficial characteristics of both parents (94). As yet, no economically important variety has been produced using this method (62).
- Transferring subcellular organelles such as nuclei or chloroplasts from one plant species to another can be accomplished by a variety of techniques. One of these, liposome transfer, involves surrounding the organelle with a lipid membrane. Because chloroplasts carry many of the genes important in photosynthesis, liposome transfer may be instrumental in improving photosynthetic efficiency.
- Vector-mediated DNA transfer (and microinjection of DNA) is the most specific, and potentially the most versatile, of the genetic manipulation techniques. Recently, foreign plant genes have been inserted and expressed in plants.

Figure 16.—

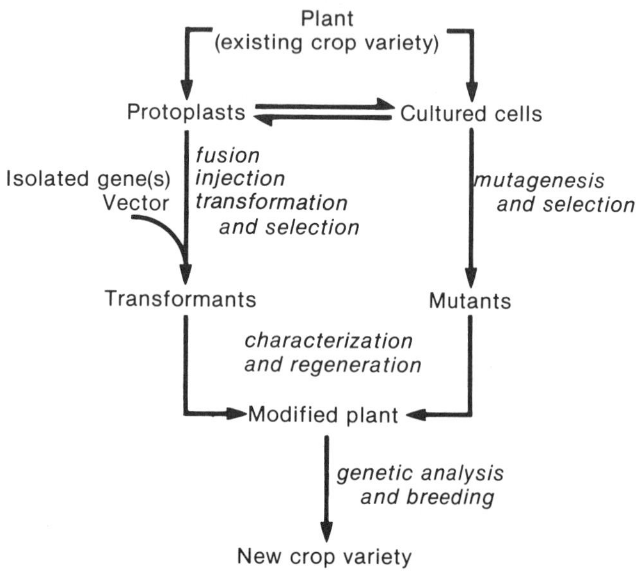

Recent advances in the methods of plant cell culture and the techniques for introducing DNA from one plant species to another are discussed in *Box C.—Methodology Important in Plant Agriculture*. The applications of these methods to specific problems in plant agriculture, such as disease resistance, photosynthetic efficiency, and nitrogen fixation and the commercial aspects of biotechnology in plant agriculture are discussed in the sections that follow.

Improvement of specific plant characteristics

Greater crop yields or a reduction in the cost of crop production would be possible if plants were resistant to disease and certain environmental factors and contained a larger amount of higher quality product. In the United States, there is great research interest in crop resistance and crop quality improvement in academic, Federal, and industrial laboratories. Unlike most other plant traits, some resistances and specific improvements may be accomplished with one or a few gene modifications. This area, therefore, is probably the most active area of industrial research, and it is likely that considerable research progress in this area will be made in the next 5 to 10 years.

PLANT RESISTANCE FACTORS

Disease and environmental resistances are important to most crops in most areas of the world. Important plant resistances are shown in table 33. Productivity losses often can be attributed to the lack of resistance to one or more factors (see tables 34 and 35). Thus, the study of resistances could lead to greatly improved productivity and an increased realization of genetic potential (42).

Numerous single gene resistance factors are known in higher plants. The most common re-

Box C.—Methodology Important in Plant Agriculture

Methods of Plant Cell Culture

Plant cell culture is important commercially today. Several species of plants can be cultured and regenerated, allowing many identical plants, including trees, to be grown. For example, cultured plant cells are used for selection of virus-free cells. In the past, it has been typical for yields in potato crops to be greatly reduced by several potato viruses. Virus-free potato plants are now obtained from regenerated cell-cultured potatoes, and, as a consequence, the yield of these plants has increased substantially (98).

In the research laboratory, plant cell culture provides a bridge between molecular genetics and plant breeding. Most schemes involving the transfer of genetic information require that cells from the recipient plant be cultured, although it should be noted that many genetic experiments can now be done by the direct manipulation of pollen or embryos, which circumvents the necessity for regeneration. If single plant cells or embryos can be cultured, selected, and regenerated, the number of cells in any experiment can be large, contributing to the overall potential success of an experiment. Many millions of plant cells can be studied in a small laboratory; if the same number of whole plants were to be used in an experiment, acres of greenhouses or fields would be required. The development of cell culture conditions is far from routine, and even though the recent successes of plant cell culture are notable, there are still only a limited number of species that are cultured routinely (53).

Plant cells are useful to geneticists, but to be useful agriculturally, whole plants must be able to be regenerated from these cells. Regeneration is accomplished by placing the cells under appropriate conditions of light, temperature, nutrition, and growth-regulating chemicals. An early stage of regeneration is the formation of an undifferentiated clump of cells known as a callus. From the callus arise the differentiated cells of roots and stems. Alternatively, the formation of embryos can be induced from the callus. The crop species for which regeneration from cells can be accomplished routinely include asparagus, rapeseed, cabbage, citrus, sunflower, carrot, cassava, alfalfa, millet, clover, endive, tomatoes, potatoes, and tobacco (62). Currently, there is considerable effort in industry and academia to regenerate corn, wheat, and soybeans from single cells routinely. There is as yet, however, no routine method published for regeneration of these major U.S. crop plants, not because there are any inherent differences in these species, but because for many years scientists worked with only a few model systems. Now that direct genetic manipulations can be done on agriculturally important plants, there is renewed interest in developing regeneration schemes for these plants.

One recent innovation is the introduction of methods for freezing plant cells and calli while retaining cell viability. Several laboratories have devised specific successful freezing regimes to preserve materials (101). One application for this technology would be to freeze germplasm or plant embryos of rare mutants for later use in breeding, tissue culture, or molecular biology programs. A side benefit of the methods of freezing plant cells would be the ability to preserve germplasm of endangered species—tropical trees, for example—until ecological niche destruction has been halted. Also, frozen embryo banks might replace or supplement some seed storage facilities. The initial applications of these methods will simplify laboratory maintenance of cell cultures.

The ultimate goal of most genetics programs utilized in conjunction with molecular biology is to create improved plant varieties (47). The process may involve the transformation of plant cells where the cells are receptive to added DNA. After treatment with DNA and selection for the desired trait, a high percentage of the cells (or embryo or pollen) must proliferate to form a whole reproductive plant. Although many pieces of the transformation, selection, and regeneration procedure have been demonstrated, the outlined scenario is far from routine. Within 5 years, however, technical innovations could provide suitable cell culture and regeneration protocols for many important crop species.

One important fact recognized in the past few years is that there sometimes is extreme variation in the phenotype and genotype of cells in tissue culture. In the past, mutants were sought after mutagenesis treatment. Beginning with work on sugarcane, however, the variability uncovered or created in culture has provided a great diversity of useful plant cell lines with altered properties. This variation found in cultured cells is called somaclonal variation (46,74). Somaclonal variation is an important source of genetic diversity for the plant breeder. Several naturally occurring disease-resistant plants have been isolated from cell culture (89).

Vector Construction and Transformation

Remarkable progress has been made in perfecting vectors for delivery of novel genetic information to plants. Most of the research excitement during the past few years has focused on understanding how certain plant pathogens transfer their DNA into a host plant. The bacterial plant pathogen genus *Agrobacterium,* for example, is able to transfer a portion of its plasmid DNA to infected dicot plant cells. *A. tumefaciens* carries the tumor-inducing (Ti) plasmid, and *A. rhizogenes* carries the root-inducing (Ri) plasmid. The portions of each of the plasmids that become stably integrated into plant chromosomal DNA are the transferred-DNA (T-DNA). Because of the relative ease of doing bacterial genetics, much progress has been made in inactivitating the disease-causing portions of the Ti and Ri plasmids to provide vectors for the introduction of novel DNA without causing plant disease during regeneration (64,75). There have been several announcements of the successful introduction and expression of bacterial genes in plants using nonpathogenic forms of T-DNA (49,70,91). The inserted genes are for antibiotic resistance, allowing the transformed plant cells to be identified easily in culture. In one case, these cells have been regenerated into whole plants, and the foreign genes are still expressed (63). One group reported the introduction and expression of a seed storage protein from beans in both sunflower and tobacco cells (73). This result is the first demonstration of the directed transfer and expression of a plant gene from one species in another plant species. The next challenge is to regenerate the transformed plant cells and show that the foreign genes are expressed in a heritable form in the mature plant and transmitted to its offspring.

Vector systems for the agriculturally important monocots do not yet exist. However, natural vectors, like the plasmids of *Agrobacterium,* are being looked for, and synthetic vectors are being constructed. More basic research in plant pathogens will be required to discover whether appropriate monocot vectors already exist in nature. The advantage of using natural vectors such as *Agrobacterium* is that the prokaryotic DNA has acquired appropriate regulatory signals for effective gene expression inside a eukaryotic cell.

Another potential vector, known as a transposable element, may prove useful for introducing DNA into monocots. A transposable element is a segment of DNA that is capable of inserting itself into the DNA of a cell. Transposable elements appear to be universal, but their natural function is unknown. An isolated transposable element can be attached to a gene to be inserted into a plant. This construct is microinjected into plant cells, and the transposable element carries the gene of interest into the cellular DNA. Experiments using transposable elements as vectors have been done with corn cells (87).

sistance genes are those that confer decreased susceptibility to disease (54,71,79). In maize, for example, there are resistance genes to several diseases such as northern corn-leaf blight (51). Because most of the single gene resistance factors confer resistance to a single pathogenic organism, it is thought that a single characteristic of the host and pathogen determine the outcome of an infection.

Most of the existing disease-resistance genes have been introduced into economically important lines of interbreeding plant species by traditional plant breeding. Currently, however, there is interest in cloning disease-resistance genes from plants in order to study the nature of resistance and to determine the possibility of transferring resistance factors among species that do not normally interbreed. It is not known in most cases

Table 33.—Plant Resistances of Economic Value

Resistance to:	Relevance in United States
Disease	All crops
Saline	Irrigated soils, particularly in California and Southwest
Alkaline earth metals	Southeastern United States and West
Anaerobic soil conditions	Areas subject to flooding
Drought	All crops
Herbicides	All crops
Pesticides	All crops
Soil pH	Low pH on acid mine tailings and soil affected by acid rain; high pH on most Western soils

SOURCE: Office of Technology Assessment.

Table 34.—U. S. Soils With Environmental Limitations

Environmental limitation	Percentage of U. S. soil affected
Drought	25.3%
Shallowness	19.6
Cold	16.5
Wet	15.7
Saline or no soil	4.5
Alkaline salts	2.9
Other	3.4
None	12.1

SOURCE: J. S. Boyer, "Plant Productivity and Environment," *Science* 218:443-448, 1982.

Table 35.—Distribution of Insurance Indemnities From Crop Losses in the United States From 1939 to 1978

Cause of crop loss	Proportion of payments (%)
Drought	40.8%
Excess water	16.4
Cold	13.8
Hail	11.3
Wind	7.0
Insect	4.5
Disease	2.7
Flood	2.1
Other	1.5

SOURCE: J. S. Boyer, "Plant Productivity and Environment," *Science* 218:443-448, 1982.

what the disease-resistance genes in plants actually do to plant metabolism or structure. By understanding how the products of plant disease-resistance genes work, better screening programs for enhanced resistance can be designed. Environmentally, it is desirable to develop pest-resistant plants, because such plants would reduce the need for spraying crops with pesticide* chemicals, and disease control would be more effective. It should be kept in mind, however, that much of the agricultural research effort is being made by the agricultural chemical industry, and this industry may see the early opportunity of developing pesticide-resistant plants rather than undertaking the longer term effort of developing pest-resistant plants.

Resistance to environmental conditions probably depends on both single and multigenic inheritance. These traits, as well as disease resistance, can be selected for in tissue culture. If analogs of the disease or detrimental environment conditions can be applied to plant cells in culture, the entire procedure can take place in a few test tubes or petri dishes in a laboratory setting. Millions of individual cells can be treated simultaneously and then examined for survivors. Stepwise selections under gradually more stringent conditions (e.g., a gradual increase in the salinity of the medium) are accomplished readily.

Some of the traits that could be selected in tissue culture are listed in table 33. Many of the traits are resistance factors that confer protection against disease and salinity. In selection schemes for these factors, the test organism is exposed either to normally lethal doses of the toxins produced by a disease organism or to high doses of salt (to mimic salinity), and the surviving cells are identified by their growth under these normally toxic conditions. This protocol holds great promise for identifying rare cells that have spontaneously acquired a novel resistance. Somaclonal variation probably supplies much of the variation seen in tissue culture (see box C) (47).

A rate-limiting step in applying selection techniques more widely is the present inability to regenerate major cereal and legume crops from individual cells or small cell clumps on a routine basis. Furthermore, some of the traits selected in tissue culture for resistance to a specific factor may not be manifested in the whole plant, because it is possible for cells to develop nongenet-

*A pesticide is an agent that prevents the growth or propagation of deleterious organisms, including weeds and insects. Both herbicides and insecticides are pesticides.

ic adaptations. Consequently, numerous regenerated plants will be required to determine if a particular selection procedure yields whole plants with important agronomic traits. On the other hand, pollen or embryo manipulation may circumvent some of these problems.

Genetic manipulations can make plants resistant to chemicals or can enhance their response to chemicals. These traits are of particular importance to the agricultural chemical industry. For instance, various plant growth regulators are produced by this industry. These chemicals can affect many stages of the growth or reproduction activities of plants to give a crop with increased yield. Enhanced response to these chemicals allows the crop to be grown at a lower cost.

Producing herbicide-resistant plants can have definite benefits, especially in crop rotation. For instance, corn is naturally resistant to triazine herbicides, whereas soybeans are not. Occasionally, soybeans do not grow well in a field the year after triazine-sprayed corn was grown there. In this case, one solution would be to introduce triazine resistance into soybeans. This particular resistance is due to a modified protein in the chloroplast membrane. Therefore, resistance between dissimilar species could be transferred by protoplast fusion or liposome-mediated chloroplast transfer (38,66). It should be noted, though, that increased use of agricultural chemicals could have serious environmental consequences.

PRIMARY PLANT PRODUCTS

The largest research effort in the modification of plant products using biotechnology is concentrated on the improvement of seeds and seed proteins. Seeds serve a dual role in agriculture. They are the major source of food for people and animals and represent an easily stored form of plant material, and they are also the material for propagating the next plant generation. The storage materials of seeds contain all of the materials necessary to nourish a plant, because each seed must support the initial phases of germination and seedling establishment until the plant is self-sufficient. During domestication, various crops have undergone an enormous exaggeration of the normal storage reserves. Today, far more material is stored in agricultural crop seeds than in the seeds of wild relatives; sometimes the increase is as much as tenfold (68).

Although the agronomic (applied) research effort is devoted primarily to increasing the amount of seeds and seed protein, current basic research efforts are devoted both to increasing the quality of the stored materials and to exploring plant gene structure. Because plants are capable of synthesizing all of the amino acids required for protein synthesis from simple carbon- and nitrogen-containing precursor molecules, the exact amino acid composition of the stored protein in seeds may not matter to a plant, and seed proteins often have an unbalanced amino acid composition. Because humans and most animals are unable to synthesize eight amino acids (the essential amino acids), the composition of ingested protein matters very much in their nutrition.

Much is known about the structure of the storage-protein products and the genes encoding these proteins in the major crop plants. In all cases studied so far, the storage-protein genes are found in small gene families with 3 to 30 members. Typically, a few genes of the multigene family contribute a significant fraction of the total protein. There is not much genetic variation among the seed storage-protein genes of a given species, although this low variation might be due to the limited diversity of the varieties currently studied. These crops may have lost much of the original diversity present in the progenitor species during the intensive plant breeding activities that have occurred throughout history.

DNA clones of storage proteins are available from several crop species: soybean, garden bean, corn, wheat, and other less significant crop plants. Changes in these genes can be made readily in vitro to improve the balance of amino acids in the protein. The difficult part is reintroducing the altered storage-protein gene back into the crop plant and ensuring that this novel gene is expressed appropriately. Most storage proteins are present only in seeds. Retention of this tissue specificity is important; storage proteins' presence in other plant cells may be detrimental. Another important consideration is that the storage-protein genes are found in families. Introduction of a new gene may change only a fraction of the total protein produced. To modify the overall amino acid composition, several genes may have to be introduced or the natural genes deleted and replaced by novel genes. In some crops such as corn, there are mutations that reduce the production of zein, the storage protein. These mutant genes can be used to reduce the zein concentration, thus allowing an introduced gene to have a greater impact on overall amino acid composition.

An alternative to the modification of existing crop species' genes is the introduction of completely novel genes isolated from other organisms. Genes whose products are very rich in the amino acids that are deficient in a particular seed type could be introduced to increase the concentration of specific amino acids. One promising example of this type is the storage protein of the Brazil nut (97). This protein is composed of 25 percent sulfur-containing amino acids (methionine and cysteine). Legume seed protein usually is deficient in these amino acids. Introduction of a few copies of the Brazil nut storage-protein gene into legume species might overcome the sulfur amino acid deficiency. Proteins of unusual composition may offer the quickest method of preparing a gene to complement deficiencies in major crop storage proteins.

SECONDARY COMPOUNDS FROM PLANTS*

Table 36 lists some of the desirable secondary products from plants. Very little research has been done on the tissue culture production of these compounds, yet it should be possible to produce important high-value plant products using culture systems instead of gathering plants from nature. Cell culture offers the advantages of reproducibility and control over production where seasonal variations, weather changes, or disease are not problems (40,57,95). On the other hand, a difficulty in the production of some products is maintaining the plant cell culture in a differentiated state such that compound production occurs.

Biotechnology offers many opportunities for the production of secondary plant compounds. The transfer of the plant metabolic pathway for a compound to a bacterial or fungal cell, for instance, could offer an opportunity for providing a steady supply of these compounds, although much more knowledge concerning the genetics and biochemistry of the pathways that produce these compounds is necessary. Another possibility is identifying and modifying the gene coding for the enzyme responsible for the rate-limiting step in product production. Overproduction of the products could result from the plants being grown in culture or in the field.

There is little current U.S. research effort to improve the yield of secondary plant compounds from cultured cells or whole plants. The Federal Republic of Germany, Canada, India, and Japan, on the other hand, have large research programs, as measured by the number of papers presented at the 1982 International Congress of Plant Tissue and Cell Culture (58). Japan, for instance, has scaled up the growth of tobacco cells to 7,000 liters, and researchers at the University of British Columbia are growing 100 liter batches of Mada-

Table 36.—Examples of Secondary Plant Products of Economic Value

Agricultural chemicals:
Pyrethrins
Rotenone
Nicotine
Allelopathic compounds
Antibiotics against soil microbes
Pharmaceutical drugs:
Codeine
Morphine
Steroids
Cardiac glycosides
Alkaloids
Reserpine
Retinoic acid
Caffeine
Cannabinoids
Antitumor compounds
Flavorings and salts:
Licorice
Coumarin
Colorings and pigments:
Anthocyanins and betacyanins
Carotenoids
Industrial intermediates:
Latex
Lignin
Dye bases
Steroid and alkaloids products

SOURCE: Office of Technology Assessment, adapted from E. A. Bell and B. V. Charlwood (eds.) *Secondary Plant Products* (New York: Springer-Verlag, 1980).

*This topic was covered by a recent OTA workshop entitled "Plants: The Potential for Extracting Protein, Medicines, and Other Useful Chemicals" (99).

gascar periwinkle cells in order to isolate anticancer compounds (58). In fact, the Japanese Government is spending $150 million over 10 years for research on obtaining secondary compounds from plants. It is argued by some, though, that plant cell culture for producing secondary products is necessary only when good farm land is not abundant (65).

PLANT GROWTH RATE

The rate at which plants grow can limit both the amount of harvestable biomass (food, fiber, secondary products) and the length of time between planting and harvesting. Traditional plant breeding has been quite successful in modifying and improving plants to respond to modern agricultural practices of herbicide, pesticide, irrigation, cultivation, and high-fertilizer application. These breeding programs have established that there is no single gene for yield. On the other hand, much is known about the genetics of harvestable products such as seed size. Additionally, there are single gene mutants, such as that for gibberellic acid, that can affect plant growth dramatically. Increased understanding of these areas of genetics may have an impact on this area of plant biology. For instance, a plant can be imagined that had a decreased amount of total biomass but an increased amount of harvestable product.

PHOTOSYNTHETIC EFFICIENCY

Photosynthesis is the basis for most life on Earth. Higher green plants, algae, and some bacteria can utilize the energy in sunlight to split water molecules; in this process, energy is generated and utilized to combine atmospheric carbon dioxide (CO_2) into an organic form as well as to drive other energy requiring processes of plants. A byproduct of this reaction is molecular oxygen (O_2). Thus, photosynthesis is not only the ultimate source of fixed carbon we use as food and fiber, but also of the oxygen we breathe.

Because photosynthesis is so important to food production, much research has focused on the mechanism of photosynthetic action. The photosynthetic system is very complex, combining enzymatic activities, key roles played by cellular organelles, and plant anatomy as well as environmental factors such as light, water, and temperature. Many proposals have been made to improve the efficiency of this system by genetic manipulation.

The critical step of the photosynthetic CO_2 fixation cycle is catalyzed by the enzyme ribulose bisphosphate carboxylase (RuBPCase), probably the most abundant protein on Earth. This enzyme is sequestered in chloroplasts, the cellular organelles where photosynthesis occurs. It is a complex molecule synthesized from both chloroplast genes and nuclear genes (50,51).

When photosynthesis originated, the Earth's atmosphere is postulated to have been nearly devoid of oxygen. The oxygen we have today is a byproduct of photosynthesis, and oxygen comprises about 20 percent of our atmosphere. RuBPCase initially evolved in a low oxygen atmosphere but now must fix CO_2 with a large excess of O_2 present. RuBPCase can utilize this O_2 in what appears to be a nonproductive enzymatic reaction. This process is called photorespiration and results in a net loss of fixed CO_2 (45). Photorespiration can decrease crop yields by as much as 50 percent (82). It is ironic that RuBPCase activity over the past millions of years has produced the O_2 that now decreases the efficiency of photosynthesis. On the other hand, it has been postulated that the ubiquitous and continued presence of photorespiratory activity implies some natural selection advantage (61).

Suggestions have been made for modifying RuBPCase or other enzymes involved in the photosynthetic system. For instance, genetic manipulations that would increase the affinity of RuBPCase for CO_2 or decrease its affinity for O_2 could substantially increase net CO_2 fixation. It has yet to be determined what effects these changes would have on the survivability of plants.

In addition to manipulating the enzymatic system, changing the plant's anatomy, such as the types of cells in leaves, might be possible. Several groups of higher plants have increased rates of CO_2 fixation that correlate with modified anatomy and physiology. Very little is known about the genetic control of leaf and cellular anatomical development, so near-term success in modifying these aspects of plant anatomy is unlikely.

Genetic manipulations to increase photosynthetic efficiency, and consequently food production, are very difficult now because of the complexity of the system. It will be several years before rDNA technology will aid in producing agriculturally important plants with increased inherent photosynthetic efficiency.

PLANT-PRODUCED PESTICIDES

Some species of plants are highly resistant to potentially damaging insects. Although not very much is understood about this phenomenon, it appears that certain plants can produce compounds that are toxic to specific species of insects or that interfere with the insects' normal reproductive or growth functions (86). An African plant, for example, produces a compound that interferes with a particular caterpillar's molting, and, as a result, the insect cannot eat (48). Other plants are known that produce chemicals that cause potentially harmful insects to avoid those plants for feeding or egg laying (59). The specificity of these plant-produced insecticides and nonpreference chemicals allows the control of pests while permitting potentially useful insects to survive. Many applied chemical pesticides do not have this specificity. It may be feasible soon to clone and transfer the genes that code for these naturally occurring chemicals, allowing them to be expressed in other plants. The result of these gene transfers could be to reduce greatly the amount of agricultural chemicals needed and, hence, the cost of production.

Investigations into chemicals released by some plants that adversely affect neighboring plants is receiving an increased amount of attention (90). These herbicides, known as allelopathic chemicals, may influence another plant directly or may act by inhibiting the micro-organisms normally associated with that plant. Allelopathic chemicals consist of a wide variety of chemical types, and their actions range from inhibiting cell division to protein synthesis to photosynthesis. Much more still is to be learned about these naturally occurring chemicals, including the factors influencing their production and how best to use them agriculturally. A goal of biotechnology is to identify the genes responsible for the synthesis and release of the plant pesticides and to transfer them to nonresistant plants. Biotechnology also could aid in the understanding of their production and possibly help develop their production in controlled laboratory culture systems.

Uses of micro-organisms for crop improvement

Applications of biotechnology in the area of crop improvement include genetic manipulations of micro-organisms that interact with plants in nitrogen fixation, for example, or that produce substances such as insecticides of potential benefit to plants. These applications are discussed further below.

NITROGEN FIXATION

Plants have a universal need for metabolically usable nitrogen in the form of ammonia (NH_3), which can originate either from the air or from applied ammonia fertilizer. Biological nitrogen fixation, the process by which living systems convert nitrogen gas in air to NH_3, is catalyzed in living systems by the enzyme nitrogenase. Nitrogenase, and consequently the capacity to fix nitrogen, is found only in prokaryotes, either bacteria or blue-green algae. Some nitrogen-fixing prokaryotes are free-living and can be either anaerobic or aerobic; other prokaryotes fix nitrogen only when they coexist symbiotically with a higher plant host. The application of biotechnology to nitrogen fixation may result in more efficient prokaryotic nitrogen fixation or the transfer of nitrogen-fixing ability to plants themselves.

Nitrogen-fixing prokaryotes share some common physiologic features. First, nitrogen fixation typically does not occur in cells already supplied with usable nitrogen. Second, nitrogenase is oxygen-sensitive, so all nitrogen-fixing organisms have mechanisms for limiting oxygen. Third, NH_3, which is toxic at high concentration, must be converted readily into organic nitrogen.

Biological nitrogen fixation is energy intensive (84,88,93), and in plant-microbe associations, this energy is derived from the plant. Estimating the energy expenditures for biological nitrogen fixation is difficult, and few reliable numbers are available. The energy cost of nitrogen fixation is an appropriate concern, but this cost should be compared with the true cost of nitrogen nutrition in field-grown plants (i.e., the cost of chemical fertilizer synthesis and other biological costs to the plant).

It may be possible to decrease the energy required for nitrogen fixation by 30 to 50 percent by preventing the evolution of hydrogen during nitrogen fixation. Some bacteria have a set of genes that allow for hydrogen recycling. These genes have been cloned and inserted into less efficient nitrogen-fixing bacteria. The recipient bacteria showed increased nitrogen-fixing efficiency (37).

Agriculturally important nitrogen-fixation systems discussed below are nonlegume nitrogen fixation and symbiotic nitrogen fixation in legumes.

Nonlegume Nitrogen Fixation.—Nitrogen fixation is performed by several groups of bacteria and blue-green algae that live free in soil or in aquatic habitats. The best studied nitrogen-fixing bacterium is the free-living *Klebsiella pneumonia,* which can easily be grown in the laboratory.* The gene complex coding for the nitrogen-fixing function in *Klebsiella pneumonia* is comprised of 17 genes, and the regulation and activities of these genes now are being studied extensively. Still, the nitrogen-fixing function is extremely complex and not well understood.

Algae have been used to fix nitrogen in Asian rice paddies for many years. Recently, research has produced strains of algae that could be used in soil to fix nitrogen for domestic crops. Algae are inexpensive compared to nitrogen fertilizer, and because they release nitrogen slowly into the soil, algae bypass the problem of nitrogen leaching (60). Furthermore, algae are being considered the botanical equivalent of yeast for genetic manipulation, and vector systems for algae transformation are in development (41).

Symbiotic Nitrogen Fixation in Legumes.—The legume-*Rhizobium* symbiosis is the most agriculturally significant biological source of fixed nitrogen. Both grain and forage legumes have large amounts of nitrogen fixed by *Rhizobium.* Recent work on legume-*Rhizobium* symbiosis has focused on several areas, including the determination of energy costs, pathways of nitrogen assimilation and transport, the biochemistry of symbiotic nodule development, and the genetics of the bacterial partner.

Symbiotic nitrogen fixation can be a significant source of nitrogen nutrition for legume crops, but its practical application can be limited by several sets of factors, some environmental, others intrinsic to the plant-bacterial partners. Soil conditions and environmental levels of fixed nitrogen have significant effects on rhizobial survival, nodule formation, and levels of nitrogen fixation. One crucial area, poorly understood at present, is the role played in symbiotic nitrogen fixation of soil micro-organisms other than *Rhizobium.* Understanding nodule formation in detail will help explain environmental effects on infection that may relate to competitiveness and effectiveness of various *Rhizobium* inocula. In addition, an understanding of why legumes, and not other plants, can nodulate would be essential for attempting to extend host range.

Another nitrogen-fixing micro-organism, the actinomycete *Frankia,* is of interest because it nodulates a number of unrelated plant genera. This ability suggests a simpler genetic symbiosis than that of *Rhizobium* and legumes. If this is true, it may be easier to extend genetically the host range of the symbiotic relationship of *Frankia* than to extend that of *Rhizobium* (41).

Specific host proteins are produced in nodules. One of these is leghemoglobin, which controls the oxygen content of the infected nodule cells. This protein is produced in high quantities in nodules. Two research groups have cloned the genes for soybean leghemoglobin (77,96), but their mechanism of action is not understood. Other new proteins appear when nodules develop (76). These are called "nodulins" and are likely to be essential for symbiotic nitrogen fixation; however, their exact role is not known. Some of these might be enzymes, such as those for ammonium assimilation (67). When nodulins and their functions are better understood, a logical extension of current research will be to move cloned nodulation genes into other plants. This may make it possible to extend nitrogen fixation to other plant species.

*Two other free-living nitrogen-fixing bacteria, *Azosporillum* and *Azotobacter,* also are important agriculturally.

Summary.—Individual nitrogen-fixing systems can be improved or extended by a knowledge of how they work and by techniques that permit the genes for nitrogen fixation to be altered and moved. One line of research will be the improvement of existing systems. Some new nitrogen-fixing systems have been proposed, as well. Proposals have been made, for example, to insert directly the genes for nitrogen fixation into the plant genome. Success of these as well as other systems in the end will be measured by the practicality of the new association. The problems of specificity, oxygen regulation, and effect on yield must be considered, and these will require broad-based knowledge of biochemistry, genetics, and physiology in a variety of nitrogen-fixing organisms (see table 37). There is a considerable amount of research being done in the area of nitrogen fixation, and genetically manipulated *Rhizobium* may be field tested soon (85).

MICROBIALLY PRODUCED INSECTICIDES

Problems or drawbacks associated with chemical insecticides, including their increasing cost and environmental hazards, their lack of specificity, and the ease with which insect resistances to such insecticides are developed, have sparked renewed interest in microbially induced insect control to improve crop yield. Microbial insecticides, because of their narrow host ranges, can control specific pests while allowing natural predators and beneficial insects to survive. Furthermore, the few characterized microbial pesticides do not appear to harm humans or animals, and they are biodegradable.

There are three natural sources of microbial insecticides: bacterial, viral, and fungal. About 100 bacteria have been reported to synthesize toxins that are insecticidal. Very few of these bacteria have been studied extensively, but in one case (*Bacillus thuringiensis kurstaki*), the gene that controls the synthesis of a toxin has been cloned using rDNA technology (69). The cellular mechanism of the toxin's insecticidal activity is not yet well understood. Genes for bacterial toxins could be put into other bacteria that normally exist on the surface of plants (48).

Viruses also can be insecticidal by virtue of their ability to cause disease in various insects. Several families of viruses have been identified as potentially pathogenic to insects, but the family *Baculoviridae* has received the most attention. The U.S. Environmental Protection Agency (EPA) has registered, or is considering registering, several baculoviruses for the treatment of such diseases as cotton bollworm, Douglas fir tussock moth, gypsy moth, and alfalfa looper (81). One particular baculovirus (*Autographa californica* nuclear polyhedrosis virus (AcNPV)) has been genetically and molecularly well characterized, making the use of rDNA techniques with this virus feasible.

In contrast to bacterial and viral insecticides, fungal pathogens need not be ingested; they can disable or kill the insect by colonizing its surface. More than 500 fungal species can infect insects,

Table 37.—Importance of Basic Research (Model Systems) on Nitrogen Fixation

Research area	Organisms used in research	Importance
Cloning nitrogenase genes	*Klebsiella pneumoniae*	Direct study of genes Introduction of nitrogen-fixing genes into other organisms
Physiology of nitrogen fixation	*Azotobacter* *Anbaena* *Klebsiella*	Improving energy efficiency of nitrogen fixation in the cell Understanding role of ammonia in nitrogen fixation
Biochemistry of nitrogenase	*Clostridium* *Azotobacter* *Klebsiella* *Rhodospirillum*	Understanding oxygen sensitivity of nitrogenase Improving energy efficiency of nitrogenase enzyme
Cell and developmental biology of nodulation	*Rhizobium*	Bacterial/plant recognition process Nodulation process

SOURCE: Office of Technology Assessment.

and there are susceptible hosts in all the major orders of insects (72). The use of fungal insecticides will require a better understanding of their pathogenesis and ecological requirements. The large-scale production of these pathogens is difficult, and, in many cases, the technology is not developed. Also, their safety with regard to higher animals and humans has not been adequately studied (43,44,81).

DISEASE-SUPPRESSIVE AND GROWTH-REGULATING MICRO-ORGANISMS

Increasing plant yields may be achieved through better understanding of the many bacteria that protect plants from naturally occurring, deleterious conditions (80,92). Some of these bacteria act by producing compounds that bind iron. Others act by altering the pH or the salinity of the soil. Still others prevent frost damage to leaves. Furthermore, there are other bacteria that produce compounds that regulate plant growth. The mechanisms by which these processes occur is not well known. With better understanding of the genetics and biochemistry of some of these bacteria and their environment, it may be possible to program them genetically to produce compounds to change any number of soil and growth conditions.

For two reasons, it probably will not be economically feasible to incorporate the useful micro-organisms directly into soil on a regular basis, because large amounts of the micro-organisms would be required and because the useful micro-organisms might not be able to compete with the well-established micro-organisms already present in the soil. Instead of being incorporated into soil, the useful micro-organisms could be given a competitive advantage by applying them to the seeds or other plant parts prior to planting. Then they would already have established a niche allowing them an advantage over naturally occurring micro-organisms (92).

The first authorized deliberate release of genetically manipulated bacteria, planned for the fall of 1983, was to prevent frost damage. The genes coding for the compounds that initiate ice crystals were identified and deleted from a bacterial strain normally found on many crop plants. The researchers intended to spray these new bacteria on field crops early in the growing season, so that they became the established strain and crowded out the natural, harmful bacteria.* It is thought that this approach could prevent up to $5 billion worth of damage to crops throughout the world (80).

Conclusion

The first successes in DNA transformation of plants to give novel characteristics have been achieved. Continued research on vectors and plant tissue culture is needed to extend these successes from model systems to major crop plants. The identification of genes that would substantially improve a crop plant requires cooperation between traditional breeders and geneticists. Plant molecular biologists need the knowledge of the more traditional plant disciplines to produce agronomically useful plants more rapidly. Interdisciplinary basic research on plant biochemistry, development, and physiology will be required to help identify important genetic traits, to define biosynthetic pathways in plants, and consequently, to develop better plants. Many single gene traits in agronomic species have been used in past breeding programs; such genes also can be studied using new biotechnology. The novel technologies also can introduce genetic material from plants that normally do not interbreed and possibly provide simultaneous introduction of many specific traits into a single breeding line.

The next 5 years will produce major breakthroughs in DNA transformation in model systems and routine regeneration of plants from our major crop species. Problems such as changing the composition of the storage proteins of cereals and legumes are difficult, because multigene families limit the impact of single gene introductions. Within the next decade, however, genes conferring resistance to stress and disease are likely to be introduced and expressed in plants.

*This experiment was indefinitely postponed pending the outcome of a lawsuit filed against the U.S. Government raising the question of the necessity of filing an Environmental Impact Statement prior to the deliberate release into the environment of a genetically manipulated micro-organism.

The production of better plants has obvious value in food production, but biotechnological developments in plants certainly will have other applications. Contributions to health care in the form of novel biological substances may result. Floriculture and the forestry industry will benefit from the development of new strains and acceleration of breeding programs. Plants could be used potentially as a source of industrial enzymes. The fiber industry is likely to see an increase in the production and quality of plant fibers, and an increased production of biomass (organic matter that grows by photosynthetic conversion of solar energy) should contribute to the generation of energy in the form of ethanol (39). The production of energy from biomass is discussed further in *Chapter 9: Commodity Chemicals and Energy Production.*

Commercial aspects of biotechnology in plant agriculture

Although the generation of new plant varieties may be important to the farmer in increased yields or decreased costs of production or to the end processor in better food products, the commercialization of plants developed using new biotechnology is in the hands of seed and live plant producers. The ability of the U.S. seed and plant producers to develop and market new plants will determine the competitive position of the United States in plant agriculture. In general, seed production is not a business where international competition plays a role. Because the climates around the world vary so greatly, researchers would have to do field trials and grow seed in other countries. Thus, each locality generally does its own research and seed production.

Excellent research programs in the applications of biotechnology to plant agriculture exist in the United States, the United Kingdom, and Australia. Because of the climatic differences among these countries, the research is concentrated on different species.

The seed market is one of the largest markets to which biotechnology is directed. In the United States, $4.5 billion of seed are sold to farmers each year. Cuttings of vegetatively propagated plants account for $500 million of this market, and the largest segment of the market, $1.2 billion, is for corn seed. The world market for seeds is estimated at $30 billion. The United States exports $250 million of seeds per year, and U.S. subsidiaries overseas contribute to world seed production (85).

Another potentially lucrative market is the market for cut flowers. This market may be one of the easiest horticultural markets to enter with novel plants produced by biotechnology because it readily accepts and depends on novel phenotypes.

It is likely that genetically manipulated plants may increase the demand for commercial seeds. Drought-resistant plants could increase the acreage planted, and other plants might be planted at higher density, both resulting in an increase in the number of seeds sold.

A phenomenon not necessarily related to biotechnology is the long-established movement by U.S. farmers toward buying new seeds every year, rather than saving and planting seeds from crops produced the previous year. The evidence is gathering that seeds from companies give better results than a farmer's own seeds. Because the cost of seeds is only 3 to 7 percent of agricultural direct costs, it behooves the farmer to get the best seeds possible (85,100). The U.S. soybean industry, for example, has moved recently from buying approximately 20 percent of its seeds a few years ago to buying approximately 40 percent of its seeds today (85). This trend could amplify the demand for seeds produced by biotechnology.

The industrial production of agricultural chemicals now produced by micro-organisms or plants could be a substantial market. These pesticides, along with pest-resistant and nitrogen-fixing plants, could begin to capture the $10 billion domestic agricultural chemical market (78).

U.S. corporate investment in agricultural research has been high in the last few years. Many of the firms that have invested in plant biotechnology are chemical firms, especially firms that produce agricultural chemicals. The investment may be a response to a potential decrease in the agricultural chemical market due to the development of plants not needing chemicals (e.g., nitrogen-fixing plants, pest-resistant plants), the development of biological pesticides, or the development of plants with enhanced responses to chemicals. Another major industrial sector investing in plant biotechnology in the United States is the petroleum industry. The firms in this sector may see plants as the next source of energy, either in the form of biomass or photosynthesis itself. Pharmaceutical and food companies also are investing in plant agriculture. How the large chemical and petroleum corporations, the existing seed companies, and the NBFs will compete for market shares is yet to be seen.

The seed and vegetative cutting market is very large, and it appears that U.S. companies are oriented mainly toward domestic markets because of the transportation costs and the expense and inconvenience of field trials in other countries. Probably because of large domestic markets, many new entrepreneurial firms are directing their efforts toward plant agriculture. In fact, the number of NBFs in plant agriculture is third only to the number in pharmaceuticals and animal agriculture (see *Chapter 4: Firms Commercializing Biotechnology*).

Priorities for future research

Animal agriculture

The prospects for the application of biotechnology in the areas of animal and plant agriculture are truly exciting. To encourage the introduction and progress of biotechnology in animal agriculture, however, several persistent problems must be overcome. These problems include the following:

- developing effective delivery systems for almost all products of new biotechnology to be used in animals;
- achieving consistent expression of polypeptides such as those used for subunit vaccines from rDNA systems;
- developing host/vector systems that yield products more closely resembling mammalian molecules (e.g., glycosylated proteins) and that secrete products for easier purification.
- demonstrating product stability under the climactic and handling conditions where these products (e.g., subunit vaccines) will be implemented; and
- achieving higher immune responses with subunit vaccines, for example, by developing delivery systems that prolong exposure to the vaccine.

More basic knowledge about biological processes in animals and about the cellular and molecular biology of pathogenic bacteria and animal parasites is required before many biotechnological applications are realistic. Advances in basic knowledge about metabolic pathways in beneficial bacteria may lead to useful growth-enhancing compounds. Finally, more basic knowledge concerning the actions of nascent products such as rDNA-produced GH is needed to discern effectiveness and safety.

Given the novelty of disciplines such as molecular genetics and cellular biology in animal science, there is some question as to whether sufficient communicative links are established yet between basic and applied scientists. The efforts of applied scientists usually are communicated to animal growers in the United States through the land grant universities' State Agricultural Experiment Stations and extension services, supported by USDA. A corollary to the productiveness of future research rests in encouraging the establishment of communication between basic and applied scientists to encourage biotechnological applications in animal agriculture.

*Plant agriculture**

Because interest in plant molecular biology is fairly recent, the most important research priority is an increased understanding of DNA structure

*Research goals similar to those outlined in this section were published recently by the National Research Council (83) and the National Academy of Sciences (82).

and gene expression in plants.* The knowledge generated from investigations of DNA sequences and their functions will be essential to the use of biotechnology in crop improvement, although the initial contributions of biotechnology will not be in crop improvement but in acquiring a better understanding of the basic biology of plants.

It is unlikely that results from laboratory "model" species can be extrapolated to agriculturally important crop plants. Therefore, research is needed for improving and understanding the laboratory culture conditions for cells from these important plants. These plants must be able to be regenerated from single cells on a routine basis before many experiments using novel biological techniques can be performed. Much more work needs to be done before any plant cell vector can be used routinely. Additionally, a continued search for vectors for monocots is necessary if rDNA technology is to have an impact on some of the most important crop species.

It also is important to develop better selection methods. For instance, it is essential to be able to determine rapidly which cells carry specific genes and whether or not those genes are acting appropriately.

Both basic and applied research efforts in improved plant characteristics are quite active. The economic impact of finding disease or environmental resistances in the near term are potentially great enough that this research area is the primary thrust of many of the new plant genetics companies in the United States. Considerable effort continues in universities, as well, although overall funding for the university effort probably is much less than that represented by the current industrial effort. For many desirable traits, the actual protein product of the gene is not known. Cloning and genetic analysis of such genes would greatly increase the knowledge of what kinds of proteins are involved in disease and other resistances. Other improvements in specific plant characteristics may be made by modifying genes in major crop plants or by the introduction of novel traits from other plants. Both approaches warrant investigation.

Plants are known to produce a variety of secondary metabolites that have either pharmaceutical or agricultural uses, yet little is known about the genetic regulation of their production or the development of culture systems for optimal production. Better understanding of these areas could lead to the production of new, improved, or less expensive drugs and compounds that attract or repel insects for controlling weeds and pests.

Goals for improved biological nitrogen fixation include extending nitrogen-fixing bacterial systems to a wider variety of plants, transferring the bacterial nitrogen-fixing genes to plants, and making existing nitrogen-fixing systems more efficient. Genetic studies will reveal how nitrogen-fixing genes are regulated, including how they respond to environmental levels of nitrogen and oxygen.

The extension of any of the nitrogen-fixing sysems depends partly on understanding more about survival and competition of nitrogen-fixing bacteria in field conditions. Temperature extremes, nutrient and pH status of soils, and presence of other micro-organisms are factors that influence colonization of host plants. Reliable, analytical descriptions of the field ecology and physiology of nitrogen-fixing organisms are needed.

Much basic biology of microbial insecticides is yet to be understood. In order to determine the appropriate strategy for their use, it is necessary to study the influence of such factors as sunlight, temperature, rain, and relative humidity on the micro-organisms. Additionally, little is known about the mode and schedule of application and dose required for effective use of micro-organisms in the field. Criteria established by EPA require an analysis of the pathogen's possible effect on human and animal health and the environment.

Even with the lack of biological knowledge currently, it is possible to apply the techniques of biotechnology to the field of microbial insecticides. Approaches include the development of more potent strains, an increase in their tolerance

*These priorities only cover the techniques discussed in this report. It should be noted, though, that genetic advances and applications are dependent on concurrent research in plant biochemistry and physiology.

to environmental stresses, and an extension of their host ranges. The cloning of the *Bacillus* toxin gene, for example, opens up possibilities for the genetic manipulation of this gene to produce a more potent toxin and for the transfer of the gene to other micro-organisms.

The virus AcNPV currently is well enough characterized that its use as a vector is now possible. Some ideas for genetic manipulation include the introduction of insect-specific toxins and broadening the host range of the virus. The use of fungal insecticides requires a better understanding of the physiology, genetics, and pathogenicity of the genes that code for these insecticides. This understanding should lead to the development of strains with increased virulence and greater ease of production in culture (81).

Plants are capable of producing insecticides, yet little is known about their biosynthesis or mode of action. Further research on this topic would allow for more specific, effective, and environmentally sound insecticides.

Because of the complexity of the photosynthetic system, more basic research is needed on the enzymatic processes of photosynthesis and their regulation and compartmentalization. Photosynthesis is used for the production of carbohydrates, and understanding how these compounds are partitioned throughout the plant may allow the ability to direct them into the harvestable parts of the plant.

Finally, knowledge concerning the ecological results of growing plants more densely or of growing plants on marginal land is scant. More research is needed on soil and water use and mineral cycling plants.

Chapter 6 references*

Animal agriculture references

1. Anderson, R. A., Krakauer, T., and Camerini-Otero, R. D., "DNA-Mediated Gene Transfer: Recombination Between Co-Transferred DNA Sequences and Recovery of Recombinants in a Plasmid," *Proc. Natl. Acad. Sci. (U.S.A.)* 79:2748-2752, 1982.
2. Bachrach, H. L., "Recombinant DNA Technology for the Preparation of Subunit Vaccines," *J. Am. Vet. Med. Assoc.* 181:992-999, 1982.
3. Bittle, J. L., Houghten, R. A., Alexander, H., et al., "Protection Against Foot-and-Mouth Disease by Immunization With a Chemically Synthesized Peptide Predicted From the Viral Nucleotide Sequence," *Nature* 298:30-33, 1982.

*4. Board of Science and Technology for International Development, et al., "Priorities in Biotechnology Research for International Development—Proceedings of a Workshop" (Washington, D.C.: National Academy Press, 1982).

5. Boothroyd, J. C., Highfield, P. E., Cross G. A. M., et al., "Molecular Cloning of Foot-and-Mouth Disease Virus Genome and Nucleotide Sequences in the Structural Protein Genes," *Nature* 209(5809): 800-802, 1981.
6. Brinster, R. L., Chen, H. Y., and Trumbauer, M. E., "Mouse Oocytes Transcribe Injected *Xenopus* 5S RNA Gene," *Science* 211:396-398, 1981.
7. Brinster, R. L., Chen, H. Y., Warren, R., et al., "Regulation of Metallothionien-Thymidine Kinase Fusion Plasmid Injected Into Mouse Eggs," *Nature* 296:39-42, 1982.
8. Campbell, W. C., Fisher, M. H., Stapley, E. O., et al., "Ivermectin: A Potent New Antiparasitic Agent," *Science* 221:823-828, 1983.
9. Della-Porta, A. J., "Current Status of Foot-and-Mouth Disease Vaccines Including the Use of Genetic Engineering," *Australian Veterinary Journal* 60:129-135, 1983.
10. Demain, A. L., "New Applications of Microbial Products," *Science* 219:709-714, 1983.
11. Goodenow, R. S., McMillan, M., Oern, A., et al., "Identification of a Balb/c H-2Ld Gene by DNA-Mediated Gene Transfer," *Science* 215:677-679, 1982.
12. Gordon, J. W., and Ruddle, F. H., "Integration of Stable Germ Line Transmission of Genes Injected Into Mouse Pronuclei," *Science* 214:1244-1246, 1981.
13. Herd, R., "Animal Health and Public Health Aspects of Bovine Parasitism," *J. Am. Vet. Med. Assoc.* 176:737-743, 1980.
14. Kleid, D. G., Yansura, D., Small, B., et al., "Cloned

*References of general interest are marked with asterisks.

Viral Protein Vaccine for Foot-and-Mouth Disease Responses in Cattle and Swine," *Science* 214 (4525):1125-1129, 1981.

15. Kupper, H., Keller, W., Kura, C., et al., "Cloning of cDNA of Major Antigen of Foot and Mouth Disease Virus and Expression in *E. coli," Nature* 289: 555-559, 1981.
16. McCauley, E. H., "Animal Diseases in Developing Countries: Technical and Economic Aspects of Their Impact and Control," *World Bank AGR Technical Note No. 7,* Washington, D.C., 1983.
17. McCauley, E. H., Big Timber, Mont., personal communication, 1983.
18. McNabb, P. C., and Tomasi, T. B., "Host Defense Mechanisms at Mucosal Surfaces," *Ann. Rev. Microb.* 33:477-496, 1981.
19. Melchers, F., Potter, M., and Warner, N. L. (eds.), "Lymphocyte Hybridomas," *Curr. Top. Microbiol. Immunol.,* vol. 81 (New York: Springer-Verlag, 1978).
20. Morris, R. S., University of Minnesota, College of Veterinary Medicine, personal communication, 1983.
21. Muir, L., Merck & Co., Rahway, N.J., personal communication, June 1983.
22. Muscoplat, C., Molecular Genetics, Inc., Minnetonka, Minn., personal communication, 1983.
23. Palmiter, R. D., Chen, H. Y., and Brinster, R. L., "Differential Regulation of Metallothionein-Thymidine Kinase Fusion Genes in Transgenic Mice and Their Offspring," *Cell* 29:701-710, 1982.
24. Peel, C. J., Bauman, D. E., Gorewit, R. C., et al., "Effect of Exogenous Growth Hormone on Lactational Performance in High Yielding Dairy Cows," *J. Nutr.* 111:1662-1671, 1981.
25. Roberts, J. M., and Axel, R., "Gene Amplification and Gene Correction in Somatic Cells," *Cell* 29:109-120, 1982.
26. Shapiro, S. Z., and Murray, M., "African Trypanosome Antigens Recognized During the Course of Infection in N'dama and Zebu Cattle," *Inf. Immunol.* 35:410-416, 1982.
27. Stewart, T. A., and Mintz, B. A., "Successive Generations of Mice Produced From an Established Cultured Line of Enploid Teratocarcinoma Cells," *Proc. Natl. Acad. Sci. (U.S.A.)* 78:6314-6318, 1981.
28. Sutcliffe, J. G., Shinnick, T. M., Green, N., et al., "Antibodies That React With Predetermined Sites on Proteins," *Science* 219:660-666, 1983.
29. Trischman, T. M., and Bloom, B. R., "Genetics of Murine Resistance to *Trypanosoma cruzi," Inf. Immunol.* 35:546-551, 1982.
30. U.S. Congress, Office of Technology Assessment, *Drugs in Livestock Feed,* OTA-F-91, Washington, D.C., June 1979.

*31. U.S. Congress, Office of Technology Assessment, *Impacts of Applied Genetics: Micro-Organisms, Plants, and Animals,* OTA-HR-132, Washington, D.C., April 1981.

32. Wagner, E. F., Stewart, T. A., and Mintz, B., "The Human -Globin Gene and a Functional Viral Thymidine Kinase Gene in Developing Mice," *Proc. Natl. Acad. Sci. (U.S.A.)* 78:5016-5020, 1981.
33. Wagner, E. F., and Mintz, B., "Transfer of Nonselectable Genes Into Mouse Teratocarcinoma Cells and Transcription of the Transferred Human β-Globin Gene," *Mol. Cell. Biol.* 2:190-198, 1982.
34. Yelverton, E., Norton, S., Obijeski, J. F., et al., "Rabies Virus Glycoprotein Analogs: Biosynthesis in *Escherichia coli," Science* 219:614-620, 1983.
35. Zimmer, S. J., and Emmitt, R. B., "Industry Report: Animal Health Products Market" (New York: F. Eberstadt & Co., Inc., 1981).
36. Zimmer, S. J., "The Impacts of Applied Genetics in Animal Agriculture," contract report prepared for the Office of Technology Assessment, U.S. Congress, August 1982.

Plant agriculture references

37. Agrigenetics Corp., press release, Boulder, Colo., Aug. 12, 1983.
38. Arntzen, C. J., and Duesing, J. H., "Chloroplast-Encoded Herbicide Resistance," abstract, *Fifteenth Miami Winter Symposium,* Miami, Jan. 17-21, 1983.

*39. Barton, K. A., and Brill, W. J., "Prospects in Plant Genetic Engineering," *Science* 219:671-676, 1983.

40. Bell, E. A., and Charlwood, B. V. (eds.), *Secondary Plant Products* (New York: Springer-Verlag, 1980).
41. Bollinger, W. H., NPI, Salt Lake City, Utah, personal communication, 1983.
42. Boyer, J. S., "Plant Productivity and Environment," *Science* 218:443-448, 1982.
43. Bulla, L. A. (ed.), "Regulation of Insect Populations by Microorganisms," *Ann. N.Y. Acad. Sci.,* vol. 217, 1973.
44. Burges, H. D. (ed.), *Microbial Control of Pests and Plant Diseases, 1970-1980* (New York: Academic Press, 1981).
45. Canvin, D. T., "Photorespiration: Comparison Between C_3 and C_4 Plants," *Photosynthesis II (Encyclopedia of Plant Physiology),* New Series, vol. 6, M. Gibbs and E. Latzko (eds.) (New York: Springer-Verlag, 1979).

46. Chaleff, R. S., *The Genetics of Higher Plants: Applications of Cell Culture* (New York: Cambridge Press, 1981).
47. Chaleff, R. S., "Isolation of Agronomically Useful Mutants From Plant Cell Cultures," *Science* 219: 676-682, 1983.
48. *Chemical Week*, "The Coming Revolution in Agricultural Chemicals," June 15, 1983, pp. 86-90.
49. Chilton, M. D., DeFramond, A., Fraley, R., et al., "Ti and Ri Plasmids as Vectors for Genetic Engineering of Higher Plants," abstract, *Fifteenth Miami Winter Symposium,* Miami, Jan. 17-21, 1983.
50. Chua, N-H., and Schmidt, C. W., "Transport of Proteins Into Mitochondria and Chloroplasts," *J. Cell Biol.* 81:461-483, 1979.
51. Coe, E. H., Jr., and Neuffer, M. G., "Corn Genetics," *Corn and Corn Improvement,* G. F. Sprague (ed.) (Madison, Wis.: American Society of Agronomy, 1977).
52. Coen, D., Bedbrook, J. R., Bogorad, L., et al., "Maize Chloroplast DNA Fragment Encoding the Large Subunit of Ribulose Bisphosphate Carboxylase," *Proc. Natl. Acad. Sci. (U.S.A.)* 74:5487-5491, 1977.
53. Conger, B. V., "Agronomic Crops," *Cloning Agricultural Plants via In Vitro Techniques,* B. V. Conger (ed.) (Boca Raton, Fla.: CRC Press, 1981).
54. Day, P. R., *Genetics of Host Parasite Interaction* (San Francisco: W. H. Freeman & Co., 1974).
55. Day, P. R., "Long Term Goals in Argicultural Plant Improvement," *Genetic Manipulation: Impact on Man and Society,* W. Arber, K. Illmensee, W. J. Peacock, et al. (eds.) (Cambridge, England: Cambridge University Press, 1983).
*56. DeYoung, H. G., "Crop Genetics: The Seeds of Revolution," *High Technology* 3:53-59, June 1983.
57. Dougall, D. K., "Tissue Culture and the Study of Secondary (Natural) Products," *The Biochemistry of Plants,* vol. 7 (New York: Academic Press, 1981).
58. Dougall, D., University of Tennessee, personal communication, 1983.
59. Dryden, R. N., "Farmer's Help," *Nature* 304:579, 1983.
60. *Economist,* "DIY Algae," July 2, 1983, pp. 76-77.
61. Ehleringer, J. R., "Photosynthesis and Photorespiration: Biochemistry, Physiology, and Ecological Implications," *HortScience* 14:217-222, 1979.
62. Evans, D. A., "Agricultural Applications of Plant Protoplast Fusion," *Bio/Technology,* May 1983, pp. 253-261.
63. Fraley, R. T., Horsch, R., and Rogers, S. G., UCLA-Keystone Meeting, Keystone, Colo., March 1983.
64. Garfinkel, D. J., Simpson, R. B., Ream, L. W., et al., "Genetic Analysis of Crown Gall: Fine Structure Map of the T-DNA by Site-Directed Mutagenesis," *Cell* 27:143-154, 1981.
65. Goldfarb, N., Calgene, Inc., Davis, Calif., personal communication, 1983.
66. Gressel, J., *Herbicide Resistance in Plants* (New York: John Wiley & Sons, 1982).
67. Groat, R. G., and Vance, C. P., "Root Nodule Enzymes of Ammonia Assimilation in Alfalfa (*Medicago Sativa*): Developmental Patterns and Response to Applied Nitrogen," *Plant Physiol.* 67:1198-1203, 1981.
68. Heiser, C. B., *Seeds to Civilization* (San Francisco: W. H. Freeman, 1973).
69. Held, G. A., Bulla, L. A., Jr., Ferrari, E., et al., "Cloning and Localization of the Lepidopteran Protoxin Gene of *Bacillus thuringiensis Subsp. Kurstaki,*" *Proc. Natl. Acad. Sci. (U.S.A.)* 79:6065-6069, 1982.
70. Herrera-Estrella, L., Depicker, A., VanMontagu, M., et al., "Expression of Chimaeric Genes Transferred Into Plant Cells Using a Ti-Plasmid-Derived Vector," *Nature* 303:209-213, 1983.
71. Holmes, F. O., "Genetics of Pathogenicity in Viruses and of Resistance in Host Plants," *Adv. Virus Res.* 11:139-161, 1965.
72. Ignoffo, C., "Possibilities of Mass-Producing Insect Pathogens," *Insect Pathology and Microbial Control* (Amsterdam: North Holland Press, 1967).
73. Kemp, J., UCLA-Keystone Meeting, Keystone, Colo., March 1983.
74. Larkin, P. J., and Scowcroft, W. R., "Somaclonal Variation: A Novel Source of Variability From Cell Cultures for Plant Improvement," *Theor. Appl. Genet.* 60:197-214, 1981.
75. Leemans, J., Shaw, C., DeBlaere, R., et al., "Site-Specific Mutagenesis of *Agrobacterium* Ti Plasmids and Transfer of Genes to Plant Cells," *J. Molec. Appl. Genet.* 1:149-164, 1981.
76. Legocki, R., and Verma, D. P. S., "Identification of 'Nodule-Specific' Host Proteins (Nodulins) Involved in the Development of *Rhizobium*-Legume Symbiosis," *Cell* 20:153-163, 1980.
77. Maier, R. J., Campbell, N. E. R., and Janus, F. J., "Expression of Hydrogenase Activity in Free-Living *Rhizobium japonicum,*" *Proc. Natl. Acad. Sci. (U.S.A.)* 75:3258-3262, 1978.
*78. Marx, J., "Can Crops Grow Without Added Fertilizer?" *High Technology,* March-April 1982, pp. 62-66.

79. Matthews, R. E. F., *Plant Virology* (New York: Academic Press, 1981).
80. Miller, J. A., "Microbial Antifreeze: Gene Splicing Takes to the Field," *Science News* 124:132, 1983.
81. Miller, L. K., Lingg, A. J., and Bulla, L. A., "Bacterial, Viral, and Fungal Insecticides," *Science* 219:715-721, 1983.
*82. National Academy of Sciences, Committee on Science, Engineering, and Public Policy, "Report of the Briefing Panel on Agricultural Research" (Washington, D.C.: National Academy Press, 1983).
*83. National Research Council (BOSTID), *Priorities in Biotechnology Research for International Development: Proceedings of a Workshop* (Washington, D.C.: National Academy Press, 1982).
84. Paau, A. D., Leps, W., and Brill, W., "Agglutinin From Alfalfa Necessary for Binding and Nodulation by *Rhizobium melilote*," *Science* 213:1513-1515, 1981.
85. Padwa, D., Agrigenetics Corp., Boulder, Colo., personal communication, 1983.
86. Patrusky, B., "Plants in Their Own Behalf," *Mosaic,* March/April 1983, pp. 33-39.
87. Peacock, W. J., "Gene Transfer in Agricultural Plants," abstract, *Fifteenth Miami Winter Symposium,* Jan. 17-21, 1983.
88. Peters, G. A., Ray, T. B., Mayne, B. C., et al., "Azolla-Anabaena Association: Morphological and Physiological Studies," *Nitrogen Fixation: Symbiotic Associations and Cyanobacteria,* W. E. Newton and W. H. Orme-Johnson (eds.), vol. 2 (Baltimore: University Park Press, 1980).
89. Pfister, K., Steinback, K. E., Gardner, G., et al., "Photoaffinity Labeling of an Herbicide Receptor Protein in Chloroplast Membranes," *Proc. Natl. Acad. Sci. (U.S.A.)* 78:981-985, 1981.
90. Putnam, A. R., "Allelopathic Chemicals," *Chem. & Eng. News,* Apr. 4, 1983, pp. 34-45.
91. Schell, J., VanMontagu, M., Hernalsteens, J. P., et al., "Ti Plasmids as Experimental Gene Vectors for Plants," abstract, *Fifteenth Miami Winter Symposium,* Miami, Jan. 17-21, 1983.
92. Schroth, M. N., and Hancock, J. G., "Disease-Suppressive Soil and Root Colonizing Bacteria," *Science* 216:1376-1381, 1982.
93. Schubert, K. R., "The Energetics of Biological Nitrogen Fixation," workshop summary, American Society of Plant Physiologists, Rockville, Md., 1982.
94. Shephard, J. F., Bidney, D., Barsby, T., et al., "Genetic Transfer in Plants Through Interspecific Protoplast Fusion," *Science* 219:683-688, 1983.
95. Shuler, M. L., "Production of Secondary Metabolites From Plant Tissue Culture—Problems and Prospects," *Biochemical Engineering II,* A. Constantinides, W. R. Vieth, K. Venkatasubramanian (eds.) *Annals of the New York Academy of Sciences,* vol. 369, 1981.
96. Sullivan, D., Brisson, N., Goodchild, B., et al., "Molecular Cloning and Organization of Two Leghaemoglobin Genomic Sequences of Soybean," *Nature* 289:516-518, 1981.
97. Sun, S., ARCO Plant Cell Research Institute, Dublin, Calif., personal communication, 1982.
98. Upham, S. K., "Tissue Culture: An Exciting New Concept in Potato Breeding," *Spudman,* November 1982 pp. 14-19.
99. U.S. Congress, Office of Technology Assessment, *Plants: The Potential for Extracting Protein, Medicines, and Other Useful Chemicals,* OTA-BP-F-23, Washington, D.C., September 1983.
100. U.S. Department of Agriculture, "Costs of Producing Selected Crops in the United States," Economic Research Service (USDA) for the Committee on Agriculture, Nutrition, and Forestry, U.S. Senate, Washington, D.C., June 1982.
101. Walbot, V., and Long, S., "Prospects for Plant Agricultural Improvement and Tissue Culture," contract report prepared for the Office of Technology Assessment, U.S. Congress, September 1982.

Chapter 7

Specialty Chemicals and Food Additives

Contents

Table

Figures

Chapter 7

Specialty Chemicals and Food Additives

Introduction

In the production of specialty chemicals, defined in this report as chemicals whose price exceeds \$1/lb (50¢/kg) in cost, there are many potential applications of biotechnology.* The nearest term applications are in the production of specialty chemicals that are already produced by processes using micro-organisms, e.g., amino acids and enzymes. Enzymes are the direct products of genes, so their production is particularly accessible with new genetic technologies.

A number of specialty chemicals are chemically synthesized. Some, including some vitamins, are synthesized chemically from petrochemicals. Others, including fatty acids and steroids, are synthesized chemically from naturally occurring compounds. Current chemical synthesis production processes often require large energy inputs, have complicated synthesis steps, and yield many byproducts. Potentially, some of the steps in current chemical synthesis processes could be replaced by biological steps catalyzed by enzymes. Enzymes that perform some of the necessary conversions in a very specific manner and with small energy inputs are already known. If appropriate microbial enzymes (or higher organism enzymes) were identified and characterized, the appropriate genetic information could be cloned and expressed fairly rapidly in well-studied micro-organisms to produce or modify compounds such as vitamins, lipids, steroids, and aromatic chemicals. Alternatively, a chemical synthesis production process might be replaced entirely by a biological process if a micro-organism were identified that performed the synthesis. Both individual enzymes and biosynthetic pathways consisting of several enzymes can be manipulated genetically to increase production.

Finally, it should be noted that there are some specialty chemicals synthesized in nature, such as complex polysaccharides, for which chemical synthesis is not feasible. Improving the syntheses of these specialty chemicals in controlled microbial processes is beginning to be investigated.

This chapter discusses the applications of biotechnology to the production of specialty chemicals. It also discusses applications to the production of animal feed and human food additives, because many of the genetic techniques applicable to the production of specialty chemicals also apply to the production of such additives. Since the main difference between specialty chemicals and food additives is in the Food and Drug Administration's (FDA's) regulatory approval process for food and food additives, food additives are discussed here as a subset of specialty chemicals.**

The several kinds of products that could be produced using biotechnology, which are discussed in this chapter, are only representative of the large range of products that could be synthesized using biotechnology. The specialty chemicals and food additives market is extremely broad, and many other applications of biotechnology to the production of such products may be evident in the future.

*The application of biotechnology to the production of commodity chemicals, defined in this report as chemicals that sell for less than \$1 per pound, is discussed in *Chapter 9: Commodity Chemicals and Energy Production.*

**FDA's regulatory approval processes are discussed in *Chapter 15: Health, Safety, and Environmental Regulation.*

Amino acids

In 1982, the worldwide sales volume of amino acids was 455,000 metric tons (tonnes) valued at \$1.15 billion (see fig. 17), and an annual growth rate of 7 to 10 percent is expected during the remainder of this decade. The world markets for amino acids are currently dominated by Japanese

Figure 17.—Uses of Amino Acids

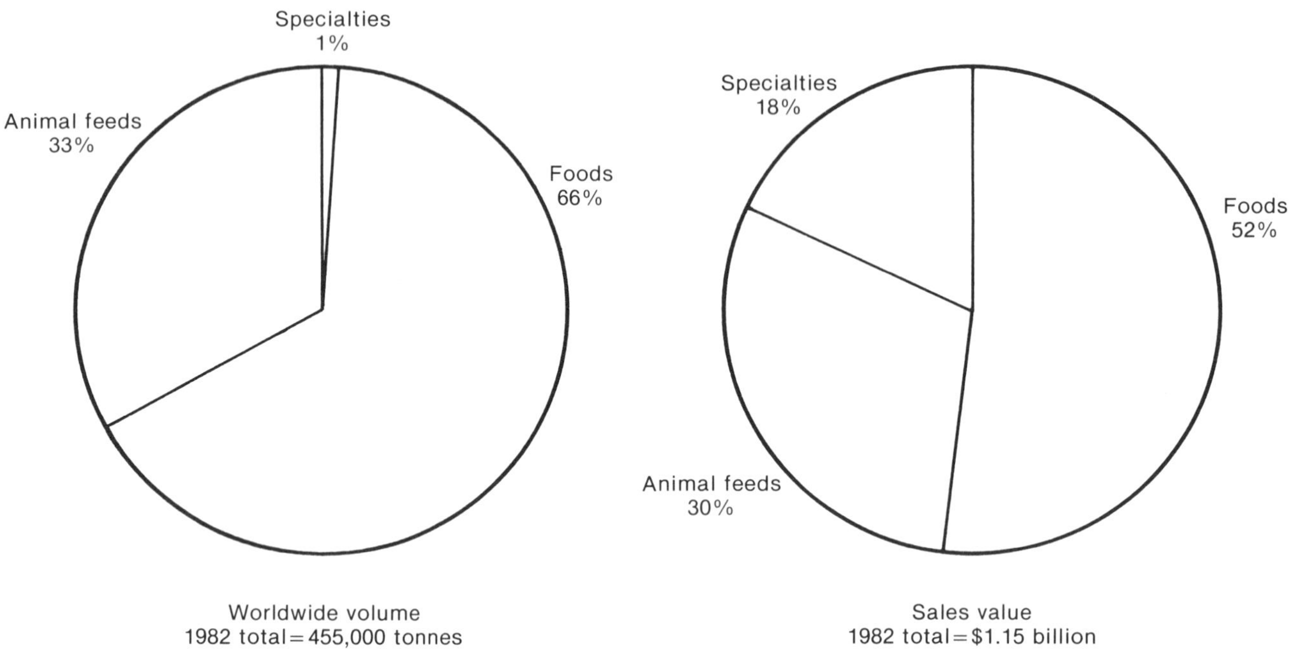

SOURCE: Office of Technology Assessment, adapted from P. L. Layman, "Capacity Jumps for Amino Acids," *Chem. & Eng. News*, Jan. 3, 1983.

producers, the largest of which is Ajinomoto. Amino acid production in the United States, however, is beginning to expand. W. R. Grace is planning to use a new plant in Maryland to produce pharmaceutical-grade amino acids, and two Japanese producers, Ajinomoto and Kyowa Hakko, are opening plants in the United States (47).

Amino acids have traditionally been used as animal feed and human food additives, and their use as animal feed additives may increase as other proteinaceous feedstuffs become more expensive. Recently, there has been increased use of pharmaceutical-grade amino acids for enteral and intravenous feeding solutions. Important constituents of these feeding solutions are the essential amino acids, those that the human body cannot make. Leading U.S. manufacturers of such solutions are Abbott Labs, Baxter Travenol, and American Hospital Supply (30). As shown in figure 17, the specialty market accounts for only 1 percent of world volume of amino acid production, but amounts to 18 percent of the sales value. The production of pharmaceutical-grade amino acids using biotechnology is receiving attention from both U.S. and Japanese companies (28,47).

Glutamic acid

The largest world market for an amino acid is the market for glutamic acid; the sodium salt of glutamic acid, monosodium glutamate (MSG), is used as a food additive. On the order of 300,000 tonnes of glutamic acid are produced annually worldwide (23,25). Approximately 30,000 tonnes are used in the United States, and about one-half of U.S. needs are met through imports at a price of about $2/kg (10).

MSG is produced by an efficient bioprocess using a strain of *Corynebacterium*. This strain was first isolated, on the basis of the microorganism's ability to synthesize and excrete glutamic acid, by the Japanese in the late 1950's. Reports through the Japanese patent literature indicate that Ajinomoto, the world's leading MSG manufacturer, is applying recombinant DNA (rDNA) techniques to *Corynebacterium* strains in an effort to improve glutamic acid production.*

*Strains of *Corynebacterium* are used extensively in Japan for synthesis of several amino acids, but the Japanese bioprocess industry did not do basic research with these bacteria until recently. However, patents and reports in the literature indicate that Japanese amino acid producing firms have begun application of rDNA tech-

Methionine

Another large market for amino acids is in animal feeds (47). Typical corn/soybean animal feeds have low concentrations of the amino acids methionine and lysine, so their nutritive value in animal diets is limited. Methionine and lysine (see below), therefore, are widely used as animal feed additives. Two companies in the United States, Monsanto and the U.S. affiliate of the West German firm Degussa, produce feed-grade methionine using a chemical process (9). Because this process is quite inexpensive, it is not likely that competitive biological routes to methionine production will be developed in the near future.

Lysine

The production of the amino acid lysine is dominated by three Japanese producers, Ajinomoto, Kyowa Hakko, and Toray Industries (11), which together account for 90 percent of the world market. Manufacturers' prices are variable, generally in the range of $3 to $4/kg for feed-grade lysine (43). The United States imports all its lysine (67) and in 1981 imported approximately 11,000 tonnes (68). A plant for lysine production is being built in Cape Girardeau, Mo., by the Japanese manufacturer Kyowa Hakko, and it is projected that the plant's initial production of lysine will be 7,500 tonnes per year (40).

Most lysine is produced in a bioprocess using mutant strains of *Corynebacterium*. A substantial increase in lysine production and a corresponding decrease in cost can be expected to result from applying rDNA techniques to these bacteria (30). In production processes, *Corynebacterium* mutants already yield large amounts of lysine from a crude carbon source such as molasses (45). Amplification of lysine biosynthetic enzymes in these bacteria through gene cloning should result in an increased synthesis rate and amount.

Tryptophan

The amino acid tryptophan is the second limiting essential amino acid in corn and the third limiting essential amino acid in combination feeds for swine and poultry (58). Although tryptophan would seem to be a prime candidate for the animal feed supplement business, marketing analyses have shown that the cost of tryptophan would have to be reduced to the $10/kg range (i.e., about three times the cost of lysine) in order to interest feed formulators in its use (47). The current cost of tryptophan, $95/kg, makes its addition to animal feeds out of the question at this time.

The development of efficient bioprocesses for tryptophan production using either modified *Corynebacterium* or enterobacteria (intestinal bacteria) such as *Escherichia coli* could potentially lower tryptophan costs. The current level of understanding of the *E. coli* aromatic amino acid pathway and sophisticated rDNA techniques that are available should facilitate strain construction in the enterobacteria. As for constructing a tryptophan-producing *Corynebacterium,* basic understanding of the synthetic pathway and development of a vector system remain to be achieved. Manipulating any micro-organism to produce tryptophan efficiently may be difficult, however, because the synthesis of tryptophan requires a greater expenditure of energy than does that of any other amino acid (1). The yield of tryptophan from a given carbon source, therefore, will be lower than the yield for other amino acids. The yield of product from glucose is an important factor in determining production cost in a bioprocess. Information concerning production cost improvements made by the Japanese companies now manufacturing tryptophan is not available.

Progress has been made in developing a two step enzymatic process for tryptophan production (32). This approach requires three substrates: glycine, formaldehyde, and indole. The high levels of the two enzymes required for this process are obtained by cloning and amplifying each of the genes for these enzymes. This process has not yet been commercialized, but is being investigated by the new biotechnology firm (NBF)* Genex (U.S.). Commercialization requires that the three substrates be priced low enough to meet the target price for tryptophan. Another enzymatic process for the production of tryptophan has been devel-

niques to *Corynebacterium.* Genex and W. R. Grace also have research programs to develop genetic techniques for these bacteria (30).

*NBFs, as defined in *Chapter 4: Firms Commercializing Biotechnology,* are firms that have been started up specifically to capitalize on new biotechnology.

oped by Mitsui Toatsu Chemicals. Commercial production of tryptophan by this Japanese firm was due to begin in January 1983 (7,12).

The relative costs of corn and soybean meal influence the use of these products as animal feed additives. As the price of soybean meal, the main source of protein, and thus amino acids, in poultry and swine feeds rises relative to feed-corn prices, as is expected during the 1980's, there will be a tendency to use less soybean meal in animal diets if less expensive feedstuffs are available. A reduction in lysine production cost and a substantial reduction in tryptophan cost could result in increased incorporation of these amino acids in animal diets as a substitute for proteinaceous soybean meal.

Aspartic acid

Innovative processes for amino acid production that involve immobilization of whole cells or enzymes for bioconversion of precursors to amino acids are being developed (30). In the case of aspartic acid, a constituent of the sugar substitute aspartame, an immobilized process has reduced the costs of production. An early process for aspartic acid production involved the enzyme aspartase in a one-step batch reaction. The life of the catalyst in this process was, at most, a few days. When the enzyme aspartase was immobilized and a continuous-flow process was developed, a 40-percent saving in aspartic acid production cost was realized (14). The life of enzymes in immobilized systems can be increased many fold, up to several months. Cost savings are due to reductions in the amount of catalyst required, in the size of equipment used, and in the labor needed to operate the system.

Phenylalanine

The demand for the amino acids aspartic acid and phenylalanine as components of the sugar substitute aspartame has spurred process development. Aspartic acid is already available at an attractive price, and the research described below will make reasonably priced phenylalanine available soon (30). Phenylalanine, like tryptophan, requires large amounts of energy for the microbial cell to make. However, it should be possible to genetically manipulate enterobacteria or *Corynebacterium* strains to overproduce phenylalanine, thereby making the process economic.

A group of Australian scientists at the University of New South Wales, Kensington, is constructing *E. coli* mutants to overproduce phenylalanine in either a batch or continuous-flow bioprocess (15). No report of the commercialization of their process has been made. Amino acid producers in Japan (Ajinomoto and Kyowa Hakko) may also be applying rDNA techniques to improve phenylalanine production by their *Corynebacterium* strains in order to reduce phenylalanine costs.

A single-step enzymatic process to produce phenylalanine for use in aspartame is being developed in the United States by Genex and in Japan by Tanabe Seiyaku (31,73). In this process, yeast cells that contain the enzyme phenylalanine ammonia lyase (PAL) are utilized. Under the appropriate conditions, PAL will catalyze the formation of phenylalanine from cinnamic acid and ammonia. The economics of the PAL process are very sensitive to the cost of the major raw material, cinnamic acid, which is currently rather expensive. Recovery of phenylalanine from the PAL process, however, will be much more straightforward than recovery from the complex broth that results from a batch bioprocess. High recovery yields in the PAL process may offset the disadvantage of a more expensive raw material.

Enzymes

Enzymes are proteins whose function in living systems is to catalyze the making and breaking of chemical bonds. They have been used commercially since the 1890's, when fungal cell extracts were first added to brewing vats to facilitate the breakdown of starch into sugar. The size of the world industrial enzyme market for 1981 was estimated to be 65,000 tonnes at a value of $400 million. A growth rate resulting in 75,000 tonnes valued at $600 million has been predicted for the end of 1985. Fewer than 20 enzymes comprise the large majority of this market. Economic sources of enzymes include a limited number of plants and animals and a few species of microorganisms (33).

The enzyme industry is dominated by two European companies, Novo Industri (Denmark) and Gist-Brocades NV (Netherlands), which together have about 65 percent of the current world market (25). Other companies marketing or planning to market large volume enzymes include CPC International (U.S.), ADM (a division of Clinton, U.S.), Miles (U.S.), Pfizer (U.S.), Dawi Kasi (Japan), Alko (Finland), Finnish Sugar (Finland), and Rohm (a division of Henkel, F.R.G.).

The leading enzymes on the world market in terms of volume are the proteases, amylases, and glucose isomerase (25). Alkaline protease is added to detergents as a cleaning aid and is widely used in Western Europe. Trypsin, another type of protease, is important in the leather industry. Two amylases, alpha-amylase and glucoamylase, and glucose isomerase are corn-processing enzymes. The reactions catalyzed by these three enzymes represent the three steps by which starch is converted into high-fructose corn syrup (see fig. 18). Fructose is sweeter than glucose and can be used in place of table sugar (sucrose) in preparation of candy, bread, carbonated beverages, and in canning. Historically, the United States imported sugar, but with the commercial development of an economic process for converting glucose to fructose in the late 1960's, corn sweeteners have decreased the amount of sugar imported. About $1.3 billion in U.S. payments for sugar imports was saved in 1980 because of the domestic use of corn sweeteners (17).

The process for converting glucose to fructose is catalyzed by the enzyme glucose isomerase. Initially, the conversion was done using a batch reaction; in 1972, however, a continuous system using immobilized glucose isomerase was initiated (36). The immobilized glucose isomerase process represents the largest immobilized enzyme process used in production in the world. A large processing plant can convert 2 million pounds of corn starch into high-fructose corn syrup per day (19).

Because of expanded sales in, for example, the detergent and high-fructose corn syrup markets, demand for enzymes will increase. The application of rDNA techniques to microbial enzyme production is expected to facilitate the expansion of the enzyme industry (25). Additionally, enzymatic activities of higher organisms could be cloned into micro-organisms, also expanding the enzyme industry. The fact that enzymes are direct gene products makes them good candidates for improved production through rDNA technology. For example, a 500-fold increase in the yield of a ligase, used for connecting DNA strands in rDNA research, was obtained by cloning the gene for that enzyme on an *E. coli* plasmid vector (25). Several research enzymes now on the market are produced by micro-organisms modified using rDNA techniques. Some are restriction endonucleases used for cutting DNA, and others are DNA-modifying enzymes. Companies that market these enzymes include Bethesda Research Laboratories (U.S.), New England Biolabs (U.S.), P L Biochemicals (U.S.), and Boehringer Mannheim (F.R.G.) (30).

Recombinant DNA technology could potentially be used to increase glucose isomerase produc-

Figure 18.—Conversion of Starch Into High Fructose Corn Syrup (HFCS)

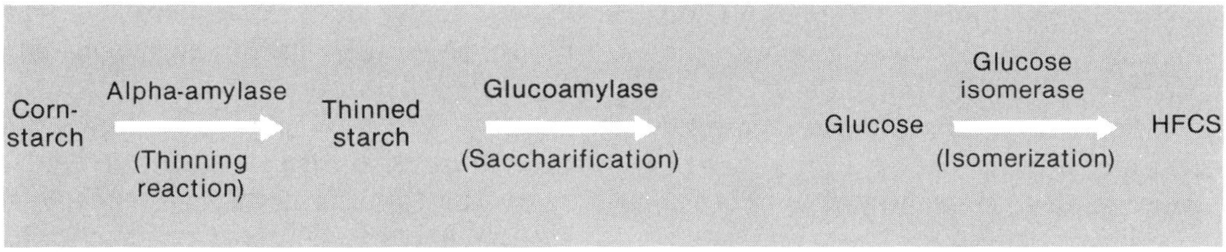

SOURCE: Office of Technology Assessment.

tion in micro-organisms and to improve the enzyme's properties. An improved glucose isomerase would have the following properties:

- a lower pH optimum to decrease the browning reaction caused by the alkaline pH now required;
- thermostability so that the reaction temperature can be raised, thus pushing the equilibrium of isomerization to a higher percentage fructose; and
- improved reaction rates to decrease production time.

Improvements in glucose isomerase will first come from the cloning of its gene into vectors and micro-organisms that have been developed for high production. It is also possible that screening a broad range of micro-organisms will yield enzymes with some improved properties. Finally, it will be possible in the future to identify the regions of the enzyme that are responsible for its various properties, such as pH optimum, and to direct changes in the gene structure to modify these properties.

Rennet is an enzyme that is essential to the cheese industry because of its milk-clotting properties. The world market for rennet from various sources is valued at approximately $64 million, over half of which is the more valuable calf rennet (25). The increasing scarcity of calf rennet has made this enzyme a very attractive candidate for gene cloning and subsequent production in a microbial bioprocess. The first announcement of the cloning of the rennet gene came from a Japanese scientist (53). Since then, it also has been cloned by four NBFs: the U.S. firms Genex, Collaborative Research (29), and Genencor (56), and the British firm Celltech (24,35). The first marketing of calf rennet produced by genetically manipulated bacteria is likely to occur in 1984 (30).

Enzymes, such as urokinase and streptokinase, are being used increasingly for treatment of human disorders. Their use and importance are discussed in *Chapter 5: Pharmaceuticals.* Many other enzymes are used for research and medical purposes in small quantities. Because rDNA technology potentially allows the construction of enzymes with improved stability and faster reaction rates, the use of enzymes industrially and medically could increase dramatically.

Vitamins

In 1981, the U.S. Department of Commerce reported that sales of vitamins for human use amounted to $1.1 billion (69). This market is expected to grow substantially over the next decade because of the current trend toward a more health- and nutrition-conscious population. A smaller but significant sector of the human vitamin market is for food processing and fortification.

Another important use of vitamins is in commercially prepared animal feeds. The vitamin content of natural feedstuffs is variable, so animal producers often supplement animal diets with vitamins. The U.S. market for vitamins as supplements in commercially prepared animal feed is large but is expected to increase an average of only 2.5 percent annually over the next decade (26) because of a decrease in the consumption of animal products.

Vitamins are either synthesized chemically or isolated from natural sources, and to date, biotechnology has had essentially no impact on vitamin production. Nevertheless, some opportunities do exist for reducing vitamin production costs using biotechnology. First, the cost of existing bioprocesses for vitamin production, such as that for vitamin B12, might be reduced by using a genetically manipulated micro-organism that synthesizes the vitamin in larger amounts at a higher rate. Second, some steps in a chemical synthesis might be replaced by biological steps, or the chemical synthesis might be replaced entirely by identifying micro-organisms able to synthesize particular vitamins. Once such microbes have been identified, vitamin synthesis can be enhanced by various biochemical, traditional genetic, and rDNA techniques. Finally, a micro-organism might be identified that produces a vitamin precursor. Such a micro-organism might then be genetically modified so that it would produce the vitamin itself by introducing a gene (or genes) that specifies an enzyme that would convert the precursor to the vitamin.

There are technical problems that introduce risks to research programs for new process developments for vitamins. One major problem is the dearth of information concerning vitamin biosynthetic pathways, especially in micro-organisms. Another problem is that any new biotechnology-based process will have to be very efficient to compete with the established chemical production methods.

Since vitamins are naturally occurring substances, they all have the potential for biotechnological production. The discussion below concentrates on vitamins B2, B12, C, and E to illustrate the range of biosynthetic pathways and potential problems for industrial production.

Vitamin B2

Riboflavin (vitamin B2) is known to be synthesized in small quantities by micro-organisms, but is manufactured primarily by chemical synthesis. The synthesis of riboflavin by the bacterium *Bacillus subtilis* has been studied extensively by a group of Soviet scientists (13), and strains of *B. subtilis* that overproduce and excrete riboflavin have been isolated (22). Because *B. subtilis* has been the subject of extensive studies by U.S. and European scientists, techniques such as DNA transformation, protoplast fusion, and gene cloning have been developed for this bacterium (21). The availability of such techniques should facilitate the construction of a strain of *B. subtilis* for the production of riboflavin.

Vitamin B12

Vitamin B12 is currently produced by a microbial bioprocess (27). The U.S. market for vitamin B12 is supplied both by U.S. and European firms. One U.S. company (Merck) supplies the major part of the feed-grade vitamin B12 market, while imports from Europe account for the major portion of the pharmaceutical grade (30). The current manufacturers' price for vitamin B12 is approximately $8,000/kg for pure material (9).*

Reducing the cost of vitamin B12 production will require genetic modifications of bacterial strains so that the micro-organisms synthesize vitamin B12 more efficiently. Vitamin B12 is one of the most complex molecules of living systems, however, and its biosynthetic pathway has not been definitively characterized.

Vitamin C

The U.S. market for vitamin C is very large, 17,500 tonnes in 1982 (30). Approximately two-thirds of this volume is supplied by U.S. producers, while the remaining third is imported. The current price of vitamin C is approximately $12/kg (9).

Although some of the synthesis of vitamin C is done microbially, efforts to replace other steps with bioconversions have not been successful (18). The synthesis of vitamin C has been reported in a few micro-organisms (50). The first step in developing a vitamin C bioprocess, therefore, will be screening for a potential production organism. Analysis of the biosynthetic pathway must be done, because little is known about microbial pathways for vitamin C synthesis. Once the rate-limiting steps of the pathway have been identified, rDNA techniques could possibly be used to increase production. A complicating factor in a vitamin C bioprocess is the fact that this vitamin, in solution, is readily oxidized when exposed to air. Controlling dissolved oxygen and complexing vitamin C with other compounds are two potential techniques for controlling the rate of vitamin breakdown during production. The wealth of unknowns makes it impossible at this time to predict a time frame for developing an improved vitamin C production process.

*The prices in this reference are for small volumes. The purchase of large quantities of these chemicals can result in a substantial price reduction.

Vitamin E

If an approach to natural vitamin E production using biotechnology could be developed, its impact would be quite significant. In 1979, approximately 3,200 tonnes of vitamin E were used in the United States (39). Of this amount, 700 tonnes were the natural form of vitamin E. The remaining 2,500 tonnes were synthetic forms. Synthetic vitamin E is a mixture of closely related compounds that vary in biological activity, whereas the natural vitamin preparation consists of only the most active compound. Demand for vitamin E as an antioxidant could increase the market for this vitamin by as much as 1,500 tonnes per year, depending on FDA's decisions concerning continued use of chemical antioxidants. The U.S. demand for natural vitamin E is met by two U.S. manufacturers, Eastman Chemicals and Henkel, and 95 percent of synthetic vitamin E is produced in the United States (30). The May 1983 price of the synthetic vitamin mixture was $27/kg (9). The price of the natural vitamin was several times that amount, depending on the activity of the preparation.

Natural vitamin E is now purified from vegetable oil by a process that involves several steps. If a one-step fermentation process could be developed based on a high-producing microbial strain, the manufacturing cost of natural vitamin E might be lowered substantially.

Blue-green algae are the only well-characterized micro-organisms that are known to produce vitamin E (20,55). It might be possible to increase vitamin E synthesis by altering the biosynthetic pathway in blue-green algae, but the biochemistry and physiology of this pathway is poorly understood, and gene cloning in these micro-organisms is at a rudimentary stage of development.

The discovery in bacteria, such as *E. coli, B. subtilis,* and *Pseudomonas,* of a compound that is potentially a vitamin E precursor suggests another route for vitamin E production (37). These bacteria are well-characterized species for which genetic transfer techniques are developed. Construction of a vitamin E-producing strain first would involve isolating mutants that overproduce the precursor. Then the genes for the enzyme that catalyzes the conversion of the precursor to vitamin E could be isolated from blue-green algae and introduced into the potential production strain. Although the savings in production cost of vitamin E could be great, this project involves a substantial amount of risk related to the lack of information concerning the biosynthesis of this vitamin. For example, it is not known if only one enzyme is needed for the conversion of precursor to vitamin, how complex such an enzyme is, how many genes encode it, and what cofactor requirements it might have.

Summary

Biotechnological techniques for improving the efficiency of vitamin production are similar to those being used in amino acid process development. The research and development (R&D) effort for vitamins will be more extensive than that for the amino acids, because vitamin biosynthetic pathways are more complex and less understood. In some instances, screening programs to identify micro-organisms with potential for producing a particular vitamin may be required. Furthermore, for some micro-organisms that have good potential for vitamin production, it will be necessary to develop techniques of genetic manipulation. In summary, the impact of biotechnology on vitamin production will be more long range than its impact on the production of either amino acids or enzymes.

Single-cell protein

The term "single-cell protein" (SCP) refers to cells, or protein extracts, of micro-organisms grown in large quantities for use as human or animal protein supplements. Although SCP has a high protein content, it also contains fats, carbohydrates, nucleic acids, vitamins, and minerals. Interest in SCP production is not new, as evidenced by the fact that Dutch, German, and British patents for SCP production were issued as early as 1920 (51). Interest in SCP has waxed and waned throughout the ensuing years, but SCP production has never achieved great significance, mostly because of economic considerations (49,64). With the advent of new biotechnology and the threat of potential world food shortages, interest in SCP may once again return (49).

SCP can be used as a protein supplement for both humans and animals. In animal feed, it is a replacement for more traditional supplements, such as soybean meal and fishmeal. For humans, SCP is used either as a protein supplement or as a food additive to improve product functionality, for example, flavor, whipping action, or fat binding (49). The use of SCP in human food presents a problem: humans have a limited capacity to degrade nucleic acids. Therefore, additional processing is necessary before SCP can be used in human food. The animal feed market is more attractive for SCP, not only because there is less processing of the product, but also because the regulatory approval process is less stringent.

Relative protein content of the various commercial sources of concentrated protein is shown in table 38. Nutritionally, the amino acid composition of SCP resembles meat, fish, and shrimp meal rather than vegetable protein. It has been shown through extensive testing both in the United States and abroad to be a suitable substitute for at least part of the former high-cost protein sources. The high protein content, good storage properties in dry form, texture, and bland odor

Table 38.—Typical 1982 Selling Prices of Selected Microbial, Plant, and Animal Protein Products

Product	Protein content (%)	1982 selling price ($/kg)
Food-grade products:		
Candida utilis (tortula yeast)	50 to 55	$1.87 to $2.24
Kluyveromyces fragilis	45 to 50	2.09 to 2.29
Soy protein concentrate	72	0.88 to 1.03
Soy protein isolate	92	2.59 to 2.68
Dried skim milk	37	1.16 to 1.21
Feed-grade products:		
Saccharomyces cerevisiae	45 to 50	$0.48 to $0.66
Soybean meal	44	0.19 to 0.20
Meat and bonemeal	50	0.19 to 0.21
Fishmeal	65	0.23 to 0.40

SOURCE: J. H. Litchfield, "Single-Cell Proteins," *Science* 219:740-746, 1983.

and taste of SCP suggest real potential in feed and food markets. In prepared aquaculture feeds where, for juvenile animals, protein content up to 50 percent and above is required, SCP appears to be an attractive product. Another application is as a calf, lamb, or kid starter, thus leaving more milk for human consumption.

Incentives for production of SCP are fourfold. First, some parts of the world, for example, the high rainfall, tropical areas, have agricultural feed and food products high in carbohydrates; in such places, there is a chronic shortage of protein, which results in deteriorated physical and mental health. SCP would raise the protein content of food. Second, the land in other regions, including the Middle East and Africa south of the Sahara, cannot produce sufficient food of any type to prevent hunger. Here also an SCP supplement would be an asset. Third, there is demand worldwide for very high protein ingredients for feeds in the aquaculture industry, i.e., in the production of shrimp, prawns, trout, salmon, and other finfish and shellfish. Finally, SCP does not rely on temperature, rainfall, or sun for survival. At least one of the variety of feedstocks is usually available in almost any country or region of the world. The security of having such an internal source of protein is attractive to many countries.

Economically feasible SCP production is dependent on the efficient use of an inexpensive feedstock by a micro-organism. A large variety of feedstocks have been used for SCP production over the years, including carbon dioxide, methane, methanol, ethanol, sugars, petroleum hydrocarbons, and industrial and agricultural wastes. These feedstocks have been used industrially with different micro-organisms, including algae, actinomycetes, bacteria, yeasts, molds, and higher fungi. The choice of a feedstock includes such considerations as cost, availability, efficient growth of the micro-organism, and requirements for pretreatment (49).

SCP has yet to become an important source of protein, mainly because of high production costs. Some SCP-production processes that were economical at one time have not remained so because of changes in prices of competitive sources of protein such as soybean meal or fishmeal. In comparison to SCP, these protein sources are quite inexpensive (see table 38). In fact, the price of most SCP processes would have to be decreased one-half to one-fifth for SCP to be competitive with soybean meal and fishmeal.

Through the years, the high cost of SCP relative to that of these other sources of concentrated protein has prevented extensive utilization of SCP, primarily in animal feeds. In the case of SCP produced from methanol, for example, the methanol represents approximately 50 percent of the cost of the product. In the United States, the cost of SCP made from methanol exceeds the average cost of fishmeal by a factor of 2 to 5. A plant in the United Kingdom (ICI) is operating at a loss because of such a situation (49,52). In some parts of the world, such as the Middle East, low-cost methanol and high shipping costs for fishmeal and other natural protein sources make the cost differential considerably less. In countries without methanol, biomass presents an option as a cheap feedstock source. However, this market has not been developed yet.

It is possible that the application of biotechnology will help to reduce the cost of production of SCP. Strains of micro-organisms could be improved using rDNA techniques. Improvements could include increasing the production of proteins with a better amino acid balance* or improving the ability of the micro-organism to utilize the feedstock efficiently. Technological improvements in the process and recovery steps would also be important. The use of automated, continuous processes could improve the efficiency of production. Recovery steps could be aided by using micro-organisms that have been genetically manipulated to excrete protein. Additionally, it is possible that an enzyme that degrades cell walls could be cloned and produced in large amounts. Its use would help in the production of a protein concentrate from cells. New technologies will probably improve the production of SCP, but widespread introduction of SCP will be governed by economic and regulatory factors.

Several companies in Western and Eastern Europe, the United States, and Japan have built SCP production plants in the last 15 years (3,5,64). Many of these are no longer operating because of high production costs and regulatory approval problems. Nevertheless, there are several companies operating plants, including Shell Chemicals (Netherlands), British Petroleum (U.K.), ICI (U.K.), Rank Hovis MacDougall (U.K.), Sosa Texaco (Mexico), Finnish Pulp and Paper Institute (Finland), Amoco (U.S.), Phillips Petroleum (U.S.), Pure Culture Products (U.S.), Rhinelander Paper Corp. (U.S.), and Amber Laboratories (U.S.). In addition, there is one plant in the German Democratic Republic, and there are several in the U.S.S.R.

The center of SCP technology is in England, especially at ICI (71). The ICI process uses aerobic bacteria with methanol and ammonia as feedstocks. The bacteria are grown in the world's largest continuous bioprocess system with computerized control and monitoring of performance. The product, Pruteen®, contains 80 percent crude protein as well as a high content of essential micronutrients, especially B group vitamins. Pruteen® is used in animal feed diets (poultry, swine, fish) and as a milk replacer (calves). In 1981, ICI had scaled up its process to produce 3,000 tons of SCP per month. It is beginning research using rDNA technology to facilitate protein harvesting (49). So far, however, the production of Pruteen® has not been economic even though it is twice as nutritious as soybean meal (52).

Two of the SCP plants in the United States (Amber and Rhinelander) use wastes produced in other parts of their plants for feedstocks, assuring a constant and inexpensive source of raw materials for SCP production (49). This type of small-scale operation using internally generated wastes as feedstocks may be the most appropriate use of SCP technology in the United States and other countries where animal- and plant-derived protein sources are abundant.

The U.S.S.R. is actively pursuing the production of SCP. The Soviets consider the construction of plants to produce SCP a high priority in order to decrease their dependency on foreign sources of protein for animal feed (5). The U.S.S.R. produces about 1 million tons of SCP per year, but production has not increased since 1976 (62). About half of the Soviets' SCP feedstock is cellulose, and the balance is petroleum. The current Five-Year Plan

*As do proteins from plants, proteins from micro-organisms often lack one or more essential amino acids. Most commercial SCP products are low in methionine (51).

calls for doubling SCP production by 1985 to 2 million tons per year, but the Soviets will have to produce a total of 3 million tons per year in order to be able to stop importing soybeans for use as a protein source.

Low-cost or waste biomass feedstocks have been cited as one means to product cost reduction. Inedible biomass can serve as an indirect feedstock for SCP processes by high-temperature conversion to synthesis gas and then to methanol (2).

Engineering improvements expected include bioreactor designs for continuous operation and high cell density. High cell densities decrease cost, because at high cell densities, the cell suspension leaving the fermentor can be dried without preconcentration of the cells by centrifugation, and because extracellular nutrients are recovered in the product.

Conventional genetic and rDNA methods for SCP production are currently being directed toward the following goals: 1) broadening the range of utilizable feedstocks; 2) increasing the optimum bioprocessing temperature and achieving a concomitant decrease in cooling requirements; 3) increasing the efficiency of utilization of the feedstock with the associated benefit of decreased generation of heat; 4) optimizing the balance of the essential amino acids in the product; and 5) producing of high-value products in conjunction with the SCP (e.g., growth stimulators) which may be either left in the SCP product or isolated from the broth.

The future of SCP depends largely on reduction in cost and improvement in quality. Means to meet these requirements involve lower cost feedstocks, improved engineering of the conversion and recovery processes, and upgrading the yield and quality of the product through conventional genetic and rDNA methods. The renewed interest in all of biotechnology, in part due to rDNA technology, is leading to increased effort in developing economically competitive SCP with improved qualities.

Complex lipids

Lipids are water-insoluble compounds found in cells whose many functions include serving as the structural components of membranes and storing of metabolic fuel. The term lipid designates a general class of compounds that includes the complex lipids (saponifiable lipids) which contain fatty acid components and simple lipids (nonsaponifiable lipids) which have no fatty acid component. The simple lipids include some vitamins, steroid hormones, and other highly specialized fat-soluble biomolecules.

Complex lipids are readily available and are extracted from natural sources. Some lipids such as sophorolipids have commercial uses. By far the most valuable attributes of lipids, however, are the products that can be derived from them, including fatty acids and fatty alcohols and the potential of lipids to replace petroleum feedstocks (48). Biotechnology could be used to develop new methods for economical production of lipid-derived products.

Fatty acids

Fatty acids are important industrial chemicals used in cosmetics, plastics, lubricating greases, rubber compounding, polymer emulsifiers, specialty household cleaners, foods, paints, varnishes, and flotation reagents (46). In the United States alone, the present consumption of fatty acids is about 1.65 billion pounds annually (46). The major sources of fatty acids are the naturally occurring fats and oils of plants and animals. The major plant sources of fatty acids in the United States are tall oils and coconut oil, and the major animal source is tallow (46). Synthesizing fatty acids from petroleum feedstocks is possible, but the process requires complex reactions and is more expensive than obtaining the acids from natural sources.

Fats and oils are composed of triglycerides, which can be broken down to free fatty acids and glycerol, a valuable coproduct. The usual decomposition method is a chemical process whereby the triglycerides are continuously hydrolyzed (16). This chemical process is efficient; 99 percent of the available triglycerides are hydrolyzed to free fatty acids and glycerol. Because the process requires both high temperatures and high pressure, however, it is also energy-intensive.

An attractive alternative to chemical hydrolysis of triglycerides is an enzymatic process that uses lipases to split the triglycerides into free fatty acids and glycerol (see fig. 19). Such a process does not require severe reaction conditions and is therefore more energy-efficient. Two Japanese companies have begun to commercialize the production of fatty acids from natural oils and fats using lipases. Miyoshi Oil and Fat Co. has reportedly constructed two plants for the lipase-catalyzed production of fatty acids. Its initial plant reportedly is producing 300 tons of fatty acids annually. Similarily, the Nippon Oil and Fat Co. has begun trial operation of a pilot plant at its Amagaski facility. It plans to produce initially about 1,000 tons of fatty acids per month. These Japanese companies report that the lipase-based production of fatty acids is both energy- and labor-efficient (38,39).

Because of their stability and lack of cofactor requirements, lipases are good candidates for use in an immobilized enzyme process. At the present time, however, the apparent requirement of lipases for an emulsified substrate represents a barrier to an immobilized enzyme process. Research on both process design and the identification of lipases that are more amenable to immobilization should result in the development of an immobilized enzyme process for the production of fatty acids. Such process development might take several years.

Figure 19.—Hydrolysis of Triglycerides

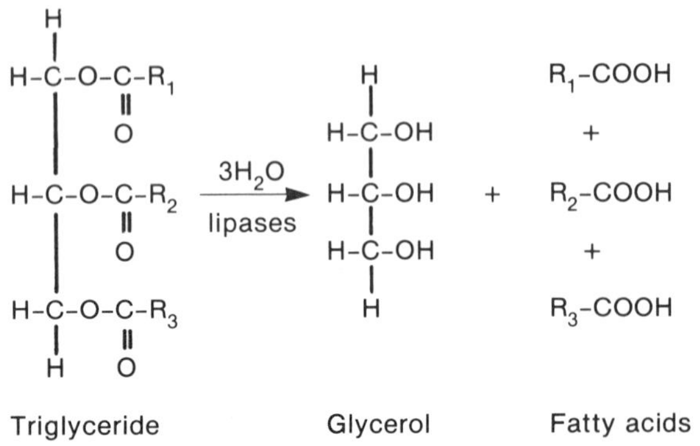

SOURCE: Office of Technology Assessment.

The cost of obtaining sufficient quantities of lipase will have a major impact on the economic viability of such processes. The application of biotechnology to develop or improve techniques for the recovery and reuse of lipases would be desirable. Supplies of specific lipases could be increased through gene cloning and amplification.

Fatty alcohols

Fatty alcohols are important industrial chemicals. The plasticizer ester industry uses large quantities of shorter chain (6 to 10 carbons) alcohols, while alcohols of longer length (11 to 18 carbons) are used to make detergents. Fatty alcohols can be synthesized chemically from ethylene, which is derived from petroleum feedstocks. Alternatively, some Japanese companies use a chemical process to convert fatty acids obtained from coconut oil into fatty alcohols (30). Although the Japanese chemical process does not rely on nonrenewable petroleum feedstocks, it does require extreme reaction conditions and therefore high energy consumption. A number of micro-organisms are capable of converting fatty acids to fatty alcohols, but these biological conversions are also energy consumptive. Furthermore, both the substrate and product are toxic to micro-organisms. Hence, the development of a biological process would require, at best, a number of years of R&D effort.

Microbial oils

Although naturally occurring fats and oils can currently be obtained cheaply from plants and animals, there is a resurgence of interest in exploiting micro-organisms for the production of oil. Israel, for example, is actively pursuing the development of a microbial source for oil (57) to reduce its dependence on imports. A number of eukaryotic oil-producing micro-organisms have already been identified, and preliminary research in developing micro-organisms as a source of oil is underway. It is impossible to predict when such processes will be commercial. The United States has sufficient plant and animal sources for fats and oils, but the supply is affected by climate. European countries, unless they develop a microbial source, will have to rely on imported materials to satisfy demands for vegetable oils and fats (57).

Sophorolipids

There is increasing interest in identifying and exploiting microbial biosurfactants (biologically derived emulsifying agents). One group of glycolipids, the sophorolipids, shows considerable promise for use as biosurfactants. Sophorolipids can be produced from vegetable oils by the yeast *Torulopsis*. These sophorolipids are comparable in activity to other surfactants, but are produced by the yeast in much higher yield and are easily separated from reaction broths, thus minimizing costs. Further characterization of the sophorolipids and their potential markets is required before applications of biotechnology to their production are likely to be considered.

Steroids

With the recognition of the therapeutic value of the natural steroid hormones and their analogs, it became necessary to develop efficient processes for producing these products. The steroids currently in therapeutic use are synthesized primarily by modifying naturally occurring steroids obtained from plants. Two commercially important modifications, 11-beta-hydroxylation and delta^1-dehydrogenation, are difficult to achieve via chemical routes, but micro-organisms have been reported to perform both reactions. Examples of a microbial 11-beta-hydroxylation and a delta^1-dehydrogenation are shown in figure 20.

Microbial reactions have been identified for the hydroxylation of virtually every position of the steroid nucleus. Because whole-cell bioconversions for introducing the 11-beta-hydroxyl group occur at low levels and are plagued by the formation of byproducts, they have not been developed for commercial use. Further study of the enzymatic process should establish whether the byproducts are the result of many steroid-metabolizing enzymes or a lack of specificity of the 11-beta-hydroxylating enzyme. If the enzyme is specific, it may be possible to obtain the desired conversion levels by cloning and expressing at high levels the genes that encode the 11-beta-hydroxylase.

Microbial delta^1-dehydrogenations are used commercially today. However, an efficient microbial process that combines delta^1-dehydrogenation and 11-beta-hydroxylation has not yet been developed. Biotechnology could make a significant contribution to the steroid industry by achieving both the delta^1-dehydrogenation and 11-beta-hydroxylation in a single biological process step. The latter reaction is catalyzed by a complex enzyme, so it is unlikely that an immobilized enzyme system could be developed for it. Therefore, the most efficient process would be to have the two reactions carried out by one cell.

The steroid market is readily accessible to biotechnology. Microbial processes are used routinely in the manufacture of steroid products. Furthermore, bioconversions with potential value to the steroid industry have been identified, and rDNA technology could be used to construct a micro-organism that more efficiently converts the steroid substrate to the desired product. The primary barriers to further biotechnological applications in the manufacture of steroids are the lack of rDNA host/vector systems for some of the micro-organisms involved and a lack of understanding of the specific enzymatic processes of steroid synthesis.

Figure 20.—Microbial Modifications of Steroid Molecules

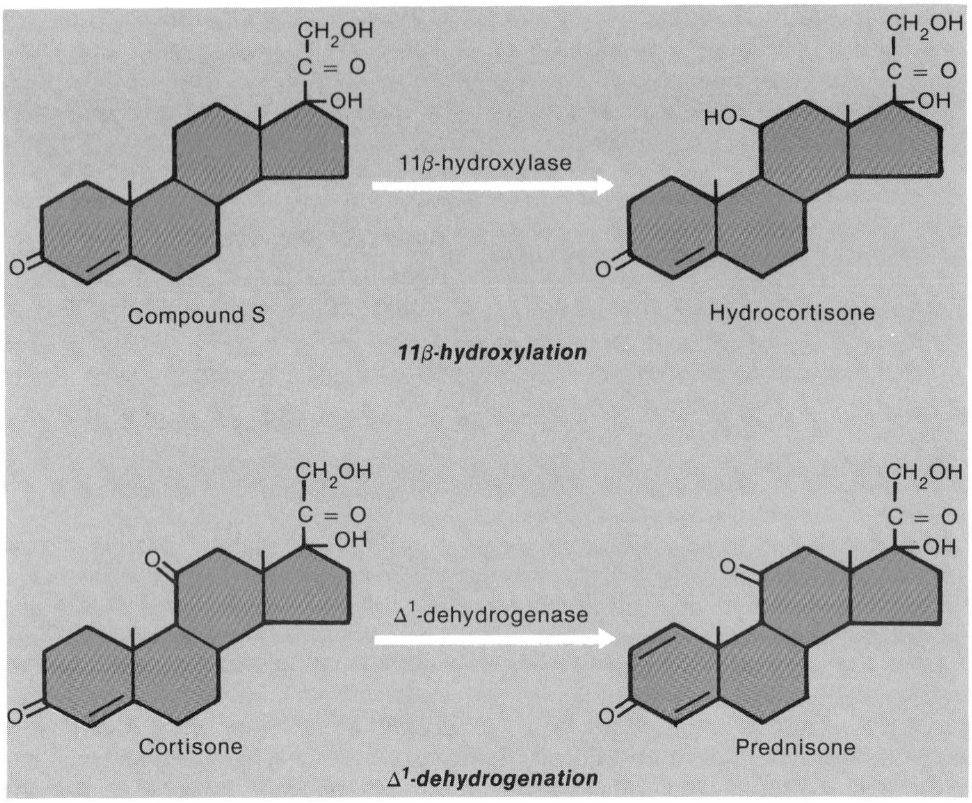

SOURCE: Genex Corp., "Impact of Biotechnology on the Specialty Chemicals Industry," contract paper prepared for the Office of Technology Assessment, U.S. Congress, April 1983.

Aromatic specialty chemicals

Aromatic compounds occur in many household products, medicines, agricultural products, pesticides, paints, cosmetics, and dyes, and their synthesis is a major component of the specialty chemical industry (6). Aromatic compounds that contain a hydroxyl group on the aromatic ring are an important group of specialty chemicals. Examples are the parabens and their esters, which are used as preservatives; 2,4-dichlorophenoxyacetic acid (2,4-D), which is the most extensively used herbicide; and N-acetylated para-aminophenol, an aspirin substitute. The synthesis of each of these compounds requires the specific hydroxylation of the aromatic ring.

The chemical hydroxylation of the aromatic ring is generally an inherently expensive step in the synthesis of an aromatic specialty chemical. This expense often results from the nonspecificity of the hydroxylation reaction, which forms unwanted byproducts and is therefore an inefficient use of the starting material. Additional processing may be required in order to remove the byproducts and to dispose of them properly. Chemical hydroxylations also require severe reaction conditions and therefore consume a large amount of energy. In addition, chemical reactions can result in the formation of undesirable contaminants. One highly publicized case is the dioxin contamination that occurs during the chemical synthesis of 2,4,5-trichlorophenoxyacetic acid (2,4,5-T), an herbicide and a component of the now banned Agent Orange.

By replacing a chemical reaction with a biological process, biotechnology has the potential to decrease the manufacturing cost of aromatic specialty chemicals, especially in processes that involve aromatic hydroxylations. Many micro-organisms are able to grow on aromatic compounds, and aromatic hydroxylations are key reactions in these growth pathways. These enzymatic reactions occur under mild conditions and result in specific hydroxylations of the aromatic ring. Furthermore, using enzymatic reactions, hydroxylations can be obtained at positions not readily hydroxylated by chemical reactions. The development of bioprocesses for aromatic hydroxylation reactions represents a valuable biotechnological opportunity for the specialty chemical industry.

Microbial aromatic hydroxylations are mediated principally by oxygenases that catalyze the direct incorporation of molecular oxygen into the aromatic ring (6,54,65,66). An example of an aromatic hydroxylation mediated by a microbial oxygenase is shown in figure 21. Many oxygenases have been studied in detail; while differences do exist among the various types of oxygenases, oxygenases generally are complex enzyme systems that require cofactors for activity.

As found in nature, the conversion efficiencies of most aromatic hydroxylations are generally too low to be commercially viable (30). However, the conversion efficiency could be improved by cloning the gene(s) encoding the oxygenase and expressing the cloned gene at high levels in an appropriate production strain. Once the oxygenase gene(s) have been cloned and expressed in an appropriate production strain, more research time and effort will be required for process development. One major consideration is how to minimize the toxic effects of the aromatic compounds on micro-organisms. One solution would be to develop an immobilized enzyme process; however, because of the complexity of the hydroxylation reaction it may not be possible to apply this technology. Toxic effects in bioprocesses have been minimized by innovative process design, and it is anticipated that there will continue to be significant advances in this area of research. Another consideration in developing an effective process is that the substrates and products are not soluble in water. Again, innovative process design could minimize this problem.

Figure 21.—An Example of a Microbial Aromatic Hydroxylation

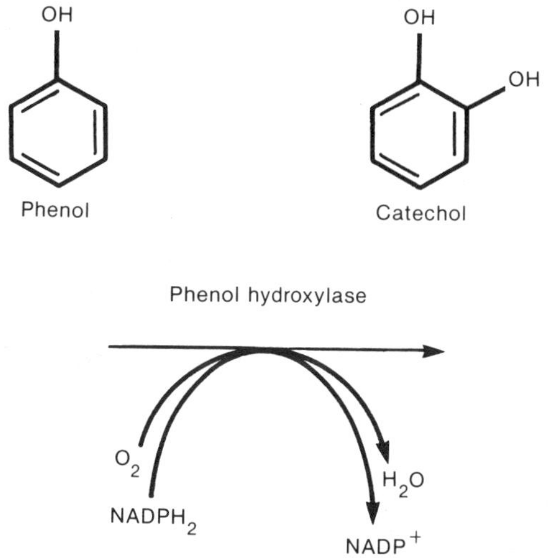

SOURCE: Office of Technology Assessment.

Polysaccharide biopolymers

Biopolymers are naturally occurring macromolecules that include proteins, nucleic acids, and polysaccharides. The discussion here will emphasize the polysaccharide biopolymers and the opportunities for the application of biotechnology to their synthesis.

The major commercially available water-soluble biopolymers are used as viscosifiers (thickening agents), flocculating agents (aggregating agents), and lubricants. Currently, there is a trend toward increased use of synthetic polymers as flocculating agents in place of natural products (70). This trend, however, is sensitive to the availability and cost of the petroleum feedstocks required for manufacturing synthetic polymers, and biopolymers will be important if the price of oil rises.

The market for viscosifiers is several times larger than that for flocculants. The currently used viscosifiers, unlike flocculants, are biopolymers obtained from plants, especially seaweed. Although these sources are not dependent on petroleum feedstocks, the use of plants as biopolymer sources has several disadvantages, including labor costs associated with extraction and purification, limited availability of the sources, and a supply that can be affected by adverse climatic conditions. Micro-organisms could provide a constant and reliable supply of these products (72). Microbial biopolymers produced in controlled processes would not suffer from the problems associated with climate, disease, and other factors that normally affect plant products. Furthermore, microbial biopolymers have relatively uniform chemical and physical properties.

These attributes have led to increasing interest in the production of biopolymers that could be used in novel applications as well as in place of commercial biopolymers that are not now microbially produced. For example, alginate is a commercially important gum obtained from kelp. The markets for alginates demand different specific characteristics such as solution viscosities and gelling qualities. The alginates obtained from kelp can vary in composition, so they must be separated, evaluated, and categorized for the different markets. Alginate is also synthesized by *Azotobacter vinelandii* (41). Because the composition of the microbial alginate can be closely controlled by bioprocessing conditions, separate microbial bioprocesses could be developed to produce specific alginates with uniform chemical and physical properties. Another microbial biopolymer that has been developed by the Kelco Co. and has recently become commercially available is gellan. Gellan is a *Pseudomonas* polysaccharide that can be used as a solidifying agent for laboratory media or food products (44).

While a number of microbial biopolymers are being developed for commercial applications as gums, plastics, and other products, only xanthan gum, dextran, polytran, and gellan are currently being produced commercially (44,72). In terms of production volume, xanthan gum is the major microbial polysaccharide. At present, over 20,000 tons of xanthan gum are manufactured in the United States annually (30). Xanthan gum's primary use is as a food additive for stabilizing liquid suspensions and for gelling soft foods, such as ice creams and cheese spreads. More recently, it has been used in the new clear-gel toothpastes. The use of xanthan gum in enhanced oil recovery is still experimental, but this appears to be the largest potential market for this product.* Xanthan gum is commercially produced in an aerobic batch bioprocess using the bacterium *Xanthomonas campestris* (30).

The importance of polysaccharide biopolymers is likely to grow. For example, the microbial polysaccharide pullulan is synthesized by *Aureobasidium pullulans* from a number of substrates (42). Pullulan has potential applications in the cosmetic industry, in diet foods, and, more importantly, as a biodegradable plastic to be used in place of wraps and plastic containers. Plastic wraps and containers are now made from petroleum-based plastics which are not biodegradable and are dependent on nonrenewable feedstocks. The Japanese are already at the pilot plant stage for the microbial production of pullulan, and pullulan has the potential to develop into a significant market.

Another microbial biopolymer that is expected to be available commercially in 1983 is emulsan. A potent hydrocarbon emulsifier, emulsan is expected to gain widespread use in cleaning oil-contaminated vessels, oil spill management, and enhanced oil recovery (4).** Like many biologically produced polymers, emulsan exhibits a specificity that generally is not observed in chemically synthesized materials; the emulsifying activity of emulsan is substrate-specific, acting only on hydrocarbons that have both aliphatic and cyclic components. Emulsan was originally discovered by researchers in Israel (34,59,60,61,75).

---

*Enhanced oil recovery is discussed in *Chapter 8: Environmental Applications.*

**See discussion in *Chapter 8: Environmental Applications.*

Emulsan was awarded patents in the United States in 1982, and Petrofirm, USA, a subsidiary of Petroleum Fermentations, N.V., headquartered in Netherlands Antilles, is developing emulsan as a commercial product (4). To date, the development has been confined to strain improvement through mutation and selection techniques. Because of the complexity surrounding the microbial biopolymer, the feasibility of applying rDNA technology for strain improvement is uncertain.

Useful microbial biopolymers can extend beyond the polysaccharides. For example, polyhydroxybutyrate (PHB), a metabolic product of the bacterium *Alcaligenes eutrophus,* has potential commercial applications as a biodegradable thermoplastic that could be used as a surgical material. The unique electrical properties of PHB are also useful in other specialty markets (8). ICI (U.K.) soon will market a PHB product known as Biopol®, made with a bioprocess using glucose as a feedstock. ICI does not know yet what Biopol® 's first markets will be. PHB has properties similar to polypropylene but costs substantially more. Its edge is its biodegradability, and ICI believes that its customers will pay the higher price for this quality (63).

There are several inherent problems in using bacteria to produce polysaccharides (30). There are probably at least 100 enzymatic steps important in the production of these biopolymers, very few if any of which have been identified. Therefore, it is much more likely that classical genetic selection techniques will be more useful than rDNA techniques initially for improving the charactertistics of the compounds. Before it is possible to predict the role that rDNA technology will play in microbial biopolymer production, the producing micro-organism will have to be characterized genetically and physiologically. It will also be important to have an understanding of the complex biochemical pathways for the production of the biopolymer and its regulation. Most biotechnology advances will only appear several years into the future, if at all.

More immediate improvements in the production of microbial biopolymers might be realized by the development of novel bioreactor designs. The polysaccharides have very large molecular weights and are viscous, two characteristics that preclude the use of most standard bioreactors. One way to generate a large quantity of polysaccharides is to maintain live cells in an immobilized cell bioreactor. The cells cannot be microencapsulated, because the product is too large to be washed away. Therefore, they need to be attached to a solid surface by a procedure that does not damage the cells. Another critical research area is improved product recovery from the broth. Current methods for the recovery of xanthan gum, for example, often result in preparations that contain water-insoluble solids such as nonviable cells and residual medium constituents. For xanthan gums to be used in enhanced oil recovery, it is important to have a product free of cells and other fine particulates because the fluid must be able to flow through porous rocks.

Another area of research is the identification of thermophilic polysaccharide producers. Development of a thermophilic micro-organism could result in substantial gains in productivity and lower process costs due to energy conservation. Screening thermophiles for polysaccharide production is an active area of research (74). To date, no thermophilic xanthan gum producers have been identified. Thermophilic *Bacillus* and *Clostridium* bacteria are being screened for the production of polymers that would be useful as biosurfactants (74).

Commercial aspects of biotechnology in specialty chemicals

Some specialty chemicals are currently made using bioprocesses, most notably amino acids and enzymes. The amino acid markets are dominated by Japanese companies, especially Ajinomoto and Kyowa Hakko, whereas the enzyme markets are dominated by two European firms, Novo and Gist-

Brocades. Japan also leads the world in the biotechnological production of fatty acids, a relatively new process.

Most of the opportunities for the use of biotechnology in the production of specialty chemicals are still in planning or early development stages. Many potential bioprocesses would replace chemical processes, necessitating a large investment in new plants. Thus, the potential of a process using biotechnology must justify this investment. On the other hand, enzymes that could withstand high temperatures and pressures could be used to replace existing chemical steps without having to change the basic chemical process. Enzymes with these characteristics are beginning to be studied.

U.S. companies are beginning to enter some specialty chemical markets with biotechnology products. Corn sweetener companies are planning to market enzymes that they have produced for inhouse use for some time. Other established firms, such as W. R. Grace, are entering markets with biotechnologically derived specialty chemicals. Several U.S. NBFs, such as Genex, Genentech, Chiron, Amgen, Ingene, Enzo, and Industrial Genetics, have stated interests in specialty chemical markets. Although 20 percent of U.S. companies using biotechnology say they are working in the specialty chemicals field, their interests are not well known and most of their research is highly proprietary.

Priorities for future research

The most glaring lack of knowledge for the successful application of biotechnology to the production of specialty chemicals is in the identification and characterization of micro-organisms that perform particular chemical conversions. Often when industrially useful reactions in micro-organisms have been identified, the micro-organism is so poorly understood that the application of new biotechnology is not possible. There are many opportunities for the specialty chemical industry to expand and improve its production capabilities using biotechnology, but before it can take advantage of these opportunities, useful micro-organisms, especially those that function at high temperature and pressure, will have to be screened and identified.

For the specialty chemical industry to take full advantage of biotechnology, sharing of information between industrial chemists and biologists is needed. The sharing of information has to proceed beyond identification of specific steps in a chemical synthesis that are inherently expensive to discussion of the total process for the manufacture of a specialty chemical. Broad discussion could suggest a bioconversion that uses a less expensive starting material and that would replace several steps of the chemical process. Processes for the manufacture of many specialty chemicals could ultimately combine chemical and biological steps, thereby resulting in more economic and energy-efficient manufacturing.

Chapter 7 references

1. Atkinson, D. E., "Adenine Nucleotides as Universal Stoichiometric Metabolic Coupling Agents," *Advances in Enzyme Regulation,* G. Weber (ed.) (New York: Pergamon Press, 1971).
2. *Biomass Digest,* "International Harvester Builds Methanol Units Around SERI Gasifier," November 1982, pp. 4-6.
3. *Biotechnology Newswatch,* "East German Fermentation Process Produces Single-Cell Protein Plus Arctic Diesel Fuel," Mar. 1, 1982, p. 3.
4. *Biotechnology Newswatch,* "Oil-Eating Bioemulsifier Passes Field Tests, Wins Patents, Federal OKs," Apr. 5, 1982, pp. 1-2.
5. Carter, G. B., "Is Biotechnology Feeding the Russians?" *New Scientist,* Apr. 23, 1981, pp. 216-218.
6. Cerniglia, C. E., "Aromatic Hydrocarbons: Metab-

olism by Bacteria, Fungi and Algae," *Reviews in Biochemical Toxicology,* E. Hodgson, J. R. Bend, and R. M. Philpot (eds.) (New York: Elsevier/North-Holland Biomedical Press, 1981).

7. *Chemical Engineering,* "Mitsui Toatsu Chemicals Announces a New Amino Acid Process," Dec. 27, 1982, pp. 9-10.
8. *Chemical and Engineering News,* "ICI Weighs Marketing of New Thermoplastic," Nov. 22, 1982, p. 6.
9. *Chemical Marketing Reporter,* "Current Prices of Chemicals and Related Materials," May 30, 1983, pp. 35-39.
10. *Chemical Purchasing,* "Chemical Briefs 2: Monosodium Glutamate," September 1982, pp. 39-43.
11. *Chemical Week,* "A Japanese Foray Into the U.S. Lysine Market," Oct. 18, 1982, p. 31.
12. *Chemical Week,* "An Enzymatic Route to L-Tryptophan," Dec. 15, 1982, p. 52.
13. Chernik, T. P., Kriviskii, A. S., Skvortsova, A. R., et al., U.S.S.R. Patent 161,184; 1969.
14. Chibata, I., *Immobilized Enzymes* (New York: John Wiley & Sons, 1978).
15. Choi, Y. J., and Tribe, D. E., "Continuous Production of Phenylalanine Using an *Escherichia coli* Regulatory Mutant," *Biotech. Letters* 4:223-228, 1982.
16. Considine, D. M. (ed.), "Soaps," *Chemical and Process Technology Encyclopedia* (New York: McGraw Hill, 1974).
17. Corn Refiners Association, Inc., "The Amazing Maize," Washington, D.C., 1981.
18. Crawford, T. C., and Crawford, S. A., "Synthesis of L-Ascorbic Acid," *Adv. Carboh. Chem. Biochem.* 37:79-155, 1980.
19. Danzig, M., CPC International, Argo, Ill., personal communication, 1983.
20. Dasilva, E. J., and Jensen, A., "Content of Alpha-Tocopherol in Some Blue-Green Algae," *Biochem. Biophys. Acta* 239:345-347, 1971.
21. Dubnau, D. (ed.), *The Molecular Biology of the Bacilli,* vol. I: *B. subtilis* (New York: Academic Press, 1982).
22. Enei, H., Sato, K., Anzai, Y., et al., Fermentative Production of Riboflavin, U.S. Patent No. 3,900,368, 1975.
23. *European Chemical News,* "MSG Demand To Double," July 15, 1981, p. 16.
24. *European Chemical News,* "Celltech Poised To Develop Recombinant Rennin Technology," Apr. 12, 1982, p. 22.
25. Eveleigh, D. E., "The Microbiological Production of Industrial Chemicals," *Scientific American* 245: 155-178, 1981.
26. *Feedstuffs,* "Trends in Agribusiness," Nov. 16, 1981, p. 8.
27. Florent, J., and Ninet, L., "Vitamin B12," *Microbial Technology,* 2d ed., vol. 1, H. J. Peppler and D. Perlman (eds.) (New York: Academic Press, 1979).
28. *Food Engineering,* "Field Report: New Amino Acid Plant," May 1982, p. 152.
29. *Genetic Technology News,* "Collaborative Research Clones Rennin Gene," February 1982, p. 1.
30. Genex Corp., "Impact of Biotechnology on the Specialty Chemicals Industry," contract paper prepared for the Office of Technology Assessment, U.S. Congress, April 1983.
31. Genex Corp., *Prospectus,* Rockville, Md., Mar. 9, 1983.
32. Glick, J. L., Peirce, M. V., Anderson, D. M., et al., "Utilization of Genetically Engineered Microorganisms for the Manufacture of Agricultural Products," *Proceedings of the Beltsville Symposium VII, Genetic Engineering: Applications to Agriculture* (Totowa, N.J.: Rowman & Allanheld, 1983).
33. Godfrey, T., and Reichelt, J. R., "Introduction to Industrial Enzymology," *Industrial Enzymology,* T. Godfrey and J. R. Reichelt (eds.) (New York: Nature Press, 1983).
34. Goldman, S., Shabtal, Y., Rubinovitz, C., et al., "Emulsan in *Acinetobacter calcoaceticus* RAG-1: Distribution of Cell-Free and Cell-Associated Cross-Reacting Material," *Appl. Environ. Microbiol.* 44:165-170, 1982.
35. Harris, T. J. R., Lowe, P. A., Lyons, A., et al., "Molecular Cloning and Nucleotide Sequence of cDNA Coding for Preprochymosin," *Nucleic Acids Research* 10:2177-2187, 1982.
36. Hebeda, R. E., "Syrups," *Kirk-Othmer: Encyclopedia of Chemical Technology,* 3d ed., vol. 22 (New York: John Wiley & Sons, 1983).
37. Hughes, P. E., and Tove, S. B., "Occurrence of Alpha-Tocopherolquinone and Alpha-Tocopherolquinol in Microorganisms," *J. Bacteriol.* 151:1397 1402, 1982.
38. *Japan Chemical Week,* "New Fatty Acid Technique Using Lipase in Spotlight," May 14, 1981.
39. *Japan Chemical Week,* "Nippon Oil Develops New Fatty Acid Technique," Feb. 4, 1982, p. 1.
40. *Japan Chemical Week,* "Kyowa Hakko Enters U.S. on Lysine for Feedstuff," Oct. 14, 1982, p. 1.
41. Jarman, I. R., "Bacterial Alginate Synthesis," *Microbial Arginate Synthesis,* R. C. W. Berkeley, G. W. Gooday, and D. C. Ellwood (eds.) (New York: Academic Press, 1979).
42. Jeans, A., "Dextrans and Pullulans: Industrially Significant Alpha-D-Glucans," *Extracellular Micro-*

bial Polysaccharides, P. A. Sanford and A. Laskin (eds.) (Washington, D.C.: American Chemical Society, 1977).
43. Jurgens, R., Miwon, USA, Hoboken, N.J., personal communication, 1983.
44. Kang, K., "Some New Bacterial Gums With Potential Industrial Applications," presented at the 83d Annual Meeting of the American Society for Microbiology, New Orleans, La., 1983.
45. Kinoshita, S., and Nakayama, K., "Amino Acids," *Economic Microbiology,* vol. 2, *Primary Products of Metabolism,* A. H. Rose (ed.) (New York: Academic Press, 1978).
46. Layman, P. L., "Fatty Acid Makers Bullish on Outlook," *Chem. & Eng. News,* Oct. 18, 1982, pp. 11-12.
47. Layman, P. L., "Capacity Jumps for Amino Acids," *Chem. & Eng. News,* Jan. 3, 1983, pp. 18-20.
48. Lipinsky, E. S., "Chemicals From Biomass: Petrochemical Substitution Options," *Science* 212:1465-1471, 1981.
49. Litchfield, J. H., "Single-Cell Proteins," *Science* 219:740-746, 1983.
50. Loewus, F. A., "L-Ascorbic Acid Metabolism, Biosynthesis, and Function," *The Biochemistry of Plants,* P. K. Stumpf and E. E. Conn (eds.) (New York: Academic Press, 1980).
51. Marstrand, P. K., "Production of Microbial Protein: A Study of the Development and Introduction of a New Technology," *Research Policy* 10:148-171, 1981.
52. *New Scientist,* "Protein Process Faces German Competition," May 20, 1982, p. 495.
53. Nishimori, K., Kawaguchi, Y., Hidaka, M., et al., "Cloning in *Escherichia coli* of the Structural Gene of Prorennin, the Precursor of Calf Milk-Clotting Enzyme Rennin," *J. Biochem.* 90:901-904, 1981.
54. Nozaki, M., "Oxygenases and Dioxygenases," *Top Curr. Chem.* 78:145-186, 1979.
55. Powls, R., and Redfearn, E. R., "The Tocopherols of the Blue-Green Algae," *Biochem. J.* 104:24c-26c, 1967.
56. Rastetter, W., Genentech Corp., South San Francisco, personal communication, 1983.
57. Ratledge, C., "Microbial Oils and Fats: An Assessment of Their Commercial Value," *Prog. Indust. Microbiol.* 16:119-206, 1982.
58. RECON Associates, "Poultry Dietary Study," Salisbury, Md., 1982.
59. Rosenberg, E., Perry, A., Gibson, D. T., et al., "Emulsifier of *Arthrobacter* RAG-1: Specificity of Hydrocarbon Substrate," *Appl. Environ. Microbiol.* 37:409-413, 1979.
60. Rosenberg, E., Zuckerberg, A., Rubinovitz, C., et al., "Emulsifier of *Arthrobacter* RAG:1: Isolation and Emulsifying Properties," *Appl. Environ. Microbiol.* 37:402-208, 1979.
61. Rubinovitz, C., Gutnick, D. L., and Rosenberg, E., "Emulsan Production by *Acinetobacter calcoaceticus* in the Presence of Chloramphenicol," *J. Bacteriol.* 152:126-132, 1982.
62. Severin, B., Central Intelligence Agency, McLean, Va., personal communication, 1982.
63. Sherwood, M., "Bacterial Plastic Comes to Market," *Bio/Technology,* July 1983. pp. 388-389.
64. Skinner, K. J., "Single-Cell Protein Moves Toward Market," *Chem. & Eng. News,* May 5, 1975, pp. 24-26.
65. Smith, R. V., and Davis, P. J., "Induction of Xenobiotic Monooxygenases," *Adv. Biochem. Eng.* 7:62-100, 1979.
66. Smith, R. V., and Rosazza, J. P., "Microbial Models of Mammalian Metabolism, Aromatic Hydroxylation," *Arch. Biochem. Biophys.* 161:551-558, 1974.
67. U.S. Congress, Office of Technology Assessment, *Impacts of Applied Genetics: Micro-Organisms, Plants, and Animals,* OTA-HR-132, Washington, D.C., April 1981.
68. U.S. Department of Commerce, Bureau of the Census, Foreign Trade #246, Washington, D.C., 1981.
69. U.S. Department of Commerce, Bureau of the Census, "Pharmaceutical Preparations, Except Biologicals," *Current Industrial Reports* MA-28G(81)-1, Washington, D.C., 1981.
70. *Water Treatment Handbook,* "Coagulation and Flocculation," 5th ed. (New York: Halsted Press, 1979).
71. Waterworth, D. G., "Single Cell Protein," *Outlook on Agriculture* 10:403-408, 1981.
72. Wells, J., "Extracellular Microbial Polysaccharides—A Critical Review," *Extracellular Microbial Polysacchrides,* P. A. Sanford and A. Laskin (eds.) (Washington, D.C.: American Chemical Society, 1977).
73. Yamada, S., Nabe, K., Izuo, N., et al., "Production of L-Phenylalanine From *Trans*-Cinnamic Acid With *Rhodotorula glutinis* Containing L-Phenylalanine Ammonia-Lyase Activity," *Appl. Environ. Microbiol.* 42:773-778, 1981.
74. Zajic, J. E., "Biosurfactants Produced by Thermophilic Bacteria," presented at the 83d Annual Meeting of the American Society for Microbiology, New Orleans, La., 1983.
75. Zuckerberg, A., Diver, A., Peeri, Z., et al., "Emulsifier of *Arthrobacter* RAG-1: Chemical and Physical Properties," *Appl. Environ. Microbiol.* 37:414-420, 1979.

Chapter 8

Environmental Applications

Contents

Figures

Chapter 8

Environmental Applications

Introduction

Micro-organisms have several uses in the environment, and new biotechnology can potentially be used to improve these micro-organisms. One application is in the control of pollution and treatment of toxic wastes. As discussed in this chapter, micro-organisms are currently used in pollution control, and the potential applications of biotechnology to treat liquid and solid wastes are numerous. Additionally, techniques are beginning to be used to select micro-organisms that can degrade extremely toxic compounds. In the mining industry, microbes are used to leach metals from mine dumps and concentrate metals from dilute solutions, and there are possibilities for using biotechnology to improve the efficiencies of these processes. A third environmental application of biotechnology is in enhanced oil recovery. About 50 percent of the world's subterranean oil is either reserves trapped in rock or is too viscous to pump. It is possible that either micro-organisms themselves or microbially produced compounds could be injected into oil wells to release the trapped oil.

None of the environmental applications of new biotechnology are ready to be marketed, and there are still many technological problems to be overcome. Nevertheless, several companies are pursuing research and development (R&D) in these environmental applications, and their development will progress over the next several years.

Pollution control and toxic waste treatment

Waste products and the pollution problems associated with such products have been part of human existence since the dawn of civilization. Troublesome wastes are of three types: those in the atmosphere, those in aqueous systems, and solids. In the treatment of both liquid and solid wastes, there are significant opportunities for the use of biotechnology. Indeed, most liquid and solid wastes have been dealt with for millennia by natural biological processes. Moreover, humans in their initial attempts to control such wastes have generally resorted to contained biological systems, particularly for the treatment of liquid wastes. The possibilities for using biological systems to control atmospheric pollution, in contrast, are rather limited. The discussion here, therefore, focuses on the applications of biotechnology in the treatment of liquid and solid wastes.

Treatment of nontoxic liquid and solid wastes

Of the conventional microbiological systems for the treatment of liquid wastes now in use, the most complex is that found in publicly owned water treatment plants. As shown in figure 22, there are four basic unit operations in a wastewater treatment plant:

- primary processing;
- secondary processing;
- tertiary processing; and
- digestion.

The primary treatment step removes solids from the wastewater. These solids (sludge) are then either disposed of or sent to a sludge digester, and the wastewater is forwarded to second-

Figure 22.—Steps in Waste Treatment

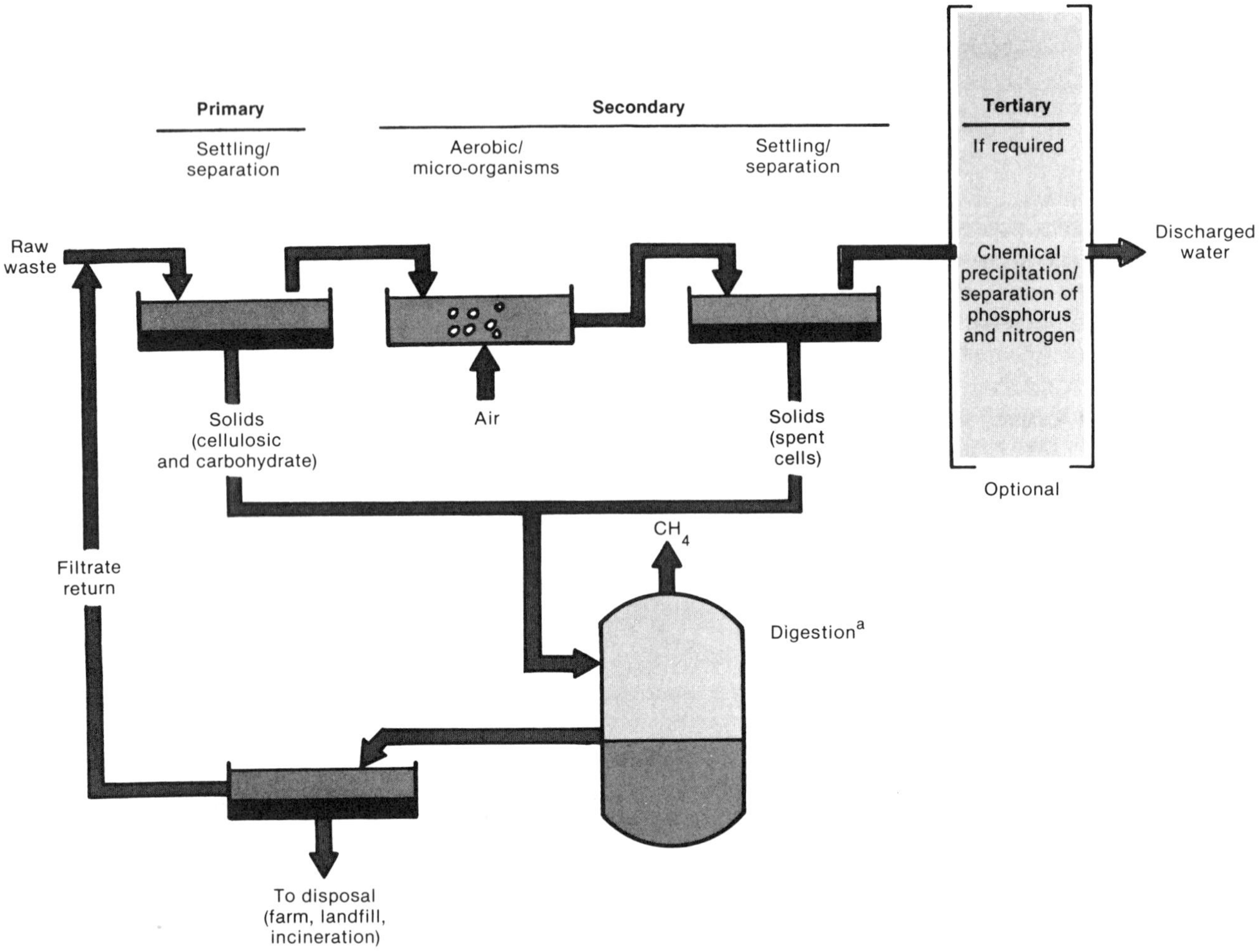

SOURCE: Office of Technology Assessment.

ary treatment. The secondary treatment system generally consists of natural aerobic microbes in a large open basin with some type of forced aeration. The purpose of this processing step is to degrade the dissolved organic compounds. The sludge resulting from this operation is primarily composed of microbial cells and is either disposed of or sent to a digester. The liquid from the secondary operation is sometimes subjected to tertiary processing, which can involve precipitation and separation of phosphorous and nitrogen, sand filtration, detention ponds, or biological filters. The water from the tertiary unit (or, in the absence of tertiary treatment, from the secondary unit) is returned to the environment.

The sludge digestion process used to treat the sludge resulting from the primary and secondary treatments is conventionally an anaerobic bioprocess. Its purpose is threefold: to reduce the total volume of solids requiring disposal, to reduce the odor, and to reduce the number of pathogenic organisms. Another potential objective of solid waste treatment can be to recover useful methane from the anaerobic bioprocess. Although the effective anaerobic treatment of solid wastes is more a problem of engineering than of biotechnology, there is a possibility that enzymes added to the waste could improve the efficiency of this treatment. Like secondary processing, sludge digestion is a classic bioprocess open to further technological improvements.

The total cost of running publicly owned water treatment systems in the United States has been estimated to be $5 billion to $6 billion per year (12). The cost of the chemicals used in these systems represents approximately 20 percent of the total operating costs (12). The biotechnology-based improvements that could be used by these treatment systems will either:

- increase the capacity of the treatment plants and therefore reduce the need for new capital expenditures,
- replace existing synthetic organic chemical additives, or
- remove newly identified, potentially harmful materials.

Processes similar to those just described for publicly owned water treatment plants are also used in the treatment of industrial wastewater, particularly wastewater from the chemical, petroleum, food processing, and pulp and paper industries. For that reason, biotechnology-based improvements in bioprocessing or solids separation procedures that are applicable to public water treatment systems will very likely be applicable to the industrial sector.

IMPROVEMENT OF CONVENTIONAL WASTEWATER TREATMENT PROCESSES

Both physical and biological processes are utilized in the treatment of wastewater. Improvements in any of these operations would be reflected in reduced capital and operating costs for wastewater treatment. Some specific opportunities for biotechnology-based improvements in wastewater treatment are discussed below.

Solids Separation: Flocculation.—The major physical operation in wastewater treatment is that of solids separation. Suspended solids must be separated during both the primary and secondary treatment steps. Quite frequently, it is also desirable to "thicken" the sludges resulting from these settling operations. The present techniques for accomplishing these separation and thickening operations generally include the use of materials known as flocculants. Because of the increased use and reuse of water, the U.S. market for flocculants is expanding (8,10).

Examples of classical flocculants are iron or aluminum salts and activated silica. In recent years, synthetic polymers have been used as flocculants, and in some cases, they have produced very promising results (8,10). Unfortunately, most of these synthetic polymers are based on acrylamide, a toxic compound. Moreover, these synthetic polymers are usually subjected to postpolymerization chemical modification, which adds to their cost. For both safety and economic reasons, therefore, biologically derived flocculants could be very desirable.

A few microbially produced polyelectrolyte polysaccharides that may prove to be effective flocculants have been identified (15). Before these potential bioflocculants can be commercially applied, micro-organisms with the potential for high-level production of effective polysaccharides at low cost will have to be identified. The potential bioflocculants will also have to be tested for their flocculating ability in waste treatment situations. Because the potential bioflocculants are polysaccharides and not proteins, improving their production through recombinant DNA (rDNA) technology may be a complex task (see discussion of polysaccharide biopolymers in *Chapter 7: Specialty Chemicals and Food Additives*). It should be noted, however, that improvements in microbial polysaccharide production have already been achieved with classical chemical mutagenesis and selection (34).

Sludge Dewatering.—For ease of handling of solid residues from water treatment processes, the water content of such residues must be reduced to a minimum to reduce their total weight. It is particularly important to reduce the water content of these residues to the smallest practical value if the sludge is to be disposed of by incineration.

The sludge dewatering operations with current technology (filtration and centrifugation, for example) result in a solids content of 15 to 40 percent, leaving a water content of 60 to 85 percent.

A significant proportion of the water that is retained is "microscopic" in nature, i.e., it is associated with microbial cells and organic debris present in the sludge. If techniques for releasing this retained water could be developed, they would find a ready and profitable market in the field of residue disposal.

Because much of the water retained in sludge is probably held in polymeric matrixes composed of cellulosics, fats, polysaccharides, and proteins (38), partial degradation of these matrixes by using some combination of cellulases, proteases, amylases, and polysaccharide hydrolases should release it. Some enzymes potentially useful for sludge dewatering may already be available in sufficient quantities and at economically attractive costs. For other potentially useful enzymes, techniques for economic, high-yield production will have to be developed. In some instances, these developments will simply involve process development using known microbial strains. In other instances, it may be necessary to construct genetically strains of micro-organisms for high-level production of specific enzymes and perhaps specifically alter the characteristics of the enzymes through directed protein modification. It may also be desirable to identify new enzymes from nature that have superior characteristics for use in sludge dewatering.

Conventional Uses of Biological Processes.—Biological processes are used in two operations of wastewater treatment plants, the secondary treatment step involving an aerobic process and the sludge digestion operation involving an anaerobic process. The performance of these standard aerobic and anaerobic biological treatment processes could conceivably be improved by the addition of specific enzymes that could augment the ability of the natural micro-organisms to degrade, for example, protein, starch, polysaccharides, and cellulosics. Such enzymes could be applied selectively at specific wastewater treatment plants where their particular substrates are present in unusually high concentrations. Enzyme "augmentation" might also help accommodate fluctuating loads on a particular treatment plant. The R&D involved in providing enzymes for this purpose would be similar to that for providing enzymes for sludge dewatering.

One potential byproduct of anaerobic bioprocesses is gas. Solid wastes, when held in sanitary landfill, very often encourage the growth of micro-organisms that produce methane. The generation of methane has become a serious problem in many sanitary landfill sites around the country. Experiments concerning the possibility of tapping this methane as an energy resource are in progress (39). Preliminary results indicate that the costs of the required anaerobic equipment are so high as to make the methane gas thus generated uneconomic as an energy source (39). Research is continuing, however, and it is conceivable that at some point in the future, improved micro-organisms or added enzymes could improve to a limited extent the economics of methane production from solid waste.

CONTROL OF ORGANIC MICROPOLLUTANTS

In recent years, significant pollution problems have arisen with regard to drinking water (27). Analyses of surface waters in the United States and Europe have demonstrated the presence at low concentrations of certain naturally arising soluble organic compounds that, when chlorinated, lead to the formation of trihalomethanes (THMs) (23,24,25,28,35,36). Increasing attention is being focused on these precursors of THM, because THMs are classified as potential carcinogens (23,24,25,28,31,35,36). In addition, there has been a series of toxic compounds discovered in ground water called volatile organic compounds (VOCs). VOCs are apparently leached from a variety of sources in the ground. Both VOCs and the precursors of THMs are potentially amenable to biological treatment methods.

Biotechnology can potentially offer improved techniques for the removal of organic micropollutants (13). It is possible, for example, that their removal could be accomplished by the use of enzymes that are capable of polymerizing aromatic compounds (e.g., fulvic acids and phenolic compounds) that often contaminate drinking water. These low molecular weight aromatic compounds are not precipitated in the traditional flocculation procedures, and they do not adsorb readily to activated carbon (26). These compounds also contribute to the formation of THMs and chlorophenols during chlorination procedures (2,23,24,25, 28,35,36).

Enzymatic polymerization should result in the removal of most of these low molecular weight aromatic compounds during flocculation procedures. Horseradish peroxidase is one enzyme that can catalyze polymerization reactions of this type (1,19,20), but it is not clear that purified or even crude horseradish peroxidase could be employed in a cost-effective manner. Other potentially useful polymerizing enzymes are synthesized by micro-organisms, but the current production levels are much too low for these enzymes to be commercially viable (5,6,11,33). Development of enzymatic polymerization to remove low molecular weight aromatic compounds will therefore require one or more of the following biotechnological developments (13):

- microbial strain improvement and process development programs using known polymerizing enzyme-producing microbial strains;
- identification of micro-organisms that produce useful polymerizing enzymes in high yield; or
- the genetic manipulation of a micro-organism to produce high levels of a polymerizing enzyme.

Another potential approach for using biotechnology to remove organic micropollutants from water is to develop micro-organisms that will better degrade these contaminating compounds. Such micro-organisms could be introduced into the water treatment cycle by seeding them onto activated carbon. When activated carbon is employed in water treatment processes, it accumulates naturally occurring microbes from the water. The goal would be to expand the degradative capacity of that microbial population. Although certain micro-organisms of various genera (*Pseudomonas, Acinetobacter, Arthrobacter, Klebsiella*) will degrade a variety of organic compounds, it will probably be necessary to identify or develop novel micro-organisms for the degradation of specific classes of pollutants. One procedure for accomplishing this, plasmid-assisted molecular breeding, is discussed below in the section on toxic waste treatment. Because micro-organisms of the genera listed above are generally present in natural populations, it should be possible to transfer genes that encode degradative enzymes from strains developed in the laboratory to the naturally occurring micro-organisms to encourage their survival in the environment.

The comments above have been made with respect to the control of organic micropollutants in drinking water. Any technology developed to solve the problems associated with drinking water, however, would most likely be applicable to similar organic contamination problems in industrial wastewater.

CONTROL OF HEAVY METAL CONTAMINATION

Heavy metals in drinking water have long been of concern (3). The concern has focused on lead, zinc, copper, and cadmium, although iron, at relatively high concentrations, can also present health risks (38). In addition to contaminating drinking water supplies, heavy metals can have detrimental effects on the operation and performance of biological processes used in wastewater treatment (3). Moreover, heavy metal contaminants in effluents from wastewater treatment plants can have potentially deleterious effects on downstream flora and fauna (3).

Micro-organisms used in metal accumulation (see section on microbiological mining below) are not useful for concentrating the heavy metals discussed here (except copper), because most metals found in contaminated water are toxic to micro-organisms. One potential approach to solving the problems of heavy metal contamination involves the use of metallothioneins (see also section on microbiological mining). These proteins, found principally in higher organisms, have a high affinity for various heavy metals (21). The economics of this process would depend on efficient release of the bound metals and reuse of the metallothionein. In fact, the gene coding for mouse metallothionein has been cloned and expressed (22,46). It is possible, therefore, that this protein could be produced in large amounts by bacteria, immobilized on a solid support, and used to extract metals from any solution passed over the immobilized protein (41). This process would be highly controlled and could be used not only for decontamination of waste streams from any industrial process, but also for concentrating metals by the mining industry.

Toxic waste treatment

The chemical and petroleum industries produce a variety of highly toxic organic wastes that are not initially amenable to conventional microbial treatment. Such wastes can be either liquid or solid. For developing biologically based processes that will degrade or otherwise detoxify them, a variety of techniques can be envisioned. A specific micro-organism or enzyme will probably have to be developed for each toxic compound.

As the number of toxic compounds that are leached or dispersed into the environment increases, the development of technologies for the treatment of toxic wastes becomes more critical. Toxic wastes are often resistant to natural biological degradation and therefore persist in the environment. Because of their toxic character, developing biotechnological approaches for effective treatment of such wastes may be difficult.

Toxic wastes are generally present in the environment in one of two forms. In some cases, they are purposefully concentrated at specific disposal sites in the form of dumps or lagoons. In other instances, the toxic compounds have already been dispersed into the environment, and they are often present at very low concentrations in soil and water over a fairly large geographical area. In general, toxic wastes in dumps or lagoons are likely to be more amenable to biological treatment than those that have been more widely dispersed. Dumps and lagoons have the advantage of presenting a reasonably high concentration of a particular type of compound or family of compounds at a specific site. Thus, the feasibility of developing a very specific treatment process tailored to both the waste to be detoxified and the environment in which it is found is increased. For more widely distributed wastes, even if biological methods for detoxification are developed, it may be impossible to apply them effectively.

It has often been observed in traditional biological waste treatment systems that the microbial population will adjust to the presence of a toxic compound and eventually achieve some degree of efficiency in its decomposition. This phenomenon, traditionally termed acclimatization, probably represents the selection of mutant micro-organisms that are able to both tolerate and degrade the toxic compound. In the case of certain toxic wastes, it may be possible to accelerate this natural mutation and selection process in the laboratory by the use of a technique called chemostat selection.

In traditional chemostat selection, the natural microbial populations present in soil or water samples collected from or near the waste disposal sites are grown continuously over several months in the presence of steadily increasing concentrations of the relevant toxic compound. This process provides steadily increasing selective pressure for the growth of mutant micro-organisms able to tolerate and potentially degrade the toxic substrate. The mutation rate in the chemostat can often be increased by the use of chemical or physical agents.

In a more modern version of chemostat selection, plasmid-assisted molecular breeding, laboratory strains of *Pseudomonas* that contain plasmids encoding enzymes involved in the degradation of toxic compounds are added to the chemostat (16). This technique is based on the observation that in nature degradative plasmids often evolve by the recruitment of genes from other plasmids in other micro-organisms. Plasmid-assisted molecular breeding has resulted in the generation of both a mixed-culture and a pure *Pseudomonas* strain that degrade the normally recalcitrant molecule, 2,4,5-T, which is a component of herbicides and Agent Orange (16,17). It has also been possible to develop micro-organisms that degrade novel substrates by introducing into a single bacterial strain plasmids specifying the degradation of different, but analogous, compounds or different portions of a single degradative pathway (32). Because degradation of a toxic compound usually involves a complex and often uncharacterized series of reactions, it has generally been preferable to let nature select for the proper genetic combination rather than to attempt to construct it de novo in the laboratory.

More recently, however, in a joint research project between the University of Geneva (Switzerland) and the University of Göttingen (F.R.G.), researchers have cloned the gene for one of the key enzymes in the degradation of 2,4,5-T. Their hope is to understand better the degradation pathways that have been naturally selected and possibly use this knowledge to develop a more capable micro-organism (4).

One or more of the techniques described above could potentially lead to the isolation of either a mixed culture or a pure strain that degrades a particular toxic compound that might be able to be used at a disposal site or in a contaminated area. The *Pseudomonas* strain that degrades 2,4,5-T has been shown to function successfully both in laboratory tests using contaminated soil and in field tests (17). The micro-organisms being investigated now are aerobic. However, if the toxic waste is present in a dump, it may be necessary to develop anaerobic micro-organisms for detoxification.

The development of micro-organisms for the degradation of both organic micropollutants and toxic wastes will require screening of natural microbial populations or chemostat selection for the appropriate degradative abilities. Once micro-organisms with the ability to degrade the offending compound(s) are available, it may be desirable to transfer that ability to a different microbial host by using rDNA technology to increase the efficiency of degradation or to increase the ability of the micro-organisms to survive in the environment in which they are utilized.

For certain toxic wastes, an alternative approach to detoxification might involve the use of specific enzymes. Enzymatic processes would not totally degrade the toxic compound but simply would convert it to a nontoxic derivative that might then be degraded through natural biological processes. Development of such enzymatic processes would probably involve an extensive research effort, and only very hazardous toxic wastes would justify this degree of effort.

Slime control

Slime can be broadly defined as an aggregation of microbial cells held together by the extracellular polysaccharides produced by the micro-organisms. Wherever water moves in significant quantities, slimes proliferate. The proliferation merely requires the presence of a nutrient, even in minute quantities. In the manufacture of paper, slime control is of major concern because slimes have a very deleterious effect on product quality (7,9,29,30). This problem arises because of the high nutrient availability and favorable temperature and pH in the paper processing environment.

The slimicides currently in use are often heavy metal-based poisons that can result in significant pollution and waste treatment problems (7,9,29, 30). However, the potential for using enzymatic methods for slime control appears quite promising. The formation of slimes is principally due to the extracellular polysaccharides produced by micro-organisms, so it should be possible to use polysaccharide hydrolases to degrade the slimes rather than toxic agents to destroy the micro-organisms.

Grease decomposition

Facilities processing meats, poultry, and certain other foods have particularly difficult problems with grease. Grease problems also appear throughout the wastewater collection and treatment cycle. Both pipe collection branches and pump stations are susceptible to the problems of grease accumulation, which include plugging of lines, accumulation of debris in wet wells, slippery working surfaces, unsightly conditions, odor, and operational problems at the facility site. Scum layers on sedimentation tanks and scum mats in digesters cause additional problems. The two basic problems are the congealing (solidifying) of the grease and the difficulty, if not an impossibility, of decomposing the grease once it arrives at the wastewater treatment plant.

Techniques that result in the emulsification and decomposition of grease would significantly improve the operation of all waste treatment facilities. Bacterial formulations have been used in the past for grease decomposition (18). Improvement of these cultures might be possible. Additionally, an enzymatic approach, such as the use of lipases, could improve the operation of waste facilities.* However, because grease contamination generally is in the form of nonaqueous, congealed deposits, substrate availability may be a significant prob-

*See *Chapter 7: Specialty Chemicals and Food Additives.*

Photo credit: David W. Taylor, Naval Ship Research and Development Center

Grease buildup in a holding tank on a U.S. Navy ship after 5 months of normal operation

Photo credit: David W. Taylor, Naval Ship Research and Development Center

Grease buildup in the same tank after 4½ months of operation with daily addition of degreasing bacteria produced through classical genetic selection techniques

lem. A mechanism for delivering the enzyme to the substrate might solve the problem, but no approaches for accomplishing this have been postulated.

Commercial aspects of biotechnology in pollution control and toxic waste treatment

In contemporary times, basic developments and improvements in water treatment have originated primarily in Western Europe and spread through the Western Hemisphere. Higher population and industrial densities coupled with fewer water resources have forced Western European countries to advance the technology at a much faster pace than required in the United States. In a sense, Western Europe has been the proving ground for new technologies used for water and wastewater treatment. This historical pattern suggests that Western Europe has probably been making initial assessments of the impact of advanced biotechnology in this area. Japan is also conducting a small amount of R&D in this area.

In the United States, there probably is more activity oriented to biotechnology, much of it financed by the U.S. Government, in the municipal solid waste treatment sector than in either the air or liquid waste treatment sectors. Additionally, R&D efforts aimed at improving the technology of wastewater treatment are concentrated in a handful of small bioprocess-oriented companies and certain academic microbiology laboratories. Only recently did interactions begin between these research groups and the plant operators involved in purifying wastewater (14). In the past, industry has relied primarily on engineering consultants, not technology-based companies, to address pollution problems; these consultants have used the most basic existing technologies for treatment of organic wastes.

Two potential barriers to the commercial application of novel approaches to the problems of pollution control and waste treatment are the performance of the products that are developed and scientific uncertainty regarding their application. For example, although the technology for high-level production of enzymes and metallothioneins certainly exists or can be developed, the performance of these products in the desired application is as yet untested. If their performance turns out to be poor, then the R&D effort for commercialization would be much more extensive and might not be worth pursuing. Furthermore, although reasonable approaches can be designed to identify or develop micro-organisms for the degradation of organic micropollutants and toxic wastes, the success of these approaches is uncertain. It is also unclear whether genetically manipulated micro-organisms or micro-organisms that have been otherwise selected in the laboratory will be able to survive in a nonlaboratory environment. Their ability to survive and function in the field will probably be greatest if the desired degradative activities can be introduced through minimal alteration of a naturally occurring micro-organism.

If the technological barriers to commercial application can be surmounted, the other areas of importance will be markets, Government policy, and regulation. Biotechnological improvements in the area of conventional wastewater treatment processes and slime control would provide economic benefits. If the performance is satisfactory, markets for these products should develop. The primary limitation to commercialization will be the rate of acceptance by the treatment plant operators.

In the case of pollution control, whether it be control of organic micropollutants, heavy metals, or toxic wastes, the primary nontechnological barrier will be Federal Government policy. Biotechnological solutions to these problems are likely to be vigorously pursued only if the Government sets goals and criteria for reducing these contaminants that must be met by both the public and private sectors. The effort for developing these biotechnological solutions will probably initially require Federal funding. However, the requirements could eventually create a demand for a commercial product, and funding might then shift partially to the private sector. At the present time, most industries will not fund biotechnological research on waste treatment problems. They are only interested in licensing or purchasing such technology if it has already been developed.

Another potential barrier to commercialization of products for pollution control is Government regulation of the products themselves. In the case of enzymes and other proteins, few significant safety problems requiring regulation are anticipated, although care must be taken in handling these products. The application of micro-organisms, in contrast, could involve significant regulatory implications. Since the micro-organisms proposed here will have the potential for being released into the environment, it will probably be necessary to establish their safety or to develop methods for their containment at the site of treatment. U.S. policy with regard to the regulation of micro-organisms, particularly genetically manipulated ones, is dynamic. The regulatory constraints that will be placed on the use of micro-organisms in the future, therefore, cannot be accurately predicted. The benefits of using micro-organisms in the area of pollution control to protect human health will have to be carefully balanced against any perceived dangers associated with their use.

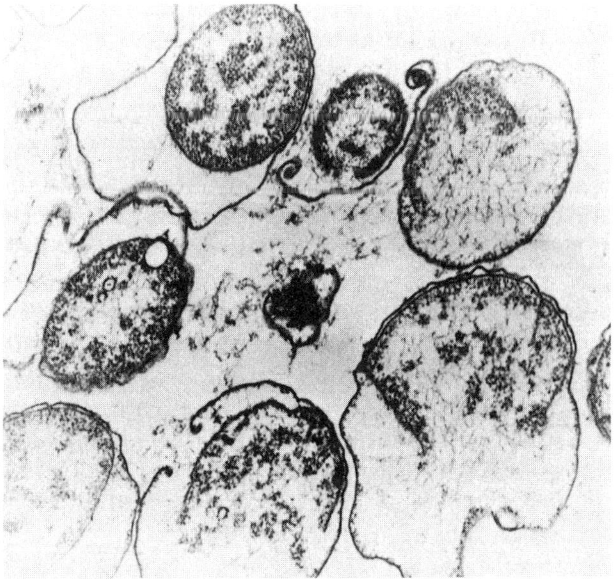

Photo credit: G. E. Pierce and M. K. Mulks

Pseudomonas putida, a bacterium capable of degrading hydrocarbons

Microbiological mining

Micro-organisms have been used to some extent in mineral leaching and metal concentration processes for many years. For the most part, these processes have been fortuitous, relying on micro-organisms found associated with mine dumps. With the recent advent of novel biological techniques, people in the mining industry and biologists have begun to think about ways to manipulate genetically some of the micro-organisms important in metal recovery processes to increase their efficiency and allow them to function on a larger variety of substrates.

Mineral leaching

More than 10 percent of the copper produced by the United States is leached from ores by micro-organisms (41,48). The micro-organisms used are found naturally associated with ores; the ores are not inoculated with selected strains. Until recently, the use of micro-organisms in the mining industry received little research attention because of the ease of mining high-grade ores and the relatively low energy cost for conventional mineral processing. The use of micro-organisms is gaining new attention, not only because of the depletion of high-grade ore and the soaring cost of energy, but also because of the possibility for genetic manipulation to increase the efficiency and broaden the application of microbial leaching.

There are many advantages to the use of micro-organisms. Besides having a low energy requirement due to their growth at ambient temperature and pressure, micro-organisms work efficiently and are less polluting than smelting techniques. It is possible they could be used for leaching in deep underground sites that are inaccessible to more traditional mining equipment. Mining with micro-organisms requires relatively low capital and operating costs, making it feasible for small-scale mining operations. The major drawback to the use of micro-organisms is that the biological processes are slow compared to the equivalent chemical ones (41).

Micro-organisms have been used mostly to leach copper and uranium (40). The organism that is most often used in these operations, and consequently the best studied, is *Thiobacillus ferrooxidans*. *T. ferrooxidans* has also been shown to effect solubilization of cobalt, nickel, zinc, and lead (43). This organism, and most of the other bacteria found in mine dump sites, are autotrophic: they use carbon dioxide from the air for their carbon source, and they generate their energy from the oxidation of inorganic matter. In other words, they need no raw material input from miners who wish to exploit them.

The solubilization (leaching) of metals from ore by bacteria occurs in two ways: indirect and direct. The *indirect method* involves the transformation of ferrous iron to ferric iron by *T. ferrooxidans*. The ferric iron is a very powerful oxidizing agent that consequently converts metal sulfide minerals into acid soluble metal sulfate compounds. For example, ferric iron reacts with copper sulfide to form soluble copper sulfate. The *direct method* involves an enzymatic attack by the bacteria on sulfide minerals to give soluble sulfates and, in the process, also oxidizes the ferrous iron to ferric iron. The result is the same, i.e., the metal is soluble in an acid solution. The metal-laden solutions are collected and the metals are removed from solution by chemical and physical processes (see fig. 23). Additionally, since the use of coal as an energy source will increase, bacteria may be used to extract the sulfur from coal, making it less polluting.

Biotechnology could be used by the mining industry to create more efficient micro-organisms. Recombinant DNA technology could be used to effect the following improvements in selected bacteria:

- an enhancement in the rate at which the bacteria regenerate the ferric iron;
- greater tolerance to acidic conditions;
- greater tolerance to saline conditions;
- a decrease in the bacteria's sensitivity to some metals, especially thorium, silver, mercury, and cadmium; and
- an increase in the bacteria's ability to withstand high temperatures for deep mine operations.

Figure 23.—One Possible Configuration for a Leaching Process

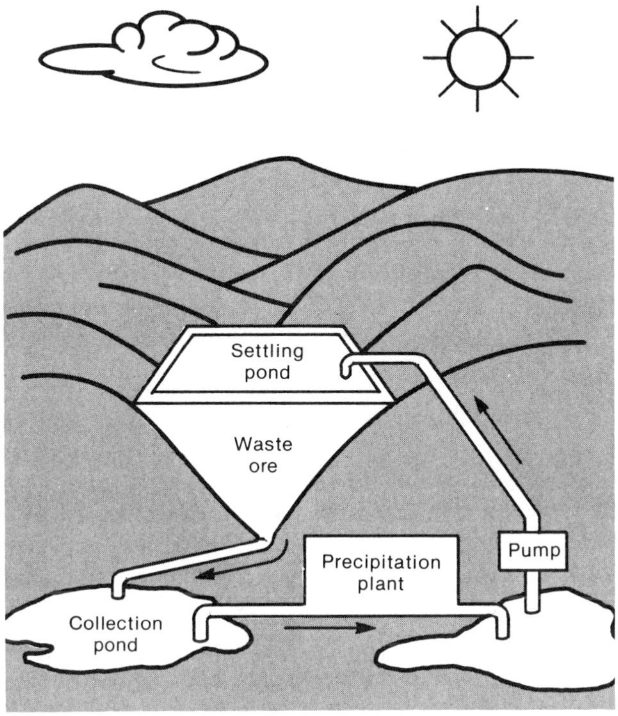

SOURCE: Office of Technology Assessment.

It is likely that rDNA technology will be able to address some of these problems in the near future (41,43,47).

Micro-organisms used in a metal-leaching operation are subjected to very different stresses than those used in a laboratory setting. These differences must be kept in mind when considering the use of rDNA technology, especially since most of the experience with the new technology has focused on well-defined laboratory strains in controlled environments. The bacteria used in leaching endure variable weather conditions, some quite inhospitable for most organisms. When a micro-organism is placed in the environment, it will most likely have to interact with other organisms, and this fact has to be taken into account when researching organisms of interest. Additionally, the mineralogy at each mine site is unique, so micro-organisms either will have to be modified for each site or will have to be able to act on varied feedstocks. It is unlikely that feedstocks will be prepared to suit the micro-organism. It seems that the most likely application for genetically manipulated micro-organisms in mineral leaching will be in the same area in which micro-organisms are used now, for the treatment of large quantities of discarded waste rock that have small quantities of valuable metals (41,43).

Because the leaching process takes place in the environment, the biological process cannot be completely controlled. Nevertheless, there are ways to optimize the reaction conditions for the micro-organisms of interest. The particle size and particle-to-solution ratio of the mine dumps can be manipulated. It is also possible to some extent to control the pH, temperature, and oxygen and carbon dioxide levels. By optimizing these conditions, the leaching organisms can be given an advantage over naturally occurring organisms.

In recent years, the search for new micro-organisms in such primeval environments as hot acid springs, volcanic regions, and deep ocean thermal vents has revealed many micro-organisms capable of metal transformations under harsh conditions. Not only are these organisms likely to have application in commercial metal recovery, but they also represent an enormous gene pool for improving existing leaching bacteria through rDNA technology.

Concentration of metals

Another area where micro-organisms could be useful to the mining industry is the concentration of metals from aqueous solutions. The R&D of this kind of process is somewhat easier than R&D of leaching because it can occur in more controlled laboratory situations, making manipulation of the organism's environment possible. There are two biological methods for concentrating metals. In one case, the metals are nonspecifically adsorbed to the surface of the organism. In the other, the metals are specifically bound and taken up by the organism. In the latter mechanism, metals can be concentrated up to 10,000 times. There is a great diversity of organisms that have been shown to concentrate metals, including bacteria, fungi, and algae. The metals they concentrate are primarily copper, uranium, silver, and the lanthanides. Recombinant DNA technology could be useful in developing organisms to expand the range of metals concentrated.

Another approach to concentrating metals involves the use of specific metal-binding proteins produced in higher organisms. One of the best studied metal-binding proteins is metallothionein, which binds cadmium, zinc, mercury, and copper. The use of these proteins is discussed earlier in the section on pollution control and toxic waste treatment.

Commercial aspects of biotechnology in microbiological mining

In the United States, there is no Federal R&D funding specifically earmarked for mining microbiology. The National Science Foundation and the U.S. Department of Energy (DOE) have funds under various programs that can be used for basic research studies on micro-organisms important in mining. In fiscal year 1984, neither agency anticipates funding at levels more than $300,000. The Bureau of Mines of the U.S. Department of the Interior did not fund any microbiology in fiscal years 1981 and 1982. In fiscal year 1983, it funded the Idaho National Engineering Laboratory at about $300,000 to study the leaching and concentrating of cobalt. The Bureau intends to continue the funding of this project at the same level in fiscal year 1984 (45).

Much of the R&D funding in this field comes from both large and small firms in the mining industry. Atlantic Richfield Co. is doing a substantial amount of research in this area. Other large companies investing in microbiological mining include General Electric, Koppers, Eastman Kodak, International Nickel Co., Chevron, W. R. Grace, and Standard Oil of California. Additionally, at least four small U.S. companies, Advanced Mineral Technologies, Inc. (Socorro, N. Mex.), Polybac (Allentown, Pa.), Genex, and Biogen S.A.* are researching mining and metal microbiology.

Two spinoff applications could derive from the work in the area of microbiological mining. One application is the recovery of expensive metals such as silver from processes such as photograph developing. In the past, the developing solutions containing the silver were disposed, but with the increased price of silver over the past few years, there has been increasing interest in silver recovery. Another application is using micro-organisms to reactivate metal catalysts, recovering metals that have been deposited on the catalyst. Both the catalyst is regenerated and the metal is recovered (48).

Several other countries, notably the United Kingdom, Australia, South Africa, and Canada, are interested in the applications of biotechnology in the mining industry. The majority of the R&D, however, is being done by private industry. Very little is funded by the Governments of these countries.

As of mid-1983, there were no genetically manipulated micro-organisms on the market (44). Yet it is possible that research efforts could yield useful, new bacteria for leaching and concentration of metals in a few years. If scale-ups and field trials (for leaching) were carried out expediently, marketable products for leaching and concentration could be available in less than 10 years (42). This research is proceeding slowly, however, because of the currently depressed state of the minerals market. Most industry experts hesitate to speculate when micro-organisms used for mining might reach the marketplace, because the worldwide availability and price of these metals will determine how fast the research will proceed. There will have to be a scarcity of the metal before much microbiological research will be done.

*Biogen is about 80-percent U.S. owned, but most of its work in microbiological mining is done by Biogen S.A. in Switzerland.

Microbial enhanced oil recovery

Conventional oil extraction technologies can recover only about 50 percent of the world's subterranean oil reserves. The balance either is trapped in rock or is too viscous to pump. The application of micro-organisms or their products possibly could be used to aid in the recovery of trapped oil. The use of microbial processes for this purpose is called microbial enhanced oil recovery (MEOR).

The interest in MEOR has increased substantially since 1975. Several conferences on the subject have brought together petroleum engineers and microbiologists to begin to analyze the roles that micro-organisms could play in the recovery of trapped oil. To date, several field tests have been done, but none have yet revealed a micro-organism that is broadly applicable in MEOR (51).

There are three general experimental approaches to MEOR (51):

- the stimulation of endogeneous micro-organisms by injection of nutrients into the well,
- the injection of laboratory-selected micro-organisms into the well, and
- the production by micro-organisms of specific biological compounds and the subsequent use of these compounds in wells.

As discussed further below, new biotechnology offers possibilities in the latter two approaches.

Uses of micro-organisms in oil wells

Various micro-organisms are now being isolated and examined for properties useful for oil extraction. Micro-organisms evolve gases, notably carbon dioxide, that could aid in repressurizing an oil well. An ideal microbe would use the less valuable parts of oil as a carbon source to produce surfactants or emulsifiers to lower the viscosity of the oil allowing it to be pumped to the surface. Several problems complicate this senario. No micro-organism has yet been found that degrades only the less useful components of oil; micro-organisms usually also degrade the compounds important to the petroleum industry. Some micro-organisms will not degrade the oil at all, but these micro-organisms need to have a carbon source, usually molasses, pumped into the well, and this increases the cost of production.

Microbes currently being studied survive only under conditions of moderate heat, salinity, and pressure (55,56). Given the wide variability in geological deposits, these micro-organisms have limited usefulness. However, there is substantial evidence that the oil reservoir is not as an untenable, restrictive environment for micro-organisms as some laboratory studies would indicate. Micro-organisms can, in fact, be isolated from deep reservoirs, and they may have developed specialized mechanisms to cope with low amounts of oxygen. Other micro-organisms have been isolated that do not need oxygen for growth. Further study of these organisms may lead to the development of micro-organisms useful to the petroleum industry (52).

Use of microbially produced compounds in oil wells

Another approach to MEOR, the use of microbially produced compounds in oil wells, could be a relatively near-term application of biotechnology. Biological compounds that could be injected into wells include surfactants and viscosity enhancers and decreasers. The search has begun for these compounds, but it is becoming increasingly obvious that little is known about these compounds and the micro-organisms that produce them.

Even with the lack of knowledge, however, two promising compounds have been isolated and studied. One substance, characterized at the University of Georgia, is a glycolipid from a bacteria named H-13. This substance reduces the viscosity of various heavy crude oils (51). Another substance, originally isolated in Israel but now studied in the United States, is called emulsan and has the property of emulsifying oil, allowing better flow and dispersal (54).* Field trials have included the cleaning of an oil tanker hold and an aircraft carrier runway (57). Emulsan proved effective at these jobs and holds promise for use in oil wells. Emulsan is being developed by Petroferm, USA (Amelia Island, Florida), and produced and marketed by Pfizer (50).

*Emulsan is discussed further in *Chapter 7: Specialty Chemicals and Food Additives*.

Commercial aspects of biotechnology in microbial enhanced oil recovery

Many of the major oil companies are thought to be investing in MEOR (49). The U.S. leader in this field appears to be Phillips Petroleum. Small U.S. firms doing R&D in MEOR include Petroferm, Genetics International (Boston), and Worne Biotechnology (Medford, N.J.). Only one company, Shell Oil Co., has stated that MEOR is too speculative for its R&D laboratories (55). Additionally, the U.S. Government, through DOE, is investigating MEOR.

Foreign companies and countries are also investigating MEOR, notably the Swiss firm Petrogenetic AG, the British Government and British Petroleum, the U.S.S.R., and the Peoples Republic of China (49,53).

The status of potential markets for MEOR is very questionable because of the lack of knowledge about MEOR's real potential. However, MEOR could potentially increase the production of oil and decrease the costs of recovery significantly.

Priorities for future research

The applications of new biotechnology in the environment are at a rudimentary stage, primarily because of the lack of knowledge about the genetics and biochemistry of the potentially useful micro-organisms and the environment in which they operate. Currently, most basic research is done with pure cultures that do not represent the real world situation. There is certainly no guarantee that a species of bacterium will perform in an outdoor environment as it does in the laboratory. Additionally, scale-up problems will be great because of the large size of the operations. Studies in all of these research areas are interdisciplinary. Unless there is close collaboration between biologists and engineers, it is unlikely that the research will be very productive.

Specific challenges for pollution control and toxic waste treatment include:

- the isolation and characterization of enzymes to polymerize low molecular weight organic compounds,
- better characterization of metallothioneins from various species,
- the identification of polysaccharides to serve as bioflocculants,
- the development of enzymes for sludge dewatering,
- the development of microbial strains or enzymes that degrade toxic compounds, and
- the development of improved polysaccharide hydrolases to degrade slimes.

Specific challenges for microbiological mining include:

- the development of micro-organisms that could leach valuable metals such as thorium, silver, mercury, gold, platinum, and cadmium;
- a better understanding of the interactions between the micro-organisms and the mineral substances; and
- the development of DNA transfer technologies for use at low pHs.

Specific challenges for MEOR include:

- better biochemical and physiological understanding of micro-organisms already present in oil reservoirs,
- the development of a micro-organism that degrades only the less useful components of oil, and
- screening of micro-organisms for the production of surfactants and viscosity enhancers and decreasers.

Chapter 8 references

Pollution control and toxic waste treatment references

1. Alberti, B. N., and Klibanov, A. M., "Enzymatic Removal of Dissolved Aromatics From Industrial Aqueous Effluents," presented at American Chemical Society Meeting, New York, 1981.
2. AWWA Research Foundation, *Workshop on VOCs,* Denver, Colo., 1982.
3. Barth, E. F., Ettinger, M. B., Sulotto, B. V., et al., "Summary Report on the Effect of Heavy Metals on the Biological Treatment Process," *J. Water Poll. Control Fed.* 37:86-96, 1965.
4. *Biotechnology Newswatch,* "To Degrade Stubborn Toxins, European Labs Recruit Transposons," Aug. 15, 1983, p. 1.
5. Blaich, R., and Esser, K., "Function of Enzymes in Wood Destroying Fungi: II. Multiple Forms of Lacasse in Whiterot Fungi," *Arch. Microbiol.* 103:271-277, 1975.
6. Bollag, J. H., Liu, S. Y., and Minard, R. H., "Asymmetric Diphenol Formation by a Fungal Laccasse," *Appl. Environ. Microbiol.* 38:90-92, 1979.
7. Buckma, S., "Water Reuse and Deposit Control," *Southern Pulp and Paper Manufacturer,* April 1973, pp. 17-22.
8. Dugan, P., *Bioflocculation and the Accumulation of Chemicals by Floc-Foaming Chemicals,* Environmental Protection Technology Series No. EPA-600-2-75-032 (Cincinnati, Ohio: Municipal Environmental Research Laboratory, Office of Research and Development, 1975).
9. Dyck, A. W. J., "What's New in Slime and Deposit Control," *American Paper Industry,* May 1974, pp. 16-17.
10. Freedman, L., "Using Chemicals for Biological Control in Cooling Water Systems; Some Practical Considerations," *Indus. Water Engineer.* 16:14-17, 1979.
11. Froehner, S. C., and Eriksson, K. E., "Purification and Properties of *Neurospora crassa* Laccasse," *J. Bacteriol.* 120:458-465, 1974.
12. Genex, Rockville, Md., internal communication, 1982.
13. Genex, "Impact of Biotechnology on Pollution Control and Toxic Waste Treatment," contract report prepared for the Office of Technology Assessment, U.S. Congress, Washington, D.C., April 1983.
14. Impact of Applied Genetics in Pollution Control, program offered by the Center for Continuing Education, University of Notre Dame: sponsored by the Department of Civil Engineering, University of Notre Dame, and Hooker Chemical Co., 1982.
15. Kang, K. S., and Cottrell, I. W., "Polysaccharides," *Microbial Technology,* vol. 1, H. J. Peppler and D. Perlman (eds.) (New York: Academic Press, 1979).
16. Kellog, S. T., Chatterjee, D. K., and Chakrabarty, A. M., "Plasmid-Assisted Molecular Breeding—New Technique for Enhanced Biodegradation of Persistent Toxic Chemistry," *Science* 214:1133-1135, 1981.
17. Kilbane, J. J., Chatterjee, D. K., Karns, J. B., et al., "Biodegradation of 2,4,5-trichlorophenoxyacetic Acid by a Pure Culture of *Pseudomonas cepacia*," *Appl. Envir. Microbiol.* 44:72-78, 1982.
18. Kirkup, R. E., and Nelson, L. R., "City Fights Grease and Odor Problems in Sewer System," *Public Works Magazine,* October 1977.
19. Klibanov, A. M., and Alberti, B. N. "Enzymatic Removal of Toxic Phenols and Aromatic Amines From Industrial Waste Waters," *Abstracts of Third Symposium on Biotechnology in Energy Production and Conservation,* Gatlinburg, Tenn., May 12-15, 1981.
20. Klibanov, A. M., and Morris, E. D., "Horseradish Peroxidase for the Removal of Carcinogenic Aromatic Amines From Water," *Enzyme Microb. Technol.* 3:119-122, 1981.
21. Kojima, Y., and Kagi, J. M. R., "Metallothionein," *Trends in Biological Science* 3:90-92, 1978.
22. Mbikay, M., Brzezinski, R., and Thirion, J., "Differential Expression of Cloned Mouse Metallothionein Sequences in *Escherichia coli*," *DNA* 2(1): 23-29, 1983.
23. Morris, J. C., and McKay, G., "Formation of Halogenated Organics by Chlorination of Water Supplies," U.S. Environmental Protection Agency, Office of Research and Development, Washington, D.C., 1975.
24. Muckerman, D. K., "Production of Trihalomethanes From the Chlorination of Cast Iron Water Pipe Bacterial Films," prepared by the Department of Engineering, Missouri University, Rolla, for the Office of Water Research and Technology, U.S. Department of the Interior, Washington, D.C., 1979.
25. O'Brien, J. D., "Formation and Control Trihalomethanes in Chlorinated Drinking Waters Containing Fulvic Acid," Project No. A-048-NH, prepared by the Water Resource Research Center, University of New Hampshire, for the U.S. Department of the Interior, Washington, D.C., 1980.
26. O'Brien, J. D., Bixby, R. L., Jewett, M. A., et al., "For-

mation and Control of Trihalomethanes in Chlorinated Drinking Waters Containing Fulvic Acid," Research Report No. 27 (Durham, N.H.: Water Resource Research Center, University of New Hampshire, 1980).
27. Page, T., Harris, R., and Bruser, J., "Waterborne Carcinogens: An Economist's View," *The Scientific Basis of Health and Safety Regulation,* R. W. Crandall and L. B. Lave (eds.) (Washington, D.C.: Brookings Institution, 1981).
28. Rook, J. J., "Haloforms in Drinking Water," *J. Amer. Water Works Assn.* 8:168-172, 1976.
29. Sandborn, J. A., "Progress in Methods of Slime Control," *Paper Trade Journal* 119:243-249, 1944.
30. Sandborn, J. A., "Four Steps To Control Troublesome Outbreaks of Slime," *Paper Trade Journal,* Oct. 18, 1965, pp. 56-59.
31. Saunders, F. M., Chairman, Subcommittee on Toxic Substances Research Committee, *Research Needs Associated With Toxic Substances in Wastewater Treatment Systems,* prepared for the Water Pollution Control Federation, Washington, D.C., 1982.
32. Schwien, V., and Schmidt, E., "Improved Degradation of Monoclorophenols by a Constructed Strain," *Appl. Environ. Microbiol.* 44:33-39, 1982.
33. Sjoblad, R. D., and Bollag, J. M., "Oxidative Coupling of Aromatic Pesticide Intermediates by a Fungal Phenol Oxidase," *Appl. Environ. Microbiol.* 33:906-910, 1977.
34. Sutherland, I., "Biosynthesis of Microbial Polysaccharides," presented at the 83rd Annual Meeting of the American Society of Microbiology, New Orleans, La., 1983.
35. Symons, J. M., "Interim Treatment Guide for Controlling Organic Contaminants in Drinking Water Using Granular Activated Carbon" (Washington, D.C.: Water Supply Research Laboratory, Office of Research and Development, U.S. Environmental Protection Agency, 1978).
36. U.S. Environmental Protection Agency, Office of Research and Development, "Formation of Halogenated Organics by Chlorination of Water Supplies," EPA-600/175-002, Environmental Health Effects Series, Washington, D.C., 1975.
37. U.S. Environmental Protection Agency, Office of Toxic Substances, *Preliminary Assessment of Suspected Carcinogens in Drinking Water,* report to the U.S. Congress, Washington, D.C., 1975.
38. *Water Treatment Handbook,* 5th ed. (New York: Halsted Press, 1979).
39. Wise, D. L., Kisdert, R. E., and Langton, E. W., "A Review of Bioconversion Systems for Energy Recovery From Municipal Solid Waste: Part II, Fuel Gas Production," *Resources and Conservation* 6:117-136, 1981.

Microbiological mining references

40. Brierley, C. L., "Bacterial Leaching," *CRC Critical Reviews in Microbiology* 6:207-262, 1978.
41. Brierley, C. L., "Microbiological Mining," *Scientific American,* August 1982, pp. 44-53.
42. Brierley, C. L., Advanced Mineral Technologies, Inc., Socorro, N. Mex., personal communication, 1983.
43. Chakrabarty, A. M., "Genetic Mechanisms in Metal-Microbe Interactions," *Metallurgical Applications of Bacterial Leaching and Related Microbiological Phenomena,* L. E. Murr, et al. (eds.) (New York: Academic Press, 1978).
44. Curtin, M. E., "Microbial Mining and Metal Recovery: Corporations Take the Long and Cautious Path," *Bio/Technology,* May 1983, pp. 229-235.
45. Kesterke, D., U.S. Department of the Interior, Bureau of Mines, personal communication, 1983.
46. Palmiter, R. D., Brinster, R. L., Hammer, R. E., et al., "Dramatic Growth of Mice That Develop From Eggs Microinjected With Metallothionein—Growth Hormone Fusion Genes," *Nature* 300:611-615, 1982.
47. Ralph, B. J., "The Implications of Genetic Engineering for Biologically Assisted Hydrometallurgical Processes," *Genetic Engineering-Commercial Opportunities in Australia* (Canberra: Australian Government Publishing Service, 1982).
48. Volger, H. C., "Microbial Leaching," *Biotechnology: a Dutch Perspective,* J. H. F. van Apeldoorn (ed.) (Delft, Netherlands: Delft University Press, 1981).

Microbial enhanced oil recovery references

49. *Biotechnology Newswatch,* "Oilmen Enlisting Microbiologists To Aid Tertiary Oil Recovery," June 7, 1982, p. 1.
50. *Biotechnology Newswatch,* "Pfizer To Scale Up 'Emulsan' Output Marketing, Research for Petroferm Patent Holder," Sept. 19, 1983, p. 1.
51. Finnerty, W. R., and Singer, M. E., "Microbial Enhancement of Oil Recovery," *Bio/Technology,* March 1983, pp. 47-54.
52. Finnerty, W. R., University of Georgia, personal communication, September 1983.
53. *Genetic Technology News,* "Recombinant DNA Could Help Microbial Enhanced Oil Recovery," March 1982, p. 1.

54. Gutnick, D. L., et al., "The Interaction of Emulsan With Hydrocarbons," *Proceedings of the International Conference on Microbial Enhancement of Oil Recovery,* Afton, Okla., in press, 1982.
55. Lancaster, H., "Scientists Turning to Microbes To Improve Oil Well Output," *Wall Street Journal,* May 28, 1982, p. 17.
56. Springham, D., "Bugs to the Oil Industry's Rescue," *New Scientist,* May 13, 1982, pp. 408-410.
57. Warren, R., Naval Air Systems Command, personal communication, 1983.

Chapter 9

Commodity Chemicals and Energy Production

Contents

Tables

Figures

Chapter 9

Commodity Chemicals and Energy Production

Introduction

In 1982, the U.S. chemical industry produced about 158 billion pounds (lb) of organic chemicals (36). About 30 commodity chemicals—defined in this report as chemicals that sell for less than $1 per lb*—constitute the majority of this market (see table 39).

*Chemicals with higher value such as vitamins, food additives, and amino acids, form the subject of *Chapter 7: Specialty Chemicals and Food Additives*. The difference between "commodity" and "specialty" chemicals is somewhat fluidly determined by price versus quantities produced. Some of the compounds described in chapter 7 are considered by some analysts to be commodity chemicals. These include vegetable oils and their derivatives, single cell protein, and fructose. Because of their predominant use as food additives, however, these compounds are considered in the earlier chapter.

Practically all commodity chemicals are currently made from petroleum and natural gas resources and are used as precursors for a variety of materials such as polymers and solvents. The United States, which now imports about 30 percent of its petroleum (34), uses about 7 to 8 percent of its total petroleum and natural gas supply for the production of commodity chemicals (10,18,22); the remainder of this supply is used as an energy source.

The chemical industry's reliance on petroleum feedstocks raises a number of problems. Two problems are the fluctuating cost and uncertain

Table 39.—Annual Production and Selling Price of the Major Organic Commodity Chemicals in the United States

Chemical	Production in 1982 (billion pounds)	Price in 1982 (¢/lb)	Major uses
Ethylene	24.7	25.5	Polyethylene derivatives
Toluene	15.3	26.7	Benzene, gas additive, solvents, polyfoams
Propylene	12.3	24.0	Polypropylene, isopropanol
Ethylene dichloride	10.0	13.7	Vinyl chloride
Benzene	7.9	21.1	Styrene, phenol, cyclohexane
Methanol	7.3	10.8	Formaldehyde
Ethylbenzene	6.6	30.0	Styrene
Vinyl chloride	6.5	22.0	Polyvinyl chloride, resins
Styrene	5.9	37.5	Polystyrenes
Xylene	5.3	18.9	p- and o-xylene, gas additive, solvent
Terephthalic acid	5.0	N.A.	Polyester fibers
Ethylene oxide	4.9	45.0	Ethylene glycol
Formaldehyde	4.7	24.4	Resins
Ethylene glycol	4.0	33.0	Antifreeze, polyesters
p-xylene	3.2	31.0	Synthetic fibers
Acetic acid	2.8	26.5	Vinyl and cellulosic acetate
Cumene	2.7	24.0	Phenol
Phenol	2.1	36.0	
Acrylonitrile	2.0	44.5	Polymers
Vinyl acetate	1.9	37.5	Polyvinyl acetates, alcohols
Butadiene	1.8	40.0	Rubber
Acetone	1.8	31.0	
Propylene oxide	1.5	40.5	Propylene glycol, urethanes
Isopropanol	1.3	32.9	Acetone, solvents
Cyclohexane	1.3	25.3	Nylon, caprolactum
Adipic acid	1.2	57.0	Nylon
Acetic anhydride	1.1	41.0	Cellulose esters
Ethanol	1.1	25.8	Detergent, solubilizer, cosmetics, solvent, fuel

SOURCE: Office of Technology Assessment, adapted from D. Webber, "Basic Chemical Output Fell Third Year in a Row," *Chem. Eng. News*, May 2, 1983, pp. 10-13; T. C. O'Brien, "Feedstock Trends for the Organic Chemical Industry," Planning Report 15, U.S. Department of Commerce, National Bureau of Standards, April 1983; and *Chemical Marketing Reporter*, "Weekly Price Report," May 31, 1982, pp. 35-39.

supplies of petroleum. Commodity chemical prices are especially sensitive to the cost of petroleum because feedstock costs typically represent 50 to 75 percent of commodity chemical manufacturing costs (6). Other problems of the commodity chemical industry include a current overcapacity of production by the capital-intensive petrochemical companies, the high costs of energy associated with "cracking" petroleum into chemical feedstocks, and environmental, safety, and ideological concerns surrounding the use of nonrenewable, fossil resources (6).

These well-publicized problems, which increase in urgency with the passing of time, have intensified the search for nonpetroleum feedstocks for chemical and energy production. The options being pursued at present include the liquification and gasification of coal, the development of synthetic fuel from natural gas, and the conversion of biomass** to fuels and a wide variety of organic chemicals.

The substitution of natural gas, coal, and other nonrenewable resources for petroleum are issues that have been discussed in several previous OTA reports (28,29,31). Despite the drawbacks outlined in those reports, coal is favored as an alternative resource by U.S. petroleum companies, which control 20 percent of U.S. coal production and 25 percent of U.S. coal reserves (3,27). Processed coal feedstocks fit readily into most petroleum feedstock schemes for the production of commodity chemicals and thus do not require large capital investments for new chemical plants. Nevertheless, at least one analyst thinks that petroleum will continue to be used as a feedstock for commodity chemicals for some time and that coal will not make a significant impact on the production of chemicals until the 21st century (22).

It appears that countries with substantial inexpensive supplies of petroleum, such as Mexico and Saudi Arabia, are turning to the production of commodity chemicals as a way of adding value to their resources. Thus, countries with petroleum may begin to control the price of these chemicals. Because such countries may be able to produce commodity chemicals at a lower price, companies in the United States, Europe, and Japan may have to develop new ways of using commodity chemicals to produce compounds of greater value or to move directly to the manufacture of higher value-added chemicals from biomass. In any case, a rapid or dramatic shift in feedstock use is unlikely; it is much more probable that there will be a slow transition to the use of biomass as a feedstock in particular instances.

Although nonrenewable resources such as coal will probably be adopted earlier, biomass—including crop and forest product wastes and municipal and agricultural wastes—may provide solutions to some of the long-term problems associated with chemical and energy production from petroleum. It is technologically possible to produce essentially all commodity chemicals from biomass feedstocks such as starch or cellulose, and most commodity chemicals can be synthesized biologically (10,24). A viable biomass feedstock for the production of commodity chemicals may be starch. Less than 1 percent of the U.S. corn crop would be required to obtain the cornstarch needed to produce a typical commodity chemical at the rate of 1 billion lb per year (18). Although a few high-volume chemicals that could be produced from biomass, such as ethanol, can be used for fuel, the volume of biomass needed to produce a nation's energy would be substantially greater than that needed to produce its commodity chemicals. Starch probably could not be used for energy production without putting a strain on food and feed uses. Thus, if biomass is to be used extensively for energy production, the biomass source will most likely be lignocellulose.

Biomass as an alternative to petroleum for U.S. energy production was described in OTA's July 1980 report *Energy From Biological Processes* (30). As emphasized in that report, substantial societal change, i.e., more public support and a higher priority for research on biomass use in the U.S. Departments of Agriculture and Energy programs, will be necessary if biomass is to become a viable alternative to petroleum as a source of energy in the near future. At present, the level of U.S. public support for biomass research is not high. Furthermore, Federal support of applied research and development (R&D) programs for al-

**Biomass is all organic matter that grows by the photosynthetic conversion of solar energy.

ternative fuel sources has been plummeting in the recent climate of intense fiscal scrutiny.

A shift from petrochemical processes to bioprocesses for the production of commodity chemicals will be difficult because of the existing infrastructure of chemical and energy production. This infrastructure allows a barrel of oil to be converted to products in a highly integrated system in which the byproduct of one reaction may form the substrate for another reaction. Most chemicals derived from biomass cannot yet compete economically with chemicals made from oil in this infrastructure.

As the costs of bioprocesses are reduced through R&D, however, a transition to biomass resources may become a more realistic proposition. This chapter examines ways in which biotechnology might improve the efficiency of biomass conversion, thus facilitating the transition to the use of biomass resources. The advances biotechnology could provide for the improved growth of plants used for biomass conversion are discussed in *Chapter 6: Agriculture.*

Since commodity chemicals represent only a small portion of today's U.S. petroleum consumption, a transition to biomass-based commodity chemical production without a concurrent transition to biomass-based energy production will not substantially reduce the country's dependence on petrochemical resources. For moving the United States toward the goal of reduced reliance on imported, nonrenewable resources, a unified approach to chemical and fuel production will be necessary.

Biomass resources

The United States has abundant biomass resources. The largest potential amount of cellulosic biomass is from cropland residues such as corn stover and cereal straw,* although the potential amount of cellulosic biomass from forest resources is also quite large. About 550 million dry tons of lignocellulose are easily collected and available for conversion to chemicals each year. In addition, some percentage of the 190 million dry tons of corn produced yearly could be converted to starch and used for chemical production (21).

Parameters used to determine the optimal kind of biomass used in microbial systems include availability of the biomass, its energy content per dry weight, the amount of energy that must be expended to achieve the desired products, the environmental impact of the process, and the amenability of the material to conversion by existing microbial systems. Ultimately, biomass resources that minimize usurpation of food sources are sought (e.g., nonfeed crops grown on extant arable land).

This chapter emphasizes the use of the two most abundant feedstocks from biomass: starch and cellulose. Starch and cellulose are both polymers of glucose units (6-carbon simple sugars) which, when hydrolyzed, yield glucose molecules (see fig. 24). These glucose sugars provide the starting point for biological chemical production, for example, the transformation of glucose to ethanol. Other derivatives of biomass, such as vegetable oils, are used in bioprocesses, and those resources are considered in *Chapter 7: Specialty Chemicals and Food Additives.*

One drawback to the use of biomass as a feedstock for commodity chemical and energy production is its relatively low energy content per unit dry weight. Dry cellulose biomass, for example, yields roughly 16 million Btu per ton and cornstarch yields 15 million Btu per ton, whereas petroleum yields 40 to 50 million Btu per ton. Thus, the energy yield per unit of weight is lower for biomass than for petroleum. Furthermore, the costs of transporting biomass to a factory may be an important economic consideration. Raw material and transportation costs are particularly important in the production of commodity chemicals, because of the low value added to the feedstock in the synthesis of final products.

*Agricultural residues left on the soil aid in the sustainability of soil. The environmental impact of the removal of these residues must be studied more thoroughly in order to determine whether agricultural wastes are, in fact, true wastes.

Figure 24.—Polysaccharides of Biomass: Starch and Cellulose

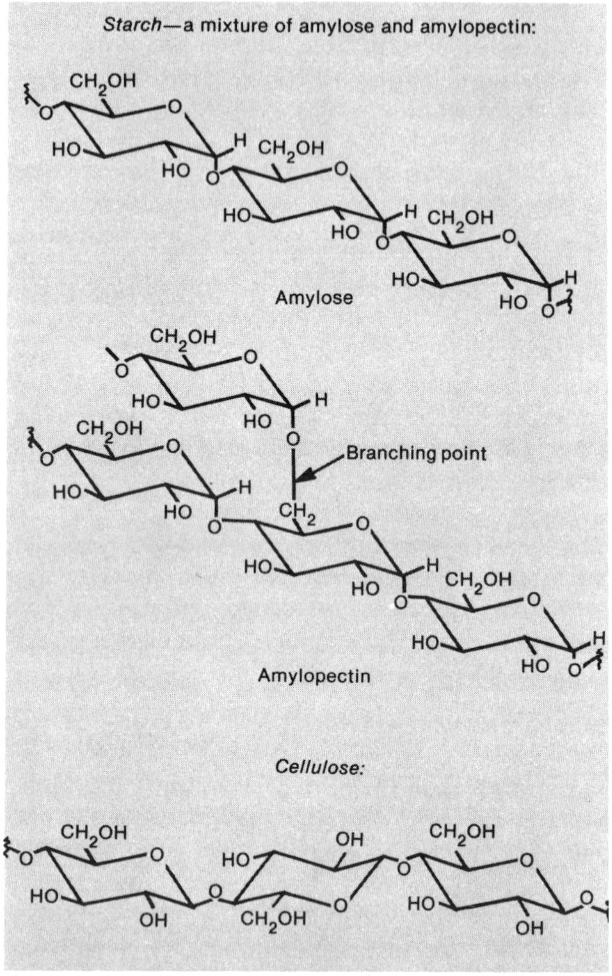

SOURCE: Office of Technology Assessment.

Starch

Starch, a molecule composed of many hundreds of glucose units bound together in branched or unbranched chains, is the principal carbohydrate storage product of higher plants and is readily available from such crops as corn and potatoes. In 1979, the United States produced about 666 billion lb of grain from six major cereal crops, and this grain contained 470 billion lb of starch. The major grain produced, corn, contained 316 billion lb of starch (10), which could provide 285 billion lb of glucose.

As shown in figure 25, world grain production has increased steadily over the past several years, and this trend is expected to continue through the end of the century as the result of yield improvement and an expansion of acreage planted (19). Furthermore, the price of corn has remained relatively constant over the past decade, especially when compared to the nearly tenfold increase in the price of oil over the same period of time.

Figure 25.—Trends in World Grains Production (million metric tons)

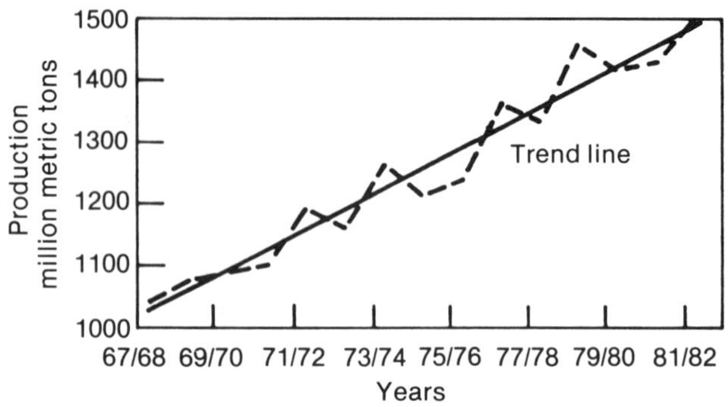

SOURCE: Office of Technology Assessment, adapted from CPC, Int., 1983.

The utilization of U.S. corn has changed over the past 10 years. A decrease in U.S. meat consumption caused a concurrent decrease in the amount of corn used for animal feed, while at the same time, technological advances increased corn yields. Consequently, the export market for U.S. corn has risen from 15 percent of the crop to 35 percent. Since U.S. corn production is expected to increase and meat consumption is expected to decrease, U.S. farmers will need new markets for their corn. Commodity chemical production from a starch feedstock could provide a market for U.S. corn. The potential for industrial use of starch from corn alone is large, and an increase in the industrial use of corn would probably aid in supporting farm prices. Currently, only about 7 percent of the corn produced in the United States is processed into cornstarch (7,19). Figure 26 suggests that 14 percent of the 1990 corn crop could go to chemical production, and enough corn would still be available for other uses.

Because of its high solubility in water and ease of hydrolysis into individual glucose units, starch is highly amenable to bioprocessing and may be an ideal feedstock for chemical production. The use of starch for both chemical and fuel production, however, might be at the expense of its use

Figure 26.— Trends in U.S. Corn Utilization

in food production. Starch may not be produced in large enough quantities to be used both as a source of food and a source of energy.*

Lignocellulose

Lignocellulose is composed of cellulose, an unbranched chain of glucose units, lignin, a linked mixture of aromatic molecules, and hemicellulose, a polymer composed mainly of 5-carbon sugars. This structure provides the rigidity necessary for cellulose's primary function, the support of plants. Because of its wide availability, lignocellulose has the potential to be the most important of all the raw materials for use in bioprocessing. Currently, however, several problems impede the use of lignocellulose on a large scale. Lignocellulose is highly insoluble in water and its rigid structure makes cellulose much more difficult than starch to hydrolyze to individual sugars. Furthermore, most micro-organisms cannot utilize lignocellulose directly without its having been pretreated either chemically or physically. Despite the considerable advances made in both chemical and enzymatic hydrolysis techniques, the cost of glucose derived from cellulose is still much higher than that derived from starch.

The inherently diffuse nature of lignocellulose resources means that very high collection costs, especially in energy and manpower, will be encountered in any attempt at large-scale utilization. These considerations have given rise to the concept that the utilization of lignocellulose for energy will be feasible only through a widespread network of smaller manufacturing facilities that draw on local resources and supply local needs. Indeed, this pattern has already been established for farm-scale alcohol production from corn. An alternative to multiple small-scale production units

*As detailed in OTA's July 1980 report *Energy From Biological Processes* (30), starch could be used to produce approximately 1 billion to 2 billion gal of ethanol in the United States each year (about 1 to 2 percent of U.S. gasoline consumption) before food prices might begin to rise.

is the concept of centralized, intensive lignocellulose production on so-called "energy plantations." The potential ecological problems and highly questionable economics have detracted from prospects for success, but the development with biotechnology of more effective biological agents for lignocellulose utilization could radically change this picture.

Conversion of biomass to commodity chemicals

As noted above, there are numerous types of biomass resources, including lignocellulosic products and feed crops such as corn. Because of the varying compositions of these raw materials, different methods are used in rendering them into useful chemicals. Nevertheless, all microbial conversion of biomass to marketable chemicals is a multistep process that includes:

- pretreatment (particularly with lignocellulosic biomass),
- hydrolysis (saccharification) to produce hexose (6-carbon) and pentose (5-carbon) sugars,
- bioprocessing of these sugars by specific micro-organisms to give commodity chemicals,
- subsequent bioprocessing or chemical reactions to produce secondary commodity chemicals, and
- separation and purification of end products.

Figure 27 is a schematic summary of the multistep processes for the conversion of starch and lignocellulosic biomass to commodity chemicals. Although figure 27 emphasizes the microbial steps that could be used for these processes and applications of biotechnology to them, it should be noted that a variety of chemical syntheses can also be used to convert the components of biomass into useful chemicals (9,10,17,18).

Figure 27.—

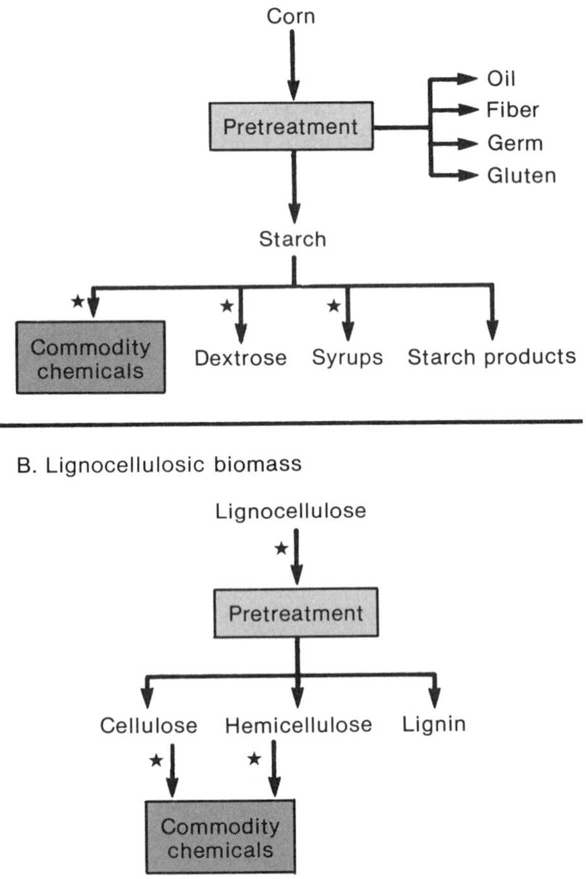

★ = Possible microbiological step

SOURCE: Office of Technology Assessment.

Pretreatment

Before either starch or lignocellulosic biomass can be used as feedstocks for bioprocesses, they must be pretreated in preparation for hydrolysis. Starch from corn requires little pretreatment. Lignocellulosic materials such as wood, however, demand extensive pretreatment to make cellulose and hemicellulose available for hydrolysis.

STARCH

The United States relies primarily on corn for starch feedstocks. About 500 million bushels are processed by corn refiners yearly to produce cornstarch products. In the production of cornstarch, refiners employ a process known as "corn wet milling" in which corn kernels are cleaned, soaked in warm, dilute acid, and ground to yield a slurry composed of starch, protein, and oil. Much of the starch is further converted to sweeteners, such as glucose and high fructose corn syrup (7). Cornstarch is the milling product that could be used to make commodity chemicals.

The pretreatment of starch requires minimum inputs of acid and heat. Energy requirements are low compared with the potential energy gained, and almost all byproducts are marketable. Combined with starch thinning and saccharification costs (see below), corn wet milling is estimated to yield monomeric sugar at a cost of 12¢/lb (at $3.40/bushel of corn) (21).

LIGNOCELLULOSE

Methods used to pretreat lignocellulosic biomass include chemical pretreatment in acids and bases, steam explosion, and mechanical grinding. These methods, described in OTA's July 1980 biomass report (30), add substantially to the costs involved in using lignocellulosic biomass as a chemical resource.

In the future, biodelignification (the biological degradation of lignin) by micro-organisms may prove useful in the pretreatment of lignocellulosic biomass (8,24). Biodelignification results in removal of lignin, exposing the crystalline cellulose and lowering the costs of mechanical pretreatment. At present, however, biodelignification is an inadequate, expensive means of pretreatment, and it is not used in the pilot projects for use of lignocellulose currently underway. As yet, there are no valuable uses for lignin. Uses must be found for lignin derivatives before these processes will be commercially viable (2).

Several groups are working toward obtaining faster biodelignification using mixed cultures of micro-organisms, but microbial reaction rates at present do not approach those needed for economic feasibility. With use of the best candidate, the degradative mold *Chrysosporium pruinosum*, 40 percent of lignin remains intact after 30 days of treatment (1). At least 20 strains of bacteria that have lignodegradative abilities have been identified, but efforts to use micro-organisms for delignification are hampered by the fact that lignin metabolites are toxic to these micro-organisms. Thus, more work remains to be done before biodelignification and other methods of biological pretreatment are competitive with the currently used chemical or mechanical pretreatment methods. Were more information available on these micro-organisms, biotechnology could be used to improve their efficiency.

Hydrolysis

STARCH

Enzymes from microbial systems are widely used industrially to catalyze hydrolysis of starch into sugars.* Batch bioprocesses are used for hydrolysis. Three enzymes, alpha-amylase, beta-amylase, and glucoamylase, are used to hydrolyze the starch chains to yield complete hydrolysis and the formation of glucose (15). The largest industrial use of enzymes is in the corn wet milling industry.

The major U.S. corn refiners have ongoing active research programs for the improvement of enzymatic degradative processes, and these manufacturers have made major advances in the areas of bioprocessing and enzyme immobilization. These manufacturers have continued their efforts toward improvement of enzymes by using new biotechnology (32).

CELLULOSE

The well-ordered crystalline structure of cellulose necessitates harsher treatments than those used for starch. Whereas hemicellulose is readily hydrolyzed into its 5-carbon sugars under mild conditions, the hydrolysis of cellulose requires strong acids, heat, and pressure. These conditions lead to the formation of byproducts which must be separated and utilized to minimize the overall costs of lignocellulose use. In addition, the acid used for the hydrolysis of cellulose must be neutralized before the mixture is used for bioprocessing, a requirement that raises the cost of hydrolysis.

*For further discussion of these enzymes, see *Chapter 7: Specialty Chemicals and Food Additives*.

The use of enzymes known as cellulases (and micro-organisms that produce cellulases) to hydrolyze cellulose, either alone or in conjunction with chemical treatment, offers an increasingly popular alternative to chemical methods of hydrolysis. Cellulose is the most abundant biological compound on earth, and a myriad of micro-organisms employ cellulases to obtain energy for growth from the resulting glucose molecules. Research efforts to improve cellulase activity by mutagenesis and selection of cellulolytic (cellulose-degrading) micro-organisms have yielded mutant strains of micro-organisms (particularly fungi) that produce cellulases with higher tolerance to glucose (the product of hydrolysis that inhibits cellulase activity), increased efficiency and reaction rate, and better functioning at the elevated temperatures and high acidities used in industrial bioprocesses (1).

The enzymatic activity of cellulases has been improving over the past several years, and in some cases, the time needed for saccharification and subsequent bioprocessing to produce ethanol from cellulose has been reduced several fold (11). Despite these improvements, however, the activity of cellulases does not begin to compare with the activities of amylases, which are about 1,000 times more catalytically efficient (5).

Although research into the molecular biology of cellulases is in its early stages, biotechnology is being used to improve the cellulase-catalyzed hydrolysis of cellulosic biomass in several ways. Two challenges for biotechnological approaches to cellulase production are increasing the low activity of the cellulase and making sure the entire cellulase gene complex is expressed. Processes that optimize cellulase activity and efficiency are prerequisite to the use of lignocellulosic biomass resources.

Researchers at the National Research Council of Ottawa, Canada, the University of British Columbia, the University of North Carolina, and Cornell are using recombinant DNA (rDNA) techniques to clone cellulase genes from several micro-organisms into bacteria that may be induced to produce cellulase in large quantities (20). Similarly, researchers at the U.S. Department of Agriculture are cloning cellulase genes from the fungus *Penicillium funiculosum* (12).

Another possibility for a biotechnological improvement is to transfer the ability to utilize the 5-carbon sugars from hemicellulose into cellulose-utilizing micro-organisms. A third possibility is improving the specificity of organisms that can utilize lignocellulose directly, e.g., *Clostridium thermocellum.* The "wild types" of these micro-organisms produce a range of products, typically ethanol and several organic acids. This varied synthesis results in low yields for each product and great difficulties in subsequent recovery and purification. Genetic mechanisms could be used to select for high production of any one of the products.

Microbial production of commodity chemicals

Some commodity chemicals, including ethanol and acetic acid, are now produced in the United States with microbial bioprocesses (9), while other chemicals, such as ethylene and propylene, will probably continue to be made from petroleum feedstocks because of lower production costs. The commodity chemicals that are attractive targets for production from biomass include ethanol, acetone, isopropanol, acetic acid, citric acid, propanoic acid, fumaric acid, butanol, 2,3-butanediol, methyl ethyl ketone, glycerin, tetrahydrofuran, and adipic acid (9,18). Additionally, some chemicals, such as lactic and levulinic acids, could be used as intermediates in the synthesis of polymers that might replace petrochemically derived polymers (18).

Because the chemical composition of biomass differs from that of petroleum and because micro-organisms are capable of a wide range of activities, it may be that the most important commodity chemicals produced from biomass will be, not chemicals that directly substitute for petrochemicals, but other chemicals that together define a new structure for the chemical industry. Micro-organisms used to produce organic chemicals could be used with micro-organisms that fix nitrogen to produce nitrogeneous chemicals, either higher value-added compounds or ammonia, a high volume commodity chemical. Other micro-organisms, such as the methanogens or the micro-organisms that metabolize hydrogen sulfide, may be used to produce sulfur-containing chemicals (14).

The aerobic and anaerobic microbial pathways leading to a number of important compounds are shown in figure 28. Some of the micro-organisms responsible for these reactions are listed in table 40. Knowledge of biochemical pathways for the synthesis of particular chemicals will lead to the

Figure 28.—

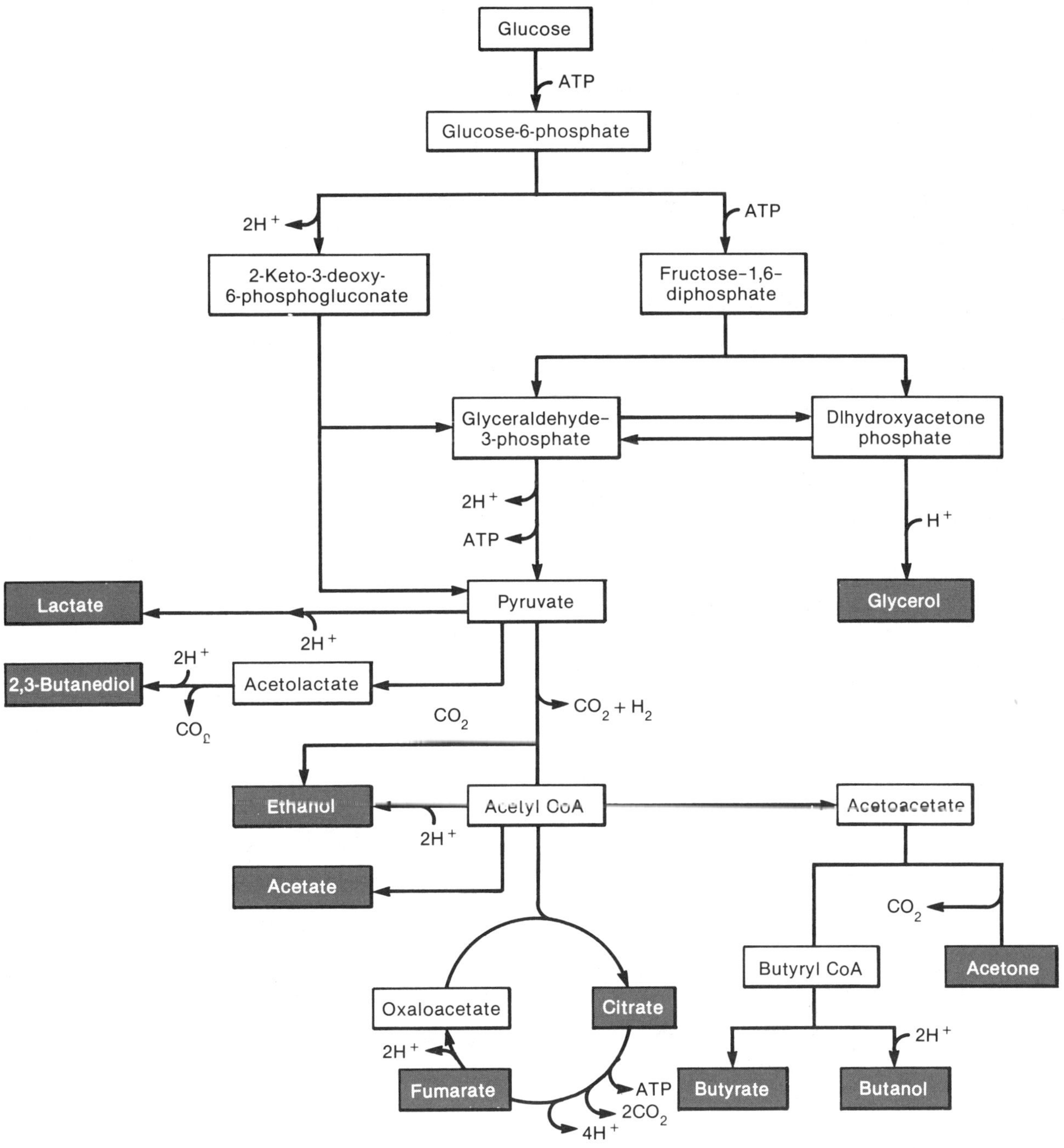

Potentially Important Bioprocessing Systems for the Production of Commodity Chemicals

Micro-organism	Carbon source(s)	Major fermentation product(s)
Saccharomyces cerevisiae	Glucose	Ethanol
Saccharomyces cerevisiae	Glucose	Glycerol
Zymomonas mobilis	Glucose	Ethanol
Clostridium thermocellum	Glucose, lactic acid	Ethanol, acetic acid
Clostridium thermosaccharolyticum	Lactic acid	Glucose, xylose, ethanol, acetic acid
Clostridium thermohydrosulfuricum	Glucose, xylose	Ethanol, acetic acid, lactic acid
Schizosaccharomyces pombe	Xylulose	Ethanol
Kluyveromyces lactis	Xylulose	Ethanol
Pachysolen tannophilus	Glucose, xylose	Ethanol
Thermobacteroides saccharolyticum	Xylose, glucose	Ethanol
Thermoanaerobacter ethanolicus	Glucose, xylose	Ethanol, acetic acid, lactic acid
Clostridium acetobutylicum	Glucose, xylose, arabinose	Acetone, butanol
Clostridium aurianticum	Glucose	Isopropanol
Clostridium thermoaceticum	Glucose, fructose, xylose	Acetic acid
Clostridium propionicum	Alanine	Propionic acid, acetic acid, acrylic acid
Aeromonas hydrophilia	Xylose	Ethanol, 2,3-butanediol
Dunaliella sp.	Carbon dioxide	Glycerol
Aspergillus niger	Glucose	Citric acid
Aerobacter aerogenes	Glucose	2,3-butanediol
Bacillus polymyxa	Glucose	2,3-butanediol

SOURCE: Office of Technology Assessment, from T. K. Ng, R. M. Busche, C. C. McDonald, et al., "Production of Feedstock Chemicals," *Science* 219:733-740, 1983; J. C. Linden and A. Moreira, "Anaerobic Production of Chemicals," *Basic Biology of New Developments in Biotechnology* (New York: Plenum Press, 1983); and D. I. C. Wang, Massachusetts Institute of Technology, personal communication, 1982.

identification of the genes that control the synthesis of these chemicals. With such knowledge, it will be possible in some instances to employ rDNA technology or cell fusion methodology to yield micro-organisms with improved bioconversion efficiencies. Improvements of these micro-organisms by genetic manipulation at present are limited to a few cases. Examples include the development of a *Pseudomonas putida* plasmid that codes for proteins that hydroxylate chemicals and the development of rDNA plasmids in *Escherichia coli* that provide the genes that code for enzymes that convert fumarate to succinate (21).

In developing commercial bioprocesses, a major need is for micro-organisms with characteristics such as tolerance to increased levels of products during bioprocess reactions;* better efficiency of sugar utilization; faster rates of production; tolerance to higher temperatures, so that separation and purification methods (which often require elevated temperatures) can be coupled with bioprocesses;** selected drug tolerance, so that growth of contaminant bacteria can be inhibited by drug treatment; and better growth on a variety of biomass sources (26). Another major need is the identification of plasmids that can be used as vectors for the transmission of useful genetic information.

*The most commonly used micro-organism for ethanol fermentation is yeast, which tolerates ethanol concentrations up to about 12 percent. Since the purification of ethanol from such dilute solutions is costly, a desirable goal is to develop organisms (and thus enzymes) whose tolerance to end products is higher. Such organisms could be used as hosts for cloned bioconversion genes.

**A combination of bioprocessing and purification could be implemented whereby products are continuously removed and collected. In this case, the high temperatures would minimize contamination by other organisms and avoid product concentrations high enough to kill the micro-organisms (13,37).

International research activities

Biomass-related research in the United States is conducted by the Department of Energy (DOE), the National Science Foundation, and private companies. Programs within DOE include the Biomass Energy Technology program, which examines the technical feasibility of innovative biomass feedstock production and conversion technology; the Alcohol Fuels program; and the Biological Energy Research program (within DOE's Office of Basic Energy Science), which funds research on genetic manipulation of plants for increased biomass production and of micro-organisms for improved bioprocessing. DOE's Energy Conversion and Utilization Technologies (ECUT) group recently started a program in biocatalysis specifically in response to the potential use of rDNA organisms in chemical production processes. The goal of this generic applied research program is to build "biocatalysis technology to enable industry to displace a significant level of nonrenewable resource requirements by [the year] 2000" (33). The ECUT program focuses on research on scale-up of bioprocesses, monitoring continuous bioprocesses, bioreactor design, and downstream product separation.

The Reagan administration's proposed fiscal year 1984 budget is not generous to biomass conversion for energy programs. The budget requests $17.3 million to support "fundamental R&D" in this area, a small increase of $1.3 million (8.1 percent) from fiscal year 1983. Alcohol fuels R&D, formerly budgeted separately, would be combined with biomass energy programs (25). Since some of this R&D relates to studies of microbial chemical production, any change in Federal support for R&D of biomass energy will effectively alter R&D for biological commodity chemical production. The only DOE program specifically directed toward the use of new biotechnology, the ECUT program, received no funding for fiscal year 1984.

Differing emphasis is placed on the biological production of chemicals and fuels by the governments of foreign countries. The United Kingdom funds biotechnological applications to chemical production processes through several governmental departments. The Canadian Development Corporation is pursuing technology development for producing ethanol from aspen wood ($21 million over 5 years), and several other Canadian Government agencies are addressing chemical and energy production from biomass. Japan, France, and Sweden also have Government-funded programs pursuing the use of biomass as a feedstock for chemicals and energy (33).

Profiles of recent U.S. patent activity indicate widespread attention by private inventors and companies in the United States and other countries to biomass conversion, particularly in areas related to hydrolysis of starch to sugar, the production of higher value-added chemicals such as amino acids from microbial systems, and improvements in bioprocess systems such as enzyme immobilization (32). Organizations with the most U.S. patents in starch hydrolysis and related bioprocesses include CPC International Inc. (U.S.), with 21; A. E. Staley Manufacturing Co. (U.S.), with 18; A. J. Reynolds Tobacco Co. (U.S.), with 8; France's National Agency for the Funding of Research (L'Agence Nationale de Valorisation de la Recherche); Anheuser-Busch, Inc. (U.S.), and Hayashibara Biochemical Laboratories, Inc. (Japan), with 7 each; and Novo Industri A/S (Denmark) and Miles Laboratories Inc. (U.S.), with 5 each (32). Even though patents in starch hydrolysis do not give a conclusive view of future biotechnological applications to the commodity chemical industry, they do indicate that U.S. companies are the predominant developers of the bioprocess technology underlying the utilization of starch biomass.

Conclusion

The production of low-value-added, high-volume commodity chemicals demands the use of the most economic production schemes available. The most economic schemes for chemical and energy production at present favor the use of petrochemical feedstocks. In the future, however, decreasing petroleum supplies, increasing oil prices, and technological advances in biomass utilization may foster a transition to the use of feedstocks derived from biomass. Such a transition is not expected to occur on an industrywide scale in the near future, but bioprocesses are being used to produce significant amounts of fuel-grade ethanol from corn and other crops economically.

Because of the potential for disruption of the existing industrial structure, the complex interrelationships that characterize the production of commodity chemicals will affect the success of the introduction of particular compounds produced by microbial bioprocesses. Projected bioprocessing costs of commodity chemicals and the structure of the chemical industry have been investigated by B. O. Palsson, et al. (23). These investigators concluded that the potential exists for a smooth introduction of four microbial products (ethanol, isopropanol, n-butanol, and 2,3-butanol) into the U.S. chemical industry, and that these products may foster other bioprocess development. In order for this transition to take place, however, either the costs of producing these products must be reduced (about 20 to 40 percent of their existing costs) or the price of petroleum must rise. Reducing the costs of production of chemicals from biomass is a prerequisite to commercial success in all case studies thus far.

U.S. Government support for applications of biotechnology to the conversion of biomass is decreasing, while high levels of government support are provided in several competitor countries, particularly Japan and the United Kingdom. U.S. companies appear to be active in developing certain biotechnological applications, but most of this activity as reflected in patents is concentrated in applications to starch conversion, with primary emphasis on higher value chemicals which are expected to be produced before biomass-based commodity chemicals are made. Some companies in the United States and other countries are active in bioprocess development, but given the current slow pace of R&D in microbial systems that perform the chemical conversions, these processes will not be applicable in the chemical industry for some years.

Priorities for future research*

Biotechnology will be a key factor in developing economic processes for the conversion of biomass to commodity chemicals. A number of priorities for research that will improve the efficiency of the conversion of biomass to useful chemicals can be identified:

- bioprocess improvements, including the use of immobilized cell and enzyme systems and improved separation and recovery methods,** an area especially important to the production of commodity chemicals because incremental improvements in bioprocess technology will be readily reflected in the price of these chemicals.
- screening programs to identify micro-organisms (and their biochemical pathways) useful to processes such as commodity chemical synthesis, cellulose hydrolysis, lignin degradation, and catalysis of reactions that utilize byproducts that are currently unmarketable;
- developing host/vector systems that facilitate increased production of commodity chemicals by gene amplification and increased gene expression of desired products and that allow

*Many of these suggestions are from Rabson and Rogers (24).
**See *Chapter 3: The Technologies* for a more extensive discussion.

the transfer of genes into industrially important micro-organisms;
- understanding the structure and function of the cellulase and ligninolytic activities of micro-organisms;
- understanding the mechanism of survival of micro-organisms in extreme environments, such as high temperature, high pressure, acid, or salt;
- understanding the mechanism of cell tolerance to alcohols, organic acids, and other organic chemicals;
- understanding the genetics and biosynthetic pathways for the production of commodity chemicals, especially for the strict anaerobic bacteria such as the methanogens and the clostridia;
- understanding microbial interactions in mixed cultures; and
- developing an efficient pretreatment system for lignocellulose.

Chapter 9 references*

*1. Bungay, H. R., *Energy, The Biomass Options* (New York: John Wiley & Sons, 1981).
2. Bungay, H. R., "Commercializing Biomass Conversion," *Environ. Sci. Technol.* 17(1):24-31, 1983.
3. *Business Week,* "The Oil Majors Bet on Coal," Sept. 24, 1979, pp. 104-122.
4. *Chemical Marketing Reporter,* Weekly Price Report, May 31, 1982, pp. 35-39.
5. Cooney, C. L., Massachusetts Institute of Technology, personal communication, 1982.
*6. Cooney, C. L., "Prospects for Chemicals and Fuels Production by Fermentation," *Basic Biology of New Developments in Biotechnology* (New York: Plenum Press, 1983).
7. Corn Refiners Association, *1982 Corn Annual,* Washington, D.C., 1982.
8. Crawford, D. L., Pometto, A. L., and Crawford, R. L., "Lignin Degradation by *Streptomyces viridosporus:* Isolation and Characterization of a New Polymeric Lignin Degradation Intermediate," *App. Env. Micro.* 45:898-904, 1983.
9. Daniel, J. R., and Whistler, R. L., "Industrial Chemicals From Cereals," *Cereals: A Renewable Resource* (St. Paul, Minn.: American Association of Cereal Chemists, 1981).
10. Doane, W. M., "Starch: Industrial Raw Material," in *Cereals: A Renewable Resource* (St. Paul, Minn.: American Association of Cereal Chemists, 1981).
11. Emert, G. H., and Katzen, R., "Chemicals From Biomass by Improved Enzyme Technology," *Biomass as a Nonfossil Fuel Source,* D. L. Klass (ed.), Symposium Series 144 (Washington, D.C.: American Chemical Society, 1981).
12. *Genetic Technology News,* "Recombinant DNA May Improve Conversion of Lignocellulose Into Ethanol," February 1982, pp. 3-4.
13. Glick, L., "Fundamentals of Genetic Technologies and Their Impact on Energy Systems," Ninth Energy Technology Conference, Washington, D.C., February 1982.
14. Haber, C. L., Allen, L. N., Zhao, S., and Hanson, R. S., "Methylotrophic Bacteria: Biochemical Diversity and Genetics," *Science* 221:1147-1153, 1983.
15. Hebeda, R. E., "Syrups," *Kirk-Othmer: Encyclopedia of Chemical Technology,* 3d ed., vol. 22 (New York: John Wiley & Sons, 1983).
16. Linden, J. C., and Moreira, A., "Anaerobic Production of Chemicals," *Basic Biology of New Developments in Biotechnology* (New York: Plenum Press, 1983).
17. Lipinsky, E. S., and Kresovich, S. "Sugar Stalk Crops for Fuels and Chemicals," *Progress in Biomass Conversion* (New York: Academic Press, 1980).
*18. Lipinsky, E. S., "Chemicals From Biomass: Petrochemical Substitution Options," *Science* 212:1465-1471, 1981.
19. McElmury, D. E., CPC Int., Chicago, Ill., personal communication, August 1983.
20. Montenecourt, B. S., "Application of Recombinant DNA Technology to Cellulose Hydrolysis," *Biotechnology* 1:166, 1983.
*21. Ng, T. K., Busche, R. M., McDonald, C. C., et al., "Production of Feedstock Chemicals," *Science* 219: 733-740, 1983.

*Note: References of general interest are marked with asterisks.

22. O'Brien, T. C., "Feedstock Trends for the Organic Chemical Industry," Planning Report 15, U.S. Department of Commerce, National Bureau of Standards, April 1983.
23. Palsson, B. O., Fathi-Afshar, S., Rudd, D. F., et al., "Biomass as a Source of Chemical Feedstocks: An Economic Evaluation," *Science* 213:513-517, 1981.
*24. Rabson, R., and Rogers, P., "The Role of Fundamental Biological Research in Developing Future Biomass Technologies," *Biomass* 1:17-37, 1981.
25. Shapley, W. H., Teich, A. A., and Weinberg, J. P., *AAAS Report VIII: Research and Development, FY 1984* (Washington, D.C.: American Association for the Advancement of Science, 1983).
26. Skotnicki, M. L., Lee, K. J., Tribe, D. E., et al., "Genetic Alteration of *Zymomonas mobilis* for Ethanol Production," *Genetic Engineering of Microorganisms for Chemicals,* A. Hollaender (ed.) (New York: Plenum Press, 1982).
27. Stobaugh, R., and Yergen, D. (eds.), *Energy Future* (New York: Random House, 1979).
28. U.S. Congress, Office of Technology Assessment, *The Direct Use of Coal,* OTA-E-86, Washington, D.C., April 1979.
29. U.S. Congress, Office of Technology Assessment, *Alternative Energy Futures, Part 1: The Future of Liquefied Natural Gas Imports,* OTA-E-110, Washington, D.C., March 1980.
*30. U.S. Congress, Office of Technology Assessment, *Energy From Biological Processes,* vols. I and II, OTA-E-124, Washington, D.C., July 1980.
31. U.S. Congress, Office of Technology Assessment, *Increased Automobile Fuel Efficiency and Synthetic Fuels: Alternatives for Reducing Oil Imports,* OTA-E-185, Washington, D.C., September 1982.
32. U.S. Department of Commerce, Patent and Trademark Office, Office of Technology Assessment and Forecast, *Patent Profiles: Biotechnology,* Washington, D.C., July 1982.
33. U.S. Department of Energy, Supplement to *ECUT Chemical Processes Project Annual Report,* 1982, Washington, D.C., March 1983.
34. U.S. Department of Energy, Energy Information Administration, *Monthly Energy Review,* July 1983, pp. 30-31.
35. Wang, D. I. C., Massachusetts Institute of Technology, personal communication, 1982.
36. Webber, D., "Basic Chemical Output Fell Third Year in a Row," *Chem. Eng. News,* May 2, 1983, pp. 10-13.
37. Wiegel, J., and Ljungdahl, L. G., "*Thermoanaerobacter ethanolicus* gen. nov., spec. nov., A New, Extreme Thermophilic, Anerobic Bacterium," *Arch. Microbiol.* 128:343-348, 1981.

Chapter 10

Bioelectronics

Contents

Figure

Chapter 10
Bioelectronics

Introduction

The potential for the use of proteins in electronic devices has received attention recently with the advent of recombinant DNA (rDNA) technology and the potential for computer-aided design of proteins (1,2,3,5,6,7,11,13,14,15,19,21). Work is focused in two areas: biosensors and biochips. Biosensors (biologically based sensors) have been used for several years, but design problems have limited their acceptance. Biotechnology is expected to increase the variety, stability, and sensitivity of these devices. Biochips (biologically based microchips) capable of logic and memory are still only speculative, and their development is many years away.

Biosensors

A potential application of biotechnology is in the development of improved sensing devices. Because of their high specificity for given substances, enzymes and monoclonal antibodies (MAbs) are particularly suited for use as sensors. Sensors using these biological molecules have the potential to be smaller and more sensitive than traditional sensors.

Biosensors using enzymes have been used to detect the presence of various organic compounds for many years (12). Most of them have used a free or immobilized enzyme and an ion-sensitive electrode that measures indirectly (e.g., by temperature or color changes produced during an enzymatic reaction) the presence of a product the formation of which is catalyzed by the enzyme. Because of the proximity of the enzyme and electrode, these biosensors are rapid and sensitive. They have not had wide application, however, because of the high cost of many enzymes, lack of particular enzymes, and temperature instability.

The use of rDNA and MAb technology and computer-aided design of enzymes and other proteins may allow the problems associated with existing biosensors to be overcome. The cloning in bacteria of genes coding for useful enzymes, for example, could allow the enzymes to be made in large amounts cost effectively. The use of MAbs, which can be made for virtually any molecule, not only could obviate the need for enzymes but also could substantially broaden the applications of biosensors. A longer term solution to the lack of particular enzymes might be to have computers design enzymes with particular catalytic functions. Finally, features of proteins that determine temperature stability could be incorporated into the genes that code for important sensing enzymes.

A new approach to fabrication is yielding biosensors with greater speed, sensitivity, and ease of operation (4). The new biosensors use a field-effect transistor that translates a chemical reaction, such as that catalyzed by an enzyme, into an electronic signal. Because the electronic response is a direct measure of the chemical reaction, the sensitivity and speed of the device is increased. (It is postulated that these sensors could use MAbs as specific detection agents.) The British Government has one of these new biosensors on the market; it detects a particular nerve gas (4).

There are many potential applications for improved biosensors in the medical, industrial, environmental, and defense fields (2,12). These are discussed in turn.

In medical diagnostics, many substances need to be measured accurately and rapidly, but the sensors now available are often expensive, slow, and insensitive. Improved biosensors could potentially solve many or all of these problems. Such biosensors could detect, for example:

- antigens associated with infectious disease,
- hormonal levels to examine endocrine function, and
- serum protein levels indicative of disease.

One particularly important medical application of improved biosensors could be in the treatment of diabetic patients for whom proper levels of insulin and glucose must be maintained. Small, implantable devices that sample blood for glucose and regulate the delivery of insulin could be developed.

As mentioned in *Chapter 3: The Technologies,* one of the hindrances to effective bioprocess monitoring and control is the need for a wide range of sterilizable sensors. Biosensors could be developed to measure levels of key reaction substances, such as reactants, intermediates, nutrients, and products. Continuous monitoring of several substances with biosensors interfaced to a computer would allow better control over the reaction process and thus increase productivity. The use of thermotolerant enzymes could potentially allow these sensors to be sterilized in place.

A potential environmental application of improved biosensors would be to monitor water and air quality. However, cost considerations limit the use of the extremely sensitive sensors now available. Additionally, very few measuring systems are portable enough for monitoring in the field. Better biosensors might circumvent these problems. Other environmental biosensors could be developed to detect exposure of workers to hazardous substances and to monitor indoor air pollution in the office or at home.

The U.S. Department of Defense (DOD), in the near future, will be the major supporter of biosensing research in the United States ($8 million over the next 4 years). DOD's aim is to develop biosensors for the detection of chemical and biological warfare agents that are small, have high sensitivity, quick response times, and no false alarms (18). If such devices were developed, they would have broad civilian applications such as those just mentioned. Companies funding research on biosensors for a number of uses include IBM, IT&T, and Johnson & Johnson in the United States and Cambridge Life Sciences in the United Kingdom (4,9,16).

Many technical barriers to developing highly reliable biosensors remain (8,17). Operating limitations (e.g., a narrow temperature range) and fabrication problems have yet to be overcome. Research is needed to identify which proteins are most appropriate for this technology. Moreover, sensors implanted in animals or used to monitor bioprocesses must be sterilized prior to implantation or use, and research is needed to develop biosensors that are not destroyed by sterilization methods. Over the next 5 to 10 years, many of these generic problems inherent in the development of biosensors will probably be solved (2,18).

Biochips

Probably the most novel potential application of biotechnology is in the production of a biomolecular electronic device. Such a device would contain a specially designed organic molecule that would act as a semiconductor surrounded and stabilized by specially designed proteins, as shown in figure 29. Researchers have studied the use of proteins as a matrix for semiconductors since the early 1970's, but the possibility of designing proteins aided by computers and producing the proteins with rDNA technology has sparked more intense interest.

Two small entrepreneurial firms in the United States are doing research on biological microchips, or biochips: Gentronix (Rockville, Md.) and Ean-Tech (San Francisco, Calif.). Furthermore, DOD will be funding biochip research beginning in fiscal year 1984 at $3 million to $4 million for 5 years. A few large electronics companies in the United States (Westinghouse, General Electric, and IBM) have small inhouse programs in this area. Japan, France, the United Kingdom, and the U.S.S.R. have indicated interest in biomolecular computers (10,20).

Figure 29.—The Use of Proteins in Constructing a Circuit

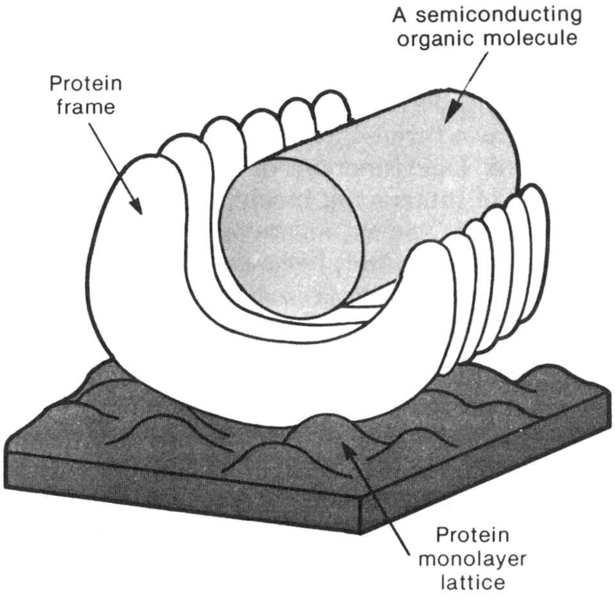

SOURCE: Office of Technology Assessment.

Bioelectronics research is in its infancy. Although most potential applications of proteins in this field are only speculative, the successful development of these applications could have a substantial effect on the electronics industry. Computers using protein-based biochips, for example, would be smaller, faster, more energy efficient, and possibly more reliable than computers using silicon chips.* The impact of such biochips would be as broad as that of present computers, from hand-held calculators to robotics.

The biological nature of biochips also raises the possibility of some exciting medical applications—they could be implanted in the body to interface with the living system. Some possibilities include:

- brain implants to circumvent damage that has caused blindness and deafness,
- cardiac implants to regulate heart beat,
- blood implants to regulate drug delivery (e.g., insulin for diabetics), and
- implants to control artificial limbs.

DOD considers biochip technology potentially very useful. Because the circuits would be the width of molecules, the resulting devices would be very small and should find use in areas where small size is essential (e.g., in missile guidance). Furthermore, because of the nonmetallic nature of biochips, it is thought that they would be immune to "electromagnetic pulse," the extraordinary electrical charge that results from a nuclear explosion and renders useless all metallic devices in a large area. In spite of the potential uses, however, it is likely to be many years before any complex biochip will be developed.

A conventional silicon chip contains a set of optically imprinted circuits on a wafer of silicon. Four factors limit the number of circuits contained on a chip. First, the lower limit of the width of a circuit is determined by the wavelength of light used for imprinting. The current limit is 1 to 10 microns; it has been postulated that by 1990 the width could be 100 times narrower (2). Second, the distance between circuits is limited by the nature of the silicon circuit construction itself. When circuits are too close together, electrons can "tunnel" between circuits. This tunneling decreases the reliability of the electronic device. The lower limit for the distance between circuits is rapidly being approached for silicon chips. The third limiting factor for conventional chips is heat dissipation. As circuits are packed more closely together, the chip becomes too hot to function effectively. Lastly, as the amount of information processing ability per chip increases, the problems with fabrication and quality control increase.

Biological and chemically synthesized molecules conceivably could solve these problems associated with conventional silicon chips as well as provide additional advantages in design. Because the molecules themselves would be the conductors, the lower limit of the circuit width would be the width of molecules, which is several orders of magnitude narrower than silicon circuits used (or even postulated) at present. Molecular circuits could be placed very close to one another without tunneling effects, because the proposed molecules conduct current without losing electrons. Furthermore, since almost no energy is required for molecules to conduct current, very little heat would be generated even when the circuits were close together. The specificity of complex interactions among proteins and the self-assembling

*Mutations that occur at a certain low level during growth of micro-organisms could affect the reliability of the final product.

processes characteristic of biological systems would facilitate the fabrication of very reliable biochips.

The fabrication of complex three-dimensional biochips with the fabrication technology now used in the electronics industry is probably impossible. An essential feature of the use of a protein matrix is that the proteins direct their own assembly and the appropriate positioning of the semiconductor molecules. There are numerous examples of self-assembling protein structures, including virus particles, and these are being studied intensely for potential applications to biochip technology.

Several proteins, including MAbs, have been suggested for constructing a biochip in three dimensions. The movement of microelectronics from two- to three-dimensional structures would allow not only for increased complexity but also for greatly reduced size. The use of a three-dimensional protein matrix necessitates the design of proteins that will interact with other proteins at correct and unique angles. The construction of these proteins will rely on computer-aided design and rDNA technology.

There are many problems to be solved before a three-dimensional biochip will become available. Biological equivalents of capacitors, transistors, and resistors are yet to be developed. Switching devices, necessary for use with the computer binary system, are only theorized. No one has determined how a three-dimensional biological structure will do logic functions or store memory. The problem of interfacing biochips so they can be programed or can assimilate other input has not been addressed. And, because the chips would use complex molecules, research needs to be done on their environmental stability.

Biochips will not be possible without computer-designed proteins and rDNA technology. Yet it will probably be several years before rDNA technology will be able to contribute substantially to biochip research, because it is first necessary to understand more about the relationship between protein structure and function, the biological self-assembly processes, and the mechanisms by which molecules could do logic functions and store memory.

Priorities for future research

Increased funding for research in the following areas could speed the development of bioelectronics:

- computer-aided design of proteins,
- temperature stability of proteins,
- field-effect transistors,
- miniaturization of sensors,
- biological self-assembly processes, and
- molecular-switching mechanisms for electronic signal propagation.

Chapter 10 references*

*1. Angier, N., "The Organic Computer," *Discover,* May 1982, pp. 76-79.
2. Clerman, R. J., "State-of-the-Art Survey: Biochips Technology" (McLean, Va.: MITRE Corp., 1982).
3. *Economist,* "Putting Organic Molecules to Work," June 4, 1983, pp. 87-88.
4. *Economist,* "Sensing Devices: A New Breed," June 25, 1983, p. 92.
5. Fox, J. L., "Molecular Electronic Devices Offer Challenging Goal," *Chem. & Eng. News,* May 23, 1983, pp. 27-29.
6. Freidman, S., "Molecular Electronics Promises 'Biochips', But Research Proceeds Behind 'Silicon Wall'," *Genetic Engineering News,* September/October 1982, p. 1.

*References of general interest are marked with asterisks.

*7. Haddon, R., and Lamola, A., "The Organic Computer," *The Sciences* 23:40-44, 1983.
8. Janati, J., and Huber, R. J., "Ion-Sensitive Field Effect Transistors," *Ion-Selective Electrode Rev.* 1:31-79, 1979.
9. Mazur, L., "Biosensor Research Forges Ahead in the U.K.," *Bio/Technology,* August 1983, pp. 468-470.
10. McAlear, J. H., Gentronix Laboratories, Rockville, Md., personal communication, 1983.
11. North Carolina Biotechnology Center, "Biotechnology: The Forging of Multidisciplinary Strategies for Research in Biomolecular Electronics, Materials Sciences, and Microecology," workshop report, Research Triangle Park, N.C., December 1982.
12. O'Brien, T., "NBS and Industrial Biotechnology: Instrumentation and Measurement Needs," Planning Report 14 (Washington, D.C.: National Bureau of Standards, U.S. Department of Commerce, March 1983).
13. Ostroff, J., "Biochips," *Venture,* February 1983, pp. 65-68.
*14. Peterson, I., "The Incredible Shrinking Computer," *Science News* 123:378-380, 1983.
15. Robinson, A. L., "Nanocomputers From Organic Molecules?" *Science* 220:940-942, 1983.
16. Scantland, P. A., and Gatz, R. L., "Progress in Bioelectronics," *Bio/Technology,* August 1983, pp. 492-493.
17. Schenck, J. F., "Technical Difficulties Remaining to the Application of ISFET Devices," *Theory, Design, and Biomedical Applications of Solid State Chemical Sensors,* P. W. Cheung, et al. (eds.) (West Palm Beach, Fla.: CRC Press, Inc., 1978).
18. Skurnick, I., Defense Advanced Research Projects Agency, Arlington, Va., personal communication, 1983.
19. Ulmer, K., "Protein Engineering," *Science* 219:666-671, 1983.
20. Woolcock, K., "Molecules Rival Integrated Circuits," *Washington Times,* Feb. 10, 1983.
21. Yanchinski, S., "And Now—The Biochip," *New Scientist,* Jan. 14, 1982, pp. 68-71.

PART IV
Analysis of U.S. Competitiveness in Biotechnology

Chapter 11

Framework for Analysis

Contents

Chapter 11
Framework for Analysis

With the increasing importance of high-technology industries in the United States and the decreasing competitiveness of U.S. goods in world markets, U.S. policymakers need to be able to assess the country's future with respect to the commercialization of emerging technologies. If the country's potential competitive position can be defined, policy analysis can suggest possible governmental steps to improve that position.

Factors influencing competitiveness in biotechnology

To analyze the future competitive position of the United States in biotechnology, OTA identified 10 factors believed to have potential influence on the international competitiveness of products resulting from an emerging technology.* Many of these factors relate to the legal system and various governmental policies, although societal and private sector factors were also identified. The 10 factors are:

- financing and tax incentives for firms;
- government funding for basic and applied research;
- personnel availability and training;
- health, safety, and environmental regulation;
- intellectual property law;
- university/industry relationships;
- antitrust law;
- international technology transfer, investment, and trade;
- targeting policies in biotechnology; and
- public perception.

These 10 factors are described in the chapters that follow. The chapters are presented, more or less, in the order of the factors' importance to competitiveness in biotechnology. Each of these factors was analyzed for the United States and five countries identified as the major potential competitors of the United States in biotechnology: Japan, the Federal Republic of Germany, the United Kingdom, Switzerland, and France.

*OTA's model for determining the future competitive position of different countries with respect to the commercialization of biotechnology could very well be useful in determining international competitiveness with respect to the commercialization of other emerging technologies. For emerging technologies other than biotechnology, however, the relative importance of specific factors would not necessarily be the same.

The three factors that OTA believes to be most important to a country's success in commercializing an emerging technology such as biotechnology are financing and tax incentives for firms, government funding of basic and applied research, and the availability of trained personnel.

The first of these factors encompasses the availability of capital both for starting new firms and for financing the growth of existing firms. It also includes tax policies that affect the formation and availability of capital as well as the strategic decisionmaking in firms.

Funding of basic, generic applied,* and applied research is necessary both to maintain a science base and to ensure the availability of the technical means to apply scientific knowledge industrially. The distinction between basic, generic applied, and applied science research is an important one, because, in establishing a competitive position, a comparative advantage in applied science may be more important than an advantage in basic research. Optimally, an analysis of funding for basic, generic applied, and applied research would include funding from both government and industry. Industry figures are usually proprietary, however, so the analysis in this report necessarily concentrates on government funding.

The third factor, availability of personnel trained in essential disciplines in a new tech-

*Generic applied research is research whose objective is to gain the understanding necessary to solve a problem common to a particular industry. Such research falls between basic research, the objective of which is to gain understanding of the basic aspects of phenomena without goals toward the development of specific processes or products; and applied research, the objective of which is to gain understanding necessary to meet a recognized and specific need, process, or product.

nology, is important to firms considering the commercialization of that technology. Furthermore, the quality of science and engineering education is a major factor in determining the future availability of personnel.

Three factors were identified as having moderate importance in the commercialization of biotechnology: health, safety, and environmental regulation, intellectual property law, and university/industry relationships.

To determine the importance of health, safety, and environmental regulation, several issues had to be weighed. On the one hand, the more stringent the regulations protecting against potential risks of the technology, the more positive the public's reaction to the development of the technology is likely to be. On the other hand, stringent regulations may discourage commercialization. Most companies will seek to enter domestic markets first, and for these companies, the domestic regulations will be of primary importance. Companies interested in developing international markets, however, must also consider the regulations of other countries. Some countries' regulations are effective nontariff trade barriers that discourage entry by foreign firms into domestic markets.

The intellectual property laws of a country partially affect whether a company will pursue a line of inquiry. If one is unlikely to reap the benefits of the discovery of an invention, then one is less likely to work on such an invention. Furthermore, if a country's patent laws are not sufficiently protective, then a company may choose to keep its inventions as trade secrets. Protection through trade secrets usually discourages technology transfer.

Active interaction between industry and academia is a factor that could promote the competitiveness of a country in an emerging technology. Usually when a technology is in the early experimental phase, most of the important research is carried out in universities. Ongoing dynamic university/industry relationships are an effective means of domestic technology transfer. Generally, therefore, such interactions promote a country's competitiveness.

Three factors were determined not to be very important to the development of biotechnology now, although these factors could increase in importance as biotechnology becomes a more mature technology. They are antitrust law; international technology transfer, investment, and trade policies; and government targeting policies in biotechnology.

U.S. and foreign antitrust laws were originally intended to stimulate competitiveness among domestic industries by prohibiting restraints of trade and monopolization. As countries have sought international markets, however, questions have been raised about whether antitrust restrictions accomplish their intended purpose. Governments of some countries have taken a relaxed attitude toward the interpretation of these laws with respect to research joint ventures and technology licensing, while the governments of some countries continue to have strict interpretations. It is possible that the strict interpretation of antitrust law with respect to joint ventures and technology licensing could decrease a country's international competitive position.

Trade policies and laws that guide the transfer of products and technology internationally could influence a country's competitive position if the laws and policies are not reciprocal among countries. Technology transfer laws are generally concerned with national security issues and transnational joint ventures. Investment control and exchange laws when applied to technology licensing or technical assistance agreements or foreign investment, can restrict the importation of foreign technology or capital into particular countries and thereby restrict foreign access to that local market. Trade policies important to biotechnology include tariffs and nontariff barriers, such as packaging requirements and nonacceptance of foreign clinical data.

Some governments target selected emerging technologies to promote rapid commercialization. In consultation with experts from academia and industry, they formulate the direction, backed by funds, that technologies should take to ensure rapid commercialization. Countries with targeting policies may have a competitive advantage in commercializing an emerging technology.

The last factor analyzed is how the public perceives the benefits and risks of the technology.

In democratic countries in particular, public perception can promote or undermine the commercialization of an emerging technology. Depending on the nature and intensity of the public's response to an emerging technology, which cannot be readily predicted, public perception could be an overriding factor in the commercialization of a new technology. In the case of biotechnology, the public's perception of an accident or perceived risk could significantly influence the development of the technology.

Firms commercializing biotechnology

In addition to analyzing the factors just discussed, it is also necessary for this competitive assessment to analyze the aggregate level of industrial activity. OTA's industrial analysis, presented in *Chapter 4: Firms Commercializing Biotechnology,* was approached from the following perspectives:

- the number and kinds of companies commercializing biotechnology,
- the commercial areas toward which industrial biotechnology R&D is being directed,
- the interrelationship among the companies applying biotechnology, and
- the overall organization of the commercial effort.

The analysis focused on the United States and then made comparisons with other countries.

U.S. efforts to commercialize biotechnology are currently the strongest in the world. The U.S. strength is in part derived from the unique complementarity that exists between small entrepreneurial firms founded specifically to develop new biotechnology and established companies in a variety of industrial sectors. While the entrepreneurial new biotechnology firms (NBFs) specializing in research-oriented phases of development have been the major force behind the commercialization of biotechnology in the United States to date, the role of established companies is expanding. Established companies have assumed a major share of the responsibility for production and marketing of, and, when necessary, obtaining regulatory approval for, some of the earliest products developed by NBFs. Through equity investments and licensing and contract agreements, these companies have also provided many of the NBFs with the necessary financial and marketing resources to remain solvent. Furthermore, many established companies are now beginning to make substantial contributions to the commercialization of biotechnology in the United States through their increasing investments in their own research and production facilities.

In European countries such as the Federal Republic of Germany, Switzerland, France, and the United Kingdom, biotechnology is being commercialized almost exclusively by large pharmaceutical and chemical companies, many of which already have significant strength in biologically produced product markets. Large established companies are critical to the development of biotechnology in Europe, and they also establish the rate at which biotechnological development takes place. Although such companies have been slow to invest in biotechnology R&D, their inherent financial, production, and marketing strengths will be important factors as the technology continues to emerge internationally.

In Japan, dozens of strong "old biotechnology" companies from several industrial sectors have extensive experience in bioprocess technology, and these large companies are using new biotechnology as a lever to enter profitable and expanding pharmaceutical markets. Japanese companies dominate biologically produced amino acid markets and are also major competitors in new antibiotic markets. They could dominate new specialty chemical markets as well.

Pharmaceutical markets will be the first proving ground for U.S. competitive strength in biotechnology. International competition will be intense. American pharmaceutical and chemical companies will be competing not only against Japanese companies, but also against the pharmaceutical and chemical companies of Western Europe, all of whom expect to recover their biotechnology investments through extensive international market penetration.

Results of the analysis

The results of the analysis of the relative importance of the factors affecting the competitive position of the United States and other countries in biotechnology both now and in the future is presented in *Chapter 1: Summary*. Also discussed is the current U.S. competitive position with respect to the other countries analyzed.

Congressional issues and options for improving the competitive position of the United States in biotechnology are discussed at the end of the following chapters. To improve the competitive position of the United States, legislation could be directed toward any of the factors discussed, although coordinated legislation directed toward all the factors might be more effective in promoting U.S. biotechnology.

The chapters that follow discuss only those congressional options that are specific to the development of biotechnology or were pointed out to OTA by U.S. firms commercializing biotechnology. Policy options in some areas are not specific to biotechnology, but to high technology or industry in general. These options are:

- to improve U.S. science and engineering education and the retraining of industrial personnel,
- to ease U.S. antitrust law to promote more research joint ventures among domestic firms,
- to regulate U.S. imports to protect domestic industries,
- to regulate the transfer of technology from the United States to other countries, and
- to target specific industries or technologies for Federal assistance.

There are many arguments for and against these options that are beyond the scope of this report. Because of their broad applicability to industry in general, these options are not discussed in the chapters that follow. It is important to note, however, that legislation in any one of these areas could affect the development of biotechnology.

Chapter 12

Financing and Tax Incentives for Firms

Contents

Tables

Figure

Chapter 12

Financing and Tax Incentives for Firms

Introduction

Two of the most important factors in the development of biotechnology in the United States have been the supply of venture capital to finance the startup and growth of new biotechnology firms (NBFs)* and the tax incentives provided by the U.S. Government to encourage capital formation and stimulate research and development (R&D) in the private sector. As noted in *Chapter 4: Firms Commercializing Biotechnology,* the types of companies commercializing biotechnology in the United States include a large number of NBFs and a smaller yet growing number of established companies from a variety of industrial sectors. In Japan and the European countries, by contrast, it is predominantly established companies that are commercializing biotechnology. A variety of reasons might explain the different nature of foreign commercialization efforts, but certainly of major importance is the fact that venture capital to fund the startup of new companies is not generally available outside the United States.

*NBFs, as defined in *Chapter 4: Firms Commercializing Biotechnology,* are firms established around 1976 or later specifically to pursue applications of biotechnology.

The first section of this chapter examines financial needs of firms commercializing biotechnology, emphasizing the needs of NBFs in the United States. It also evaluates the sources and availability of capital for firms in the United States and other countries. The second section examines tax incentives for firms. Tax incentives are an indirect source of government funding.** Such incentives can expand or contract the supply of funds available to companies engaged in biotechnology and can thereby affect the overall rate at which biotechnology develops. They also can affect the financial decisionmaking and thus the methods of financing used by companies applying biotechnology.

**Direct government funding of basic and applied research is treated in *Chapter 13: Government Funding of Basic and Applied Research*. Direct government funding in the United States is provided exclusively for research. In some countries, notably Japan, the Government provides direct funding to industry. Such funding is discussed in this chapter.

Financing in firms commercializing biotechnology

Starting a new company, expanding the product line of an existing company, and manufacturing an existing product in a new way all require some form of financing. The discussion below outlines the financial needs of U.S. companies applying biotechnology. It also examines the sources and availability of private sector funds to meet these needs. Brief comparisons are made with the five countries likely to be the major competitors of the United States in the commercialization of biotechnology—Japan, the Federal Republic of Germany, the United Kingdom, Switzerland, and France.

Financial needs of firms commercializing biotechnology

As discussed in chapter 4, a distinction can be made in the United States between two types of firms that are active in the commercialization of biotechnology: NBFs and established companies. NBFs, as defined in this report, are firms established around 1976 or later specifically to pursue applications of biotechnology.* Established

*Cetus (U.S.) and Agrigenetics (U.S.), though established before 1976, are included in the NBF category. Cetus was founded to capitalize on classical genetic techniques for product development.
(footnote continued on next page)

companies have considerably longer corporate histories than NBFs and are generally much larger. In Japan, the Federal Republic of Germany, the United Kingdom, Switzerland, and France, efforts to commercialize biotechnology are led by established companies, although the United Kingdom and France do have a few NBFs. Because of their large financial assets, established companies generally do not need external sources of funds for R&D in new areas such as biotechnology. Furthermore, if they do need such funds, established companies are generally able to obtain debt financing. Debt financing, a traditional means to fund corporate growth, is not available to NBFs, because they lack both collateral to secure a loan and sufficient means to repay the lender (27). The discussion in this section, therefore, focuses on the financial needs of NBFs.

(continued)
It showed early interest in biotechnology and began aggressively pursuing product development with the new techniques. Agrigenetics was formed in 1975 to link new genetic research with the seed business. Thus, the behavior and research focus of both Cetus and Agrigenetics places them in the new firm category despite their early founding dates.

Even the most mature NBFs at present have only a few products to generate revenues that can be used to cover operating expenses and provide capital for future growth. In order to generate revenue, as described in chapter 4, NBFs in the United States are currently relying heavily on research contracts. The reliance of entrepreneurial firms on research contracts to generate revenue is almost without parallel, except perhaps for the small firms that do defense contracts.

Table 41 shows profit/loss figures for 18 NBFs in the United States, all of which are publicly held. Of these firms, only three, Cetus, Genentech, and International Genetic Engineering (INGENE), have shown earnings in the most recent fiscal year for which data are available. The favorable financial position of Cetus and Genentech is mostly due to earned interest income from funds obtained in public offerings. However, revenues from sales (including contract research) fall far short of expenses for all three of these companies, and all three are losing money on an operating basis.

As shown in table 42, NBFs' investment in R&D is currently very large in comparison to their op-

Table 41.—Breakdown of Revenues and Net Income/Losses for 18 New Biotechnology Firms in the United States, Fiscal Year 1982 (millions of dollars)

	Operating revenues						
New biotechnology firm	Revenues from research	Contract revenue as a percent of total revenues	Revenues from product sales or royalties	Total	Interest income	Total revenues	Net income/loss[a]
Amgen[b]	$ 0.13	9.4 %		$ 0.13	$ 1.38	$ 1.5	($7.0)
Biogen	12.1	58.8		12.1	8.5	20.6	(3.9)
Biotechnica International	0.031	34		0.031	0.059	0.09	(1.6)
Bio-Technology General	0.15	93		0.15	0.16	0.011	(2.3)
Centocor	2.4	84.2		2.4	0.45	2.85	(2.76)
Cetus	15.2	46.5	$0.79	15.99	16.7	32.7	4.5
Chiron[b]	1.58	92		1.58	0.14	1.72	(2.2)
Damon Biotech	0.81	48		0.81		1.7	(1.38)
Enzo Biochem[c]	0.10	11.2	0.17	0.27	0.62	0.89	(1.25)
Genentech[c]	28.8	88.3		28.8	3.76	32.6	0.625
Genetic Systems	2.2	71.66		2.2	0.87	3.70	(1.0)
Genex	5.2	85.3		5.2	0.67	6.1	(5.6)
Hybridoma Sciences[d]	0.07	73		0.07	0.024	0.095	(0.186)
Hybritech	1.3	27.4	1.8	3.1	1.6	4.75	(7.26)
Integrated Genetics	0.6	60		0.6	0.46	1.0	(1.76)
International Genetic Engineering (Ingene)	1.78	90		1.78	0.211	1.98	0.13
Molecular Genetics	0.66	61		0.66	0.42	1.08	(3.75)
Monoclonal Antibodies[b]	0.10	1.5	0.16	0.26	0.39	6.5	(2.7)

[a] Losses are shown in parentheses.
[b] Fiscal year 1983.
[c] Stock split.
[d] Units offered (one unit=three shares of common stock and three Class A Warrants).

SOURCE: Office of Technology Assessment, based on information from E.F. Hutton & Co., company annual reports, and company prospectuses.

Table 42.—Capital Expenditures, R&D Budgets, and Operating Revenues of Nine New Biotechnology Firms in the United States, Fiscal Year 1982 (millions of dollars)

New biotechnology firm	Capital expenditures	R&D budget	Operating revenues	R&D as a percent of operating revenues
Biogen	$ 8.7	$ 8.7	$12.1	72%
Cetus	22.9	25.9	16.0	143
Enzo Biochem	0.09	1.2	0.3	400
Genentech	31.8	31.9	28.8	111
Genetic Systems	0.46	3.9	2.2	177
Genex	1.8	8.3	5.2	160
Hybritech	1.44	5.0	3.1	161
Molecular Genetics	1.4	2.8	0.66	424
Monoclonal Antibodies	0.57	1.1	0.26	423

SOURCE: Office of Technology Assessment based on information from company annual reports.

erating revenues. Furthermore, NBFs that are incurring large R&D costs to develop products are sustaining large losses relative to their earnings (see table 41).* These losses, which will likely continue for several years, are eroding the capital bases of many NBFs and increasing their need for additional sources of funds. NBFs such as Biogen N.V.** do not expect operating revenues to meet R&D expenses, and consequently do not expect to operate at a profit for at least several years (2). For the next several years, expenditures by NBFs for R&D will probably equal 20 percent or more of sales (27).

*The cumulative losses shown in table 41 understate the level of funding required to sustain these companies because they do not fully reflect capital outlays. Only the depreciated portion of capital outlays shows up in a profit and loss statement and, hence, in cumulative loss (27).

**Biogen N.V., the parent company of the Biogen group, is registered in the Netherlands Antilles but is about 80-percent U.S. owned. Biogen's principal executive offices are located in Switzerland. Biogen N.V. has four principal operating subsidiaries. Biogen Research Corp. (a Massachusetts corporation) and Biogen S.A. (a Swiss corporation) conduct research and development under contract with Biogen N.V. Biogen B.V. (a Dutch corporation) and Biogen Inc. (a Delaware corporation) conduct marketing and licensing operations. Available figures pertaining to Biogen refer to Biogen N.V. and its subsidiaries.

Because of the emphasis on R&D in biotechnology, skilled labor for firms applying biotechnology is relatively more important than labor for firms in other areas. Such labor is also quite expensive. The average Ph. D., supported by two technicians, costs on the order of $150,000 to $175,000 per year with overhead (27). As a result, labor may initially constitute a large percentage of a new firm's operating expenses.

The most revealing indicator of the NBFs' potential need for cash is the *rate* at which such firms are consuming funds. Table 43 shows decreases in working capital for six NBFs. Except for Cetus, which raised an exceptional amount of money in its initial public offering, the drop in working capital for these firms is large compared to their equity capital. In 1981, Genentech used up 21 percent of its ending equity capital, while Molecular Genetics used up 10 percent, and Cetus 12 percent (27). Hybritech increased its working capital by 72 percent of beginning equity in 1981 by means of a public stock offering; by October 1982, however, Hybritech had returned to the public markets to raise additional equity because its

Table 43.—Cash Drain Relative to Equity for Six New Biotechnology Firms in the United States, Fiscal Year 1982 (millions of dollars)

New biotechnology firm	Equity capital	Cash flow[a]	Yearly change in working capital	Cumulative deficit
Biogen	$ 61.9	($3.0)	($12.1)	$10.0
Cetus	128.3	5.7	(15.7)	(0.3)
Genex	13.3	0.6	(9.4)	(2.3)
Genentech	53.1	1.0	(11.4)	(0.03)
Hybritech	17.6	(4.3)	6.3	(12.8)
Molecular Genetics	1.5	(3.6)	(1.6)	(4.0)

[a] Cash flow is sum of net income or loss plus noncash expenses such as depreciation.

SOURCE: Office of Technology Assessment, based on information from company annual reports.

working capital had dropped to 43 percent of stockholder's equity by the end of 1981 (27). Other NBFs, including Monoclonal Antibodies, Genex, and Molecular Genetics, have also had to return to the public market not long after their initial or second public offerings.

The financial needs of NBFs are largely dependent on which market they are trying to enter. To enter each of the markets described below, increasing amounts of funds are necessary.

Contract Research and Development Market. The funding needed to support entry into the contract R&D market is generally less than that required for entering product markets, because research that a firm does for another company, university, or government agency is funded by that organization, often through progress* or advance payments. Most NBFs perform contract R&D to generate revenues to fund their own proprietary research, although the costs of proprietary research generally exceeds their contract research fees (27).

*In Vitro Monoclonal Antibody Diagnostic Products Market.*** The funding needed to support entry into the market for in vitro (used outside the body) monoclonal antibody (MAb) products is more than funding needed to support entry into the contract research market. Because of the small amount of plant and equipment required to develop such products and because of the comparatively low cost of complying with the Food and Drug Administration's (FDA's) testing requirements for in vitro diagnostic products for humans, the financial requirements are relatively low.*** A number of NBFs, including Hybritech, Monoclonal Antibodies, Molecular Genetics, Centocor, and Genetic Systems, have developed in vitro MAb diagnostic products for humans that are "substantially equivalent" to products that FDA has already approved and thus do not require rigorous testing. Other MAb products being developed by these firms are intended for research or production (e.g., separation and purification) purposes and thus do not require FDA approval. Several of these NBFs are within a few quarters of achieving operational profitability for these product lines (27).

*Specialty Chemicals Market.** Specialty chemicals are defined in this report as chemicals whose price exceeds $1 per pound (50¢ per kilogram). These include substances such as enzymes, amino acids, vitamins, fatty acids, and steroids. Most specialty chemicals do not need regulatory approval. For specialty chemicals considered foods or food additives, however, FDA approval is required, and significant funds may be expended to meet FDA requirements. Thus, the amount needed to enter the specialty chemicals market varies depending on the product. In general, though, the amount of funds needed to enter the specialty chemicals market is more than the amount required to enter the contract research market but less than that needed to enter the commodity chemicals market.

*Agricultural Products Market.*** For the animal agricultural market, the R&D cost are very similar to those for pharmaceuticals (in vitro and in vivo products), because many of the products, such as diagnostics, vaccines, and hormones, are essentially the same. However, the regulatory requirements promulgated by the U.S. Department of Agriculture (USDA) and FDA for animal health products are much less stingent than the requirements for pharmaceuticals. Some animal agriculture products (e.g., vaccine for colibacillosis) have received approval and are already reaching the market.

The R&D costs for applications of biotechnology to plant agriculture vary over a broad range. The genetic manipulation of micro-organisms important to plant agriculture, for the most part, is less costly than the genetic manipulation of the plants themselves. Furthermore, the various traits being investigated are at different stages of research. For instance, plants with traits conferring resistance to drought or saline stress are more near term than those with improved photosynthesis or nitrogen fixation. The financial requirements for developing the latter plants are much greater

*Progress payments are received when the contracting company reaches certain milestones in the research project.

**In vitro MAb products are discussed in *Chapter 5: Pharmaceuticals*.

***For a discussion of FDA's regulatory processes, see *Chapter 15: Health, Safety, and Environmental Regulation*.

*Applications of biotechnology to specialty chemicals are discussed in *Chapter 7: Specialty Chemicals and Food Additives*.

**Applications of biotechnology to animal and plant agriculture are discussed in *Chapter 6: Agriculture*.

than those for developing simpler genetic applications. Firms doing research in the more complex agricultural applications of rDNA technology are unlikely to have commercial products available until the late 1980's or 1990's. Some companies, such as Plant Genetics (U.S.), hope to finance themselves through the research period by developing commercial products using conventional plant genetics (27).

*In Vivo Diagnostic and Therapeutic Products Market.** The financial requirements for entering the market for in vivo (inside the body) diagnostic and therapeutic products for human use are very large, in part because such products require extensive clinical testing to meet FDA regulatory requirements. Taking a pharmaceutical product from research to market in the United States generally requires 7 to 10 years and costs $70 million or more (14). To date, no NBFs attempting to enter this field are operationally profitable, nor are they likely to be in the near future (27). Hybritech, for example, does not anticipate profitability for its therapeutic line until about 1988 (26).

*Commodity Chemicals Market.*** For several reasons, the financial requirements for entering the commodity chemicals market are the largest. Currently, practically all commodity chemicals, defined in this report as chemicals that sell for less than $1 per pound, are made from petroleum feedstocks. Although it is theoretically possible to produce essentially all commodity chemicals from biomass feedstocks such as starch or cellulose, and most commodity chemicals can be synthesized biologically, most commodity chemicals derived from biomass cannot yet compete economically with chemicals made from petroleum in the highly integrated production infrastructure that now exists. Furthermore, profitability in commodity chemicals requires the achievement of economies of scale in production plants costing hundreds of millions of dollars (27).

*Applications of biotechnology to in vivo diagnostic and therapeutic products intended for human use are discussed in *Chapter 5: Pharmaceuticals*.

**Applications of biotechnology to commodity chemical production are discussed in *Chapter 9: Commodity Chemicals and Energy Production*.

With the exception of firms developing in vitro MAb assays and diagnostic products, it will be some time before NBFs, most of which are U.S. companies, can be self-financing; some estimate that NBFs cannot be self-financing before the late 1980's (27). The new firms must finance not only losses due to operating expenses but also expenditures needed for capital assets. For some NBFs, meeting FDA regulatory requirements will also require substantial funds. Because, as noted earlier, debt financing may not be available to many NBFs, the financial needs of these firms must for the most part be met by additions to equity capital (27). Thus, in many cases, the receptivity of the public market to NBF stock issues and the use of R&D limited partnerships is a matter of great importance.

Sources and availability of financing for U.S. firms

The sources and availability of financing for the two types of firms that are important to the development of biotechnology in the United States—i.e., NBFs and established companies—are quite different. The discussion below, therefore, treats each type of firm separately.

NEW BIOTECHNOLOGY FIRMS

The main sources of financing for NBFs, the small, new firms specializing in biotechnology, are the following:

- revenues from contract research and interest on cash previously obtained from public or private offerings;
- various sources of venture capital; and
- public stock offerings.

Revenues From Contract Research, Product Sales or Royalties, and Interest.—Research and product development agreements between NBFs and established companies are generally cost reimbursement contracts with additional fees and incentives for reaching agreed on milestones. The NBF generally retains the patent rights to any technology involved and grants the contracting company an exclusive license to that technology. Thus, such agreements usually provide for royalty payments to the NBF by the established company on the future sales of the product that results from the R&D work; these royalties may range from 2 to 10 percent of total sales, depending on the size of the product market.

Table 41 breaks down total fiscal year 1982 revenues for 18 NBFs into operating revenues received from contract research or product sales or royalties and interest income. In most NBFs, no income or very limited income was obtained from the sale or licensing of products. Most revenue, even for the larger NBFs such as Genentech, Cetus, and Biogen, was contract revenue and interest on cash raised through public offerings and private investment. Genentech reports, for example, that 88 percent of its total $32.6 million revenue in 1982 was derived from contracts and the balance derived from interest income. Cetus reports that, in fiscal year 1982 (which ended in June 1982), income from contracts accounted for almost 47 percent of its total revenues and interest income for most of the remainder. Similarly, Biogen reports that 59 percent of its revenue comes from contract sales with the balance being interest income.

Biogen and Genentech are concentrating on product development using rDNA technology. Some NBFs, including Genetic Systems, Monoclonal Antibodies, Centocor, and Hybritech, are developing MAbs for in vitro assays, diagnostics, and research products. These firms will probably achieve an income stream from product sales more quickly. In fiscal year 1982, however, these firms also show primarily interest income. Currently, Hybritech has the greatest percentage of total revenue coming from product sales, 38 percent. In the near future, product sales should contribute more substantially to revenues for Hybritech as well as other diagnostic product companies.

Venture Capital.—In the United States, there are several sources of venture capital. These are:

- corporate venture capital,
- R&D limited partnerships,
- venture capital funds, and
- Small Business Investment Corporations (SBICs).

Each of these is discussed further below.

From 1969 to 1977, the total venture capital pool in the United States remained relatively unchanged, at the level of about $2.5 billion to $3 billion each year (27). Since then, however, the venture capital pool has increased sharply, reaching between $3.5 billion and $4 billion in 1979 (45), $5.8 billion in 1981 (46), and an estimated $7.5 billion as of the end of 1982 (48).

Variability in the amount of venture capital in the United States is influenced by many factors. These include general macroeconomic variables (e.g., interest rates and inflation), changes in capital gains tax laws, and changes in pension fund investment rules. In 1969, the U.S. capital gains tax was increased from 29 to 49 percent. In addition, the U.S. inflation rate increased sharply in 1972, causing investors to seek a much higher rate of return on their investments. In 1973-74, the price index of the National Association of Security Dealers Quotation of over-the-counter securities, which represents smaller companies, declined more than did the Dow-Jones industrial price index, which represents larger companies (27), indicating a decline in investor interest in newer, smaller firms relative to larger, more established companies.

Recent changes in U.S. laws and regulations affecting the formation of venture capital have led to a resurgence in the supply of venture capital in this country. In 1979, Employee Retirement Income Security Act pension fund regulations were interpreted to allow some pension fund money to flow into venture capital investments. Around the same time, the Securities and Exchange Commission adopted Rule 144 allowing founders of companies to liquidate their "restricted" stock holdings sooner than previously allowed. The opportunity to liquidate sooner provides investors with a stronger incentive to invest. Especially important to the supply of venture capital in the United States have been decreases in the rate at which long-term capital gains are taxed. The current long-term capital gains tax rate for individuals, established under the Economic Recovery Act of 1981, is 20 percent (28 percent for corporations), making venture investments even more attractive than they were under the pre-1969 rate of 29 percent.

Table 44 shows the distribution of venture capital disbursements in the United States by industry for 1980 and 1981. In 1980, investments in "genetic engineering"* accounted for 4.2 percent of the total number of investments but 7.6 percent of the dollars invested. In 1981, "genetic engineering" accounted for 6.2 percent of the number of investments but absorbed 11.2 percent of venture dollars. The disproportionately large average size of "genetic engineering" investments reflects the fact that a large amount of funds must be dedicated to R&D before a concept is proven. In other high-technology industries, "seed money" is usually sought to prove a concept and averages around $1 million per project. But in biotechnology, seed money and startup financing from venture capitalists is generally combined to obtain enough money for product development and initial marketing. Financing for biotechnology projects averaged about $2.2 million per project in 1982 (27). As shown in table 45, seed money is a very small percentage of total venture capital disbursements in the United States. In biotechnology, venture investments have tended to combine both seed and startup financing, making the average disbursement disproportionately high.

*A definition of "genetic engineering" was not given by the *Venture Capital Journal*.

The peak period for raising venture capital in biotechnology in the United States occurred in 1980. That year, the valuations of NBFs ranged from $5 million to $25 million for 25 percent of the company (41). The stock market decline of 1981-82 was accompanied by changes in the venture capital market with respect to biotechnology ventures. Valuations of NBFs ranging from $2 mil-

Table 44.—Distribution of Venture Capital Disbursements in the United States by Industry, 1980 and 1981

	Percent of total number of investments		Percent of dollar amount invested	
	1980	1981	1980	1981
Communications	11.5%	11.4%	10.9%	11.2%
Computer related	27.4	30.0	25.7	34.3
Other electronics related	9.6	14.5	9.6	13.1
Genetic engineering	4.2	6.2	7.6	11.2
Medical/Health related	10.5	7.0	9.3	5.8
Energy	8.3	4.9	19.9	5.8
Consumer related	7.5	4.9	3.7	1.9
Industrial automation	4.5	6.2	2.7	5.3
Industrial products	3.6	4.4	2.0	3.4
Other	12.9	10.5	8.6	8.0
Total	100.0%	100.0%	100.0%	100.0%

SOURCE: *Venture Capital Journal* 22(6):8, June 1982.

Table 45.—Distribution of Venture Capital Disbursements in the United States by Stage of Investment, 1981

	Percent of number of investments		Percent of dollar amount of investments		Average size of venture financing
State of investment	Venture development	Total activity	Venture development	Total activity	($000)
Seed	4%		2%		$1,000
Startup	26		31		2,200
Other early stage	19		19		2,000
Total early stage	49%	39%	52%	46%	$2,000
Expansion	51	40	48	41	$1,750
Total	100%	79%	100%	87%	$1,900
Other		10%		8%	1,850
Stage unrecorded		11		5	900
Total		100%		100%	

SOURCE: *Venture Capital Journal* 22(6):9, June 1982.

lion to $4 million for 40 to 50 percent of the company became more common. The following two factors may have accounted for the decrease in the valuation of NBFs in 1981 and 1982:

- increased investor knowledge of the time that would be required for commercializing applications of biotechnology, and
- decreased investor interest in biotechnology because most venture capitalists who desired to invest in an NBF had already done so.

At least one venture capitalist stated that the number of new proposals based on biotechnology decreased substantially from 1981 to 1982 (27). Possible reasons for the decrease in proposals include the following:

- the existence of many competing companies in each of the major application areas discouraged additional entrants, and
- the fact that many of the scientist/entrepreneurs who wanted to form a new firm had already done so.

Table 46 shows the cost of venture capital for selected NBFs in the United States, although it should be noted that few general rules can be determined from this table. Genentech and Hybritech, which the venture capital firm Kleiner, Perkins, Caulfield, and Byer partly organized as well as financed, turned out to be particularly good investments. For Hybritech, a $300,000 investment initially purchased 72 percent of the company at a price of $0.20 per share. At the time of the public offering at $26.75 per share, Kleiner, Perkins, Caulfield, and Byer held 29.3 percent of Hybritech worth $1.7 million. For Genentech, a $200,000 investment eventually equated to 14.3 percent of the common stock ($0.21 share cost) worth around $33 million at the time of the public offering. Wilmington Securities, a later investor in Genentech, purchased 6.2 percent of the company for $500,000 or $2 per share of a stock that went public at $35. Lubrizol, a still later investor in Genentech, paid $10 million for 24 percent of the company or $6.43 per share.

Table 47 contrasts the private valuations and public (market) valuations of some recently offered NBF issues. Hybridoma Sciences exhibits the greatest increase in valuation (and thus the highest rate of return to original investors) in the shortest period of time—over 1,100 percent in just over 2 years.

The four sources of venture capital in the United States, which were mentioned at the beginning of this section, are discussed further below. Independent private venture capital funds have accounted for an increasing share of total venture capital relative to that provided by corporate investors and SBICs, as shown in table 48.

Corporate venture capital. A number of major corporations provide revenue to NBFs through R&D contracts as well as equity investments and joint ventures. Contractual relationships provide benefits to the corporate investors as well as the NBF. *Chapter 4: Firms Commercializing Biotech-*

Table 46.—Cost of Venture Capital for Selected New Biotechnology Firms in the United States

	Private venture capital			
New biotechnology firm	Venture capital invested	Percent of company purchased	Price per share	Price per share in public offering
Cetus:				$23.00
1st stage	$ 1,999,600	16.5%	$ 0.91	
SOCal—2d round	5,000,000	10.4	3.60	
Genentech:				35.00
Wilmington Securities, early stage	500,000	6.2	2.00	
Lubrizol	10,000,000	24.0	6.43	
Genetic Systems	200,000	9.7	0.51	6.00
Hybritech	300,000	72.0	0.20	26.75
Molecular Genetics:				9.00
Founders	40,560	59.9	0.02	
Sale of 632,366 shares to American Cyanamid	2,750,000	18.7	4.35	
Monoclonal Antibodies	825,116	29.2	0.52	10.00

SOURCE: Office of Technology Assessment, based on information from company prospectuses.

Table 47.—Comparison of Private and Public (Market) Valuations of Eight New Biotechnology Firms With Initial Public Offering in 1983 (millions of dollars except offering price per share)

New biotechnology firm	Date company founded	Private valuation[a]	Date of initial public offering	Public valuation: Offering price and millions of shares offered	Public valuation: Total market valuation and millions of shares outstanding	Ratio of private valuation to public valuation
Amgen	1980	$54—Feb. 1981	6/17/83	$18 2.35	$187 10.4	1:3.5
Advanced Genetic Sciences	3/79	$48—April 1981	7/83	$20[b] 2.0	$242[b] 12.2	1:5
Biogen	1978	$100—April 1981	3/83	$23 2.5	$425 18.5	1:4.25
Cambridge Bioscience	3/81	$8.3—July 1982	3/31/83	$ 5 1.0	$ 20.3 4.08	1:2.4
Chiron	5/81	$29—April 1983	8/83	$12 1.5	$ 87.6 7.3	1:3
Hybridoma Sciences	4/81	$2.2—Feb. 1983	8/83[b]	$ 6[c] 0.70	$ 25.7 4.29	1:11.7
Immunex	7/81	$10.5—July 1982	7/1/83	$11 1.65	$ 64.3 5.85	1:6.1
Integrated Genetics	1981	$33.3—Dec. 1982	7/19/83	$13 1.6	$107.9 8.3	1:3.2

[a]Based on most recent transaction.
[b]Estimated
[c]One unit. One unit = three shares common stock and three Class A Warrants.

SOURCE: Office Technology Assessment, adapted from E. F. Hutton, prepared July 18, 1983.

Table 48.—U.S. Venture Capital Pool, 1977 and 1982 (millions of dollars)

	1977	Percent of total	1982	Percent of total
Independent private funds and venture capital partnerships	$ 887	35%	$4,400	58%
SBICs (exclusive of nonventure capital related SBICs)	612	24	1,300	17
Corporate (financial and industrial subsidiaries and non-SBIC public funds)	1,022	41	1,900	25
Total pool	$2,521	100%	$7,600	100%

SOURCE: *Venture Capital Journal* 22(10):7, October 1982.

nology provides a discussion of these joint ventures and the costs and benefits accruing to both parties. Table 13 in chapter 4, entitled "Equity Investments in New Biotechnology Firms by U.S. Established Companies, 1977-1983," summarizes established U.S. firm equity investments in and joint equity ventures with NBFs.

R&D limited partnerships. R&D limited partnerships, consisting of at least one general and one limited partner, are a financing mechanism that allows businesses to engage in research activities without paying for the activities out of retained earnings or borrowed capital.* Most of the 300 to 400 R&D limited partnerships that exist in the United States have been formed since 1980 (29).

From August 1982 to May 1983, over $200 million was raised through R&D limited partnerships by NBFs alone (4). One analyst estimates that R&D limited partnerships will raise a total of $500 million in 1983 (3). In R&D limited partnerships in biotechnology, the NBF typically serves as the general partner and assumes liability. The limited partners are the investors whose money buys a share of the partnership's future profits or losses. The liability of the limited partners is limited to the loss of their investment. More than 10 R&D limited partnerships in biotechnology have been formed since 1980, and 10 to 20 more are now being formed (40).

Such partnerships have enabled NBFs to reduce their reliance for financing on established companies and venture capital firms and to reduce their costs of capital. They have also provided many NBFs with a stable source of financing for the next 4 to 5 years—the time frame written into most of the partnerships. In other words, R&D limited partnerships are providing NBFs with the financial ability to undertake their own proprietary research and early product development and in some cases clinical testing without relying on established companies and venture capital firms.

As shown in table 49, the total amount raised by 12 NBFs for R&D limited partnerships in biotechnology exceeds $400 million. The amount raised for each partnership ranged from just under $1 million (Neogen) to $80 million (Cetus). The first NBF to raise a fairly large amount of money ($55 million) through an R&D limited partnership was Agrigenetics. Genentech, which is using an R&D limited partnership as a novel approach to financing clinical trials of human growth hormone and gamma interferon, raised

*The U.S. Supreme Court decision in the 1974 precendent-setting *Snow* v. *Commissioner* (416 U.S. 500) held that limited partners could offset their other income with partnership research or other experimental expenditures. It also extended the reach of section 174 (Title 26 U.S.C. IRS §174) to include businesses that had not yet offered any products for sale.

Table 49.—R&D Limited Partnerships Used by 12 New Biotechnology Firms in the United States

New biotechnology firm	Partnership formation date	Amount (millions of dollars)
Agrigenetics	1981	$ 55.0
Genetic Systems	1982	3.4
Cetus	1982	80.0
California Biotechnology	1982	27.5
Genentech	1982	55.0
Molecular Genetics	1982	11.1
Neogen	1982	0.96
Hybritech	1982	7.5
Cetus	1983	78.0
Genentech	1983	34.0
Genetics Institute	1983[a]	25.0
Serono Labs	1983	29.0
Total		$405.46

[a] As of 8/83 not yet closed.

SOURCE: Office of Technology Assessment, based on information from the trade press and company reports.

$34 million (27). R&D limited partnerships can provide more financing than the average amount raised by NBFs in the most recent initial public stock offerings (see below).

One advantage to the general partner in an R&D limited partnership is the fact that partnership funds appear on the corporate balance sheet as contract revenue rather than as debt or equity, thus enhancing future investment prospects. Another advantage for the general partner is that the limited partners do not participate in the management of the partnership; in this respect, an R&D limited partnership is unlike other forms of equity financing where investors may sit on the board of directors and shareholders vote on major management decisions.

The limited partner (investor) in an R&D limited partnership is generally interested in investing in such a partnership because R&D limited partnerships, unlike corporations, are treated under the U.S. Internal Revenue Service (IRS) Code as nontaxable entities, meaning that partnership profits and losses are "passed through" to the individual partners who then combine them with their other items of income and expense.* Since an R&D project typically generates tax losses in its initial years (because of large R&D expenditures), limited partners can use those losses immediately to offset other income which might be taxable at rates as high as 70 percent. Furthermore, partners can deduct as much as 85 to 95 percent of their initial investment, immediately decreasing their after-tax cost (and risk) and more than doubling the potential rate of return.

Venture capital funds. Venture capital funds are professionally managed funds dedicated to investment in one or more industries. Sources of capital for these funds include pension funds (e.g., John Deere, General Electric, and Ohio Public Employees Retirement Fund), insurance companies (e.g., Wausau Insurance, Prudential Life, and Metropolitan Life), trust departments of commercial banks such as Morgan Stanley or City Bank, and corporate investors interested in potential profit from discoveries arising from the fund's support.

Of interest is the fact that a few independent private venture capital funds have been formed to invest a significant percentage of their funds in biotechnology. One example is Plant Resources Venture Fund, a $15 million to $20 million fund that invests in companies doing plant-related R&D. In the first 18 months of its operation, this fund invested in three companies, taking all the outside equity in each. Two of the companies are engaged in tissue culture research and the other is a plant genetics company. The strategy of the Plant Resources Venture Fund is to invest $500,000 to $1.75 million in each company in several stages. In first-stage financing, the fund expects to assume the major share of investment. In subsequent financings, the fund will take progressively smaller amounts as other investors are brought in. Plant Resources Venture Fund anticipates financing another seven to nine companies by 1984 (10).

Small Business Investment Corporations. In 1982, approximately 17 percent of the venture capital funds in the United States were raised by SBICs. SBICs are private companies licensed by the Small Business Administration (SBA) that must invest their funds in U.S. small businesses. There are three major groups of SBICs: 1) bank affiliates, 2) subsidiaries of venture capital and other financial companies, and 3) independent SBICs and units of nonfinancial companies. Each SBIC must have paid-in equity capital contributed by shareholders of at least $500,000. After the paid-in capital requirement is met, SBA will loan up to three times the paid-in amount of capital, thus extending the resources of the SBIC. In effect, SBICs leverage their paid-in capital by four times with SBA's assistance. SBICs obtain funds from SBA at very favorable interest rates, several points below the prime rate. They then lend the money to small businesses at a rate that is higher than the rate at which they have obtained it but still less than the prevailing rate.

An SBIC provides at least three kinds of tax advantages for shareholders (34). First, a loss on the sale or exchange of the stock can be treated by stockholders as an ordinary loss, i.e., such loss does not have to be offset against gains from sales of stock, and it can be regarded as a business loss for net operating loss deduction purposes. Second, a loss on the sale or exchange of convertible debentures purchased from small businesses (or stock obtained through conversion) can be

*Corporate profits, by contrast, are taxed both at the corporate and the shareholder level, and deductions for losses incurred by the corporation are not available to the individual shareholders.

treated by the company as an ordinary loss. Third, rather than the normal 85-percent deduction for dividends received from domestic corporations, the company gets a 100-percent dividends received deduction.*

For NBFs that might want to use funds from SBICs, there are two problems. First, because SBICs obtain much of their money as loans from SBA and must repay the SBA in a prescribed period of time, SBICs lend their money rather than use it to buy stock in small businesses. However, an increasing number of equity investments are being made by SBIC bank affiliates such as First Capital Corp. of Chicago. Most NBFs do not seek money from SBICs, because such firms need to retain dollars internally rather than use them to pay interest on debt to an SBIC. Second, SBICs do not generally commit public funds guaranteed by public institutions to high-risk ventures, which is exactly what NBFs are. However, in spite of the interest risks associated with investments in new high-technology firms, some SBICs have invested in NBFs. SBICs raised $4,108,197 in capital for NBFs in 1981 and $3,383,333 in 1982 (50).* They invested in 15 NBFs in 1981 and 9 NBFs in 1982. Thus, although the total amount of capital invested by SBICs decreased from 1981 to 1982, the average amount of capital invested per company increased.

Public Stock Offerings.—Public offerings can be divided into initial public offerings, the first time a firm attempts to raise money by offering shares in itself to the public, and subsequent public offerings, when the firm returns to the market to raise additional funds. As a way to obtain funds, the initial public offering differs in an important way from the other methods for raising funds that have already been discussed. The initial public offering is the first time that the firm must publicly disclose its financial and product development status. Going public also requires registration with an oversight organization, the Securities and Exchange Commission, and commits the firm to continued public scrutiny through publicly available reports to shareholders and annual statements to the Securities and Exchange Commission (Form 10-K). Meeting the requirements for public accountability is expensive, both in time and money, and meeting the earnings expectations of the investors can inhibit long-term R&D. In confirmation, Gabriel Schmergel of Genetics Institute in Boston says "reasons why companies haven't gone public is because sometimes they are under great pressure to produce earnings" (18). Thus, although a great deal of money can be raised in a public offering, its costs, both fiscal and otherwise, must also be considered.

The amount that a firm can raise through a public offering depends not only on the performance of the firm itself but also on the stock market and the receptiveness of investors. In times of recession, institutional investors tend to undervalue high-technology stocks because they are interested in short-term gains (16). Yet, during the early 1980's, despite the recession, high-technology issues were fairly successful, with the peak years for biotechnology stocks being 1980 and 1981. In 1982, some NBFs that made public offerings were not able to raise as much as they had expected. Until September of 1982, the performance of biotechnology stocks paraleled that of Standard and Poor stocks. After September, however, the biotechnology stocks outperformed the Standard and Poor stocks. Thus far, the 1983 bull market has been accompanied by a boom in new issues, greater in magnitude and scale than ever before. For biotechnology issues, 1983 is a banner year. Between March and July of 1983, 23 NBFs raised about $450 million (18). Figure 30 provides a comparative market performance of some MAb, rDNA, and biotechnology support companies with the Standard and Poor 500 for the period April 1982 through April 1983..

During the 1970's, venture capitalists were accustomed to waiting 5 to 7 years before seeing their investments achieve liquidity in the public markets. With the advent of the microprocessor, a number of electronic companies developed applications that became profitable quickly. In some cases, these companies were able to achieve profitability in 18 months and a public offering within 2 to 3 years from founding, in part because of better capital markets after 1978 (8). As a result,

*A corporation pays tax on dividends distributed. The dividend is also taxed as part of income of the distributee. To partially compensate for this double taxation, if the distributee is a corporation, 85 percent of dividends received is excluded from this second taxation. However, if the corporation is an SBIC, 100 percent of dividends received is excluded.

*The 1982 figures available from the SBA did not include November and December figures.

Figure 30.—Comparative Market Performance:
Companies Using Biotechnology vs. Standard and Poor's 500 Companies, April 1982 through April 1983

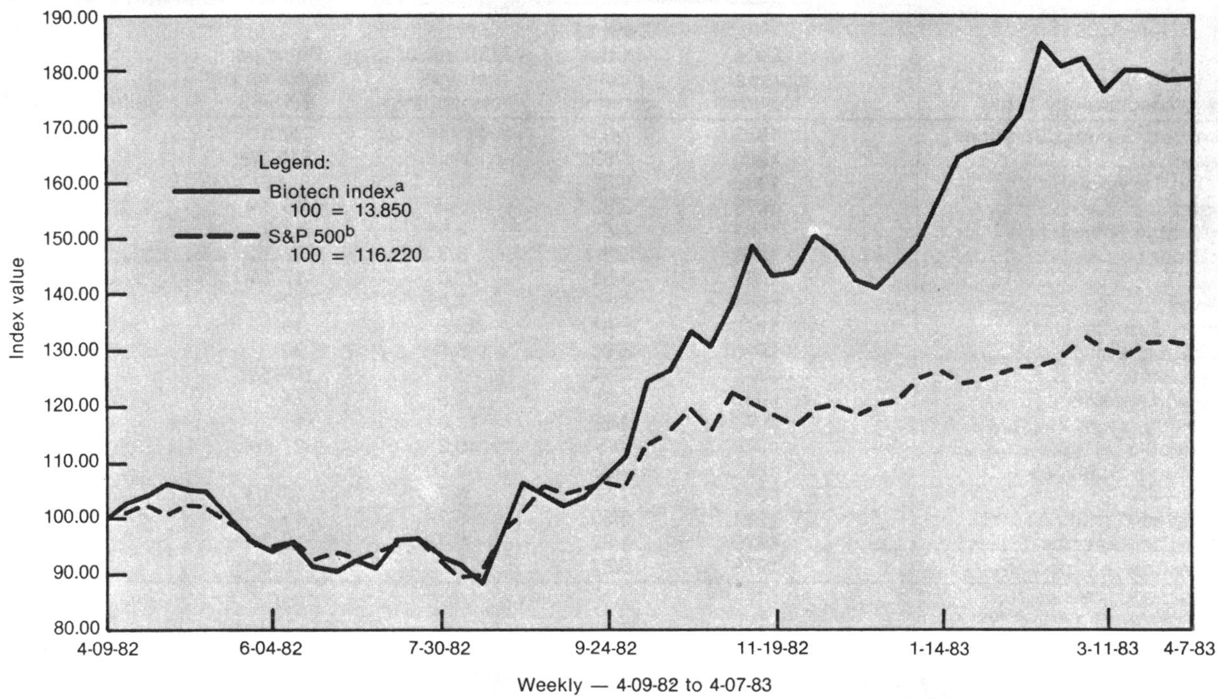

[a]Biotech index includes A. B. Fortia, Bioresponse, Cetus, Damon, Enzo-Biochem, Flow-General, Genentech, Genetic Systems, Hybritech, Monoclonal Antibodies, Novo Industri A/S. OTA did not include A. B. Fortia or Flow General as companies using biotechnology.
[b]Standard and Poor's 500 is an index of a broad cross section of companies traded on American stock exchanges.

SOURCE: Office of Technology Assessment, adapted from E. F. Hutton.

some venture capitalists may have shortened their investment time horizons (41), a development that now might be affecting the time taken to bring NBFs to the public market. Table 50 shows the elapsed time between company founding date and initial public offering for 19 NBFs.

The number of, and the amount of money raised in, initial public offerings in all industrial sectors in the United States over the past 10 years is shown in table 51. As can be seen, both the number of offerings and the amount raised first decreased and then increased dramatically. The years 1981 and 1982 were record years for new stock offerings, both in the number of offerings and in the amount raised (though the total amount raised in 1982 was 25 percent less than the amount raised in 1981). Not since the boom of the late 1960's, however, has the new issues market been as active as in 1983.

The initial public offering history and market valuations as of July 1983 for 19 NBFs is shown in table 50. No NBFs made offerings prior to 1980. Two firms went public in 1980, five in 1981, and three in 1982; as of August 1, nine had gone public in 1983. The drop in the number of biotechnology public offerings between 1981 and 1982 parallels the drop in initial public offerings in all sectors during the same period (table 51).

The first recognized "biotechnology firm" to go public, in October 1979, was BioResponse,* with an offering of 1,320,000 units** at $2.50 per share. Thus, the total raised was $3.3 million. It is interesting to note that at the time of the initial public offering, BioResponse had no revenues and a negative net worth of more than $600,000.

*BioResponse was founded in 1972 and is not included here as an NBF.

**One unit = one share of common stock plus one warrant.

Table 50.—Initial Public Offering History and Market Valuations as of July 1983 for 19 New Biotechnology Firms in the United States

New biotechnology firm	Date company founded	Date of initial public offering	Market valuation as of July 1983: Millions of shares outstanding	Price per share as of 7/15/83	Market value (millions of dollars)
Advanced Genetic Sciences	1979	7/83	12.2	N/A[a]	N/A
Amgen	1980	7/83	10.0	$13 3/8	133.75
BioCell Technology	1980	8/81	N/A	1/2	N/A
Biogen	1978	3/83	18.5	15 3/4	291.375
Cambridge Bioscience	1981	4/83	4.08	11 3/4	48.175
Centocor	1979	12/82	5.3	17 1/2	92.75
Cetus	1971	3/81	22.0	17 1/4	379.5
Chiron[a]	1981	8/83	7.28	12[b]	87.4
Damon Biotech	1983	6/83	19.5	16	312
Enzo Biochem[c]	1976	6/80	5.8	30	174
Genentech[c]	1976	10/80	1.4	46 3/4	65.45
Genetic Systems	1980	6/81	1.8	14	25.2
Genex	1977	9/82	12.6	19	239.4
Hybritech	1978	10/81	10.3	27 1/4	280.67
Hybridoma Sciences	1981	8/83	4.29	6[b,d]	25.7
Immunex	1981	7/83	5.7	13 1/4	7.5
Integrated Genetics	1981	6/83	8.3	13	107.9
Molecular Genetics	1979	4/82	6.13	18 3/4	114.94
Monoclonal Antibodies	1979	8/81	2.4	18 3/4	45

[a]N/A—Information not available.
[b]After public offering August 1983.
[c]Stock split.
[d]One unit = 3 shares common stock + 3 Class A Warrants.

SOURCE: Office of Technology Assessment, adapted from E.F. Hutton & Co., Inc., Washington, D.C., personal communication, August 1983.

Table 51.—Number of Initial Public Offerings and Amount Raised in All Industrial Sectors in the United States, 1972-83

Year	Number of initial public offerings	Amount raised[a] (millions of dollars)
1972	568	$2,700
1973	100	330
1974	15	51
1975	15	265
1976	34	234
1977	40	153
1978	45	249
1979	81	506
1980	237	1,400
1981	448	3,200
1982[b]	222	1,470
1983[a,b]	516	7,900

[a]Through August 1983.
[b]Howard & Co., Philadelphia, personal communication 1983.

SOURCE: Office of Technology Assessment, adapted from K. Farrell, "Going Public 1982," *Venture*, April 1982, p. 30.

No revenues had been recorded by September 1982 (27), yet stock in BioResponse is trading in 1983 at about $13 per share. The successful experience of BioResponse established a precedent for bringing NBFs with similar financial characteristics to the market.

The history of the initial public offering of BioResponse illustrates the extraordinary investor interest in firms commercializing biotechnology. Indeed, biotechnology has produced two "firsts" on Wall Street. In 1980, Genentech set a new record with a price rise from $35 to $89 per share in the first 20 minutes of trading in its initial public offering. In 1981, Cetus set a new high for an initial public offering—$120 million (net amount was $107 million). Even in 1983, the best year ever for raising money for biotechnology, few products had been introduced.

Public offerings in 1982 were less successful than had been hoped for, probably because of an increasing realization by the public that the fruits of biotechnology R&D might be more distant than was first anticipated and also because the stock market was depressed in 1982. Thus, Collaborative Research in February of 1982 raised less than half of the $28.5 million it had hoped to raise in its initial public offering, while Molecular Genetics obtained only $3.3 million, less than one-third of its goal. Genex, in a 2.5 million share initial offering, sought to raise about $30 million to support scale-up of its research products, but first day over-the-counter sales totaled only about 1 million shares, and the closing price was $9 rather than the $10 to $12 initially predicted.

The boom in the 1983 public offerings market has provided many new firms including NBFs, with capital. Venture capital for NBFs increasingly difficult to obtain, the result being that public offerings in 1983 are supplying second- and third-round financing. NBFs that are either seeking or already have raised second- and third-round financing in 1983 include Cambridge BioScience, Damon Biotech, Molecular Genetics, Biotechnica, Genetics Institute, Biogen, Integrated Genetics, Applied BioSystems, California Biotechnology/Synergen, DNA Plant Technology, Amgen, Hybridoma Sciences, INGENE, Advanced Genetic Sciences, Biotechnology General, Immunex, and Chiron. Table 52 lists some recent initial public offerings by NBFs and the amounts raised.

The price/earnings ratios for NBFs appear high in 1983, given their negative or low earnings records. Continued reliance on the public market for funds will place increased pressure on public NBFs to earn a profitable income stream quickly. If products are not manufactured and income generated within the time frame demanded by investors in the stock market, NBFs will face additional financial constraints. If they have to rely on the stock market and R&D limited partnerships for funds, NBFs might face problems in financing the long-term risky research in scale-up processes that is needed to commercialize biotechnology products.

ESTABLISHED COMPANIES

Established U.S. companies like Eli Lilly, DuPont, and Monsanto can finance their entry into biotechnology using internal funds generated from a variety of sources, (e.g., the sale of products, interest income on capital, and other sources). Such companies also have ready access to debt financing (e.g., loans) or through debt offerings and the sale of bonds. The cost of borrowing is less for established companies than for new companies, because financing is available to established companies at or near the prime rate. Those NBFs that are able to qualify for loans may pay 2 or 3 percentage points over the prime rate (27). In sum, for established U.S. companies considering commercial applications of biotechnology, the question is not whether financing is available, but whether or not to spend their sizable resources (or those that they borrow) on the new commercial pursuits of biotechnology.

To illustrate the magnitude of established company resources to enter biotechnology, a few examples can be noted. In 1981, DuPont budgeted $120 million for biotechnology R&D out of a total R&D budget of $570 million (19). In 1982, DuPont began construction of a new $85 million life sciences center, and it acquired New England Nuclear (U.S.) for $340 million, in part to expand its capability in the life sciences. As another example, in 1984, Eli Lilly expects to complete a $60 million research center that will emphasize rDNA and immunological applications of biotechnology (13). The annual R&D budgets of established U.S. companies such as DuPont and Eli Lilly dwarf the

Table 52.—Amounts Raised in Recent Initial Public Offerings by Six New Biotechnology Firms

New biotechnology firm	Date of initial public offering	Shares offered (in millions)	Offering price per share	Amount raised (millions of dollars)
Amgen	6/83	2.35	$ 18	$ 42.3
Biogen	3/83	2.5	23	57.5
Cambridge Biosciences	3/83	1.0	5	5.00
Chiron	8/83	1.5	12	18.0
Immunex	7/83	1.65	11	18.15
Integrated Genetics	7/83	1.6	13	20.8

SOURCE: Office of Technology Assessment, adapted from E.F. Hutton & Co., Inc., Washington, D.C., personal communication, July 18, 1983.

amounts that have been raised by NBFs in the United States in even the most successful public stock offerings. In 1981, for example, the NBF Cetus raised a record breaking $120 million in its initial public offering—a little more than 20 percent of DuPont's annual R&D budget.

Sources and availability of financing for firms in other countries

The sources and availability of financing for companies commercializing biotechnology in Japan, the Federal Republic of Germany, the United Kingdom, Switzerland, and France—the five countries considered the major competitors of the United States in the area of biotechnology—are outlined in the discussion below.

JAPAN

As noted in *Chapter 4: Firms Commercializing Biotechnology,* predominantly large established companies are developing biotechnology in Japan. Established companies in Japan, like those in the United States, are able to rely on debt financing or revenues generated from the sale of products and other internal sources of funds to finance their entry in the field of biotechnology.

The industrial and financial structures of Japan are very different from those of the United States and most European countries. In Japan, equity markets are relatively unimportant for allocating capital. Instead of raising capital by sharing equity, Japanese companies continue to favor debt financing.* The emphasis on personal savings by Japanese families has produced a large pool of funds in banks and postal savings accounts, and these funds are lent to Japanese corporations. Thus, private sector financing of biotechnology in Japan is usually mediated through the banking system.

NBFs, especially prevalent in the United States, and to a lesser extent in the United Kingdom and France, are not found in Japan because of the low level of equity funds there (39).* Public offerings, venture capital, and other equity instruments are of relatively minor importance there. The low level of equity funding available in Japan is illustrated by comparing the over-the-counter securities markets in Japan with those in the United States. About 111 companies are traded on the Japanese market, compared to 13,000 in the United States. Differences in venture capital investments are also indicative of the relative importance of venture capital in the two countries. In 1982, venture capital investments in Japan amounted to about $84 million, whereas those in the United States amounted to $5.8 billion (6). The low level of interest by Japanese investors in venture capital is further shown by the fact that a venture capital firm established in July 1982 by the Daiwa Securities and Long Term Credit Bank was the first venture capital company to be started in 8 years (6).

The Japanese Government has made two efforts to encourage the development of a venture capital industry in Japan. One effort was made by the Ministry of International Trade and Industry (MITI) in the early 1970's but yielded little in the way of results (22). In a resurgence of interest in this area, in 1982, MITI set up an Office of Venture Enterprise Promotion in parallel with the creation of the Office of Biotechnology Promotion (32).

Japan's private sector has recently taken some initiative in developing a source of "venture capital" by pooling corporate resources. The Japan Associated Finance Corporation (JAFCO) is a private venture capital fund that was organized by Nomura Securities Company. One French, three Hong Kong, and 10 Japanese firms are involved in JAFCO, which plans to offer financial help to new businesses until they qualify for listing as a joint stock company. When the firm reaches this stage of maturity, its income gains will be distributed among the partners of the fund accord-

*A majority of Japanese companies commercializing in biotechnology have debt to equity ratios that exceed 3 (39), as compared to U.S. ratios that are generally closer to 1. Although the Japanese figures are biased upwards because of differences in land values and because off-sheet financing is used more frequently in the United States than in Japan, the differences in debt to equity ratios are significant.

*Other reasons for the scarcity of NBFs in Japan are cultural attitudes that discourage entrepreneurism, the rigid separation in Japan between university basic research departments and industry, and Japan's weak basic science base in molecular biology (39). Some of these subjects are addressed in *Chapter 17: University/Industry Relationships*.

ing to the ratio of the capital contribution of the fund (22).

These new sources of venture capital may or may not succeed in increasing the supply of venture capital in Japan. In any case, the amount of venture capital these sources currently provide is very small when compared to the amount available in the United States.

The one source of "venture capital" that has been very important to the development of biotechnology in Japan is personal loans of sizable amounts by wealthy individuals who are the managers of progressive Japanese companies such as Hayashibara, Suntory, and Green Cross. As entrepreneurial managers, these individuals are very unusual in Japanese history. A venture by Hayashibara for producing interferon with hamsters was possible only because the owner, who owns or controls 12 institutions (hotels, gas stations, and candy manufacturing firms) and does about $150 million worth of business a year, put his capital behind it (51). The diversification by Suntory (a whiskey company) into rDNA research to produce pharmaceuticals was similarly supported. Significantly, Japan's giant pharmaceutical companies were far slower and more bureaucratic in their response to the potential of biotechnology than these newer Japanese more progressive firms.

In fiscal year 1981, a Government-related organization called the Center for Promoting R&D Type Corporations guaranteed approximately $3.7 million (¥ 750 million) in loans (a total of 24 loans). Beginning in 1982, the center was to begin making loans as well as guaranteeing other lender's loans. Up until now, however, the Japanese Government has not been a major source of financing for Japanese companies developing biotechnology.

There is no indication that significant funds are being channeled into biotechnology by financial institutions connected with the Japanese Government to make up for the shortage of venture capital. In the past, Government-funded banks like the Japan Development Bank (JDB) lent only to projects that fit into articulated Government policy and were located in Japan. In the past decade, however, private bank loans have expanded to such an extent that they are competitive commercially with the Government financial institutions (39). Certain funds within the JDB loan portfolio are targeted for technology promotion. For the past 4 years, this fund has remained fairly constant at the level of $500 million (¥ 100 billion), approximately 10 percent of the total loan portfolio. Loans from the JDB are made at interest rates between 7.5 and 8.4 percent. There is no indication that any of these funds are being channeled into biotechnology.

FEDERAL REPUBLIC OF GERMANY

In the Federal Republic of Germany, nearly all private sector investment in biotechnology has been made by the established pharmaceutical and chemical companies. There is no parallel in the Federal Republic of Germany to the U.S. venture capital industry. Commercial banks provide most of the funds used for industrial expansion, and it is common for such banks in Germany, unlike those in the United States, to have equity participation in companies in which they invest. The West German commercial banking sector is dominated by three banks, and the linkages between the banking and corporate structures are so close that the Monopoly Commission concluded in 1976 that the banks effectively utilize management functions to the detriment of competition (23).

In 1975, a consortium of 28 banks recognized that the German banking system is not conducive to high-risk, innovative, startup firms and formed a venture capital concern called Risk Financing Society (WFG, Deutsche Wagnisfinanzierungs-Gesellschaft) (7). The principal objective of this organization is to aid small and medium-sized firms in commercializing their products. So far, the electronics industry has been the major recipient of WGF funds; biotechnology firms have not yet been of great interest to WFG. Since 1980, WFG has been looking for innovations that could achieve commercial success within 24 months. If this continues to be the criterion for any firm receiving funds from WFG, then it would be surprising if many startup firms in biotechnology were established in the Federal Republic of Germany with WFG funds (23).

UNITED KINGDOM

The present Government of the United Kingdom believes that the successful industrial development of biotechnology depends on private industry. The main source of funds will be the retained earnings of established companies and the capital provided by private financial institutions. The United Kingdom does not have a well-developed venture capital market, and the tax structure in the United Kingdom is not conducive to the formation of risk capital (the capital gains tax rate there is higher than in the United States, as are the marginal income tax rates for higher incomes).

Despite the little direct availability of venture capital, the United Kingdom is providing public and institutional support to encourage the formation of small firms. The Unlisted Securities Market (USM), for example, was formed in 1980 primarily to raise capital for small companies. At the time of its opening, USM had 6 firms; 2 years later, it had a membership of 115 firms and was capitalized at a total of $2 billion. Most of the trading volume in this market is accounted for by small investors. The value of the shares of USM's 20 largest companies has increased 45 percent over the past 2 years, excluding dividends (43). Before USM was established, companies could be listed only on the London Stock Exchange, and listing there required profits of at least $1 million. In addition, until 1977, the London Stock Exchange required a company to sell off at least 35 percent of its equity for listing (the requirement has since been scaled down to 25 percent).

The British Government has introduced two new measures to encourage the formation of small firms. The first measure is designed to encourage the private sector to make equity investments in startup firms by offering tax relief at the top marginal rate to investors in new (up to 5 years old) qualifying businesses. As a result of this measure, a number of professionally managed funds have been established wherein individuals have pooled their money allowing the professional managers of the fund to make their investments. Cambridge Life Sciences, the first British biotechnology firm to go public, used this measure in April 1982 (43). The second Government measure is to guarantee loans made by banks and other financial institutions for qualifying projects that are considered to be viable (in the institution's judgment) but are not backed by personal securities. This measure means that individuals need not have substantial income in order to form a company.

Views on whether there is a shortage of funds available for biotechnology firms in the United Kingdom vary depending on the source of information. Financial institutions say funds are not in short supply; rather, the shortage is in well-presented ideas with commercial value that are capable of earning the relatively high rates of return desired by investors with risk capital. Entrepreneurs say that there is a shortage of funds because institutions demand more evidence than they can supply to prove that their products are capable of earning high profits.

Several institutions in the United Kingdom are supplying funds for the development of biotechnology, including Biotechnology Investments Ltd., Prutec, Advent Eurofund, Cogent, and Technical Development Capital (43). Biotechnology Investments Ltd., a branch of N.M. Rothschild Asset Management, is the largest, with an initial capital pool of $55 million (17). Although Rothschild has invested mostly in U.S. NBFs and other foreign companies, it recently purchased equity in Celltech (U.K.) and is considering several proposals from other British firms. Another fund, Technical Development Capital (TDC), provides equity financing in addition to loans and has a policy of becoming actively involved in management teams. TDC has an annual budget of $5.7 million (£10 million) of which $1.4 million (£2.5 million) is devoted to biosciences, one of three priority areas. The time scale of investments required depends on the industrial sector (e.g., in the medical field, the time horizon is 5 to 7 years; in agriculture, it is 15 to 20 years). TDC has investments in Celltech, Imperial Biotechnology, and three other NBFs in the United Kingdom. Prutec, a wholly owned subsidiary of the Prudential Assurance Co., Ltd., was established in 1980 and makes investments in technology-based firms. Prutec has identified biotechnology as one of 10 strategic areas for investment.

A public institution, the British Technology Group (BTG), is sponsored by the Department of Industry and is the major public source of venture capital in the United Kingdom. BTG invests a certain percentage of its funds in high-risk, long-term investments. The aim of BTG's investment group is to invest on commercial terms in minority partnership with private industry. The best known example of this policy is BTG's investment in Celltech.

Although the number of NBFs forming in the United Kingdom is increasing, the established firm sector is largely responsible for the development of biotechnology there.

SWITZERLAND

Funding for new, high-risk enterprises in Switzerland is not readily available. Analysts attribute this situation to many factors. The Swiss banking industry is oriented to large-scale international financial transactions in areas such as securities, foreign exchange, and precious metals. The banking expertise to evaluate and finance new technologies is lacking. Some argue that the structure of the savings system is changing, with private savings declining and pension funds, traditionally more conservative in investment policies, increasing. Added to these factors is the national reluctance to take risks. The NBF Biogen S.A., for example, has relied heavily on U.S. venture capital and the U.S. stock market to obtain needed capital to finance operations (24).

All of the established Swiss chemical and pharmaceutical companies have substantial capital investments in the United States. Because of the small size of Switzerland's domestic market, most Swiss companies are multinational. The Swiss companies spend a substantial fraction of their R&D costs abroad (this fraction varies among companies). Ciba-Geigy, for example, traditionally spends about 60 percent of its research expenditures in Switzerland and 40 percent in other countries; in 1981, Ciba-Geigy's expenditures on R&D in the United States rose to 23 percent of its total research expenditures, and expenditures on R&D in Europe and in Asia accounted for 20 percent (24).

FRANCE

The number of companies involved in commercializing biotechnology in France is fairly small, and the Government expects this situation to continue. The French Government, which generally believes that only large companies have the necessary resources to undertake biotechnology, has identified three centers of development in the private sector: Rhone Poulenc, Elf Aquitaine, and Roussel Uclaf. Rhone Poulenc and Elf Aquitaine are now nationalized, and Roussel Uclaf is 40-percent Government owned (44).

The venture capital market is poorly developed in France. Banks are the major source of financing. Banks in France, like their counterparts in the United Kingdom but unlike those in West Germany, have always hesitated to take equity positions in industry. The Government of France would like to change this attitude (28). A mutual guarantee company, INODEV, was established by the French Government to guarantee bank credit for the purpose of innovation (33). Since French banks do provide long-term financing, French firms do not have to worry as much about second- and third-round financing as do firms in the United States (44).

Tax incentives relevant to firms commercializing biotechnology

The various tax provisions in the United States, Japan, and Western Europe that are potentially important to companies commercializing biotechnology* are those pertaining to R&D expenditures, capital formation, corporate taxation, and tax treatment of small businesses.** A summary of the tax provisions described for the United States, Japan, the Federal Republic of Germany, the United Kingdom, and France is presented in table 53. Switzerland is excluded from the table, because Swiss tax rates vary among cantons, and the Federal tax system is less important.***

U.S. tax provisions affect NBFs and established companies differentially. In order for corporate tax rates to make a difference in the decisionmaking process of firms, taxable income, the base on which taxes are figured, must be present. Since the NBFs are not experiencing substantial profits, and because there are loss carry-forward provisions in the tax code (for the United States, the period that a company can carry forward losses is 7 years), most NBFs are not now focusing a lot of attention on tax incentives.* Established companies earning taxable income from a number of product lines, by contrast, are interested in current tax benefits.

*The tax codes of various countries change frequently. The discussion here is based on the latest information available in existing sources. The intent of this section is to sketch the major provisions, not to detail specifics of each tax code.

**Local or regional taxes are not included, except in the case of Switzerland, which taxes primarily on a cantonal level. Value-added taxes are also not included, since not all countries have this tax.

***In Switzerland, taxes are governed by Federal law and the tax laws of 26 cantons. While the Federal Government collects practically all indirect taxes, it receives only a small portion of direct taxes levied. The 26 Swiss cantons have a number of obligations, which in other countries would be the responsibility of the Central Government, such as education, road construction, health, police, and justice expenses. To be able to meet these obligations, tax revenue is collected from taxes on income and net assets of individuals and business entities by each canton.

*For this analysis, OTA solicited the views of the following companies engaged in biotechnology: Biogen, Cetus, Genex, Genentech, DuPont, Hybritech, and Monoclonal Antibodies. Industrial Biotechnology Asssociation and the U.S. Department of Commerce were also contacted. Most stated that tax incentives are of secondary importance to other tax provisions (e.g., loss carry forward provisions, R&D limited partnership, and capital gains treatment) given the stage of the company's development.

In a recent study of California biotechnology companies, few participants in the survey stated that tax abatement programs would be useful to their companies (16). Tax abatement programs were rated on a scale of possible utility to the company; evaluations of these programs by the executives responding to the survey ranged from "possible" (at best) to "unlikely." This pattern may reflect the essentially entrepreneurial nature of the NBFs included in the survey. The more established firms with a diversity of product lines would be more interested in tax incentives not primarily focused towards capital formation. It may happen that as established companies become more important in the field, tax incentive programs will be viewed with more interest.

It is important to note that some countries rely more on tax provisions to stimulate capital formation or industrial development than others that use grants or subsidies to assist specific industrial projects. The United States, Switzerland, and to a lesser extent the United Kingdom, for example, tend to rely more on tax incentives to encourage overall capital formation than, for example, the Federal Republic of Germany or France, which use grants or subsidies for specific projects. Other countries (e.g., the United Kingdom, France, and Japan) use tax incentives to encourage investment in R&D or plant and equipment required for scale-up or scientific research. Furthermore, some countries (e.g., the United States and Japan) favor formation of small businesses by tax provisions that are specifically aimed at smaller establishments. Japan targets particular industries and uses both tax incentives and grants.

Some analysts state that the tax incentives in the United States, when compared to those in Western Europe, are not a major factor in decisions about the location of foreign subsidiaries of biotechnology companies (26). However, others argue that sharp differences in the corporate tax rate between countries such as the Netherlands Antilles (whose nominal corporate tax rate is 3 percent) and the United States (whose nominal corporate tax rate is 46 percent) have led some

Table 53.—Tax Treatment of Innovation Activities in the United States and Other Countries

Capital expenditures for R&D	Current expenditures for R&D	Venture capital investments in new technology-based firms	Small business tax treatment	R&D tax credits/ investment grants[a]
United States:				
Treated in same manner as other depreciable assets	Immediately expensed	R&D limited tax partnerships allow investors to write off current expenses as losses and treat future gains as capital gains Investors can pool funds in a regulated investment company of which venture capital corporations are a member, and the company can avoid taxes if the company distributes all its income	SBIC treatment: 1) dividends-received deduction of 10% is allowed to SBICs for dividends received from taxable domestic corporations; 2) loss on stock is treated as an ordinary loss and does not have to be offset against gains from sales of stocks; 3) gains are treated as capital gains Subchapter S corporations: A sub S company gives owners of closely held corporations the advantage of limited liability for depts while taxing the corporation's income at shareholder's income rates. Number of shareholders permitted is 35	Can deduct 25% of the difference between the current year's R&D expenditures and the moving average of a 3-year period.
Japan:				
Firms that are members of Research Association can take 100% depreciation allowance on all fixed assets used in connection with Research Association activities	Immediately expensed	No special provisions	The corporate tax rate for small-and medium-sized corporations on the first ¥ 7 million ($28,107) is 22% (as opposed to regular rate of 30%). A small business can add each year to the ordinary depreciation allowance up to 14% of the original value of new equipment and machinery acquired between Apr. 1, 1972 and Mar. 31, 1983 Additional depreciation allowances are allowed for small businesses that are entering new industries.	Can deduct each year from its income tax 20% of the difference between the current year's R&D expenditures and the highest R&D expenditures in a year before the current year if the difference is positive
Federal Republic of Germany:				
Depreciated in same way as other assets. For expenditures of plant and equipment embodying new technology, the depreciation allowance includes reasonable allowance for obsolescence	Immediately expensed	No special corporate tax treatment for venture capital investments	There is no special corporate tax treatment apart from a provision applicable to foundations and associations. For these organizations, there is a deductible tax free amount of DM5,000 (U.S. $2,060). If corporate income exceeds DM10,000 ($4,120), the tax-free amount is reduced by half the excess	Investment grant of 20% of cost can be claimed for the first DM500,000 (U.S. $206,049) of the costs of assets used in R&D. The excess of cost DM500,000 qualifies for an investment grant of 7.5%
United Kingdom:				
For scientific research assets, a 100% tax allowance (or deduction) is given. Allowances are given for capital expenditures (e.g., labs) and current expenditures (e.g., research workers' salaries)	Immediately expensed.	No special tax provisions for venture capital investments	A closely held company's investment income is apportioned, provided it is surplus to the requirements of the business. Corporation tax rate is 40% if profits do not exceed £ 70,000 (U.S. $122,527)	

Table 53.—Tax Treatment of Innovation Activities in the United States and Other Countries (Continued)

Capital expenditures for R&D	Current expenditures for R&D	Venture capital investments in new technology-based firms	Small business tax treatment	R&D tax credits/ investment grants[a]
France:				
Can depreciate 50% of the cost in first year with the balance depreciable over useful life	Current expenditures are immediately expensed—carry-backs are not allowed	Businesses which purchase shares in Qualified Research Companies and shares in Innovation Finance Companies may deduct 50% of the cost of the shares in the year of acquisition. If shares are sold, the additional gain attributable to this 50% deduction is eligible for capital gains tax treatment. If shares are held for 3 years or more, no capital gains tax is assessed	Small and medium-sized businesses (fewer than 2,000 employees, not legally dependent on a larger business and having less than 50% of their shares held by quoted companies) are entitled to an exceptional deduction of 50% of the cost of equipment and tools used for R&D Tax allowance amounting to one-third of the firm's taxable profits in the fiscal years of its establishment and in the 3 subsequent tax years	—

[a] Information on the tax rules of foreign countries obtained from tax services and other secondary sources, not from the foreign statutes themselves. While efforts were made to obtain accurate and up-to-date information, it should be noted that reliance on secondary sources does increase the potential for error.

SOURCE: Office of Technology Assessment, based on information from National Science Foundation, *Corporation Income Tax Treatment of Investment and Innovation Activities in Six Countries*, Washington, D.C., 1981; Price Waterhouse & Co., *Price Waterhouse Information Guide: Doing Business in Germany*, September 1978; Price Waterhouse & Co., *Price Waterhouse Information Guide: Doing Business in France*, 1979; Price Waterhouse & Co., *Price Waterhouse Information Guide: Doing Business in the United Kingdom*, 1980; and Price Waterhouse & Co., *Price Waterhouse Information Guide: Doing Business in Switzerland*, 1982.

biotechnology companies to incorporate in the Netherlands Antilles and then form a subsidiary in the United States (20). Generally, tax incentives aimed at capital formation, such as the R&D limited tax partnership or capital gains tax rate, are viewed with much more interest in the short term by U.S. NBFs than tax incentives because NBFs need taxable income to use them.

Tax incentives relevant to new biotechnology firms in the United States and other countries

Tax incentives beneficial to NBFs include R&D tax incentives, capital formation tax incentives, and tax treatment of small businesses.

R&D TAX INCENTIVES

The U.S. tax code offers no special incentives for R&D beyond those available for investment generally and for investment in depreciable structures or equipment used for research and experimental design. The buildings used for R&D are not given preferential tax treatment in the United States as they are in Western European nations. Thus, the United States has no special tax incentive for construction of plant or equipment used in biotechnology. Such an incentive may be, depending on the importance of the costs of depreciable assets in the total production costs, an important factor in determining cost competitiveness in biotechnology products. As products move from research to scale-up stages of production, these costs become more important.

Companies in the United Kingdom are entitled to a 100-percent first year writeoff on capital expenditures for scientific research, the most rapid allowance offered by any country (1). Tax provisions allowing the immediate deduction of capital expenditures for assets used in R&D provide a current tax benefit* rather than a deferred tax benefit, because the capital expenditures for R&D may be offset against income earned in the year of the capital asset's acquisition rather than offset against income earned over the useful life of the asset. Accelerated depreciation provides a tax benefit in that it permits a much faster recovery of the cost of an R&D asset; however, the immediate deduction of the total cost of the asset provides an even faster recovery of costs. The Federal Republic of Germany allows accelerated depreciation for R&D assets in the form of additional depreciation taken in the first few years the assets are used. For investments of less than $234,750 (DM570,000), there is an investment grant of 20 percent of the cost of the assets used in R&D (9,30). France allows 50 percent of the cost of buildings used for scientific or technical research to be written off in the first year.

The United States, Japan, the Federal Republic of Germany, the United Kingdom, and France allow deductibility of current R&D expenditures, but only the United States and Japan give a tax credit for incremental R&D. The Japanese tax credit allows a company to deduct each year from its income tax 20 percent of the difference between the current year's R&D expenditures and the highest R&D expenditures in a base year before the current year. The U.S. tax credit allows a company to deduct 25 percent of the difference between the current year's R&D expenditures and the moving average of a 3-year period's R&D expenditures. In order to qualify for the credit, a company must be carrying on a trade or business. The U.S. Treasury was given leeway in defining the trade or business, and it was widely hoped that the newest proposed regulations would give small firms, primarily engaging in research but not yet selling products, an advantage. Some have stated that Treasury's position is inflexible towards the small firms not yet able to produce products (5).

Some analysts argue that the U.S. tax credit for incremental R&D encourages more R&D than Japan's tax credit, because the base used in the United States (the moving 3-year average) may be lower than the base used in Japan (the highest R&D expenditure in a previous year); the lower base in the United States may allow a higher tax credit given the same rate of increase in R&D expenditures. The U.S. tax credit is currently scheduled to expire in 1985, and many are urging an automatic extension of the credit, especially since the planning and implementation of R&D is a long-term process. Legislation introduced in

*This current benefit is of immediate benefit only to firms with sufficient current taxable income to use the tax benefit.

the 98th Congress, H.R. 3031, sponsored by Representative Fortney Stark, and S. 738, sponsored by Senator John Danforth, would amend the IRS Code by making the R&D credit permanent in the United States. France is considering a 25-percent tax credit for R&D expenditures, thus encouraging through the tax system an increase in R&D expenditures (49). Whether the implementation of additional tax credits will affect the amount of money devoted to R&D expenditures will depend in part upon the permanency of the tax provision in each country.

The treatment of income derived from the sale or license of technology differs among countries. In the United States, proceeds from the sale of patents are treated as long-term capital gains (taxed at the long-term corporate capital gains tax rate of 28 percent). Royalties are taxed as ordinary income (30). In Japan, both proceeds and royalties are treated as ordinary income. Sales of patent rights, technical and manufacturing processes, and know-how are taxable in France at the reduced 15 percent long-term capital gains tax rate (1). Royalties are taxed at the standard 50-percent corporation tax rate unless industrial property rights have the characteristics of fixed assets or the license is granted for 8 years and for exclusive use within a geographical area. In the latter instances, royalties are taxed as long-term capital gains. In the United Kingdom, any capital sum received on the sale of a patent by a U.K. resident is charged as if it were a corporation (at a tax rate of 52 percent); the sum is generally spread over 6 years, so that one-sixth of the sum is liable to tax in each year. Royalties received are treated as ordinary income (30). Overall, the United Kingdom has the most adverse tax treatment of income resulting from the sale of technology (whether involving the sale of patents or licensing).

CAPITAL FORMATION TAX INCENTIVES

Tax incentives designed to stimulate capital formation are of special importance to the formation and growth of NBFs, because few NBFs have enough income derived from product sales or contract revenue to sustain high costs for both R&D and scale-up production. In affecting the amount of capital available to smaller firms, the tax treatment of individual capital gains and R&D limited partnerships are important.

Tax Treatment of Individual Capital Gains.—The long-term capital gains tax rate for individuals in the United States is 20 percent, down from 49 percent in 1976. Industry analysts suggest that this decrease in the individual capital gains tax rate is the primary reason for the substantial increase in venture capital available in the United States (27).

In Japan, capital gains on the sale of securities are exempt from tax, unless the sales are habitual or in the course of business. For nonexempt gains, the first $2,232 (¥ 500,000) is exempt, and the remainder of gain is either taxed as short-term capital gains (treated as ordinary income) or long-term gains (50 percent taxed at ordinary income tax rates) (42).

In the Federal Republic of Germany, no capital gains tax is payable by individuals on assets held longer than 6 months. If an asset is held less than 6 months, the capital gains income is taxed as ordinary income. Capital gains arising from the sale of business assets by an individual are liable to tax at normal rates where the assets form part of the business property. Extraordinary income arising as a result of a gain from the sale of an entire unincorporated business or from the sale of shares by a substantial shareholder are taxed at half the individual's marginal tax rate, i.e., at a maximum of 28 percent (35).

In the United Kingdom, capital gains income is subject to a tax rate of 30 percent (42). The tax treatment of capital gains in France depends on the length of time the asset is held. Short-term capital gains (on assets held for less than 2 years) are included in operating profit and are taxable at a 50-percent tax rate (37). The taxpayer may elect to spread the capital gains tax over 3 years. Long-term capital gains (on assets held for 2 years or more) are taxable at a 15-percent tax rate. Long-term capital gains and losses of the same fiscal year are offset against each other.

Tax Treatment of R&D Limited Partnerships.—As discussed in the section of this chapter on "Financing in Firms Commercializing Biotechnology," an important tax tool used for risk capital formation in the smaller companies engaged in biotechnology in the United States is the R&D limited partnership. Some NBFs using R&D limited partnerships as a method of raising capital have stated that they prefer the partnerships as a method of financing, because the revenues from a partnership are treated as revenues and allow a company to show a profit even if it has few or no products to sell (27). By using R&D limited partnerships, NBFs have postponed issuing stock, selling equity to established firms, or searching for venture capital, thereby keeping more control over their company. Neither Japan nor West European countries use a similar type of tax treatment.

An R&D limited partnership is formed to support R&D that will result in something that is marketable and patentable. As discussed below, financial advantages accrue to the limited partners (investors) at both the R&D phase and the marketing phase, provided certain conditions are met.

Turning attention first to advantages at the R&D phase, the applicable part of the Internal Revenue Service (IRS) Code is section 174 (Title 26 U.S.C. IRS §174). Section 174 allows each limited partner to deduct all expenses for research (generally, the amount the limited partner invested in the partnership) from income in the year the expenses were incurred, provided the limited partners were at risk.* If the limited partners are *not* at risk, such deduction is not allowed. The challenge, therefore, is to write the agreement establishing the partnership so that the limited partner is at risk. This is generally done by structuring the agreement so that the general partner does not *automatically* buy the results of the research from the limited partners. An automatic purchase provision in the agreement would presume the research would be successful and imply that there was no risk. Similarly, agreements usually base any financial return to the limited partners that may arise from the partnership on sales rather than profits, because the term "profits" in the agreement implies success and hence a no-risk situation.

*To ascertain whether the partners bear the required risk, one asks, "Who loses if the research effort is a complete failure?"

Upon successful completion of an R&D project supported by an R&D limited partnership, the limited partners may realize economic returns either through royalties or license fees derived from the sale or transfer of a patent or by sale of the product back to the general partner or to a third party. Both of these may qualify for favorable tax treatment. If the research results in a patent, the patent may be sold or transferred by the limited partners to the general partner, generally in return for royalties or license fees. Under section 1235 of the IRS Code (Title 26 U.S.C. IRS §1235), any royalties received as a result of transfer of a patent qualify as long-term capital gain rather than ordinary income. The current tax rate on long-term capital gains for individuals is 20 percent, whereas the tax rate on ordinary income can be as much as 50 percent. The usual 1-year period necessary for the sale of a capital asset to qualify for capital gains treatment does not apply.

Generally, section 1235 treatment applies to a transfer of property consisting of all substantial rights to a patent by any holder. A holder is defined as any individual whose efforts created the patentable property or any other individual who has acquired interest in the patentable property in exchange for money paid to the creator prior to the actual reduction to practice of the invention.* This definition of holder makes it difficult for R&D limited partnerships to acquire rights to a patent when a university has the rights to the patent through employment agreements with its university scientists. Universities that have obtained patent rights through employment agreements with university scientists are excluded from the present definition of holder. As a result, relatively few universities have formed R&D limited partnerships as a means for helping to commercialize their research results.

If the research results in nonpatentable knowhow or technology, the sale of the property must meet the requirements of sections 1221-1223 or section 1231 of the same IRS code for the proceeds to be taxed to the limited partners as capital gains rather than ordinary income. Under this

*Reduction to practice is a term used in patent law referring to when the invention has been tested under operating conditions.

section, capital assets must be held for at least 1 year before they are sold to qualify for long-term capital gains treatment. Another challenge then is to write R&D limited partnerships so that they result in a patent.

Two recent changes in the U.S. tax code have increased investor interest in economic return from tax shelters, rather than just a tax deduction. First, the maximum tax rate on unearned income has been reduced from 70 percent to 50 percent. This reduction in the maximum tax rate makes unearned income for individuals in high tax brackets more valuable than it used to be and also reduces their need to shelter it. Second, investors may no longer deduct more than the amount they are actually at risk; thus, they can no longer recoup more than their full cash investment in tax savings.

There are two potential disadvantages of R&D limited partnerships for the limited partner. The first is low liquidity: the only way for a limited partner to get out of the agreement is to convince the general partner to buy his or her interest in the partnership. The second is that patents are the only assets that qualify for tax treatment under section 1235. Other types of intellectual property, such as plant variety protection certificates and trade secrets, do not qualify.*

R&D limited partnerships permit the partners to deduct partnership expenses for R&D activities from their individual incomes and then allow any income from the sale of the successfully developed invention to be treated as capital gains income, which is taxed at lower individual tax rates.

Because the financial markets are so dissimilar amoung countries, it is difficult to compare the effect on investments of different capital gains tax treatment. However, the United States has a more developed capital market than its competitors in biotechnology and also has more options for financing smaller firms. If the NBFs continue to serve as an important source of innovation, the expanded financing options for these firms will help the competitive position of the United States. The ability of firms to commercialize innovations will serve as a better indicator of a country's competitiveness than the ability of firms to serve as a source of innovation.

*Patent law and plant breeders' rights statutes are discussed in *Chapter 16: Intellectual Property Law*.

TAX TREATMENT OF SMALL BUSINESSES

Some countries have special tax incentives to promote the growth of small businesses. Studies suggest that small businesses serve as an important source of innovation as well as of the diffusion of technology.

The most favorable tax treatment for smaller businesses is provided by the United States. Subchapter S corporations* give the owners the advantage of limited liability for debts, while the corporation's income is taxed at the shareholder's tax rate rather than at the corporation's tax rate. A key advantage of subchapter S is that if a company generates operating losses, these can be "passed through" to the individual shareholders. The shareholders can use the losses to offset other taxable income. If the owners of a small company have incorporated as a "Sub-S" and they are in the 50-percent tax bracket, then the effect is that the U.S. Treasury is financing 50 percent of the new company expansion. Most NBFs are experiencing losses, so this form of corporation is attractive.

Japan also has special tax treatment for small businesses. A small business can add each year to the ordinary depreciation allowance up to 14 percent of the original value of new machines and equipment. In addition, there is a special depreciation allowance for encouraging small businesses to enter new industrial sectors. A small business that plans to change its business can treat its old machines and equipment as ones newly acquired when it calculates depreciation allowance. Special first-year depreciation credits are now allowed on this machinery (39).

A recent study by the Organisation for Economic Co-Operation and Development (OECD) outlined member government policy towards small businesses and concluded that European countries had fewer policies aimed at small firms than did either the United States or Japan (33).

*Any corporation satisfying requirements described in the Subchapter S Act and Subchapter S Revision Act of 1982 is known as a Subchapter S corporation.

The French Government has been giving increasing attention to startup firms since 1976. Three problems for smaller businesses have been addressed: self-financing, external capital financing, and access to medium- and long-term bank credit (33). The first problem is being addressed through a tax allowance for startup firms equal to one-third of the firm's taxable profits in the fiscal year of their establishment and in the 3 subsequent tax years. The usefulness of this incentive for the small firms using biotechnology in its present stage of development is questionable. Few NBFs are experiencing profits, so few would be able to use the tax allowance. The second problem, external capital financing, is addressed in France through the establishment of regional financing companies (Societés de Financement Regional) and incentives for these financing companies to acquire holding in new firms. The last problem, access to bank credit, has been and still continues to be a problem for smaller companies in France. As noted earlier, the Government of France has established a mutual guarantee company, INODEV, to guarantee bank credit for the purposes of innovation (33). In addition, small and medium-sized businesses (i.e., businesses that have fewer than 2,000 employees, are not legally dependent on a larger business, and have less than 50 percent of their shares held by quoted companies) are entitled to an additional deduction of 50 percent of the cost of equipment and tools used in R&D. However, the small firm sector is not expected to play as innovative a role in France as it has in the United States.

In the Federal Republic of Germany, there is no special tax treatment of small businesses other than a provision applicable to research foundations and associations (30).

The United Kingdom has few tax provisions available to investors or owners of small businesses that would encourage the formation of startup firms. To the extent that the NBFs are important in determining a country's ability to capture world market share in biotechnology products, the United Kingdom would be at a disadvantage. A U.K. resident company which is controlled by five or fewer persons (a person is defined as an individual and near relatives) or by its directors is known as a close company. There is exemption for certain companies which, although closely controlled, have a 35-percent public shareholding and are quoted on a recognized stock exchange. A close company is subject to special tax provisions, of which the most important before March 26, 1980, was that all or part of the company's undistributed after-tax income, after allowing for certain business requirements, could be apportioned (i.e., attributed to its shareholders according to their respective interests in the company and treated as their income). For accounting periods ending after March 26, 1980, only a close company's investment income can be apportioned (37). Therefore, the income of a close company is, to the extent attributed to shareholders under these provisions, subject to the progressive rates of personal income tax and investment income surcharge. Companies whose pretax profits do not exceed $40,000 (£22,900) pay a corporate tax rate of 40 percent instead of the usual 52 percent (37).

Various countries have national programs of regional tax incentives to encourage industries to develop in particular geographical locations. France is divided into four zones, Zones A through D, for incentive purposes. Zone D is the Paris Basin area and the Lyon region, and for this area, there exist no incentives. The other areas have varying amounts of grants and other incentives available (33). In the United Kingdom, enterprise zones are to be designated to encourage the creation of new businesses in economically declining areas. Generous depreciation allowances will be granted in these areas on the cost of certain new buildings in these zones. There also exist regional tax incentives in the Federal Republic of Germany, but the incentives only apply to the West Berlin area. In the United States, there are no Federal programs to encourage industry development in certain sections of the country. Increasingly, however, local and State governments are offering their own tax incentive programs.

Tax incentives relevant to established companies in the United States and other countries

Tax incentives for established companies include R&D tax incentives, capital formation tax incentives, and corporate taxation.

R&D TAX INCENTIVES

The depreciation allowances that apply to the capital assets used in R&D by established companies are the same as those discussed in the R&D tax incentives section for small firms above. Additional tax incentives for established companies are noted below.

Large established companies in the United States can utilize the same R&D tax credits as those used by small firms. An early assessment of the recent U.S. R&D tax credit suggests that it is not likely to induce significant increases in the growth rate of R&D in the short run, but the tax credit may have been one of a number of factors helping to *maintain* R&D budgets in the tight financial situation of 1980-82 (31).

Table 54 shows initial calculations relating U.S. firm size to tax credits earned in 1981. The assumptions underlying this table are: 1) about 63 percent of total R&D budgets is actually eligible for inclusion as R&D expenses for the credit; and 2) half of 1981 eligible expenditures occurred in the second half of the year (because only the second half of 1981 is covered by the credit). The tax credit as a percent of total 1981 R&D falls from about 2 percent on average for firms with fewer than 1,000 employees to about 1 percent for firms with 25,000 employees or more. The inverse relationship between firm size and tax credit as a percentage of R&D reflects the inverse relationship between firm size and rate of growth of R&D. The initial results tend to suggest that the tax credit for R&D is relatively more important to small than large companies.

Japan allows companies that are members of a Government Research Association* such as the one formed for biotechnology research to take a 100-percent depreciation allowance on all fixed assets used in connection with their Research Association activities. Only established companies are members of Research Associations. The Federal Republic of Germany provides a 7.5 percent tax-free cash subsidy for investment in R&D facilities for investments exceeding $206,050 (DM500,000).

Some countries allow businesses to deduct payments to research institutes for contract research. The United Kingdom allows deduction for payments made to research institutes approved by the Secretary of State or the Minister of Technology (1). The United States allows corporations to deduct the cost of equipment given to universities. Also, a manufacturer of new R&D equipment in the United States can donate equipment to universities and obtain a deduction of cost plus one-half the difference between price and cost, up to a limit of twice cost. Payments to universities for contract research or basic research by firms may be included in eligible expenditures for computing R&D tax credit.

CAPITAL FORMATION TAX INCENTIVES

The corporate capital gains tax rate and investment tax credits are discussed below as they relate to capital formation for established compa-

*Research Associations are government-sponsored groups of established companies in Japan performing joint research in specified fields.

Table 54.—Estimated Relationship Between Tax Credit Earned and U.S. Firm Size[a]

Number of employees in company	Number of companies	R&D expenditures (millions of dollars) 1980	1981	Change 1980 to 1981	Tax credit as a percent of R&D expenditures
Not available	24	$ 102	$ 130	$ 28	1.46%
Under 1,000	113	185	240	55	1.91
1,000 to 4,999	286	1,260	1,563	302	1.56
5,000 to 9,999	99	872	1,031	158	1.28
10,000 to 24,999	108	2,781	3,282	500	1.25
25,000 and over	147	22,686	25,862	3,176	0.99
Total	777	$27,886	$32,107	$4,221	1.06%

[a] Based on figures published in *Business Week's* "R&D Scoreboard 1981."

SOURCE: National Science Foundation, *An Early Assessment of Three R&D Tax Incentives Provided by the Economic Recovery Act of 1981*, PRA report 83-7, Washington, D.C., April 1983.

nies. In a broader sense, all of the tax incentives discussed in this chapter have some influence on companies' decisions concerning investment.

Corporate long-term capital gains are taxed in the United States at a maximum rate of 28 percent. In the Federal Republic of Germany and Japan, corporate capital gains are taxed at ordinary corporate income tax rates. In the United Kingdom, corporate long-term capital gains are effectively taxed at 30 percent (30,37). France allows long-term capital gains and losses of the same fiscal year to be offset against each other. Any remaining net after-tax gain (after off-setting) is credited to a special reserve, where it is allowed to remain for an indefinite period of time. If capital gains in the special reserve are distributed as cash dividends, a complementary tax equal to the difference between the long-term capital gains tax and the corporate tax is assessed. If the amount is a loss (after off-setting), it may be carried forward for 10 years to offset future long-term capital gains (36).

The United States and Japan have investment tax credits. In the United States, the credit is equal to 10 percent of qualified investment in depreciable property up to 70 to 100 percent of the tax liability for the year the equipment was placed in service; the excess may be carried over. In Japan, the credit is equal to 10 percent of the purchase price up to 20 percent of total corporate tax liability in the year of purchase for certain industries; the excess may be carried over for 3 years.

CORPORATE TAXATION

The top-bracket corporate tax rate on retained earnings or distributed earnings in the United States for established companies is 46 percent. The corporate tax rate in Japan is 40 percent on retained earnings and 30 percent on distributed earnings. In the Federal Republic of Germany, the corporate tax rate is 56 percent on retained earnings and 36 percent on distributed earnings. In United Kingdom, the corporate tax rate on retained earnings is 52 percent. In France, the corporate tax rate is 50 percent (42).

For international comparisons, effective corporate tax rates should be used rather than the statutory rates just cited. The effective rates take into account different definitions of taxable income and treatments of depreciation. Available studies suggest that effective corporate tax treatment in the Federal Republic of Germany, France, and the United States is relatively equal, with Japan and the United Kingdom having lower effective corporate tax rates; however, these studies need to be updated.

In Switzerland, different cantons have different corporate tax rates: some allow taxes that are paid to other tax authorities as a deduction; others have different loss carry-forward provisions; still others will tax capital gains at a separate rate or not tax the gains at all. The effective corporate tax rates (including Federal defense taxes) in Switzerland range from 8.85 percent to 36.89 percent, depending on the size of profits and the particular canton (38). These tax rates are among the lowest in Europe, and Switzerland is favorable in its treatment of established companies. Switzerland does not have any special treatment for small businesses, only for companies that invest in the equity of other companies and derive most of their income from dividends.

Findings

As a factor determining competitiveness in the commercial development of biotechnology, financial resources to support entry into this new field are of critical importance in all countries, especially now when the technology is new and its applications are just being developed. Financial resources available to commercialize biotechnology are greatest in the United States and Japan and somewhat less in the four other countries examined: the Federal Republic of Germany, United Kingdom, Switzerland, and France.

In the United States, a variety of funding sources are available to support the commercialization of biotechnology in both NBFs and established companies. Most major U.S. corporations have sizable internal sources of funds and are therefore less likely than NBFs to use external sources of funds to support R&D efforts in biotechnology. If external funds are needed, however, they are most likely to be obtained through debt financing.

Funding needs of NBFs depend on the market selected for entry. Funding needed to support entry into the contract research market is very low. Higher, but still quite low, are the funds needed to manufacture in vitro MAb diagnostic products; indeed, such product lines should be profitable within 2 to 3 years. Greater financial resources are required to enter the pharmaceutical market involving products for internal human use because of the expense of testing and clinical trials to obtain FDA approval. Nevertheless, about 55 percent of the NBFs in the United States plan to enter this market.* The amount of financial resources needed to enter the specialty chemicals market varies depending on the product. Most specialty chemicals do not require regulatory approval; however, FDA approval is required for specialty chemicals considered foods or food additives. Because research is near term for many of the products, 3 to 5 years, and most do not require approval, the financial costs of entering this market fall betwen those for the contract research and commodity chemicals markets. Very great financial resources are needed if an NBF wishes to enter the market for applications to plant agriculture requiring the manipulation of many genes, such as nitrogen fixation or photosynthesis, because a great deal of basic science remains to be done before commercial applications can be achieved, so a firm must plan on many years of research without financial return. Entry into the commodity chemicals market also requires major financial resources, because economies of scale are essential for economic production, and production plants for commodity chemicals cost millions of dollars. The commodity chemicals market is a risky one to select because it involves competition over a few cents difference in price. Additionally, the biotechnology that would be used needs substantial basic research.

*The commercial applications of biotechnology being pursued by NBFs are discussed in *Chapter 4: Firms Commercializing Biotechnology*.

The major sources of financing available to NBFs in the United States may be broadly categorized as:

- revenues from contract research and interest on cash previously obtained from public or private offerings,
- various sources of venture capital, and
- public stock offerings.

Research and product development agreements between NBFs and established companies are generally cost reimbursement contracts with additional incentives for reaching agreed upon milestones. Prepayments and advance payments may be obtained, and licensing agreements may bring royalties to the NBF from marketable products of the research. The funding that NBFs receive from research contracts is likely to diminish in the future as large corporations establish greater in-house capabilities in biotechnology. The funds available from corporate sponsors will increasingly be for truly innovative research, which historically has been done by small firms. As contract research funds decrease, however, many NBFs may find themselves in financial jeopardy.

Venture capital sources include venture capital from major corporations, R&D limited partnerships, venture capital funds, and SBICs. SBICs have provided relatively little venture capital to NBFs, although recently an increasing number of equity investments in new firms including NBFs have been made by SBIC bank affiliates. Many equity investments have also been made by major corporations in NBFs. Such investments appear to be motivated more by the corporations' desire to gain "a window on the technology" than by the hope of financial gain from their investments.

Some venture capital firms are set up by major corporations to invest corporate funds in new ventures. Because the firms are independent entities, the corporation is protected from loss. If successful, the venture firm returns some profits to the parent corporation. Other venture firms have no connection to major corporations. Venture capital firms can provide seed money (used to write business plans for new firms), but most often, they fund startups, underwrite public offerings, and invest in R&D limited partnerships as limited partners. A few of these firms have invested a significant amount of their money in NBFs.

R&D limited partnerships are a very important source of funds for NBFs; next to public offerings, R&D limited partnerships have so far provided the most funds for NBFs. Although such partnerships have been available for some time, NBFs are responsible for popularizing their use. Such partnerships have enabled NBFs to attract the substantial funding needed to fund research and early product development and have also been formed for novel purposes, such as supporting the cost of clinical trials.

The number of public stock offerings in biotechnology in 1982 declined to about half the number in 1981, paralleling a similar decline in the number of public offerings in all U.S. industrial sectors. Furthermore, the amounts raised by NBFs in 1982 public offerings were less than NBFs had hoped for. The disappointing return on public offerings probably reflected increased public knowledge about biotechnology and more realistic appraisals of the time necessary before investments in biotechnology are likely to pay off. Thus far in 1983, there is a boom in the new issues market and a large number of NBFs are using the market as a means to finance expansion. Between March and July of 1983, 23 NBFs raised about $450 million (18). The stock market is also providing newly public NBFs with second- and third-round financing. Some of these firms, however, may encounter future financial constraints if they continue to rely on the stock market, because many investors are interested in relatively short-term returns.

In European countries and Japan, there is significantly less venture capital available than there is in the United States, and venture capital has therefore not been a major funding source for biotechnology R&D. Furthermore, because of a lack of venture capital in these countries, the number of NBFs in Europe and Japan is tiny compared to the number of NBFs in the United States. Some governments, such as those in France, Japan, and the United Kingdom, have attempted to stimulate the formation of venture capital, but the results have been disappointing. Outside the United States, direct government funding of industry is proportionately a far more important funding source for the commercial development of biotechnology than it is within the United States. In Japan, corporate funds supply most of the financing for biotechnology.

The United States tends to use tax incentives more than direct government funding to encourage industrial development. In the United States, the tax measures aimed at capital formation and R&D are important to NBFs in their present stage of development. As scale-up proceeds, tax measures aimed at R&D capital assets will become more important. The United States tax code offers no special incentives, beyond those available for investment generally, for investment in depreciable structures or for equipment used for research and experimental design. Currently, France and the United Kingdom have accelerated write-offs for R&D capital assets, and West Germany has an investment grant allowing a company to recover up to 20 percent of the cost of R&D capital expenditures. Japan also has extremely favorable depreciation allowances for capital assets used in R&D for members of Government Research Associations such as the one formed for biotechnology.

Available studies suggest that Switzerland, followed by Japan and the United Kingdom, have the lowest effective corporate tax rates. The effective rates in the United States, the Federal Republic of Germany, and France are higher and about equal.

In most countries, proceeds from patents are treated as either capital gains income or ordinary income. In the United Kingdom, however, proceeds from patents are taxed as corporate income (at a rate of 52 percent). Royalties are taxed as ordinary income, except in France under certain circumstances. From a tax viewpoint, the United Kingdom has the most adverse treatment of income derived from innovational activity, because proceeds from patents are taxed at corporation tax rates and the long-term capital gains tax rate in the United Kingdom is the highest of the competitor countries.

The United States has the most favorable tax treatment for raising capital for smaller firms. This is an important advantage in fostering the growth of startup and small expanding firms. The people contacted in NBFs agreed that this feature of the U.S. tax system aided the formation of their companies, especially compared to the tax treatment abroad. Recently, OECD published a study comparing the treatment of small businesses among its members and concluded that the European governments had few policies directly aimed at small businesses (33). The European governments are trying to develop policies to encourage entrepreneurs, but there are cultural as well as economic obstacles to be overcome.

Issues and policy options

ISSUE 1: How could Congress help new biotechnology firms obtain the financing necessary for production scale-up?

Many NBFs in the United States are currently sustaining large losses because of the very large investment in R&D relative to operating revenues required to develop a biotechnology product. Most NBFs at present have few or no products to generate revenues and will have difficulty financing production scale-up. Furthermore, as more and more NBFs carrying large losses approach production stages in the future, financing difficulties are expected to increase. If NBFs do not have the financing necessary for production scale-up, the commercialization of biotechnology in the United States may be hindered.

Although many NBFs are currently using public stock offerings and R&D limited partnerships to obtain funds for scale-up, it is not at all certain that these sources of financing will remain available to them. The public market is not generally considered a reliable source of funds for investments characterized by long time horizons and high risk; and R&D limited partnerships may not be a reliable source of funds given current legal uncertainties and uncertain IRS interpretations which affect the tax status of the partnership. If future returns on investments are lower than expected by current investors or if the time horizons for biotechnology scale-up are longer than expected, these sources of financing might become less available.

It might be argued that sufficient investment capital is available to commercialize biotechnology in the United States and that the Government need not intervene with specially targeted guaranteed loans or special tax provisions to further stimulate the U.S. biotechnology effort. However, the commercialization of biological technologies appears more costly both in time and investment than other high technologies. For this reason, Government support may be necessary to maintain the current competitive status of the United States. To help NBFs obtain the financing necessary for production scale-up, Congress could adopt one or more of the following options.

Option 1: Provide guaranteed loans for production scale-up.

A guaranteed loan program, much the same as the 1950 V-loan program that supplied working capital for U.S. semiconductor firms,* could be formulated for biotechnology. Under a V-loan program for biotechnology, the Federal agency guaranteeing a loan would be obliged to purchase a stated percentage of the loan if the borrower defaulted. The loans would be granted at less than

*The development of the semiconductor industry is discussed in *Appendix C: A Comparison of the U.S. Semiconductor Industry and Biotechnology*.

prevailing interest rates and would thus decrease the cost of capital for the individual firm. Because the guarantees would not be tied to a particular loan but to a particular level of debt, they would serve as a system of revolving credit. As periodic repayments reduce the outstanding debt, additional loans could be taken out as long as repayment kept the debt within the face amount of the authorization. The V-loan program of 1950 authorized a total of $2.9 billion over its life, which permitted loans totaling about $11.6 billion. It also returned a profit to the Federal Government of about $24.5 million, because the Federal guaranteeing agent was entitled to a portion of the interest paid on the loan.

Funds for biotechnology earmarked for scale-up projects could be placed in a "Biotechnology Development Bank" or allocated to an interested agency such as the National Institutes of Health, National Science Foundation, or the SBA. The funds could be authorized for a specific amount and aimed at a particular level of debt, thus allowing *successful* biotechnology firms to pay back the loans to the level of debt only. Once the level of debt was paid back, the firms could obtain additional funds from the agency/Bank.

Option 2: Allow rapid depreciation for capital assets required for production scale-up.

The current depreciation schedule for plant and equipment assets in the United States is a set of statutorily provided depreciation periods: 15 years for most structures, 5 years for most equipment, and 3 years for R&D equipment. This schedule is faster than earlier schedules and provides a greater incentive than was provided before for the purchase of long-lived equipment such as bioreactors. A depreciation schedule that would allow an even more rapid recovery of capital costs incurred in production scale-up would help alleviate some of the financial constraints faced by NBFs in production scale-up. The increased write-offs could be made available to investors through equipment partnership agreements or leasing arrangements. Such agreements would allow NBFs to obtain additional money instead of relying on tax provisions alone.

The Defense Procurement Act of 1950, which allowed participating firms to write off their capital expenditures in a 6-month period, could be used as a model for new legislation that would similarly benefit firms using biotechnology. The new legislation could allow NBFs to write off 100 percent of their expenditures for pilot plant equipment.

Currently, the United Kingdom and France have tax provisions applicable to scientific R&D equipment, allowing up to 100-percent write-offs in the first year. Congress could allow similar write-offs or accelerated depreciation for equipment used in biotechnology pilot plants.

Option 3: Refund the R&D tax credit to NBFs not earning enough taxable income on which to apply the R&D tax credit.

The R&D tax credit legislation currently allows unused tax credits to be carried over to each of the 15 taxable years following the unused credit year. For NBFs experiencing cash flow problems while scaling-up production, a tax credit refundable in the year sustained would help alleviate these financial constraints. In addition, in present value terms, a refundable tax credit would be more valuable to NBFs in the year earned than a tax credit carried forward to the years in which enough taxable income would be earned to take advantage of the credit.

The major disadvantage of this option would be the loss of revenue to the U.S. Treasury in times of high deficits. In addition, political and equity-related objections might be raised concerning Government rebates to businesses.

ISSUE 2: How could Congress encourage broader use of R&D limited partnerships in biotechnology?

R&D limited partnerships have been an important source of financing for NBFs. As noted above, NBFs incur high R&D costs relative to their revenues and have few marketable products. NBFs have found R&D limited partnerships useful vehicles by which to attract the substantial funding needed to fund research, early product development, and in the case of some pharmaceutical products, clinical trials required by FDA. Such partnerships may allow more NBFs to enter markets such as that for pharmaceuticals, where extensive regulation makes the costs of entry high.

Given the very large amounts of capital which will be required to support the further commercial development of biotechnology and the variability of the stock market as a source of funds through public offerings, R&D limited partnerships are probably critical to the survival and growth of NBFs. To encourage broader use of R&D limited partnerships and increase their role in providing financing for NBFs, Congress might consider the following options.

Option 1A: Amend section 1235 of the IRS code so that it applies to plant variety protection certificates.

The most favorable tax treatment of income for R&D limited partnerships is provided under section 1235 of the IRS Code. Section 1235 treatment applies to a transfer of property consisting of all substantial rights to a patent by any holder. Under section 1235, any royalties received as a result of transfer of a patent qualify as long-term capital gains rather than ordinary income. Because they are legally distinct from patents, plant variety protection certificates are currently excluded from section 1235 treatment. Their exclusion from section 1235 treatment may have limited the use of R&D limited partnerships for biotechnology research in plant agriculture—an area where some of the most important applications of biotechnology are likely to occur. Adopting this option would very likely encourage the formation of R&D limited partnerships for plant-related biotechnology.

Option 1B: Amend section 1235 of the IRS code so that universities are included in the definition of holder.

Under section 1235, a holder is defined as any individual whose efforts created the patentable property or any other individual who has acquired interest in the patentable property in exchange for money paid to the creator prior to the actual reduction to practice of the invention. A holder cannot be the employer of the creator. This definition of holder may discourage some university/industry R&D limited partnerships. The present definition makes it difficult for an R&D limited partnership to acquire rights to a patent directly from the inventor when a university has such rights through its employment agreement with its scientists. Amending section 1235 to include universities in the definition of holder, in addition to allowing universities to obtain additional money, would enable wider use of R&D limited partnerships.

Option 2: Allow R&D limited partnerships to qualify for tax credits under the Economic Recovery Act of 1981.

Under the Economic Recovery Act of 1981, tax credits are provided for any incremental R&D expenses incurred above a 3-year moving average. The language as it is currently written and statements in the legislative history, suggest that R&D limited partnerships do not qualify for these credits. If they did, the credits could be passed on to the limited partners, thus making R&D limited partnerships more attractive to investors.

On the other hand, it can be argued that R&D limited partnerships are already attractive to investors. The additional incentive to investors that would be provided by enabling limited partnerships to qualify for the R&D tax credits might be small. The loss to the U.S. Treasury must also be considered.

Chapter 12 references

1. Arthur Anderson & Co., "National Research and Development Study," unpublished study, Washington, D.C., January 1983.
2. Biogen, N.V., Biogen Prospectus, Curacao, Netherlands Antilles, 1982.
3. *Biotechnology,* "Predicted Bankruptcies Fail To Materialize," October 1983, p. 646.
4. *Biotechnology News,* "Genentech Makes Monoclonals via rDNA, Forms New RDLP," 3(10):7, May 15, 1983.

5. *Business Week,* "A Furor Over Business Tax Breaks," Mar. 8, 1982, p. 28.
6. *Business Week,* "Tokyo Tries To Encourage Equity Financing for Startups," Oct. 11, 1982, p. 54.
7. *Business Week,* "High Tech Tries an End Run Around Germany's Banks," July 18, 1983, p. 74.
8. Byers, B., Kleiner, Perkins, Cayfield & Byers, interview cited in L. W. Borgman & Co, "Financial Issues in Biotechnology," contract report prepared for the Office of Technology Assessment, U.S. Congress, March 1983.
9. Committee for Economic Development, Research and Policy Committee, *Stimulating Technological Progress* (New York: Committee for Economic Development, 1980).
10. *Corporate Financing Week,* "New Fund Narrows Focus," 8(6):6, Nov. 1, 1982.
11. E. F. Hutton & Co., Inc., Washington, D.C., personal communication, July 18, 1983.
12. E. F. Hutton & Co., Inc., Washington, D.C., personal communication, August 1983.
13. Eli Lilly & Co., Eli Lilly Report, South Bend, Ind., Feb. 9, 1982.
14. Erickson, T., Arthur D. Little, Inc., "Worldwide Pharmaceutical Industry Overview: The Context for Shifting R&D," at seminar on Research and Development in the Global Economy, Institute for Alternative Futures, Alexandria, Va., Apr. 11, 1983.
15. Farrell, K., "Going Public 1982," *Venture,* April 1982, p. 30.
16. Feldman, M., and O'Malley, E., *The Biotechnology Industry in California,* submitted to the California Commission on Industrial Innovation, Sacramento, August 1982.
17. Fishlock, D., "Biotechnology Fund Still Seeks British Investment," *Financial Times,* Jan. 11, 1983.
18. *Genetic Technology News,* "Public Invests Nearly Half Billion in Genetic Engineering," 3(8):5, August 1983.
19. Hardy, R., "Biotechnology: What, How, Where, and When," *Japan Techno-Economics Society Forum Technical Feature* 11, Oct. 30, 1981.
20. Harsanyi, Z., E. F. Hutton & Co., Inc., New York, personal communication, March 1983.
21. Howard & Co., Philadelphia, personal communication, 1983.
22. *Japan Economic Journal,* "Japan Associated Finance Forms U.S.-Type Venture Capital Funds," Mar. 23, 1982.
23. Jasanoff, S., "Public and Private Sector Activities in Biotechnology: Federal Republic of Germany," contract report prepared for the Office of Technology Assessment, U.S. Congress, January 1983.
24. Jasanoff, S., "Public and Private Sector Activities in Biotechnology: Switzerland," contract report prepared for the Office of Technology Assessment, U.S. Congress, January 1983.
25. Jungwirth, J., Hybritech, San Diego, Calif., personal communication, February 1983.
26. Jungwirth, J., Vice President, Finance, Hybritech, cited in L. W. Borgman & Co., "Financial Issues in Biotechnology," contract report prepared for the Office of Technology Assessment, U.S. Congress, March 1983.
27. L. W. Borgman & Co., "Financial Issues in Biotechnology," contract report prepared for the Office of Technology Assessment, U.S. Congress, March 1983.
28. Marsh, D., "Looking for Closer Relationship," *Financial Times,* Nov. 12, 1982.
29. Michaud, S., "The Dark Side of R&D Shelters," *Venture,* November 1982, p. 28.
30. National Science Foundation, *Corporation Income Tax Treatment of Investment and Innovation Activities in Six Countries,* Washington, D.C., 1981.
31. National Science Foundation, *An Early Assessment of Three R&D Tax Incentives Provided by the Economic Recovery Tax Act of 1981,* PRA report 83-7, Washington, D.C, April 1983.
32. *Nihon Keizai Shimbun* (Japan Economic Journal), "Dainiji bencha boomu torai (The Second Venture Capital Boom is Coming)," Sept. 9, 1982.
33. Organisation for Economic Co-Operation and Development, *Innovation in Small and Medium Firms,* Paris, 1982.
34. Pratt, S. (ed.), *Guide to Venture Capital Sources* (Wellesley Hills, Mass.: Capital Publishing Corp., 1981).
35. Price Waterhouse & Co., *Price Waterhouse Information Guide: Doing Business in Germany,* September 1978.
36. Price Waterhouse & Co., *Price Waterhouse Information Guide: Doing Business in France,* 1979.
37. Price Waterhouse & Co., *Price Waterhouse Information Guide: Doing Business in the United Kingdom,* 1980.
38. Price Waterhouse & Co., *Price Waterhouse Information Guide: Doing Business in Switzerland,* 1982.
39. Saxonhouse, G., "Biotechnology in Japan," contract report prepared for the Office of Technology Assessment, U.S. Congress, June 1983.
40. Sladovich, H., E. F. Hutton, Washington, D.C., personal communication, August 1983.
41. Strogen, D., Princeton Montrose Partners, interview cited in L. W. Borgman & Co., "Financial Issues in Biotechnology," contract report prepared for the Office of Technology Assessment, U.S. Congress, March 1983.

42. U.S. Congress, Congressional Research Service, *Tax Rates in Major Industrial Countries: A Brief Comparison,* No. 80-224E, Washington, D.C., December 1980.
43. Vaquin, M., "Biotechnology in Great Britain," contract report prepared for the Office of Technology Assessment, U.S. Congress, February 1983.
44. Vaquin, M., "Biotechnology in France," contract report prepared for the Office of Technology Assessment, U.S. Congress, February 1983.
45. *Venture Capital Journal,* "Evolution of an Industry: Venture Capital Redefined for the 1980's," January 1980.
46. *Venture Capital Journal,* "Venture Capital Disbursements 1981: A Statistical Overview," 22(6):7, June 1982.
47. *Venture Capital Journal,* "Special Report—The Growth of An Industry, Venture Capital 1977-1982," 22(10):7, October 1982.
48. *Venture Capital Journal,* "Capital Transfusion 1982," 23(1):6, January 1983.
49. Virolleaud, P., "Research Finally a Tax Break," *L'usine Nouvelle,* Oct. 21, 1982, p. 75. Translated by the Foreign Broadcast Information Service in *West Europe Report, Science and Technology* (Joint Publications Research Service, Arlington, Va.).
50. Werner, J., Director, Program Policy and Evaluation Staff, U.S. Small Business Administration, Washington, D.C., personal communication, January 1983.
51. Yamamura, K., University of Washington, personal communication, July 1982.

Chapter 13

Government Funding of Basic and Applied Research

Contents

Tables

Figure

Chapter 13

Government Funding of Basic and Applied Research

Introduction

Federally funded basic research in the United States has been essential to the development of biotechnology. The United States currently has a strong and diversified basic research capability, the foundation for which was laid during World War II by the Office of Scientific Research and Development (OSRD). The National Institutes of Health (NIH) was established to succeed OSRD's Commmittee on Medical Research in 1930.

Within a few years after World War II, several patterns of U.S. Government funding for basic research had been established. First, funding of scientific research would further the broad aims and priorities of the U.S. Government as defined by Congress and the President. Second, nongovernmental laboratories (e.g., research universities) would perform much of the research of interest to the Federal Government; in-house Government laboratories would also perform such research. Third, direct relationships between Federal agencies and university researchers would be established; funds for university research would be awarded to individual investigators or small teams of investigators rather than to the institutions themselves (legally, funds are administered through institutions in the name of investigators). Fourth, university research and graduate training in the United States would be closely related functions. These patterns, with elaboration, have persisted until the present (21).

The launching of Sputnik in 1957 triggered a spectacular increase in the U.S. research effort. From 1953 to 1967, national expenditures in current dollars for research and development (R&D) increased by more than 350 percent, and current dollar R&D expenditures by the Federal Government increased almost 425 percent. In 1967, Federal Government expenditures represented 62 percent of total national expenditures for R&D. After 1967, the rate of growth in R&D expenditures declined, and by 1976, the Federal Government's contribution had dropped to an estimated 53 percent of total national R&D expenditures (21).

National basic research expenditures by the Federal Government have decreased more sharply in constant dollars than in total R&D outlays. Between 1968 and 1976, basic research expenditures declined in constant dollars by an estimated 15 percent. Since universities perform the greatest share of basic research, they have suffered the most from constraints on Federal research funding. In real dollars, fewer basic research funds were spent in universities in 1976 than in 1968 (21). In spite of this leveling off of Federal support, the basic research effort of the United States is prodigious and led to the recent developments in biotechnology.

One aspect of the development of biotechnology demonstates the unanticipated results of a long-term commitment by the U.S. Federal Government to basic research. The "war on cancer" stimulated investigators to study the properties of viruses that cause tumors.* A great deal of work was done to locate the genes in several tumor viruses, such as SV40 virus, that cause tumors in hamsters and mice. These viruses are particularly recalcitrant to classical genetic procedures for mapping genes. This problem led to the use of bacterial restriction enzymes—enzymes that cut DNA at specific locations—to construct physical maps of genes. Physical mapping of an entire genome (a complete set of genes of an organism) using restriction enzymes was first accomplished on SV40 DNA. It was the knowledge of the mechanism of action of these restriction enzymes, generated originally from cancer research, that led to the cloning of genes.

*See *Appendix C: A Comparison of the U.S. Semiconductor Industry and Biotechnology*.

As biotechnology is commercialized, different emphases will be placed on various aspects of the continuum that stretches from basic to applied research. The objective of basic research is to gain a better understanding of the fundamental aspects of phenomena without goals toward the development of specific processes or products. The objective of applied research is to gain the understanding necessary to meet a recognized and specific need, process, or product (13). Bridging the gap between basic and applied research is "generic applied" research, which is more specific than basic research, but longer term and more risky than most applied research.* The Federal commitment to basic and generic applied research in the United States will be a necessary element in the commercialization of biotechnology in the coming years.

Donald Kennedy has characterized the process that moves from basic, to generic applied, to applied research as the "trajectory of innovation" (10). Within this trajectory, particular kinds of institutional sponsors play defined roles:

- *Phase One* (Basic Research). Characterized by loose, informal organization, open communication, quick publication of all the details of an experiment. Usually takes place in university departments or laboratories such as those at NIH, or sometimes in a special organization such as Bell Laboratories. Most often publicly funded, oriented toward the discovery and explanation of phenomena.
- *Phase Two* (Generic Applied Research). Focused on processes, the application phase. Takes place in various settings: applied institutes, some university departments, nonprofit organizations (e.g., Stanford Research Institute, Battelle). Mixed public and private funding. Environments variable with respect to proprietary secrecy.
- *Phase Three* (Applied Research). Innovative emphasis on products, the development stage, attention given to practical application. Funding by private risk capital, environment tends to be closed for proprietary reasons, essentially all work takes place in private laboratories.

Biotechnology is moving rapidly along the trajectory of innovation. The role of Federal funding in the process has been and will continue to be critical to the U.S. competitive position in biotechnology.

Assessing the U.S. competitive position in biotechnology research is difficult for several reasons. First, the definition of biotechnology used in this report is a definition specific to the commercialization of biotechnology, and thus is more likely to fit traditional definitions of applied research. Second, basic or fundamental research in biotechnology can include research on topics as diverse as cancer, developing new vectors to improve recombinant DNA (rDNA) techniques, increasing oxygen solubility in aqueous systems, understanding immune function, and neurobiology. Basic research by its very nature is wide ranging; many elements drawn from basic research of various kinds go into the innovation and development of a particular patentable product. Third, the use of rDNA techniques or rDNA research may be but a small component of a particular research project, or the description of the particular research may not have contained key words that warranted its inclusion in an agency classification of biotechnology research. In addition, as rDNA techniques are more widely used, much of basic research at the cellular and subcellular level will use these techniques; thus, much of basic biomedical research will use the techniques of biotechnology. Fourth, even in the United States, biotechnology is defined differently among funding agencies. Added to problems of definitions are differences in granting procedures by various agencies, as well as different accounting procedures for indirect costs (indirect costs are part of the cost of doing research and therefore must be included). And, finally, overall funding levels give some indication of the total research effort but do not reveal the quality of the research. Nevertheless, most experts would agree

*Generic applied research is a part of the continuum between the two poles of basic and applied. This research may be characterized as follows: 1) it is not committed to open-ended expansion of knowledge as university basic research typically is, but is less specific (more widely applicable or "generic") than the typical industrial product or process development effort; 2) it has more well-defined objectives than basic research, but is longer term than typical product and process development efforts; and 3) it is high risk, in the sense that the stated objectives may fail and the resources committed may be lost for practical purposes.

that the two are closely correlated and that the United States leads the world both in its investment in science and in the quality of its science. The totals for Federal funding for biotechnology research are shown in table 56 and will be discussed in the sections to follow.

Since the focus of this chapter is an assessment of the relative strengths of basic, generic applied, and applied research in biotechnology in the United States, Japan, the Federal Republic of Germany, the United Kingdom, Switzerland and France, the estimates of government funding for biotechnology research in other countries that are available have been included in this chapter. Given problems with respect to definitions, currency exchange fluctuations, and lack of complete data, these figures must be interpreted with caution. For detailed analysis of agency budgets within the United States, the reader is referred to the American Association for the Advancement of Science and National Science Foundation (NSF) documents listed in the references (1,13).

The three sections of this chapter that follow are intended to provide a perspective on the U.S. commitment to biotechnology research by discussing basic, generic applied, and applied biotechnology research, respectively, within individual U.S. Government agencies. A separate section considers instrumentation initiatives by the U.S. Government that have bearing on biotechnology research. Near the end of the chapter, research expenditures in biotechnology and channels of research funding in Japan, the Federal Republic of Germany, the United Kingdom, Switzerland, and France are presented in a comparative overview. The final section of the chapter identifies issues and congressional policy options pertaining to U.S. Government funding of biotechnology research and instrumentation initiatives.

Table 56.—U.S. Federally Funded Research in Biotechnology[a]

		Amount of funding (millions of dollars)		
		Basic	Generic applied	Applied
NIH:				
Molecular biology, generic manipulation, hybridoma, monoclonal antibodies	FY 1982	$378.0	—	—
Immobilized enzymes	FY 1982	—	$ 2.0	—
NSF:				
rDNA research	FY 1982	12.8	—	—
Bioprocess engineering	FY 1982	—	1.7	—
Other biotechnology-related research (broadly defined)		38.6	—	—
USDA:				
ARS plant biotechnology	FY 1983	7.2[b]	—	—
ARS animal biotechnology	FY 1983	6.4[b]	—	—
CSRS competitive grants (CRGO)	FY 1982	5.0	—	—
SAES	1081 82	15.6[b]	—	—
DOD:				
DARPA	FY1983	—	2.2	—
Army/Navy/Air Force rDNA research	FY 1983	3.3	—	—
Other biotechnology	FY 1983	2.0	—	—
DOE:				
Photosynthesis, stress mechanisms of plants and micro-organisms, genetic mechanisms, methanogenesis, etc.[c]	FY 1983	9.9[b]	—	—
Conservation & Renewable Energy Program	FY 1983	23.7[b]	—	—
Other	FY 1983	2.0[b]	—	—
Biocatalysis research	FY 1983	—	0.5	—
Total		$510.9	$ 6.4	$5.0(SBIR)

[a]Unless otherwise specified, see text for explanation of figures.
[b]Some of this research may be generic applied research.
[c]Biotechnology, broadly defined.

SOURCE: Office of Technology Assessment.

U.S. Government funding of basic research in biotechnology

U.S Government agencies funding basic research in biotechnology are NIH, NSF, the U.S. Department of Agriculture (USDA), the Department of Energy (DOE), and the Department of Defense (DOD).

National Institutes of Health

In November 1983, the fiscal year 1984 budget of NIH was appropriated at $4.3 billion with some of the unauthorized programs still under continuing resolution. The number of new and competing project grants will be maintained at 5,000.* The 16,560 research project grants—5,000 competing and 11,560 noncompeting—will be the largest number of research project grants supported in the history of NIH. Budget estimates indicate that direct costs for noncompeting continuation grants will be reduced by about 1 to 2 percent and those for competing grants by 2 to 4 percent. A 4-percent reduction in average costs was applied to these grants in both 1982 and 1983.

Most of the basic research that has been and is done in biotechnology is NIH-funded research. Despite the budget pressures on NIH funding as a whole, the number of extramural projects using rDNA techniques has increased. Funding figures for NIH projects in biotechnology for the fiscal years 1978 to 1982 are shown in table 57. Since data are catalogued by NIH staff on the basis of grant applications or progress reports and indexed by staff who looked for key words such as "genetic manipulation," "hybridoma," "monoclonal antibodies," and "immobilized enzymes," the figures may be slightly misleading. For example, the term "genetic manipulation" includes some projects that do not involve rDNA techniques. Also, the figures are the total costs associated with the awards, including direct and indirect costs, and are not related to the proportion of rDNA research in the total research effort. With the exception of generic applied research on immobilized enzymes, the work is primarily basic research, so many of the industrial applications associated with new biotechnology may be in the distant future. Despite these classification problems, it is evident from the figures in table 57 that research using rDNA techniques is becoming more widespread and comprises a larger proportion of the total grants awarded each year.

Table 57.—NIH Projects in Biotechnology, Fiscal Years 1978-82

Fiscal year	Number of projects	Dollars awarded (millions of dollars)
Genetic manipulation:		
1978	546	$ 61
1979	847	103
1980	1,061	131
1981	1,400	164
1982	1,588	185
Hybridomas (term not created until 1980):		
1980	256	$ 22[a]
1981	479	49
1982	654	64
Monoclonal antibodies (term not created until 1980):		
1980	268	$ 22
1981	768	78
1982	1,274	129
Immobilized enzymes:		
1978	25	$ 1
1979	33	2
1980	26	2
1981	27	2
1982	25	2

[a]Data are probably not complete.

SOURCE: U.S. Department of Health and Human Services, National Institutes of Health, Office of Recombinant DNA Activities, 1983.

Funding figures for biotechnology research in NIH intramural programs are unavailable; however, this research is a much smaller portion of all NIH-sponsored research.

National Science Foundation

The total fiscal year 1984 budget request for NSF is $1.2 million, a 17.4-percent increase over fiscal year 1983. Research instrumentation and support for graduate students are high priorities. Within NSF's Biological, Behavioral, and Social Sci-

*New projects are those competing for first time; competing projects are those that are competing but have been funded before by NIH (a competitive renewal); and noncompeting projects are ongoing projects awarded for more than 1 year.

ences program, the physiology, cellular, and molecular biology program is increased 20 percent over fiscal year 1983. The Chemical and Process Engineering Division budget in NSF's Engineering program is also up 21.5 percent; this may have some effect on biotechnology (4).

The total NSF expenditure for grants having some rDNA component from 1975 through October 1982 was just over $57 million. From fiscal year 1975 through fiscal year 1980, about $35.3 million was spent. Funding for grants having some rDNA component in fiscal year 1981 was $9.8 million and in fiscal year 1982, $12.8 million.

U.S. Department of Agriculture

The fiscal year 1984 budget proposal calls for USDA's agricultural research programs to get along with essentially the same amount of money in 1984 as in 1983 (19).

The division of funds among USDA's bureaus—the Agricultural Research Service (ARS), the Cooperative State Research Service (CSRS), and the Forest Service—and the USDA research agenda have been the subject of several reports and studies. The latest, from the White House Office of Science and Technology Policy (5), has caused considerable debate. The findings from that report indicate that research at the land-grant colleges and universities lags far behind current developments in plant biology, that agricultural research funds should be more widely distributed, that much of the research conducted by ARS is duplicative, and that the agriculture system overall is no longer energy- nor resource-efficient. In addition, this and other reports have suggested that the competitive grants program within CSRS funds high-quality basic research within USDA and should be expanded in order to create a critical mass of long-term high-quality research. Hearings on this issue are expected in the next year.

In fiscal year 1984, there will be an increase of $4.6 million for the competitive research grants within CSRS in order to initiate a program in animal science. Some of these grants may include biotechnology research. In fiscal year 1981 (latest year for which data are available), of the $15.8 million total being spent for competitive research grants, approximately $5 million was spent on biotechnology research (17).

The Agriculture Committee on Biotechnology of the National Association of State Universities and Land-Grant Colleges (12) has estimated that during 1981-82, $34.7 million was committed to biotechnological research by State Agricultural Experiment Stations (SAES). (This estimate was derived from a survey of SAES that totaled the number of persons plus full-time equivalents working on biotechnological research.) The distribution of this total is 42 percent State, 45 percent Federal, and 14 percent private funding.

ARS has funded a total of $13.6 million in biotechnology research in fiscal year 1983; $7.2 million of this was devoted to plant biotechnology and $6.4 million to animal biotechnology (27).

Department of Energy

DOE has several programs involved in biotechnology research. DOE's Office of Basic Energy Sciences, which funds fundamental research in plant sciences and microbiology (photosynthesis, stress mechanisms of plants and micro-organisms, genetic mechanisms, methanogenesis, genetics of anaerobic micro-organisms, and regulatory aspects of metabolic pathways), had a budget of $9.9 million in fiscal year 1983 and will have $11.0 million in fiscal year 1984. Work on anaerobic digestion, algal production, and genetic manipulation is funded through DOE's Conservation and Renewable Energy programs (including DOE's Solar Energy Research Institute); the budget for these programs is $23.7 million. Other programs support biotechnology research relating to pollutant control, beneficiation of coal, and microbial enhanced oil recovery. The aggregate of these latter activities totaled between $1.5 million and $2.0 million in fiscal year 1983 (14).

Department of Defense

The Federal agency with the greatest increase in the fiscal year 1984 budget proposal for R&D funding is DOD—up 29.7 percent over fiscal year 1983 in current dollars. Although most of this increase will fall in the development areas of research, a 9-percent increase in basic research is also proposed (18). Within this framework, there are some data available on biotechnology R&D.

The total funding for rDNA basic research over all three military services for fiscal year 1983 is $3.3 million; $2.9 million of this is funded with $0.4 million obligated but not yet funded (2). DOD is currently amassing data on fiscal year 1983 funding for biotechnology activities (research on cell culture, monoclonal antibodies, etc.). DOD estimates that in-house research is probably at a level of $1 million per year and that contract research in biotechnology is at least as great as that. More accurate figures should be available in fiscal year 1984 (2). These figures represent a very small proportion of the total military basic research budget ($787.5 million for basic research in fiscal year 1983) (19).

U.S. Government funding of generic applied research in biotechnology

NSF, DOE, DOD, and NIH are the only U.S. Government agencies funding generic applied research in bioprocess engineering. Because of limited Federal support, bioprocess engineering could prove to be a critical bottleneck in the United States as biotechnology moves toward production scale-up. Not only is bioprocess engineering research underfunded relative to other types of engineering research, but trained bioprocess engineers are in short supply.*

The major U.S. Government funding group for generic applied research in bioprocess engineering is NSF's Chemical and Process Engineering Division. In fiscal year 1983, $1.7 million of its $4.5 million budget was used to fund projects in bioprocess engineering. In fiscal year 1984, there is no increase in its budget, but more of the budget, $2.7 million, is being allocated to bioprocess engineering (29).

DOE has a Biocatalysis Research Activity within its Energy Conversion and Utilization Technologies Program. Although this activity was funded up to $525,000 through fiscal year 1983, the administration's fiscal year 1984 budget request implies that biocatalysis research activities will be terminated. This research project, begun in 1981 at $130,000 was a generic applied research project designed specifically to capitalize on basic research conducted at universities. Its goal was to build the technical and engineering base of biocatalysis technology to enable U.S. industry to displace a significant level of nonrenewable resource requirements by the year 2000. The project supported applied research and exploratory development to help establish the technology base that the chemical process industry will need to develop cost-competitive products from genetically manipulated organisms based on renewable energy feedstocks. Unfortunately, this beginning toward a federally funded generic applied research base in bioprocess engineering has been terminated. Currently, however, discussions are underway in DOE's Office of Energy Research to begin a broader bioengineering initiative.

DOD's Defense Advanced Research Projects Agency (DARPA), with an overall budget for fiscal year 1983 of $719.5 million (projected to increase 9.7 percent in fiscal year 1984), has two program areas in biotechnology, one underway and one beginning in fiscal year 1984. Thie first program, a research effort in chemical and biological ultrasensors, began in fiscal year 1982 with a budget of $888,000. Funding for this program is expected to increase to $2.2 million in fiscal year 1983, stay level at about $2.2 million in fiscal year 1984, and increase to $2.9 million in fiscal year 1985. The research is being done through contracts with four universities, two private companies, and three Federal laboratories. The purpose of the second initiative, which is to begin in fiscal year 1984, is to study the mechanical properties of bio-

*See *Chapter 14: Personnel Availability and Training* for a discussion of the shortage of bioprocess engineers.

polymers. Funding in fiscal year 1984 will be $1.4 million, rising to $2 million in fiscal year 1985 and 1986, $2.7 million in fiscal year 1987, and decreasing to $1 million for phaseout in fiscal year 1988.

Projects are undertaken in DARPA if there is a perception that there will be downstream applications of interest to the military. Thus, the research DARPA funds is generic applied. If a particular initiative appears to be fruitful, additional funding will be targeted to basic research in the area. Programs are viewed as successful if the technology is transferred to secondary agencies within 5 years. Thus, most research initiatives are for 5 years, at which time they are phased out. New initiatives are continually being phased in as projects demonstrate merit (20).

Although most NIH research is basic research, NIH research on immobilized enzymes, which totaled about $2 million in 1982, could be characterized as generic applied.

U.S. Government funding of applied research in biotechnology

U.S. Government funding of applied research in biotechnology is provided principally through the Small Business Innovation Research (SBIR) program, a program that was established to promote research by small businesses because only about 1 to 2 percent of the total research budgets of Federal funding agencies were set aside for research by small businesses. The Small Business Innovation Development Act establishing this program was passed in 1982, so it is too early to evaluate it. Furthermore, each Federal agency is implementing the program slightly differently. In several of the agencies, however, there is potential for some funding of applied biotechnology research. The status of the SBIR program with regard to biotechnology in specific Federal agencies is detailed below. Also discussed is the Small Business Set Aside program.

Small Business Innovation Research program

The findings of both Government and private studies on technological innovation in small firms convinced the U.S. Congress of the need to increase the share of Federal R&D dollars going to small businesses. The new Federal SBIR program was created to meet this objective. The SBIR program provides a source of nonequity capital to small businesses in the United States. The SBIR program is designed as an expanded version of continuing smaller programs in DOD and NSF. When the program is fully phased in, nearly $430 million annually will be set aside for small high-technology firms, including many new biotechnology firms (NBFs).*

On July 22, the Small Business Innovation Development Act of 1982 was signed into law by President Reagan. The purposes of this act are to: 1) stimulate technological innovation from Government-funded R&D, 2) use small businesses to meet Federal R&D needs, and 3) increase private sector innovation derived from Federal R&D by coupling the SBIR to venture capital. In the first NSF SBIR solicitation, NSF awards totaled $5.3 million. Approximately $42 million in follow-on funding was awarded to the first recipients.

In order to accomplish the three objectives of the law, the SBIR program is structured in three phases. Phase I is a screening phase to evaluate the technical and commercial feasibility of proposals. Usually, the period of performance is months. The awards given in Phase I are up to $50,000. This money is most effectively used for either out-of-pocket expenses and the salary of a technician or for financial sustenance while developing a business plan and looking for venture capital. Only winners of Phase I awards can compete for Phase II awards, and only about 50

*NBFs, as defined in *Chapter 4: Firms Commercializing Biotechnology*, are small firms that have been started-up in recent years specifically to capitalize on new biotechnology.

percent of the Phase I winners receive Phase II awards.

Phase II provides funds for the projects found most promising after Phase I. These awards are generally used for the principal research effort. The period of performance is up to 2 years and the awards given are up to $500,000. In Phase II, the law requests (but does not require) the proposer to obtain a follow-on funding commitment from a third party, usually a large corporation or a venture capital firm. The third party is used not only because the small firms tend to be undercapitalized but also to provide an objective look at the management, market, technology, and long-term financial requirements.

Phase III consists of private investments to stimulate commercial production. This phase is not funded by the Federal Government.

The SBIR law requires that each Federal agency for the next 6 years set aside a specific percentage of its R&D budget for awards to small businesses. Federal agencies with external R&D budgets exceeding $100 million—i.e., the National Aeronautics and Space Administration (NASA), the Department of Health and Human Services (of which NIH is a part), NSF, DOE, USDA, the Department of Transportation, the Department of the Interior, the Environmental Protection Agency, and the Nuclear Regulatory Commission—must set aside 0.2 percent of their external R&D budget for small businesses in fiscal year 1983, 0.6 percent in 1984, 1 percent in 1985, and 1.25 percent in 1986-88. In those agencies with external R&D budgets exceeding $10 billion (DOD), the set aside begins at 0.1 percent and increases to 1.25 percent in the fifth year. Each agency sets its own guidelines for implementation and its own R&D areas for solicitations.

Because the SBIR law is so new, it is difficult to determine the extent to which it might affect technological innovation and the overall competitiveness of NBFs. Nevertheless, it is clear that the SBIR program gives the U.S. Federal Government an opportunity to influence technological innovation in the U.S. private sector. If biotechnology research areas are given adequate support by Federal agencies, innovations in biotechnology might very well be fostered.

U.S. DEPARTMENT OF AGRICULTURE

USDA has reserved almost $550,000 for its SBIR program in fiscal year 1983. There are five project areas. The two most likely to initiate biotechnology proposals are animal production and protection and plant production and protection. Solicitations were sent May 1, 1983. USDA anticipates making 10 to 14 awards.

DEPARTMENT OF ENERGY

DOE has set aside $5.5 million for the SBIR program for fiscal year 1983. One topic of the 25 in the solicitation schedule deals with bioprocess technology and applied microbiology. Of the 1,700 proposals DOE received, 100 were on this topic. Traditionally, DOE's relationships with small businesses have been through subcontracting of funds allocated to the National Laboratories and contractors in universities and elsewhere. The work has usually involved procurement of materials, construction, and fabrication rather than research. The SBIR program will provide DOE with another means of supporting applied research in small R&D firms (14).

DEPARTMENT OF DEFENSE

For fiscal year 1983, DOD has almost $17 million set aside in its SBIR program. Unlike all other Federal agencies, with the exception of NSF, DOD already relies on the small business sector for R&D contracts. In fiscal year 1981, DOD awarded 7.4 percent ($679 million) of its external budget to small businesses—almost twice the small business share of total Federal R&D. Because DOD does not classify R&D projects by industrial application or research area, the amount awarded to small businesses for biotechnology R&D is unknown.*

Because of the important contribution small firms have made to DOD's R&D effort, the Department designed its own SBIR program in 1981—the Defense Business Advanced Technology (DESAT) program—and has made awards to small businesses through that program as well as through regular procurement channels. In fiscal

*DOD's classification system is as follows: 6.1-Basic Research, 6.2-Exploratory Research, 6.3-Advanced Research, 6.4-Engineering, 6.5-Support, 6.6-Major Systems. These headings are not immediately recognizable as biotechnology.

year 1982, 1,103 proposals were received from the first solicitation under the DESAT program and 100 awards were made. The DESAT program will in all likelihood be replaced by the SBIR program.

All three military services plus DARPA participate in DESAT.

- *Air Force.* The Air Force is not pursuing any biotechnology-related R&D with small business or otherwise.
- *Navy.* In fiscal year 1982, the Navy granted 36 awards under the DESAT program; few if any of which were in biotechnology-related areas. Other awards were made to small business, but no agency or service is able to break down biotechnology-related contracts for small businesses only, unless they fall under a specific small business program. Most contract research carried out by the Office of Naval Research and the Naval Research Laboratory in the past has been unsolicited. Of the unsolicited business in the past, 48 percent was done by small business and 50 percent was done by universities.
- *Army.* Under the SBIR program, biotechnology and chemical defense "correspond to the U.S. Army's 'New Thrust' program designed to take advantage of U.S. technology unmatched by Soviet capabilities that can provide the leverage technologies needed for the future battlefield" (23). The Army's R&D efforts under the SBIR program will emphasize the application of novel technologies such as rDNA and hybridoma technology in the development of vaccines, antidotes, analgesics, and blood substitutes (mostly for casualties). About 3,000 proposals are expected to be received for this topic alone.
- *Defense Advanced Research Projects Agency.* In fiscal year 1983, DARPA has set aside $750,000 for its SBIR program. It is unlikely that more than one biotechnology-related contract will be awarded under the program this year, because there are 14 research areas to be covered and the average contract price is about $50,000. In fiscal year 1982, about 12 percent of all awards went to small businesses. Most proposals that come into DARPA are unsolicited. Earlier in fiscal year 1983, when the schedule for solicitations was being formulated, biotechnology R&D was given the highest ranking for research areas to be pursued. As the schedule went through the review process, however, the specificity of the proposals was changed and the proposals were broadened. A biotechnology effort will, however, be funded in DARPA, in the area of biopolymers. Some of the contract awards will no doubt go to NBFs.

DEPARMENT OF HEALTH AND HUMAN SERVICES

The fiscal year 1983 SBIR budget for the Public Health Service, of which NIH is a part, is $5.6 million. Within NIH, it is difficult to speculate about the amount of R&D money to go to NBFs. NIH uses what it refers to as an omnibus solicitation. This approach is designed to generate new business. NIH has little experience awarding applied research contracts to small for-profit companies. In fiscal year 1981, contracts totaling $40 million went to small businesses, mostly for research support (e.g., building animal cages). In fiscal year 1982, the amount increased to $70 million. However, only since January 1982 has NIH been making awards to other types of profitmaking organizations. Most of the forthcoming NIH research solicitations under the SBIR program are in the field of biotechnology.

NATIONAL AERONAUTICS AND SPACE ADMINISTRATION

NASA's fiscal year 1983 SBIR program has a budget of $11 million. However, biotechnology as defined in this report does not fall within the mission of NASA and is therefore not a NASA research area.

NATIONAL SCIENCE FOUNDATION

NSF's SBIR budget for fiscal year 1983 is $5.5 million, approximately the same as the SBIR budget for the Public Health Service. In fiscal year 1982, NSF did not give any awards in biotechnology, and few good proposals were received by NIH. Congressman Don Fuqua sent a letter to NSF and NIH asking why so few proposals for biotechnology research topics were received (6). The response given was that many of the NBFs had received funding from private sources for their first-round financing needs (6). Such firms were ineligible to receive Phase II funding without having participated in Phase I.

Small Business Set Aside program

The Small Business Set Aside program was created to help small businesses obtain Federal Government contracts and subcontracts by setting aside "suitable" Government purchases or competitive awards to small businesses. The set aside contracts (not grants) reserve an entire procurement or a portion of a procurement for the exclusive bidding of small business concerns. The program was designed to give small businesses equal opportunity to compete for Government contracts and subcontracts. It was not designed specifically with R&D contracts in mind and has had limited significance in stimulating technological innovation in small businesses (22).

U.S. Government instrumentation initiatives

The obsolescence of analytical instruments is an increasingly severe problem for U.S. universities. As instrumentation becomes more sophisticated, it also becomes more costly; furthermore, obsolescence occurs more rapidly. DOD has estimated that upgrading all qualified laboratories to "world class" status in instrumentation would take an infusion of $1.5 billion to $2 billion. Instrumentation is needed not only to carry out research but also to teach the next generation of researchers and industrial personnel.

Since reduced funding levels have caused universities to cut back purchases of necessary technical equipment, a special fund totaling $150 million over 5 years for the purchase of equipment has been set up in DOD. The purpose of the special DOD fund is to upgrade the equipment of universities. Each of the three military services contributes equally to DOD's special fund, and the Office of Naval Research coordinates its administration. The solicitations sent out by DOD stipulate that the requests are to be for major pieces of equipment that cannot be purchased with other funding. One goal of DOD's fund is to stimulate program projects, i.e., to encourage several researchers to work together. The research they would undertake would necessitate the purchase of equipment costing a minimum of $50,000 (this may be raised to $100,000). The primary criterion for evaluating proposals is the relevance of the proposed research to DOD's interests. The second criterion is the scientific merit of the research to be performed with the equipment. By the closing date of November 30, 1982, 2,478 proposals totaling $645 million had been received. The announcement of 204 awards was made in late April 1983, with awards averaging $148,000. The large response to the DOD initiative is one index of the need for updating instrumentation in universities (15).

For fiscal year 1984, major increases in NSF's R&D equipment and instrumentation initiative are proposed (see table 58). Rather than taking the form of a single dedicated line-item, the funding is distributed among the regular disciplinary elements of the budget. NSF stresses that a few manufacturers of equipment recently have agreed to provide substantial discounts for equipment purchased by NSF grantees. Efforts to broaden participation by manufacturers in this program are continuing.

DOE has a $4 million university equipment initiative in fiscal year 1984 for DOE contractors who need equipment costing more than that allowed in the DOD instrumentation initiative; these requests can have a minimum of about $100,000 (14).

Table 58.—NSF R&D Equipment and Instrumentation, Fiscal Year 1984 Request (obligations in millions of dollars)

	Actual FY 1982	Estimate FY 1983	Estimate FY 1984	Increase (percent) FY 84/82	FY84/83
Mathematical and Physical Sciences	$41.7	$ 56.4	$ 86.3	107.0%	53.0%
Engineering	6.4	8.7	18.3	184.4	109.2
Biological, Behavioral, and Social Sciences	14.3	16.2	24.6	72.0	51.9
Astronomical, Atmospheric, Earth and Ocean Sciences	19.6	22.1	36.7	87.2	66.1
U.S. Antarctic Program	6.0	6.6	12.1	101.7	83.3
Scientific, Technological and International Affairs	2.1	2.3	2.3	9.5	0.0
Total, NSF	$90.1	$112.3	$180.2	100.0%	60.5%

SOURCE: American Association for the Advancement of Science, *R&D in the FY 1982 Budget: A Preliminary Analysis*, Washington, D.C., March 1983.

International comparisons

A brief overview of Government research funding in the foreign countries expected to be the major competitors of the United States in biotechnology—Japan, the Federal Republic of Germany, the United Kingdom, Switzerland, and France—is presented below.

Government funding of biotechnology research in other countries

The amounts spent by foreign governments on biotechnology research (including basic, generic applied, and applied) are extremely difficult to estimate. Any estimate is at best a guess, and, except where indicated, breakdowns by basic or generic applied cannot be made. Currently available estimates for the countries identified as the major competitors of the United States in the area of biotechnology are as follows:

- *Japan.* Funding for biotechnology research in Japan is divided among the Ministry of International Trade and Industry (MITI), the Science and Technology Agency, the Ministry of Agriculture, Forestry and Fisheries, and three other Government agencies. This research is a mix of basic, generic applied, and applied. The figures are shown in table 59.
- *Federal Republic of Germany.* Estimates of spending for projects funded by the Federal Ministry for Research and Technology (BMFT, Bundesministerium für Forschung and Technologie) range from $49 million to $70 million (DM120 million to DM170 million). A large proportion of this research is generic applied.
- *United Kingdom.* The British Government is spending about $43.8 million to $52.5 million (£25 million to £30 million) per year on generic applied and applied research in biotechnology. If basic research is included, the figure probably ranges upward toward $60 million.
- *France.* Estimates for Government expenditures for biotechnology range from $35 mil-

Table 59.—Government Funding for Biotechnology Research in Japan, 1982 and 1983 (in millions)

	1982		1983	
	Yen	Dollars	Yen	Dollars
Ministry of International Trade and Industry	¥ 2,381	$ 9.56	¥ 2,503	$10.04
Science and Technology Agency	2,172	8.72	2,338	9.40
Ministry of Agriculture, Forestry, and Fisheries	1,874	7.53	2,017	8.10
Ministry of Education, Ministry of Welfare, and Environmental Protection Agency	7,557	30.35	7,906	31.75
Total	¥ 13,984	$56.16	¥ 14,761	$59.29

SOURCE: G. Saxonhouse, "Biotechnology in Japan" contract report prepared for the Office of Technology Assessment, U.S. Congress, June 1983.

lion to $60 million (F230 million to F395 million).

Organization of basic and applied research in other countries

The organization of basic research in the United States and other countries competing in biotechnology is described in *Chapter 17: University/Industry Relationships* and in *Appendix B: Country Summaries*. The organization of generic applied and applied research efforts in countries likely to compete with the United States in biotechnology is outlined below.

JAPAN

Because of Japan's continuing interest in bioprocess engineering and because MITI has identified biotechnology as a "next-generation" project, there is a great deal of activity in biotechnology research in Japan. Much of the research is carried out by MITI's Agency for Industrial Science and Technology. Some biotechnology projects that MITI is sponsoring are listed in table 60. This agency oversees several research institutes, including the Fermentation Research Institute (FRI). FRI was founded in 1940 to develop fermentation technology and has expanded to include any microbial application in industry and environmental protection. Additionally, FRI has a depository for patented micro-organisms. Its fiscal year 1982 budget was $4.4 million (¥ 1.1 billion). FRI and other institutes in Japan meet many of industry's needs for generic applied research in biotechnology. Their equivalent does not exist in the United States (16).*

FEDERAL REPUBLIC OF GERMANY

The Society for Biotechnological Research (GBF, Gesellschaft für Biotechnologische Forschung) is without doubt the most important of the federally owned research centers for biotechnology in West Germany and perhaps the most ambitious governmentally operated institution of its kind in the world. In 1982, GBF's operating expenses were $13.1 million (DM32 million). Generously funded by the West German Government, GBF is one of the best equipped facilities of its kind in Europe. Its bioprocess laboratory, for example, permits considerable experimentation with bioprocess technology as well as scale-up of biotechnological processes to the pilot-plant stage.

GBF was set up to perform a variety of substantive research tasks as well as to cooperate with other researchers working in the field of biotechnology. GBF's major functions include the following (9):

- to develop environmentally sound biotechnological processes in order to assure a suf-

*For further details, see *Chapter 17: University/Industry Relationships*.

Table 60.—Some Biotechnology Projects in Japan

Title of R&D project	Ministry with jurisdiction	Institutions conducting projects	Project period
Utilization of biomass	Ministry of Agriculture, Forestry, and Fisheries	Business Office, Agriculture, Fishery and Forestry Technology Council National Institute of Agricultural Sciences Forestry Experiment Station National Agricultural Experiment Station National Research Institute of Agriculture University and private research institutes	1980-90
Enzymatic reactors	MITI	National Chemical Laboratory Agency for Industrial Science and Technology	1979-83
Industrial enzyme use	MITI	Fermentation Research Institute Agency for Industrial Science and Technology	1980-84
Physiologically active macromolecules and production processes	MITI	Research Institute for Polymers and Textiles Agency for Industrial Science and Technology	1978-82
Biochemical pulp technology	MITI	Government Industrial Research Institute, Shikoku Agency for Industrial Science and Technology	1980-83

SOURCE: G. Saxonhouse, "Biotechnology in Japan," contract report prepared for the Office of Technology Assessment, U.S. Congress, June 1983.

ficient supply of chemicals, pharmaceuticals, and foodstuffs;

- to scale-up biotechnological processes from the laboratory to the pilot-plant stage, this being the basis for the development of full-scale industrial processes;
- to make new sources of raw materials available for the manufacturing of natural products by micro-organisms and to make plant and tissue cultures available;
- to make new pharmacologically significant natural products available and to investigate their modes of action;
- to make its scientific facilities available to non-GBF research groups, provided that their projects fit within the R&D program of GBF;
- to support other research groups in the fields of biology, chemistry, and medicine by supplying noncommercial natural products;
- to participate in joint projects, provided they are within the framework of BMFT's Biotechnology Program; and
- to provide advanced interdisciplinary training for scientists, engineers, and technicians.

In keeping with its overall mission, GBF is involved in a number of cooperative arrangements with industry and with academic institutions. GBF's resources and expertise are used by industrial and academic researchers, and GBF relies on other institutions, usually private industry, for services such as toxicological and pharmacological testing of new products. GBF is also engaged in joint activities with academic and international research centers. GBF fosters international scientific exchanges by receiving temporary visitors from other countries. An acknowledged objective of BMFT is to strengthen existing ties between GBF and private industry in order to facilitate technology transfer in the field of biotechnology (9).

Since 1979, the German Collection of Micro-Organisms (DSM, Deutsche Sammlung von Mikroorganismen) has been incorporated into GBF. DSM has served since October 1981 as an international depository of patented or patent-related micro-organisms pursuant to the Budapest Treaty.* More generally, DSM's mission is to collect micro-organisms of scientific and technological significance, to conserve them unchanged, and to make them available for R&D and teaching purposes. The proposed budget for operating DSM in 1982 was $833,000 (DM2 million).

UNITED KINGDOM

The United Kingdom has several Government-sponsored research centers that are involved in biotechnology development projects (see table 61). Some of the centers are entirely Government owned, whereas others have significant industrial commitments.

*See *Chapter 16: Intellectual Property Law.*

Table 61.—Government-Sponsored Applied Biotechnology Centers in the United Kingdom

Name of center	Funding (in millions)	Source of funds
Center for Applied Microbiology and Research (CAMR)	£ 2 ($3.5)	Department of Health and Social Security, sales of products, industry contracts
British Technology Group (BTG)	£13 ($22.8)	Government
Celltech	£ 12 ($20)	BTG (44%)[a] Technical Development Capital (14%) Prudential Assurance (14%) Midland Bank (14%) British & Commonwealth Shipping Co. (14%)
Agricultural Genetics	about £40 ($70)	BTG (about one-third), Ultramar, Advent Eurofund
Biotechnology Institute and Studies Centre Trust	N.A.[b]	Government through: Polytechnic of Central London, University of College London, University of Kent at Canterbury No committed industries

[a]BTG recently released 14 percent of its equity to the Rothschild Biotechnology Investments Group and Boots Co.
[b]N.A. = Information not available.

SOURCE: M. Vaquin, "Biotechnology in Great Britain," contract report prepared for the Office of Technology Assessment, U.S. Congress, December 1982.

One Government-sponsored center is the Center for Applied Microbiology and Research (CAMR). As shown in table 61, CAMR is financed in part through the Department of Health and Social Security and in part from sales of products and contract research. Its current operating budget is $3.5 million (£2 million), and there are plans for expansion. CAMR has been singled out by the British Government to play a special role in the development of biotechnology. It has well-developed and established contacts with both universities and industry and sees itself as an intermediary between basic university research and production on an industrial scale. CAMR's major commercial contract in biotechnology is with KabiVitrum (Sweden) to scale-up and develop a process for manufacture of human growth hormone using rDNA bacteria developed for Kabi by the U.S. firm Genentech. CAMR also has contracts with Cadbury Schweppes (U.K.), Unilever (U.K.), Technofirm Development, Ltd. (U.K.), and Celltech (U.K.).

The British Technology Group (BTG) is a public corporation sponsored by the Department of Industry with the aim of supporting the development of biotechnology by facilitating the transfer of technology from the laboratory to the marketplace (see fig. 31). BTG has committed about $22.8 million (£13 million) for biotechnology projects to date, with annual increases of $6.5 million projected. BTG has four major investment areas: research support, joint venture funding, startup financing of small firms, and equity and loan financing. It is not clear what portion, if any, of BTG's funds is being used for scale-up and development processes. In addition to and separate from BTG activities, the Department of Industry has initiated a 3-year $30 million "Biotechnology in Industry" program.

Celltech was founded in 1980 by the National Enterprise Board (now BTG), Technical Development Capital, Prudential Assurance, Midland Bank, and British and Commonwealth Shipping Co., with an initial outlay of $20 million (£ 12). Recently, the BTG and Technical Development Capital released 14 percent of their equity to the Rothschild Biotechnology Investments Group and Boots Co. The establishment of Celltech represented one of the first steps initiated by the British Government to involve industry in commercializing the results of research in public sector laboratories. While the company was being formed, it successfully negotiated exclusive access to all work in the Medical Research Council, where monoclonal antibodies (MAbs) were discovered in 1975. Although the firm, which intends to concentrate on the development of MAbs for human diagnostic and therapeutic applications, has yet to make a profit on its limited product sales, it has extensive plans for the future, including the development of a continuous cell culture bioreactor that would produce MAbs in higher volumes than current bioprocessing technologies permit.

Agricultural Genetics is a company similar in design to Celltech that will commercialize research of the Agricultural Research Council. BTG will provide about one-third of the capital ($8.6 million; £5 million). The industry sponsors are Ultramar and Advent Eurofund.

The Biotechnology Institute and Studies Centre Trust (BISCT) is a recently established organization that draws on the expertise of some of United Kingdom's foremost biotechnologists. Currently, BISCT is offering continuing education in the form of a 1-year postgraduate degree in biotechnology, short courses, and an advisory service for industry. It hopes to undertake research programs sponsored by industry in bioprocess engineering and applied microbiology (26).

SWITZERLAND

Switzerland has no publicly owned research institute specifically for biotechnology comparable to GBF in West Germany. Outside industry, research related to biotechnology, both basic and applied, is carried out primarily in the university system, which at present includes 10 institutions of higher learning.

The leading Swiss center for research on the generic applied and applied aspects of biotechnology is at the Federal Institute of Technology (ETH, Eidgenössiche Technische Hoschschule) in Zurich, one of the two polytechnic universities managed by the Federal Government through the Swiss School Council (Schweizerischer Schulrat). Headed by a former research director of the Swiss pharmaceutical company Hoffmann-La

Figure 31.—British Technology Group Support for Biotechnology

BTG Support for Biotechnology

- Support research
- Joint venture funding
- Minimum-fuss finance for small firms
- Equity and loan finance

Total funds committed by BTG: £13 million

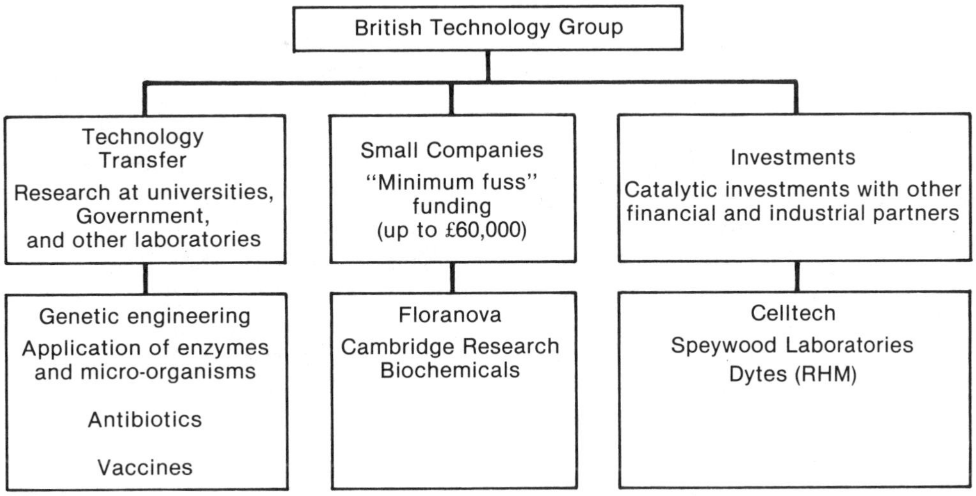

Biotechnology Subject Areas Being Supported

Agricultural applications	8 projects
Industrial applications	8 projects
Medicinal applications	17 projects
Veterinary applications	4 projects
Enabling technology	2 projects

Strategy for Further Investment

Seek out and promote:

- Opportunities for industrial investment in downstream applications of genetic engineering and cell fusion
  - — Low volume, high margin products
  - — Healthcare, food production and fine chemicals

Respond positively to:

- Technology transfer opportunities from universities and public sector laboratories
  - — Back a lot of starters
  - — Involve potential industrial partners as early as possible
- Opportunities for industrial investments in "biotechnology infrastructure":
  - — Laboratory reagents and equipment
  - — Fermentation hardware

Avoid:

- Early investment in "big biotechnology" projects
  - — e.g., heavy organic chemicals, bioenergy, and waste recovery

SOURCE: British Technology Group, Prutec Ltd., and Technical Development Capital, "Minutes of Evidence to Education, Science, and Arts Committee on Biotechnology," H. M. Stationery Office 289iii, April 26, 1982.

Roche, ETH proved receptive to the idea of biotechnology at a fairly early date, and its department of biotechnology was established in 1976. One of the department's achievements to date is the development of a new bioreactor design, which is being tested along with more conventional models in the ETH bioprocessing facility.

The channels for transfer of knowledge from the universities to industry appear well established in the area of biotechnology, although the large pharmaceutical companies may not yet be major beneficiaries of this exchange. The president of ETH, for example, has endorsed the practice of industrial contracts with professors in the biotechnology department. Joint funding by industry and the Commission for the Encouragement of Scientific Research provides another avenue for collaboration with the private sector, one that has been actively utilized by the ETH biotechnology group. The Swiss firm Biogen S.A.* is not only closely linked to the Swiss university research system, but has built an important share of its competitive strength on the productivity of these ties (8).

FRANCE

France has no Government-sponsored applied research centers like GBF in West Germany and the ETH-Zurich in Switzerland. The Institut Pasteur, a nonprofit organization jointly sponsored by the Government and industry, is the single most important facility in biotechnological research in France, but is primarily concerned with basic research. The Institut Pasteur receives 47 percent of its income from the French Government (Directorate General for Research). The rest of its income comes from the sale of services: royalties from Institut Pasteur Production (13 percent), industrial contracts (33 percent), and donations collected by the Association for the Development of Institut Pasteur (7 percent). Although the Institut Pasteur is mostly concerned with basic research (e.g., projects on vaccines and monoclonal antibodies), it does support the development aspects of biotechnology (e.g., projects on the use of cellulose for alcohol production and biological insecticides) with industrial contracts.

The Institut Pasteur has plans to open a new biotechnology building in 1985 or 1986. This building, which will have 3,000 square meters of new laboratory space, will be used partly to rehouse existing projects and partly for new projects. It will also contain bioprocess scale-up facilities (at present, the Institut Pasteur cannot do any scale-up work itself). The new biotechnology building is to be financed by the Government, but the Institut Pasteur will have to cover the operating costs, probably by increased industrial contracts.

An organization within the Institut Pasteur, G3, was started several years ago to encourage applied research in rDNA technology. G3 is funded by a set of Government groups: Institut Pasteur, the National Center for Scientific Research (CNRS, Centre National de la Recherche Scientifique), the National Institute for Agricultural Research (Institut National de la Recherche Agronomique), and the National Institute of Health and Medical Research (INSERM, Institut National de la Santé et de la Recherche Médicale). G3 has no capital, cannot employ directly, and does not own any laboratory space. It only has an operating budget. Now working with a staff of only 10, G3 plans to expand into the new biotechology building. The work program is proposed in part by the Government partners and in part undertaken at the request of industries. It is too early to predict whether G3 will contribute significantly to a generic applied research program in bioprocess technology (25).

*Biogen N.V., the parent company of the Biogen group, is registered in the Netherlands Antilles. Biogen S.A., one of Biogen N.V.'s four principal operating subsidiaries, is located in Switzerland, along with Biogen N.V.'s principal executive offices.

Findings

U.S. Government expenditures for basic research in biotechnology—the largest in the world—amount to approximately $511 million per year (mix of data from fiscal years 1982 and 1983). U.S. Government expenditures for generic applied research in bioprocess engineering and applied microbiology are estimated to be approximately $6.4 million (see table 56), although the amount could possibly range as high as $20 million or $30 million if the portions of USDA and DOE expenditures devoted to generic applied biotechnology research were known. U.S. Government funded applied research in biotechnology is virtually nonexistent, except for the SBIR program and some work being done in the National Laboratories. Most of NIH's solicitations for the SBIR program and about 5 percent of DOE's are for biotechnology; if all solicitations are funded, this could total about $5 million plus. The U.S. Army has also included a major initiative for biotechnology under its SBIR program. Since none of these grants has been funded, it is too early to estimate the amounts that will be devoted to applied biotechnology research.

Data on Government expenditures on biotechnology research in Japan are the best for purposes of international comparisons. The total amount being spent by the Japanese Government for biotechnology research in Japan is about $60 million, but Japan's definition of biotechnology is a broad one. A significant proportion of the Japanese Government's funding is for generic applied research in bioprocess engineering. The Federal Republic of Germany, United Kingdom, and France are probably spending similar amounts for biotechnology research (approximately $60 million to $100 million each), probably with relatively equal portions of basic and generic applied research.

The current pattern of U.S. Government funding for basic and generic applied research in biotechnology in the United States may compromise the U.S. competitive position in the commercialization of biotechnology. There is no doubt that past Federal support for basic research has produced a scientific infrastructure and knowledge base in the United States that is the best in the world. Furthermore, continued Federal support of basic research is critical for future innovation in a high-technology society. Because the U.S. Government has provided comparatively little funding for generic applied research, however, Americans may not be as efficient as the Japanese in applying the scientific base to the development of marketable goods and services. The Japanese Government's funding for generic applied research allows companies in Japan to make optimal use of the basic scientific knowledge of the United States and other countries and very efficiently develop this knowledge into marketable products. U.S. industry draws on the basic science knowledge base also, but the speed of the diffusion and development of this knowledge may be slower and ultimately more costly than it would be if more generic applied research were funded by the U.S. Government.

In comparison with other types of engineering research, as well as with molecular biology research, bioprocess engineering research in the U.S. is severely underfunded by the Federal Government. The personnel and academic research needs are enormous. If current funding levels for bioprocess engineering research are not increased, the United States' competitive position in biotechnology may not be as strong in the future as it is now. Bioprocessing expertise currently rests in private industry (chiefly in the pharmaceutical industry). Because private industry's bioprocessing research is proprietary, the diffusion of generic applied knowledge in this area is not as rapid as it might be. Industrialists generally agree that roughly 20 person-years of engineering research are required to go from the test-tube stage to the point where the design of a plant can begin. (Each person-year costs from $80,000 to $120,000). If existing processes or engineering techniques can be used, then about 8 instead of 20 person-years of engineering research are required. The 12 person-year difference is partially attributable to generic applied research that is now duplicated among companies at great cost. Generic applied research in bioprocess engineering could, at least partially, be supported by Federal funds. Federal support could ensure more rapid diffusion of generic applied knowledge, thus enhancing U.S. competitiveness in biotechnology.

In Japanese universities, there is a clear separation between basic and generic applied research. In addition, the Japanese Government supports generic applied research through institutes such as FRI. Japan currently is increasing its funds to basic research, although it relies to some extent on the basic research of the United States.

GBF, generously funded by the West German Government, is one of the best equipped applied biotechnology research centers in Europe. Its bioprocess laboratory, for example, is excellent. In its various activities, GBF also serves as a bridge between academia and industry.

The United Kingdom has a high standard of excellence and a cadre of highly trained basic research personnel. Recently, the British Government funded either wholly or in part several institutes and organizations to carry out generic applied and applied research and to train researchers in industry in the new techniques. These include CAMR, a center to carry out generic applied microbiology research and diffuse it to industry; Celltech, a company formed to exploit public sector microbiology research; Agricultural Genetics, a company formed to exploit public sector agricultural research; and BISCT, a biotechnology institute and studies center trust to offer continuing education, especially to industrialists.

Switzerland has an excellent basic research base in molecular biology, especially considering its small size. In addition, ETH in Zurich undertakes applied research and in 1976 established a biotechnology department. ETH and faculty at universities have a tradition of close interaction with industry in Switzerland.

In France, universities are regarded as teaching rather than research institutions. The Government funds its own laboratories through CNRS or INSERM. These laboratories are attached to several universities. The most important center for biotechnology research is the Institut Pasteur which is funded jointly by the Government and industry and carries out primarily basic research. G3, an organization established several years ago within the Institut Pasteur, was specifically mandated to encourage applied research in rDNA technology. It is too early to predict whether G3 will contribute significantly to the development of the field. The major lack in French biotechnology is a supply of trained researchers, because the biological disciplines have not traditionally been favored in France.

Basic, generic applied, and applied research are necessary for any country's competitive position in biotechnology. In terms of funding of basic research, the United States is clearly the leader with the largest and most extensive basic research enterprise in the world. The United Kingdom, West Germany, and Switzerland follow, and Japan is slightly behind them. France is sixth because it only now is beginning to exert a concerted effort to study molecular biology.

In contrast, the Japanese Government leads all countries in its commitment to generic applied and applied research. The West German Government also has an extensive commitment to generic applied research with the best equipped generic applied research laboratory in Europe. The United Kingdom and Switzerland follow. The United Kingdom is beginning to fund applied efforts with its support, for instance, of Celltech and Agricultural Genetics, and Switzerland, with ETH, has had a biotechnology effort since 1976. The United States ranks behind these four countries in its relative commitment to generic applied research as opposed to basic research, and is followed by France, which ranks sixth in all three categories of research.

Issues and options

ISSUE 1: How could Congress improve U.S. competitiveness in biotechnology by promoting generic applied research?

With its continual support of basic research, Congress has endorsed a Federal commitment to long-term funding of basic research that is essential to technological development and innovation in this country. It is crucial to the U.S. competitive position in biotechnology that this commitment to basic research continue.

Over the last three decades, the Federal commitment to generic applied research in biology and bioprocess engineering has declined relative to the commitment to basic research. Researchers in the United States have not been attracted to fields such as applied microbiology or bioprocess engineering because only small amounts of Federal funding have been available. Two critical factors underlie this decline: 1) there is no flexible, broad-based Federal system for carrying out such work; and 2) there has been a steady erosion of these generic applied science efforts in U.S. universities.

The governments of the major industrial countries of Western Europe and Japan all possess generally effective and sometimes extensive mechanisms for funding generic applied R&D. Furthermore, the university systems of these countries have not become as unaware of the needs of industrial technology as have the universities in the United States (7). To improve U.S. competitiveness in biotechnology by promoting generic applied research, Congress could adopt one or more of the following options.*

Option 1: Fund one or more biotechnology institutes within universities.

The interdisciplinary nature of biotechnology requires interaction among people with backgrounds in biology and engineering, but most American universities are not structured to facilitate this interaction. The creation at selected campuses of biotechnology institutes, in which faculty in both biology and engineering could be located in the same physical structure and work on common research projects, could facilitate this interaction. These institutes could carry out basic and generic applied research. Funding could come from Federal and State Governments and from industrial sources. Several States have already begun development of biotechnology centers; Federal funding might help leverage State funding to bring in more industrial support. Industrialists as well as academicians could work in the institutes; this arrangement would foster domestic technology transfer. In addition, students could be trained in both academic and industrial environments and industry personnel could be retrained at the institutes.

Option 2: Increase funding for university-industry cooperative programs within NSF.

NSF currently has two university/industry cooperative programs. One, the Industry/University Cooperative Research Projects program, encourages industry/university cooperation for basic research because it will fund up to half of the cost of a grant for basic research projects involving the cooperation of investigators from industry and universities. The program is advantageous to industry, because it allows industry to leverage its research funding effort, and, through cooperation, to gain a competitive edge in the innovation process. University researchers benefit from the program as well, because they improve their awareness of industrial problems and applications of basic research work.

The other NSF program, the Industry/University Cooperative Centers program, provides seed money for a university to set up a center in cooperation with industrial partners. Federal funding is phased out after 3 to 5 years. This program allows the establishment of settings that encourage university/industry cooperative research, while market demand helps to determine the type of research to be undertaken. Government funding adds incentive for industry to fund long-term, generic applied research. The infrastructure for

*See *Chapter 12: Financing and Tax Incentives for Firms* for indirect funding options for R&D.

the continued implementation of the program already exists within NSF.

The peer review system for reviewing university/industry cooperative research projects at NSF is separate from the system for reviewing other research projects. Thus, the generic applied nature of these cooperative research projects is taken into consideration, while high standards of research are assured.

Although increased Federal funding for university/industry cooperative programs within NSF could promote generic applied research, if the funding is not supplemental to needed increases in basic research in bioprocess engineering, the cooperative program could be damaging to the extension of fundamental knowledge in bioprocess engineering and applied microbiology.

Option 3: Establish special grants for interdepartmental cooperative research in biotechnology.

Currently, there is little communication between bioprocess or chemical engineers and basic biologists in universities. Special grants stipulating that a bioprocess engineer and a biologist be coprinciple investigators on a cooperative research project could make researchers in these disciplines more likely to conduct research on bioengineering or applied microbiology research relevant to the commercial development of biotechnology. The grants could be administered by NSF, since it has the technical personnel to administer such a program.

One potential problem with special grants, given current difficulties in obtaining funding, is that the researchers might cooperate in order to write the proposal then do essentially separate pieces of research once funding is obtained. Thus, the research conducted might not be truly cooperative. Avoiding this problem would necessitate carefully stated requests for proposals and careful monitoring of research.

Option 4: Develop generic applied research capability for biotechnology in the National Laboratories.

The National Laboratories are an existing resource, both in terms of physical plant and personnel, that would be expensive to duplicate. Currently, the National Laboratories do not have a great deal of expertise in biotechnology. Nevertheless, there would be several advantages to developing their generic applied research capability. These laboratories have a commitment to research, facilities to conduct research, an objective attitude towards industrial development, an array of personnel trained in relevant disciplines, and unique instrumentation development capabilities that could have a major impact on biotechnology development. DOE's Energy Research Advisory Board has just assessed the laboratories, and the White House Office of Science and Technology Policy is currently reviewing them. An assessment of the capability of the National Laboratory system to carry out generic applied research in biotechnology has not been a part of this report. This is an option for further study by Congress.

Option 5: Increase funding for the SBIR program.

Increased funding for the SBIR program would foster applied research not only in biotechnology but also in other high-technology areas. Furthermore, this program maintains the traditional philosophy of keeping much of applied research in industry and fostering entrepreneurship.

Two counterarguments to this option should be mentioned. First, although DOD and NSF have had programs similar to the SBIR program, the SBIR program has not been in existence long enough in other agencies to be evaluated. Second, because SBIR-funded research must have commercial potential within 3 years, it is too short term for problems that are generic applied, i.e., studies that fall between fundamental research and applied research. The SBIR program, as it is structured, is funding research that is further on the continuum toward product development than generic applied research. Although the program is important for biotechnology because it could help support small businesses that are doing biotechnological research, it may not be a viable option for increasing support of generic applied research in biotechnology.

ISSUE 2: Should the U.S. Government fund a germplasm screening program?

USDA (under ARS) has a network of centers for accession, storage, screening, and research on germplasm. The work at most of the centers is devoted to study of plants (the center at Fort Collins, Colo., being the largest). The center in Peoria, Ill., however, also includes micro-organisms in its collection. The Peoria center currently houses about 80,000 accessions of micro-organisms (pathogens are not included in the program) of potential interest to bioprocesses, especially for foods and drugs. It also houses 15,000 accessions of wild plant species and is screening these for industrial and medical potential. Of these, 8,000 wild species have been analyzed. Since the Peoria center is a repository for patented and industrially important micro-organisms, there is no specific program to screen these or other micro-organisms for potentially useful genes. The National Academy of Sciences is currently reviewing the USDA germplasm storage program in order to evaluate the relative efforts spent on accession and storage versus screening and analysis for potentially useful genes. A germplasm screening program might be an oversight issue for Congress as biotechnology develops.

ISSUE 3: How could Congress help U.S. academic institutions meet their needs for modern equipment and instrumentation?

There is an enormous need for modern equipment and instrumentation at universities, colleges, and secondary schools. Instrumentation is needed for teaching as well as research purposes, because teaching and research institutions have not been able to meet the needs for rapidly changing technology in instrumentation. In addition, as research grows more sophisticated and specialized, the instrumentation also grows more costly. To enable academic institutions to meet their needs for equipment and instrumentation, Congress could adopt one or more of the following options.

Option 1: Increase the special DOD fund for upgrading university equipment.

The purpose of DOD's fund, obligated in fiscal year 1982 and totaling $150 million over 5 years, is to upgrade university equipment. The solicitations stipulated that the equipment must be for basic research, must be multiuser, and must cost more than $50,000. By the closing date, proposals totaling $645 million had been received from U.S. universities. An increase in funding would help to alleviate the huge need manifested by the $645 million in proposals.

One disadvantage of relying exclusively on the instrumentation fund in DOD is that DOD awards are granted only to projects that are of interest to DOD. A second problem is that DOD's fund does not address equipment needs in the $10,000 to $50,000 range.

Option 2: Increase the instrumentation fund within NSF.

The NSF research instrumentation initiative is slated for major increases in fiscal year 1984, with the biological sciences component up 51.9 percent and engineering up 109 percent (some portion of which will be spent on bioprocess engineering). The NSF funds will concentrate on multiuser equipment. Various manufacturers of equipment have agreed to give NSF grantees reduced prices for purchase of this equipment.

The NSF research instrumentation initiative, although it moves in the right direction toward reducing instrumentation needs, is a part of the awards process. That is, more money will be available only for NSF grantees to use for instrumentation needs for NSF-funded research projects. Instrumentation initiatives similar in amount to DOD's but without the defense-related restrictions do not exist in the United States. An instrumentation initiative within NSF or some other agency could be steadily increased over the next several years to begin addressing the instrumentation needs of teaching and research institutions. Some funds could be earmarked for instrumentation needs primarily for teaching purposes.

Option 3: Legislate tax deductions for the installation and servicing of new or used equipment that companies have donated to universities

Tax deductions to encourage industry to donate equipment to universities and colleges already exist. Often, however, because they cannot afford the installation and service costs, universities are unable to use the equipment that is donated. A change in the tax law to stipulate that installation charges and even service maintenance charges would also be tax deductible would increase the university research benefit of the measure.

Chapter 13 references

1. American Association for the Advancement of Science, *R&D in the FY 1984 Budget: A Preliminary Analysis,* Washington, D.C., March 1983.
2. Beisel, W. R., U.S. Army, personal communication, 1983.
3. British Technology Group, Prutec, Ltd., and Technical Development Capital, "Minutes of Evidence to Education, Science, and Arts Committee on Biotechnology," H. M. Stationery Office, 289-iii, Apr. 26, 1982.
4. Crowley, J. C., "National Science Foundation, FY 1984," *R&D in the FY 1984 Budget: A Preliminary Analysis* (Washington, D.C.: American Association for the Advancement of Science, March 1983).
5. Executive Office of the President, Office of Science and Technology Policy, *Science for Agriculture,* Washington, D.C., 1982.
6. Fish, M., National Institutes of Health, personal communication, November 1982.
7. Gaden, E., University of Virginia, personal communication, 1983.
8. Jasanoff, S., "Public and Private Sector in Biotechnology: Switzerland," contract report prepared for the Office of Technology Assessment, U.S. Congress, January 1983.
9. Jasanoff, S., "Public and Private Sector Activities in Biotechnology: The Federal Republic of Germany," contract report prepared for the Office of Technology Assessment, U.S. Congress, January 1983.
10. Kennedy, D., "The Social Sponsorship of Innovation," *Technology in Society,* vol. 4, in press, 1982.
11. Lewis, H., National Science Foundation, personal communication, 1982.
12. National Association of State Universities and Land-Grant Colleges, Division of Agriculture Committee on Biotechnology, *Emerging Biotechnologies in Agriculture: Issues and Policies,* Progress Report, November 1982.
13. National Science Foundation, *Science Indicators,* 1980, Report of the National Science Board, Washington, D.C., 1981.
14. Rabson, R., U.S. Department of Energy, personal communication, 1983.
15. Ryan, R., Office of Naval Research, personal communication, 1983.
16. Saxonhouse, G., "Biotechnology in Japan," contract report prepared for the Office of Technology Assessment, U.S. Congress, June 1983.
17. Schauer, A. H., U.S. Department of Agriculture, personal communication, 1983.
18. Shapley, W. H., "R&D in the FY84 Department of Defense Budget," *R&D in the FY 1984 Budget: A Preliminary Analysis* (Washington, D. C.: American Association for the Advancement of Science, March 1983).
19. Shapley, W. H., Teich, A. A., and Weinberg, J. P., *AAAS Report VIII: Research and Development, FY 1984* (Washington, D.C.: American Association for the Advancement of Science, 1983).
20. Skurnick, I., Defense Advanced Research Projects Agency, personal communication, 1983.
21. Smith, B., and Karlesky, J., *The State of Academic Science* (New York: Change Magazine Press, 1977).
22. Templeman, D., Small Business Administration, November 1982.
23. U.S. Army, memorandum through Chief of Staff, "Estimate of Set Aside Funds for the Small Business Innovative Research Program FY 83," Sept. 29, 1982.
24. U.S. Department of Health and Human Services, National Institutes of Health, Office of Recombinant DNA Activities, 1983.
25. Vaquin, M., "Biotechnology in France," contract report prepared for the Office of Technology Assessment, U.S. Congress, February 1983.
26. Vaquin, M., "Biotechnology in Great Britain," contract report prepared for the Office of Technology Assessment, U.S. Congress, February 1983.
27. Victor, J. R., U.S. Department of Agriculture, personal communication, 1983.
28. Walsh, J., "Panel Pans DOE Management of Labs," *Science* 217:1015, 1982.
29. Zaborsky, O., National Science Foundation, personal communication, 1983.

Chapter 14

Personnel Availability and Training

Contents

Tables

Chapter 14

Personnel Availability and Training

Introduction

Adequately trained scientific and technical personnel are vital to industrial competitiveness in biotechnology. Countries lacking highly skilled personnel cannot have companies that compete internationally in highly technical operations such as the design and manufacture of a computer-controlled bioreactor, the discovery of a new biochemical pathway for the production of a specialty chemical, or the development of a microorganism that produces a desired protein.

An important factor in the success of companies attempting to commercialize biotechnology is the degree of sophistication of their research and development (R&D) personnel with respect to state-of-the-art developments in the field. Despite the fact that there is no "typical" firm or organizational structure among the firms using biotechnology, most corporate activity in new biotechnology at present is dedicated to R&D.* Thus, for example, a July 1982 report on a survey of California firms using new biotechnology estimated that 63 percent of the employees in these companies were professional and technical personnel involved in R&D (11).** The other employees were clerical workers (17 percent), managers (15 percent), and floor-level production and maintenance workers (5 percent).

An indication that the commercial development of biotechnology is highly dependent on skilled personnel is the fact that companies are offering special inducements to highly qualified personnel. Many companies have given their scientists and engineers considerable freedom with respect to the pace and direction of their work. U.S. firms using biotechnology stress the independence and flexibility of the work environment in order to attract qualified personnel from academic environments (11). In Japan, companies that persuade Japanese doing academic research abroad to return promise them a flexible research environment (35).

As background for the analysis that follows, the first section of this chapter discusses the quantity and types of scientific and technical personnel needed for the commercial development of biotechnology. The second section compares and contrasts the availability of especially important categories of personnel in the United States and four other countries commercializing biotechnology—Japan, the Federal Republic of Germany, United Kingdom, and France—while the third section compares the training systems in biotechnology-related areas in these countries. Also presented is the information that is available on Switzerland. In the concluding section, congressional issues and policy options with respect to the training and retraining of U.S. personnel in biotechnology are outlined. Because the amount of government funding of specific research areas can attract or discourage students from entering those areas, the reader may wish to review *Chapter 13: Government Funding of Basic and Applied Research*.

*See *Chapter 4: Firms Commercializing Biotechnology* for a description of the firms involved in the development of biotechnology in the United States and other countries.

**This survey identified 50 companies and interviewed a simple random sample of 10 firms (20 percent). All were new biotechnology firms, as defined in *Chapter 4: Firms Commercializing Biotechnology*. The survey's definition of biotechnology was "the use of living organisms or their components in industrial processes."

Size and future growth of the biotechnology labor force

It is very difficult to estimate the size of the biotechnology labor force. Theoretically, the number of personnel in supply and technological support firms, which is approximately four to five times that of firms commercializing biotechnology (11), should be included in the estimate. This chapter, however, focuses exclusively on the personnel requirements for professional and technical personnel of firms commercializing biotechnology. It does not consider the requirements of supply and technological support firms, the vast majority of which market products not only to companies commercializing biotechnology but to other companies as well.

A July 1982 report estimated total U.S. private sector employment in "synthetic genetics" to be 3,278,* including about 2,000 "professional and technical" employees (11). The same report estimated that U.S. private sector employment in "synthetic genetics" had grown at a rate of 54 percent annually since 1976 and projected that total employment would reach about 40,000 in 1992.

OTA estimates that about 5,000 employees are employed by companies in the United States in biotechnology R&D. In April 1983, OTA and the National Academy of Sciences (NAS)** conducted a survey to determine the personnel needs in biotechnology of companies in the United States. The questionnaire, reproduced in *Appendix E: OTA/NAS Survey of Personnel Needs of Firms in the United States,* was sent to 286 companies. Of the 133 that responded, 18 indicated that they were not engaged in biotechnology activities, and 20 others were determined not to be engaged in biotechnology activities from their answers to the questionnaire. To estimate the total number of firms engaged in biotechnology in the United States, OTA determined which of the 153 nonresponding companies were engaged in biotechnology by telephoning the companies, examining annual reports, reading newspaper reports, etc. OTA's estimate of the total number of companies engaged in biotechnology activities in the United States is 219.*

As of April 1983, the 95 companies that responded to the OTA/NAS survey employed 2,591 individuals in industrial biotechnology R&D. These 95 firms represent 43 percent of the 219 firms in the United States estimated to be engaged in biotechnology activity. Extrapolation of this number suggests that the number of individuals employed in biotechnology R&D in all 219 companies using biotechnology could be about 5,000.

The 95 firms that responded to the survey indicated plans to hire an additional 1,167 technically trained employees over the next 18 months.** No company indicated plans to reduce the number of technically trained employees in the next 18 months, so this figure represents an annual employment growth rate approaching 30 percent (not including any new companies formed in the next 18 months). A 30-percent annual growth rate in the number of R&D personnel probably will not be sustained over any length of time, so it is unlikely that the commercialization of biotechnology will lead directly to large increases in employment in the R&D sector. The need for marketing and sales personnel and the potential for spinoff industries are difficult to assess at this time. However, these sectors could be high-growth sectors for biotechnology.

*This number was arrived at by taking estimates of total worldwide shipment of biotechnology products estimated for the target year from OTA's 1981 report *Impacts of Applied Genetics: Micro-Organisms, Plants, and Animals* (40). This estimate was converted to employment of production workers by using case study data from the same OTA report. Next, this estimate was converted into total employment, including nonproduction workers, by utilizing data for established industries. Finally, total worldwide employment was subdivided and a weighted allocation made to the United States.

**NAS Committee on National Needs for Biomedical and Behavioral Research Personnel; Robert Barker, Cornell University, Chair of Panel on Basic Biomedical Personnel.

*For a list of companies engaged in biotechnology in the United States, see *Appendix D: Index of Firms Commercializing Biotechnology in the United States.*

**For a tabulation of the numbers and types of employees these companies indicated they planned to hire, see question 4 in *Appendix E: OTA/NAS Survey of Personnel Needs of Firms in the United States.*

One reason that commercialization of biotechnology will not directly contribute to a rapidly expanding U.S. work force is that bioprocess technology is not labor intensive (11).* It is estimated that personnel requirements for bioprocessing, even after firms enter mass production, will be only 10 to 15 percent of the total biotechnology work force. Furthermore, with more sophisticated, computer-controlled continuous bioprocesses, the labor intensity of bioprocesses could decrease (11).

*Feldman cites a 1980 report by the National Institute for Occupational Safety and Health (NIOSH), in which NIOSH reported on a Schering-Plough (U.S.) process for producing human leukocyte interferon. Only six people were assigned to production, and probably all six were not needed to monitor the bioprocess (11).

The demands for biotechnology R&D personnel are estimated to be fairly small in foreign country calculations as well. Britain's Royal Society has estimated that about 100 graduate biotechnologists per year will be needed over the next 10 years to commercialize biotechnology in the United Kingdom; about four times that number of technicians and technical support staff will be needed (45). The French Biotechnology Commission has forecast a need for about 1,830 researchers and engineers in biotechnology in France over the next 5 years (44).

Availability of biotechnology personnel

Categories of technical expertise

The industrial development of biotechnology will require several specific categories of technical personnel, many of which are listed in *Appendix E: OTA/NAS Survey of Personnel Needs of Firms in the United States.* Especially important categories include specialists in genetic manipulation such as molecular biologists and immunologists, specialists in scale-up and downstream processing such as bioprocess engineers, biochemists, and microbiologists. Generalizations with regard to the relative importance of these various categories of technical specialization in the development of biotechnology can be drawn from the responses of the 95 companies that responded to the OTA/NAS survey.

SPECIALISTS IN GENETIC MANIPULATION: MOLECULAR BIOLOGISTS AND IMMUNOLOGISTS

The development of hybridoma and recombinant DNA (rDNA) technologies brought molecular biology into the marketplace. A sufficient supply of molecular biologists and immunologists who are specialists in genetic manipulation has been critical to the development of corporate biotechnology R&D in the United States. As shown in table 62, about one-third of the technical personnel employed by the 95 companies responding to the OTA/NAS survey are specialists in rDNA/molecular genetics or hybridoma/monoclonal antibody (MAb) technology (there are twice as many specialists in rDNA as in hybridoma technology). These specialists in genetic manipulation are expected to become increasingly important in the next 18 months, constituting 37 percent of new hirees.

Most molecular biologists trained in the United States at present are specialists in animal molecular biology. The development of agricultural applications of biotechnology will require specialists in plant molecular biology with knowledge of both plant physiology and molecular genetics. According to the OTA/NAS survey, specialists in plant molecular biology currently constitute only 3 percent of the U.S. biotechnology R&D labor force and will constitute 5 percent of all new hirees in biotechnology in the next 18 months.

SPECIALISTS IN SCALE-UP AND DOWNSTREAM PROCESSING: BIOPROCESS ENGINEERS, BIOCHEMISTS, AND MICROBIOLOGISTS

Specialists in scaling-up the production of genetically manipulated micro-organisms (and higher organism cells) and in separation and purification

Table 62.—Major Categories of Biotechnology R&D Personnel in Firms in the United States (OTA/NAS Survey)

	Present employees		Employees to be hired in the next 18 months	
Area of technical expertise	Number	Percent of total[a]	Number	Percent of total[a]
Areas related to genetic manipulation:				
rDNA/molecular genetics	586	23%	302	25%
Hybridoma/monoclonal antibodies	247	10	146	12
Plant molecular biology	76	3	63	5
Areas related to scale-up/downstream processing:				
Microbiology[b]	334	13	160	13
Biochemistry[c]	326	13	125	10
Bioprocess engineering	186	7	100	8
Areas related to all aspects of biotechnology:				
Enzymology/immobilized systems	219	9	59	5
Cell culture	187	7	66	5

[a]The total number of industrial personnel (currently engaged in R&D in new biotechnology) identified in the OTA/NAS survey was 2,591. The total number of personnel to be hired in the next 18 months, according to the survey responses, was 1,167 (see app. E).
[b]Microbiology, as used in this table, combines the OTA/NAS survey responses to industrial microbiology and general microbiology (categories g and s of the survey questionnaire reproduced in app. E).
[c]Biochemistry, as used in this table, combines the OTA/NAS survey responses to analytical biochemistry and general biochemistry (categories j and k of survey questionnaire reproduced in app. E).
SOURCE: Office of Technology Assessment.

of products will become increasingly important as companies developing commercial applications of biotechnology move into production. Although few companies have reached the scale-up stage for new biotechnology products to date,* a substantial amount of R&D in companies developing commercial applications of biotechnology is related to scale-up.

As shown in table 62, about one-third of the biotechnology R&D technical personnel at the 95 companies responding to the OTA/NAS survey are specialists in areas related primarily to scale-up and downstream processing: bioprocess engineering, biochemistry, and microbiology. Bioprocess engineers are needed to design, construct, and maintain scale-up equipment and bioprocesses. Biochemists (apart from enzymologists, discussed below) are involved in the recovery, purification, and quality control of protein products. Microbiologists are needed for the isolation, screening, and selection of micro-organisms having particular catalytic properties. Such specialists are also needed to determine the optimal growth and production conditions for micro-organisms in order to facilitate the design of environments that maximize the micro-organisms' productivity. In the context of the commercialization of biotechnology, bioprocess engineering, biochemistry, and microbiology are generally considered to be more applied science disciplines than are molecular biology and immunology.

*In 1982, about 2 percent of all biotechnology workers in California were production workers (11).

As shown in table 62, the OTA/NAS survey of firms in the United States found that bioprocess engineers constitute approximately 7 percent of the current biotechnology R&D work force and will constitute 8 percent of all new hirees over the next 18 months. Specialists in microbiology constitute 13 percent of current employees and 13 percent of the employees to be hired in the next 18 months. Biochemists constitute 13 percent of current employees and will constitute 10 percent of new hirees in the next 18 months.

SPECIALISTS IN ALL ASPECTS OF BIOTECHNOLOGY: ENZYMOLOGISTS AND CELL CULTURE SPECIALISTS

Enzymologists and cell culture specialists are important for many aspects of biotechnology. Advances in the understanding of enzyme structure and function are important in developing the potential of biocatalysts for product formation. Cell culture is used at early R&D stages, but it is becoming increasingly important for the large-scale growth of higher organism cells, especially hybridomas. As shown in table 62, according to the OTA/NAS survey, enzymologists constitute 9 percent of current biotechnology employment in R&D; cell culture specialists constitute 7 percent of current biotechnology employment. Both categories of specialists constitute a smaller fraction of future biotechnology hirees (5 percent each) than they do of current employees.

Availability of biotechnology personnel in the United States

Of the countries studied, the United States has the largest number of specialists in genetic manipulation. The large supply of well-trained molecular biologists and immunologists in the United States is one reason for the rush of small company startups and the initial American lead in biotechnology. A primary reason for the large number of basic life science specialists in the United States is that for the past three decades, there has been substantial support from the U.S. Government, primarily from the National Institutes of Health (NIH), of basic research in the life sciences (26). In 1978, for instance, while the governments of most other developed countries were putting 2 to 4 percent of their R&D expenditures into health-related basic research, the United States was putting 11 percent of a much larger R&D base into health research (26). U.S. Government funds have strengthened the foundation of basic life science research, produced trained graduates, and generated an infrastructure for U.S. industrial growth in molecular biology (12). The dominance of the United States in the life sciences is supported by scientific and technical article publishing data. In 1979, U.S. authors published 40 percent of the world's articles in biology and 43 percent of the world's articles in biomedicine (26).

The results of the OTA/NAS survey of U.S. industrial biotechnology personnel needs reflect, with few exceptions, the United States' abundance of personnel trained in basic biological science. Relatively few of the 95 companies responding to the survey indicated that they were experiencing shortages of biochemists, pharmacologists, and toxicologists, who will be needed for the purification, recovery, and testing of biotechnology products. Furthermore, relatively few companies cited shortages of personnel in the areas of hybridoma and cell fusion technology.*

*For a tabulation of responses, see question 1 in *Appendix E: OTA/NAS Survey of Personnel Needs of Firms in the United States.*

Despite the abundance of personnel in the basic biological sciences in the United States, participants at two recent National Science Foundation (NSF) workshops* expressed concern that the United States currently may not have enough well-trained bioprocess engineers necessary for design and monitoring of biological scale-up processes (27). A shortage of highly trained bioprocess engineers in the United States, workshop participants suggested, could be a bottleneck to the rapid commercialization of biotechnology in the United States. The NSF workshop participants also pointed to an insufficient supply of industrial microbiologists. Between 1979 and 1981, the number of industrial microbiology positions listed in the United States nearly doubled, while the number of doctorates in "microbiology and bacteriology" has remained constant for the past 15 years (4). As shown in table 63, the results of the OTA/NAS survey also suggest that the United States may be experiencing shortages of bioprocess engineers: 11 of the 26 U.S. companies planning to hire Ph. D. bioprocess engineers in the next 18 months are experiencing shortages. The OTA/NAS survey results with respect to shortages of microbiologists are more equivocal.**

Shortages in bioprocess engineers, and possibly, industrial microbiologists, may be due in part to the fact that in the past three decades, there has been relatively less Federal support for applied microbiology, applied biochemistry, and bioprocess engineering research than for basic research in molecular biology, biochemistry, and immunology. Thus, university research activities have been guided by Federal funding toward basic biological research and away from these applied disciplines. The shortages may also reflect the fact that U.S. industrial support for university R&D in applied biology and bioprocess engineering has declined in the past three decades (12). After World War

*"Prospects for Biotechnology," University of Virginia, Apr. 5-6, 1982; "Developing the Biotechnology Component of Engineering," North Carolina Biotechnology Center, Apr. 24-25, 1983.

**Results concerning personnel shortages from the OTA/NAS survey are equivocal because the responses of the firms that indicated that they were not experiencing personnel shortages could indicate merely that the firms have not begun a search for personnel or instead indicate that they are not having any difficulty finding trained personnel. Furthermore, the 95 firms that responded to the survey represent less than half of the total number of companies commercializing biotechnology in the United States and may not be representative of the level of scale-up taking place as a whole.

Table 63.—Shortages in Major Categories of Ph. D. Biotechnology R&D Personnel in Firms in the United States (OTA/NAS Survey)

Area of technical expertise (Ph. D.)	Number of firms: Experiencing shortages and plan to hire in the next 18 months	Experiencing shortages and do not plan to hire in the next 18 months	Not experiencing shortages but plan to hire in the next 18 months
Bioprocess engineering	11	1	15
Recombinant DNA	10	1	29
Gene synthesis	7	3	7
Plant molecular biology	4	4	15
Industrial microbiology	3	4	14

SOURCE: Office of Technology Assessment.

II, U.S. chemical companies switched from biomass to petroleum feedstocks and consequently decreased their demand for bioprocess engineering and applied biology programs. Conditions in Japan, the Federal Republic of Germany, and the United Kingdom have differed markedly from those in the United States; in these countries, both public and industrial support have helped maintain a strong academic base for the microbial and bioprocess industries over the past several years (12).

The late David Perlmann wrote in 1973 (8):

> The interest in the U.S. has shifted in the past 20 years toward molecular biology. Few students are being trained for the fermentation industries. In the long run, this has worked to the disadvantage of the industries. Unless present trends in the U.S. are reversed, we can expect that in the future it will be desirable to send our students to Japan to learn the techniques that will assure the continuation of the fermentation industries in the United States.

This situation does not appear to have changed much in the last 10 years.

The OTA/NAS survey also showed that 10 of 39 companies planning to hire Ph. D. specialists in rDNA in the next 18 months are experiencing shortages. Much of the R&D activity now in the commercialization of biotechnology is in this area, and, thus, the demand for these specialists is high. However, as companies move toward production, the demand for scale-up and downstream processing specialists will increase, while the demand for the more basic scientists will not. Thus, the current shortages of bioprocess engineers and industrial microbiologists are considered to be more serious.

The shortages in biotechnology personnel in the United States may be partially counteracted by a flow of skilled foreign personnel into the United States.* A representative of one U.S. company stated that of the company's R&D staff of 130, 13 were foreign nationals (9 Ph. D.s). The foreign nationals were from Taiwan, India, Canada, and Hong Kong, and had expertise in nucleotide chemistry, applied microbiology, and bioprocess engineering. U.S. companies using biotechnology might be hiring an even greater percentage of foreign technical personnel if cumbersome and strict immigration regulations did not exist.

Availability of biotechnology personnel in other countries

The number of scientists and engineers engaged in R&D activities in the United States, Japan, the Federal Republic of Germany, the United Kingdom, and France is shown in table 64. As can be seen from that table, in 1977, the United States had more R&D scientists and engineers than any of its principal competitors in biotechnology. Japan had the second largest number, with half that of the United States. The size of a country's R&D labor force is one measure of a nation's R&D capacity. It is only an approximate measure, however, because it does not take into account such factors as the level of sophistication or specialization, utilization, or productivity of a country's R&D personnel. Furthermore, these data cannot

*Reliance on foreign R&D personnel has been common in other U.S. high-technology industries. Many semiconductor and computer companies hire foreigners in order to compensate for shortages of U.S. electrical engineers. At Intel (U.S.), for instance, 50 percent of the engineers holding M.S. degrees and 64 percent of the engineers with Ph. D.s are foreign (15).

Table 64.—Number of Scientists and Engineers Engaged in R&D by Country, 1977

Country	Number of scientists and engineers	Scientists and engineers as percentage of work force
United States	573,900	0.58%
Japan	272,000	0.50
Federal Republic of Germany	111,000	0.44
United Kingdom	80,700[a]	0.31
France	68,000	0.30

[a]1975.

SOURCE: National Science Foundation, *Science Indicators, 1980*, Report of the National Science Board, Washington, D.C., 1981.

be dissected into the percentage of biological personnel.

There are few statistics documenting numbers of specific types of biotechnology personnel in countries other than the United States. For that reason, shortages and surpluses in foreign countries are difficult to identify. Nevertheless, distinct patterns with respect to the availability of biotechnology personnel in foreign countries can be discerned through an examination of available government policy documents and other supporting evidence.

JAPAN

Several experts noted that in the early 1980's, Japan experienced a shortage of experts in genetic manipulation. This shortage was undoubtedly due to the inadequacy of the basic biological sciences in the universities.* Japanese universities have received limited Government support for basic research, so most Japanese universities have not developed extensive research programs in the basic biological sciences. Japan's public universities have been a relatively minor source of highly trained personnel in rDNA and hybridoma techniques (35). Thus, Japanese companies have had to look to other sources of trained basic biological scientists. Some companies have started in-house training programs. Japanese companies have also hired Japanese researchers from abroad, sent employees to be trained abroad and at Japanese universities, and recruited midcareer researchers from other Japanese companies (35). The last option is particulary unique for Japan, a country noted for a lack of personnel mobility. The extensive effort exhibited by Japanese companies seems to have overcome the personnel shortages documented a few years ago.

The supply of bioprocess engineers and industrial microbiologists is larger in Japan than in any of the competitor countries. Japanese Government officials monitoring biotechnology have indicated that the supply of personnel to handle the challenges of scale-up in Japan is not an area of concern (19,35). In fact, a major proportion of biotechnologists in Japan have their background in microbial physiology, an area of neglect in every country examined here except Japan (29).

The specialties of bioprocess engineering and industrial microbiology are strong in Japanese universities in part because the specialty chemical and other industries using traditional bioprocesses in Japan have kept the demand for graduates in these specialties high. After World War II, when chemical companies throughout the world largely switched to processes using petroleum feedstocks, Japanese chemical companies retained some processes using biomass feedstocks and came to dominate the international amino acid market. Furthermore, applied biology departments at Japanese universities have kept in close contact with industry representatives. Each year, 75 students in applied biochemistry graduate from Tokyo University alone; half go on to graduate studies, and half of these go beyond their M.S. degrees. Most are employed by Japan's leading bioprocess companies (35).

FEDERAL REPUBLIC OF GERMANY

The Federal Republic of Germany has sufficient personnel to compete with the United States and

*There is little communication between the basic and applied science departments in Japanese universities. Only the applied science departments have traditionally maintained close relationships with industry. For a more extensive description of the Japanese university system and its relationship with industry, see *Chapter 17: University/Industry Relationships*.

other countries in biotechnology. It is possible that there are some shortages of molecular biologists with expertise in rDNA and hybridoma research. However, according to Norman Binder, the cabinet head of the German Ministry of Science and Technology (BMFT, Bundesministerium für Forschung und Technologie), the training of people in rDNA and hybridoma technology is now a high priority in West Germany (21).

The Federal Republic of Germany's supply of personnel in specialties related to scale-up and bioprocessing appears to be adequate. Like Japan, the Federal Republic of Germany maintained a steady supply of both industrial and government funding for applied microbiology and bioprocess engineering after World War II. According to BMFT, however, the number of both bioprocess engineers and industrial microbiologists in Japan surpasses the number in West Germany (21).

Like the United Kingdom (see below), the Federal Republic of Germany is concerned about a brain drain of biotechnology R&D personnel to other countries. According to the Max Planck Society's senate and the present Minister of Research and Technology, shortages of suitably qualified workers in West Germany are partially due to a brain drain to the United States (9,37). The brain drain of scientists from West Germany, however, appears to be less serious than that from the United Kingdom.

UNITED KINGDOM

Like the United States, the United Kingdom boasts both qualified personnel and excellent training and education programs for personnel in the basic life sciences. In the 1950's and 1960's, there was considerable expansion of basic life science research in British universities. By 1972-73, health-related R&D, supported mostly by the Medical Research Council (MRC), had risen to 5 percent of the British Government's R&D budget, nearly twice the percentage of Japan, the Federal Republic of Germany, or France (26).* MRC's past investment in biology is now paying off. Molecular biologists and immunologists supported by MRC are internationally prominent in the development of rDNA and hybridoma technologies. Nevertheless, there may be shortages of molecular biologists if the industrial development of biotechnology expands rapidly (2).

Like Japan and the Federal Republic of Germany, the United Kingdom has a good academic base for training bioprocess engineers. Nevertheless, the United Kingdom appears to be experiencing a shortage of bioprocess engineers (2). A brain drain from the United Kingdom is viewed as partially responsible for this shortage. Many British biotechnologists are leaving for the United States, Switzerland, and other countries of the European Economic Community, because sufficient posts do not exist in the United Kingdom at present and salaries in the United Kingdom are not competitive with those in other countries (45). When the Swiss company Biogen S.A.* advertised for 30 molecular biologists, half of the 600 applications they received were well-qualified British (45).

Analysts estimate that a total of between 100 and 1,500 experts in some aspect of biotechnology have left the United Kingdom over the past several years (30). Governmental institutions are taking active measures to counteract the brain drain. The Research Councils, the United Kingdom's public research institutes, have adopted an active policy of encouraging scientists from the United Kingdom who have spent time in industry abroad to return home. The Science and Engineering Research Council (SERC) maintains a list of British biotechnologists outside the United Kingdom and may be taking measures to encourage them to return (30), and MRC has announced publicly that it will provide laboratory space and allow reentry into the career structure without penalty for scientists who return to the United Kingdom (45).

SWITZERLAND

The access to distinctive universities and the high standard of living in Switzerland attract highly qualified personnel from around the world to participate in Swiss biotechnology. Although the availability of personnel may not be impor-

*Since 1973, Government expenditures in the United Kingdom for health-related research have dropped and are now equivalent to those of the other foreign countries studied here (26).

*Biogen S.A. is one of the four principal operating subsidiaries of Biogen N.V., which is registered in the Netherlands Antilles. Biogen N.V. is about 80-percent U.S.-owned.

tant for the large pharmaceutical companies which conduct a large proportion of their R&D in other countries, it is crucial to the Swiss advancement of biotechnology in other sectors. The attraction of talent from other industrialized countries may help the competitive efforts of Swiss companies in biotechnology in the future.

FRANCE

France has a serious shortage of qualified personnel that could well undermine the country's basic and applied science base and prevent France and its industries from competing successfully in the world biotechnology marketplace. Specialists in the fields of general and industrial microbiology, rDNA and hybridoma technologies, enzymology, plant and animal cell culture, and bioprocess engineering are few (3). Although some French research centers boast internationally recognized teams, such as the enzymology and bioprocess technology teams at the technical University of Compiegne or the immunology groups at the Institut Pasteur (44), these are isolated clusters of expertise. Thus, France will have difficulty matching the total output of the large and balanced national research bases of other competitor countries.

The scarcity of personnel in France cuts across several sectors of R&D in these technologies and applies equally to different categories of personnel, from scientists and bioprocess engineers with advanced degrees to skilled laboratory and production technicians. In order to correct this situation, the French Government has given special attention to the education and training of qualified personnel. The research law passed in July of 1982 called for the active involvement in the educational process of public sector researchers outside universities (46). And the Programme Mobilisateur presents educational guidelines for all stages of schooling from secondary to postdoctoral levels, placing special emphasis on an interdisciplinary approach within the universities (24). The education of a specialist in rDNA technology, nonetheless, takes many years, as does the implementation of such training programs. As a short-term solution to its present lack of personnel, therefore, France imports foreign experts (24).

Personnel training

The availability of the scientific and technical personnel necessary for the commercialization of biotechnology is highly dependent on a country's educational infrastructure. The discussion here compares various aspects of training, all of which are important to the development of biotechnology: 1) secondary school education, 2) biotechnology-related undergraduate and graduate education, 3) transnational training opportunities, and 4) mid-career retraining opportunities.*

Secondary school education in the United States and other countries

Secondary school education in science and mathematics in the United States trails that in Japan and many European countries. High school students in Japan are required to complete 2 years of mathematics and 2 years of science before graduating (42). Secondary school students in many European countries, even students specializing in classics or languages, similarly get far more extensive training in mathematics and science than do students in the United States (6).

Several recent studies have identified a decline in the quality of science and mathematics education in U.S. secondary schools, attributing it to a lack of good teachers, instrumentation, Federal support, and local community support in the form of bonds and taxes (6,10,16,31,47). Furthermore, many leading scientists, engineers, and politicians in the United States fear that the decline is leading the United States to become a nation of technological illiterates and is compromising the U.S.

*For general information on science and engineering education and personnel internationally, see (39).

position in international competition in high-technology areas (1,38,49).

Undergraduate and graduate education in the United States and other countries

There is near unanimous agreement that the development of biotechnology will require personnel capable of operating in an interdisciplinary environment with various levels of expertise in both biology and engineering (29). Because of traditional barriers between basic biological science and engineering departments in most higher educational institutions, the challenge of providing interdisciplinary undergraduate and graduate education for personnel in biotechnology is a challenge common to all industrialized countries.

UNITED STATES

The United States has an adequate supply of personnel in nearly all the fields of basic biological sciences relevant to biotechnology, with the possible exception of plant molecular biology. For the training of plant molecular biologists, new and modified curriculum offerings may be needed. Most classical plant breeders in the United States are trained at agricultural research stations and land-grant colleges; thus, their training does not traditionally include molecular biology. Because the new genetic technologies grew out of biomedical research at universities and NIH, few traditional plant breeders have the training that would allow them to do experiments using rDNA technology. Nevertheless, interest in plant molecular biology is increasing dramatically. Botanists are learning the new techniques, and biomedically trained researchers are applying their expertise to plants. Because of the separation of agricultural researchers and plant molecular biologists in the United States, however, there are problems of communication between these groups which may slow research advances (34).

There is a growing concern that a shortage of plant molecular biology professors in the United States could result from a drain of Ph. D. plant molecular biologists from U.S. universities to industry (25). As numerous companies have started efforts in plant molecular biology and existing companies have expanded into plant molecular biology, industry has been competitively recruiting university researchers. As shown in table 65, according to the OTA/NAS survey, all of the companies wanting to employ Ph. D. plant molecular biologists intend to hire from academia, and half intend to hire from industry as well.

Bioprocess engineering education in the United States, now almost exclusively provided in university chemical engineering departments at the graduate level,* is closely tied to training opportunities in chemical engineering (12). Between 1970 and 1980, the number of Ph. D.s graduating in chemical engineering declined by nearly 25 percent, and the bioprocess subset of the chemical

*At the undergraduate level, there are only two accredited bioengineering (distinct from biomedical engineering) programs in the United States, one at the University of Illinois at Chicago and one at Texas A&M.

Table 65.—Sources of New Ph. D. Biotechnology R&D Personnel In Selected Categories in Firms in the United States (OTA/NAS Survey)

Area of technical expertise (Ph. D.)	Companies planning to hire from industry		Companies planning to hire from academia		Companies planning to retrain current staff	
	Number	Percent of total[a]	Number	Percent of total[a]	Number	Percent of total[a]
Recombinant DNA	15	38%	35	84%	3	7%
Gene synthesis	9	64	13	93	3	21
Industrial microbiology	11	67	13	81	2	13
Bioprocess engineering	19	86	11	50	2	9
Plant molecular biology	9	50	18	100	3	17

[a]Refers to percent of companies that both indicated plans to hire in the specialty area *and* revealed the sources from which they would hire new personnel. Many companies indicated more than one hiring source for each specialty area.

SOURCE: Office of Technology Assessment.

engineer category probably declined proportionally. At most, only about 10 percent of the recent M.S.s and Ph. D.s in chemical engineering are ready to enter the bioprocess industry without additional formal training (13).

The decline in the number of Ph. D.s graduating in chemical engineering in the United States in part reflects declining graduate student enrollment. Because industry salaries are quite high for bachelor's degree engineers, fewer and fewer people have gone to graduate school. Another reason for the decline is a shortage of engineering professors. Most American universities do not pay salaries commensurate with industry. Currently, there are 1,600 faculty vacancies at U.S. engineering schools in all disciplines (43). Participants at a 1982 workshop on biotechnology sponsored by the University of Virginia and NSF agreed that the shortage of faculty in engineering is a more pressing problem for the long-term educational stability of the United States than the declining engineering graduate student enrollment (28).

According to the OTA/NAS survey of firms using biotechnology in the United States, Ph. D. bioprocess engineers are in high demand by industry (see table 63). If incentives for Ph. D. bioprocess engineers to remain in the academic field are not improved, the loss of these Ph. D.s to the private sector may reach the point that the American Society for Engineering Education refers to as "industry eating their seed corn" (32). If the United States is to produce high-quality Ph. D. engineers, salary money and research funding for engineering faculty, as well as a restructuring of bioprocess engineering education emphasizing interdisciplinary training may be necessary.

JAPAN

In Japan, training in basic biology research is relatively weak. The director of the new Bioindustry Office of Japan's Ministry of International Trade and Industry (MITI) has listed as one of his primary concerns the state of basic biology research in Japan. However increased Japanese Government funding for such research is not apparent. The University of Tsukuba, the heart of a new $5 billion "science city" 37 miles north of Tokyo, has the largest budget of Japan's 95 national universities, but has no plans to expand its graduate enrollment in biology (22).

The distinction between basic and applied science departments at Japanese universities is great. At Tokyo University, for example, basic and applied science departments are located on separate campuses and have little interaction. Furthermore, professors in pure science areas such as biology are proud of their independence from industry (35). There is little direct correlation in Japan between university basic sciences curricula and corporate personnel needs. Special interdisciplinary biotechnology programs combining basic and applied sciences have not been instituted at Japanese universities.*

Because of Japan's need to generate and transfer basic science to industry more rapidly, the Japanese Government is attempting to end the isolation of Japan's basic research. Japan's Science and Technology Agency (STA) funds "Leading Technology" (Senatsu Gijutsu) projects, that allocate research responsibilities between university and corporate laboratories, but this funding has not yet been applied to the biotechnology field. STA is also funding a new program called the New Technology Development Fund (Shingijutsu Kaihatso Jigyodan) that was established to help companies commercialize university-generated research. The Government has also proposed building two new biotechnology centers open to private sector corporations through universities. Each researcher will conduct research in his or her own laboratory, but exchange of information between the corporate and academic researchers will take place on a regular basis (35).

National laboratories supported by the Agency for Industrial Science and Technology of MITI encourage the flow of personnel into interdisciplinary generic applied research. The national laboratories provide a place for university professors, Government researchers, and corporate researchers to work together. These laboratories have been especially important in the development of agricultural sciences and applied microbiology, because there are few private institutes

*See *Chapter 17: University/Industry Relationships.*

carrying on significant research in these areas (35).

FEDERAL REPUBLIC OF GERMANY

In the Federal Republic of Germany, three types of nonindustry laboratories conduct basic research in biotechnology: 1) laboratories belonging to universities, 2) laboratories dependent on BMFT for operating expenses and on the German Research Society (DFG, Deutsche Forschungsgemeinschaft) for project support,* and 3) laboratories in institutes supported by the Max Planck Society (Max-Planck Gesellschaft zür Forderung der Wissenschaften), which in turn receives support from BMFT.

Although laboratories supported by BMFT and DFG, such as the Cancer Research Center at Heidelberg, carry out important biotechnology-related work, institutes funded by the Max Planck Society are responsible for the bulk of basic research advances in biotechnology. The Max Planck Institute for Plant Breeding Research in Cologne, which recently received an unrestricted grant from Bayer, boasts some of the best plant genetics teams in the world. BMFT would like to see closer cooperation between the Max Planck institutes and industry (21).

The center for generic applied research in biotechnology in the Federal Republic of Germany is the Society for Biotechnological Research (GBF, Gesellschaft für Biotechnologische Forschung). GBF is a Government-supported private institution that was founded to conduct generic bioprocessing research to meet the needs of industries (23). In 1972, 89 percent of its $13 million (DM31.6 million) came from BMFT (14).

Among the factors cited to explain Germany's slow entry into biotechnology is an educational system that prevents the kind of interdisciplinary cooperation that is viewed by most experts as essential to the development of this field (21). Because of the traditional separation of technical faculties from arts and science faculties in West Germany, bioprocess technicians, usually located in technical schools, rarely come into contact with colleagues holding university appointments in biochemistry or microbiology (21). In August 1981, BMFT policy called for greater interdisciplinary cooperation among biologists, chemists, medical experts, and engineers (21).

*See *Chapter 13: Government Funding of Basic and Applied Research.*

UNITED KINGDOM

The United Kingdom's system of funding research in biology and the medical sciences at universities has produced highly trained personnel in rDNA and hybridoma technology for industry. Furthermore, the country's Plant Breeding Institute is considered a model for interdisciplinary research on plants. Unlike the United States, therefore, the United Kingdom is probably not suffering interdisciplinary training problems in plant molecular biology.

Many British universities have programs in bioprocess engineering. Bioprocess engineering has been taught at the postgraduate level at University College in London and Birmingham to biologists and biochemists for nearly 20 years. Furthermore, at least 10 to 15 university centers are now involved in postgraduate biotechnology education, and these centers are receiving extra money from the University Grants Committee. One of these, the Centre for Biochemical Engineering and Biotechnology, was set up by three universities both to acquire new laboratory space and to launch new courses. Imperial College in London set up the Centre of Biotechnology with four new faculty positions. This center will work with other departments of the college involved in biotechnology to launch a biotechnology masters course. Funding for bioprocess graduate research and training in Britain's universities is also being provided by SERC. SERC has plans to fund four new specialized biotechnology courses in universities, which will all contain elements of bioprocess engineering. SERC will fund a maximum of 60 places for graduate students, and industry is encouraged by the Government to finance more places (45).

British universities have 30 to 40 teaching staff who teach biotechnology (including bioprocess engineering) on a full-time basis and a much greater number of teaching staff who devote varying proportions of their time to teaching biotechnology. According to bioprocess expert Malcolm Lilly, the United Kingdom has more teaching biotechnologists than the United States and is also ahead of other European countries. Thus, there appears to be no current shortage of biotechnology faculty at British universities. Nevertheless, Government officials are worried that a lack of bioprocess engineering faculty may be a problem for the United Kingdom in the future because of the fairly small numbers of chemical engineers getting higher degrees in bioprocess engineering in recent years (45). To counter any shortage in teaching capabilities, the United Kingdom plans to involve industrialists in teaching bioprocess engineering courses at the universities.

FRANCE

In France, those pursuing higher education in scientific and engineering education go either to universities, to the more prestigious grandes écoles, or to Government-funded laboratories. French universities do not have graduate interdisciplinary courses in microbiology, rDNA technology, enzyme engineering, or bioprocessing techniques (33), and their creation will be difficult because of the lack of funds and a shortage of faculty. Four grandes écoles have interdisciplinary courses in biotechnology, but they produce only about 40 graduates a year total. However, other grandes écoles are now introducing courses in biotechnology (44). The Institut Pasteur, which is 49-percent Government-owned, regularly accepts doctorate students in biotechnology fields.

Other important loci of graduate training for biotechnology personnel in France, apart from the grandes écoles, are public research centers (grandes organismes), a very important part of the French research establishment. The grandes organismes have approximately 600 technical workers in biotechnology-related fields (nearly one-half of all of France's personnel in biotechnology), but they will probably find it difficult to create interdisciplinary training programs. At the largest and most significant organisme, the National Center for Scientific Research (CNRS, Centre National de la Recherche Scientifique), for example, there are communication problems between the scientific and engineering departments (44).

Transnational training in the United States and other countries

A trend evident in many scientific and technical fields, including biotechnology, is the training of increasing numbers of foreign students in the United States. In 1982, foreign students constituted 2.6 percent of the total U.S. university enrollment, and 23 percent of the foreign students enrolled at U.S. universities were studying engineering. In 1981, for the first time, more foreigners than Americans received doctoral degrees in engineering in U.S. graduate programs (15). The proportion of foreign students in American postdoctoral engineering programs was more than 60 percent. Furthermore, foreign students constituted a third of all postdoctoral students in American science and engineering programs (26). These numbers illustrate the esteem with which U.S. science and engineering education is held throughout the world (43).

In the areas of molecular biology and immunology, foreign nationals are actively seeking training at U.S. institutions. Hoechst's (F.R.G.) 10-year, $70 million contract with Massachusetts General Hospital, for example, was, in part, established to train Hoechst's personnel at Harvard Medical School (21).*

NIH has several programs that sponsor research by foreign nationals in NIH laboratories. Under the "visiting program," NIH sponsors and pays visiting scientists studying at NIH labs. In 1983, 810 foreign nationals were enrolled in this program. Of these visiting scientists, 158 were from Japan, 97 from India, 62 from Italy, 27 from France, and 6 from the United Kingdom. Under the "guest researchers program," foreign nationals are sponsored by their native country. In 1983, 32 Japanese were enrolled, 23 Italians, 21 French, 10 Indians, and 4 British (36).

Japanese personnel trained in the United States are now being actively recruited by Japanese

*This arrangement is discussed in *Appendix H: Selected Aspects of U.S. University/Industry Relationships.*

firms. In a 1982 Keidanren survey* of 60 Japanese companies using biotechnology, 35 percent of the companies were active in recruiting researchers already studying or working abroad (35). When the Japanese company Suntory hired new employees for about one-third of the 126 research positions in its Biomedical Research Institute established in 1979, for example, many of the new employees were Japanese who had been working abroad (35).

The larger more established Japanese companies sponsor transnational training of their employees. Sixty-two percent of Japanese companies responding to the 1982 Keidanren survey indicated that some scientific and engineering personnel would be sent abroad for training in specialized technologies (35).

Foreign nationals are being trained not only at university and government centers in the United States, but at U.S. companies looking for supplemental sources of revenue. Five corporate researchers from Japan recently attended a 3-month course at Genex in rDNA technology offered at $120,000 per person. According to the Japanese companies, they learned "highly specific knowledge . . . and key points for developing specific products by using the rDNA technology" (18).

Amid all the evidence that foreign countries are making use of U.S. training facilities, data show that U.S. doctoral graduates are going abroad for postdoctoral study less frequently. During the decade of the 1970's, postdoctoral training abroad decreased by nearly 50 percent (26). In biotechnology especially, postgraduate training abroad appears to be an area poorly funded by the United States. Professor Arnold Demain, for example, has indicated that 8 of the 11 students currently enrolled in his graduate program in industrial microbiology at Massachusetts Institute of Technology (MIT) are foreigners, all sponsored either by their government or company. Money to send Americans overseas to do postdoctoral work in industrial microbiology, however, is not available (7).

*Keidanren, the Japan Federation of Economic Organizations, is a national organization composed of about 700 of the largest Japanese companies. It enjoys the regular and active participation of the top business leaders working closely with a large professional staff to forge agreements on behalf of business as a whole. It often surveys its members on issues of economic importance.

Midcareer retraining in the United States and other countries

To address the challenges of biotechnology, industrial scientists and engineers can probably be retrained. Retraining in the United States is often viewed as the responsibility of the individual scientist or engineer and not that of the employer, with some exceptions (see below). A problem is that it is very difficult for a scientist or engineer in midcareer to take a year off to go back to school.

Reflecting concern over this situation, four senior professors at MIT recently published a report advocating "lifelong cooperative education" (48). The report's major recommendation was that engineering schools and neighboring industries collaborate in making off-campus graduate programs available to working engineers. Although the report was addressed specifically to the electrical engineering department of MIT, it could also be addressed to a larger community, and many of its recommendations may apply to biotechnology. For example, MIT Professor Daniel Wang recently stated that chemical engineers who "don't know the faintest thing about how proteins are isolated," if taught some basic protein chemistry, could develop new techniques for large-scale purification (17). Historically, chemical engineers in the United States have been retrained by pharmaceutical companies to be bioprocess engineers (7).

As shown in table 65, a relatively small percentage of the 95 companies responding to the OTA/NAS survey intend to retrain their workers to fill vacancies in areas of biotechnology personnel shortages. For most categories of Ph. D. personnel, hiring from academia is considered the optimal choice. In the case of Ph. D.s in bioprocess engineering, however, 86 percent of the companies planning to hire Ph. D. bioprocess engineers intend to hire them away from other companies, 50 percent plan to hire from academia; only 9 percent of the companies plan to retrain.* One

*These percentages exceed 100, because some companies indicated more than one hiring source.

reason for the very small amount of retraining in biotechnology may be the small size of many of the U.S. companies using biotechnology. The small companies that account for much of the biotechnology research activity in the United States probably do not have the resources to retrain personnel in-house.

Some foreign countries are pursuing the retraining of personnel more actively than the United States. The retraining of workers in Japan, more than in any other industrialized country, is viewed as the responsibility of the corporation. The Japanese permanent employment programs, prevalent in a majority of companies in the Japanese biotechnology-related industries, make it economically feasible for a firm's employees to be optimally trained at company expense (35). Japanese employees' salaries are in part based on the number of years they have been employed by the firm, so employees have strong incentives not to leave the firm for which they are working. Because employees in Japan are more likely to stay with their firms than employees in the United States, a far larger proportion of total training is sponsored by the Japanese private sector than in the United States (35).

The provision of corporate funding for worker retraining in biotechnology is common in Japan. According to the 1982 Keidanren survey, 53 companies indicated that they planned to use in-house training to meet, at least partially, their personnel needs (35). Some Japanese corporations, by commissioning research on a particular topic, are able to send their researchers to train at a university laboratory with a professor and his or her staff. At national universities, each professor is limited to approximately six or seven corporate trainees a year, but at private universities, there is no such restriction. As discussed above, training of Japanese workers at institutions in other countries is also common.

Japan's ability to overcome weaknesses in its labor force rapidly, due largely to corporate financing of worker retraining, is truly extraordinary. In 1981, for example, no more than 10 private Japanese firms had more than 10 researchers working on rDNA projects. A year later, the Keidanren survey in March 1982 revealed that 52 out of the 60 leading Japanese firms surveyed had 10 or more research workers in the area (35). It is partly because of the large-scale retraining of industrial personnel that Japan has been able to overcome a weak biological science base to remain a leading international competitor in the commercial development of biotechnology.

The European leader in the industrial retraining of its biotechnology work force is the Federal Republic of Germany. The German chemical industry association, DECHEMA, has an expert group on biotechnology, a standing body to bring academics and industrial scientists into regular contact. It organizes continuing education courses in various aspects of biotechnology (e.g., the use of immobilized enzymes, measurement needs, and control of bioreactors) (21).

The British and French Governments are adopting active policies to encourage retraining. In the United Kingdom, some Research Councils are offering short courses for midcareer scientists. Currently, MRC establishments are providing training in cell fusion and rDNA technology to the employees of Celltech and some larger companies, including Glaxo, ICI, and Seralab (45). In France, the Institut Pasteur runs postgraduate courses in biotechnology, long courses in both microbiology and immunology, and short specialized training courses (44).

Findings

The OTA/NAS survey of 95 companies using biotechnology in the United States suggests that approximately 5,000 workers are now doing biotechnology R&D in the 219 companies using biotechnology in the United States. Though the number is expected to increase about 30 percent over the next year, it is unlikely that a 30-percent annual growth rate can be maintained over the next decade. The commercialization of biotechnology is unlikely to contribute directly to large increases in employment. Bioprocess technology, an essential part of industrial biotechnology activities, is not labor intensive.

About one-third of the technical personnel currently employed in 95 surveyed companies using biotechnology are specialists in basic science areas related to genetic manipulation: rDNA/molecular genetics and hybridoma/MAb technology. Specialists in these categories will continue to be important to biotechnology R&D, and more hirees are expected. Another third of the technical personnel currently employed by the U.S. companies using biotechnology are specialists in areas of applied science related to scale-up and downstream processing: microbiology, biochemistry, and bioprocess engineering. Of these categories, only hirees in bioprocess engineering will increase over the next 18 months. About one-fifth of the biotechnology work force are specialists in areas important to all aspects of biotechnology: enzymology and cell culture. The balance of people are specialists in such fields as pharmacology and toxicology.

The United States currently has a competitive edge in the supply of scientific personnel able to meet corporate needs for R&D in rDNA and hybridoma technology. This edge is primarily due to generous Federal support for university life science research since World War II. Nevertheless, the supply of Ph. D. specialists in plant molecular biology and in applied disciplines such as bioprocess engineering and industrial microbiology may be inadequate for U.S. corporate needs. It may be difficult to alleviate rapidly the shortage of engineers because of the shortage of Ph. D. engineers serving as university faculty and the lack of governmental training programs. To an extent, foreign technical personnel are alleviating some of the industrial shortages.

With the exception of France, the other competitor countries have adequate supplies of basic biological scientists. French companies are importing foreign specialists. German and Japanese companies, where slight shortages do exist, are making efforts to train some of their personnel abroad and to retrain workers. Some Japanese companies are making successful efforts to repatriate Japanese workers trained overseas.

Japan, the Federal Republic of Germany, and the United Kingdom, unlike the United States, maintained a steady supply of both industrial and government funding for applied microbiology and bioprocess engineering after World War II. Japan's supply of scale-up personnel appears to be sufficient. However, the United Kingdom and West Germany are suffering from a brain drain to foreign countries (in particular to the United States), and shortages of scale-up personnel may occur.

The United States has very few undergraduate or graduate interdisciplinary programs in biotechnology. Consequently, in the agricultural fields, for example, there are communication barriers between classical plant breeders and plant molecular biologists. Bioprocess engineering education in the United States is provided almost exclusively at the graduate level and is closely tied to training opportunities in chemical engineering with few interactions occurring between biologists and engineers. Funds for Ph. D. and postgraduate education in bioprocess engineering in the United States have been inadequate for the training of sufficient numbers of specialists for industry and academia. Furthermore, the high industrial demand for Ph. D. bioprocess engineers is likely to create a shortage of university faculty in the field.

Universities in the United Kingdom, in contrast to their counterparts in the United States, have long had interdisciplinary programs in biotechnology, and the British Government is encouraging the formation of overarching biotechnology programs in those universities where they do not already exist. Though France, the Federal Republic of Germany, and Japan have systematic barriers to interdisciplinary programs, their governments are utilizing national research institutes to facilitate interdisciplinary research in biotechnology.

The funding by foreign governments and companies for the training of domestic workers overseas is far more extensive than that of organizations within the United States. In fact, in biotechnology-related areas, the U.S. Government appears to fund more the training of overseas nationals in the United States than the training of U.S. nationals abroad.

Switzerland, which has not been extensively discussed in this chapter, appears to have no trouble meeting the personnel needs in either its universities or companies developing biotechnology. Particularly in relation to the size of the country, Swiss academic institutions show unusual strength in both basic and applied research relevant to biotechnology. Swiss companies seeking to develop and expand their expertise in these technologies may choose to work with the qualified Swiss researchers in the university or may recruit foreign scientists, with apparently little difficulty, to work in Switzerland (20).

Retraining of corporate workers in biotechnology is being pursued more actively in foreign countries than in the United States. Japanese companies, in particular, make a regular practice of sending their workers to be retrained at Japanese and foreign universities and research institutions. Only a very small percentage of companies using biotechnology in the United States intend to retrain their workers in areas of personnel scarcity.

Issues and options

ISSUE 1: How could training for biotechnology at the graduate and postdoctoral levels be improved?

The United States appears to be suffering shortages of Ph. D. plant molecular biologists, applied microbiologists, and bioprocess engineers in its biotechnology-related industries. Although improved science education at the secondary school and undergraduate level could enhance the development of biotechnology in the future, the graduate level seems to be the best place to address the shortages of certain types of personnel.

For the past several years, U.S. Government funding for research in the areas of plant molecular biology, applied microbiology, and bioprocess engineering has been far less than funding for research in animal and bacterial molecular biology and immunology. Increasing Federal funding for research grants in plant molecular biology, applied microbiology, and bioprocess engineering, by encouraging more investigators to enter these fields, could help alleviate shortages of personnel. Since fields of faculty endeavor are at least partially determined by the availability of research grants, increased funding for research might encourage training and indirectly prevent future shortages of faculty. Options for directing research funds toward areas of personnel shortages are discussed in *Chapter 13: Government Funding of Basic and Applied Research.*

Another area where more Federal research funding could potentially reduce personnel shortages is that of interdisciplinary research. The interdisciplinary nature of biotechnology requires research collaboration among people with backgrounds in biology, engineering, and chemistry. Options that Congress could take to encourage interdisciplinary research are discussed below.

OPTION 1: Authorize increased funding for USDA, NIH, and NSF graduate and postdoctoral training grants in plant molecular biology, applied microbiology, and bioprocess engineering.

The lack of training grants is probably the single most outstanding reason for U.S. shortages in selected areas of biotechnology personnel. There are no NIH or NSF training grants for industrial microbiology or process engineering. The U.S. Department of Agriculture (USDA) this past year gave only five training fellowships in plant science. NSF until recently had no training grants at all in plant science, although in May of 1983, NSF's Biological and Behavioral Directorate approved 24 postdoctoral fellowships for study in plant cell biology.

In fields such as molecular biology, competitive training grants have been one of the most effective uses of Government funds for graduate and postdoctoral education. Training grants encourage university departments to carry on a cohesive training program and allow money from faculty research grants to be used for research instead of salaries. The institution of adequate training grants in the areas of plant molecular biology, applied microbiology, and bioprocess engineering would be a long-term strategy to counter personnel shortages in these areas. Such grants could be administered by NIH (for applied microbiology and plant biology), USDA (for plant biology), and NSF (for all three).

OPTION 2: Continue to support special incentives to encourage young engineers to stay in academia.

The shortage of engineering faculty at U.S. universities could seriously hamper efforts to increase the number of qualified engineers, including bioprocess engineers, in the United States. The recently instituted Presidential Young Investigator Awards to be administered by NSF is an example of the sort of special incentives program that Congress could continue to support to counteract the shortage of engineering faculty. Two hundred of these awards, 100 of which are to go to engineers, are to be awarded each year for 5 years to scientists and engineers in academia who have fewer than 7 years postdoctoral experience. Each award could total up to $100,000 per year for 5 years. The first $25,000 per year is to come from NSF. Industry funding for the engineers, of up to $37,500 per year, is matched by NSF, giving the total amount of $100,000.

OPTION 3: Specify that a certain percentage of NSF graduate and postdoctoral grants be used for training in other countries and authorize NIH and other relevant agencies to initiate researcher exchanges with other industrialized countries.

Increasingly fewer U.S. Ph. D.s are doing postdoctoral work abroad, while the number of foreign Ph. D.s doing postdoctoral work in the United States is increasing. The U.S. Government supports the training of its nationals overseas far less than its industrialized competitors.

Foreign countries have many significant and growing research programs in biotechnology that U.S. researchers could fruitfully be visiting—e.g., Japan's Fermentation Research Institute and University of Tokyo; the Society for Biotechnological Research (GBF) in Braunschweig, Federal Republic of Germany; and the John Innes Institute and Plant Breeding Institute in the United Kingdom. Few Americans are studying at those institutions. Though NSF's Science and Engineering Directorates can give grants to students studying overseas, such grants are not generally given because they are usually more costly than regular grants.

NSF's Science, Technology, and International Affairs Directorate has an International Cooperation and Scientific Activities program that provides special funds for researchers to study abroad—funds that can supplement the grants of other programs within NSF. One advantage of authorizing more money for this program is that this program has had experience negotiating standards of bilateral student exchange with foreign governments, having negotiated a successful bilateral agreement with France. In most foreign countries, American students cannot study at the best institutions (usually national) without the proper contacts and encouragement of the domestic government.

Congress could also specify that the NSF international grants that are given have a clearer training component. Currently, even the international fellowship grants are evaluated on the basis of their proposed research, rather than the quality of training for the U.S. nationals. It should be noted, however, that setting aside a part of NSF international grants for graduate and postdoctoral training would probably reduce the current percentage of international grants given to junior professors.

NIH's unilateral programs to support the study and research of foreign postdoctoral personnel in the United States could also be expanded to support the study of American nationals overseas. Since the United States is not the sole source of advanced R&D capability, Congress could authorize NIH to formulate programs that result in reciprocal exchanges and postdoctoral research opportunities for American scientists and engineers in areas of foreign expertise.

ISSUE 2: How could Congress improve interactions between classical plant biologists and plant molecular biologists?

Many people would argue that the agricultural research system in the United States does not need to be improved because the United States has the most productive agricultural system in the world. Nevertheless, there are specific areas where some advances in plant science, aided by new biotechnology, may be crucial to feeding the world's population in the coming years. These advances can be made only with the interaction of classical plant breeders and plant molecular biologists. Yet, because of the historical separation of agricultural researchers and plant molecular biologists in the United States, these groups do not have established communication networks. Most of the classical plant breeders are trained at agricultural research stations and land grant colleges, whereas most of the plant molecular biologists were originally trained in biochemistry, bacterial genetics, and animal biology (funded extensively by NIH) and are now working at the universities where much of the molecular biology is done. The lack of interaction between these two disciplines puts the United States at a disadvantage in modern agricultural research.*

The agricultural surpluses that the United States has today could vanish in a single year and probably are temporary. Greater productivity will be necessary as we move into the 21st century. The United States is also depleting its water resources and its topsoil. Advances in biotechnology can contribute to the solution of these problems with the development of plants that need less water, have greater nutritive value, and are more resistant to the high saline content of irrigation water. The costs of production can be lowered if plants are pest-resistant, and fewer fertilizers will be needed if plants can fix their own nitrogen. These advances cannot be made without greater interaction between classical plant breeders and plant molecular biologists. The Federal Government is spending about $20 billion on an acreage diversion program. This money subsidizes the market price, but does not address the central agricultural production issue, the farmer's low profit margin. Diverting a portion of this money to research on plant genetics could go a long way toward reducing agricultural production costs.

OPTION 1: Legislate the creation of one or more plant research institutes.

A plant research institute was established under the Department of Energy's (DOE's) management and with cooperation from the State of Michigan in 1965. DOE's contribution to this effort was $1.65 million in fiscal year 1983 and will be $1.7 million in fiscal year 1984. This is a beginning toward solving some of the problems of communication among biologists of different disciplines, but it is only one effort.

The creation of several more plant research institutes could facilitate interdisciplinary research between classical plant biologists and plant molecular biologists, although there could be some problems. First, a large amount of money would be required. Second, scientists to work in the institute would have to be drawn from other institutions, thereby possibly causing a shortage of teaching faculty. Faculty shortages could be partially alleviated if the institute were located near a major research university or land grant college. Third, it is not obvious what agency would administer the institute. DOE is one choice because it already has experience with one institute. USDA is another choice, but recent studies (see preceding footnote) have suggested that the research stations it already administers have not kept up-to-date with the latest molecular techniques being applied to plants. NIH, which is well versed in molecular biology, is not an ideal agency to administer an essentially agricultural program. NSF might be a candidate to administer a new plant research institute because of its interdisciplinary staff.

OPTION 2: Establish grants for cooperative research between classical and molecular plant biologists from different institutions.

*The administration of basic research in agriculture has recently been reviewed by several agencies (5,34,41). Changes in the administration of USDA research will be extremely important to the direction of development of biotechnology in agriculture. A proposal within USDA to significantly increase the competitive grants in plant biology has recently been published (25). However, an assessment of the USDA technical and administrative infrastructure is beyond the scope of this report.

An increase in funding alone would facilitate interaction between classical and molecular plant biologists. Because of its interdisciplinary focus, NSF might be the agency to administer these grants.

Careful specification of requests for proposals and monitoring of the grants by technically qualified staff would be needed to ensure that the research that is funded is truly cooperative. Otherwise, some researchers experiencing difficulties in obtaining research funding might be tempted to cooperate in proposal writing in order to obtain a grant and then carry out independent research.

ISSUE 3: How could the retraining of industrial personnel in biotechnology be improved?

The OTA/NAS survey of companies using biotechnology in the United States shows that there is little retraining of personnel in this field. This situation is probably due, in part, to the fact that many of the U.S. companies using biotechnology are small and have neither the resources nor incentives to retrain personnel. These small companies depend on their ability to attract already highly qualified personnel. However, the pharmaceutical industry has shown that chemical engineers can be retrained in bioprocess engineering.

Continuing education sponsored by large U.S. companies, in general, takes place through short courses or joint research performed at universities. University/industry training and research agreements in biotechnology are being developed without the assistance of the Federal Government. But the Government could further encourage retraining in biotechnology by increasing funding for NSF's Industry/University Cooperative Centers Program, which provides seed money for a university to set up a research center with industrial partners. This option is discussed in *Chapter 13: Government Funding of Basic and Applied Research.*

Whether human resources in the United States are used and retrained adequately is a larger, national question that addresses the transition of the U.S. labor force from declining to growing industrial sectors. Suggestions to encourage more retraining have included revision of the tax code to encourage business loans to employees for retraining and an extension of unemployment insurance to include payment for retraining. The comparative evaluation of measures such as these that include other disciplines is beyond the scope of this study.

Chapter 14 references

1. Abelson, P. H., "Science and Engineering Education," *Science* 210:965, 1980.
2. Advisory Council for Applied Research and Development, Advisory Board for the Research Councils, and The Royal Society, "Biotechnology: Report of a Joint Working Party" (Spinks' Report) (London: H. M. Stationery Office, March 1980).
3. *AFP Sciences,* "Scientific Research Policy and Organization," July 22, 1982, pp. 4-5.
4. *ASM News,* "Shockman Testifies on Manpower Needs . . . ," 48(8):349-350, 1982.
5. Bentley, O., "Perspective on Agricultural Research," *Science* 219:4584, 1983.
6. *Chemical and Engineering News,* "Foreign Precollege Curricula Gives Strong Dose of Science," Jan. 25, 1982, p. 42.
7. Demain, A., Professor of Industrial Microbiology, Massachusetts Institute of Technology, personal communication, 1983.
8. Demain, A. L., and Solomon, N. A., "Responsibilities in the Development of Biotechnology," *Bioscience* 33(4):239, 1983.
9. Dunnet, J. S., "Brain Drain Threatens Progress," *Nature* 302:644, 1983.
10. Feinberg, L., "Panel Urges Measures To Halt Decline of Education in America," *Washington Post,* Apr. 27, 1983, p. 1.
11. Feldman, M., and O'Malley, E. P., *The Biotechnology Industry in California,* contract paper prepared for the California Commission on Industrial Innovation, Sacramento, Calif., August 1982.
12. Gaden, E., "Bioprocess Technology: An Analysis

and Assessment," contract report prepared for the Office of Technology Assessment, U.S. Congress, July 1982.
13. Gaden, E., University of Virginia, personal communication, 1983.
14. Gesellschaft für Biotechnologische Forschung (Society for Biotechnological Research) *Programmbudget* (Program Budget), Braunschweig, 1982.
15. Greer, W. R., "Foreign Students: Boon or a Threat?" *New York Times,* Mar. 27, 1983, sect. 12, p. 72.
16. Herman, R., "Science Slips in the Land of Opportunity," *New Scientist,* July 14, 1983, pp. 110-113.
17. Hochhauser, S. J., "Bringing Biotechnology to Market," *High Technology,* February 1983, pp. 55-60.
18. *Japan Economic Journal,* "Researchers From 4 Companies Study at Genex," Jan. 26, 1982, p. 10.
19. Japanese External Trade Organization (JETRO) mission, meeting of Japanese corporate, university, and government officials in biotechnology-related areas with the Biological Applications Program, Office of Technology Assessment, U.S. Congress, Washington, D.C., November 1982.
20. Jasanoff, S., "Public and Private Sector Activities in Biotechnology in Switzerland," contract report prepared for the Office of Technology Assessment, U.S. Congress, January 1983.
21. Jasanoff, S., "Public and Private Sector Activities in Biotechnology in the Federal Republic of Germany," contract report prepared for the Office of Technology Assessment, U.S. Congress, January 1983.
22. Knox, R., "Basic Research a Biotech Barrier," *Boston Globe,* p. 6 of Special Report, Dec. 5-10, 1982.
23. Maddox, J., "Research Labs Galore," *Nature* 297:279, 1982.
24. Ministère de la Recherche et de l'Industrie (Ministry of Research and Industry), Mission des Biotechnologies (Biotechnology Mission), "Programme Mobilisateur: l'Essor des Biotechnologies" (Mobilizing Program: The Expansion of Biotechnology), Paris, July 1982.
25. National Association of State Universities and Land-Grant Colleges, "Emerging Biotechnologies in Agriculture: Issues and Policies," Progress Report II, November 1983.
26. National Science Foundation, *Science Indicators, 1980,* Report of the National Science Board, Washington, D.C., 1981.
27. National Science Foundation, "Special Initiative: Biotechnology," Washington, D.C., July 15, 1982.
28. National Science Foundation, "Biotechnology Initiative," Washington, D.C., Apr. 30, 1982.
29. Organisation for Economic Co-Operation and Development, *Biotechnology: International Trends and Perspectives,* Paris, 1982.
30. Pearson, R., Institute of Manpower Studies, University of Sussex, personal communication, June 1983.
31. Peterson, B., "Academic Gap Found Narrowing," *Washington Post,* Feb. 8, 1983, p. 1.
32. Pierce, K., "Bull Market For Engineers," *Time,* Nov. 10, 1980, p. 69.
33. Pinon, J. C., Scientific Attaché to Embassy of France, Washington, D.C., May 1983.
34. Rockefeller Foundation and Executive Office of the President, Office of Science and Technology Policy, *Science for Agriculture* (New York: Rockefeller Foundation, October 1982).
35. Saxonhouse, G., "Biotechnology in Japan," contract report prepared for the Office of Technology Assessment, U.S. Congress, June 1983.
36. Schambra, P., National Institutes of Health, Bethesda, Md., personal communication, 1983.
37. Strack, G., "Bonn Entices Bioscientists," *Frankfurter Rundschau,* Frankfurt/Main, February 1983, translated by the Foreign Broadcast Information Service in *West Europe Report: Science and Technology* (Joint Publications Research Service, Arlington, Va.)
38. Tangley, L., "Science Education in U.S. Criticized," *Science News,* May 22, 1982, p. 341.
39. U.S. Congress, Library of Congress, Congressional Research Service, "U.S. Science and Engineering Education and Manpower: Background; Supply and Demand; and Comparison with Japan, the Soviet Union, and West Germany," report prepared for the Subcommittee on Science Research and Technology, Committee on Science and Technology, U.S. House of Representatives, April 1983 (Washington, D.C.: U.S. Government Printing Office, 1983).
40. U.S. Congress, Office of Technology Assessment, *Impacts of Applied Genetics: Micro-Organisms, Plants, and Animals,* OTA-HR-132, Washington, D.C., April 1981.
41. U.S. Congress, Office of Technology Assessment, *An Assessment of the U.S. Food and Agricultural Research System,* OTA-F-155, Washington, D.C., December 1981.
42. U.S. Congress, Office of Technology Assessment, *International Competitiveness in Electronics,* OTA-ISC-200, Washington, D.C., November 1983.
43. U.S. Department of Commerce, *An Assessment of U.S. Competitiveness in High-Technology Industries,* Washington, D.C., February 1983.
44. Vaquin, M., "Biotechnology in France," contract report prepared for the Office of Technology Assessment, U.S. Congress, February 1983.
45. Vaquin, M. "Biotechnology in Great Britain," contract report prepared for the Office of Technology Assessment, U.S. Congress, February 1983.

46. Walgate, R., "La Loi at Last," *Nature* 298:110, 1982.
47. Walsh, J., "Trouble in Science and Engineering Education," *Science* 210:215, 1980.
48. Walsh, J., "New Slant on Engineering Training," *Science* 218:269, 1982.
49. Worthy, W., "Classroom Crisis in Science and Math," *Chemical and Engineering News,* July 19, 1982, pp. 9-16.

Chapter 15

Health, Safety, and Environmental Regulation

Contents

Chapter 15

Health, Safety, and Environmental Regulation

Introduction

Regulation has been and will continue to be a factor in the development of biotechnology, especially for recombinant DNA (rDNA) processes and products. When the rDNA technique was first developed, its novelty and tremendous power to manipulate organisms raised the specter of potentially drastic consequences to human health and the environment through the creation and proliferation of organisms with unknown but potentially hazardous traits. In the United States, therefore, Congress moved to develop stringent regulation of rDNA. This movement was forestalled in part by the adoption in 1976 of fairly restrictive self-regulatory guidelines by the scientists (27).

As time passed, however, concern and fears diminished greatly. As scientists learned more about molecular genetics, perceived risks associated with probing the unknown diminished, and no evidence was discovered to support many of the early risk scenarios. Formal risk assessment studies also led to downward evaluation of potential risk. Molecular biologists gained the confidence of the public by bringing other experts and the public into the decisionmaking process that established the system of voluntary self-regulation. And, most importantly, there has been no evidence of any harm to human health or the environment from rDNA. Consequently, the requirements of the rDNA guidelines in the United States have been substantially relaxed.

Today, most experts believe that the potential risks of rDNA research were drastically overstated and that rDNA technology generally does not involve a risk beyond that already inherent in the host, vector, DNA, solvents, and physical apparatus being used (35). This is not to say, however, that biotechnology—like most new technologies—does not continue to raise special concerns or present special risks. In particular, questions have been raised about the long-term effects on workers' health from exposure to novel organisms and products and about the risks of deliberately releasing genetically manipulated organisms into the environment. In addition, some of the products that will be made by biotechnology may present special risks. For example, the U.S. Food and Drug Administration (FDA) has been concerned about bacterial endotoxins found in drugs produced by *Escherichia coli* (28).

Regulation will have a moderately important effect on the development of biotechnology and, consequently, on U.S. competitiveness in biotechnology. Special risks may lead to limited new regulation that could direct commercial efforts away from certain areas or at least slow advancements in those areas. In addition, most of the products that could be made by biotechnology and associated processes are already subject to considerable regulation, pharmaceuticals and chemicals being the best examples. This existing regulation also will affect corporate strategies and patterns of industrial development.

The costs and time involved in complying with regulatory requirements are the price society pays for safety. However, unreasonable restrictions and unnecessary burdens may delay or prevent important products from reaching the market or may increase the business risks of developing those products. Uncertainties, for example, about what the regulatory requirements will be or which agencies have jurisdiction, will also affect the risk, time, and cost of product development. Those countries that have the most favorable regulatory environment in terms of least restrictions and uncertainties will have a competitive advantage in the commercialization of biotechnology.

This chapter evaluates the regulatory environment for the commercialization of biotechnology in the United States and five competitor countries being examined in this assessment—the Federal Republic of Germany, the United Kingdom, France, Switzerland, and Japan. Two specific factors are considered in the evaluation: 1) the restrictiveness of the regulation, and 2) the uncertainties with respect to possible agency jurisdiction or requirements. Congressional options for improving U.S. competitiveness in biotechnology through changes in the regulatory environment are presented at the end of the chapter.

In the analysis that follows four areas of regulation are considered:

- regulation directed specifically toward biotechnology;
- existing regulation that would apply to biotechnology products;
- environmental regulation relevant to biotechnology; and
- worker health and safety regulation.

The chapter concentrates on the guidelines for rDNA research adopted by the competitor countries and the approval requirements for pharmaceuticals (human drugs and biologics) and for veterinary medicines (animal drugs and biologics). The guidelines for rDNA research merit significant attention because they are the only type of governmental oversight developed specifically for biotechnology. The approval requirements for pharmaceuticals and veterinary medicines also merit attention because those products are subject to the most restrictive regulation, even when made by conventional means,* and because so much of the current activity in biotechnology is directed toward those types of products. In addition, with respect to regulation of products in other countries, most of the information OTA was able to obtain related to the approval process for pharmaceuticals and veterinary medicines. Sufficient information on foreign regulation of food, food additives, medical devices, and chemicals was not available for meaningful international comparisons; however, this information is included for the United States because of its availability and because of the interest in it.

Two inherent limitations could qualify the analysis in this chapter. The first results from the difficulty of determining and interpreting foreign laws and especially the rules and policies of the foreign agencies. Much of this material is not readily available in English or even in the native language. In addition, enforcement of laws and regulations in other countries generally is much more discretionary than in the United States.* Thus, there may be a wide gap between the written laws and regulations and the actual regulatory environment in which foreign companies operate. The second limitation results from the fact that the analysis does not consider the positive effects of regulation and a country's track record for safety. In other words, the restrictiveness of regulation theoretically should be balanced against some measure of the harm avoided. However, the necessary data are generally not available, and such an analysis is beyond the scope of the chapter.

*Significant regulation also exists for commodity and specialty chemicals (including herbicides and pesticides), but it is generally not as restrictive as for pharmaceuticals and some types of veterinary medicines. The use of genetically modified organisms in the environment will probably face some moderate degree of regulation. Agricultural products currently face little health, safety, or environmental regulation, but this situation could change in the case of genetically modified plants and animals.

*In fact, this discretion has led to claims of selective enforcement against U.S. companies, thus creating a nontariff trade barrier. For discussion of other nontariff trade barriers, see *Chapter 19: International Technology Transfer, Investment, and Trade.*

Regulation directed specifically toward biotechnology: rDNA research guidelines

The only oversight mechanism directed specifically toward biotechnology is the rDNA research guidelines. These guidelines grew out of the concerns in the mid-1970's about potential risks of rDNA research and the desire to proceed cautiously in the face of the uncertainties. Guidelines similar to the National Institutes of Health Guidelines for Research Involving Recombinant DNA Molecules (NIH Guidelines) in the United States have been adopted by Japan, the Federal Republic of Germany, the United Kingdom, France and Switzerland. Over time, they have been substantially relaxed worldwide in a series of revisions that reflect decreasing concern about the risk. In fact, many types of experiments involving rDNA are now exempt from the guidelines. The guidelines are essentially self-regulatory.

The guidelines for rDNA research reflect the decision by experts and policymakers that rDNA research presents some special risks and uncertainties that require special attention. They are based on two underlying concepts:

- rDNA research should be conducted at increasing levels of physical and biological containment related to the degree of possible hazard, and
- the degree of oversight should relate to the degree of possible hazard.

The implementation of these concepts is fairly similar in the competitor countries, because the worldwide scientific community was involved in their development and because most countries followed the lead of the United States. Nevertheless, there are some important differences among the guidelines adopted in the various countries, and different countries are at different stages in the process of relaxing them.

This section surveys the rDNA research guidelines of the six competitor countries with respect to their scope, containment requirements, approval requirements, and enforcement mechanisms in order to assess their impact on competitiveness in biotechnology.* The commercial development of biotechnology in many of these countries, however, will depend less on the specific biological and physical containment measures required by their rDNA research guidelines than on the scope of activities reached by the guidelines (i.e., whether they cover large-scale research) and the structure set up for implementing and enforcing the guidelines. The analysis presented here is based on the more detailed description of the rDNA research guidelines of the six countries and the European Economic Community found in *Appendix F: Recombinant DNA Research Guidelines, Environmental Laws, and Regulation of Worker Health and Safety,* which the reader is urged to examine.

*Provisions relating specifically to worker health and safety are discussed in the section of this chapter entitled "Regulation of Worker Health and Safety."

Scope

In the United States, Japan, and France, the guidelines technically apply only to government-funded rDNA research, while in Switzerland, the Federal Republic of Germany, and the United Kingdom, they apply to all rDNA research. (Actually, all the guidelines also apply to large-scale rDNA work to varying degrees, as discussed below.) While U.S., Japanese, and French private laboratories might seem to have some advantage over private laboratories in the other countries because they could dispense with safety measures perceived to be unnecessary, this "advantage" is probably illusory. Industry perceives compliance with the guidelines to be in its best interest, and there has been no publicized evidence of noncompliance.

Perhaps the single most important issue for companies using biotechnology is the rDNA guidelines' treatment of large-scale research (i.e., work with cell cultures in volumes exceeding 10 or 20 liters), which is a necessary step in successful commercial development. The guidelines in Japan are easily the least favorable in this regard. Recombinant DNA research with volumes exceeding 20 liters can be conducted in Japan only after Government permission, and that permission has been quite difficult to obtain.* It should be noted, however, the situation in Japan is expected to change shortly.** Under the U.S. guidelines, the large-scale work need only be reviewed by each

*Six companies have obtained permission for large-scale work (14).

**The Council for Science and Technology, which advises the Prime Minister and oversees rDNA work by private institutions in Japan, is expected to recommend the elimination of the prohibition of large-scale work without special Government approval. Instead, large-scale bioprocess facilities would classify into two categories, LS1 and LS2. LS1 facilities would be covered by rules similar to those for conventional microbiological laboratories. LS2 facilities, which would involve work with more hazardous micro-organisms, would be covered by more stringent rules. The Prime Minister is expected to act favorably on the recommendation in August 1983.

Institutional Biosafety Committee (IBC), although NIH has made specific recommendations regarding physical containment, which were recently incorporated into the U.S. guidelines. Large-scale research in the United Kingdom is treated on a case-by-case basis by the supervising authority, the Genetic Manipulation Advisory Group (GMAG).* But in explaining the need for a different kind of review of large-scale research, GMAG has suggested that large-scale research will not be subject to as stringent containment measures as smaller scale research. The French rDNA guidelines exclude large-scale research from their coverage, but the Government's oversight agency will apparently consider such activity on a case-by-case basis. The West German guidelines do not mention large-scale research. The Swiss guidelines permit scaling-up without special approval; it is unclear whether the small-scale rules continue to apply or whether, as with the NIH guidelines, large-scale research is subject to the safety measures decided on by the IBC.

Containment requirements

Each country's rDNA guidelines specify requirements for physical and biological containment of the research organisms. Except for the United Kingdom, each country assesses risk in the same manner—according to the source of the DNA used in the experiment and the pathogenicity of the host-vector system. The United Kingdom determines risk by considering the survivability and likely harm of the organism containing rDNA. Whether this risk assessment method gives the United Kingdom an advantage or disadvantage depends on the particular experiment. The United Kingdom does have an advantage with respect to rDNA production of insulin and interferon, which are classified at a lower containment level there than in the United States (8). Each country uses four levels of physical containment. Most research is now conducted at the lowest physical containment level.

*GMAG's status was recently reviewed by the Health and Safety Executive, and the subsequent report recommended relocation of the group from the Department of Education and Science to the Department of Health and Social Security. GMAG has been moved and is now called the Health and Safety Commission Advisory Group on Genetic Manipulation.

The physical and biological containment measures required for an experiment vary slightly from country to country, but it is difficult to determine what effect on a country's competitive position any one requirement might have. It is difficult to determine, for example, what effect will come from the fact that at the United Kingdom's physical containment level II, a continuous air flow into the laboratory is required, while it is not required in other countries until the third containment level. The measures with the greatest impact are probably the biological containment rules in Japan, which severely restrict the types of organisms that can be used in host-vector systems. These restrictions may prevent commercially promising rDNA research from going forward.

Approval requirements

Notice and approval requirements depend on the risk of the experiment. Research in the United States at the highest risk level is subject to the approval of NIH and the appropriate IBC; at the next level, only IBC approval before initiation is necessary. IBC notification at the time of initiation is required for some lower level risk experiments, while many are exempt entirely. More than 85 percent of all rDNA work in the United States is done at the lowest containment levels (23), and virtually all monitoring of rDNA work is done by IBCs.

The recommendation of the European Economic Community (EEC) on rDNA research suggests that notice of experiments be given to the central authority in each member state, usually before the work begins. For some types of research, notice would not have to be made before work is begun. The United Kingdom, France, and the Federal Republic of Germany are members of the EEC.

In the United Kingdom, the Health and Safety Executive (HSE) is directed to inspect the facilities for rDNA research at the two higher containment levels, categories III and IV. For research at these levels, GMAG also must have notice and an opportunity to give advice. Advance notice is required for research at the category II level but not approval. Activities at the category I level can go forward provided only that the local safety committee notifies the central authorities once a year of new research. Companies in the United Kingdom also have to deal with two separate agencies: GMAG, which promulgates and monitors the rules, and the HSE, which enforces them.

Scientists in France must notify the French Control Commission (Commission de Contrôle) of planned research. This commission must approve certain high-risk research. Local safety committees monitor the research.

In the Federal Republic of Germany, the Central Commission for Biological Safety (Zentrale Kommission für die Biologische Sicherheit) must be notified of all research except that at the lowest level of containment. This requirement makes for one of the most restrictive approval processes in the countries surveyed. Experiments at the two high-risk levels require the entire commission's approval, while those at the second lowest containment level must be approved by one or two individual members of the commission. The Commission for Biological Safety must also authorize the use of host-vector systems not enumerated in the rDNA research guidelines and may approve reductions in levels of containment employed.

Switzerland, where rDNA research is now conducted under guidelines that are essentially equivalent to the April 1982 NIH Guidelines (34), differs from the United States in an important respect. The research is overseen by a commission created by the Swiss Academy of Medical Sciences. The commission, as a private entity, may be more willing than NIH to modify requirements for projects with which it is familiar.

In Japan, two different bodies monitor rDNA research, the Council for Science and Technology, which supervises activities by private institutions, and the Science Council (in the Ministry of Education), which monitors the activities of public institutions such as universities. The Science Council is not required to approve university experiments, which may go forward simply on the approval of the president of the university and the university safety committee. However, it must approve the use of hosts other than those specified in the guidelines. Only a limited number of hosts and vectors have been approved for use, which puts Japan at a competitive disadvantage.

Enforcement

In all of the countries except the United Kingdom, the only direct sanction for noncompliance with the rDNA research guidelines is the ability of the government to restrict or withdraw funding for an institution's or a scientist's rDNA research. The guidelines in the United Kingdom are promulgated under the Health and Safety at Work Act of 1974 and are backed-up by the general legal sanctions created by that act.

Effect on competitiveness

The commercial effect of the rDNA research guidelines is difficult to assess, because their effect depends on the specific research done and because commercial exploitation of rDNA research has only recently begun. With the exception of Japan and possibly the Federal Republic of Germany, no country's rDNA research guidelines place it in a noticeably disadvantageous position. However, the U.S. rDNA research guidelines are probably the least restrictive of the six competitor countries. The European countries and Japan have generally followed the U.S. guidelines but are often following earlier, more restrictive versions.

Existing regulation of biotechnology products

A comparative assessment of the regulation of biotechnology products in the competitor countries involves two stages. Since biotechnology products generally will be subject to existing regulation for generic products, it is first necessary to compare these general regulatory regimes. In other words, biotechnologically made pharmaceuticals, for example, will be subject to the general regulations covering pharmaceuticals, regardless of how they are made; thus, comparing the pharmaceutical laws of the different countries will provide information about competitiveness. In this context, the following questions are particularly relevant:

- How much time and effort does it take to get products through the approval process?
- What is the usual or average cost for securing regulatory approvals?
- What are the import and export restrictions on approved and unapproved products?
- Will the regulatory authorities accept foreign test data in the approval process?

The second stage of the analysis involves looking at specific issues raised by biotechnology. Some of these are the following:

- Will new biotechnology products chemically identical to approved products made by other means still be required to go through the full regulatory review process?
- Will the classification of a pharmaceutical as a drug or biologic affect the time or cost of securing regulatory approval?

United States

Three Federal agencies will be most involved in regulating biotechnology products. They are the Food and Drug Administration (FDA), the U.S. Department of Agriculture (USDA), and the Environmental Protection Agency (EPA).

FOOD AND DRUG ADMINISTRATION

FDA regulates drugs, biologics, food, food additives, and diagnostics pursuant to the Federal Food, Drug, and Cosmetic Act (FFDCA) (21 U.S.C. §§301-392) and section 351 of the Public Health Service Act (21 U.S.C. §262).

Since the first commercial applications of biotechnology (i.e., pharmaceuticals) have been in areas subject to FDA jurisdiction, FDA is the agency having the most experience with biotechnology products. FDA has approached rDNA-produced products on an agencywide basis by creating a Recombinant DNA Coordinating Committee, composed of representatives of its centers and bureaus, the Office of General Counsel, and Office of Regulatory Affairs. FDA's Recombinant DNA Coordinating Committee has determined that rDNA products whose active ingredients are identical to ones already approved or to natural substances will still have to go through the new product approval process. Data requirements may be modified and often abbreviated, however, and each case will be handled on an ad hoc basis.* (In the case of many conventionally produced products, abbreviated review procedures are available when the active ingredient of the new product is identical to one already approved or to natural substances.) FDA will not require compliance with the NIH Guidelines as a condition of approval. For monoclonal antibody (MAb) products, no coordinating body similar to the Recombinant DNA Coordinating Committee exists; FDA's policy for these products has been set by the National Center for Devices and Radiologic Health (NCDRH) and the Office of Biologics. Actual product regulation will occur at the individual bureaus or offices as discussed below.

Human Drugs.—FDA's Office of New Drug Evaluation has taken the position that drugs made by rDNA technology, even if identical to currently approved drugs, are "new drugs."** Therefore, such drugs cannot be marketed until approved by FDA as safe and effective.

FDA's approval process for a new drug can take several years because it requires a series of animal and human tests. Clinical investigations can be carried on only after a drug's sponsor files a Notice of Claimed Investigational Exemption for a New Drug (IND). The IND contains the results of animal testing, a description of the planned clinical investigations, and other information. The preclinical investigations generally last from 1 to 2 years (20). The human studies then go through

*FDA has been concerned about bacterial endotoxins and immunogens contaminating the products and about the genetic stability of the rDNA organism. In the latter case, the product might be affected if the DNA underwent changes.

**A new drug is a drug whose composition is not generally recognized by qualified experts as safe and effective under the conditions of use set forth in its labeling or, even if so recognized, has not been used to a material extent or for a material time (sec. 201(p) of the Federal Food, Drug, and Cosmetic Act; 21 U.S.C. §321(p)). A drug is a substance intended for use in the diagnosis, treatment, or prevention of disease or which is intended to affect the structure or function of the body (sec. 201(g) of the Federal Food, Drug, and Cosmetic Act; 21 U.S.C. §321(g)).

three phases to establish safety, set dosage levels, and establish efficacy. This clinical testing often takes 5 to 6 years (20). During or after the clinical studies, the sponsor files a New Drug Application (NDA), which contains the results of animal and human testing, a statement of the drug's composition, a description of the methods and controls used in its manufacture, and other information. The time required for processing an NDA depends on the completeness of the data, the drug's performance, and the speed of FDA review. In 1980, the duration of the NDA phase for new chemical entities varied from about 1 to 7 years and averaged slightly less than 3 years (20).* Taking into account the research and development (R&D) costs of drugs that fail to reach the market, various economic analyses indicate that the R&D costs per marketed new chemical entity range from $54 million to over $70 million (11).

There are abbreviated approval procedures that FDA might eventually permit sponsors to use after it gains more familiarity with rDNA technology and if warranted by the risks. One is the Supplemental New Drug Application (SNDA), which is required when an NDA holder intends to market the drug under conditions materially different from those approved in the NDA. An SNDA could become available in the case where the manufacturer of an approved drug made by chemical synthesis decides to make the drug by using rDNA and bioprocess techniques. A second procedure is the Abbreviated New Drug Application (ANDA), which is available for generic versions of drugs first marketed between 1938 and 1962. An ANDA might be used by a manufacturer using rDNA techniques to make an approved drug made by conventional techniques by another manufacturer. The final procedure is a "paper" NDA, available for generic copies of drugs marketed after 1962. Such drugs require an NDA, but FDA is willing to accept published reports demonstrating safety and efficacy, thus saving the new sponsor the time and costs of clinical trials. A "paper" NDA could become available in the case where a manufacturer wants to make an rDNA-produced drug whose NDA is held by another manufacturer, if adequate data are available in the published literature to establish safety and effectiveness.

*A General Accounting Office study of the U.S. drug approval process found that for 132 NDAs submitted to FDA in 1975, the average approval time was about 20 months (20).

Human Biologics.—A biologic is a vaccine, therapeutic serum, toxin, antitoxin, or analogous product for the prevention, treatment, or care of diseases or injuries. The distinction between a drug and a biologic is largely historical and bureaucratic and is becoming even more blurred with the advent of biotechnology.

Although biologics also come within the definition of drugs in section 201(g) of the Federal Food, Drug, and Cosmetic Act (FFDCA), they primarily are regulated under section 351 of the Public Health Service Act and by FDA's Office of Biologics rather than the Office of New Drug Evaluation.* Section 351 creates a regulatory structure for biologics similar to that for drugs. However, it is a licensing procedure; both the product and the establishment where it is produced must be licensed. At the investigational stage, the Office of Biologics follows the requirements for INDs. After clinical trials, the procedure involves a license application for the establishment and for the product; together they provide essentially the same information as required by an NDA. Differences, however, occur in practice. The Office of Biologics generally has been perceived to be more flexible than the Office of New Drug Evaluation. It often uses informal, unpublished guidelines, or "regulatory memoranda."** On the other hand, it is the administrative practice of the Office of Biologics to require lot by lot approval of many biologics before they are released by the manufacturer, which is not usually required by the Office of New Drug Evaluation (1).

Biologics made by biotechnology will have to go through the approval process outlined above. In accordance with announced policy, rDNA-produced biologics, even if chemically identical to approved biologics, will have to go through the

*The Office of Biologics also regulates diagnostics related to blood bank products. All other diagnostics, including most of those incorporating monoclonal antibodies (MAbs), are regulated by FDA's National Center for Drugs and Radiologic Health (NCDRH). The first MAb diagnostic kits related to blood products and were approved by the Office of Biologics.

**It has published three about biotechnology. One covers MAb diagnostic kits for blood bank related products (31). Another covers MAbs for use in human therapy (33). A third covers the production and testing of interferon (32).

full approval process, but data requirements may be lessened. For MAbs, there has been no announced policy, but virtually all of those that would be used for therapeutic purposes would be truly new and therefore have to go through the full review process.

Medical Devices.—Medical devices are regulated by FDA's National Center for Devices and Radiologic Health (NCDRH), except for those in vitro diagnostic products used in connection with blood banking activities such as tests for hepatitis B surface antigen. Those products are regulated by the Office of Biologics.

The Medical Device Amendments to FFDCA in May 1976 required that all devices for human use marketed before the amendments be classified by FDA into one of three categories on the basis of recommendations by expert panels. Class I products are subject to general controls, such as good manufacturing practice regulations. Class II devices are required to meet performance standards in addition to the general controls. Class III devices require FDA premarket approval for safety and effectiveness. For devices marketed after May 1976, those that are "substantially equivalent" to a preamendment device are classified with that product, and those that are not substantially equivalent are placed in Class III. Under section 510(k) of the act, manufacturers are required to give FDA a 90-day notice before they can market a device, during which period FDA determines whether the device is substantially equivalent to a preamendment device.

Manufacturers of MAb diagnostic kits generally have been successful in using the 510(k) notice procedure to get their products to the market quickly. Although MAbs are different from and generally superior to polyclonal antibodies for diagnostic purposes, applicants have been successful in showing that MAbs are "substantially equivalent" to polyclonal antibodies marketed before May 1976. That is, the applicants have demonstrated to the satisfaction of NCDRH that the MAbs provide essentially the same (or better) results as polyclonal antibodies used for the same diagnostic purposes (1). Since the review panels of experts required by the statute have placed most preamendment diagnostic kits in Class II (1), the new MAb kits have been placed in Class II, which requires certain performance standards to be met, rather than Class III, which would require the manufacturer to demonstrate safety and efficacy.* The availability of the 510(k) application is highly desirable from a company's perspective because NCDRH must respond within 90 days.

Food and Food Ingredients.**—The distinction between food and food ingredients (substances added to food) is important in terms of the regulatory approval process. Food can be marketed without FDA clearance, but food ingredients are subject to the food additives provisions of FFDCA, which may require premarketing approval. FFDCA defines food broadly and circularly as food or any component thereof (sec. 201(f)). A food additive is defined as a substance that may, by its intended use, become a component of food or affect the characteristics of food (sec. 201(s)). This definition excludes, among other things, substances generally recognized as safe (GRAS) by qualified experts and certain prior-sanctioned (previously approved) substances. A new food additive requires premarketing clearance by FDA, and its sponsor has the burden of demonstrating its safety. Favorable action by FDA results in a published regulation stipulating the concentration and other conditions under which the additive may be used. GRAS substances technically can be marketed without prior approval by FDA, but also can be the subject of published FDA regulations.***

FDA's Bureau of Foods has not been confronted with any foods, food additives, or GRAS substances produced by rDNA techniques; however, on the basis of the announced policy of FDA's Recombinant DNA Coordinating Committee and discussions with the staff, the Bureau appears likely to take the following positions. If FDA were concerned about the safety of such a food, high lysine corn, for example, it could take various

*If a MAb kit were placed in Class III, the sponsor could petition for a reclassification to Class II; however, such reclassifications are supposedly difficult to obtain.

**This section uses the term food ingredient instead of the term food additive used in other chapters, because the term food additive has a particular meaning under FFDCA. As explained in this section, under FFDCA a food additive is one type of food ingredient (substance added to food).

***FDA publishes lists of what it considered to be GRAS substances and sometimes it will consider a substance GRAS only when used under certain conditions.

steps to prevent its sale or remove it from the market by proving it was "ordinarily . . . injurious to health" and, therefore, was adulterated within the meaning of section 402(a) of FFDCA. It might be able to require premarketing clearance if the corn were used as an ingredient in other foods, such as stew, because then it would be subject to the food additive requirements (21 C.F.R. §170.30 (f)). Recombinant DNA products that are similar or chemically identical to GRAS substances or food additives already approved for use will be required to go through the approval process by FDA's Bureau of Foods, although the Bureau will be flexible on data requirements.

Animal Feeds, Feed Additives, and Devices.—These products are regulated in a way similar to the way in which human foods, food additives, and medical devices are regulated; however, the regulation for animal products is less rigorous than that for human products. For animal feeds and feed additives, the requirements for demonstrating safety are less than for the comparable case of human food and food additives. In the case of animal feed additives, however, there is an additional requirement that they be shown to be safe to people consuming edible products from animals receiving the additive. For animal devices, there is no premarket approval requirement as there is for many human devices. At this time, there is no reason to expect any particular regulatory problems if these products are made by biotechnology.

Veterinary Medicines.—For veterinary medicines (animal drugs and biologics), FDA's authority is similar to its authority for human drugs or biologics with two exceptions. First, there is an additional requirement in the animal drug approval process, i.e., animal drugs must not leave unsafe residues or metabolites in edible tissues or other food products. Second, FDA does not have the primary regulatory authority over animal biologics; USDA regulates them under the Virus, Serum, Toxin Act of 1913 (VST Act) (21 U.S.C. §§151-158), even though they are also technically drugs under FFDCA. USDA's authority applies only to interstate marketing. According to a recent case, FDA has jurisdiction over intrastate marketing (10).

These jurisdictional distinctions have been blurred by rDNA and MAb technology. An FDA/USDA memorandum of understanding creates a standing committee to sort out regulatory responsibilities in this area (29).* The memo says FDA will regulate where the VST Act does not apply or does not offer an appropriate remedy.

The first product to be considered by the standing committee is bovine interferon. Both agencies claimed jurisdiction, and the committee has split along agency lines. Several attempts to resolve the impasse on scientific grounds have failed; however, efforts are continuing. In the meantime, the manufacturer has encountered additional costs and burdens by attempting to meet the requirements of both agencies (6).

Control Over Exports.—Under section 801(d) of FFDCA, unapproved food additives and medical devices can be exported if certain conditions are met.** Unapproved new human drugs or biologics and unapproved new animal drugs, however, cannot be exported except in the following two cases: 1) if the products are subject to an IND, providing the importing country's government has approved such imports; or 2) if the importing country's government formally requests through the U.S. Department of State that the product be exported (for purposes of clinical trials only) (21 C.F.R. §312.1(a)). As to unapproved animal biologics, there is some question about whether the VST Act applies to exports. Nevertheless, it is clear that FDA has authority over such exports, and, as indicated in the previously

*The FDA/USDA memorandum of understanding defines animal biologic products as those that "generally act through a specific immune process and are intended for use in the treatment (including prevention, diagnosis, or cure) of diseases in animals. Such products include but are not limited to vaccines, bacterins, sera, antisera, antitoxins, toxoids, allergens, diagnostic antigens prepared from, derived from, or prepared with micro-organisms, or growth products of micro-organisms, animal tissues, animal fluids, or other substances of natural or synthetic origin."

**An approved food additive can be exported if the exporter determines, without any need to inform or petition FDA, that the four conditions in sec. 801(d)(1) of FFDCA are met. The same is true for a Class I medical device, but an unapproved Class II or Class III medical device cannot be exported unless a petition has been submitted to FDA and FDA has found that the exportation is not contrary to the public health and safety of the importing country and has the approval of the importing country, under sec. 801(d)(2) of FFDCA.

discussed memorandum of understanding, FDA could exercise that authority.

The U.S. policy of restricting the export of unapproved drugs and biologics is essentially based on paternalism. Many countries do not have the mechanisms either to evaluate or to regulate the quality of the drugs they import. In addition, there have been cases of drug dumping—situations where drugs deemed unsafe or ineffective by the United States or other developed countries have been marketed in less developed countries (25).

This policy has several implications for U.S. companies using biotechnology, and the implications may differ depending on the size of the company. In part because of the export restrictions, several of the large U.S. pharmaceutical companies have established manufacturing facilities in foreign countries, where their products are approved or where the law permits the export of unapproved products. These actions result in the transfer of technology, lost employment opportunities for U.S. workers, and lost opportunity to help the U.S. international balance of payments. These consequences can be expected to continue with respect to biotechnology products. The existence of such facilities in foreign countries may provide the large companies with at least a short-term competitive advantage over small, new biotechnology firms (NBFs).* The vast majority of the latter companies do not have and probably cannot afford to establish foreign facilities.

The export restrictions will also have an important implication for NBFs and for U.S. competitiveness in general because they may foster technology transfer to foreign companies with which they have joint ventures. In their joint ventures with large foreign companies, some NBFs in the United States are required to provide bulk product produced by the micro-organism to the foreign partner, which would secure necessary approvals and purify, package, and market the drug in foreign markets. If the U.S. firm is unable to provide bulk product, the foreign partner then has the right to obtain the organism for its own use. The U.S. prohibitions on the export of unapproved drugs and biologics might be one reason why an NBF could not fulfill its agreement to supply bulk product to its foreign partner, thereby being required to transfer the organism and the technology.

In proposed revisions to the regulations governing the approval of new drugs, FDA has taken the position that bulk products, which it calls "drug substances," can be exported only if they are used in the manufacture of approved drugs and if certain labeling requirements are followed (30). FDA has proposed to define "drug substance" as "an active ingredient that is intended to furnish pharmacological activity or other direct effect in the diagnosis . . . treatment or prevention of disease. . . ." (30). This definition would cover drug products produced by biotechnology, even if they required purification, packaging, and labeling, because such products usually will be active. At least one NBF in the United States has argued that section 801(d) of FFDCA should not be interpreted to prohibit the export of such substances for purposes of clinical trials (if the conditions of sec. 801(d) are met) and that such an interpretation will require it to transfer technology for the reasons mentioned in the preceding paragraph (9).

This entire problem concerning the export of unapproved drugs can be avoided in the future, however, without changes in the law or regulations. As mentioned previously, the current U.S. regulations allow the export of unapproved drug substances upon the formal request of the importing country's government. NBFs in the United States rightly point out that such requests are unlikely in cases where the government is actively seeking to encourage inward technology transfer. However, the NBFs' licensing agreements with foreign companies could be written so that the NBFs would not have to transfer the technology if the foreign company's government did not make the necessary request.

Imported Pharmaceuticals and Foreign Test Data.—Imported pharmaceuticals must meet FDA's IND and NDA requirements, even if approved for clinical testing or marketing in a foreign country. A question naturally arises regarding the acceptability of foreign test data.

*NBFs, as defined in *Chapter 4: Firms Commercializing Biotechnology,* are firms that have started up specifically to capitalize on new biotechnology.

Currently, FDA will accept foreign clinical data in support of an NDA, but it very seldom approves an NDA solely on the basis of foreign data, even if the study that produced the data meets FDA requirements for well-conducted studies. Under proposed revisions to its regulations, FDA would consider approving NDAs based solely on foreign clinical trials on a case-by-case basis if: 1) the data are applicable to the U.S. population and U.S. medical practice; 2) the studies have been performed by investigators of recognized competence; and 3) FDA is able to assure itself of the validity of the data (30).

If adopted, the revised data requirements would have at least two implications for this country's competitiveness in biotechnology. First, they would allow large U.S. drug companies to continue their practice of conducting much of their clinical work in foreign countries where drug approval has been quicker than in the United States, but also to secure quicker drug approvals in the United States. Second, they would lessen a U.S. nontariff trade barrier faced by foreign firms.

U.S. DEPARTMENT OF AGRICULTURE

Under the VST Act, the manufacturer of an animal biologic to be sold interstate needs premarket clearance by getting licenses for the product and the factory from USDA. The agency has broad authority to require any data it thinks necessary to judge product identity, purity, safety, and efficacy. USDA regulation is generally seen as significantly less costly and time-consuming than FDA regulation. However, USDA's position on biotechnological products appears to be consistent with FDA's, i.e., such products will need a new license, even if identical to other licensed products, although data requirements may be lessened.

ENVIRONMENTAL PROTECTION AGENCY

EPA has extremely broad authority over chemicals, herbicides, and pesticides. Chemicals are covered by the Toxic Substance Control Act (TSCA) (15 U.S.C. §§2601-2629). TSCA is intended to fill gaps in other environmental laws. It authorizes EPA to acquire information on "chemical substances" in order to identify and evaluate potential hazards and then to regulate the production, use, distribution, and disposal of those substances. Commodity and specialty chemicals made by biotechnology (except those regulated under FFDCA) will face the same kind of regulation under TSCA as those chemicals made by conventional means. TSCA will also be applied to organisms used in the environment, as noted in the "Environmental Regulation" section below.

Pesticides, herbicides, and related products are covered by the Federal Insecticide, Fungicide, and Rodenticide Act (FIFRA) (47 U.S.C. §§136(a)-(y)). FIFRA creates a premarketing clearance procedure under which EPA reviews data on safety and then registers the pesticide, provided it will not generally cause unreasonable adverse effects on the environment. EPA has proposed a rule on data requirements for such registration (36). Sections 158.65 and 158.165 of the proposed rule cover biological pest control agents, including genetically manipulated ones.*

European Economic Community countries

The Federal Republic of Germany, the United Kingdom, and France are members of the EEC, or Common Market, which was established by the Treaty of Rome in 1958.** The regulations of the EEC and the national regulatory processes of these three countries that are relevant to biotechnology products are discussed further below.

EUROPEAN ECONOMIC COMMUNITY

Since 1965, the EEC has issued a series of directives aimed at harmonizing the member states' testing and approval processes for proprietary medicinal products and veterinary medicinal products. None of these directives specifically deals with biotechnological products. The directives are important for the development of biotechnology because, to the extent biotechnological products are proprietary or veterinary medicinal products,*** their approval for manufacture or

*These sections set extensive data requirements on product performance, toxicology, residue analysis, hazards to nontarget organisms, and environmental fate and expression.

**The other members of the EEC are Belgium, Denmark, Greece, Ireland, Italy, Luxembourg, and the Netherlands.

***Proprietary medicinal products are drugs, biologics, or similar products sold under brand or trade names. In practice, most member states regulate biologics differently from chemically synthesized drugs and the European Community directives have not been used to try to harmonize those regulations.

marketing will be governed by national procedures conforming to the directives.

Although the ultimate aim of the EEC directives is to replace national drug approval processes with a Community-wide system, such a system is unlikely in the near future. The speed with which the EEC does achieve a Community-wide drug approval system, however, will have a significant impact on the development of biotechnology, because such a system could cut costs, provide uniform regulation, speed up the approval process, and open access to new markets.

Currently, the existing directives deal only with drugs and veterinary medicines, not biologics. The directives also deal only with some aspects of the pharmaceutical approval process—marketing authorizations and certain testing requirements. A system has been set up for obtaining multiple authorizations for marketing in EEC member states, but control over exports outside the EEC is entirely up to member states.

Council Directive 65/65/EEC established the basic regulatory framework with respect to drugs (4). It requires an authorization from the competent authority of a member state before a drug can be marketed in that state. It sets forth the required information that must be submitted to the authorizing agency and provides that authorization of the product shall be based on a finding of safety, efficacy, and quality. Licenses are to be granted for a 5-year period, subject to extension. A similar directive exists for veterinary medicine (5).

Two questions that will be important to biotechnology companies that manufacture drugs and that seek EEC marketing authorizations remain unanswered. The first concerns the so-called paper NDA issue. The EEC permits a new manufacturer of an already approved product to rely on published data to establish the safety, efficacy, and quality of its version. It is unclear, however, whether this policy will apply to biotechnological products. Under most member states' existing regulations, a change in manufacturing process from chemical to biotechnological synthesis requires either a new market authorization or an amendment to an existing one. Since the EEC has not addressed the issue, the individual member states will determine whether published tests results can be relied on or whether new tests must be undertaken.

A variation of this same issue involves unpublished test results. Under current regulatory policies for drug approvals in both Europe and the United States, the documents submitted in support of an application for approval of a drug (the "dossier") are treated as confidential. Proposals are being considered in Europe, particularly in the Federal Republic of Germany, to change the scope of the confidentiality of the dossier. One proposal is to retain the confidentiality of the dossier for a certain number of years, and then allow access to the information after the payment of compensation to the original manufacturer who performed the tests.

FEDERAL REPUBLIC OF GERMANY

The Law on the Reform of Drug Legislation of 1976 sets forth the approval process for drugs, biologics, and veterinary medicines (7). It is designed to conform with the relevant EEC directives, and responsibility for its administration lies with the Federal Health Office (BGA, Bundesgesundheitsamt).

The licensing procedure for new drugs and biologics produced through biotechnological processes will be the same as for more traditional products. A manufacturer of pharmaceuticals must obtain individual marketing authorizations to distribute each drug or biologic that it manufactures and separate manufacturing authorizations for each of its production plants. Generally, the drug approval process takes 4 to 6 months from the time the application is filed. In the case of biologics, BGA defers to the Paul Ehrlich Institute, which provides authorizations for the manufacture of sera, vaccines, test allergens, test sera, and test antigens. Before deciding to approve a new drug or biologic, BGA must consult an independent commission of experts composed of physicians and representatives of the pharmaceutical industry. After an authorization for a drug or a biologic is given, BGA continues to monitor the competence of the managers and the adequacy of the facilities. An authorization may be withdrawn, revoked, or suspended if satisfactory standards are not maintained.

BGA regulations governing clinical testing of drugs and veterinary medicines track the applicable EEC directives. No specific prior approval of clinical testing is required, but BGA guidelines for such trials must be followed. The process for obtaining marketing approval and the information required in the application follows the EEC directives on proprietary medicinal products and on veterinary medicinal products.* In addition, the manufacturer must show that it holds a manufacturing license.

Anyone seeking to market an imported product must show that the product's foreign manufacturer has the equivalent of a manufacturing license and a marketing license in the country of manufacture; otherwise an explanation of why such authorization has not been granted must be supplied.

With respect to exports, it appears that a manufacturer intending to produce an item solely for export must comply with the requirements and obtain a manufacturing license but need not obtain a marketing license.

Certain biologics, specifically sera, vaccines, or test allergens, may only be marketed if each batch is approved by the Paul Ehrlich Institute. Approval is given only if a test shows that the batch possesses the required safety, efficacy, and quality, and has been manufactured and tested by methods which conform to the standard set by scientific knowledge currently prevailing.

Several aspects of the Federal Republic of Germany's pharmaceutical approval process are of particular significance to pharmaceuticals produced by biotechnology, because a change in manufacturing process from chemical synthesis to biotechnology would necessitate a reauthorization of these products. In certain cases, a manufacturer must apply for reauthorization of a drug despite an existing authorization. The circumstances in which such a reauthorization must be sought include a change in the composition of the active constituents either in type or quantity, a change in dosage form, or an extension in the field of application. For biologics such as sera, vaccines, and test allergens, a change in the manufacturing process also requires a reauthorization.

Two regulatory issues currently being debated in the Federal Republic of Germany are also relevant. The first is a regulation now in force that requires any person who markets a drug in the country to maintain a legal presence in the Federal Republic of Germany. The EEC has recently ruled that this requirement is illegal and has asked that it be abolished. Whether the Federal Republic of Germany will do so remains to be seen.

The second issue involves current proposals to modify the confidentiality of drug authorization dossiers. As in most of Europe, no manufacturer in West Germany has access to confidential information in another manufacturer's dossier unless it specifically receives permission from the original manufacturer, permission which is usually granted, if at all, only after the payment of substantial compensation. A second manufacturer of a drug that has already been approved may also rely on published material in lieu of relying on the dossier or conducting its own tests, but most important drugs are not the subject of published studies. Almost any scientifically reliable material will be contained in the confidential dossier that the first manufacturer submitted. Under active consideration are proposals that would maintain absolute confidentiality of the dossier for a given number of years, but then allow for access to the dossier with a statutorily prescribed compensation system. It will probably be some time before any such system is enacted (8).

UNITED KINGDOM

Because the United Kingdom is a member state of the EEC, its regulations conform to the basic requirements of the EEC pharmaceutical directives. Its current standards are embodied in the Medicines Act of 1968 and in the regulations adopted under this statute. No specific regulations governing the approval of biotechnologically pro-

*The application data must contain data showing: 1) the toxicological effects and pharmacological properties of the drug; 2) its effectiveness in the given indications; 3) the propriety of the suggested dosage; 4) side effects; 5) the drug is of appropriate quality; and 6) the production control methods correspond to scientific knowledge currently prevailing and are suitable for quality assessment. An application for an authorization for veterinary medicines and medicated foodstuffs must include residue tests and indicate how long it takes for residues to occur in edible tissues and how such residues are to be assessed.

duced pharmaceuticals have yet been adopted, so such products are subject to the general approval process set forth in the Medicines Act. The approval process for pharmaceuticals and related substances is similar to the U.S. system in several respects, but it is somewhat less restrictive and much more efficient in terms of the time for approval.

The Medicines Act of 1968 provides a comprehensive framework for the regulation of "medicinal products" which include drugs, biologics, and veterinary medicines. Its provisions are administered by the Health and the Agriculture Ministers of the United Kingdom, acting with the advice of the Medicines Commission. The day-to-day operation of the act is the responsibility of the Medicines Division of the Department of Health and Social Security.

The regulations governing the use of medicinal products focus on the safety, efficacy, and quality of the product. The system utilizes five types of licenses: licenses as of right, clinical trial certificates,* product licenses, manufacturers' licenses, and wholesale dealer licenses. These licenses apply to the manufacture, sale, storage, import, or export of any medicinal product. The requirements for the issuance of clinical trial certificates are considered to be among the strictest in Europe. Before a certificate can be granted, an applicant must present animal pharmacokinetic data, acute and chronic toxicity data, and information on potential reproductive toxicity. The basic documentation required to obtain a product license is similar to that required by the relevant EEC directives. Trial certificates valid for up to 2 years and product licenses valid for 5 years are issued for drugs on which clinical testing and production began after September 4, 1971. Either may be renewed.

Additional requirements are imposed with respect to "biologicals," which include vaccines, toxins, antigens, sera, and enzymes. Such biological medicines are licensed on a batch release system. The manufacturing license requires that each batch of product be subject to certain tests and that samples and the results of the tests be submitted to the National Institute of Biological Standards and Controls (NIBSC). The basic tests administered by NIBSC, which may have to be modified in the case of new biotechnological applications, include potency, purity, toxicity, pyrogenicity, and immunogenicity.

NIBSC has begun considering how its testing requirements may have to be modified for biotechnological products but has not formally adopted new requirements (2). Among the issues which NIBSC has identified as requiring modification of its procedures for biotechnologically produced products are establishment of the identity of large proteins produced by rDNA technology, adaptation of bioassay techniques, biological potency, contamination of biotechnologically produced products with macromolecules of bacterial origin, and chemical modification of the required products.

Several aspects of the pharmaceutical approval process in the United Kingdom will be particularly relevant to the development of biotechnology. The batch release system for testing biologics will apply to many biotechnologically produced products, but that will be the case in many countries. Also of importance for biotechnology is the treatment of already licensed drugs produced with new methodologies. In the United Kingdom, such drugs require product and manufacturing licenses. However, the full documentation that would be required for a completely new drug need not be provided. The precise amount of documentation will vary with the particular drug. In general, the United Kingdom will allow the substitution of published references for actual test results in those situations permitted under the EEC Council Directive 65/65/EEC (4). However, a second manufacturer is not permitted to rely on the confidential information submitted in the dossier of a first manufacturer. Thus, a new manufacturer of an already approved drug is required independently to demonstrate the safety, efficacy, and quality of the drug through its own research or that of independent researchers.

Imported drugs also require a product license. The manufacturer may be required to declare that any requirements imposed by the law of the country in which the drugs are manufactured

*Licenses of right and clinical trial certificates are self-limiting. The trial certificates terminate automatically once the trial process has ended. Licenses of right are transformed into product licenses once the drug has been reviewed by the Medicines Review Commission and found safe, effective, and of proper quality.

have been complied with and to permit the licensing authority to inspect his premises to ensure that they comply with any prescribed conditions of the license. Drugs produced solely for export also must be licensed, but the licensing authority is required to consider only quality, not the safety and efficacy of the drug.

FRANCE

The French approval processes for pharmaceuticals includes many of the same steps as the processes in the United States. The basic standards for approval are quality, safety, and efficacy of the pharmaceuticals, and the necessary tests are largely the same.

The authority responsible for the registration of new drug products is the Directorate of Pharmacy and Medicaments of the Ministry of Health, which administers the requirements of the Public Health Code, Book V, and the EEC protocols for analytical, toxicological, and pharmacological tests and clinical trials. The Ministry of Health uses the same basic standards of quality, safety, and efficacy required by the EEC.

A manufacturer must notify the Ministry of Health before commencement of clinical trials of a new product or for a new indication of an established product. The trials must be carried out under the supervision of an "approved expert"* and must follow procedures and present data in the format established by the Ministry of Health. Toxicological and pharmacological data must be submitted to the approved expert prior to commencement of the trials. Except for the analytical data, the information does not have to be generated by local French studies; however, the foreign data can only be accepted if it is justified by approved experts and conforms to EEC protocols. These rules apply also to clinical data generated by studies conducted abroad. Clinical trials must be performed in hospitals as controlled experiments.

Prior to obtaining a marketing license, a manufacturer is also required to request authorization from the Directorate of Pharmacy and Medicaments to manufacture the new drug product. If the product is to be manufactured abroad, the manufacturer must attach to the French marketing application the document granting it authority to manufacture the product in the foreign country. The marketing authorization itself is subject to the documentary requirements established by the EEC directives.

Once a manufacturer has submitted all relevant data to the Ministry of Health, the Minister must announce a decision on the application for marketing registration within 120 days. This period may be extended for another 90 days in exceptional cases. In practice, however, the processing time for an application averages 6 to 8 months. A second manufacturer cannot rely on the dossier of a first manufacturer to qualify its drug, so a new manufacturer making an already approved drug by biotechnological processes would have to show the drug's safety, efficacy, and quality all over again. However, as in other EEC countries, a manufacturer may rely on some published data to support its application.

Once registration has been approved, as in the rest of the EEC, the marketing license is valid for 5 years. It may be renewed for additional 5-year periods only if the manufacturer formally declares that no modification has occurred in the scientific data submitted in support of the original application. The Ministry of Health must therefore be notified of any new data.

A drug may be imported from another EEC country and, in exceptional circumstances, from a non-EEC country, provided that a marketing license has been obtained in France. A certificate is required proving authorization for sale or distribution within the exporting country. Authorization for the marketing of an imported drug is only valid for 6 months, but presumably may be renewed.

Drug products designed for animal consumption are also regulated by the Ministry of Health. The application procedures for obtaining authorization to market veterinary drugs are basically the same as those for human drugs.

*"Approved experts" are scientists with expertise in various aspects of pharmaceutical testing who are approved by the Minister of Health. The Minister maintains lists of these experts from among whom an applicant may select experts to review his or her data and supervise further testing. Approved experts need not be French.

Switzerland

The Intercantonal Convention for the Control of Medicaments is the authority for the regulation of drugs and related products. Under the Convention, the Intercantonal Office for the Control of Medicaments (IOCM, Interkantonale Kontrollstelle für Heilmittel) administers the drug regulatory system. IOCM has four principal tasks: quality control of marketed drugs, quality control of manufacturing, the licensing of new drugs, and the review and relicensing of existing drugs. IOCM has responsibility for pharmaceuticals,* veterinary medicines, and medical devices. Food and cosmetics are controlled by the Federal Office of Public Health under separate Federal authority. The quality control functions of IOCM are exercised through sampling of drugs at the time of their registration and periodically thereafter and through periodic inspections of pharmaceutical facilities.

The licensing of pharmaceuticals is much more streamlined than in other countries. There is no requirement for government approval before initiation of clinical trials. This is due both to the small size of IOCM and to greater reliance on the good faith of manufacturers and the common sense of medical practitioners participating in the clinical trials of new drugs.

Approval of the marketing of a drug is based on its efficacy and safety, which are judged by an independent board of university scientists. Approval can be refused not only if the drug is found not to be safe and effective, but also if its price is excessive. Licenses are issued for a 5-year period and may be renewed by the same board.** The drug approval process generally takes 6 to 10 months.

Of particular importance for biotechnology is the fact that less documentation is required for drugs that are not new chemical entities. Switzerland's streamlined drug approval process should mean faster action on new drug applications and on old drugs being produced through biotechnology.

For imports, it is necessary to have a certification that the drug is authorized for sale or distribution in the country of manufacture and that the manufacturer is subject to regular inspection. Drugs intended solely for export are exempt from registration, but voluntary registration can be made.

*This includes in vivo diagnostics, contraceptives, narcotics, anesthetics, antibiotics, some industrially produced homeopathic medicines, herbal remedies, radiopharmaceuticals, and certain blood products.

**In special cases, up-to-date analytical, preclinical, and chemical data as well as samples may be required if requested by IOCM.

Japan

The approval process for drugs, biologics, and veterinary medicines in Japan is set forth in the Pharmaceutical Affairs Law (17). The law generally requires each manufacturer or importer to obtain a license for each manufacturing plant or business office and a separate approval for each drug manufactured or imported.* The manufacturer's or importer's license must be renewed every 3 years. The product approval has no set duration, but in practice many drugs are reviewed again after 6 years. The approval process is quite drawn out and complex because many agencies are involved. The time from submission of an application to approval is supposed to take 1 to 3 years but in practice takes longer (13).

The information that must be filed with the application for the approval of a new drug in Japan include data on origin, discovery, use in foreign countries, physical and chemical structure and properties, stability, various forms of toxicity and other dangerous side effects, pharmacological action, how the drug will be used in the body, and results of clinical trials (15). Most of the data is required to be published as an original article in a Japanese scientific journal. Data on animal tests for toxicity must meet certain special requirements. The application will be denied if the drug has no effect, efficacy, or efficiency as indicated in the application, if the drug is "remarkably dangerous" in comparison to its effect, or if the drug has been designated improper under the Ministry of Health and Welfare Ordinance (17).

An application to import a new drug must meet these standards. It must also contain a document

*The separate approval for each drug is unnecessary if the drug is listed in the Japanese Pharmacopoeia and has been exempted by the Minister for Health and Welfare.

certifying that the exporting country approves its manufacture and copies of the import contract or similar document (16). The import or manufacture of biologics is prohibited unless special requirements concerning their processing, properties, quality, and storage are met (16). Each batch of biologics must be tested and approved by the National Institute of Health.

New drugs must be reexamined about 4 to 6 years after approval, largely so that the safety of the drug can be assessed in light of post-approval clinical tests and other scientific research. The reexamination is to determine whether the drug now displays any condition that would, if a new drug application were now filed, require its rejection, i.e., that the drug is not efficacious, is more dangerous than efficacious, or has been designated improper (17). The approval for a drug may be canceled if the drug cannot pass reexamination, if health or sanitation reasons so require, if the licensee fails to submit accurate reexamination material, or if the licensee has not produced the drug for 3 years (17). How this will affect drugs produced with biotechnology is unclear.

Environmental regulation*

Protection of the environment is one aim of the rDNA guidelines in each of the competitor countries; none of them have any other rules specifically directed to the environmental effects of biotechnology. Nevertheless, the more general environmental laws will apply to biotechnological processes, products, and waste products. The extent to which these general laws will apply to genetically modified organisms used in the environment is uncertain in all of the countries except the United States, where EPA has asserted jurisdiction under TSCA.

The environmental requirements in the rDNA guidelines are likely to have little effect on the competitive position of any country. The specific measures required for any physical containment level vary little from country to country. Moreover, most rDNA activities are now conducted at low containment levels that require essentially only that good microbiological practices be followed. Deliberate release of genetically modified organisms is generally prohibited, although procedures exist for exceptions from the prohibition. In the United States, deliberate release is not prohibited as such, but one who would do so under the guidelines must have the approval of IBC and NIH, after consultation with the Recombinant DNA Advisory Committee.*

It is difficult to determine what effect, if any, the more general environmental regulations of each country dealing with air and water pollution and waste disposal will have on biotechnology in that country. Since much of the environmental regulation in any country is performed on the local level, generalizations about national environmental controls can be misleading. States (Länder) in the Federal Republic of Germany, for example, are about to enact specific legislation to fill in the framework set up by Federal laws. Certain environmental legislation in Japan, though enacted at the national level, applies only to certain areas. Local authorities in France and the United Kingdom possess considerable responsibility for administering and enforcing environmental rules. Switzerland leaves most decisions on environmental regulation to the cantons, as it does decisions on other subjects. The United States has one of the more centralized systems for environmental control, but even Federal statutes allow for responsibility to be transferred to the States.

*For specific information regarding the six countries, see the section on environmental regulation in *Appendix F: Recombinant DNA Research Guidelines, Environmental Laws, and Regulation of Worker Health and Safety*.

*A lawsuit has been filed against NIH claiming that approval by the Recombinant DNA Advisory Committee is not consistent with the National Environmental Policy Act and claiming that an Environmental Impact Statement must be prepared (*Foundation on Economic Trends* v. *Heckler*, No. 83 Civ. 2714 (D.D.C. Sept. 14, 1983)).

All of the countries except Switzerland have fairly comprehensive and stringent environmental regulation. Switzerland's national regulation is directed only toward water pollution. Thus, its biotechnology companies may have a competitive advantage over those in the other countries because of less restrictive environmental regulation. Yet even the more stringent regulation in other countries would not necessarily handicap companies because the regulation is directed mainly toward toxic chemicals. The degree of traditional environmental problems that companies using biotechnology might create—air and water pollution and hazardous waste—does not now appear to be so great that environmental controls will significantly affect the commercialization of biotechnology. However, increasing commercialization of biotechnology eventually will require more consideration about the disposal of waste byproducts. All countries are now about equal in this area, but those who undertake to resolve uncertainties about the specifics of that regulation should enhance the competitive positions of their biotechnology companies.

The United States seems to be the farthest ahead in considering the risks and regulation of the deliberate release of genetically modified organisms. This may simply be the result of the fact that this area of biotechnology is further along in the United States than in the other countries. In any event, NIH recently has reviewed and approved several proposals to release organisms into the environment. Also, on June 22, 1983, two congressional subcommittees held a joint hearing on the topic of regulating such releases (22).

At the hearing, EPA took the position that such organisms are "chemical substances" as defined by TSCA* and therefore subject to regulation by EPA under TSCA (3). Although the matter is not free from doubt, a consensus has been developing among the experts that TSCA would apply (18).

TSCA gives EPA broad authority to regulate the products of biotechnology, and, assuming EPA's interpretation of the definition of "chemical substance" survives any subsequent legal challenge, TSCA would have great potential for regulating the deliberate release of genetically modified organisms. Under section 4 of TSCA, EPA can adopt rules requiring testing of chemical substances that "may present an unreasonable risk of injury to health or the environment" or will be produced in substantial quantities (and enter the environment in substantial quantitites or result in substantial human exposure) when existing data are insufficient to make a determination and testing is necessary to develop adequate data. Section 5 requires the manufacturer of a new chemical substance to notify EPA 90 days before beginning production and submit any test data it may have on the chemical's health or environmental effects. If the agency decides that the data are insufficient for evaluating the chemical's effects and that it "may present an unreasonable risk of injury to health or the environment" or will be produced in substantial quantities (and enter the environment in substantial quantities or result in substantial human exposure), it can propose an order to restrict or prohibit the chemical substance's manufacture or use. Under section 6, EPA can prohibit or regulate the manufacture or use of any chemical substance that "presents, or will present an unreasonable risk of injury to health or the environment." TSCA also provides for recordkeeping and information gathering about the environmental and health effects of chemical substances.

Despite its theoretical applicability, TCSA may leave much to be desired in terms of a practical program to regulate the use of genetically manipulated organisms in the environment. First, EPA has little expertise or experience in the area of genetic manipulation. Second, its toxic substances program has been significantly understaffed, according to a 1980 study by the U.S. General Accounting Office study (21). Third, TSCA may not give EPA sufficient regulatory power, if the risks presented by deliberate release are viewed as substantial. For example, section 5, which creates the premanufacturing notice requirement, does not require the generation of toxicological data. A recent OTA background paper found that nearly half of the premanufacturing notices submitted

*A "chemical substance" is defined in the relevant part under sec. 3(2)(A) of TSCA as "any organic or inorganic substance of a particular molecular identity," including "any combination of such substances occurring in whole or in part as a result of a chemical reaction or occurring in nature. . . ."

to EPA do not have information about the chemical's toxicity (26).* Moreover, the burden is on EPA to take legal action if it believes that insufficient data exists for a new chemical substance.

USDA also has an environmental role to play with respect to biotechnology. It regulates importation and interstate shipment of plants, animals, and their pathogens (21 U.S.C. §§101-135; 7 U.S.C. §§151-167; 7 U.S.C. §150aa et seq.). Thus, some of the "raw materials" of interest to biotechnologists in the agriculture field are subject to USDA restrictions. For example, two potential mechanisms for transferring genes into plants are the bacterium *Agrobacterium tumefaciens* with its integrating Ti plasmid and the cauliflower mosaic virus. Both the bacterium and the virus are subject to the restrictions. Similarly, work with particularly dangerous animal viruses may be prohibited or severely restricted. For example, work on foot and mouth disease virus can only be performed at Plum Island, a high containment USDA laboratory located off the coast of Long Island, N.Y. USDA also bars entry into the United States of 22 other pathogens that might be of interest to companies desiring to produce animal vaccines.

*As to the importance of such information, OTA's background paper stated (26): "Certainly, the absence of toxicity data complicates EPA's efforts to decide whether a new chemical may present an unreasonable risk to health or the environment. But the importance of toxicity data for making decisions about particular chemicals varies. Those data are less important for chemicals that closely resemble others for which there is much information and experience. They are critical for unusual chemicals or chemicals for which there is limited information."

Regulation of worker health and safety*

The rDNA research guidelines in each of the six countries (but not those of the EEC itself) contain provisions for the safety and health of laboratory workers. Each country also has more widely applicable laws and regulations, but it is the rDNA guidelines that will have the most immediate impact on the biotechnology companies.

The substance of the various worker health and safety provisions in the national rDNA guidelines varies among the six countries studied, although most set forth rules to ensure that laboratory workers are knowledgeable about laboratory procedures, that emergency procedures are known and safety equipment is available, and that worker health is monitored for certain types of work. It seems fair to infer that the costs and burdens associated with these requirements are modest, because there has been little criticism or complaints about them from academia or industry (8).

The more general worker health and safety laws in the United States and in each of the five foreign countries have had no measurable effect as yet on the industries using biotechnology in each country. Each country imposes general duties on employers to maintain safe workplaces and to eliminate or control hazardous substances (although when these substances are specified, they do not include materials likely to be found in a biotechnology laboratory). The most that can be said is that each country has at least one authority able to impose further requirements to protect worker health and safety, but none has yet done so. Such requirements would be primarily process rather than product oriented.

The United States has studied the question of the possible risks posed to workers from long-term exposure to novel organisms and products. The Centers for Disease Control and the National Institute of Occupational Safety and Health (NIOSH) created an ad hoc working group on medical surveillance for industrial applications of rDNA. The group concluded that, while physical containment of rDNA-containing organisms and their products is the first line of defense, medical surveillance of industrial workers can play a valuable auxiliary role in protecting their health (19). Others have disagreed with this finding, questioning the need for surveillance and the ability to construct a meaningful program.

*For specific information regarding the six countries, see section on regulation of worker health and safety in *Appendix F: Recombinent DNA Research Guidelines, Environmental Laws, and Regulation of Worker Health and Safety*.

The NIOSH findings have not been implemented by the Occupational Safety and Health Administration (OSHA), the U.S. agency primarily responsible for worker safety and health. Under the Occupational Safety and Health Act of 1970, OSHA can promulgate workplace standards to protect workers from toxic substances or harmful physical agents. Under a recent decision by the U.S. Supreme Court (12), such standards must be "reasonably necessary to remedy a significant risk of material health impairment." Although this requirement would appear to prevent OSHA from acting on those purely conjectural risks associated with biotechnology, the agency could act on known biological risks (e.g., those presented by known pathogens), or physical risks (e.g., those presented by the use pressurized containment vessels). In any event, OSHA has not promulgated any standards for bioprocesses in general, nor has it taken any position on regulating biotechnology.

At this point and for the foreseeable future, worker health and safety regulation of biotechnology is minimal. Thus, it will give neither an advantage nor a disadvantage to any of the competitor countries.

Findings

Health, safety, and environmental regulation can affect the cost, time, and financial risks of getting products to market. Thus, such regulation can be expected to affect international competitiveness in biotechnology.

The only government controls directed specifically toward biotechnology are the rDNA guidelines adopted by the EEC and the six competitor countries. They are essentially voluntary and directed primarily at research, although they do apply to large-scale work to varying degrees. Their containment and oversight provisions have been substantially relaxed since they were originally adopted, and this trend is expected to continue.

The rDNA guidelines in the competitor countries are quite similar in their regulatory goals, requirements, and implementation because they are generally patterned after the U.S. guidelines, which were initially developed through the efforts of the international scientific community. Nevertheless, there are differences that allow the guidelines to be ranked in terms of their restrictiveness and potential impact on the competitiveness of the various countries.

The rDNA guidelines of the United States are the least restrictive of the guidelines in any of the competitor countries. The vast majority of the experiments that are done with the most commonly used host-vector systems are either exempt or can be done at the lowest containment levels. Prior approval, even by the IBCs, is required only for a limited category of experiments. The rDNA research guidelines of Japan and the European countries are more restrictive than the U.S. guidelines in one or more of the following ways:

- they require more stringent containment;
- they require more time-consuming approval procedures;
- they have fewer categories of approved host-vector systems; or
- they severely restrict large-scale work.

Japan has the most restrictive rDNA guidelines. A limited number of host-vector systems have been approved for use. More importantly, companies have had extreme difficulty in obtaining approval to do work with more than 20 liters of culture, but this is expected to change soon.

Of the remaining countries, Switzerland appears to have the least restrictive guidelines. Its Government has played no role in the guidelines, and there are no requirements covering large-scale work. However, Switzerland follows an earlier, and thus more restrictive, version of the U.S. guidelines. The guidelines in France and the United Kingdom appear to be roughly equivalent with regard to their impact on biotechnology. The Federal Republic of Germany appears to be slightly more restrictive, primarily because Government approval must be obtained before even moderate risk experiments can be started.

It is the existing regulation that will most affect biotechnology: product approval laws, environmental laws, and worker health and safety laws. The most important of these for biotechnology will be the product approval requirements, especially for pharmaceuticals and veterinary medicines because those products are the most stringently regulated or the subject of much of the current effort in product development. For this reason, and because of insufficient information on foreign regulation of the other products, the analysis for product approval in this chapter concentrated on the regulation of pharmacuticals and veterinary medicines.

With respect to the product approval process, particularly for pharmaceuticals and animal drugs, the United States appears to be at a competitive disadvantage with respect to all of the other countries except Japan. The competitive disadvantage for the United States results mainly from the time and cost necessary to secure premarketing approval. In contrast, the United Kingdom has the most expedited pharmaceutical approval process, even though its substantive requirements are quite similar to those of the United States. Switzerland is the least restrictive of the countries in terms of substantive requirements. For example, it does not require Government approval before initiation of clinical trials. In contrast to pharmaceuticals and animal drugs, the regulatory requirements for animal biologics are less restrictive in the United States and roughly on par with those in other countries.

Another reason the United States is at a competitive disadvantage is that the United States, in contrast with the other countries, does not allow the export of unapproved pharmaceuticals. In addition, bulk drug products may also not be able to be exported. Given certain provisions in joint venture agreements between U.S. NBFs and their foreign partners, these requirements could enhance the transfer of biotechnology to foreign companies.

Specific requirements regarding biotechnology products are or will be set at the agency level within the existing statutory framework. In the United States, FDA has taken the lead in developing and publishing informal statements. Since these statements help dispel uncertainties, they will help product development. In its policy statements, however, FDA has taken the position that rDNA products whose active ingredients are identical to ones already approved or to natural substances will still need to go through the new product approval process. However, data requirements may be modified and abbreviated. This appears not to be the situation in other countries, although there have not been definitive pronouncements by the regulatory agencies.

One area of uncertainty that could hinder U.S. competitiveness in biotechnology to some degree is the question of jurisdiction over animal biologics. FDA and USDA are engaged in a jurisdictional dispute that could delay product approvals.

Environmental and occupational safety and health regulations are not likely to give any of the countries a significant competitive advantage in biotechnology. This regulation is likely to play a minor role, except in the area of deliberate release of genetically manipulated organisms into the environment. For that application of biotechnology, uncertainties exist as to what, if any, kind of special regulation will develop. The United States appears to be the farthest along in considering the problem; thus, to the extent that decisions are made and the regulatory picture clarified for corporate planners, the United States may have a slight advantage.

Issue and options

ISSUE: How could Congress improve U.S. competitiveness in biotechnology through changes in the regulatory environment?

Regulation imposes costs, constraints, and delays on biotechnology companies that are justified when they promote such general goals as the enhancement of human health or quality of the environment. To the extent that such regulation is inefficient or unnecessarily restrictive or creates uncertainties that impede business planning, however, it will restrict biotechnological innovation and U.S. competitiveness in biotechnology without achieving the other goals.

OTA has identified several options that could improve U.S. competitiveness in biotechnology through changes in laws, regulations, and administrative policies regarding health and safety. Many of these are not specific or limited to biotechnology but nevertheless could significantly affect this technology. Furthermore, many of the actions could be taken by executive agencies, and, in fact, are being considered. Nevertheless, Congress may decide legislative action is necessary or more appropriate.

Option 1: Amend the Federal Food, Drug, and Cosmetic Act (FFDCA) to permit the export of unapproved drugs and biologics.

Of the six competitor countries identified in this assessment, the United States is the most restrictive regarding the export of unapproved drugs and biologics. The relevant provision of FFDCA is designed to prevent "drug dumping"—situations where drugs deemed unsafe or ineffective by the United States or other developed countries have been marketed in developing countries.

Those who advocate eliminating this provision of FFDCA argue that a U.S. company can have ethical reasons for wanting to export a drug that is unapproved by FDA. For example, it may be supplying a company that sells the drug in a country that has approved the drug for sale. Advocates of eliminating this provision also argue that the provision simply embodies U.S. paternalism toward other countries, which are capable of making their own health and safety decisions. Partly to avoid the U.S. ban on the export of unapproved drugs, the multinational drug companies have established foreign manufacturing facilities. This practice results in the transfer of technology and jobs from the United States and has an adverse effect on the U.S. balance of payments. For NBFs, which may not have the money to establish foreign facilities or the time before contract revenues and capital run out, the export restriction may be especially burdensome.

FDA has taken the position that bulk pharmaceutical products made by biotechnology are drugs because such products are biologically active; thus, the export prohibition of FFDCA applies. One U.S. company, Genentech, has asserted that its inability to sell bulk pharmaceutical products to its foreign joint venturers will result in its being required to transfer the technology to produce that bulk product to its foreign partners. This company has argued that bulk pharmaceutical products produced by biotechnology and not labeled as drugs should not be considered drugs under FFDCA and FDA regulations. Clearly, this question of interpretation could be resolved on the administrative level without congressional action. To change the general prohibition in FFDCA against the export of unapproved human and animal drugs and biologics, however, legislation would be necessary.

The arguments against amending FFDCA to permit the export of unapproved drugs and biologics are essentially moral ones. There have been documented cases of drug dumping in developing countries. Supporters of the existing restrictions argue that the United States has a moral duty to try to prevent such actions and that the developing countries are unofficially in favor of these export restrictions.

There are several different ways that legislation to permit the export of unapproved human and animal drugs and biologics could address these moral arguments. First, the legislation could exclude products that have actually been barred by FDA. Second, it could permit the export of unapproved drugs and biologics only if they have been approved by at least one other developed country. Third, it could permit the export of unapproved drugs and biologics only to countries where the products has been approved. Finally, the legislation could be drafted so that unapproved drugs and biologics can be exported only to developed countries. The potential diplomatic problems that could arise by having to decide which countries are "developed" could be avoided or lessened by using the definitions of various international organizations, such as the International Monetary Fund.

Option 2: Pass legislation to merge the Virus, Serum, Toxin Act of 1913 into the Federal Food, Drug, and Cosmetic Act.

The reasons for the different statutes are primarily historical, and the distinctions between animal drugs and biologics, if they were not already anachronistic, have virtually been made so by rDNA and hybridoma technology. Nevertheless, USDA and FDA were engaged in a jurisdictional dispute over bovine interferon and may well continue to engage in disputes over future products. By trying to satisfy both agencies, U.S. companies using biotechnology are likely to incur additional costs and delays. In addition, the uncertainties over regulatory authority may hinder corporate planning for what product areas to pursue or may steer firms away from pursuing these kinds of products. As a result, U.S. firms may be at a competitive disadvantage with respect to foreign firms.

Although combining the regulatory jurisdiction into one agency, FDA, may make sense conceptually, there will be substantial institutional barriers to doing so. If USDA is unwilling to give up its jurisdiction, as it appears to be, it can count on substantial political support from inside and outside of government. In addition, despite the adverse consequences of this jurisdictional dispute, the biotechnology companies themselves may well prefer USDA to retain or enhance its jurisdiction over animal biologics because USDA regulation is viewed as substantially less burdensome and costly than FDA regulation. This option has been proposed several times in past years, but there has been little progress toward its implementation.

Option 3: Amend the patent law to extend the term of patents on products or processes that need regulatory approvals before marketing.

This option was considered extensively by the 97th Congress, in which legislation passed the Senate and failed to pass the House by a few votes. It was also the subject of an OTA report, *Patent-Term Extension and the Pharmaceutical Industry* (24). Legislation to accomplish this option (S. 1306, H.R. 3502) has been introduced in the 98th Congress.

Firms that are heavily involved in basic research support patent-term extension. They claim that R&D costs and risks are rising, yet the effective life of patents on the products resulting from the R&D is declining because of the increasing time necessary for securing regulatory approvals before marketing. Since this may cause returns on R&D investments to decrease, the firms assert that innovation will suffer. Several biotechnology firms have supported this option publicly.

Generic drug producing firms and consumer groups oppose patent-term extension. The generic drug firms, which derive most of their revenues from drugs equivalent to the pioneering ones whose patents have expired, assert that patent-term extension will delay their entry into the market or not make that entry worthwhile because of limited product life remaining. They also assert that patented products often maintain an exclusive market position after their patents expire because of nonpatent barriers to market acceptance of generically equivalent products. As a result, patent-term extension would cause competition to decline and prices to increase. The consumer groups support this position and also note that the pharmaceutical industry has been extremely and consistently profitable for a great many years, even while the regulatory burdens have been increasing.

The OTA report mentioned above found that "[t]he evidence that is available neither supports nor refutes the position that innovation will increase significantly because of patent-term extension." It did note, however, that the incentives provided by patents for pharmaceutical R&D would be enhanced.

Option 4: Address the uncertainties and concerns about the deliberate release of genetically manipulated organisms into the environment by passing new legislation or amending the Toxic Substance Control Act to clarify its applicability to living organisms.

There are risks associated with releasing nonindigenous organisms into the environment. Although most nonindigenous such organisms do not establish an ecological niche, many have done so with disastrous consequences. For example, over half of the insect pests in the United States today came from abroad; similarly, the microorganism causing Chestnut blight was not indigenous to the United States.

The risks of releasing genetically manipulated organisms into the environment are not known. On one hand, changing the genetic makeup of an organism usually decreases its ability to survive. On the other hand, many of these organisms, such as microbes used for enhanced oil recovery, will have to be manipulated so as to be competitive with indigenous micro-organisms and to be able to withstand extreme environments in order to be able to accomplish the task. Some industry spokespeople, who believe that rDNA-containing micro-organisms do not present any special risks when properly contained in bioreactors, have expressed concern about the deliberate release of such micro-organisms into the environment.

The concern about releasing genetically manipulated organisms into the environment and the uncertainties about the Federal Government's authority to regulate such activities may impede developments in the use of biotechnology in areas such as microbial enhanced oil recovery, pollution control, and mineral leaching. It may even hinder genetic manipulation of plants,* although the risks involved are seen as much less than those for micro-organisms. Given the concern about risk and the uncertainty over the Federal Government's possible regulatory response, U.S. companies may have difficulty planning where to place limited resources for research and product development.

Opponents of this option are likely to question whether legislative action is needed to accomplish the goal of environmental protection. Although most experts acknowledge that there is uncertainty about whether TSCA covers organisms, a consensus seems to be developing that it does. More importantly, EPA has taken the position that TSCA applies. In addition, voluntary oversight is being exercised by the Recombinant DNA Advisory Committee, although the quality of that oversight is the subject of litigation.

*The U.S. Recombinant DNA Advisory Committee (RAC) recently approved a change in the guidelines that would permit field tests with plants containing rDNA with the prior approval of the local Institutional Biosafety Committee and a working group of the RAC under certain conditions.

Chapter 15 references*

1. Bozeman, N. H., "The Regulation of Hybridoma Products," *The Impact of Hybridoma Technology on the Medical Device and Diagnostic Product Industry,* T. J. Henry (ed.) (Washington, D.C.: Health Industry Manufacturers Association, 1982).
2. Bristow, A. F., and Bangham, D. R., "The Control of Medicines Produced by Recombinant DNA Technology," *Trends in Biological Science* (Amsterdam: Elsevier/North-Holland Biomedical Press, August 1981), 6:VI-VIII.
3. Clay, D. R., Acting Assistant Administrator, Office of Pesticides and Toxic Substances, Environmental Protection Agency, statement in *Environmental Implications of Genetic Engineering,* hearings before the Subcommittee on Science, Research, and Technology and the Subcommittee on Investigations and Oversight, House Committee on Science and Technology, U.S. Congress, June 22, 1983 (Washington, D.C.: U.S. Government Printing Office, 1983).
4. Council of the European Communities, Council Directive of 26 January 1965, No. 65/65/EEC, on the Approximation of Provisions Laid Down by

*Note: F.2d = Federal Reporter, Second Series.
S. Ct. = Supreme Court Reporter.

Law, Regulation or Administrative Action Relating to Proprietary Medical Products, *Official Journal of the European Communities,* 22, Feb. 9, 1965, reprinted in Commission of the European Communities, *The Rule Governing Medicaments in the European Community,* Catalogue No. CB-24-78-120-EN-C (Brussels: Office for Official Publications of the European Communities, May 1978).

5. Council of the European Communities, Council Directive of 28 September 1981, No. 81/851/EEC, on the Approximation of the Laws of the Member States Relating to Veterinary Medicine Products, *Official Journal of the European Communities,* 24 (L317):1-15, Nov. 6, 1981.
6. Cunningham, B., General Counsel, Genentech, Inc., South San Francisco, Calif., personal communication, May 5, 1983.
7. Federal Republic of Germany, Arzneimittelgesetz (Law on the Reform of Drug Legislation) of 24 August 1976 Bundesgesetzblatt [BGB1] 2445 (W. Ger.), nonofficial translation edited by Federal Ministry for Youth, Family Affairs, and Health, Bonn, 1977.
8. Fox, A., and Sciortino, J. (of Kaye, Scholer, Fierman, Hays, & Handler), "International Health and Safety Regulation of Biotechnology," contract report prepared for the Office of Technology Assessment, U.S. Congress, July 1983.
9. Genentech, Inc., "Supplemental Comments by Genentech, Inc. on Proposed New Drug and Antibiotic Regulations," U.S. Food and Drug Administration Docket No. 82N-0293 (Jan. 5, 1983). [comments on file at FDA]
10. *Grand Laboratories, Inc.* v. *Patricia Harris,* 660 F.2d 1288 (8th Cir. 1981).
11. Hayes, A. H., Jr., Commissioner, Food and Drug Administration, statement in *Patent Term Extension and Pharmaceutical Innovation,* hearings before the Subcommittee on Investigations and Oversight, House Committee on Science and Technology, U.S. Congress, Feb. 4, 1982 (Washington, D.C.: U.S. Government Printing Office, 1982).
12. *Industrial Union Department, AFL-CIO* v. *American Petroleum Institute,* 100 S. Ct. 2844 (1980).
13. International Federation of Pharmaceutical Manufacturers Association, *Legal and Practical Requirements for the Registration of Drugs (Medicinal Products) for Human Use,* 2d ed., 1980.
14. *Japan Economic Journal,* "Council Helps Biotech Firms by Easing 20-liter Regulation," 21:16, June 21, 1983.
15. Japan, Enforcement Regulations, Article, 18-3; *Requirements for the Registration of Drugs in Japan* (Yakuji Nippo, Ltd., 1982).
16. Japan, Ministry of Health and Welfare, Pharmaceutical Affairs Bureau, *Pharmaceutical Administration in Japan,* 2d ed., Tokyo, Japan, February 1980.
17. Japan, Pharmaceutical Affairs Law, Law No. 145 of August 1960.
18. Karny, G. M., Senior Analyst, Biological Applications Program, Office of Technology Assessment, statement in *Environmental Implications of Genetic Engineering,* hearings before Subcommittee on Science, Research, and Technology and the Subcommittee on Investigations and Oversight, House Committee on Science and Technology, U.S. Congress, June 22, 1983 (Washington, D.C.: U.S. Government Printing Office, 1983).
19. Landrigan, P., Cohen, M., and Dowdle, W., "Medical Surveillance of Biotechnology Workers: Report of the CDC/NIOSH Ad Hoc Working Group on Medical Surveillance for Industrial Applications of Biotechnology," *Recombinant DNA Technical Bulletin* 5(3):133-138, 1982.
20. U.S. Congress, General Accounting Office, *FDA Drug Approval—A Lengthy Process That Delays the Availability of Important New Drugs,* Washington, D.C., 1980.
21. U.S. Congress, General Accounting Office, *Report to the Congress: EPA Is Slow To Carry Out Its Responsibility To Control Harmful Chemicals,* Washington, D.C., Oct. 28, 1980.
22. U.S. Congress, House Committee on Science and Technology, *Environmental Implications of Genetic Engineering,* hearings before the Subcommittee on Science, Research and Technology and the Subcommittee on Investigations and Oversight, House Committee on Science and Technology, U.S. Congress, June 22, 1983 (Washington, D.C.: U.S. Government Printing Office, 1983).
23. U.S. Congress, Office of Technology Assessment, *Impacts of Applied Genetics: Micro-Organisms, Plants, and Animals,* OTA-HR-132, Washington, D.C., April 1981.
24. U.S. Congress, Office of Technology Assessment, *Patent-Term Extension and the Pharmaceutical Industry,* OTA-CIT-143, Washington, D.C., August 1981.
25. U.S. Congress, Office of Technology Assessment, *World Population and Fertility Planning Technologies: The Next 20 Years,* OTA-HR-157, Washington, D.C., February 1982.
26. U.S. Congress, Office of Technology Assessment, *The Information Content of Premanufacture Notices: A Background Paper,* OTA-BP-H-17, Washington, D.C., April 1983.
27. U.S. Department of Health, Education, and Welfare, National Institutes of Health, "Guidelines for

Research Involving Recombinant DNA Molecules," *Federal Register* 41:27911-27922, July 7, 1976.
28. U.S. Department of Health and Human Services, Food and Drug Administration, "Minutes of the Industrial Practices Subcommittee Meeting," (Sept. 4, 1980), reprinted in *Recombinant DNA Research* 6:492-496, 1981.
29. U.S. Department of Health and Human Services, Food and Drug Administration, "Notice of Memorandum for Understanding with the United States Department of Agriculture Animals and Plant Health Inspection Service," *Federal Register* 47:26458, June 18, 1982.
30. U.S. Department of Health and Human Services, Food and Drug Administration, "Proposed New Drug and Antibiotic Regulations," *Federal Register* 47:46622-46663, Oct. 19, 1982.
31. U.S. Department of Health and Human Services, Food and Drug Administration, National Center for Drugs and Biologics, Office of Biologics, "Points to Consider in the Manufacture of *In Vitro* Monoclonal Antibody Products Subject to Licensure," Bethesda, Md., June 1983.
32. U.S. Department of Health and Human Services, Food and Drug Administration, National Center for Drugs and Biologics, Office of Biologics, "Interferon Testing Procedures: Points to Consider in the Production and Testing of Interferon Intended for Investigational Use in Humans," Bethesda, Md., July 1983.
33. U.S. Department of Health and Human Services, Food and Drug Administration, National Center for Drugs and Biologics, Office of Biologics, "Points to Consider in the Manufacture of Monoclonal Antibodies for Human Use," Bethesda, Md., July 1983.
34. U.S. Department of Health and Human Services, National Institutes of Health, "Guidelines for Research Involving Recombinant DNA Molecules," *Federal Register* 47:17180-17198, Apr. 21, 1982.
35. U.S. Department of Health and Human Services, National Institutes of Health, "Program To Assess the Risks of Recombinant DNA Research; Proposed Second Annual Update," *Federal Register* 47:55104-55109, Dec. 7, 1982.
36. U.S. Environmental Protection Agency, "Proposed Data Requirements for Pesticide Registration," *Federal Register* 47:53192, Nov. 24, 1982.

Chapter 16
Intellectual Property Law

Contents

Chapter 16
Intellectual Property Law

Introduction

Biotechnology will give rise to a vast array of new inventions. The inventions may be placed into two general categories: products and processes. Products will include organisms, such as genetically modified micro-organisms, cell lines, hybridomas, plants, and possibly even animals. Products also include parts of organisms and related material such as high expression plasmids, viral vectors, synthetic genes, probes, and restriction enzymes. Finally, there will be products of organisms, such as drugs, chemicals, biologics, and monoclonal antibodies (MAbs). Processes will include various ways to make new organisms or parts thereof or to use an organism to make some product such as insulin. Other examples of processes include various bioprocessing techniques, regeneration of plant tissue culture, breeding techniques, and methods of treating the human body. In addition, research and development (R&D) will give rise to new knowledge, which will be of value to whoever possesses it.

The ability to secure a property interest in an invention and to protect related know-how generally is perceived as providing an extremely important incentive for a private company to spend time and money to carry out research, development, and scale-up for the commercialization of new processes and products. Without the ability to prevent other companies from taking the results of this effort, many new and risky projects that could lead to important new products would not be undertaken. Empirically proving this notion, however, is difficult (47). It is beyond the scope of this chapter to delve into the debates among experts on that problem. This chapter will assume—as our society does—that the ability to secure property interests in or otherwise protect technological processes, products, and know-how will encourage development of technology. Therefore, one factor to evaluate in assessing U.S. competitiveness in biotechnology is how well the law of intellectual property of the United States and the five other major competitor countries—Japan, the Federal Republic of Germany, the United Kingdom, France, and Switzerland—allows inventors, private companies, and others to protect the results of their efforts.

The three categories of intellectual law most relevant to biotechnology are those dealing with trade secrets, patents, and plant breeders' rights. These are the focus of this chapter.* Copyright may also be relevant, because it protects the tangible expression of information, and a gene may be viewed as the tangible expression of information (36). Because this idea has not been widely accepted, and several commentators have criticized its usefulness (16,40,52), here it will not be discussed any further.

The categories of intellectual property law work together as a system. If one has disadvantages, a company can look to another. To the extent that a country has available many alternative ways for companies to protect biotechnological inventions, it is more likely to be competitive in biotechnology.

This chapter compares and contrasts the law relating to the protection of biotechnological inventions and related know-how in the United States, the United Kingdom, the Federal Republic of Germany, Switzerland, France, and Japan. The chapter begins by examining U.S. law in order to provide a basis for comparisons, raise the relevant issues, and explain some basic legal concepts.

*Two other areas of law are also relevant to biotechnology but will not be considered in this chapter: personal property law and contract law. Traditional personal property law will apply to cell lines and many other biological inventions because they are physical objects—just like cars and jewelry. Contracts create legally enforceable rights and duties between the contracting parties. Thus, biotechnological inventions can be protected by contract, and in view of some of the uncertainties in the intellectual property law regarding biotechnology, contracts can be important to biotechnology companies in many instances. These topics will not be considered further in this chapter, because OTA was unable to obtain information on how they would apply to biotechnology in other countries. Some commentators have addressed their applicability to biotechnology in the United States (10,40,42).

Foreign intellectual property laws are considered after the discussion of the U.S. law and also in appendix G. The strengths and weaknesses of the laws of the six countries are then analyzed by considering three basic questions: 1) what interests will the law protect; 2) how well will they be protected; and 3) what questions are unanswered? Policy options for Congress addressing the issue of how to improve U.S. competitiveness in biotechnology by strengthening U.S. intellectual property law are identified and discussed at the end of the chapter.

Intellectual property law of the United States

As noted above, three categories of intellectual property law are particularly relevant to biotechnology: trade secrets, patents, and plant breeders' rights.

Law of trade secrets

An inventor is regarded in the United States as having a natural right to keep an invention secret. This right is recognized by the law of trade secrecy. A trade secret is generally viewed as "any formula, pattern, device, or compilation of information which is used in one's business, and which gives him (sic) an opportunity to obtain an advantage over competitors who do not know or use it" (1).* Examples of trade secrets in biotechnology are a method for genetically manipulating an organism, a method for selecting among the organisms for those particular characteristics, and the organism itself.

The holder of a trade secret in the United States can enforce his or her interests in State courts by securing either an injunction or monetary damages against a person who takes or otherwise acquires the secret through improper means, or even against a person who acquired it through mistaken disclosure by the owner.** Criminal penalties may also be available in egregious cases in the majority of States. The underlying policy is that a person should not benefit by unfairly using another's efforts.

*In recognizing the existence of a trade secret, the courts do not use a hard and fast definition, but look at numerous factors, such as the extent to which the information is known outside of the business, the effort involved in developing and guarding the information, and the difficulty with which the information could be properly acquired by others (see 34).

**The cases also recognize secret information that does not qualify as a trade secret, but a person acquiring or using that information is liable only if he does so by "improper means" (1).

In the United States, virtually any biological invention, including cells and their components, or related information would be protectable by the law of trade secrets.*

It should be noted, however, there are some limitations on its scope. One important limitation arises from the fact that a trade secret must be continuously used in a business. This requirement raises questions about the results of basic research. Generally, the courts have held that if information is merely a preliminary idea, it does not qualify as a trade secret (41,51). Some degree of commercial value must be established if the information is to be considered a trade secret. A few States have taken a more expanded view of the concept of trade secret and protect information that also has only potential economic value. In those States—Arkansas, Idaho, Kansas, Minnesota, and Washington—the results of basic research clearly would be protected.

Another possible limitation on the scope of the law of trade secrets arises from the fact that the holder of a trade secret must know the information and attempt to keep it secret from others. In the well-known case involving disputed ownership of an interferon-producing cell line, *Hoffmann-La Roche, Inc.* v. *Golde* (28). Genentech (U.S.) and Hoffmann-La Roche (Switzerland) apparently argued that the University of California had no trade secret interest in the cell line because the university did not know about its ability to produce interferon (10).

The advantages of a trade secret to its holder are several. First, there is no time limit on trade

*Misappropriation of an organism or other tangible biological material constitutes misappropriation of the information it contains (see 53).

secret protection. It should be noted, however, that in a fast moving area like biotechnology, the "useful life" of a trade secret may actually be quite short. Second, a trade secret does not have to be a patentable invention. Third, maintenance and enforcement are generally less expensive for trade secret rights than for patents. Fourth, competitors are not apprised of the information, in contrast to the situation with patents (see below). Fifth, trade secret protection is valuable for certain inventions that would be hard to police if patented. For example, if a product is capable of being made by many different processes, keeping secret a new process for making the product might be preferable to patenting it. Sixth, if there is doubt as to the patentability of an invention, trade secrecy is a viable alternative. Finally, certain organisms and parts thereof, such as high-expression plasmids, may be better off held as trade secrets, since they could not be reverse engineered from the products that they produce, but, if patented, would be placed in the public domain.

Disadvantages of relying on trade secrecy include the following. First, the protection exists only as long as secrecy exists. The holder of a trade secret has no rights against someone who independently discovers and uses the trade secret and has no rights against someone who may have innocently learned the secret from someone who originally obtained it improperly. Second, reverse engineering (the examination of a product by experts to discover how it was made) is a legitimate way to discover a trade secret. The structure of a gene, for example, may be determined by reverse engineering a polypeptide that is on the market. Because of the complexity of biological processes and organisms, however, most of these will not be capable of being discovered by reverse engineering of their products. Third, trade secrecy is, by definition, incompatible with the desire of most scientists to publish the results of their research. If a company wishes to attract and retain good scientists, it may not be able to rely on trade secrecy to protect their work. Fourth, there is always the chance that a trade secret will be independently discovered by another, who then obtains a patent on it. The patent holder may then prevent the holder of the trade secret from using it. Finally, the acquisition of a trade secret by a competitor through misappropriation or breach of a confidential agreement may be difficult to prevent, discover, or prove. Micro-organisms are especially easy to steal, once one gains access to them, because of their small size and self-replicating nature. Further, the thief would not even have to understand exactly the valuable information contained in the micro-organisms; he or she has acquired the factory (i.e., the micro-organism) and the ability to grow it in any amount desired.

Patent law

U.S. patent law, Title 35 of the United States Code, is designed to encourage invention by granting inventors a limited property right in their inventions. A U.S. patent gives the inventor the right to *exclude* all others from making, using, or selling the invention within the United States without the inventor's consent for 17 years. In return, the inventor must make full public disclosure of the invention.

The policy behind U.S. patent law is twofold. First, by rewarding successful efforts, a patent provides inventors and their backers with an incentive to risk time and money in R&D. Second, and more importantly, the patent system encourages public disclosure of technical information, which may otherwise have remained secret, so that others are able to use it. The inducement in both cases is the potential for economic gain through exploitation of the patent right.

To qualify for patent protection in the United States, an invention must meet the following requirements:

- it must be capable of being classified as a process, machine, manufacture, or composition of matter;*
- it must be new, useful, and not obvious; and
- it must be disclosed to the public in sufficient detail to enable a person skilled in the same or the most closely related area of technology to construct and operate it.

*These categories are set out in §101 of Title 35 of the United States Code (35 U.S.C. §101). Sec. 101 is the basic section under which most inventions are patented. Patents under 35 U.S.C. §101 are often called utility patents.

Plants that reproduce asexually may also be patented under slightly different criteria.

The criteria for obtaining and enforcing patents on biotechnological inventions are quite similar in the six countries being examined in this report. The following eight subsections discuss the criteria of patentable subject matter, novelty, utility, nonobviousness, disclosure requirements, deposit requirements, claims, and enforcement in the United States in order to provide a basis for a comparative analysis of how each country's patent law will affect its competitiveness in biotechnology.

PATENTABLE SUBJECT MATTER

The categories of patentable subject matter under 35 U.S.C. §101—process, machine, manufacture, or composition of matter—are quite broad but they are not unlimited. The courts have held scientific principles, mathematical formulas, and products of nature to be unpatentable on the grounds that they are only discoveries of preexisting things—not the result of the inventive, creative action of human beings, which is what the patent laws are designed to encourage.

One of the major patent law questions arising with respect to biotechnology is whether living organisms are patentable subject matter. The U.S. Supreme Court addressed this question in 1980 in the landmark case *Diamond* v. *Chakrabarty* (21). In a five to four decision, the Court held that the inventor of a new micro-organism, whose invention otherwise met the legal requirements for obtaining a patent, could not be denied a patent solely because the invention was alive. The Court ruled that Congress had not intended to distinguish between unpatentable and patentable subject matter on the basis of living v. nonliving, but on the basis of "products of nature, whether or living or not, and human-made inventions" (22).

The U.S. Supreme Court stated that its decision in the *Chakrabarty* case was limited to a human-made micro-organism, leaving unresolved questions of whether eukaryotic cells or other higher organisms would be patentable subject matter. In theory, however, the *Chakrabarty* decision stands for the proposition that any organism is potentially patentable, because the crucial test used by the Court was whether or not the organism is human-made. As a result, eukaryotic cells, cell lines, tissue culture, and even plants are generally viewed as being patentable under 35 U.S.C. §101. The harder question is whether the U.S. Patent and Trademark Office or the courts would permit patents on higher organisms such as animals.*

There is no question, however, that virtually any other biotechnological invention would be patentable subject matter, providing that it meets the other requirements. Such inventions would include processes using micro-organisms, recombinant DNA (rDNA) molecules, subcellular units such as plasmids, methods for making these inventions, and biotechnological methods for treating human or animal disease (29).**

NOVELTY

The statutory requirement of novelty signifies that an invention must differ from the "prior art," which is publicly known technology. Novelty is not considered to exist, for example, if: 1) the applicant for a patent is not the inventor; 2) the invention was previously known or used publicly by others in the United States; or 3) the invention was previously described in a U.S. or foreign publication or patent (35 U.S.C. §102). The inability to meet novelty requirement is another reason why products of nature are unpatentable.

Two questions are particularly relevant to biotechnology. First, how can naturally occurring substances, such as genes, plasmids, and even organisms, be patentable? Second, what actions on the part of an inventor, such as discussing the invention with colleagues or publishing a paper about the results of research, can place the invention in a public domain, thus barring patentability because the invention will not be novel?

*The U.S. Patent and Trademark Office has stated that it will determine questions as to patentable subject matter on a case-by-case basis following the test set forth in *Chakrabarty* (49).

**A U.S. Patent and Trademark Office official estimated that there are currently 500 genetic manipulation related patent applications pending, that the office is receiving applications at the rate of 200 per year, and that the rate is increasing (46). These applications are classified in Class 435, Subclass 172 in the U.S. Patent Classification System (46). This classification is *not* coextensive with OTA's definition of biotechnology.

As to the first question, the crucial element of patentability for most biological inventions in the United States, as shown in the *Chakrabarty* case, will be the fact that the substance was in some way changed from the naturally occurring substance by human intervention. For example, although genes and regulatory sequences may be obtained from natural sources, it is the removal of the DNA sequences from their natural habitat and their joining to other DNA sequences that provides the human-made requirement of the *Chakrabarty* case. Thus, it is not the sequence that is new, but the environment, such as the host or flanking DNA regions (44).*

As to the second question, it should be noted that U.S. law, in contrast to the laws of most foreign countries, provides a 1-year grace period between the date of any publication by the inventor relating to the invention and the filing of a patent application. This grace period in the United States is generally viewed as favorable to the rapid dissemination of new scientific knowledge, because knowledge pertaining to an invention can be published without the inventor's foregoing the opportunity to file for a patent. Most countries other than the United States require the patent application to have been filed before the invention is disclosed, for example, in a scientific paper. This requirement is known as "absolute novelty" and will be discussed in greater detail in the section comparing and contrasting U.S. and foreign law.**

UTILITY

The utility standard in the United States is generally not a difficult standard for an invention to meet to qualify for a U.S. patent. There is one potential problem, however, with regard to biological inventions. Since the courts have held that an invention must show some practical or commercial utility (12,32,33), certain results of research that may be very important for research purposes (e.g., a new DNA probe or even certain organisms) may not meet the utility standard. This problem can generally be avoided by describing some practical use of the invention in the patent application, even if that use will not be the one that is of ultimate commercial value to the company.

NONOBVIOUSNESS

The nonobviousness standard that inventions must meet to qualify for a U.S. patent pertains to the degree of difference between the invention and the "prior art." An invention that would have been obvious at the time it was made to a person with ordinary skill in the relevant field of technology is not patentable (35 U.S.C. §103). The U.S. patent law requirements for nonobviousness and novelty together represent a policy that a patent should not take from the public something that it already enjoys or potentially enjoys as an obvious extension of current knowledge.

Given the fact that many of the basic techniques in biotechnology are well known and straightforward to competent scientists, how can the various inventions meet the nonobviousness standard? The answer is that biotechnology is still in many respects a very inexact science. Many of the various manipulations of genetic material, for example, will give unexpected results. Difficulty in the isolation or preparation of materials and the unexpected or superior nature of results are some of the criteria that would be used to show nonobviousness.

It is interesting to note that some scientists view hybridoma technology as more straightforward than rDNA technology. If this is true, patents may be more difficult to obtain for hybridoma technology than for rDNA inventions, necessitating a greater reliance on trade secrets. However, there are still many problems associated with human-human hybridomas, so broad patents may be able to be secured for inventions in that area. (See *Box D.—Patents on Hybridoma Inventions* for further information on patenting hybridoma technology.)

The nonobviousness requirement may present another problem for biotechnology. The rapid

*In a companion case to *Chakrabarty*, a lower court, the Court of Customs and Patent Appeals (now the Court of Appeals for the Federal Circuit), held that a purified culture of naturally occurring bacteria was patentable subject matter (3). For procedural reasons, the Supreme Court did not rule on this issue.

**Japan provides for a limited 6-month grace period for: 1) experimentation, publication, and papers presented before scientific organizations by the applicant; 2) unauthorized disclosure by third parties; and 3) displays at authorized exhibits. Otherwise, it is considered an absolute novelty country.

Box D.—Patents on Hybridoma Inventions*

Many scientists and others unfamiliar with the patent law have questioned how a technology invented by Kohler and Milstein in the mid-1970's and well known to practitioners in the field could give rise to patentable inventions. It is important to remember, however, that hybridoma technology has many technical problems associated with it—anyone who solves any one or more of those problems will likely be able to obtain a patent on that improvement in the state of the art. The following improvements, for example, would be potentially patentable (17):

- new myeloma cell lines that offer improvements over existing myeloma cell lines,
- new culture media that offer improved growth,
- new methods of fusion that offer significant improvement over those currently employed,
- new and improved selection procedures,
- new hybridomas that are more stable and consistent in the production of MAbs,
- new MAbs that react to antigens different from those that prior patented MAbs react to, or
- new methods of using MAbs whether for diagnostic kits, cell sorting, tissue typing, purification, or other uses.

There may be some problems with respect to patenting hybridoma technology. One relates to the perception that the U.S. Patent and Trademark Office is allowing fairly narrow claims with respect to hybridoma technology, particularly with regard to MAbs themselves. If this turns out to be true, it may be easy to "invent around" the patented invention. Furthermore, because hybridoma cell lines are often unstable and may change over time, there may be a problem with regard to enablement. However, this problem may be solved by freezing the cell line so that it is available to the public when desired yet not continuously replicating and possibly changing.

*See *Chapter 3: The Technologies* for a discussion of hybridoma/MAb technology.

development and complexity of the field will make it difficult to determine as of a given point in time what is ordinary skill or what is obvious.

DISCLOSURE REQUIREMENTS

The requirement for adequate public disclosure of an invention is designed to ensure that the public receives the full benefit of the new knowledge in return for the granting a limited monopoly to the patent holder. Thus, a U.S. patent, which is a public document, must contain a sufficiently detailed description of the invention to enable others in that field of technology to build and use the invention without "undue experimentation." This is known as the enablement requirement. The patent also must disclose the best mode known to the inventor for carrying out the invention at the time the patent application is filed.

In the case of biological inventions, satisfying the enablement requirement is a major hurdle. Because of their complex and unknown nature, many biological inventions, especially organisms, cannot be sufficiently described in writing to allow their predictable reproducibility on the basis of that description alone. Even with fairly precise techniques such as rDNA, random events provide uncertainty as to predicting the exact nature of the final product. There is always the possibility during the manipulation of DNA fragments, plasmids, and transformed organisms that random changes have occurred. The final product may in fact be quite different from the description provided by the experimenter, even though the experimentation process itself may have been accurately described.

This problem has been dealt with for patent applications on new micro-organisms or processes involving them by permitting the micro-organisms to be placed in culture depositories, where they are available to the public (31). The depository and the culture catalog number are then referenced in the patent application, and if the patent issues, the public gains access to the culture.* There is some debate over whether such things as plasmids must be deposited, because there is some question as to the reproducibility of the plasmids on the basis of a written description alone.**

*The case law has left open the possibility of satisfying enablement in ways other than through a deposit (25,31).

**One of the questions raised by the patent examiner in the pending Cohen-Boyer patent application on the products of rDNA technique, e.g., plasmids, was whether the application disclosed a reproducible way to make a certain plasmid (5).

In any event, the enablement requirement will be one major hurdle to the patentability of higher organisms because of the logistical problems associated with depositing those organisms.

DEPOSIT REQUIREMENTS

Deposit requirements in the United States have developed by court decision and administrative action. The practice of the U.S. Patent and Trademark Office has been to require a deposit to be made at a recognized depository no later than the patent application filing date (50). The office further requires that deposits be maintained for the life of the patent (50).

Along with the other five countries being considered in this report, the United States is party to the Budapest Treaty on the International Recognition of the Deposit of Microorganisms for the Purpose of Patent Procedure (14), which attempts to harmonize the deposit requirements of the signatory countries. Under the treaty, the signatory states recognize in their own patent procedures a micro-organism deposit made in another country if the deposit is made in a depository meeting the requirements of the treaty.* Thus, if the patent applicant is filing applications in several countries, only one deposit need be made. Deposits made under the treaty must be maintained for at least 30 years.

A potential problem that arises with respect to deposits should be noted. Although any valid patent must describe an invention with sufficient specificity so as to enable a person of ordinary skill in that technology to make the invention, there is a significant difference between describing an invention and actually turning it over to the other person. The know-how that is associated with the actual making and subsequent perfection of an invention clearly provides the inventor with an advantage over a competitor who must construct the invention from the description in the patent. Yet in the case of a micro-organism, the invention must actually be turned over to any competitor who desires it. In essence, therefore, the holder of a patent on a micro-organism that produces a commercially useful polypeptide such as insulin must turn his or her "factory" (i.e., the micro-organism) over to competitors. Given the current state of the technology, this situation is probably unavoidable. Possibly, however, consideration could be given to allowing various restrictions to be placed on access to the deposits.

CLAIMS

Claims are the precise language that define the boundaries of an invention protected by a patent. U.S. law permits a series of claims, ranging from broad to narrow, to be made with respect to an invention, so that if one or more of the claims are subsequently held invalid (e.g., for covering some of the prior art or being indefinite), the inventor may still be able to rely on a narrower invention. Of course, all of the claims could be held invalid.

The scope of permitted claims will be important for biotechnology. The scope is initially determined by what the U.S. Patent and Trademark Office will accept. In any new technology, the initial inventions tend to be broad and pioneering, so broad claims are usually permitted. As time passes, however, prior art develops and new extensions of the art become more obvious. Then, the claims permitted by the Patent and Trademark Office will be narrower. The Cohen-Boyer patent on the basic rDNA technique (U.S. Patent 4,237,224) is an excellent example of a broad, pioneering invention, although some commentators have questioned its validity (7). In the case of hybridomas and MAbs, however, there is some indication that the Patent and Trademark Office is being fairly conservative from the start. The data supporting this perception are largely anecdotal, because there have been few patents issued on hybridoma technology. If the claims being allowed are more narrow, however, the value of patents on this technology would be lessened.

A recent decision by the U.S. Patent and Trademark Office, *Ex parte Jackson* (24), has important implications for the scope of permitted claims on micro-organisms, cell lines, and processes for producing or using them (6). The case involved the isolation and purification of three strains of bacteria that made a new antibiotic. All three strains had been deposited and referenced in the patent application. Although the Board of Appeals

*The American Type Culture Collection in Rockville, Md., and USDA's Northern Regional Research Laboratory in Peoria, Ill., together with five foreign institutions, currently meet the requirements (45).

of the U.S. Patent and Trademark Office upheld a claim to producing the antibiotic by using a micro-organism selected from the deposited strains (or mutants thereof), it rejected a claim to producing the antibiotic by using any micro-organism of the same species on the grounds that the claim was not enabling. Thus, the scope of the patent on the applicant's process for producing the antibiotic will be limited, and others may be able to legally practice the invention by using other strains. This case, if broadly applied, may have a significant adverse impact on the incentive to patent many kinds of biotechnological inventions, because inventors may see the scope of patent protection as being too narrow.

Subsequent to patenting, the scope of the claims will be determined by Federal courts ruling in patent infringement suits. If the patent is upheld, the court has some discretion on how broadly to interpret the written claims. It will tend to interpret the scope more broadly for fundamental inventions. Sometimes the scope of the literal wording of the claims can be extended, if the infringing invention does substantially the same thing, by substantially the same means, and in substantially the same way, as does the patented invention, yet the literal wording of the claims in the patent for the invention does not cover the infringing invention (26). In such cases, the courts will interpret the claim as covering the infringing invention. This is known in patent law as the "doctrine of equivalents."

The fact that the claims define a new invention does not mean that the new invention does not infringe on a previously patented invention. For example, consider the Cohen-Boyer patent on the fundamental rDNA technique. Its existence will not prevent new applications of the rDNA technique from being patented (providing they also meet the other requirements of the patent law); however, the new inventions may infringe the Cohen-Boyer patent. Thus, for a holder of the new patent to make use of that invention, he or she may have to pay royalties to the owners of the Cohen-Boyer patent.

ENFORCEMENT

Patent infringement in the United States is defined as the unauthorized making, using, or selling of any patented invention within the United States (35 U.S.C. §271(a)). No liability for infringement exists prior to the date the patent is issued.

With respect to enforcing a patent, certain problems arise. One problem, generally not a problem for products but potentially a very serious problem for processes, is knowing whether or not an infringer is using the patent. If an unpatented product can be made by many different processes, the owner of a patent on one of those processes may have no way of knowing whether a product made by a competitor has been made by a different process or by the patent owner's process. This is a special problem for any process involving a micro-organism or cell line. To get a patent on such a process, a deposit must be made, making the micro-organism or cell line available to anybody who desires to use it. For this reason, processes using such organisms are likely to be held as trade secrets unless the process is truly a major advance.

Another problem with respect to enforcing process patents granted in the United States is the fact that the patented process may be used in other countries to make the same product, which can then be imported into the United States and compete with the product made by the owner of the U.S. process patent. Although many countries would define this action as infringement of that process patent, the United States does not. A remedy for the owner of the process patent is available through an action before the U.S. International Trade Commission. If the owner of the patent can prove that the foreign activity infringes the U.S. process patent and that importation of the product would injure an efficiently conducted U.S. industry (or prevent its establishment), the product can be excluded from the United States (19 U.S.C. §1337, §1337(a)). This remedy has been criticized as leaving much to be desired (39). However, one commentator has pointed out many substantial advantages of going this route as compared to an action in Federal district court (13). The requirement for proving injury to an industry is not as problematical as it might seem because the International Trade Commission has held that the domestic industry may consist of only one company, the U.S. patent owner (13). Thus, the issues of whether biotechnology is an industry or whether one imported product could injure that whole "industry" would not be relevant. In fact, an International Trade Commission action is one way the owners of the Cohen-Boyer patent might enforce it against foreign users of the rDNA process.

Another problem area relevant to biological inventions has been the general attitude of the courts in the United States toward patents. Despite a statutory presumption of validity, about one-half of all litigated patents are held invalid by the courts (48). There has been a certain judicial hostility toward patents because they are "monopolies," even though permitted by the U.S. Constitution and Title 35 of the U.S. Code (29). Certain language in U.S. Supreme Court decisions, for example, refers to such "monopolies" and states that patents must be construed very narrowly and must not be upheld on "mere gadgets" (27). In the 15 years before *Chakrabarty,* the Supreme Court had not ruled in favor of a single patent applicant or patentee (29).

On the other hand, this judicial hostility appears to be changing. In some recent U.S. Supreme Court decisions, including the *Chakrabarty* case, the Court has upheld the patents and has used broad language to do so (20,23).

PATENT V. TRADE SECRET PROTECTION*

Patents and trade secrets are alternative and not necessarily mutually exclusive ways to protect biotechnological inventions. Companies are likely to choose between them on a case-by-case basis. In choosing, they would evaluate the following factors:

- whether there is any significant doubt that the invention can meet the legal requirements for patenting,
- whether there is the likelihood of others discovering the invention independently or through reverse engineering,
- what the invention's projected commercial life is and how readily others could improve on it if it were disclosed in a patent,
- how easily the patent could be "policed,"
- whether it is a pioneer invention,
- the cost of the related R&D and regulatory approvals,
- whether there are any plans for scientific publication, and
- what the costs of patenting are versus reliance on trade secrecy.

The first factor speaks for itself. The next two factors require difficult decisions to be made on the basis of the characteristics of the invention and the competitive environment. If research to develop a particular product is widespread and intense (as is the case with interferon), the risk of a competitor developing the invention independently provides a significant incentive for patenting. On the other hand, reverse engineering by competitors is virtually impossible for most products of micro-organisms because of the variability and biochemical complexity of microbiological processes.

The fourth factor, how easily the patent could be policed, is especially relevant for processes. Greater protection may lie in keeping a process secret, even if the microbe and the process could be patented. This is especially true for a process that is only a minor improvement in the state of the art or that produces an unpatentable product already made by many competitors. The commercial life of the process might be limited if it were patented, beause infringement would be difficult to detect and not worth the time and money to prosecute. Reliance on trade secrecy might then extend its commercial life.

Most companies would patent truly pioneer inventions, which often provide the opportunity for developing large markets. Moreover, patents of this sort tend to have long commercial lives, since it is difficult to circumvent a pioneer invention and since any improvements are still subject to the pioneer patent. Furthermore, infringement is easy to detect because of the invention's trailblazing nature. This would be true for processes also.

*This section draws on the analogous section in OTA's report *Impact of Applied Genetics: Micro-Organisms, Plants, and Animals* (47).

High costs for research, development, and regulatory approval of products is a factor in favor of patenting because a company will want to protect its investment. The research-oriented pharmaceutical companies have traditionally relied on patents for this reason.

The last two factors involve considerations secondary to a product and its market. Obviously, any publication of the experiments leading to an invention forecloses the option of trade secrecy. Also, a company must evaluate the options of protection via either patenting or trade secrecy in terms of their respective cost effectiveness.

Plant breeders' rights statutes

Ownership rights in new varieties of plants are specifically granted by two Federal statutes: 1) the Plant Patent Act of 1930 (35 U.S.C. §§161-164) and 2) the Plant Variety Protection Act (PVPA) of 1970 (7 U.S.C. §2321 et seq.).

The Plant Patent Act, which covers new and distinct asexually reproduced varieties other than tuber-propagated plants or those found in nature, confers the right on the patent holder to exclude others from asexually reproducing the plant or from using or selling any plants so reproduced, for a period of 17 years. Because of the impossibility of describing plants with the same degree of specificity as machines and the inability to recreate a new plant solely from a written description, this law also liberalized the enablement requirement; the description need be only as complete as "reasonably possible."

PVPA provides for patent-like protection to new, distinct, uniform, and stable varieties of plants that are reproduced sexually, excluding fungi, bacteria, and first-generation hybrids. The breeder may exclude others from selling, offering for sale, reproducing (sexually or asexually), importing, or exporting the protected variety. In addition, others cannot use it to produce a hybrid or a different variety for sale. However, saving seed for crop production and for the use and reproduction of protected varieties for research is expressly permitted. The period of exclusion is 18 years for woody plants and 17 years for other varieties.

These acts are basically consistent with an international treaty designed to provide consistency in the international protection of plant breeders' rights—the International Union for the Protection of New Varieties and Plants—known as UPOV.* UPOV has been signed by 16 countries, including all those discussed in this chapter, but not all of those countries have yet conformed their laws to it.

Until the *Chakrabarty* decision, the Plant Patent Act and PVPA were generally viewed as the sole source of plant breeders' rights in the United States. The *Chakrabarty* decision raises the possibility of protecting plants under 35 U.S.C. §101, because the essential point of the decision is that a human-made organism is a "manufacture" or "composition of matter" as those terms are used in §101. Further, there is no indication in the decision that the Plant Patent Act and PVPA preempt protection for plants.

There would be certain advantages and disadvantages of securing protection of sexually and asexually reproduced plant varieties through §101. One advantage is that more than one claim could be presented, as opposed to the single claim permitted under the Rules of Practice relating to plant patent applications (37 C.F.R. §1.164) This would allow parts of the plant to be covered as well as the whole plant. Further, a patent grant under 35 U.S.C. §101 for a new variety would provide more comprehensive protection against infringement in certain situations.

The disadvantages of proceeding under 35 U.S.C. §101 are that other currently irrelevant sections of the patent law would come into play. For example, the Plant Patent Act (35 U.S.C. §162) significantly modifies the disclosure requirements of 35 U.S.C. §112, simply requiring that the description be as complete as reasonably possible. This would at least theoretically no longer be true. However, the use of depositories for plant material, as required for micro-organisms, could satisfy the enablement requirement. A further potential factor is the applicability of the nonobviousness

*The Plant Patent Act conforms, but PVPA does not. Since the United States is a party to UPOV, some changes in PVPA may be necessary. At this time, however, it is hoped that conformity can be achieved through administrative practices (45).

requirement of 35 U.S.C. §103. This test is inherently difficult for plant material.

On balance, the *Chakrabarty* decision is likely to provide yet another protection option which can, in certain circumstances, be very useful. For example, tuber-propagated plants such as potatoes, which are not patentable under the Plant Patent Act, would appear to be patentable under 35 U.S.C. §101.

Comparison of U.S. and foreign intellectual property law

Much of the analysis in this section is based on the more detailed description of intellectual property law of the Federal Republic of Germany, the United Kingdom, France, Switzerland, and Japan found in *Appendix G: Intellectual Property Law.*

Patent law

The Federal Republic of Germany, the United Kingdom, France, and Switzerland, along with seven other Western European countries, are signatories to a treaty that creates a European patent system. That treaty, known as the European Patent Convention (EPC), went into force on October 7, 1977. The EPC establishes a legal system for granting European patents through a single supranational European Patent Office and a uniform procedural system with respect to patent applications. The single European patent application, if granted, become a bundle of individual European patents, one for each of the countries designated by the applicant.* The EPC system and the resulting patents exist in parallel with the patent systems of the member countries. Enforcement, however, is handled by the individual member countries. The ultimate goal is for each of the member countries to adopt in its national law the same substantive law of patents set forth in the EPC. The following discussion compares the patent law of the EPC countries and Japan with that of the United States.

*A proposed European Community Patent Convention would take the EPC one step further by providing for a single patent covering the entire European Economic Community.

PATENTABLE SUBJECT MATTER

One of the most difficult problems facing the owners of biological inventions is the inability of the law to respond rapidly enough to keep pace with the development of the technology. This is especially a problem in the case of the law's definition of patentable subject matter. Questions about what constitutes patentable subject matter create a significant degree of uncertainty for owners of inventions.

One of the basic decisions to be made by owners of inventions is whether to maintain their inventions as trade secrets or to attempt to protect them by patents. An intelligent decision is nearly impossible when one does not even know which basic subject matter is patentable under the laws of particular countries. In the United States, the trade secret route can still be selected in the event that no patent protection is ultimately secured. In most foreign countries, including the United Kingdom, France, the Federal Republic of Germany, and Japan, however, pending applications are published before it is known whether patenting will be possible, thereby providing complete and enabling disclosure to the public, including samples of any deposited micro-organisms necessary to carry out the invention. Such publication usually occurs 18 months after the application is filed. This situation effectively precludes reliance on trade secrecy once a patent application is filed. As a result, there exists in many foreign countries today considerable disincentive to seek patent protection for certain types of biological inventions, particularly those involving basic genetic procedures and the resulting products. However, with respect to the five foreign countries under study here, much of the uncertainty surrounding subject matter patentability of biotechnological inventions has been resolved.

This uncertainty in many foreign countries may indirectly discourage U.S inventors from filing for patent protection in the United States, since there is no way available at present to confine within the United States the culture deposit samples which must be made available once a U.S. patent issues. While enabling disclosure theoretically is communicated upon issuance of a U.S. patent to all countries, regardless of whether corresponding protection is available or is actually sought in those countries, it is only in connection with many biological inventions that an applicant is required to provide also the physical means to carry out the invention, i.e., a self-replicating organism, which in many instances is a "factory" capable of carrying out the invention.

One important aspect of this problem of uncertainty in the definition of patentable subject matter is the uncertainty of classification of certain types of biological inventions. It is not clear in the case of certain lower organisms, for example, whether they are to be classified as plants, animals, or something else (e.g., protista) (see, e.g., 15,19). Fortunately, in the United States, it seems to be a matter of choosing between multiple options for protecting such subject matter by either utility patents or plant patents, but in most other countries, plants and animals are explicitly excluded from patentability. Thus, a definition may be determinative of patentability.

As a result of the 1980 U.S. Supreme Court's decision in the *Diamond v. Chakrabarty* case, the U.S. definition of patentable subject matter is very broad. It is broader than that under the EPC or any of the national laws of the five other countries being examined in this assessment. In contrast to the United States, the EPC, which has a very liberal definition of patentable subject matter, excludes methods for treatment of the human or animal body by surgery or therapy and diagnostic methods. Also, the EPC excludes plant and animal varieties and biological methods for producing them, which are apparently not excluded by *Chakrabarty*. In all other respects pertaining to biological inventions, the United States and EPC appear to permit patenting of the same general classes of subject matter. France, Switzerland, the United Kingdom, and the Federal Republic of Germany follow the EPC, except Switzerland does not allow patents on micro-organism themselves.

Japan's definition of patentable subject matter is essentially coextensive with the definition of the EPC, excluding processes in the fields of medicine, diagnosis, therapy, and pharmacology in which the human body is an indispensible element. However, certain microbiological inventions could be excluded from patentability in Japan if they are "likely to injure the public health." The situation with respect to plants and animals in Japan is unclear.

NOVELTY

U.S. law requires the patent application to be filed by the inventor. If two different applicants happen to have the same invention, the patent will issue to the one who invented it first. Hence, the U.S. system is called a "first-to-invent" system. The laws of the other five countries, in contrast to U.S. law, permit someone other than the inventor (e.g., the employer) to file the patent application. If there are two applications for the same invention, the patent will issue to the applicant who filed first. These countries thus have what is called a "first-to-file" system. The combination in the United States of a first-to-invent system with the provision of a 1-year grace period between the date of any publication relating to an invention and the filing of a patent application makes the U.S. system fundamentally different from nearly all foreign systems, which are generally first-to-file systems are characterized by absolute novelty (i.e., allow no grace periods).

This difference manifests itself in connection with prior disclosures by the applicant. Under U.S. law, the general rule is that a disclosure of an applicant's own invention cannot be used to prevent the applicant from obtaining a patent, unless the disclosure satisfies the requirements of one of the statutory bars under 35 U.S.C. §102 (18). For example, consider the following types of possible disclosure by an inventor of his or her own work:

1. Communicating with colleagues by telephone, letter or in person;

a. under expressed confidentiality;
b. with no indication as to confidentiality; or
c. under expressed nonconfidentiality.
2. Delivering a paper at a conference or seminar, orally only.
3. Delivering a paper at a conference or seminar, both orally and with a disseminated written text.
4. Submitting a paper for publication.
5. Submitting an abstract prior to a conference to the conference promoting organization.

Under U.S. law, items 1, 2, 4, and 5 would not bar patentability.* Item 3 will become a statutory bar 1 year after the paper is disseminated in some tangible form, assuming the disclosure was enabling.

Under the laws of the four Western European countries, items 2 and 3 would prevent the granting of a patent if they occurred before the earliest effective filing date (e.g., before a U.S. applicant filed a patent application in the United States which will later serve as a basis for claiming the right of priority in corresponding foreign applications).** Items 4 and 5 would normally not bar a patent, assuming that the paper and/or abstract were not disseminated to members of the public, (e.g., conference attendees) prior to the actual date the patent application was filed. This is based on the implied confidentiality under which submissions of this type are usually handled by publishers. Similarly, the concept of expressed or implied confidentiality prevents items 1(a) and 1(b) from constituting prior art under German law concepts, which commentators believe will apply to the EPC and other European countries (11). It appears that even item 1(c), in and of itself, does not necessarily constitute prior art under German principles, inasmuch as such a nonconfidential disclosure must be available to an *unlimited* number of persons (43). If the disclosure were limited to the colleagues contacted and not otherwise made freely available, it would not defeat novelty of a subsequently filed application. It is too early to tell how EPC law will develop on this issue. The same can be said for the United Kingdom, where introduction of the EPC novelty standards represents a significant change from prior law and practice.

The Japanese law provides a limited 6-month grace period for publications and papers presented before scientific organizations. Thus, items 1, 4, and 5 would not bar patentability, and items 2 and 3 would bar patentability after 6 months.

It must be noted that the above discussion regarding bars to patents because of lack of novelty is predicated on the assumption that the disclosure is enabling. If the disclosure is *not* enabling, even a published paper about the invention would not bar patentability.

Because of the different approaches with respect to novelty, the U.S. patent law provides a competitive advantage in that scientific information can be quickly disseminated in the United States without forgoing patent rights, if the application for a patent is filed within a year. This advantage is qualified by the fact that the inventor who also wishes to file abroad cannot publicly disclose the invention until the priority application is filed. The case of the Cohen-Boyer patent on the rDNA technique is a well-known example of a case in which the inventors were able to obtain a U.S. patent, even though they had published papers about the techniques, but were unable to file for foreign patents because of the absolute novelty requirement in other countries. The probable result will be a substantial loss of income from foreign royalties.

UTILITY

The U.S. patent law's requirement for practical utility differs slightly from the requirement of European and Japanese law for industrial applicability. The U.S. utility doctrine has been criticized by the American patent bar, but has not proved to be a major obstacle for industry (45). It has undoubtedly disadvantaged some researchers and simultaneously deprived the public of

*If a paper or proceedings of conference were published, however, then the inventor would be barred if he or she filed a patent application more than 1 year after the date the proceedings or paper were published. Also if the invention were sufficiently disseminated so that it was deemed to be "in public use," then the inventor would be barred by sec. 102(b) from patenting it after the expiration of the 1-year grace period.

**Under the Paris Union Convention, to which all six competitor countries subscribe, applications filed in any country within 12 months of the first filing in a member country have, as their *effective* filing date, the filing date of the first application. This is known as the "right of priority."

prompt disclosure of research on, for instance, new pharmacological compounds and processes that do not yet have an established utility (45). In some cases, effort has undoubtedly been wasted in establishing trivial or unimportant yet "practical" utilities for such inventions in order to satisfy the U.S. Supreme Court's definition (45). This problem will affect researchers in biotechnology to some extent, particularly those working with pharmaceuticals.

On the other hand, the foreign systems present a different problem of "utility." They exclude method inventions in the field of therapeutic or diagnostic treatment, at least those involving treatment of humans, as not being part of "industry." Thus, certain types of biological inventions (e.g., monoclonal antibody diagnostic assays) will not be patentable in EPC member countries or possibly in Japan, although patent protection can be obtained for them in the United States. This is, in most cases, not a serious obstacle, since patent protection is not precluded for the materials that are used in the excluded methods or the products of those methods.

DISCLOSURE REQUIREMENTS

U.S. disclosure requirements are stricter than those of the EPC and Japan. The U.S. law requires (35 U.S.C. §112):

- a written description of the invention,
- enablement both with respect to "how to make" the invention and also with respect to "how to use" the invention, and
- a disclosure of the best mode known to the inventor for carrying out the invention as of the time of filing.

As to the basic enablement standard, however, U.S. law does not differ substantially from the foreign laws. Under the U.S. law, the test of enablement is whether the invention can be carried out by a person of ordinary skill in the art without "undue experimentation" (30). This is another way of stating the requirement for "reproducibility" which is fundamental to European law.

As previously mentioned, compliance with the enablement requirement creates serious difficulties for many biological inventions, because such inventions may have been produced by random mutation and selection or another procedure that cannot be repeated with the certainty of obtaining the same results. The solution that has been adopted essentially worldwide is to permit a deposit of the appropriate biological material in a depository, from which samples will be made available to the public.

The Federal Republic of Germany's requirement for reproducibility raises additional obstacles to patenting a micro-organism itself. It requires that a patent application describe a repeatable procedure for reproducing with certainty the deposited organism apart from the deposit itself (i.e., "from scratch" so to speak) before a patent can be granted on the organism per se. This is not required if one claims only a method of using such a deposited organism. Thus, this requirement, in effect, could preclude patents on many microorganisms.

Neither the EPC countries nor Japan specify a best mode requirement in their respective laws. In the United States, the best mode requirement arguably requires the best producing microorganism strain to be deposited, but this issue is not resolved.

The written description of the invention requirement under U.S. law is not articulated as such in foreign laws, but a requirement similar in principle is applied in some situations under the laws of most countries.

DEPOSIT REQUIREMENTS

At present, uncertainty regarding the deposit requirements exists in many countries. The circumstances under which a deposit is necessary are not clearly spelled out. Moreover, before receiving a substantive examination on this question in the EPC, for example, the patent applicant must take action that has the effect of making the deposit, and also access thereto upon publication of the application, irreversible. In the United States, the same basic uncertainty exists, but the applicant need not make a commitment until after substantive examination is completed.*

*As a practical matter, however, if patent protection is sought in other countries, this irreversible effect will have taken place already, prior to conclusion of the examination in the United States because of the 18-month publication practice in other countries.

The United States does not have any explicit deposit requirements in the patent statute or rules thereunder. For deposits necessary in order to comply with the enablement requirement, however, certain requirements for the deposit have been developed by administrative action (50) and court decisions.

As far as timing and location of deposit, the U.S. practice is basically consistent with the practice most countries, i.e., the deposit is to be made no later than the patent application filing date and at a recognized depository (50). The United States does not have a specific list of recognized depositories and therefore maintains more flexibility than the EPC and certain national offices that do have such lists. Of course, the United States also recognizes deposits meeting the requirements of the Budapest Treaty.

The U.S. Patent and Trademark Office has required only that deposited cultures be maintained for the life of the U.S. patent (although any deposit made under the Budapest Treaty must be maintained for a minimum of 30 years). The EPC and many European countries have opted to apply the longer period of the Budapest Treaty to any deposit made in accordance with national law. This will require additional costs for the applicants in those countries.

Samples of deposited micro-organisms become available to the public under U.S. practice at the time the patent issues, after which time no restrictions on access are permitted. The situation in the United States is quite different than that in the EPC countries and Japan. In the EPC countries (except for Switzerland) and Japan, patent applications are published approximately 18 months after the effective filing date. Such publication, which also makes the deposit publicly available, may place foreign applicants at a disadvantage.

On the other hand, under many foreign systems, including the EPC, the patentee is entitled to maintain certain limited restrictions on those receiving samples of the deposited culture throughout the life of the patent. The restrictions also apply to cultures derived from the original one (EPC Rule 28(6)). The Federal Republic of Germany also allows territorial restrictions to be placed on deposited micro-organisms.

Potential problems exist in the present deposit system as a result of import/export restrictions imposed by countries. In one case, a German applicant was unable to perfect a deposit in a U.S. depository (one of two in the world which accepted his type of cell line) within the 12-month priority period because of health-oriented import restrictions imposed by the United States (9). It is also possible that a patentee could lose his or her rights entirely in a given country if that country imposed restrictions on the import of samples of a culture in a foreign depository that is otherwise recognized by its patent office. The same result could occur if the country in which the depository is located refuses to permit export of samples of the deposited culture. In the latter instance, however, the Budapest Treaty permits a second deposit to be made in another depository without loss of deposit date.

CLAIM PRACTICE

Claim practice in the United States is extremely liberal and is regulated primarily by the requirement for definiteness contained in the second paragraph of 35 U.S.C. §112. This fact, together with the fact that patentable subject matter in the United States is generally less restricted than in most other countries, results in an very broad freedom for an applicant to claim his or her invention in a U.S. patent application.

There is a dearth of experience with claims directed to the relatively new inventions of biotechnology, and the EPC itself is too new for any significant precedent. Existing precedent primarily involves processes for the use of micro-organisms.

Under U.S. practice, biological inventions can be claimed in many different ways. In addition to process claims directed to methods of genetic manipulation, the products thereof can be claimed with regard to their structure, or if their structure is not known, with regard to their chemical and/or physical characteristics or in terms of the process steps for preparing them. Despite this flexibility, however, the previously discussed *Jackson* case (24) indicates that the U.S. Patent and Trademark Office may impose significant limitations on the breadth of claims.

Some of the patent offices in foreign countries have taken positions similar to that taken in the *Jackson* case. Switzerland and Japan have refused to grant claims that are broader than the specific micro-organisms disclosed in the application and deposited (Swiss Patent Ordinance, Section 15.15.3, May 12, 1980; Japanese Examination Guidelines).

There is little reported precedent regarding judicial interpretation of claims pertaining to biological inventions in infringement cases. Nevertheless, one can extrapolate from general principles of claim interpretation in the various foreign patent systems. The law in most countries provides for application of the doctrine of equivalents in some form, although in some countries, including Japan, the scope of equivalents is apparently very limited. As a general rule, it can be said that the scope of equivalents must be determined on a case-by-case basis, depending on factors such as the degree of unpredictability of the technology (i.e., equivalents must be obvious to persons of ordinary skill) and the degree of advance which the claimed invention exhibits over the "prior art." The more unpredictable the subject matter, the smaller the scope of equivalents, whereas the more pioneering the invention, the broader the scope of equivalents. Biological inventions typically involve highly unpredictable phenomena; thus, claims are likely to be narrowly interpreted.

Even if it is assumed that a reasonable degree of equivalents will be given for biological inventions, the next problem is to determine what constitutes an equivalent. No precedent is available, and, of course, the determination will be made on a case-by-case basis. It would seem that good arguments can be made to the effect that closely related strains of the same species can be looked on as equivalents, that different species normally would not constitute equivalents, and that mutants of the basic strain would, in most instances, be expected to have equivalent properties to the basic strain (see 8).

ENFORCEMENT

The United States, the four European countries, and Japan define patent infringement in similar ways. The major difference is that, unlike the other countries, the United States does not grant extraterritorial effect to process patents by defining as infringement the importation of a product made by the patented process without the authorization of the patent owner.

The United States grants the basic remedies of injunction and monetary damages for infringement (35 U.S.C. §283, §284), as well as reasonable attorneys' fees to the prevailing party in exceptional cases (35 U.S.C. §285). The foreign countries provide for similar remedies. There are no criminal penalties provided under the U.S. patent statute, contrary to many foreign patent laws.

Enforcement of patents claiming biological inventions involves unique problems. The first is simply identification of infringing activity. Many of the products will be unpatentable for lack of novelty and will be manufactured in small quantities. Thus, it will be difficult to determine if a competing product infringes one's patented process. In addition, strains of micro-organisms can be altered through mutation and other modification techniques to produce different organisms that possess the same basic characteristics of the protected organism.

It may prove to be an essential, or at least important, element of the case for the patentee to establish that the alleged infringer actually derived his or her organism from a sample obtained directly or indirectly from the culture deposit of the patentee's organism. Without adequate controls on the access to samples of deposited strains, proof of this fact will be extremely difficult.

Proving the identity and equivalence of the patented micro-organism with an allegedly infringing micro-organism can also present difficult problems for the present state of this technology. The technology is still sufficiently undeveloped that much room exists for honest differences of opinion among experts. Most questions of infringement will probably turn out to be a battle between the respective parties' expert witnesses, until more objective criteria are established.

Trade secret law

Of the countries considered in this assessment, the Federal Republic of Germany seems to have the strongest statutory system for the protection of proprietary information, and its courts are most consistent in enforcement of those statutes. Switzerland's system, which closely resembles West Germany's, has also been very effective in protecting such information. However, Swiss law does not recognize as trade secrets the secrets held by professors, scientists, and others not engaged in a business (45). This could affect the exploitation of commercial rights by educational institutions in Switzerland.

The United States and the United Kingdom appear to be slightly less effective than the countries just mentioned in protecting proprietary information. The British courts emphasize the "confidential" over the "secret" aspects of such information. Breaches of confidence are therefore not tolerated, regardless of whether the particular information misappropriated fits within a pre-established "trade secret" category. The U.S. courts often overlook the breach of obligation aspect of misappropriation and concentrate on determining whether or not the information qualifies as a "trade secret." As a result, misappropriators of confidential information are sometimes held not liable in the United States, whereas they would be held liable for the same activity in the United Kingdom (45). Nevertheless, U.S. courts have shown much greater flexibility than their British counterparts in fashioning remedies that prevent the use of misappropriated information. Furthermore, U.S. law provides for criminal penalties in addition to the usual civil remedies provided for under U.K. law. Finally, the sheer mass of successful trade secret cases, including favorable rulings from the U.S. Supreme Court in the *Kewanee* case (38) and in *Aronson* v. *Quick Point* (4), indicates that the United States is probably more effective than the United Kingdom in safeguarding such information (45).

France does not have as strong a system for protection of proprietary information as the United Kingdom or the United States. French courts have been rather restrictive in defining the types of information that may receive protection and more protective of the employee who leaves with the employer's confidential information than the courts in other industrialized countries (45).

The protection of proprietary information in Japan has been improving over the last two decades, but still is not on a level with the protection in the United States or the major European countries. As Japan continues its development from a technology-importing country to a technology-generating country, further progress in this area may be expected (45).

Plant breeders' rights

CHOICE OF TYPE OF PROTECTION

A breeder of asexually reproduced varieties of plants in the United States will normally proceed under the Plant Patent Act. However, 35 U.S.C. §101 may provide a viable option. Although additional disclosure requirements for asexually reproduced plant material will be required (e.g., the deposit of plant material in a satisfactory depository), this is not an onerous burden. Moreover, with the depository, there is the additional advantage that the patented plant material will be available during the life of the patent for comparison purposes with any alleged infringing varieties. The public would also be able to practice the invention when the patent expired.

For sexually reproduced plant varieties, the principal advantages of proceeding under 35 U.S.C. §101, as opposed to PVPA, are the substantially reduced costs of filing a patent application (as opposed to an application under PVPA)* and the possible increased protection afforded by the patent as opposed to the protection certificate issued pursuant to PVPA. Moreover, whereas numerous judicial decisions have been rendered under the patent statutes, judicial interpretation of PVPA is relatively limited.

In the United Kingdom, the Federal Republic of Germany, Switzerland, France, and Japan, a single statute covers both sexually and asexually reproduced plant varieties. As previously noted, protection is in the form of protection certificates rather than patents. Therefore, there is no choice of the type of protection obtained in these countries.

*The cost of filing an application under PVPA is $1,000, as compared with the cost of filing a utility patent application ($150 for small entities and $300 for others).

LIMITATION ON PROTECTABLE VARIETIES

In the United States, only tuber-propagated plants or plants found in an uncultivated state are excluded from protection under the U.S. Plant Patent Act. As a practical matter, this exclusion affects only the Irish potato and the Jerusalem artichoke. All other plant varieties that can be propagated true to type through asexual reproduction can be protected. Similarly, under PVPA, only first-generation hybrids are excluded, and all other varieties otherwise meeting the act's requirements can be protected.

In most countries other than the United States, by contrast, the number of specific genera or species that can be protected is restricted. The 1978 UPOV Text requires only a very limited number of designated genera or species for a country to comply with the provisions of the text. Thus, the protection provided in the European countries and Japan is relatively limited when compared with the all-encompassing protection provided by the U.S. Plant Patent Act and PVPA.

EFFECT ON COMPETITIVENESS

With respect to plant breeders' rights, U.S. law provides a competitive advantage over the other countries. The scope of protection is much broader in terms of the types of varieties than can be protected, and U.S. law provides the additional option of using §101 of the patent law (35 U.S.C. §101).

Evaluation of effectiveness of intellectual property law to promote the development of biotechnology

United States

U.S. patent law embodies a number of pro-innovation features: a "first-to-invent" system coupled with a 1-year grace period; secrecy of the invention subject matter until grant of the patent; and, as a result of the latter, no requirement for owners of biological inventions to grant access to deposited cultures until after protective rights have been established. These features provide incentive for owners of biological inventions to utilize the patent system, thereby making their inventions known to the public to aid further development. They also provide a sufficient period of time for the patentee to develop a leading position in the technology before being forced to hand over his or her enabling disclosure (including means for immediately practicing the invention, in the case of culture deposit samples) to competitors, both domestic and foreign. The "first-to-file" systems in the other competitor countries do not provide these advantages to applicants.

Another strength of the U.S. system is the choice of protection routes it now offers to inventors. Developers of new varieties of plants can now choose between the special plant protection provisions of the law and the possibility of obtaining a utility patent.

The 1980 *Chakrabarty* decision has far greater significance than merely holding that living organisms constitute patentable subject matter under U.S. law. It, together with other recent cases, represents the first truly positive pronouncement in many decades from the U.S. Supreme Court regarding the role and value of the patent system in promoting and maintaining technological competitiveness of U.S. industry (37,45).* This should have an effect on the way in which the lower courts will treat patents in the future. In addition, creation of the new Court of Appeals for the Federal Circuit should provide uniformity and consistency at the appellate level, as well as a body of law that is well informed and respected by those whom the patent laws serve. The important role of trade secret protection has been reaffirmed by the Supreme Court in its 1974

*Justice Jackson was prompted to state in his dissenting opinion in *Jungerson* v. *Ostby & Barton Co.* (35) that: "The only patent that is valid is one which this Court has not been able to get its hands on."

Kewanee decision (38). Finally, the United States has responded to the needs of plant breeders of asexually reproduced varieties by adhering to UPOV, and conformity between UPOV and the Plant Variety Protection Act of 1970 involves only a matter of the time to necessary reconcile minor language differences. With these positive developments, the intellectual property law of the United States may be viewed as entering a period of unprecedented strength and vitality (45). It should play an important, positive role in the development of biotechnology in the United States and thereby aid the international competitiveness of U.S. companies.

There are also several weaknesses in the U.S. system. One is that the patentee is not permitted to maintain sufficient control over samples of deposited cultures. A second is that the U.S. system provides less protection for process inventions than foreign systems, because the U.S. system allows competitors to practice a patented process invention outside the United States (e.g., in a jurisdiction where patent protection may not be available) and import the product into the United States, thereby lowering the value of the U.S. process patent. This may prove to be particularly relevant to the field of biological process inventions, especially those inventions in connection with which the patentee is obliged to provide to competitors with a culture sample. The U.S. process patent holder has a remedy in the form of a proceeding before the U.S. International Trado Commission, but its usefulness has been questioned.

Foreign countries

It would appear that the United Kingdom, the Federal Republic of Germany, Switzerland, France, and Japan have provided adequate incentives under their intellectual property laws for development of biotechnology. All provide reasonably broad definitions of patentable subject matter, and most protect plant varieties, even though these are generally excluded from patent protection. Animal husbandry does not enjoy such widespread possibilities for protection. Trade secrets are adequately protected.

Some disadvantages or disincentives for the development of biotechnology can be seen in the rigid manner in which many of these countries approach the subjects of disclosure requirements, reproducibility, and culture deposits. In Switzerland and, to a large extent in the Federal Republic of Germany, micro-organisms per se are not protectable. This may not be a serious problem, at least not at this stage of the technology, in view of the other ways in which an invention can be claimed (e.g., as a process using the micro-organism). The practice in Europe and Japan of requiring access to deposited cultures upon the publication of unexamined applications can be viewed as a disincentive, and it may foster a greater reliance on trade secret protection. This could restrict the flow of information and thereby retard the development of the technology.

Findings

Although there is a large degree of uncertainty in most countries over what kinds of biotechnological inventions can be patented, much of this uncertainty has been resolved in the United States, the United Kingdom, the Federal Republic of Germany, Switzerland, France, and Japan. Of the six countries, the United States has the broadest interpretation of patentable subject matter for biotechnology. The EPC has adopted a broad interpretation of patentable subject matter in the field of microbiology, even though plant and animal varieties are excluded from patentability. This broad interpretation will make it possible to patent under the EPC most of the technology dealing with the techniques of genetic manipulation. The EPC has affected or will ultimately affect the law of the Federal Republic of Germany, the United Kingdom, France, and Switzerland. Switzerland now diverges from EPC practice, however, by not permitting micro-organisms per se to be patented. A major departure from U.S. law under the EPC and in Japan is the exclusion from patentability of therapeutic and diagnostic methods.

Japan appears to be moving in the direction of providing significant patent protection for biotechnology products and processes. One possible obstacle, however, is that Japan has strict health and safety guidelines regarding genetic research, which may bar patenting of genetically manipulated organisms viewed as hazardous.

The concept of utility in the patent laws of most foreign countries is based on industrial (including agriculture) applicability, which differs in interpretation from the utility standard in the United States. In countries with the former concept, including the five foreign countries discussed in this chapter, even products and processes of scientific research satisfy the utility requirement, as long as the basic endeavor falls into the broad category of industry; however, therapeutic and diagnostic methods do not. In the United States, certain chemical products and processes of research interest only are considered not to satisfy the utility requirement. The fact that utility under U.S. law includes utility in the therapeutic and diagnostic fields, however, helps U.S. competitiveness in biotechnology.

The four European countries studied here have an absolute novelty standard, with no grace period for either oral or written disclosures of an invention by the inventor before the date he or she files an initial patent application covering the invention. The United States has a 1-year grace period, and Japan has a limited 6-month grace period for presenting scientific papers before filing a patent application. In all of the countries, the novelty defeating disclosure must be enabling. Thus, the notion that *any* disclosure before filing a patent application will bar patentability is incorrect.

Most countries have a disclosure standard for inventions based on the concept of enablement. This standard typically includes an aspect of reproducibility, i.e., an invention must be repeatable with a fair degree of certainty and the results must not be merely randomly achievable. Particular problems in satisfying the disclosure standard have been encountered up until now in connection with many biological inventions. This situation has led to the practice of requiring a culture deposit of new micro-organisms used to carry out an invention or forming the subject matter of the invention. The Federal Republic of Germany has refused to grant patent protection on micro-organisms themselves in those cases where disclosure of a reproducible method for producing the micro-organism cannot be given apart from a culture of the micro-organism itself.

In those countries that publish unexamined patent applications (all but the United States and Switzerland of the six competitor countries), a serious problem for owners of biological inventions is the fact that deposited cultures can become publicly available before any patent rights are granted. Although the access to deposited cultures usually is granted with some safeguards in the form of assurances given by the recipient, these safeguards often do not adequately protect the valid interests of the technology owner (e.g., they usually are not geographically limited or do not restrict the activities of the recipient to only experimental use). In fact, it may be desirable to have some restrictions on access even after the patent grant, in view of the fact that the patentee must furnish a "working model" of the invention, which patentees in other fields are not required to do.

Because of the nature of biotechnology, special problems are faced by patentees in the enforcement of their rights. Apart from the general problems of policing for infringement, the possibilities for disguising the use of a biological invention by genetic manipulation will present difficult questions of law and fact. The law and practice of claim interpretation in this field are in their infancy. In the present state of the technology, it is likely that patent-granting authorities generally will limit claims to the specific organisms or parts thereof disclosed in patent applications.

All of the countries studied provide some element of legal protection for trade secrets. Most aspects of biotechnology lend themselves to protection via the trade secret route, and owners of such technology may rely on trade secrets when patent rights are uncertain or when they judge trade secrecy to be more advantageous in a particular case.

With the major international and national efforts regarding plant variety protection, culminating in the 1978 UPOV treaty, there is a trend toward providing such protection without requiring satisfaction of any enablement standard. The nature of the protection for plant varieties is different from traditional patent protection in that it protects basically against derivation and copying.

The U.S. intellectual property law system appears to offer the best protection for biotechnology of any system in the world. In general, it appears that the United States offers protection for broadest scope of biological subject matter, especially because of the 1980 ruling by the U.S. Supreme Court in the *Diamond* v. *Chakrabarty* case (21) that the inventor of a micro-organism could not be denied a patent solely because the invention was alive. The United States also offers some of the best procedural safeguards for inventors, including the 1-year grace period and no publication of the patent application before patent grant. In addition, the United States offers a choice of protection to plant breeders. Finally, the trade secrecy protection offered in the United States is as good as that offered in most countries, with the exception of the Federal Republic of Germany and Switzerland.

Issue and options

ISSUE: How could Congress improve U.S. competitiveness in biotechnology by strengthening U.S. intellectual property law?

Option 1: Pass a statute specifically covering living organisms and related biological inventions.

The advent of the new biotechnology has raised questions in the United States regarding what inventions will be patentable, under what conditions, and what the scope of protection will be. Although the *Diamond* v. *Chakrabarty* case in 1980 answered in the affirmative the basic question of whether living organisms would be patentable, other questions remain.

A statute specifically covering living organisms and related materials could help resolve this uncertainty. Greater certainty would allow companies to plan their R&D and marketing strategies better and in some cases would lower the financial risks involved. The result should be increased innovation. The alternative is to rely on case-by-case developments in the U.S. Patent and Trademark Office and the courts. Patent litigation is extremely expensive and may be unaffordable for small, new biotechnology firms.

Another argument in favor of a special statute is that it could help patentees to secure better ownership rights in biological inventions. The existing U.S. patent law was developed primarily for inanimate objects and processes. Living organisms are fundamentally different. Unlike a machine, a living organism reproduces itself and occasionally mutates during its lifetime. Furthermore, a living organism is extraordinarily more complex than any machine. Although the inventor of the most complex machine knows all of its parts and understands completely how it functions, no one knows all of the components of the simplest micro-organism or understands completely how it functions. Finally, many biochemical pathways in an organism are not unique to that organism; because there are many different ways to produce a product, a patent on one of the ways may provide only limited protection. In the case of biological inventions, therefore, there may be problems in meeting the enablement and written description requirements, in securing an adequate scope of protection for inventions, and in policing for infringement.

The complexity of living inventions will make it difficult to fully describe them. Although depositing a micro-organism in a culture collection may circumvent these difficulties with regard to enablement, it may be of little help in establishing novelty and the bounds of patent protection. Microbial taxonomy is an imprecise art. Micro-organisms have different characteristics in different environments, and taxonomists often disagree on their classification and description. Thus, it may be impossible to distinguish sufficiently a micro-organism for patent law purposes from similar ones created by other inventors or from ones existing in nature.

The fact that organisms reproduce may require a change in the definition of infringement. The law currently defines infringement as the unauthorized making, using, or selling of the patented invention. If someone took a patented organism from a public depository, reproduced it, and gave it away to many users, would this be infringement? One could argue that the person did not "make" the invention.

The fact that organisms mutate may cause problems with respect to the scope of the claims and infringement. For example, if a patented organism subsequently mutated, it might no longer be within the scope of the claims. Also, if the deposited organism is the standard against which infringement is measured, a patent holder may have difficulty enforcing the patent if the organism mutated after it had been deposited. On the other hand, culture deposits generally are preserved by freezing, so mutations may not be much of a problem.

Finally, there is the problem of adequately protecting a product that can be made many different ways, only some of which may be known at the time the patent application is filed. For example, because of the degeneracy of the genetic code, a particular protein can be made by various base sequences. Claiming a particular sequence will provide insufficient protection, and claiming the protein will not help if the protein is not novel. Claiming the novel organism is one solution, but others can easily construct different organisms to produce the same product.

These problems have been addressed in PVPA (and to a lesser extent in the Plant Patent Act), which could be used as a model. For a plant variety to be protected under PVPA, for example, it must be distinct, uniform, and stable. The definitions of these terms embody the concept that it is necessary only to know the important characteristics of a new plant variety in order to distinguish it from others and that only these characteristics need to be stable through succeeding generations. In addition, PVPA defines infringement to include unauthorized reproduction. If this approach were taken, the plant acts could be subsumed in the new statute.

There are several arguments against this option. First, any new technology raises questions about the scope and nature of patent protection, and many of these will only be able to be resolved on a case-by-case basis rather than by statute. Second, most patent attorneys argue that the patent laws are flexible enough to accommodate any new technology, including biotechnology. Third, despite the possible limitations in applying the patent law to living organisms, utility patents actually may provide the patentee with the greatest degree of protection when compared to the protection provided by a statute like PVPA. One of the principal reasons is that a multiplicity of claims is permitted for utility patents, which could cover components of organisms, whereas just the plant itself (and its seeds) is covered by a plant variety protection certificate. Fourth, many experts would argue that since the *Chakrabarty* case resolved the fundamental issue—the patentability of living organisms—there is no need to undertake the major effort needed to pass legislation to solve more minor problems. In addition, since there is some degree of public sentiment against patenting living organisms, the fundamental issue also would be raised again. Finally, a new statute would create its own new issues and questions of interpretation.

Option 2: Allow patentees to place restrictions on micro-organism cultures supplied to third parties.

U.S. patent law requires complete and enabling disclosure of an invention in order to place it in the public domain. In the case of patented micro-organisms, the patentee is in effect required to turn over more than his or her invention—the micro-organism is virtually a complete "factory" ready to begin production. For this reason, inventors may be more inclined to rely on trade secrets than on patents, and the public will not gain the benefits of the new knowledge embodied in their inventions. This problem is even greater for process patents involving micro-organisms, which are difficult to police. Reasonable restrictions on micro-organism cultures supplied to third parties, designed not to prevent public access to the culture deposits but to prevent patent infringement, would be consistent with the aims of the U.S. patent system. For example, restrictions might be placed on how the organism is used and subsequent transfers of the organism. The other countries surveyed, particularly the Federal Republic of Germany permit certain restrictions on culture deposits.

A drawback to the option might be that the restrictions could hinder the dissemination of information, one of the fundamental goals of the patent law. However, those who support such restrictions argue that they can be devised so as not to hinder the dissemination of information yet prevent infringement, another important goal of the law.

Option 3: Amend section 271 of Title 35 of the United States Code to define as infringement the importation of a product made outside the United States by a process patented in the United States without the authorization of the patent owner.

The four Western European countries and Japan grant extraterritorial effect to their process patents in the way that is envisioned by this option. Although U.S. law provides a different remedy to the situation—an action for unfair competition before the International Trade Commission—many attorneys believe this remedy leaves much to be desired. This option would strengthen the patent system by providing an additional way for patentees to protect their rights. Although the effect of this option would not be limited to biotechnology, it would be important to this technology because of the ease with which micro-organisms used in patented processes can be acquired and used by overseas competitors. Many companies using biotechnology and their patent attorneys see this option as a potentially important part of their program to protect the results of their R&D efforts.

A bill to implement this option (H.R. 3577) was introduced on July 14, 1983.

Chapter 16 references*

1. American Law Institute, *Restatement of Torts,* sec. 757 comment (b) (Chicago: 1939).
2. American Law Institute, *Restatement of Torts,* sec. 759 (Chicago: 1939).
3. *Application of Bergy,* 596 F.2d 952 (C.C.P.A. 1979).
4. *Aronson* v. *Quick Point,* 440 U.S. 257 (1979).
5. *Biotechnology Law Report,* "Patent Office Withdraws Cohen-Boyer Plasmid Patent From Scheduled July 13th Issue and Rejects All Claims," 1:76, 1982.
6. *Biotechnology Law Report,* "Board of Appeals Decision on Abbott Laboratories Patent Application Takes Tough Stance on Enablement of Microbiological Inventions," 2:66, 1983.
7. *Biotechnology Newswatch,* "Rejection of Cohen-Boyer Patent No. 2 Exposes No. 1 to Possible Invalidation," Aug. 16, 1982, p. 4.
8. Blum, U. D. E., "Der Patentschutz für Mikrobiologische Erzeugnisse" (Patent Requirements for Microbiological Inventions), Dissertation *Verlag Ostschweiz AG,* University of Zurich, 1979, pp. 114-120.
9. Boeters, H. D., "Erfahrungen bei der Hinterlegung von Gewebekulturen," Mitteilungen der deutschen Patentwälte, 1982, pp. 73-74.
10. Bomse, S., "Rights in the Products of Biotechnology: Issues and Strategies," *Industrial Applications of Genetics Engineering: The Legal and Regulatory Regime* (Philadelphia: American Law Institute, American Bar Association, 1983).
11. Bossung, V., "Stand der Technik and eigene Vorverlautbarung in internationalen, europaeischen und nationalen Patentrecht," *Gewerblicher Rechtsschutz und Urheberrecht International Part,* 1978, pp. 381-398.

*Notes: C.C.P.A. = Court of Customs and Patent Appeals.
F.2d = Federal Reporter, Second Series.
Ill. App. 2d = Illinois Appellate Court Reports, Second Series.
N.E.2d = North Eastern Reporter, Second Series.
U.S. = U.S. Supreme Court Reports.
U.S.P.Q. = U.S. Patent Quarterly.

12. *Brenner* v. *Manson,* 383 U.S. 519, 153 U.S.P.Q. 45 (1966).
13. Brunswold, B., "Analysis of the United States International Trade Commission as a Forum for Intellectual Property Disputes," *J. Pat. Off. Society* 60:505, 1978.
14. Budapest Treaty on the International Recognition of the Deposit of Microorganisms for the Purpose of Patent Procedure, *Official Gazette* 961:21-16, Aug. 23, 1977.
15. Cooper, I. P., "Arzberger Under the Microscope: A Critical Reexamination of the Exclusion of Bacteria From Plant Patent Protection," *Pat. & T.M. Rev.* 78:59, 1980.
16. Cooper, I. P., *Biotechnology and the Law,* sec. 11.02 (New York: Clark Boardman & Co., 1983).
17. Cooper, I. P., "Patent Strategies for Biotechnology Products," presented at a conference on Biotechnology: Drugs, Biologics, Medical Devices, Institute for Applied Pharmaceutical Sciences, East Brunswick, N.J., 1983.
18. *Corning Glass Works* v. *Brenner,* 175 U.S.P.Q. 516 (D.C. Cir. 1972).
19. Daus, D. G., Bond, R. T., and Rose, S. K., "Microbiological Plant Patents," *IDEA* 10:87, 1966.
20. *Dawson Chemical Co.* v. *Rohm & Haas Co.,* 448 U.S. 176 (1980).
21. *Diamond* v. *Chakrabarty,* 447 U.S. 303 (1980).
22. *Diamond* v. *Chakrabarty,* 447 U.S. 310 (1980).
23. *Diamond* v. *Diehr,* 450 U.S. 175 (1981).
24. *Ex parte Jackson,* 217 U.S.P.Q. 804 (1983).
25. *Feldman* v. *Aunstrup,* 517 F.2d 1351, 186 U.S.P.Q. 108 (C.C.P.A. 1975).
26. *Graver Tank & Mfg. Co.* v. *Linde Air Products Co.,* 339 U.S. 605, 85 U.S.P.Q. 328 (1950).
27. *Great A&P Tea Co.* v. *Supermarket Equipment Corp.,* 340 U.S. 147 (1950).
28. *Hoffmann-La Roche, Inc.* v. *Golde,* Civ. No. 80-3601 A.J.Z. (N.D. Cal.).
29. Halluin, A., "Patenting the Results of Genetic Engineering Research: An Overview," *Banbury Report 10: Patenting of Life Forms* (New York: Cold Spring Harbor Laboratory, 1982).
30. *In re Angstadt,* 537 F.2d 498, 190 U.S.P.Q. 214 (C.C.P.A. 1976).
31. *In re Argoudelis,* 434 F.2d 1390, 168 U.S.P.Q. 99 (C.C.P.A. 1970).
32. *In re Joly,* 576 F.2d 909, 153 U.S.P.Q. 45 (C.C.P.A. 1967).
33. *In re Kirk,* 376 F.2d, 936, 153 U.S.P.Q. 48 (C.C.P.A. 1967).
34. *Jet Spray Cooler, Inc.* v. *Crampton,* 282 N.E.2d 921, 925 (1972).
35. *Jungerson* v. *Ostby & Barton Co.,* 335 U.S. 560 (1949).
36. Kayton, I., "Copyright in Living Genetically Engineered Works," *Geo. Wash. L. Rev.* 50:191, 1982.
37. Kayton, I., *The Crisis of Law in Patents* (Washington, D.C.: Patent Resources Group, 1970).
38. *Kewanee Oil Co.* v. *Bicron Corp.,* 416 U.S. 470 (1974).
39. Kiley, T., "Learning To Live With the Living Invention," *APLA Q.J.* 220:228, 1979.
40. Kiley, T., "Speculations on Proprietary Rights and Biotechnology," *Banbury Report 10: Patenting of Life Forms* (New York: Cold Spring Harbor Laboratory, 1982).
41. Milgrim, R. M., *Trade Secrets,* sec. 2.02(2) (New York: Mathew Bender, 1981).
42. Ostrach, M., and Nyari, L., "Arrangements With Respect to Transfer of Biological Materials." Industrial Applications of Genetic Engineering: The Legal and Regulatory Regime (Philadelphia: American Law Institute, American Bar Association, 1983).
43. Rauh, P. A., Patent Attorney, Munich, personal communication to Richard Schwaab, OTA contractor, Aug. 3, 1982.
44. Rowland, B., "Are the Fruits of Genetic Engineering Patentable?" *Banbury Report 10: Patenting of Life Forms* (New York: Cold Spring Harbor Laboratory, 1982).
45. Schwaab, R., Jeffery, D., and Conlin, D., "U.S. and Foreign Intellectual Property Law Relating to Biological Inventions," contract report prepared for the Office of Technology Assessment, U.S. Congress, 1983.
46. Tegtmeyer, R., "Patenting of Genetic Engineering Technology," speech presented to the National Academy of Sciences' Convocation on Genetic Engineering of Plants, Washington, D.C., 1983.
47. U.S. Congress, Office of Technology Assessment, *Impacts of Applied Genetics: Micro-Organisms, Plants, and Animals,* OTA-HR-132, Washington, D.C., April 1981.
48. U.S. Department of Commerce, "Industrial Advisory Subcommittee Report on Patent Policy," *Advisory Committee on Industrial Innovation: Final Report,* Washington, D.C., 1979.
49. U.S. Department of Commerce, Patent and Trademark Office, *Manual of Patent Examining Procedure,* sec. 2105, Washington, D.C., 1980.

50. U.S. Department of Commerce, Patent and Trademark Office, *Manual of Patent Examining Procedure,* sec. 608.01(p), Washington, D.C., 1980.
51. *Vendo Co.* v. *Stoner,* 105 Ill. App. 2d 261, 245 N.E.2d 263 (1969).
52. Whale, A., "Patents and Genetic Engineering," *Intellectual Property L. Rev.* 14:93, 1982.
53. *Yoder Brothers, Inc.* v. *California-Florida Plant Corp.,* 537 F.2d 1347, 1365 (5th Cir. 1976).

Chapter 17

University/Industry Relationships

Contents

Table

Chapter 17

University/Industry Relationships

Introduction

The recent spectacular advances in molecular biology in the United States have arisen from basic research, most of which is federally funded and carried out in university laboratories. Led by the promise of biotechnology's commercial potential and the need for technical expertise, U.S. and foreign companies have been developing closer ties with universities, thus intensifying the process of university/industry technology transfer. At least in the United States, concerns have been raised about industrial sponsorship of university research (1,4,8,13,25,26). Some of these concerns are actually not new. What is new is that biology, rather than chemistry or engineering, is suddenly commercially promising.

This chapter focuses on university/industry relationships as a factor influencing the competitive position of the United States vis-a-vis other countries in the commercialization of biotechnology. Issues in university/industry relationships are not confined to relationships in biotechnology, so the chapter also includes some discussion of broader university/industry issues that have implications for competitiveness in biotechnology. The resolution of issues in U.S. university/industry relationships in biotechnology is extremely important, because the manner in which these issues are resolved will help determine the pattern of basic and applied research in the field for the next decade or so. Furthermore, research is likely to be critical to the development of biotechnology for some time.

Closer ties between universities and industry can be advantageous to the institutions involved and are important for the national innovative process. Industrial research questions can enrich the university research process, and there are financial benefits from increased industrial funding of university research. Industrial support of university research and development (R&D) in the United States currently represents about 6 to 7 percent of the total research budget of universities, although the percentage of industrial funding in some departments of universities may be much higher or lower (19). It is unlikely that industrial support will ever equal Federal support of university research, but increases in industrial funding could have significant effects on the types of research performed, especially in high-technology areas such as biotechnology.

American universities can expect some financial benefit from royalties derived from the licensing of patents, although it is unlikely that royalty income will ever be a significant portion of support. The Wisconsin Alumni Research Foundation (WARF), for example, has been instrumental in generating royalty income for the University of Wisconsin. It should be noted, however, that 39 of the 58 income-producing inventions assigned to WARF since 1925 have earned less than $100,000, and only 7 have earned more than $1 million (3,9). As shown in table 66, royalty income as a proportion of total Federal support is far less at nine other leading research universities in the United States than at the University of Wisconsin. If Public Law 96-517, the 1980 law that allows universities and small businesses to retain patent rights for federally funded research, encourages the development and marketing of products, U.S. universities' royalty income may increase. Stanford University and the University of California at Berkeley have already benefited from royalties (approximately $2 million) for the Cohen-Boyer patent for the basic recombinant DNA (rDNA) process. However, university income from biotechnology may be more dependent on whether the firms developing and marketing biotechnological products or processes rely primarily on patents or on in-house research.* If the more usual operating mode becomes in-house industrial research, then royalty income to universities may not be significant.

*The advantages and disadvantages of relying on patents or trade secrets to protect intellectual property are discussed in *Chapter 16: Intellectual Property Law*.

Table 66.—License and Patent Activity at 10 Leading U.S. Research Institutions

Institution	Fiscal year 1980 Federal R&D support: Total ($000)	Fiscal year 1980 Federal R&D support: Life sciences ($000)	Type of activity	Current annual number of disclosures	Current annual royalty income (thousands of dollars)
1. Johns Hopkins	$239,869	$60,275	Licensing program	20	$ 90
2. MIT	141,011	24,200	Licensing program	164	1,500
3. Stanford University	104,011	43,712	Licensing program	140	2,500
4. University of Washington	100,567	54,968	Research foundation	28	120
5. University of California, San Diego	90,703	37,327	Licensing program[a]	320[a]	1,700[a]
6. University of California Los Angeles	87,073	52,606	Licensing program[a]		
7. Harvard University	83,997	53,962	Licensing program	60	50
8. Columbia University	81,361	49,383	—	20	Minimal
9. University of Wisconsin	80,460	43,342	Research foundation	75	6,000 (with investment income)[b]
10. Cornell University	74,761	37,900	Research foundation	50	1,300

[a]University of California system.
[b]Investment income is the substantial portion.

SOURCE: G. S. Omenn, "University-Corporate Relations in Science and Technology: An Analysis of Specific Models," *Partners in the Research Enterprise*, T. W. Langfitt, S. Hackney, A. P. Fishman, et al. (eds.) (Philadelphia: University of Pennsylvania Press, 1983).

There are potential disadvantages to closer university/industry relationships, but some problems can be avoided if participants are aware of potential difficulties and adequate safeguards are in place. One potential disadvantage of closer relationships might be a tendency to increased secrecy on the part of university faculty; it should be noted, however, that some secrecy has always existed when a particular faculty member is close to a new discovery. A second potential disadvantage is the danger that basic research in universities will be directed toward profitable lines of inquiry instead of toward interesting questions raised by past or recent findings. This might occur if there were a precipitous decline in Federal support for research in universities and universities had to turn increasingly to industry for financial support. A third potential problem is that some universities might be associated with products and processes linking them to lawsuits for damages, causing subsequent impairment of the universities' impartiality and credibility. Finally, there is the danger that universities that traditionally have competed for the best faculty might compete instead for the most lucrative industrial contracts.

In general, the purposes of universities in the United States are education, the conservation of knowledge, and the pursuit of unrestricted knowledge. The ends of a university and its faculty are pursued in a relatively open environment that allows the exchange of ideas and unrestricted publication of research findings. This does not mean that there is no competition among scholars, nor is it to deny that secrecy can accompany the desire to be first to announce a discovery (31). Similarly, it does not mean that the pursuit of knowledge for its own sake cannot be diverted by the funds currently available for particular kinds of endeavors (e.g., a "war on cancer" or secret government research). Generally, however, the pursuit by universities of the principles of openness, aided by generous Federal funding for basic research, has enabled the United States to build the greatest research capability in the world.

In contrast to the purposes of universities, the goal of industry is to make a profit, and the mode of achieving this goal is competition. Industry is output oriented, i.e., industry aspires to the efficient production of goods and services. When a company pays for research, it may expect ownership of the results long enough to justify the investment to bring the product to market. In an industrial setting, there is less willingness than there is in a university setting to share research materials; such materials are often kept as trade secrets. The reason for greater secrecy in industry is that development of a product is often risky, costly, and fraught with many obstacles along the route to success. Although the costs of developing and marketing a product vary among industries and products, the development of a pharmaceutical product in the United States can cost from $50 million to $75 million, with no guarantee of profit (27). Thus, in industry, achieving a competitive edge in a market necessitates guarding communication and intellectual property, an operating mode quite opposite from that of universities (6,13).

Industries and universities undertake partnerships in biotechnology for a variety of reasons, ranging from the desire by industry to gain access to new technology, to gain a lead time in basic knowledge, or to obtain trained personnel, to the need by universities to fill shortfalls in funding. Ultimately, it is hoped, the effect of the partnerships in the United States will be to facilitate and speed up the process of domestic technology transfer, since this is critical to the maintenance of a competitive position by the United States. Examining U.S. university/industry relationships in biotechnology is necessary in order to gain insight into the process of technology transfer and to determine if technology is being transferred in a spirit of cooperation and without compromising the goals of two very different institutions.

E. David has described the fundamental characteristic of optimal technology transfer between universities and industry as a two-way synergistic process between equal partners (6). Basic research, usually carried out in universities, is essential to the process. It is important to note, however, that basic science itself cannot progress without advances in technology, which often is developed by industry. Just as, for example, Galileo and Newton could not have made their contributions to astronomy without the invention of the telescope, the recent advances in molecular biology could not have been made without advances such as the electron microscope, X-ray crystallography, radioisotope labeling, and chromatography. Thus, universities and industry alike must accept the requirements of the other institution and enter into agreements that maximize the ability of each to maintain its standards and goals.

The effectiveness of university/industry relationships in biotechnology transfer

Since most of the university/industry relationships in biotechnology are new, it is difficult to ascertain how effective the relationships in the United States will be in transferring the technology between universities and industry. An estimate of their effectiveness can be made however, by considering the following questions:

- Why are university/industry relationships in biotechnology being formed?
- Are the relationships working smoothly?
- Has the way research is done in university laboratories or the quality of university research been affected by the relationships?
- Has collaboration among university researchers been affected?
- Has the quality of education been affected?
- Are there lessons to be learned from university/industry relationships in fields such as microelectronics?
- What forms are university/industry relationships in biotechnology taking and what are the associated issues?
- Are university policies with respect to university/industry relationships (e.g., patent and royalty agreements, handling of tangible research property, and conflicts of interest) being formulated?
- What is the likely future of university/industry relationships in biotechnology?
- And finally, how effective is university/industry technology transfer in countries likely to compete with the United States in biotechnology?

At the request of OTA, two contractors interviewed university administrators, faculty, and graduate students (principally in biotechnology) from the University of California, Berkeley, the University of California, San Francisco (UCSF),

Stanford University, Harvard University, Massachusetts Institute of Technology (MIT), and Johns Hopkins University, and representatives from 15 companies (a mix of new biotechnology firms and other companies moving into the biotechnology area) to obtain their opinions. Although this sample was not statistically representative, it included some of the major U.S. companies and research institutions working in biotechnology; thus, the opinions came from individuals active and knowledgeable in the field.

Why are university/industry relationships in biotechnology being formed?

OTA found an almost unanimous consensus among both university and industry representatives in the United States that universities are seeking money from their relationships with industry, motivated in part by a reduction or fear of reduction in Federal funding. Industry representatives believe that universities want to gain more real-world exposure for faculty and students and offer them a look at "economic reality" (18). In addition, some faculty stated that industrial funding requires less administrative work and is longer term than Government-funded renewable grants.

Are the relationships working smoothly?

The perception of most of the respondents in OTA's survey is that university/industry arrangements in the United States are working well. The initial administration of agreements between universities and industry in the area of biotechnology was inefficient, because new policies were being formulated and new players (biologists, in contrast to engineers or chemists) are now involved in interactions with industry. In addition, some research administrators have had to learn how to administer technology transfer agreements (18). Some individuals have speculated that agreements are working well because there are almost no biotechnology products yet. Disagreements may arise, especially in limited partnerships, when product sales revenues are generated (18).

Has the way university research is done or the quality of university research been affected by the relationships?

Respondents in OTA's survey were asked to consider two potential effects of university/industry relationships on U.S. university research: effects on the way research is done (its focus or methodology) and effects on the quality of the research. Nearly 85 percent of those responding believed that university/industry relationships in biotechnology have had no effect on the way research is done, and virtually all believed there has been no change in the quality of research.

Has collaboration among university researchers been affected by the relationships?

Almost 85 percent of the respondents in OTA's survey who had an opinion about this issue believed that university/industry relationships in biotechnology have had no substantial effect on the exchange of information or the collaboration that has existed among U.S. university researchers. Most respondents believed that there is only limited collaboration in rapidly evolving areas of science anyway and that levels of communication vary among faculty. Industry representatives commented that faculty having consulting arrangements should keep proprietary information confidential (18).

Has the quality of education students receive been adversely affected by the relationships?

Slightly more than half of those who responded to this question said there has been no change in the quality of education students receive. The others said that if there has been any effect, it has been to enhance the quality. Two forces will probably keep the quality of education at American universities unaffected by university/industry relationships in biotechnology. First, the goal of the faculty and university administrators to protect and maintain standards of academic excellence will continue to influence the arrangements that universities make with industry. Second, students themselves can be expected to monitor the situation and act to prevent any deterioration in the quality of education they receive. Some students have encountered problems at the University of California, Davis and Stanford University campuses, for example, and seminars and meetings have been held to address them. Faculty and university administrators have been involved in efforts to address the problems and to ensure that students' education is not compromised.

Are there lessons to be learned from university/industry relationships in fields such as microelectronics?

The development of the U.S. semiconductor industry is often suggested as a comparison for the development of biotechnology (see *Appendix C: A Comparison of the U.S. Semiconductor Industry and Biotechnology*). Virtually all of the basic research in electronic engineering carried out by U.S. universities during the 1950's and 1960's was supported by the Federal Government. In addition, however, a specific program in electronics research was funded by the Joint Services of the U.S. Department of Defense (DOD). DOD's program had four specific aims:

- extending basic knowledge in electronics;
- strengthening the scientific qualifications of electrical engineering faculty;
- training students to enter research positions at industrial, government, and university laboratories; and
- developing new ideas that could be exploited in the development of new systems and devices in applied research and development labs.

Because of the infusion of capital from DOD's program, there developed at U.S. universities a research and training infrastructure that facilitated the growth of the U.S. semiconductor industry. From the mid-1950's on, this infrastructure generated increasingly open cooperative ties between university electrical engineering departments and private companies. By 1961, nearly half of the 400 graduate students in Stanford's electronics program were employees of local industry who attended Stanford on a part-time basis and whose education was paid for by private company contributions. Moreover, members of Stanford's electrical engineering faculty sat as directors on the boards of 13 corporations (including one board chairman and one half-time company president). Nearly all of the 30-odd electrical engineering faculty members spent one-half to 1 day per week consulting for private industry. Moreover, four or five faculty members were virtual millionaires as a result of equity participation in companies with which they were associated as either board members or consultants. During the intensifying Cold War atmosphere surrounding the launching of Sputnik in the late 1950's, most individuals in academia, government, or industry were not troubled by these overt cooperative ties between the semiconductor industry and university electrical engineering departments. Neither the quality of the education nor academic freedom appeared to suffer substantially; in fact, all were probably enhanced (2).

The impact of Federal research funding at universities during the 1950's and 1960's thus had intended and unintended effects. Federal moneys purposefully developed the research and training infrastructure at universities necessary to feed industrial growth, and, in turn, laid the basis for widespread but largely unintended collaborative ties between American universities and the U.S. semiconductor industry. Major universities seized on Federal funds to become the concentrated locational foci for the rapid growth of the dynamic, new U.S. semiconductor industry. However, few semiconductor innovations emerged directly from federally funded university research.

The potential industrial applications of biotechnology, by contrast, have emerged directly from publicly funded academic biomedical research. As biotechnology has been moving to the market, universities have been buffers in commercializing the fruits of public funding, because they are virtually the *sole source* of basic know-how. Many of the new firms in the field of biotechnology have sprung out of academia, whereas in the semiconductor field, ample DOD procurement helped to create *industrial* know-how and encouraged *industrial* spinoff. In the area of biotechnology, the traditionally distinct roles of the university as source of research and training and of industry as source of commercialization are blurred. Though the consulting arrangements, equity arrangements, and research contracts between U.S. universities and industry in the field of biotechnology resemble in form the cooperative ties that emerged between U.S. universities and industry in the field of semiconductors, their timing, substance, and scale are significantly different (2).

What forms are university/industry relationships in biotechnology taking and what are the associated issues?

The major issues in university/industry relationships, though derived in part from the differences between the two institutions, are also set in a context of broader social and economic issues. Thus, the discussion of types of university/industry arrangements below is set in this context of broader issues. First a caveat: industry and universities are not monolithic institutions. The variability within each of these two institutions is as least as great as, if not greater than, the variability between them. This diversity is essential to the health of both and must be borne in mind in any discussion of university/industry arrangements, because no two arrangements are identical.

In the following discussion, five broad types of university/industry arrangements in biotechnology are considered:

- consulting arrangements,
- industrial associates programs,
- research contracts,
- research partnerships, and
- private corporations.

Additional information about specific university/industry relationships in biotechnology is presented in *Appendix H: Selected Aspects of U.S. University/Industry Relationships in Biotechnology.*

By far the most common form of interaction is personal interaction among scientists. Personal interactions can include consulting arrangements, personnel and publication exchanges, seminars, and speaker programs. Issues arise most often with regard to consulting arrangements.

CONSULTING ARRANGEMENTS

Consulting is important for several reasons. It allows direct technology transfer between universities and industry that goes in both directions. Academicians agree that consulting keeps them apprised of new innovations in industrial R&D and that their knowledge can be applied to new kinds of problems related to, but outside of, their on-campus research. University faculty who consult publish more than faculty who do not consult (this may be a chicken and egg situation); they also do more research and participate as actively in their administrative duties as faculty who do not consult (17). Furthermore, consulting plays a significant role in faculty salary supplementation: 44 percent of calendar year faculty at doctoral granting institutions in the United States report that consulting is their first or second largest source of supplemental income (17). Consulting relationships have led to longer term industrial support of U.S. university research such as that provided by Monsanto to Washington University (see below) and Harvard and that provided by Exxon to MIT.

Industry views consulting arrangements with university faculty essentially as having an expert on retainer. Most U.S. universities have policies on consulting, but the policies vary. Some examples of university policies on consulting are presented in appendix H.

University consulting policies typically have provisions regarding conflict of interest, time regulation, disclosure, and policy enforcement. In most cases, policy enforcement is based on an honor system; each faculty member who consults is personally responsible for adhering to this. Although some faculty members may not always observe the rules, with incentives to carry on good research, train graduate students, and publish findings, most university faculty are not motivated to pursue consulting activities to the point where conflicts of interest occur on a regular basis. Disclosure policies are of interest for public access to objective scientific information. An argument could be made that because the public has supported research in universities, it has a right to know whether a particular university faculty member who is giving testimony, for example, has a consulting relationship with a company that manufactures a particular potentially harmful chemical. The negative side of disclosure policies is that "objective" information may be judged "subjective" because of guilt by association. If a faculty member's consulting arrangement with industry is declared openly, it is not necessarily the case that his or her testimony is biased. In fact, the expert may have a more objective view because he or she understands both the research and development aspects of the technology.

INDUSTRIAL ASSOCIATES PROGRAMS

Industrial associates programs usually involve entire university departments or groups of specialists within a department. Companies pay a set annual fee that allows them to participate in seminars, interact with graduate students and faculty, and preview publications.

Industrial associates programs allow university/industry contacts and at the same time avoid conflict of interest problems and patent agreements. These programs exist at several U.S. universities, and some ongoing programs now include biotechnology. At MIT, for example, the Industrial Liaison Program has begun to include biotechnology as a subject of its symposia and seminar series. One of Stanford's 19 industrial affiliates programs is a program in biochemistry. And Pennsylvania State University has just initiated a Cooperative Program in Recombinant DNA Technology.

Industrial associates programs facilitate technology transfer between universities and industry, open up opportunities for further consulting and contract arrangements, provide funding for graduate students and faculty research, and give industry access to graduate students for future employment. Industrialists generally view these programs as useful. However, some industrialists believe that a few university programs tend to give the impression that research results are being sold to members only. Exclusivity is not the purpose of these programs; rather, their purposes are support of research activities and continuing open lines of communication of research results.

RESEARCH CONTRACTS

University research contracts with industrial sponsors have been and continue to be an important type of university/industry relationship in biotechnology. Research contracts differ from consulting arrangements in that the industrial sponsor is usually paying for a specific piece of research or supporting general research activities. Contractual arrangements often grow out of consulting or industrial associates programs and are usually motivated by industry's need for research that complements research being done in-house or for some expertise in a new area.

Several of the university research contracts with industrial sponsors in biotechnology have been large and have elicited questions regarding issues such as commingling of funds, patent rights, and disclosure of equity or other financial arrangements between the industrial sponsor and the principal investigator. The larger agreements have received extensive press coverage.

Issues of conflict of interest, invention rights, commingling of funds, and university policies regarding the processing of contractual arrangements are all important. It is interesting to note that MIT, which traditionally has had a close relationship with industry and has a relatively larger (7 percent) share of industrial sponsorship than other American universities, has the most explicit guidelines for consulting, disclosure, and processing of industry-sponsored contracts. Other universities, notably, Johns Hopkins, Harvard, and the University of California, are moving toward more explicitly stated policies. See appendix H for descriptions of selected university policies on sponsored research and patents.

RESEARCH PARTNERSHIPS

Another type of university/industry arrangement taking place in biotechnology is the joint establishment of a research foundation, institute, or long-term collaborative arrangement by an industrial sponsor and a university. Three recent ones, further described in appendix H, are well known: the Hoechst/Massachussetts General Hospital agreement, the Monsanto/Washington University agreement, and the Whitehead Institute/MIT agreement. These arrangements raise several issues, some of which are pertinent to only one or two of them, others to all three.

The agreement between the West German company Hoechst and Massachusetts General Hospital, for example, raises the issue of foreign investment in and foreign benefit from U.S. Government-funded research. This agreement also raised the issue of commingling of funds (see below). For both the Hoechst and Whitehead Institute agreements, faculty selection is an issue because of the need for balance in subdisciplines in biology in Massachusetts General Hospital's medical school and MIT's biology department, respectively. Other issues raised by these agreements are external peer review of projects and controls on rights to publish. Another issue is the terms of termination of the agreements and whether adequate notification provisions have been made for the university to seek other support.

In the Hoechst/Massachusetts General Hospital agreement, the company will pay for all equipment and other expenses in order to ensure that there will be no Federal support of the research. Questions will arise if faculty cooperate with other researchers who are funded, for example, by the National Institutes of Health. Provisions have been made in both the Hoechst and Whitehead agreements to separate faculty selection and consulting. Choice of directions of research is the responsibility of the Whitehead Institute's directors and scientific board, all of whom have high academic reputations. Provisions for termination of the agreements vary, but they have been clearly stated.

Several States have established institutes for biotechnology development that encourage interactions between industry and universities. The North Carolina Biotechnology Center and the Molecular Biology Institute in Michigan have already been established; other States are in the process of establishing such centers.*

*For a list of State government initiatives for high-technology industrial development, see *Technology, Innovation, and Regional Economic Development: Census of State Government Initiatives for High-Technology Industrial Development—Background Paper (28).*

PRIVATE CORPORATIONS

Innovative approaches to connecting university research to commercial developments in biotechnology are being initiated. The establishment of Engenics (with Stanford and the University of California, Berkeley) and the establishment of Neogen (with Michigan State) are examples of two different approaches. For descriptions of these arrangements, see appendix H.

The Engenics arrangement is funded by six corporations, with money flowing through the simultaneously established nonprofit Center for Biotechnology at Stanford. The Center for Biotechnology funds contract research on the campuses of the University of California, Berkeley, and Stanford (and also funds one contract at MIT) and will funnel royalty income back into the university to fund more research. Neogen was established to utilize limited partnerships and tax benefits as a vehicle to allow Michigan State University faculty to remain on campus and simultaneously allow entreprenurial ideas to flourish. Neogen's royalties are funneled back to the university through the nonprofit Michigan State University Foundation.

It is too early to evaluate the Engenics and Neogen arrangements. It should be noted, however, that potential challenges could arise with respect to adequate mechanisms for peer review of projects, applied research being done on campus, conflicts of interest of professors, and a private company doing the same type of work as is being done on campus with the on campus principal investigator having ties (equity, consulting, board membership) to the company.

Are university policies with respect to university/industry relationships being formulated?

The control of intellectual property, commingling of funds, tangible research property, and conflicts of interest are issues that cut across all university/industry arrangements and ultimately affect technology transfer and the U.S. competitive position in biotechnology. University policies with respect to these issues are addressed in the discussion that follows, and additional information about university policies is presented in appendix H.

INTELLECTUAL PROPERTY

Different traditions have developed in the United States to deal with different kinds of property. Although some U.S. universities allow the faculty member who developed the invention to retain any patent rights, most require those rights to be transferred to the institution. Created works are subject to copyright laws. Most institutions assert that ownership, but universities do not assert rights to books written by faculty (14).

Patents.—Issues relating to patent agreements can be divided into two kinds: those dealing with retention of rights to an invention and those dealing with decisions regarding exclusive or nonexclusive licenses.

The rights of small businesses, universities, and other nonprofit organizations to inventions made under research sponsored by the U.S. Government are addressed in the 1980 U.S. patent law, Public Law 96-517. An Office of Management and Budget (OMB) circular, Circular A-124, "establishes a standard Patent Rights clause to be included in all Government grants and contracts with such organizations, which gives these inventing organizations the right to retain title to the inventions. The Circular also requires agencies to modify existing regulations to bring them into conformity with the Circular" (7). Public Law 96-517 was passed with the recognition that the public interest can in most instances be promoted by allowing exclusive licenses under those circumstances. In a competitive economy, private enterprise will not invest funds to develop ideas that can be duplicated with impunity. Without exclusive licenses, important investigations made at Government expense would remain undeveloped because development costs are high. Thus, these inventions would never be available to the public (10).

The consensus expressed in recently developed university guidelines for industrial sponsorship of academic research is that granting of exclusive or nonexclusive licenses will be on a case-by-case basis to the corporate sponsors of research. Summaries of State Agricultural Experiment Stations (SAES) and Pajaro Dunes Conference guidelines are presented in appendix H. In some cases, an exclusive license is given to allow time for development of the product. There is a division of opinion on whether exclusive licenses should be granted on all discoveries that result from university research funded by an industrial sponsor: some university representatives believe that an exclusive license should be granted, while others believe that the university should provide a nonexclusive royalty free license (see Pajaro Dunes Conference guidelines in appendix H). Most agree, however, that if a faculty member's research is being sponsored by a company in which the faculty member has substantial interest and/or equity, the university should grant only a nonexclusive license. In most of the major multimillion dollar university/industry agreements being struck in biotechnology, the corporate sponsor is receiving some exclusive rights to inventions developed as a result of the funding. In all arrangements between industry and universities, it is essential that the patent issues be carefully thought out in advance.*

COMMINGLING OF FUNDS

Since one of the purposes of the 1980 U.S. patent law (Public Law 96-517) is to foster cooperative research arrangements among Federal Government agencies, universities, and private industry, one question that immediately arises is the potential for commingling of funds. Currently, for agreements struck after 96-517 became law, no exemption for Government de minimus provisions has been made. Where the Government has funded a small percentage—even 1 or 2 percent of direct costs—then the provisions of Public Law 96-517 and OMB Circular A-124 apply.

The Comptroller General of the United States, in his reply to Congressman Albert Gore concerning the possibility for commingling of funds in the Hoechst/Massachusetts General Hospital agreement stated, "MGH must account separately for all expenses leading to an invention, including the cost of research itself as well as indirect or overhead costs, to be able to show that the

*For a discussion of patent issues in such agreements, see P. Hutt, "University/Corporate Research Agreements" (10).

expenses were paid with funds provided by Hoechst" (23). After reviewing the terms of the contract, the Comptroller General ruled that it is possible for Massachusetts General Hospital to separate the funds properly.

TANGIBLE RESEARCH PROPERTY

A basic principle among scientists is that research findings should be communicated promptly to the scientific community by written and oral means. Written and oral processes, however, are not sufficient to disseminate tangible products of research such as the antibody-producing cell lines and plasmids used in biotechnology.

Stanford University developed in March 1982 a specific policy on tangible research property (TRP), defined as "tangible (or corporeal) items produced in the course of research projects," including items such as "biological materials, computer software, computer data bases, circuit diagrams, engineering drawings, integrated circuit chips, prototype devices and equipment, etc." (16). Stanford's policy was promulgated to protect the university's ownership in such property consistent with the policy of promoting the prompt and open exchange of TRP and associated research data with scientific colleagues outside the investigator's immediate laboratory. Controlling the distribution of TRP, subject to provisions of applicable grants or contracts and university policy, is the responsibility of the principal investigator. Such control includes determining if and when distribution of the TRP is to be made beyond the laboratory for others' scientific use.

WARF has developed a confidential disclosure agreement in order to disseminate or license intellectual property, tangible or intangible property, and products arising from work conducted at the University of Wisconsin. In order for the receiver to obtain the materials, the following conditions must be met (3):

- The materials will be received and held in confidence by the receiver. Only persons within the receiver's organization and only those essential in the evaluation of the materials will be permitted access to the materials.
- If opinions or services of other persons outside the receiver's organization are needed, then the receiver will notify WARF and the confidential disclosure agreement will be executed with that person.
- The receiver will not commercially utilize the material or any part thereof without written consent of WARF or prior to entering into a licensing arrangement with WARF.

Recently, a dispute over the ownership of a cell line that produces interferon arose between the University of California and the Swiss company Hoffmann-La Roche. The University of California, as the institutional home of the scientists who created the cell line, claimed ownership of the cell line and the right to future royalties. Hoffmann-La Roche also claimed ownership on the grounds that it had funded the university research that increased interferon production by the cell line and filed a patent application covering this interferon production process. Lawyers from the university sued the company, arguing among other things that the firm had made unauthorized use of the material, taking commercial advantage of the open exchange of information and material among academic scientists. This suit was settled out of court, but the settlement has not been made public.

Another recent case has left unresolved the issue of ownership of a cell line (24). H. Hagiwara, a visting Japanese researcher at the University of California, San Diego, took, without permission, a hybridoma fused from cancer cells taken from his mother and used the resulting monoclonal antibodies to treat her for cancer. Although the usefulness of the treatment has not yet been evaluated, the cell line may have commercial potential, so the issue of ownership is important. The University of California sued Hagiwara, stating that, as the research institution, it owned the cell line. This case has been settled with the Hagiwara Institute of Health (Hagiwara's father is the director) obtaining exclusive license in Japan and other Asian countries and the patent rights assigned to the University of California. Some argue that a hybridoma is a newly created entity, so the donor has no rights of ownership; others contend that cell donors should automatically be given a share of any subsequent profits (24).

CONFLICT OF INTEREST

Conflict of interest situations have both financial and intellectual components. A potential conflict of interest could arise if a university held equity in a company in which a faculty member of the university also held equity interest as a line officer. This situation arose in a Harvard proposal to establish a firm to commercialize the research of one of its professors. The proposal was subsequently withdrawn, and Harvard President Derek Bok described the potential problems with the arrangement (1):

- The administration could be exposed to disagreements not only with the faculty partners but also with other nonpartner faculty who might also want support.
- Commercial ventures could impose responsibilities on the university it doesn't have when its endowment invests only in shares of many companies.
- Conflicts could arise if the university were associated with particular products and a public that expects high standards from the university were dissatisfied with the standards of marketing or the products.
- The arrangement could inevitably change and confuse the relationship of the university to its professors. A faculty member who joins with the administration in founding a new company is no longer valued merely as a teacher and scholar; he becomes a source of potential income to the institution.
- There could be more doubt concerning decisions made with respect to qualifications for tenure, extra leave, larger laboratory space, more graduate students, salaries, etc.
- Professors might become so involved with the challenges of seeing an enterprise grow and develop that their work commitment to university duties might be diminished. The university would be in a prejudiced position regarding assessment of that person's performance of work since that person's commercial success would be linked to that of the university.
- Participation would increase the risk of secrecy, and the university could have a stake in supporting that secrecy.

A potential conflict of interest and commitment arises when a professor holds equity within a company that engages in the same type of activities as the university. This issue has been raised in the activities of Calgene. The State Agricultural Experiment Station (SAES) at Davis helps develop new varieties of plants for California growers. University of California professors, including Ray Valentine, have part-time employment at the station. Calgene, a company Valentine founded, is undertaking for profit the same kinds of activities as the SAES undertakes for growers. Thus, there is a potential source for conflict and for taking the results of work off campus and marketing them through the company.

University conflict of interest policies and consulting policies vary, but university policies are becoming more explicit, in part because universities are responding to developments in biotechnology. It is interesting to note that MIT, which has traditionally had extensive contacts with industry, has explicit policies on industrially sponsored research. In addition, several organizations are setting guidelines for industrial sponsorship of academic research (see app. H).

At the request of Representative Albert Gore, the American Association of Universities (AAU) reviewed the ethical dilemmas posed by increases in industrial support for research. AAU suggested that it serve as a clearinghouse and monitor activities at the major American universities with regard to the formulation of policies. AAU decided not to develop policies, because "conditions exist [in the universities] for intelligent and thoughtful decisionmaking on these issues and policies that are informed by wide experience and that are tailored to individual circumstances are preferable to injunctions broadly enough cast to cover the multitude of local circumstances that exist among many universities" (15).

Also, representatives from universities, industry, and the legal community are now meeting to review issues and communicate more effectively. Recent meetings have been hosted by Columbia University, the Gulf Universities Research Consortium, the Industrial Biotechnology Association, Florida State University, Harvard, and the Bar of the City of New York, and a meeting in Philadelphia in December 1982 was hosted by eight major research universities (15).

It is clear that recent activities to formulate explicit policies are advantageous in helping to define the role of the university and ultimately facilitating effective technology transfer between universities and industry. Technology transfer will be most effective if both sides are strong, vibrant, creative, and have something to offer each other.

What is the likely future of university/industry relationships in biotechnology?

A comparison of the responses to OTA's survey of university and industry groups in the United States shows that both groups see the future of university/industry relationships in biotechnology as depending largely on the success of biotechnology companies in getting products into markets. There was divergence of opinion among the respondents, however, on what kind of research assistance—broad basic research or more specialized research—industry would seek from universities. In biotechnology as in other fields, an increase in the actual number of industry/university relationships and an increase in the total amount of funding made available by industry can be expected to develop over the short term (18).

U.S. university/industry relationships in biotechnology will most likely follow the same pattern that they have in other high-technology areas. First, scientific breakthroughs generate a period of hyperactivity in university/industry relationships. This hyperactivity phase is characterized by the promise of "big bucks," which leads to a short-term faculty and post-graduate drain. After the industry goes through its initial phases, an equilibrium state is reached and a fairly healthy symbiotic relationship emerges.

How effective is university/industry technology transfer in countries likely to compete with the United States in biotechnology?

The countries identified in this assessment as being the most likely major competitors of the United States in biotechnology are Japan, the Federal Republic of Germany, United Kingdom, Switzerland, and France.

JAPAN

Japan has a mixed situation with regard to university/industry relationships in biotechnology. First, a distinction should be made between institutions and individuals. At the national universities, which are at the top of the Japanese university hierarchy, institutional ties are very strictly regulated. At the level of individual professors, however, there is considerable opportunity for interaction. A second distinction is between the basic and applied sciences. The distinction and separation of basic and applied science departments at Japanese universities is strong. Japanese professors in disciplines such as biology, biochemistry, and medical science are proud of their independence from industry. Professors in applied disciplines such as bioprocess engineering, on the other hand, have ongoing contacts with industry. Japanese professors in applied science departments are considered to have a moral obligation to place their students as employees. Consequently, they tend to maintain good relationships with industry. Furthermore, because former students are members of industry, informal contacts continue.

Even though Japanese professors in applied sciences have contacts with industry, the level of exchange of information between universities and industry in Japan is not as high as that in the United States. Japanese professors at the national universities are forbidden to take other positions simultaneously with their university work, and all donations toward their research must be made through formal university channels. No fees for consulting can be accepted, and offers of stock options are unheard of. Japanese professors were not allowed to hold patents or collect royalties until 1981. Because of the restrictions on both professors and industry, Japanese professors often quit their posts to work in industry or private laboratories where facilities and salaries are better than in the universities. They do this in spite of the fact that there is a great deal of social prestige attached to being a professor.

There are only two mechanisms through which Japanese industry can channel funds to a university. One of these, the "itakuhi," is commissioned research on a particular topic. The company supplies a researcher and some funds, usually only $500 to $1,000 (¥ 125,000 to ¥ 250,000), to a university professor. This mechanism allows the company to have its researchers trained by the professor and the professor's staff; the professor, in turn, gets extra help in doing his research. The second mechanism, the "shogaku kifukin," is a scholarship grant donated to a specific university researcher but not for a specific topic. The grant money must be used only for equipment and other direct costs, not personnel costs; no overhead is charged by the university, and there is no limit on the amount. Money for these grants must be channeled through the Ministry of Education. Within this framework, there are a number of administrative obstacles in terms of hierarchy of approvals necessary, different policies on patents among departments, etc. In spite of the tight control of channels of funds and lack of consulting opportunities for Japanese professors, about 10 percent of all university funding for research in Japan does come from industry. Most of this is probably channeled to applied research (22).

The Science and Technology Agency (STA) has established the New Technology Development Fund in order to subsidize Japanese companies that develop and commercialize university research findings. The fund will probably be used to transfer technology between applied science departments and industry, which already have good relations, rather than between basic science departments and industry.

Another STA program is designed to cross traditional barriers between university basic science departments and industry and between the Ministry of Education and the Ministry of International Trade and Industry (MITI). Research responsibilities in STA's program are allocated between university and corporate laboratories. The success of the program will depend in part on getting MITI, the Ministry of Education, and basic research departments to work together. Basic researchers have already asked the Ministry of Education to supervise the project, so there is some doubt as to whether the program will be successful. If supervision stays in Ministry of Education, the link with industry will be weakened.

Research in Government institutes makes up for the lack of technology transfer from the heavily regulated Japanese universities. In almost every major industrial sector in Japan, there are a number of governmental research institutes that do background research for MITI policymakers and where professors, industry representatives, and Government officials meet for discussion. Mitsui Information Development and Normura Research Institute have been used for background research in biotechnology.

In addition, the Japanese Government is building two biotechnology centers, each with a P4-level laboratory facility: one in Tsukuba (a new university research community) and one at Osaka University. The P4 facilities will be available to private sector corporations via contacts with university professors. Four other universities were designated by the Ministry of Education to have P3 level facilities and received $640,000 (¥ 160 million) in fiscal year 1981 to help build them: 1) Tokyo University Medical Research Institute; 2) Kyoto University Chemistry Research Institute; 3) Osaka University Microbial Disease Research Institutes; and 4) Kyushu University Medical Department. These large-scale biotechnology facilities administered by the Japanese Government will provide a place for university professors, Government researchers, and company researchers to work together.

The applied science departments of Japanese universities have been instrumental in Japan in providing training and information exchange in biotechnology. At present, university basic research in Japan is peripheral to Japanese industrial activities. If Japan intends to develop a greater basic research capacity that industry can draw on, funding for basic research will have to be increased and mechanisms to increase communication between researchers and industry will have to be implemented. Japanese companies look to other countries to make up for the weaknesses in the technology transfer from Japanese universities. Whether the Japanese will have a competitive edge in biotechnology will rest, in part, on the differences in industrial relationships in applied and basic research. If biotechnology develops such that most research moves into industry, then the present system will be adaptive. If strong basic research and effective domestic technology transfer by universities is important to the development of biotechnology and if international technology transfer proves ineffective, then the Japanese system will have to change (22).

FEDERAL REPUBLIC OF GERMANY

Biotechnology research in the Federal Republic of Germany is carried out at the federally supported, private Max Planck institutes as well as in German universities.* Critics have charged that the Max Planck institutes may be depriving the universities of talent by drawing away promising researchers and that they are "ivory towers" conducting research of little relevance to the nation's technological well-being. The Federal Ministry for Research and Technology (BMFT, Bundesministerium für Forschung und Technologie) would like to see closer connections between the Max Planck institutes and industry. One successful outcome of its strategy is the recent cooperative agreement between the Max Planck Institute for Plant Research and Bayer Leverkusen.

The university system is in flux. Beginning in the 1960's, West German universities were subjected to a series of reforms that left the system in turmoil. According to one recent analysis (11):

> The underlying trouble is that West Germany has sought to reconcile several irreconcilables—the principle of open access to any university in the country, the doctrine that all universities are equal, the practice that the universities are run by the ministries of culture in the Länder in which they happen to be sited, and the phenomenal increase in the demand for higher education in the past twenty years.

The result is a university system in which litigation about the rights and duties of students and faculty sometimes seems to take precedence over research and teaching.

A lack of money has recently added to the administrative and legal conflicts created by the West German university reforms. Biotechnology in the universities, both because of financial cutbacks and because it is a new discipline, depends on outside sources of funding. Probably the largest single source of funding is the German Research Society (DFG, Deutsche Forschungsgemeinshaft), a nonprofit institution that serves as a German National Academy of Sciences and as a conduit for Government funding of basic research. The approval by DFG of a "special collaborative project" on bioconversion in Munich will give a boost to academic work in this area.

Other sources of funding for biotechnology in universities include the Fraunhofer-Gesellschaft and the Volkswagen Foundation, as well as private industry. Relations with industry in the past have largely taken the form of contracts or consulting agreements between individual professors and interested firms. Hoechst's arrangement with Massachusetts General Hospital (see app. H), however, has prompted BMFT to seek more systematic university/industry collaborations within West Germany. One product of BMFT's initiatives in this area is an agreement between the German chemical company BASF and the University of Heidelberg whereby the chemical company will give the university $450,000 per year for research over a 5-year period. BASF's commitment is more modest than Hoechst's support for Massachusetts General Hospital. Nevertheless, it marks an important step in the German Government's effort to engage the private sector in building up fundamental research in biotechnology inside Germany.

Among the factors cited to explain West Germany's slow entry into commercial biotechnology is an educational system that prevents the kind of interdisciplinary cooperation that is viewed by most experts as essential to the development of this field. In particular, the traditional separation of technical faculties from their arts and sciences counterparts means that process technicians, usually located in the technical schools, rarely come into contact with colleagues holding university appointments in biochemistry or microbiology.

One of BMFT's professed aims since the adoption of its performance plan for biotechnology has been to bridge this institutional gap.** A significant contribution toward meeting this objective is made by the German Society for Chemical Engineering (DECHEMA, Deutsche Gesellschaft für chemisches Apparatewesen). DECHEMA played a crucial role in the original formulation of BMFT's biotechnology program. Its working group on "technical biochemistry" was charged in 1971 with preparing a study on biotechnology that established the framework for the BMFT program.

*For a description of Federal applied research carried out through the Society for Biotechnological Research, GBF, see *Chapter 13: Government Funding of Basic and Applied Research.*

**BMFT's performance plan, *Leistungsplan: Biotechnologie,* is discussed in *Chapter 20: Targeting Policies in Biotechnology.*

DECHEMA continues to further interdisciplinary exchanges through a variety of means. Its expert group on biotechnology is a standing body that brings academics and industrial scientists into regular contact at seminars on biotechnology for small groups of experts. Attendance at these is by invitation, and one of their functions is to further a fruitful dialog between industry and academia. The confidential character of these meetings permits research scientists to discuss their results at prepublication stages. At the same time, industry representatives can present their own problems with the hope of interesting academic groups in their resolution. Finally, DECHEMA also organizes continuing education courses in various aspects of biotechnology, such as the use of immobilized enzymes or measurement and control of bioreactors (11).

UNITED KINGDOM

A traditional weakness in the United Kingdom has been a gap between university research and industry. This gap in the area of strategic applied research has been termed the "pre-development gap." There is consensus that the National Enterprise Board set up to foster university/industry relationships failed. The National Enterprise Board is now called the British Technology Group (BTG), and measures have been taken to improve its efficiency. Also, new institutions for biotechnology have been developed. Furthermore, direct contacts between British universities and industry have recently increased, in part because of economic conditions.

To stimulate the transfer of university basic science research in health-related fields to industrial applications, the British Government and four industrial partners created a new company, Celltech, Ltd., in 1981. In the original agreement, Celltech received the right of first refusal* on all work in hybridoma technology conducted by the Medical Research Council. Celltech also plans to commercialize the results of basic research in rDNA technology. Currently, the British Government owns 28 percent of the company and private companies hold the other 72 percent. Celltech's initial capitalization was $20 million (£11.4) million). Celltech currently maintains a staff of 130 persons, two-thirds of whom are scientists. It is likely to increase this number to 180 persons in the near future.

In an arrangement similar to that of Celltech, the British Government, through BTG and two private concerns (Ultramar and Advent Eurofund), recently established the company Agricultural Genetics. The objective of this company is to commercialize basic research results in nonconventional plant breeding, microbial resistance factors, and biological control products originating from research in the Agricultural Research Council. Agricultural Genetics has the right of first refusal on all work in the Agricultural Research Council. Though only about $1.2 million (£685,000) has been invested to date, the total initial capital promised approaches $28 million (£ 16 million).

To encourage direct links between academia and manufacturers, the Cooperative Research Grant Scheme has been initiated under the Science and Economic Research Council (SERC). SERC will support the academic side provided that the company in a particular arrangement makes substantial contributions in effort, materials, and expertise. Patent rights, subject to a small royalty, will be assigned to BTG. The number of projects in biotechnology under this scheme increased from 3 to 14 in the last 6 months of 1982.

Industrialists are also beginning to invest in university centers. At the University of Leicester, four companies have put up $1.7 million (£1 million) to establish a new biotechnology center. SERC is granting $316,000 (£180,000) for capital equipment.

Another program has been started by industry to help academics in British universities develop their ideas into commercial realities. At the initiative of Monsanto (U.S.) and including the Universities of Oxford, Cambridge, and St. Andrew, the Imperial College in London, and the Nuffield Foundation, $17.5 million (£10 million) initial capital has been raised (Monsanto contributed half). The program will include most fields of high technology as well as biotechnology.

Imperial College in London, in order to transfer its technology to industry, has launched a private company to exploit the pilot plant built at Imperial

*This is the right to choose whether to produce any good or service without having to bid competitively.

College in the 1960's. The plant is in good condition, but has been underused because of lack of funds. Imperial College has reserved 20 percent of the time of the plant for its own use in lieu of shares. Thus, there will be a continuing contact between research workers, the associated Imperial Biotechnology Center, and industrialists. Imperial Biotechnology's first major contract is with the Swiss firm Biogen S.A. to scale-up the firm's interferon production to 3,000 liters.

Engineers in British universities have traditionally done consulting for industry; biologists in British universities are now adopting the same practice. The extent of this phenomenon is not known, but all the large British companies involved in biotechnology are using the services of consultants in the universities. No general rules apply to consultancies in British universities; arrangements are left to individual institutions and to the consciences of the individuals involved. There is some concern on the part of the British Government, however, that foreign companies are making more use than domestic companies of British biotechnology experts.

Most authorities agree that the United Kingdom has an excellent basic research base with well-trained researchers. Traditionally, however, the United Kingdom has had a problem translating this expertise to industry; the next 5 years will determine how effective the new British measures to effect domestic technology transfer are (30).

SWITZERLAND

The field of molecular biology is highly developed in Swiss universities, particularly in relation to the size of the population. Centers of excellence include the universities of Basel, Geneva, and Zurich. The quality of research in these institutions is all the more remarkable in view of the fact that they are under cantonal jurisdiction and thus derive support primarily from local revenues.

The channels for transfer of knowledge from the Swiss universities to industry appear well established in the area of biotechnology. The president of the Federal Institute of Technology (ETH, Eidgenössiche Technische Hochschule), which established a department of biotechnology in 1976, for example, has endorsed the practice of direct contracts between professors in the biotechnology department and industry. Joint funding by industry and the Commission for the Encouragement of Scientific Research provides another avenue for collaboration with the private sector, one that has been actively utilized by the ETH biotechnology group.

The Swiss firm Biogen S.A. is closely linked to the Swiss university research system and has built an important share of its competitive strength on the productivity of these ties. Two members of the company's scientific board, Drs. Weissmann and Mach, have done seminal work for the company in the Universities of Zurich and Geneva, respectively. Finally, the city of Basel, as the fountainhead of Swiss research in the chemical and biological areas, provides unique opportunities for communication and collaboration between the academic and industrial sectors, a potential that the Basel-based pharmaceutical corporations clearly recognize and are prepared to exploit (12).

FRANCE

Universities in France are generally regarded as teaching institutions and not looked to for their research capabilities. Highly regarded research in France is usually funded by the National Center for Scientific Research (CNRS, Centre National de la Recherche Scientifique) or the National Institute of Health and Medical Research (INSERM, Institut National de la Santé et de la Recherche Médicale). Both of these are Government research institutes (grands organismes). CNRS operates its own laboratories, which are usually associated with universities. INSERM is concerned with the applied aspects of medical research. French universities that are important in biotechnology are those at Toulouse, Strasbourg, Marseille, Lille, Montepellier, Paris-Orsay, Grenoble, and Nancy. A newer university, the Technical University at Compiègne (modeled after the American university structure), which is an important center for enzymology and bioprocess technology, has concentrated on some of the disciplines underlying biotechnology and has developed good relations with industry.

There is divided opinion in France as to whether relationships between academics and industrialists should be encouraged. The December

1980 Pelissolo report concluded that relations formerly were poor; this situation has changed, however, and the problem now is not whether university research results should be transferred to industry, but how best to accomplish the transfer (29).

There are no formal constraints in France on relationships between academics and industrialists. The National Agency for the Funding of Research (ANVAR, L'Agence Nationale de la Valorisation de la Recherche) is an independent organization that stimulates the transfer of public research results to industry and encourages applied research in industry. ANVAR does not have any rights of first refusal on the results of public research, which includes university research, and apparently encourages direct links between universities and industry by offering general advice on suitable contract terms and on patenting problems. Large French companies such as Elf Aquitaine and Rhone Poulenc have organizations to keep in touch with and seek out public sector university research of interest and appear to have no problems developing and maintaining these links (except for occasional conflicts between the firms' desire for secrecy and the researchers' legitimate desire to publish). In addition, some companies are locating their new biotechnology facilities near universities in order to benefit from proximity (e.g., Elf Aquitaine at Toulouse and Transgene at Strasbourg). The University of Compiègne is situated in an agricultural region and works closely with the local foodstuffs industry. Also, according to ANVAR, a phenomenon similar to the involvement of American professors in U.S. venture capital firms is developing in France, although along more traditional lines—top French scientists are either acting as consultants to private firms or leaving the public sector for industry (this kind of transfer is generally much more common in France than in, for example, the United Kingdom (29).

Despite the absence of formal constraints on relationships between academics and industrialists, there remains a problem in France in the exploitation of the results of public sector research by industry. Except for large companies, industry has an insufficient number of qualified personnel to seek out opportunities, and French research scientists as a whole have not been very active in the pursuit of commercial opportunities (29).

The French Government has recently taken steps to encourage cooperation between the grands organismes and French industry. The institutes participated through the Committee for the Organization of Strategic Industries (CODIS, Comité d'Orientation des Industries Stratégiques)* in the establishment of the French pharmaceutical firm, Immunotech. More generally, in the recent law reforming these institutes, there are provisions for them to form profitable liaisons with industry (up until now they have been limited to contract research). This change is very recent, so it is impossible to judge its practical effect. But CNRS, in a change of statutes published on November 25, 1982, has for the first time appointed a scientific director of "Funding and Application of Research."

Fields related to biotechnology have not attracted large numbers of French researchers in the past. Government policies and funding changes have been promulgated to change and to foster university/industry relationships. It is too early to determine the effectiveness of these policies.

*CODIS mobilizes State funds from multiple sources to produce packages for project development in strategic industries.

Findings

The United States has the most effective and dynamic university/industry technology transfer process of the six countries being examined in this assessment. The process in the United States is facilitated by the openness of the university system and the freedom of faculty to pursue research. It is also facilitated by the many mechanisms by which the process can occur. These include dissemination of publications, professional meetings, consulting arrangements, contract research, cooperative research agreements, and institutes within universities funded by industry. All these mechanisms are being exploited in biotechnology. U.S. universities and industry are benefiting from the present arrangements, and diffusion of knowledge is occurring.

University and industry representatives in the United States agree that Federal support of basic research is essential. Even if industrial support of university R&D were to rise to 15 percent of universities' R&D budget, it could never replace Federal funding. Furthermore, since the goals and philosophy of industry are different from those of universities, the focus of research in industry is different from that of research in universities. Of necessity, industry is mission oriented; the emphasis in industry is on applied research leading to products. By contrast, the purpose of university basic research is to extend knowledge itself.

Universities in the United States are formulating and implementing policies that are more consistent across disciplines and more specific with regard to consulting, conflict of interest, and disclosure than policies formulated in the past. There have been some cases of potential conflict of interests with researchers who have consulting or contract arrangements with firms in which they hold equity. University administrators, faculty, and students appear to be taking measures to reduce the potential for conflicts of interest and ensure quality research and education.

Although funding of large agreements between U.S. universities and industry in biotechnology has occurred, the consensus is that, after the initial excitement has dissipated and companies have developed in-house capabilities, most of the university/industry arrangements in biotechnology will be consulting and contract research as in other fields with close university/industry ties.

Universities are looking for financial support, but the promise of patent royalties from biotechnology may be premature. Especially if biotechnology becomes a rapidly moving process field where research is carried out primarily in industry, research in biotechnology will have to move off campus and royalty income to universities may not be significant.

Biotechnology has spawned a new kind of arrangement in university/industry relationships: for-profit companies established with nonprofit buffers to funnel contract research money and royalty payments between the university and the company. One arrangement (Neogen) takes advantage of new U.S. tax laws that permit funding of R&D through limited partnerships. The other (Engenics) is built on the support of six major corporations that are funding the research and have invested in the company. It is too early to predict whether these approaches will be viable.

Biotechnology is being transferred between industry and universities in the United States; most of the arrangements are working well. Some individuals have noted potential problems and administrative bottlenecks; these should lessen as individuals on both sides gain more experience and policies are formulated to standardize administrative procedures within universities. Some individuals believe that problems may arise when sales revenues are generated as a result of some of the limited partnership agreements.

The early history of the U.S. microelectronics industry can serve as a comparison for the commercialization of biotechnology. The U.S. semiconductor industry was fueled by and developed in a milieu of DOD support for basic research and training at universities, DOD procurement of the industry's products, and DOD's need for increasingly more sophisticated products from that industry. In the history of the U.S. semiconductor industry, relationships between universities and industry were very close. Many professors had equity in companies located close to campuses, and consulting was extensive. It appears that education in this field did not suffer; in fact, it was probably enhanced, and students gained an understanding of industrial career paths. The current leveling of Federal support for biology combined with the lack of consensus that biotechnology is a strategic industry (as was microelectronics in the instance of the space race) leads to the perception of more "potential conflicts" in industry/university relations in biotechnology than actually exist.

In countries other than the United States, there are varying degrees of cooperation between universities and industry. In Japan, the ties between university applied science departments and industry have always been close. Most people acknowledge that Japan already leads the world in bioprocess engineering research, and the close relationships that already exist between Japanese industry and university applied research departments benefit the commercialization of biotechnology in that country. Currently, the Japanese Government is implementing new policies to encourage closer ties between university basic researchers and industry.

In the Federal Republic of Germany, BMFT is encouraging domestic university/industry contacts, especially in light of Hoescht's agreement with Massachusetts General Hospital. After that agreement was announced, some West Germans were concerned because they felt that research money was being funneled into American universities instead of into German universities.

The United Kingdom has an excellent basic research base. University ties with industry have been few in the past, but are now being encouraged by the Government. The British Government helped to establish two firms, Celltech and Agricultural Genetics, to capitalize on British university research in animal and plant molecular biology.

In Switzerland, the field of molecular biology is highly developed, and patterns of interaction between individuals in universities and industry are well established. ETH established a department of biotechnology in 1976 and endorses the practice of direct contracts between professors in the biotechnology department and industry.

In France, an ambitious program is underway to tie universities and industry closer together. One problem in France is that the country lacks a cadre of experts in molecular biology, because this field has not been considered an important one.

Issue

ISSUE 1: Should Congress set guidelines for university policies on industry-sponsored research?

At the request of Representative Albert Gore, the American Association of Universities (AAU) reviewed ethical dilemmas posed in the United States by increases in industrial support of university research. A select committee drawn from the AAU membership suggested that the AAU could serve as a clearinghouse and monitor of activities at major universities with regard to the formulation of university policies on industrially sponsored research. Because one policy formulated by the AAU or Congress would have to be broad enough to cover all circumstances, it might be too general to be useful. Furthermore, as the committee noted, informed decisionmakers within universities are formulating policies that fit each university's needs.

In addition, in a report of joint hearings on university/industry relationships in biotechnology, the Subcommittee on Investigations and Oversight and the Subcommittee on Science, Research, and Technology of the Committee on Science and Technology of the U.S. House of Representatives made the following recommendations: 1) universities should prepare guidelines for industrially sponsored research that require open disclosure of all faculty consulting and contractual agreements; and 2) full-time faculty should be discouraged from holding equity or directing such firms. The subcommittees further recommended that there be continued review by universities, industry, and the Federal Government of the benefits and problems resulting from large-scale corporate support for and involvement in university research programs in biotechnology.

Chapter 17 references

1. Bok, D., "The President's Report: Business and the Academy," *Harvard Magazine,* May/June 1981, p. 26.
2. Borrus, M., and Millstein, J., "Technological Innovation and Industrial Growth: A Comparative Assessment of Biotechnology and Semiconductors," contract report prepared for the Office of Technology Assessment, U.S. Congress, March 1983.
3. Bremer, H., Wisconsin Alumni Research Fund, Madison, Wis., personal communication, 1983.
4. Culliton, B. J., "Biomedical Research Enters the Marketplace," *N. Eng. J. Med.* 1195:304, 1981.
5. Dashen, A., Washington Research Foundation, personal communication, 1982.
6. David, E. E., "The University-Academic Connection in Research: Corporate Purposes and Social Responsibilities," *Journal of the Patent Office Society* 64:209-219, 1982.
7. Executive Office of the President, Office of Management and Budget, OMB 82-5, Public Affairs Cover Memo for OMB Circular A-124 (Patents—Small Firms and Nonprofit Organizations), Washington, D.C., Feb. 12, 1982.
8. Giamatti, A. B., "The University, Industry, and Cooperative Research," *Science* 1278:218, 1982.
9. Fred, E. B., "The Role of Wisconsin Alumni Research Foundation in the Support of Research at the University of Wisconsin," Wisconsin Alumni Research Foundation, Madison, Wis., 1973.
10. Hutt, P. B., "University/Corporate Research Agreements," *Technology in Society* 5(2), in press, 1983.
11. Jasanoff, S., "Public and Private Sector Activities in Biotechnology: The Federal Republic of Germany," contract report prepared for the Office of Technology Assessment, U.S. Congress, January 1983.
12. Jasanoff, S., "Public and Private Sector Activities in Biotechnology: Switzerland," contract report prepared for the Office of Technology Assessment, U.S. Congress, January 1983.
13. Kalergis, D., "The Role of the University in the Commercialization of Biotechnology," University of Virginia Law School, December 1981.
14. Kennedy, D., "The Social Sponsorship of Innovation," *Technology in Society,* vol. 4(4), in press.
15. Langfitt, T. W., Hackney, S., Fishman, A. P., et al., *Partners in the Research Enterprise, University-Corporate Relations in Science and Technology* (Philadelphia: University of Pennsylvania Press, 1983).
16. Lee, P. R., Levinson, W., Butler, L. H., et al., "Industrial-Academic Relationships in Biotechnology at Stanford University, University of California Berkeley, and University of California, San Francisco," contract report prepared for the Office of Technology Assessment, U.S. Congress, July 1982.
17. Linnell, R. A. (ed.), *Dollars and Scholars: An Inquiry Into the Impact of Faculty Income Upon the Function and Future of the Academy* (Los Angeles: University of Southern California Press, 1982).
18. Management Analysis Center, Inc., "Study of University/Industry Relationships in Biotechnology," contract report prepared for the Office of Technology Assessment, U.S. Congress, January 1983.
19. National Science Foundation, *University/Industry Research Relationships,* Fourteenth Annual Report of the National Science Board (Washington, D.C.: U.S. Government Printing Office, 1982).
20. Omenn, G. S., "University-Corporate Relations in Science and Technology: An Analysis of Specific Models," *Partners in the Research Enterprise,* T.W. Langfitt, S. Hackney, A. P. Fishman, et al. (eds.) (Philadelphia: University of Pennsylvania Press, 1983).
21. Rosenzweig, R. M., letter addressed to the Hon. Don Fuqua for the Association of American Universities, Oct. 28, 1982.
22. Saxonhouse, G., "Biotechnology in Japan," contract report prepared for the Office of Technology Assessment, U.S. Congress, June 1983.
23. Socolar, M. J., Comptroller General of the United States, letter addressed to the Hon. Albert Gore, Jr., Oct. 16, 1981, ref. B-204687.
24. Sun, M. A., "Scientists Settle Cell Line Dispute," *Science* 220:393-394, 1983.
25. U.S. Congress, House Committee on Science and Technology, *Commercialization of Academic Biomedical Research,* hearings before the Subcommittee on Investigations and Oversight and the Subcommittee on Science, Research, and Technology, June 8-9, 1981 (Washington, D.C.: U.S. Government Printing Office, 1981).
26. U.S. Congress, House Committee on Science and Technology, *University/Industry Cooperation in Biotechnology,* hearings before the Subcommittee on Investigations and Oversight and the Sub-

committee on Science, Research and Technology, House Committee on Science and Technology, June 16-17, 1982 (Washington, D.C.: U.S. Government Printing Office, 1982).

27. U.S. Congress, Office of Technology Assessment, *World Population and Fertility Planning Technologies: The Next 20 Years,* OTA-HR-157, Washington, D.C., February 1982.
28. U.S. Congress, Office of Technology Assessment, *Technology, Innovation, and Regional Economic Development: Census of State Government Initiatives for High-Technology Industrial Development—Background Paper,* OTA-BP-STI-21, Washington, D.C., May 1983.
29. Vaquin, M., "Biotechnology in France," contract report prepared for the Office of Technology Assessment, U.S. Congress, February 1983.
30. Vaquin, M., "Biotechnology in Great Britain," contract report prepared for the Office of Technology Assessment, U.S. Congress, February 1983.
31. Watson, J., *The Double Helix* (New York: Atheneum, 1968).

Chapter 18

Antitrust Law

Contents

Chapter 18
Antitrust Law

Introduction

Antitrust laws in the United States date back to 1890, when they were first passed to counter the concentration of industrial power. Their fundamental goal is to prevent the distortion of competitive market forces, and thus ensure more productivity, innovation, and lower prices. The assumption underlying the laws is that competition between industrial units generates more consumer benefits than a cartelized or managed industry.*

Today, there is much public debate about whether U.S. antitrust laws do, in fact, accomplish these goals in all cases. Some commentators have claimed, for example, that U.S. antitrust restrictions, uncertainties about their scope and applicability, and substantial penalties for violations serve to discourage research and development (R&D) joint ventures that could actually stimulate rather than retard innovation.** In addition, there are claims that antitrust restrictions have hampered the ability of many U.S. companies to compete in world markets against foreign companies that face significantly less stringent restrictions under the antitrust laws of their countries (see, e.g., ref. 21).

Antitrust law creates no issues or problems unique to biotechnology; it embodies broad economic principles and affects or potentially affects virtually any business enterprise. Much of the debate on antitrust is essentially a general debate on economic policy and high technology that is beyond the scope of this report. Two issues in the debate, however, are particularly relevant to biotechnology. One is whether current U.S. antitrust law discourages the formation of R&D joint ventures and thereby retards innovation and the competitiveness of U.S. firms in world markets. The second issue is whether current U.S. antitrust law inhibits the legitimate exploitation through licensing arrangements of the technology created by R&D efforts.

*See *Northern Pacific Railway Co.* v. *United States*, 356 U.S. 1, 4 (1958), where Mr. Justice Black wrote that "[t]he Sherman Act was designed to be a comprehensive charter of economic liberty aimed at preserving free and unfettered competition as the rule of trade."

**See, for example, the testimony of Peter McCloskey, Malcolm Baldridge, William Norris, and Admiral B. R. Inman in hearings before the Senate Judiciary Committee, June 29, 1983 (34).

These issues are relevant because biotechnology is very much in the R&D phase of its development, despite some well-known examples of products being marketed. That phase is likely to continue for some time, and R&D will always be important for many new and established companies using biotechnology, even when they are engaged in production and marketing. Similarly, technology licensing is and will continue to be important for many of these companies. As discussed in *Chapter 4: Firms Commercializing Biotechnology*, much of the current research in biotechnology is being funded by large, established companies with well-developed marketing capabilities. In return for their funds, these companies have received, among other things, rights to market the fruits of the research being conducted by new biotechnology firms (NBFs).* Moreover, even if a new or established company were to develop certain technology on its own, it might be in the company's best interest for various reasons to license the technology to others rather than to exploit the technology itself.

This chapter will first examine why and how research joint ventures and technology licensing agreements come within the scope of antitrust law. Second, the chapter will compare and contrast the relevant antitrust laws and policies of the United States, the European Economic Community (EEC), the Federal Republic of Germany, the United Kingdom, France, Switzerland, and Japan (the "competitor countries"). Third, the

*NBFs, as defined in *Chapter 4: Firms Commercializing Biotechnology*, are firms that have been started up specifically to capitalize on biotechnology. The relationship between NBFs and established companies is further explored in that chapter.

chapter will examine the current impact of these laws on biotechnology-related R&D and the licensing of the results of that R&D. Finally, the issue of whether congressional action on antitrust law is needed to promote U.S. competitiveness in biotechnology will be addressed.

Antitrust implications of research joint ventures

A joint venture is a form of association between separate business entities that falls short of a formal merger, but that unites certain agreed on resources of each entity for a limited purpose. The form of a joint venture may range from a purely contractual agreement to take joint action, to an agreement where any participant acquires certain assets of another, to the creation of a separate entity in which at least one participant acquires an equity interest. Joint venturers often agree that they will share the management and control of the joint activity's results.

Reasons for entering an R&D joint venture are as varied as the companies and individuals involved. The reasons must be strong enough to overcome the powerful disincentives among individual companies of sharing management and profits. Three reasons stand out in particular:

- *Small firm limitations.* Often small firms have the capability of inventing a process and obtaining a patent but are unable to develop or market the product without the assistance of a larger company.
- *Interdisciplinary technological areas.* Companies of any size may need to draw on expertise outside their own. It may be cheaper and faster to tie up with another company than to develop the new expertise themselves.
- *Economies of scale in R&D.* On certain large and complex technological problems, even large companies may not be able to achieve economies of scale in research if they undertake the R&D themselves.

From the perspective of antitrust policy, the last reason is the most important, since one goal of the antitrust laws is to enhance economic efficiency. In addition, joint ventures could allow certain high-risk, costly R&D to be undertaken that might not be undertaken otherwise by individual firms. Thus, research joint ventures can increase R&D and promote innovation. It is precisely because of these potential benefits to society that the antitrust authorities in both the United States and Europe have set forth official policy statements assuring companies that research joint ventures are viewed very favorably under the antitrust law and rarely raise significant questions.

Despite the general encouraging attitude that antitrust authorities have taken towards joint R&D activity, there are potentially anticompetitive effects of R&D joint ventures. Because R&D joint ventures may involve market-dominating technology, may be conducted by competitors or potential competitors, or may involve restrictive agreements concerning the use of the results, such ventures can give rise to antitrust concerns (36). In its *Antitrust Guide Concerning Research Joint Ventures,* the U.S. Department of Justice identified three kinds of effects on competition (36):

- when the association itself would lessen existing or potential competition between the participating firms,
- when the joint venture agreement or related agreements contain restrictions that restrain competition, and
- when limitations on participation or access to the results of research create or abuse market power.

The first concern is straightforward. When research ventures include most or all of the major competitors in an industry, they could reduce the competitors' separate efforts and thereby reduce innovation. The incentive to finance research and rapidly develop the results is diminished when the participants know that any invention is available for everyone to use. As Assistant Attorney General William Baxter stated, "Rivalry, in short, is important in research as it is in any other commercial activity" (4). There may be cases, however, where an industrywide effort is clearly the most efficient means to perform the research successfully (36).

In practice, the second antitrust concern is more common. Joint ventures in R&D often contain restrictions on the use of the technology once it is developed. Such restrictions may have anticompetitive effects.

Finally, a joint venture may create an important or even revolutionary new technology that would allow the participants to dominate the market. Such domination could create significant anticompetitive effects. Market domination itself, however, is not necessarily illegal; what is important is how that market power is exercised. In any event, the antitrust law must balance these anticompetitive effects with the reasonable desire of the participants to be rewarded for the risks and costs incurred by entering the joint venture.

Antitrust aspects of technology licensing

An inventor's ability to protect his or her invention long enough to reap sufficient benefits to make the inventor's investment of time and capital worthwhile will have a major impact on the inventor's decision to undertake R&D in the first place. Both the patent laws and laws permitting an inventor to license* a product, process, or discovery serve the social goal of promoting R&D. By protecting the inventor from interlopers who would otherwise benefit at little or no cost from the inventor's labor, ingenuity, or financial investment, these "legal monopolies" help ensure that invention is both encouraged and sufficiently rewarded.

Although they may at times appear to conflict, the U.S. patent laws (see *Chapter 16: Intellectual Property Law*) and the U.S. antitrust laws have virtually identical goals—the fostering of competition and innovation. Competition and innovation improve the allocation of scarce resources so that the maximum type and quantity of goods are produced at the lowest cost. The patent "monopoly," which is expressly recognized by the U.S. Constitution, is essentially a property right—the right to exclude others from making, using, or selling an invention for a limited period of time. A patent may or may not provide an economic monopoly. But even the existence of an economic monopoly based on a lawfully acquired patent is not of concern under the antitrust laws, because a patent is granted to encourage inventions that might not occur if a patent were not available. Inventions benefit the public by creating new products or more efficient means of making old products. Thus, the creation and introduction of inventions is an important form of competition.

The exploitation of the patent right involves its use by the owner or its use by other parties via a licensing agreement whereby these parties pay royalties to the owner. The antitrust laws do limit the exploitation of the market power resulting from patents. The patent owner is naturally interested in obtaining the greatest possible economic return from that market power. In patent licensing agreements, therefore, the owner/licensor may attempt to place certain restrictions on the licensee that are designed to enhance that economic return. (For example, the licensor may want the licensee to use a patented process only with materials supplied by the licensor.) However, these restrictions are not always compatible with society's goal of maximum production of goods at the lowest cost. Thus, patent licensing agreements may violate the antitrust laws.*

*A license is a contractual right granted by the owner of the technology to another party to use the technology. It is one way the owner can exploit the invention.

*In addition to the antitrust laws, the doctrine of patent misuse also serves to limit the patent owner's exploitation of the patent. It is available as a defense in a patent infringement case, and, if established, it renders the patent unenforceable. It is established by facts that do not establish an antitrust violation and is available even to a defendant who is not affected by the misuse (27). The doctrine has been criticized as vague, subjective, and mostly detrimental to innovation (5). An extended discussion of the doctrine is beyond the scope of this chapter.

For similar reasons reflecting both the concept of proprietary interest and the concept of rewarding invention, trade secrets and other forms of know-how may receive protection against improper disclosure.* And, like patents, they may be exploited through licensing agreements. Under appropriate circumstances, then, know-how licensing is a legitimate procompetitive action that promotes research and product development. Know-how licensing, however, will be subject to antitrust scrutiny.

Whether a particular form of patent or know-how licensing is anticompetitive is a determination that is fact specific and requires a detailed analysis of the terms of the agreement and the markets involved. The courts have developed various principles to guide the analysis, which will be discussed in greater detail in the next section.

*Know-how may be defined as technological information relating to manufacturing processes not protected by a patent, not generally known or accessible, and of competitive advantage to its owner (20). Legal protection of know-how is based on a theory of breach of trust and misappropriation. To the extent know-how is known only by its owner, the owner holds a limited monopoly.

A review of relevant U.S. and foreign antitrust laws

Antitrust laws and policies relevant to biotechnology in the United States are described below. Also discussed are the laws and policies of the EEC, the Federal Republic of Germany, the United Kingdom, France, Switzerland, and Japan.

United States

Four provisions of the U.S. antitrust laws are most relevant to this discussion. Section 1 of the Sherman Act (15 U.S.C. §1) prohibits "[e]very contract, combination . . . or conspiracy, in restraint of trade or commerce among the several States or with foreign nations . . ." Section 2 of the Sherman Act (15 U.S.C. §2) condemns monopolization, attempts to monopolize, or any combination or conspiracy to monopolize "any part of the trade or commerce among the several States, or with foreign nations . . ." Section 7 of the Clayton Act (15 U.S.C. §18), as amended in 1980, prohibits partial or entire corporate acquisitions " . . . by any person engaged in commerce or in any activity affecting commerce . . ." where "the effect of such acquisition may be to substantially lessen competition or to tend to create a monopoly." Section 5 of the Federal Trade Commission Act (15 U.S.C. §45 prohibits unfair methods of competition.

Taken together, these four statutory provisions prohibit any behavior that results in a substantial lessening of competition. The U.S. Department of Justice and the Federal Trade Commission have the power to investigate agreements or actions for anticompetitive effects. Violators of the antitrust laws face criminal penalties or injunctions. In addition, "injured" private parties can sue for violations of the law, which supplements Government enforcement. Under section 4 of the Clayton Act (15 U.S.C. §15), a private plaintiff* may sue for treble damages or seek injunctive relief. While in many instances private antitrust lawsuits follow successful Government litigation, private lawsuits can be the sole action challenging a given practice (32). The threat of private antitrust enforcement, coupled with the treble damages remedy, is a significant adjunct to U.S. Government enforcement and an important deterrent to anticompetitive behavior.

The U.S. antitrust laws are very different from most other statutes because they do not provide a checklist of specific, detailed statutory requirements, but instead set forth very broad principles. This approach requires private parties, Government prosecutors, and the courts to consider the

*Since enactment of the Clayton Act in 1914, Congress has twice amended §4 to qualify the rights of certain plaintiffs bringing actions under its provisions. New subsection (b) of 15 U.S.C. §15 limits monetary recovery in successful actions brought by foreign corporations to actual damages unless the plaintiff meets each of four specified tests. Other additions limit the time in which lawsuits may be filed to 4 years and establish rights and procedures governing parens patriae actions and instituted by Federal and State attorneys general. *See* 15 U.S.C.A. §§15, 15a-c,f,g (1983 Supp.).

overall purpose and effect of a business arrangement. Most arrangements are evaluated under a "rule of reason" test first enunciated by the U.S. Supreme Court in 1911 (28). Under this test, restraints on competition are evaluated by a full factual inquiry as to whether they will have a significant adverse effect on competition, what their justification is, and whether that justification could be achieved in a less anticompetitive way. Terms of an agreement may restrict some competition, yet be permitted, provided the restriction is clearly ancillary to some legitimate purpose and is appropriately limited in scope (35). The necessary vagueness of this test can create uncertainty about the legality of business arrangements, and this uncertainty may dissuade some types of arrangements.

Some types of agreements are not evaluated by the rule of reason test; instead they are considered illegal per se. Agreements between existing or potential competitors to fix prices or to allocate markets or customers, for example, are considered illegal per se. For such agreements, experience has established that their "pernicious effect on competition and lack of any redeeming virtue" makes an "elaborate inquiry as to the precise harm . . . or the business excuse" generally not worth the effort (22).

To assess the competitive impact of R&D joint ventures, the U.S. courts generally have used the rule of reason test. Under this test, a fact-intensive analysis is undertaken in which numerous factors are considered and their pro- and anticompetitive effects are balanced to assess the legality of certain behavior. The number of factors that must be assessed is often large. In *United States* v. *Penn-Olin Chemical Co.* (38), for example, the Supreme Court listed 15 factors to be considered in determining whether a joint venture violated section 7 of the Clayton Act.

In assessing the legitimacy of research joint ventures under the antitrust laws, the U.S. Department of Justice indicated in its *Guide to Research Joint Ventures* the most relevant considerations to be the following (36):

- *Whether the individual joint venturers would have undertaken the same or similar R&D on their own.* "If the cost and risk of the research in relation to its potential rewards are such that the participants could not or would not have undertaken the project individually, then the venture will have the effect of increasing rather than decreasing innovation."
- *The number and size of competitors in the relevant market, as well as the level of existing competition.* "The greater the number of actual and potential competitors in an industry, the more likely that a joint research project will not unreasonably restrain competition." The Justice Department has stated a preference for a series of several competing joint research projects, rather than industry-wide joint ventures, though the latter may be justified due to necessity.
- *The nature of the research.* "In general, the closer the joint activity is to the basic research end of the spectrum—i.e., the farther removed it is from substantial market effects and developmental issues—the more likely it is to be acceptable under the antitrust laws."
- *The scope of the research joint venture (how it is limited in time and subject matter).* "The narrower the field of joint activity and the more limited the collateral restraints involved, the greater the chances that the project will not offend the antitrust laws." Any ancillary restraining agreement is viewed more favorably if it is an important additional factor necessary to assure the venture's success.

The U.S. Department of Justice has procedures for reviewing and giving advice on the proposed business joint ventures before they are undertaken (28 C.F.R. §50.6). Though the grant of immunity is not guaranteed, approval through this procedure almost always is an effective grant of immunity from subsequent Government prosecution. From 1968 to 1978, the Department of Justice considered 29 specific requests for advice concerning proposed research joint ventures. Utilizing the procedure, the Department fully cleared 90 percent of the research joint ventures considered (14). Of all ventures granted clearance, none have been subsequently sued by private plaintiffs.

There have been few Justice Department enforcement actions with respect to R&D joint ventures. In fact, a pure research joint venture without ancillary restraints has never been challenged by the Antitrust Division (9). In the past 15 years, the Justice Department has formally challenged only one joint research arrangement, and only because it involved patent pooling and ancillary restraints that hindered the coventurers from undertaking the R&D themselves (8,24).*

Of the few private suits in the United States attacking R&D joint ventures, one recent case is the most significant. In *Berkey Photo, Inc.* v. *Eastman Kodak Co.* (7), the plaintiff, Berkey, contended that Kodak had extracted secrecy agreements from General Electric (GE) and Sylvania, its coventurers, that precluded other camera manufacturers from competing to produce cameras that could be used together with the certain new flash devices made by GE and Sylvania. The court noted that Kodak and GE were not direct competitors and that Kodak and Sylvania were potential competitors at best. However, because of Kodak's market power over cameras in general, the court found an exclusionary potential. The court recognized that if several substantial companies in an industry undertake joint research on a scale unattainable by the remaining companies and those remaining companies are not permitted to join the group, the coventurers might gain a decisive and unjustified advantage over the others. While the court had found market power to be a significant factor in assessing the joint venture's legality, it had been necessary for the plaintiff also to demonstrate that Kodak was gaining competitive advantages which were not the pure products of technological improvement (30).

Like joint ventures, technology licensing agreements are generally evaluated by the rule of reason when they contain terms that may restrict competition. Examples of license provisions that have raised antitrust concerns are limitations on how much the licensee can charge or sell, restrictions on the licensee's dealing in competing products, restrictions on the resale of the patented product, and tying arrangements.* Restraints may take several other forms, such as territorial restraints, field-of-use restrictions, and grantbacks.** Similar restraints also exist for know-how licensing. Factors relevant to assessing the legitimacy of such restraints are as follows: whether they are ancillary to a lawful main purpose of the agreement, have a scope and duration no greater than that reasonably required to achieve that purpose, and are not part of some larger pattern of anticompetitive restriction (36).

There is relatively little case law on the subject of know-how restrictions, but the existing cases state that the same type of ancillary-restraints analysis will be followed in this area as well. This is not to say that the outcome will be the same as for patents, since there are differences between patent and know-how licensing.*** Recognizing these differences, particularly the fact that know-how lacks the legislative status of the patent system, the U.S. Department of Justice at one time took the position that "know-how licenses will in general be subject to antitrust standards which, if anything, are stricter than those applied to patent licenses" (36). Further, the Justice Department took the position that restraints in know-how licenses should not last longer than the time necessary for the licensee to develop equivalent know-how for itself, "a reverse engineering

*The challenged R&D venture involved an alleged agreement between auto manufacturers to delay installation of existing emission control devices or stall the improvement of such devices. The case was ultimately settled and it enjoined the defendants from preventing or delaying the development or installation of these devices (37).

*A tying arrangement requires the licensee to purchase unpatented materials from the licensor.

**Territorial restraints are restraints that limit the licensee's use of the invention to specified geographical areas. Field-of-use restrictions limit the use of the invention to something less than all of its potential applications. For example, if Stanford licensed the Cohen-Boyer recombinant DNA process patent to a company only for making specialty chemicals but not for making pharmaceuticals, that would be a field-of-use restriction. A grantback is an agreement by the licensee to give back to the licensor (the owner of the basic patent) rights to any improvement patent.

***Some of these differences are the following: 1) all the patent claims must be definite in scope while know-how is usually of an amorphous character and cannot be described precisely; 2) patent protection is limited to the territory of the country granting the patent, while know-how could be protected, at least in theory, wherever the domestic law of the forum protects trade secrets; 3) patents are limited to the 17-year period of protection, while know-how is protected for as long as it does not become generally known; 4) a patent grant protects its owner from a duplicative independent invention, but the character of know-how can be destroyed by an independent invention; and 5) know-how content changes as new information is incorporated, and old information becomes publicly known (29).

period" (23). The rationale for the concept of the reverse engineering period appears to be that a restraint limited to the length of time necessary to invent around the licensed know-how "does not eliminate competition which would have taken place in the absence of the licensing agreement" (12). The current policy is to be more flexible on these restraints (2).

European Economic Community

The Federal Republic of Germany, United Kingdom, and France are members of the European Economic Community (EEC). The EEC, or Common Market, was created in 1958 by the Treaty of Rome. One of the goals of the treaty was the "establishment of a system ensuring that competition in the common market is not distorted." The result has been a two-tiered system of antitrust law in the Common Market. EEC law coexists with the national systems of antitrust law and is considered part of the national law of each member state. If there is any conflict between the national law and the law of the EEC, the latter prevails. Responsibility for enforcement of EEC law rests primarily with the Commission of European Communities ("Commission"). The Court of Justice, located in Luxembourg, reviews the formal decisions of the Commission.

Articles 85 and 86 of the Treaty of Rome govern anticompetitive practices. article 85(1) prohibits "all agreements . . . and concerted practices . . . which have as their object or effect the prevention, restriction or distortion of competition within the common market . . . ."* Article 86 prohibits abuses by one or more enterprises "of a dominant position within the common market," such as "limiting of production, markets, or technical development . . . ."

Article 85(3) of the Treaty of Rome provides for exemptions from article 85(1) for certain agreements or practices such as those that promote economic and technical progress and do not impose ancillary restrictions or afford the possibility of eliminating competition. A notification procedure has been created which allows the Commission to review agreements for which an article 85(3) exemption is claimed. The grant of an exemption by the Commission is binding on the national authorities and courts of the member states.* Thus, clearance by the enforcing agency is much more important in the EEC than in the United States.

The articles in the Treaty of Rome give the Commission of European Communities broad authority to prohibit: 1) R&D joint ventures that have the potential to eliminate competition between major companies, and 2) ancillary restrictions of R&D joint ventures that could restrain competition. The criteria that the Commission has shown to be important in judging whether a venture comes under the first category have generally been similar to those of the U.S. Department of Justice, i.e., the market share of the relevant companies, the ability of other enterprises to perform the research, and the extent to which the research is applied as opposed to basic. In the second category, restraints ruled illegal usually have been restrictions on the ability of the participants to compete with the joint venture itself and restrictions concerning distribution of the joint venture's end results.

Though 15 years ago the Commission published an official notice intended to reassure enterprises of the legality of most R&D agreements (in particular ventures with R&D as the "sole object"), later decisions of the Commission have showed some of its statements of leniency to be unreliable (6). For example, in 1972, two of the largest manufacturers in the oligopolistic European soap industry created a joint, equally owned subsidiary in Switzerland to conduct research into soap products.

---

*Article 85 will apply to an agreement only if it "may affect trade between Member States." Thus, if a contract only affects internal trade of one nation, trade between nonmember nations, or trade between a member and a nonmember nation, it is not covered by article 85 regardless of its impact on competition (40).

*In addition to the ability to petition for article 85(3) exemptions, an enterprise can request the Commission to rule that, based on the information supplied, it will not challenge the agreement under article 85(1). Such a ruling is provided for under article 2 of regulation 17 and is called a negative clearance. The grant of a negative clearance means that article 85(1) does not apply to the agreement at all. In practice, applications for negative clearance are often accompanied by requests for an exemption, so that if the commission finds a violation of article 85(1), it can consider whether to grant an exemption. Failure to disclose all pertinent facts or a subsequent change in the factual situation may result in cancellation of an exemption or a negative clearance.

The Commission found that the agreement eliminated competition in research and therefore violated article 85(1) (18).

Since the Commission may not grant an exemption in the absence of a notification of the agreement and its provisions, the EEC legal system has ensured that most major research ventures among European companies of different nationality are filed with the Commission.* The soap case mentioned above was in fact notified and granted an exemption because the commission ruled that the joint research would promote economic and technical progress. The exemption was subject to the condition that the companies inform the Commission of all license agreements emanating from the results of the joint research.

The Commission will also exempt anticompetitive collateral restraints on the basis of article 85(3). In one case, an agreement between two enterprises for joint R&D work on a new type of electrically powered bus was granted an exemption, even though its provisions prohibited cooperation with third parties within the field covered by the agreement (19).

The Commission's decisionmaking process differs substantially from the U.S. adjudicatory process in the sense that it is much less formal and less procedurally oriented. Before giving approval, the Commission is willing to negotiate and, wherever necessary, mandate conditions that will guarantee that the parties will remain competitive once the joint research venture has terminated.** It is rather frequent that harmful collateral restrictions are found, which usually can be eliminated or redrafted without prohibiting the joint venture itself. Although there have been no Commission decisions to prohibit research joint ventures, many recent decisions have in some way limited or controlled joint research agreements, in most cases with respect to their collateral restrictions. Since 1968, the Commission has modified at least eight cases involving joint research and subjected others to reporting obligations or otherwise limited the exemption granted in time or scope of coverage.*

Considering the list of cases that have been modified and the mandatory notification requirement, it appears that in practice the EEC is at least as tough as, and probably tougher than, the United States on joint research, particularly with respect to collateral restraints. The Commission has not hesitated to impose reporting obligations and to review periodically whether a joint venture may become anticompetitive in future years.

Patent and know-how licenses are agreements that may come within the scope of article 85. EEC law and the law of the member countries generally follow the traditional doctrine that restrictions on the licensee are valid if they do not expand the scope of the patent. A body of law has developed, based mainly on Commission decisions, with regard to what kinds of restrictions in licensing agreements are legal and what kinds are not.** The Commission has also issued a proposed exemption from article 85(1) for two-party patent licensing agreements (10). The proposed exemption is very narrow and has received substantial criticism (40).

Federal Republic of Germany

In the Federal Republic of Germany, the Act Against Restraints of Competition (GWB, Gesetzgegen Wettbewerbsbeschrankungen) prohibits restrictive business practices and is concerned with maintaining competitive market structures.

---

*Article 4(2) of regulation 17 provides that certain classes of agreement need not be notified to the commission in order to obtain an exemption. This means merely that they are eligible to be considered for the grant of an exemption under article 85(3) even if notification has not been filed. Though agreements which have as their "sole object . . . joint research and development" do not have to be notified [(Article 4)(2)(iii)(b)], R&D agreements with any sort of ancillary restraints must be.

**An example of this was the *ICI/Montedison* case (17) where the Commission proposed to mandate an obligation that would insure that "on the termination of the agreement, Montedison should be in a position to continue as an independent producer of a line if it wished, thereby increasing competition in an oligopolistic market."

*See ACEL/Berliet, 1968 C.M.L.R. D35 (1968) (modification); Henkel/Colgate, J.O. 1972, L. 14/14 (1972) (reporting obligation, exemption limited to 5 years); SOPER:EM/Rank, 1975-1 C.M.L.R. D72, (1974) (modification, reporting obligation, exemption limited to 10 years); Vacuum Interrupters, 1977-1 C.M.L.R. D67 (1977) (reporting obligation, exemption limited to 8 years); General Electric/Weir, 1978-1 C.M.L.R. D42 (1977) (modification, reporting obligation, exemption limited to 12 years); SOPELEM/Vickers, 1978-2 C.M.L.R. 146 (1977) (reporting obligaton, exemption limited to 5 years) *modified*, 1982-3 C.M.L.R. 443 (1981) (exemption extended until 1991); Beecham/Parke Davis, 1979-2 C.M.L.R. 157 (1979) (modification, reporting obligation, exemption limited to 12 years).

**For information on particular kinds of clauses, see (40).

This law is intended expressly to promote "competition based on efficiency" and is regarded as the "constitution" of the German social market economy (31). Section 1 of the law establishes a general prohibition against agreements made for a common purpose by enterprises that restrain competition, production, or market conditions. Thus, this section can preclude a research joint venture having anticompetitive market effects.

Section 5b permits small- and medium-sized firms to form rationalization cartels,* assuming no substantially adverse effect on competition and assuming that the result promotes the firms' overall efficiency. Such cartels may include cooperative R&D ventures.

The application of German law by Government authorities appears to have been at least as tough as in the United States in regard to research joint ventures. Between 1979 and 1980, the German Cartel Office caused the abandonment of two agreements involving joint research. A proposed venture between Siemans AG and VDO Adolph Schindling to develop, produce, and market liquid crystal gages for use in automobile instrument panels was prohibited, because the arrangement already jointly held 80 percent of the market for automobile instruments (13). Another proposed joint venture between Takeda Chemical of Japan and Bayer AG of Germany to develop, test, and market pharmaceutical products in the Federal Republic of Germany was prohibited because it would have represented a combination of two of the world's eight largest pharmaceutical companies and eliminated Takeda as an independent potential competitive force in West Germany (13).

With respect to technology licensing agreements, GWB section 20(1) is relevant. It nullifies agreements covering the acquisition or use of patents or protected seed varieties to the extent they impose restrictions on the business conduct of the acquirer or licensee that go beyond the scope of the protected right. However, German cartel authorities may grant an exemption to this provision under GWB section 20(3) if the economic freedom of the licensee or other company is not unfairly hurt and market competition is not substantially impaired. Thus, the approach of West Germany is similar to that of the United States in terms of having a general prohibition against agreements that extend the scope of the patent, but German law gives the antitrust authorities discretion to exempt agreements on a case-by-case basis, which makes the German system more flexible.

*A rationalization cartel is one formed to improve efficiency of production through concerted action.

United Kingdom

The U.K. antitrust law is contained in several statutes. The ones most relevant for R&D joint ventures and technology licensing are the Fair Trading Act of 1973 and the Competition Act of 1980.

Under section 76 of the Fair Trading Act, the Director General of Fair Trading has the duty to be generally informed about all mergers and to decide whether to recommend to the Secretary of State referral to the Monopolies and Merger Commission. Not all joint ventures are affected by the legislation. The Fair Trading Act does not apply if the joint venture is merely the result of an investment of capital by the coventurers in a jointly owned company. In most instances, a research joint venture will not involve the type of agreement constituting a merger under the Fair Trading Act.

Should a "merged" R&D venture be referred to the Monopolies and Mergers Commission, its legality is assessed on the basis of whether it will operate in the public interest. The five factors considered are whether the merger will promote: 1) effective competition, 2) the interests of consumers, 3) reduced costs and the development of new techniques and products, 4) a balanced distribution of industry and employment, and 5) competitive activity in British markets. Even if a proposed research joint venture were subject to the Fair Trading Act's reporting provisions, it is likely to be characterized as activity helping to develop "new techniques and products" and therefore not violate the Fair Trading Act.

The Competition Act was designed to provide a comprehensive approach to anticompetitive practices not already covered by existing statutes. Generally, the act applies to all activities that prevent, restrict, or distort competition. Thus, it would apply to R&D joint ventures and to technology licensing agreements.

Generally, the antitrust regime in the United Kingdom is relatively loose, and enforcement actions on joint R&D ventures and licensing agreements have been few. But U.K. companies formulating agreements with companies of other European countries must take into account the EEC laws.

France

The relevant statutes in France are Title II of Act No. 77-806 and Articles 50 and 51 of Price Ordinance No. 15-1483. Under title II, the Minister of Economic Affairs may act against a "concentration"* that is "of such a nature as to prevent adequate competition in the market." Articles 50 and 51 apply to concerted actions or agreements that prevent, restrain, or distort competition.

R&D joint ventures could be prohibited under title II if they involved major French companies. However, an anticompetitive concentration may be exempted under section 4 when the concentration is found to make "a sufficient contribution to economic and social progress" to compensate for its restraints on competition. A determination on this point considers the international competitiveness of the companies concerned. A biotechnology R&D joint venture among large companies would likely be exempted under this provision, and such a joint venture among small firms is unlikely to raise problems in the first place.

Ancillary restraints which accompany many joint R&D agreements would come under paragraph one of article 50, which prohibits concerted actions or agreements that may prevent, restrain, or distort competition and specifically mentions impeding technological advance. However, article 51 provides for an exception where the anticompetitive practices further contribute to economic progress, particularly through enhanced productivity.

There is no French antitrust law that applies specifically to technology licensing, but the Competition Commission has taken the position that articles 50 and 51 apply to intellectual property rights. However, there is very little case law in this area (25).

French antitrust law is of recent origin and is still developing. It is unlikely to be applied to a biotechnology R&D joint venture. How it will be applied to biotechnology licensing agreements is somewhat unclear at this point.

Switzerland

Joint ventures and licensing agreements in Switzerland are governed under the provisions of the Federal Cartels Act. The mere creation of a joint venture would not trigger liability under this act. If the venture dominated or exercised a determining influence on a product market, however, the act would apply. Unless major companies joined a biotechnology R&D joint venture, the act would not appear to apply.

Exemptions to the Federal Cartels Act are outlined in article 5. Activities that are otherwise prohibited by the act may be permitted on the "grounds of overriding legitimate interests" if competition is not prevented "to a degree that is excessive." "Overriding legitimate interests" include those aimed at: 1) establishing reasonable requirements as to training, skill, or technical knowledge for a particular occupation or industry; and 2) promoting an economic or occupational structure that is desirable in the general interest. Thus, even if a biotechnology research venture interfered with competition to a limited degree, it would appear to be exempt under article 5.

Swiss law appears to favor R&D joint ventures. There apparently have been no specific cases dealing with R&D joint ventures, and there has been no general treatment of the subject in Swiss legal periodicals (9).

The Federal Cartels Act would apply to licensing agreements in situations involving market

*A "concentration" is defined as, "the result of any legal act or transaction . . . having the object or effect of enabling one enterprise or a group of enterprises to exercise an influence, directly or indirectly, on one or more other enterprises which is of such a nature as to direct or even orientate the management or workings of the latter."

dominance. For example, a requirement that a licensee undertake no research in the same area as a patented invention would be objectionable under the act. Similar objections would be raised if a licensee were obligated to assign any improvements on the licensed technology to the licensor. However, cooperative agreements to exchange research results appear to be lawful.

Japan

Japan's antimonopoly law—the Act Concerning Prohibition of Private Monopoly and Maintenance of Fair Trade (Japanese Law 54 of 1947)—was first enacted in 1947 during the U.S. occupation and was revised three times subsequently, in 1949, 1952, and 1977. Enforcement procedures were established in the Japanese Fair Trade Commission (JFTC), an independent five-person regulatory body modeled after the U.S. Federal Trade Commission. Section 25 of the law allows private companies the right to sue for damages, but only after JFTC has found a violation.

The basic provisions of Japan's antimonopoly law are quite rigorous. Article 1 explains that the purpose of the law is to "eliminate unreasonable restraint of production, sale, price, technology, and the like . . . ." Revisions in 1977 reflected a concern for controlling large corporations so that the revitalized market structure could function more efficiently. Sections 3 and 6 of the 1977 revisions preclude entrepreneurs from engaging in any unreasonable restraints of trade or entering into international agreements with terms that might be unreasonable trade restraints. Research joint ventures could qualify, since section 2(6) defines "unreasonable restraints of trade" as: "business activities by which entrepreneurs . . . mutually restrict or conduct their business activities in such a manner as to fix, maintain, or enhance prices, or to limit production, technology, products, facilities, or customers or suppliers." The act also prohibits private monopolization.

Several provisions in articles 21 through 24 of the antimonopoly law specifically permit certain types of legal cartels, including research joint ventures. In total, there are 39 laws permitting businesses to form legal cartels exempt from the antimonopoly laws (26).

With the end of the occupation in 1951, Japan's antimonopoly law was ineffectively enforced by JFTC; its relatively severe antimonopoly restrictions and prohibitions against cartels drew considerable hostility from the Japanese Government, and JFTC virtually languished between 1952 and 1969 (15). In the meantime, Japan's Ministry of International Trade and Industry (MITI) often implemented a procedure known as "administrative guidance" in which persuasion would be used by MITI to influence businessmen within its oversight. In some instances, administrative guidance functioned to foster cartelization either by restricting production or investment, or otherwise influencing prices—all circumventing the antimonopoly law.

The last decade, however, has seen a marked increase in JFTC's enforcement activities. In 1980, for example, JFTC completed 62 cases, 24 of which involved price-fixing. It has also ordered 279 businesses to pay a total of $10 million in fines and has prosecuted a wide variety of unfair business practices (33).

Despite the increase in enforcement activity, the Japanese Government has to date not prosecuted any R&D joint ventures. The Research Association Law, passed in 1961 and amended in 1963, provides an important perspective on the Japanese Government's competition policy as opposed to its enforcement of its antimonopoly laws. This law allows several companies to undertake long-term R&D or to pool financial, personnel, and capital resources. In almost all such instances, the approved association involves R&D in which some Japanese Government ministry or agency participates. Rather than being anticompetitive, these research associations often serve to undermine collusive behavior by increasing entry into advanced industries and helping to diffuse new technology (26). Pursuant to the Research Association Law, the Ministry of Finance has specifically recognized the recently created Biotechnology Research Association.

There is one significant difference between the Japanese and U.S. antitrust perspective on research joint ventures. In Japan, there would be no objection in the case of a new technology if all the companies involved were to join in the same venture. In the United States, such a venture would raise serious antitrust problems. However, if the Japanese joint venture restricted entry into or subsequent use of the technology by competitors, then it would probably violate the antimonopoly law.

Japan's antimonopoly law creates a mechanism for Government oversight of international technology transfer. Section 6 of the law prohibits a firm or entity from "enter[ing] into an international agreement or contract which contains such matters as constitute unreasonable restraints of trade or unfair business practice." On July 23, 1982, section 6 was amended to require that international agreements that *may* constitute unreasonable restraints of trade or unfair business practices be filed with JFTC. Technology licensing and joint venture agreements are among those required to file. JFTC has promulgated a regulation covering patent licensing agreements (3). Thus, JFTC can monitor these agreements for restraints on competition.

Applicability of antitrust law to biotechnology research joint ventures

The use of joint ventures in biotechnology, as discussed in *Chapter 4: Firms Commercializing Biotechnology,* is prevalent. The capital markets have not funded all the long-term, high-risk R&D that NBFs wish to undertake. Joint ventures have been used as an important source of revenue by NBFs until they can develop the production and marketing capabilities to distribute their own products. Large, established companies have entered into several different kinds of joint ventures with NBFs, in most cases to obtain access to the new technology until their own in-house capabilities can be developed.

Joint research ventures in biotechnology currently run very little risk of violating either the U.S. or foreign antitrust laws. Two factors in particular support this assertion. One is the very high risk of biotechnology R&D. For example, total sales of biotechnology products reached $20 million in 1982 and are projected to range from $150 million to $3 billion in 1987 (16). This huge range reflects the considerable uncertainty and risk at this time over the size of future markets, a factor that depends on the number of commercially available products (16).

The track record of the first rDNA product, the human insulin product Humulin®, provides an instructive example of the risks involved in commercializing biotechnology. The micro-organisms used to produce Eli Lilly's (U.S.) product Humulin® were first provided by the NBF Genentech (U.S.) over 5 years ago. Lilly sponsored both the research and the marketing and agreed to pay Genentech royalties (see *Chapter 5: Pharmaceuticals*). The commercial success of this product, however, remains uncertain. In clinical trials, Humulin® has not shown any special advantages over naturally produced porcine insulin and has been found to cause immune reactions similar to the porcine product. Furthermore, production difficulties may have caused Eli Lilly to have run short of the drug during clinical trials. Finally, according to some critics, a newer and cheaper method of producing human insulin may already be available (11).

Eli Lilly's experience with Humulin® demonstrates that the commercial development of biotechnology products may take several years and may generate products that may become rapidly outdated. Combined, these factors indicate a very high level of risk. When the risks are as substantial as they currently are in biotechnology, enforcement authorities are far more tolerant of joint ventures.

The second reason joint research ventures in biotechnology do not currently raise antitrust concerns is the decentralization of biotechnology R&D. At the end of 1983, there were about 200 companies using biotechnology in the United States. The major thrust of all systems of antitrust law is to prevent dangerous trends towards concentration and monopolization—conditions that could signal a downturn in innovation. Although the point where dangerous concentration in R&D occurs varies from case to case, the biotechnology field remains far from that point today.

Because of the deconcentration of biotechnology R&D, research joint ventures are essentially procompetitive, assuming the absence of ancillary restraints. Most established companies that have participated in joint ventures with NBFs are also undertaking in-house R&D. The revenue earned by joint ventures for NBFs is sustaining the viability of a larger number of competitors.

Thus, joint ventures in biotechnology R&D in the United States (and most likely foreign countries as well) currently face virtually no significant antitrust restraints. The absence of measurable product markets and the lack of R&D concentration means that research joint ventures are not reducing competition. Only when successful products and measurable market shares become apparent will joint research activities among major companies invite major antitrust challenge.

Antitrust law has come under much scrutiny recently, and the trend in the U.S. Department of Justice is toward a policy that an action is unlawful only if the injury to competition outweighs the benefits. For instance, the Department of Justice recently gave preliminary approval to the formation of one of the largest cooperative research arrangements in U.S. industrial history—an amalgam of 12 major computer firms called the Microelectronics Computer Corp. (MCC) (39). Although the Department of Justice press release gave little guidance on the antitrust aspects, the decision not to challenge MCC's formation at least demonstrates that a carefully structured R&D joint venture can include most of the U.S. competitors without being considered anticompetitive.

Application of antitrust law to biotechnology licensing agreements

The preceding survey of the antitrust laws of the competitor countries in biotechnology indicates that most restraints on competition in licensing agreements will be evaluated by a rule of reason test. The authorities of the various countries have applied this test to various types of provisions in licensing agreements, including grantbacks and field of use restrictions. Other provisions, such as tying arrangements, are generally treated under per se rules. It is not useful to examine these in detail, since virtually none of them raises any unique issues with respect to biotechnology.

One type of factor relating to restrictions may have unusual significance for biotechnology. As a general rule, restrictions extending beyond the life of the technology being licensed are considered suspect. For U.S. patents, the life of the technology is arguably no more than 17 years, i.e., the term of the patent. For know-how, however, the useful life is not so easily defined. At least two commentators have suggested that most know-how can be reverse-engineered in 3 to 5 years and that restrictions exceeding 5 years should therefore be considered in the United States per se unreasonable unless the licensor can provide a special justification (1). On one hand, this view may make sense for biotechnology know-how, given the pace of technological development. On the other hand, many, if not most, production processes for biological products, i.e., the organisms themselves, are not capable of being reverse-engineered because of their complexity. Thus, the rigid and unthinking application of a 5-year rule would unfairly and unnecessarily hinder licensors in their ability to exploit their technology.

Findings

U.S. companies using biotechnology face no major antitrust compliance problems. Nevertheless, there is some degree of uncertainty about the scope and applicability of the antitrust laws to R&D joint ventures and to licensing agreements. This uncertainty, plus the expense of litigation and the threat of treble damages, could discourage some activities that might lead to innovation in biotechnology and could limit the ability of U.S. companies to commercially exploit their technology. Furthermore, the rigid application of certain per se rules in the area of licensing may actually lead to anticompetitive results. Thus, the current antitrust laws in the United States may have some modest adverse effect on biotechnology.

The antitrust laws of the United States, the Federal Republic of Germany, the United Kingdom, France, Switzerland, and Japan are generally similar in that they all prohibit restraints of trade and monopolization. Unlike the U.S. laws, however, the foreign laws generally provide for exemptions and vest much discretion with the enforcement authorities. Most of the kinds of arrangements that would be of interest to firms using biotechnology would be evaluated under a rule of reason test, but others are now per se illegal.

Under U.S. antitrust law, the legality of a research joint venture is judged on the basis of a balancing of its procompetitive v. anticompetitive effects. Factors considered important are whether the individual joint venturers would have undertaken the same or similar R&D on their own, the number and size of competitors in the relevant market, the scope of the research (basic v. applied), and the scope of the research joint venture (how it is limited in time and subject matter).

It is by no means clear that more favorable treatment is given to R&D joint ventures by the laws and enforcement authorities of European countries. Authorities in the EEC and the Federal Republic of Germany in particular have caused the abandonment or modification of a larger number of joint R&D ventures than their U.S. counterparts have. Though Switzerland, France, and the United Kingdom appear to have less stringent antitrust enforcement than the United States, European company activity across national boundaries of member states comes under EEC law.

Japan has probably been more tolerant than the United States toward anticompetitive aspects of R&D joint ventures. The highly publicized research associations sponsored by the Japanese Government best exemplify this attitude. However, this attitude may be changing, as indicated by the growing number of antitrust enforcement actions in general.

At the present time, companies applying biotechnology both in the United States and foreign countries face virtually no significant antitrust compliance problems with research joint ventures, excluding blatantly anticompetitive activities like price-fixing. In biotechnology, there is a lack of concentration of industry research and an absence of measurable markets. Only when biotechnology-related industries develop increasing concentration, successful products, and measurable market shares will R&D joint ventures be exposed to the antitrust statutes.

Technology licensing agreements are reviewed by the governmental authorities under the general principle that the agreements should not extend the scope of the patent or know-how in ways that are on balance anticompetitive. The only issue unique to biotechnology raised by the application of the antitrust laws to these agreements relates to the length of time of permissible restrictions on competition. The rule that such restrictions should not extend beyond an arbitrarily determined useful life of the licensed technology may not be especially relevant to biotechnology, and its application may hinder the exploitation of inventions through licensing.

Despite the fact that U.S. antitrust law is not likely to be a major concern with respect to biotechnology R&D joint ventures or licensing, there will be some degree of uncertainty regarding the antitrust implications of any corporate activity in this area. The degree of uncertainty is less in foreign countries than in the United States because these countries have more well-defined procedures for prior review of transactions by government authorities and less onerous penalties for violations. Lessening the uncertainties under U.S. law could be expected to have a positive effect on the development of biotechnology in the United States.

Issue

ISSUE: Should Congress change U.S. antitrust laws to encourage more joint research in biotechnology or to facilitate biotechnology licensing?

U.S. companies using biotechnology face no major antitrust compliance problems. For the reasons discussed in the findings of this chapter, however, current U.S. antitrust laws could have some modest adverse effect on U.S. competitiveness in biotechnology. The impact of these laws is not particularly unique to biotechnology, as distinguished from other areas of technology. In fact, the impact will probably be less for biotechnology than for more mature technologies, given the current lack of concentration in commercial R&D in biotechnology and the absence of measurable markets for products. Therefore, despite the many proposals to change the law and enforcement procedures now being discussed, no policy options are discussed here, because their broad applicability to technology in general is beyond the scope of this report.

Chapter 18 references*

1. Adelman, M. T., and Brooks, E. L., "Territorial Restraints in International Technology Agreements After *Topco," Antitrust Bull.* 17:763, 1972.
2. Andewelt, R., Chief, Intellectual Property Section, Antitrust Division, U.S. Department of Justice, "Competition Policy and the Patent Misuse Doctrine," remarks before the Bar Association for the District of Columbia, Patent, Trademark, and Copyright Section, Washington, D.C., Nov. 3, 1982.
3. "Antimonopoly Act Guidelines for International Licensing Agreements (Fair Trade Commission, 1968)," app. H to G. Rose, "EEC and Other Foreign Antitrust Law as Applied to Licensing," *Technology Licensing*, vol. 1, T. Arnold and T. Smegal, Jr. (eds.) (New York: Practicing Law Institute, 1982).
4. Baxter, W. F., Assistant Attorney General, Antitrust Division, U.S. Department of Justice, speech to the National Association of Manufacturers, Washington, D.C., May 20, 1983.
5. Baxter, W. F., Assistant Attorney General, Antitrust Division, U.S. Department of Justice, statement concerning S. 737, S. 568, and S. 1383, Bills Related to Joint Research and Development, in *Industry Joint Research and Development Ventures,* hearings before the Senate Committee on the Judiciary, U.S. Congress, June 29, 1983 (Washington, D.C.: U.S. Government Printing Office, 1983).
6. Bellamy, C., and Child, G. D., *Common Market Law of Competition* (New York: Mathew Bender, 1973).
7. *Berkey Photo, Inc.* v. *Eastman Kodak*, 603 F.2d 263, *cert. denied*, 444 U.S. 1093 (1980).
8. Brodley, J., "Joint Ventures and Antitrust Policy," *Harv. L. Rev.* 95:1521, 1982.
9. Cira, C., and Kolb, C., of Foreman and Dyess, "Report on Antitrust and Biotechnology," contract report prepared for the Office of Technology Assessment, U.S. Congress, April 1983.
10. Commission of European Communities, "Proposal for Commission Regulations on Article 85(3) of the Treaty to Certain Categories of Patent Licensing Agreements," *Official Journal of the European Communities* 22(C58):12, Mar. 3, 1979.
11. *Economist,* "A Shaky Start for Genetic Engineering's First Drug," Sept. 25, 1982, pp. 103-104.
12. Farmer, D. A., Jr., Special Assistant to the Assistant Attorney General, Antitrust Division, U.S. De-

*Note: C. D. Cal. = Central District of California.
F. Supp. = Federal Supplement.
U.S. = U.S. Supreme Court Reports.
F.2d = Federal Reporter, Second Series.

partment of Justice, "An Overview of the Justice Department's Guide to International Operations," speech to the Council of the Americas, June 8, 1977, reprinted in *Trade Regulation Transfer Binder: Current Comment, 1969-1983,* 12th ed. (CCH) 50, 325, p. 55, 690.
13. Federal Republic of Germany, *Tatigkutsbericht des Bundeskartellamt* (Report of the German Cartel Office), N. 9/565 at 63, 68 (1979).
14. Ginsburg, D., *Antitrust Uncertainty and Technological Innovation*, 28 (Washington, D.C.: National Academy of Sciences, 1980).
15. Hadley, E., *Antitrust in Japan* (Princeton, N.J.: Princeton University Press, 1970).
16. Hochhauser, S., "Bringing Biotechnology to Market," *High Technology,* February 1983, p. 60.
17. *ICI/Montedison, 7th Report on Competition* (EEC), point 156-159 at pp. 117-119 (1978).
18. *In re Henkel/Colgate,* decision of the Commission of European Communities of 23 December 1971, J.O. 114:14, Jan. 18, 1972.
19. Kintner, E., and Joelson, M., *An International Antitrust Primer* (New York: Macmillan Publishing Co., 1974).
20. Kirkpatrick, A., and Mahinka, T., "Antitrust and the International Licensing of Trade Secrets and Know-How: A Need for Guidelines," *Law & Policy International Business* 9:725, 1977.
21. National Research Council, "International Competition in Advanced Technology: Decisions for America," Washington, D.C., 1983.
22. *Northern Pacific Railway Co.* v. *United States*, 356 U.S. 1,5 (1958), cited in U.S. Department of Justice, Antitrust Division, *Antitrust Guide for International Operations,* Washington, D.C., Jan. 26, 1977.
23. Oppenheim, S. C., and Weston, G. E., *Federal Antitrust Laws: Cases and Comments* (St. Paul, Minn.: West Publishing Co., 1968).
24. Pfeffer, J., and Nowak, P., "Patterns of Joint Venture Activity: Complications for Antitrust Policy," *Antitrust Bull.* 21:315, 1976.
25. Rose, G., "EEC and Other Foreign Antitrust Law as Applied to Licensing," *Technology Licensing,* vol. 1, T. Arnold and T. Smegal, Jr. (eds.) (New York: Practicing Law Institute, 1982).
26. Saxonhouse, G., "Biotechnology in Japan," contract report prepared for the Office of Technology Assessment, U.S. Congress, June 1983.
27. Schlicher, J., "An Introduction to the Antitrust and Misuse Limits on Commercial Exploitation of a Patent," *Banbury Report 10: Patenting of Life Forms* (New York: Cold Spring Harbor Laboratory, 1982).
28. *Standard Oil Co. of N.J.* v. *United States,* 221 U.S. 1 (1911).
29. Stedman, A., "Legal Problems in the International and Domestic Licensing of Know-How," *A.B.A. Antitrust Section Rep.* 29:247, 1965.
30. Stein A., statement in *Japanese Technological Advancement and Possible U.S. Response Using Research Joint Ventures,* hearings before the Subcommittees on Investigation and Oversight and Science, Research, and Technology, House Committee on Science and Technology, June 30, 1983, p. 3, U.S. Congress (Washington, D.C.: U.S. Government Printing Office, 1983).
31. Stockman, K., and Strauck, V., "Germany," *World Competition Law,* J. O. von Kalinowski (ed.) (New York: Mathew Bender, 1979), 5:202[1], at pp. 2-5.
32. Sullivan, L., *Handbook of the Law of Antitrust* §246 (St. Paul, Minn.: West Publishing Co., 1977).
33. Uesugi, A., "Japanese Antimonopoly Policy—Its Past and Future," *Antitrust L. J.* 50:709, 1981.
34. U.S. Congress, Senate Judiciary Committee, *Industry Joint Research and Development Ventures,* hearings before the Senate Judiciary Committee, U.S. Congress, June 29, 1983 (Washington, D.C.: U.S. Government Printing Office, 1983).
35. U.S. Department of Justice, Antitrust Division, *Antitrust Guide for International Operations,* Washington, D.C., Jan. 26, 1977.
36. U.S. Department of Justice, Antitrust Division, *Antitrust Guide Concerning Research Joint Ventures*, Washington D.C., 1980.
37. *United States* v. *Automobile Manufacturers Association, Inc.,* 307 F. Supp. 617 (C.D. Cal. 1969), aff'd sub nom., *City of New York* v. *United States,* 397 U.S. 248 (1970) (consent decree), modified sub nom. *United States* v. *Motor Vehicle Manufacturers Association of the United States, Inc.,* 1982-83 Trade Cas. ¶65,175 (C.D. Cal. 1982).
38. *United States* v. *Penn-Olin Chemical Co.,* 378 U.S. 158 (1964).
39. Walsh, J., "MCC Moves Out of the Idea State," *Science* 220:1256, 1983.
40. Wiesen, J., "EEC Antitrust Considerations in Patent Licenses and Other Technology Transfer Agreements," *Industrial Applications of Genetic Engineering: The Legal and Regulatory Regime* (Philadelphia: American Law Institute, American Bar Association, 1983).

Chapter 19

International Technology Transfer, Investment, and Trade

Contents

Table

Chapter 19

International Technology Transfer, Investment, and Trade

Introduction

Intense international research and development (R&D) activities in biotechnology have stimulated equally intense efforts to diffuse the resulting knowledge and to sell the products of the research, both in the United States and abroad. Academic scientists are racing to publish their research results or to share them with colleagues at international conferences. Established companies and new biotechnology firms (NBFs)* are funding university research programs to gain access to potentially valuable new developments. U.S. and foreign patents arising from the increase in biotechnology R&D and international licensing agreements formulated to exploit the patents are diffusing the technology and promoting worldwide commercialization. Finally, NBFs in the United States and large U.S. and foreign companies have undertaken many R&D joint ventures to develop and market new products.

Several other chapters of this report have examined factors basic to the commercialization of biotechnology research (e.g., venture finance and patent rights). This chapter focuses on the legal environment surrounding the international exploitation of biotechnology and access to foreign markets through international technology transfer, investment, and product trade.

Although most companies are not yet marketing biotechnology products, the legal environment surrounding licensing, investment, and trade is already influencing the strategic decisionmaking of companies commercializing biotechnology—strategic decisions, such as negotiations on licensing, locational decisions for R&D, production, and clinical trials.

This chapter considers laws that can be employed directly by governments to control or influence access to foreign or domestic markets: export controls, patent laws, compulsory licensing provisions, investment and exchange controls, and trade laws. *Export controls* on technology and on products have a direct effect on potential demand and may affect the price that technology will fetch. Controls also bring delay and red tape into export transactions. *Patent laws* may contain secrecy provisions that restrict outward technology transfer for security, economic, or foreign policy reasons. They are similar in purpose and effect to export control laws. *Compulsory licensing* can be used to force inward technology transfer and can diminish return on R&D; where aggressively used, it may simply deter foreign and domestic investment in local R&D. *Investment and exchange controls* as well as *technology transfer controls* can be used to reserve national markets for locally owned firms and to force inward technology transfer. *Nontariff barriers* to trade such as product certification systems that discriminate against imported products, may block access to important markets abroad. However, trade remedy statutes such as *section 301* of the Trade Act of 1974 offer a means of negotiation for opening markets.

Any company generally has three means of exploiting its technology in a worldwide market:

- it may license the technology directly to a foreign company,
- it may invest in a foreign manufacturing subsidiary or joint venture, or
- it may make the product in its home market and export it.

Companies may also combine these alternatives, for instance, by licensing technology tied to sales of raw materials, or licensing to a joint venture abroad.

*NBFs, as defined in *Chapter 4: Firms Commercializing Biotechnology*, are firms that have been started specifically to commercialize new biotechnology. Most NBFs are U.S. firms.

At present, most NBFs in the United States have licensed at least some of their technology to established U.S. or foreign companies, the reason being that these NBFs lack the capital and expertise for full-scale manufacturing and marketing, much less manufacturing or marketing abroad. Typically, NBFs that license technology to established U.S. companies surrender their rights to the U.S. market in exchange for future royalties from sales. But a number of NBFs have preferred to reserve the U.S. market for themselves and have made licensing agreements with foreign companies, especially Japanese companies, whose production and marketing expertise resides in foreign markets. These NBFs and the licensors of their technology are interested in export controls on technology and other laws that can be employed directly by governments to control or influence biotechnology transfer.

For some NBFs and most established U.S. companies, domestic or foreign manufacturing are viable options and are particularly desirable to the extent that the alternative, licensing of technology, confers long-term benefits on foreign competitors that are not recouped by the royalties and other provisions of the licensing contract. However, in a situation in which, for instance, a foreign government makes it difficult or impossible to import biotechnology products into that country or to manufacture them there through a wholly owned subsidiary, a U.S. firm seeking to work a patent in that foreign market may find it necessary to license its technology to a local company or to enter a joint venture with a local company on terms that reflect the U.S. firm's lack of market access and therefore favor the local licensee. One NBF has expressed concerns about this issue concerning its licensing negotiations with a Japanese company (14). The short-term consequence of forced technology transfer is that part of the potential return from the technology is transferred from the U.S. licensor to the foreign licensee. The foreign licensee may also receive a valuable technological boost in the short and long term.

Even though there is already substantial diffusion of biotechnology itself, via licensing, joint ventures, and scientific journals, it is difficult to quantify and assess the present amount of biotechnology transfer,* investment, and trade and their potential impact on U.S. competitiveness in biotechnology. Studies of more mature technologies only emphasize the difficulties associated with estimating the level or direction of technology flow (13). Most observers would agree that the net flow of biotechnology is outward from the United States, but such judgments are impressionistic at this time. Also, the net flow of technology outward from the United States is likely to diminish as foreign companies enter the U.S. market (via subsidiaries or foreign manufacturing operations) bringing with them foreign technology. It is not possible to provide reliable estimates of the size of the net outflow, nor is it possible to compare biotechnology with other more advanced technologies in this respect.

In examining the effects of international technology transfer, investment, and product trade, on competitiveness in biotechnology, the first question to be asked is whether biotechnology raises any new issues at all in these areas. For instance, will the growing application of biotechnology in many industries create any new problems in these areas, problems that the existing U.S. or international legal framework does not adequately cover? The answer to this question depends largely on whether there will be relevant significant differences between:

- transfer of biotechnology and transfer of existing chemical or biological technology;
- biotechnology investment and other technology investment; or
- trade in biotechnologically-produced products and trade in the chemical or biological products they supplement or replace.

OTA concludes that biotechnology will raise no such significant novel issues for the regulation of international biotechnology transfer, product trade, or investment. Even without truly novel issues, however, the existing legal framework bears examining, because it will affect access to foreign markets and ultimately competition in industries applying biotechnology. Furthermore, the laws embodying government practices can be changed.

*Ways technology is transferred include: 1) scientific and technical literature, 2) construction of industrial plants, 3) joint ventures, 4) licensing, 5) training, 6) technical exchanges, 7) sale of processing equipment, 8) engineering documents, 9) consulting, 10) documented proposals, and 11) trade exhibits.

Export controls and biotechnology

Export control laws restrict international technology transfer, as well as trade in products, for reasons of national security, foreign policy, or economic policy. From a biotechnology standpoint, the relevant questions are:

- What technologies or products are under what types of controls?
- What is the framework for controls and how are decisions made on controls?
- What is the potential impact of export controls on U.S. competitiveness?

Like the United States, Japan, and the four European countries being considered in this assessment all have some export controls. All but Switzerland belong to the Coordinating Committee for Multilateral Export Controls (CoCom)* and are subject to its multilateral export controls. Although current CoCom control lists do not include biotechnology products as such, CoCom lists on toxicological products and commercial chemicals could become relevant to biotechnology. However, there is no indication that companies would find CoCom requirements restrictive.

United States

In the United States, biotechnology products and data are subject to a number of export controls. The most significant are under the Export Administration Act (EAA) of 1979 (50 U.S.C. App Sec. 2401, et seq.).** Under the EAA, export restrictions depend on the type of commodity or data and its destination. Exporters of any item on the Commodity Control List of the EAA regulations must have a "general license" or a "validated license" for all exports except most shipments to Canada. A general license is essentially an exemption because no application is required. A validated license, on the other hand, requires an application. The Office of Export Administration in the U.S. Department of Commerce, which administers the EAA, may deny permission to export or take a long time to issue the validated license.

With regard to biotechnology products, two groups on the Commodity Control List are especially relevant: Group 7 and Group 9. Group 7 includes chemicals, metalloids, petroleum products, and related materials, including industrial chemicals obtainable by bioprocessing, such as citric and lactic acids. The compounds in Group 7 require a validated license for export to communist countries with the exception of those compounds in a subgroup of Group 7 called "Interpretation No. 24 compounds" (CCL Category 6799G). These latter compounds include DNA, many enzymes, nucleosides, nucleotides, "protein substances," "prepared culture media," and pharmaceutical products for humans and animals. These can be exported to most countries without a validated license.

Group 9 ("Miscellaneous") of the Commodity Control List includes four pertinent categories: 1) "viruses or viroids for human, veterinary, plant, or laboratory use, except hog cholera and attenuated or inactivated systems" (CCL Category 4997B); 2) bacteria, fungi, and protozoa (except those listed in supplement No. 1 to sec. 399.2, Interpretation No. 28) (CCL Category 4998B); 3) bacteria and protozoa listed in Interpretation No. 28 (which basically covers inactivated, attenuated, or relatively harmless organisms); and 4) a catch-all category (CCL Category 6999G), which includes some medicines. Exports in the first category require a validated license for virtually every country except Canada. This category would include viral cloning vectors such as cauliflower mosaic virus, SV40, and bacteriophage lambda. Similarly, the second category requires, with certain exceptions, a validated license for export to any country other than Canada. The third and fourth categories have few restrictions unless the export is being made to certain countries like North Korea, Cuba, Vietnam, or Libya.

Certain bacteria of major industrial importance, such as those of the family Streptomycetaceae and of the genus *Lactobacillus*, fall into Interpretation No. 28 and are therefore exempted from validated

*CoCom is composed of the NATO nations, minus Iceland and Spain, plus Japan, and was formed to deny the Communist countries access to military technology and strategic materials.

**The following discussion is based on the EAA that expired on Sept. 30, 1983, but continues in effect indefinitely under the authority of the International Emergency Economic Powers Act.

license requirements. However, several other types of bacteria commonly used in industry and research, such as the genera *Escherichia, Bacillus,* and *Pseudomonas*, do not come within Interpretation No. 28 and therefore require a validated license for export to all countries except Canada (3). For a summary of the controls on biotechnology products under the EAA of 1979, see table 67.

One commentator has criticized the way in which Interpretation 28 (which will provide major exemptions for biotechnology products) was developed (6). First, the Office of Export Administration did not seek comments from the scientific community before issuing it. Second, the Office has not clarified the basis on which it decides if an organism should be placed on the list. Finally, the Office must formally amend Interpretation 28 by rulemaking before it can place new, nonpathogenic species of commercially important microorganisms on the list.

Data exports* or reexports** to certain countries are also subject to licensing under the ex-

*Export of data occurs whenever data are transmitted out of the United States, released in the United States with the knowledge that they will be transmitted abroad, or released abroad (15 C.F.R. §379.1(b)(1)).

**Reexport of data is the release of data of U.S. origin in a foreign country with the knowledge that the data will be transmitted to another foreign country (15 C.F.R. §370.2). The recipient of technical data must provide written assurances that the data will not be reexported.

Table 67.—Controls on Biotechnology Products Under the Export Administration Act of 1979

Commodity	Commodity Control List (CCL) category	Countries for which a validated license is required[a]
Organisms:		
Viruses	CCL 4997B	All except Canada
Bacteria	CCL 4998B	All except Canada
Human and animal bacterial vaccines	Interpretation No. 24 (CCL 6799G) or Interpretation No. 28	S. Z
Human and animal viral vaccines	CCL 4997B or CCL 6999G	All except Canada S, Z
Human and animal peptides and proteins (pharmaceuticals)	Interpretation No. 24 (CCL 6799G)	S, Z
Human and animal peptides and proteins (miscellaneous)	Interpretation No. 24 or CCL 6999G or CCL 5799D	S, Z P, Q, S, W, Y, Z
Recombinant DNA and related compounds (DNA nucleosides, nucleotides)	Interpretation No. 24	S, Z
Human and animal antibiotics	Interpretation No. 24	S, Z
Human and animal diagnostic agents	Interpretation No. 24	S, Z
Amino acids	Interpretation No. 24 or CCL 6999G or CCL 5799D	S, Z P, Q, S, W, Y, Z
Vitamins	Interpretation No. 24	S, Z
Enzymes	Interpretation No. 24 or CCL 57919D	S, Z P, Q, S, W, Y, A
Pesticides and herbicides (excluding microbial agents)	Interpretation No. 24 or CCL 4707B or CCL 5799D	S, Z All except Canada P,Q,S,W,Y,Z
Seeds	CCL6999G	S, Z

[a]The countries are grouped as follows: P - People's Republic of China; Q - Romania; T - essentially the Western Hemisphere, except Cuba and Canada; V - Southern Rhodesia and countries not in any other group (except Canada); W - Hungary and Poland; Y - Albania, Bulgaria, Czechoslovakia, Estonia, G.D.R., Laos, Latvia, Lithuania, Outer Mongolia, and the U.S.S.R.; Z - North Korea, Vietnam, Cambodia, Cuba; S - Libya.

Under a recent amendment to the Commodity Control List, the export of "medicine and medical products" to Libya does not require a validated license.

SOURCE: Office of Technology Assessment, 1983.

port control regulations. There are three categories of technical data that may be exported to any country under a general license (i.e., an exemption): 1) data already generally available without restriction and at nominal cost, such as in publications or through conferences; 2) scientific or educational data not directly and significantly related to industrial applications; and 3) data contained in foreign patent applications (15 C.F.R. § 379). However, if companies using biotechnology choose to protect information as trade secrets or if information has commercial value, these exceptions will not apply.

The U.S. export control regulations do provide another limited exemption of greater practical use for biotechnology data exports, depending on the destination and the nature of the exported data. Broadly speaking, exports of technical data to virtually all non-Communist countries and, under more restricted circumstances, to the eastern bloc or the Peoples Republic of China, may take place under a general license rather than a validated license.* However, a validated license is required for technical data related to Group 9 commodities, if the data is exported to Communist countries.

Controls on the export of "dual-use" technical data (data with both military and civilian uses) may become more important to the international commercialization of biotechnology in the future. In 1976, the Defense Science Board Task Force on Export of U.S. Technology issued a report (the Bucy report) which concluded that U.S. export controls should focus on design and manufacturing know-how for critical technologies rather than on products (7). In the EAA, Congress directed the U.S. Secretary of Defense to develop a "Militarily Critical Technologies List" (MCTL) and to incorporate it into the export control system after review by the U.S. Department of Commerce. The U.S. Department of Defense has developed a broad MCTL, most of which is classified (19). This list covers many technologies, including ones with primarily nonmilitary applications and has been criticized as covering virtually all of modern technology (19). The MCTL is being revised and has not yet been incorporated into the export control regulations.

Section 16.8 of the Defense Department's MCTL is most pertinent to biotechnology because it covers "technology for manufacture and dissemination of biological and toxic materials." It would cover know-how for: 1) design and production of bacterial, viral, and fungal products, including vaccines, specialized high containment facilities, and special instrumentation; and 2) design, production, and use of dissemination equipment. It would also cover related equipment, materials, and goods accompanied by sophisticated know-how. Although the MCTL has not yet been implemented, it appears that such a concept will be incorporated into the EAA renewal.

In addition, "biological agents adapted for use in war" are subject to controls under the Arms Export Control Act, as are technical data related to biological warfare agents, including "any technology which advances the state-of-the-art or establishes a new art in an area of significant military applicability in the United States" (22 C.F.R. § 125.01). Many pathogenic organisms could be viewed as biological warfare agents, yet their export could be for peaceful purposes such as for research to develop a vaccine. Ultimately, the decision on what products are "adapted for use in war" is left to the discretion of the U.S. Department of State. In addition, the broad definition of technical data could include even information indirectly related to military applications, such as information relating to cloning of genes for human neurotransmitters, because many chemical and some biological warfare agents act by affecting these neurotransmitters (4). On the other hand, a fairly recent case indicates that the courts will interpret the definition of technical data much more narrowly (17).

To sum up, the current impact of U.S. export control requirements is minimal except in the case of micro-organisms where the Commerce Department sees the need for broad controls on national security grounds. Exports of most products and technical information to non-Communist countries should be possible without need for a validated license under the EAA regulations. However, the export of most micro-organisms to all

*In many instances, the availability of this general license for exports to non-Communist countries is conditioned on assurances against unauthorized reexport to a controlled destination.

countries except Canada will require a validated license unless the micro-organisms are inactivated, attenuated, or fall within Interpretation No. 28 *E. coli* and some other micro-organisms of interest to biotechnologists do not fall within Interpretation 28 and therefore require a validated license for export (unless inactivated or attenuated). Controls over micro-organism shipments and data transfers will have most impact on those companies that do research abroad.

Although the impact of the current U.S. export controls on biotechnology companies appears to be fairly modest, the future impact is unclear. The EAA expired on September 30, 1983. Although U.S. export controls continue in effect under the International Emergency Economic Powers Act, it is not clear what form the EAA's successor will take.* Many different bills are pending. Some would strengthen U.S. export controls in general, while others would liberalize them. Furthermore, even if the broad framework of export controls does not change significantly, it is possible that controls could be tightened at the administrative level. The Undersecretary of Defense for Policy testified before Congress in 1982 that "microbiology" is one of the technologies that "pose the greatest risk to U.S. security" (11). Similarly, the April 1982 Central Intelligence Agency publication, *Soviet Acquisition of Western Technology,* identified microbiology, and especially "genetic engineering," as one of the major fields of interest to Soviet and Eastern European visitors to the United States (11). A recent interagency discussion paper for the White House Office of Science and Technology Policy (OSTP), on the other hand, concluded that more restrictive measures to control the transfer of biotechnology are not warranted and may be counterproductive (8). It also noted that existing export control regulations could be clarified and better administered. How much impact this latter report will have in the administration is unknown. OSTP has taken the position that the report is a draft only and will be part of a larger review of technology transfer and national security (4).

*For a complete discussion of the major bills and the various congressional options on export control, see the May 1983 OTA report *Technology and East-West Trade: An Update* (19).

Japan

Japanese export controls combine trade concerns with defense and foreign policy objectives. In addition, Japan cooperates with CoCom controls (18). Under the Foreign Exchange and Foreign Trade Control Law of 1949 (most recently revised in 1979) and the implementing Export Trade Control Order, Japan's Ministry of International Trade and Industry (MITI) may require export licenses on the basis of domestic short supply, export restraints for orderly marketing reasons, defense, and harm to public order or morals. The list of controlled items in the Export Trade Control Order includes blood derivatives, fertilizers, and bacterial agents for military use. (The policy of the Japanese Government is to ban all arms exports.) An export license from MITI is required to export these commodities to any foreign destination. The licensing process, in practice, involves extensive preliminary consultations resulting in informal advance clearances (18).

Federal Republic of Germany

Export controls in the Federal Republic of Germany are limited to commodities and information directly related to "implements of war," are limited in nature and scope, and must interfere as little as possible with freedom of economic activity (1). Except for data and documents concerning goods controlled multilaterally by CoCom, technical data are unrestricted. Certain biological and chemical warfare materials, including some micro-organisms, are controlled. Thus, export controls in the Federal Republic of Germany are much less restrictive than the controls in the United States. West Germany's export controls should have little or no impact on data or product exports by companies using biotechnology in the Federal Republic of Germany that wish to trade internationally.

United Kingdom

The United Kingdom controls the export of goods but not technical information under the Import, Export, and Customs Powers (Defense) Act. Export licensing decisions are national-security-based. No biotechnology products are on the Board of Trade's list of controlled commodities.

Switzerland

Swiss law formerly provided for export controls in the "national interest" on two categories of biotechnology products: serums and vaccines, and pharmaceuticals (16). Currently, however, there are no Swiss controls on biotechnology products or data.

France

The French export controls appear to be quite informal and a product of administrative action rather than statutory decree. The French Ministry of Economics and Finance's list of products requiring export licenses includes biotechnology materials usable in biological warfare and their related technical data. The controlled list does not include antibiotics, other medicinal products, or cultures of nonpathogenic organisms.

Comparative analysis

U.S. export controls in general are more restrictive than those of Japan or the four European competitor countries, and they are more restrictive with regard to biotechnology. The United States is the only country that controls exports of pharmaceuticals for foreign policy reasons and is the only nation that has perceived a national security interest in controlling the export of microbial cultures generally. The other nations only embargo shipments of biological warfare agents.

U.S. export controls could cause problems for U.S. firms using biotechnology due to delays in the export licensing process or uncertainties in the application of controls. These problems will occur primarily in the export of micro-organisms, many of which will require a validated U.S. export license. In contrast, exports of most biotechnology products and data will not require a validated license. If export controls are a significant handicap to U.S. firms' competitiveness in biotechnology, these controls may lead U.S. firms using biotechnology to source their exports from affiliates abroad, to first introduce new products abroad, or to site their R&D abroad.

Patent law provisions affecting international technology transfer

Patent laws of many countries, including the United States, contain secrecy provisions that restrict outward technology transfer for security or foreign policy reasons. On the other hand, compulsory licensing provisions can be used to force inward technology transfer. This section discusses these two types of provisions in the patent laws of the competitor countries.

National security restrictions on patent applications

The U.S. patent law provides a waiting period after filing for a patent in the United States during which the U.S. Patent and Trademark Office and the U.S. defense agencies may screen the invention on national security grounds and withhold the grant of a patent. In addition, procedures exist for the review of applications in foreign countries by U.S. parties, and secrecy orders can be issued in certain instances. The review period results in an effective prohibition against foreign filings within 3 months of the U.S. filing. French, United Kingdom, and West German patent laws have similar provisions. However, the Federal Republic of Germany will issue a secret patent instead of a secrecy order. Swiss patent law provides for expropriation with compensation; Japanese patent law does not place any national security restrictions on the application process.

National security provisions create delay in filing foreign patents for all patent applicants. It is too early to tell whether military uses of biotechnology will make patent secrecy orders a significant problem for biotechnology.

Compulsory licensing of patents

In most countries, patent owners who fail to put their inventions into practice in the country within a prescribed period may have their patent rights reduced or revoked. Failure to exploit a patented invention in the country is regarded as an abuse of the patent monopoly rights and may subject the patent to compulsory licensing, revocation, or automatic lapse (2). Compulsory licensing is the normal remedy employed in these situations. Proponents of compulsory licensing argue that it ensures early applications of a technology and diffuses control over technology. Its opponents argue that it discourages public disclosure of new technology through the patent system, expropriates property rights, and decreases incentives to innovate. In the United States, compulsory licensing is generally viewed as inconsistent with the patent owner's right to exclude others from making, using, or selling the patented invention, and U.S. law provides for compulsory licensing only in limited instances.

Countries with compulsory licensing recognize that it may be very difficult for a licensee to practice a patent without the benefit of the patent owner's continued technical assistance and that this assistance is unlikely to be forthcoming when unfavorable terms are imposed on the patent owner. Thus, compulsory licensing can discourage the transfer of know-how in conjunction with the license. This may be less of a problem in cases where an organism has been deposited in support of a patent. Since the organism is publicly available and is in esence a "factory" for the product, a licensee that obtained a compulsory license may not need the know-how. In this situation, compulsory licensing could be a threat to U.S. biotechnology companies because sufficient technology transfer could occur for the compulsory licensee to use the invention competitively without any assistance from the patent owner.

An international patent treaty known as the Paris Convention permits any of its member countries to require compulsory licensing of its patents after 3 years from the date of issuance, if the patent is not sufficiently worked. However, the Convention provides exceptions for reasons such as compliance with national safety requirements (15). All of the competitor countries are signatories to the Convention, and all but the United States have general compulsory licensing statutes consistent with the Convention.

In some cases, in the interests of free trade and regional cooperation, the requirement that an invention be worked in the country is waived when the demand for the patented product in the country is being met by manufacturing in a cooperating country. This is the case for the member states of the EEC. Bilateral agreements also exist between Switzerland and the United States and between the Federal Republic of Germany and the United States whereby the working of a patent in the territory of one of the parties is considered equivalent to its working in the territory of the other party.

Specialized compulsory licensing provisions of interest include United Kingdom and French provisions for compulsory licensing of pharmaceuticals in certain circumstances.

Although the U.S. system generally allows the patentee to use or not use the patented technology at will, certain statutes and judicially created legal doctrines provide for compulsory licensing in limited cases. For example, statutory compulsory licensing exists under the Plant Variety Protection Act* and the recent statute on ownership of federally funded inventions (Public Law 96-517). Compulsory licensing also exists de facto where courts do not enjoin patent infringement on grounds of patent misuse, antitrust violation, or public policy.

Assessing the impact of compulsory licensing laws on U.S. competitiveness in biotechnology is necessarily speculative at this time. Compulsory licensing of patents could result in transfer of biotechnology and could adversely affect U.S. competitiveness in biotechnology. Although compulsory licenses apply in theory equally to any company, foreign or domestic, in practice they could be used discriminatorily against U.S. companies; standards that provide for licenses "in the public interest" grant wide discretion to the governmental body that decides such cases.

*The act permits the Secretary of Agriculture to declare a protected variety open for use for up to 2 years at a reasonable royalty in order to ensure an adequate supply of food, fiber, or feed in this country when the owner is unwilling or unable to meet the need at a fair price (47 U.S.C. §2404).

Regulation of technology imports and foreign investment

Foreign exchange and investment control laws are sometimes applied to technology licensing or technical assistance agreements or to foreign investment, with the effect of restricting the importation of foreign technology or foreign capital and helping locally controlled firms retain control of the local market.* Such restrictions have two rationales. First, a nation in a precarious balance of payments position may look askance at what it views as the payment of exorbitant sums for foreign technology. Second, a nation might act to prevent or modify a transaction for political reasons in instances where imported technology or foreign investment might result in increased control of a local firm bv a foreign firm.

The United States, the United Kingdom, the Federal Republic of Germany, and Switzerland currently have no significant formal exchange or investment control laws. Although these countries lack statutory and administrative mechanisms for direct control over private international technology transfer agreements, de facto means exist in the United Kingdom, the Federal Republic of Germany, and Switzerland under which these governments could block foreign investments in those exceptional cases in which it might be deemed necessary to do so for screening important investments* (14). France and Japan have investment or exchange control mechanisms that do affect technology transfers and foreign investment.

*Trade and investment restrictions, together with compulsory licensing provisions can act like pincers to extract a foreign technology owner's industrial property rights. The foreign technology owner may patent the product in an important market, be blocked from using the patent himself, and have to license the patent on pain of losing its benefits.

*For example, in the Federal Republic of Germany, any enterprise whether domestic or foreign that acquires 25 percent or more of the shares of stock in a German corporation must notify the provincial banking authorities and the target company when it buys the shares. Section 23 of the Foreign Trade Law authorizes the German Federal Government to ban the sale of a company to nonresidents on national security grounds. While the Federal Government has never had to use this power, its existence makes possible an informal but well-known agreement between the Federal Government and the major banks (which often are major shareholders of companies) that no company nor block of stock be sold without prior consultation with the Government.

France moved in 1970 from a system requiring prior review of technology transfer agreements to a system requiring notification after the fact. Currently, the French party to an international "industrial property" or "scientific and technical assistance" agreement must notify and submit a copy to the Industrial Property Service of the Ministry of Industrial and Scientific Development within 1 month after the agreement is concluded (9). The French party must also submit yearly reports of payments made and reciprocal transfers of technology. The submissions are confidential, and compliance is a prerequisite to being able to transfer royalty payments (10).

This mechanism appears to be one primarily designed to gain statistical information, but one source indicates that it may have further ramifications (5). The French Ministry of Economy may express reservations if it considers the royalty payments to be too high. Such an action could result in the excess amount of royalties being prohibited from being deducted for tax purposes. Most of the reservations expressed by ministry officials have involved contracts in the chemical, pharmaceutical, and petroleum sectors (5). Thus, the ministry officials may be inclined to express reservations for biotechnology licensing agreements, if those agreements are viewed as not being sufficiently favorable to the French party.

France's investment control laws are relevant both to biotechnology transfer and to the ability to invest in the French market. Nonresidents of the European Economic Community (EEC) that plan to invest in France must submit a declaration to the Ministry of Economics and Finance. The declaration includes information on the identity of the investor, the business to be invested in, the forms, conditions, rationale, and consequences of the investment, and financial information on the companies involved. Within 2 months following the receipt of the declaration, the Ministry may order the suspension of the proposed action.

Direct foreign investment in certain industries is not encouraged in France. And in France as in all countries other than the United States, takeovers of local companies are not favored, particularly takeovers resisted by the local management (14). On the other hand, investments that provide for capital transfer or technology transfer into France are favored. Given the French Government's concerted efforts to stimulate biotechnology, investments by foreign companies in French companies using biotechnology are likely to be carefully scrutinized.

Of the countries under study, Japan has been most restrictive regarding technical assistance and licensing agreements between foreign parties and Japanese companies and direct foreign investment. In the period 1949-68, all licensing agreements and all foreign investments in Japan had to be reviewed in advance by the Japanese Government. Over the years, an increasing range of agreements and investments were given "automatic approval." Finally, the revised Foreign Exchange and Foreign Trade Control Law (effective Dec. 1, 1980) provided that foreign trade and investment is to be free in principle and restrictions are to be exceptional.

Under Japan's revised Foreign Exchange and Foreign Trade Control Law, the Japanese Government has the power to screen investments. Before a foreign investor can conduct a transaction characterized as "direct foreign investment," the investor must give notice to the Japanese Government. The foreign investor must then wait 30 days before proceeding with the transaction.* The Minister of Finance and the minister in charge of the industry concerned also have the power to designate specific companies for special controls on foreign ownership. Eleven companies have been so designated, including Sankyo Pharmaceuticals (25-percent ceiling on foreign ownership).

Articles 29 and 30 of Japan's Foreign Exchange and Foreign Trade Control Law deal specifically with "agreements for importation of technology." The parties to such an agreement must first notify the Minister of Finance and the minister in charge of the industry involved of the terms of the agreement whenever they intend to enter into, renew, or amend such an agreement. The agreement cannot be concluded until a 30-day waiting period has elapsed. (Normally, the ministries exercise their power to shorten this period for transactions not deemed "harmful.") The ministries review the agreement with respect to a number of criteria, ranging from national security to competition with other Japanese business. The Japanese Government has a fair degree of control over technology transfer agreements, although it is not clear whether the control is used to secure better contractual terms for Japanese companies, particularly terms that encourage biotechnology transfer to Japan.

The greatest significance of Japanese investment controls for biotechnology products is the lingering effect of past controls. In strategic industries where foreign companies' technology position was strong, liberalization of investment controls came late. In pharmaceuticals, for instance, 100-percent foreign ownership was not permitted in Japan until 1975, so non-Japanese drug companies either had to enter a joint venture with a Japanese firm (or license to a Japanese firm) or had to forgo the world's second largest drug market (22). Late liberalization of investment controls retarded foreign firms' establishment of their own marketing and distribution networks in Japan. Nevertheless, the international pharmaceutical companies have a strong and increasing presence in Japan, and some foreign pharmaceutical companies have even acquired smaller Japanese pharmaceutical firms. Merck's recent acquisition of Banyu Pharmaceutical, the number three firm in the Japanese industry, puts Merck in an extremely strong position in the Japanese market. Still, the waiting period for investments and for licensing contracts is at the least a nuisance to the inward investment or licensing transaction, although other factors such as interlocking directorships, cross-holding of stock, and labor resistance to foreign management may be very significant in discouraging investment entry into the Japanese market through a hostile takeover.

*If certain circumstances are found to exist, then within an extended waiting period, the Government may recommend that the agreement be altered; this power has seldom been used in recent years. If this recommendation is not accepted, the Government may suspend the transaction indefinitely by Cabinet Order.

Trade barriers affecting biotechnology products

While firms using biotechnology now trade mostly in technology through licensing and have in a few cases invested abroad, trade in the products of biotechnology is just beginning. The tariffs on biotechnology products are generally low in the competitor countries and are getting lower (as Tokyo Round tariff cuts are phased in).* Thus, it is nontariff barriers that are most likely to be important to trade in biotechnologically produced products.

Nontariff barriers to trade include any government intervention affecting competition between imported and domestic goods. The barriers most significant for biotechnology products will be those that affect technology development and technology transfer:

- health and safety standards and certification systems;
- subsidies;
- price regulation;
- to a minor degree, government procurement; and
- least significant, customs classification of new products.

Rather than addressing health and safety regulation per se, the discussion here addresses how such regulation applies specifically to imports. Similarly, rather than considering the specific production and R&D subsidies, it considers how these programs fit in with U.S. rights under trade agreements. For instance, Japan maintained until very recently a dual safety certification system that discriminated against imports, including imported drugs, medical devices (e.g., monoclonal antibodies), chemicals, and animal drugs (20).

Standards and certification systems

Product standard systems are a particularly thorny problem for exporters of health care products, because such products are extensively regulated and subject to the regulator's discretionary determination of whether imported products meet applicable standards. Product standards can affect the activities of both exporting and importing companies. Biologically produced pharmaceuticals, vaccines, foods, chemicals, and veterinary products will all be subject in some degree to inspection, approval, and/or certification of whether they meet local standards of safety and efficacy.

For a foreign manufacturer, registration and approval of a product in a certification process involves inevitable leakage of technology. Any manufacturer must explain its technology to local regulators to the extent necessary to get its products approved. In those countries where marketing approval for an imported product can only be given to a locally resident importer, as has been the case in Japan, the technology (including trade secrets and nonpatentable know-how) that is required for an application for approval must be transmitted to the regulating authority by the importer, whose possession of this information could provide the resident importer with leverage over the foreign manufacturer. This generalization applies equally to foreign manufacturers in the United States and to U.S. manufacturers abroad. Leakage of technology may also occur where a registration scheme involves disclosure of trade secrets, as in the case of disclosure of chemical identities for registration in the European Inventory of Existing Chemical Substances under the EEC's Sixth Amendment regulation scheme for toxic chemicals.

The Agreement on Technical Barriers to Trade ("Standards Code") addresses these problems. The Standards Code, negotiated in the Tokyo Round of multilateral trade negotiations, came into effect January 1, 1980 and covers all six countries discussed in this report. This code requires the following: 1) national or regional certification systems must treat products of code signatories no less favorably than domestic products, 2) imported products must be treated in a nondiscriminatory manner with regard to product testing and certification, and 3) signatories must use the same test methods and administrative procedures

*All of the competitor countries belong to the General Agreement on Tariffs and Trade (GATT). GATT is a multilateral agreement signed by 87 governments accounting for over 80 percent of world trade. GATT serves as a code of rules for international trade and as an international trade organization. A primary goal of GATT is to discourage the use of nontariff barriers to trade and then to reduce tariff levels through a series of multilateral trade negotiating rounds of which the most recent was the Tokyo Round (1973-78).

for imports and domestic products and charge comparable fees. Test results must be made available to the exporter, importer, or their agents, and confidentiality of information must be respected equally for foreign and domestic suppliers. The Standards Code does not implement a transnational standards system. It merely provides international rules for how individual national systems treat products of other code signatories, provides a forum for negotiations, and provides redress against foreign violations of the code (20).

JAPAN

Until recently, one of the most wide-ranging barriers to foreign market access in Japan was discriminatory certification systems (20). While various product standards were administered under different laws, the framework was remarkably uniform. Each law would provide two tracks: 1) an approval adapted to high-volume production and sales, requiring factory inspection and product-type approval; and 2) a low-volume approval, involving lot-by-lot inspection. The first track was legally foreclosed to foreign manufacturers. Because the person holding the product approval had to be subject to potential sanctions under Japanese law, that person had to be present in Japan. Furthermore, the product approval (and all data to obtain it) was the property of the approval holder, who under the second track had to be the Japanese importer. Transfer of the approval to another importer (even transfer of the approval from a joint venture to a wholly owned subsidiary) meant regenerating the data.

In response to foreign complaints, the Japanese Diet, on May 18, 1983, passed legislation amending 16 Japanese standards and certification laws, including the Pharmaceutical Affairs Law (drugs, medical devices), the Agricultural Chemicals Law, and the Toxic Chemicals Law. The amendments, together with their implementing regulations issued soon thereafter, are designed to give foreign producers direct access to certification systems, including direct ownership of approvals. Foreign regulated products—such as drugs or monoclonal antibody kits—still (as of fall 1983) must be unpacked, sampled, and tested, lot by lot, as they pass Japanese customs. Foreign manufacturers may now apply for, and be granted, factory inspection and U.S. product type approval. U.S. trade negotiators are now working for Japanese acceptance of factory inspections carried out by U.S. testing firms for this purpose.

The Japanese Ministry of Health and Welfare, which previously refused to accept entirely foreign clinical test data because of racial and dietary differences, agreed in January 1983 to work toward acceptance of foreign clinical data and to undertake objective studies of racial and dietary differences. However, as of December 1983 no such studies had been undertaken. In addition, the Ministry promised to clarify the line between (regulated) pharmaceuticals and (unregulated) foodstuffs and to shorten the approval period for in vitro diagnostics used as medical devices. The Ministry has also promised to allow approvals to be transferred between importers of drugs and importers of medical devices.

EUROPE

U.S. chemical exporters have been concerned about inadequate protection of proprietary data in the European registration process, in particular, the requirements for disclosure of chemical identities of substances. Another long-term concern of U.S. pesticide exporters has been pesticide registration procedures abroad, which may diminish the proprietary value of registration data by allowing national authorities to use data submitted by pioneer registrants in determining the safety of "me-too"* pesticides (20).

Subsidies

Subsidies (e.g., loans, grants, tax preferences) are a form of government intervention which, in some cases, can provide competitive advantages to domestic producers. There is basic disagreement between the United States and its trading partners both on how to define a subsidy and on how to measure its effect. The position of the United States is that a measure of a subsidy is the benefit conferred on the recipient; the position of the EEC is that the measure should be the cost to the government or the benefit to the recipient,

*"Me-too" products are generic products equivalent to an already existing product.

whichever is lower. In any case, subsidies used by governments may be important in international competition to commercialize biotechnology.

One of the most controversial agreements of the Tokyo Round was the Subsidies Code, which attempts to expand international discipline over subsidies. Three aspects of the Subsidies Code are important to firms using biotechnology. First, the code prohibits any export subsidies on industrial products. This means that neither the United States nor its competitor countries can grant export subsidies on biotechnology products without violating the code. Second, the code recognizes that domestic subsidies, which include all existing subsidies that affect biotechnology, can be used except in situations where the subsidies: 1) cause or threaten injury to another signatory's industry, 2) cause or threaten "serious prejudice,"* or 3) nullify or impair GATT benefits of another signatory. Third, the code provides for remedies. Two methods of obtaining remedies are available: countervailing duties (described under the discussion of U.S. trade law below) and multilateral dispute settlement.

The Subsidies Code sets limits on both the export subsidy behavior of our trading partners and on what the United States (and other signatories) can do to promote industry. The code also authorizes national governments to unilaterally impose countervailing duties** on subsidized imports to offset subsidies, where the importing country's government has found that there are subsidies and that injury to domestic industries is caused or threatened by reason of the subsidized imports.

All presently known government promotion measures affecting the commercialization of biotechnology are either domestic subsidies or other promotional measures that legally do not qualify as subsidies at all. Under U.S. subsidy and countervailing duty practice, R&D grants and preferential loans awarded by a government to finance research that has broad application and yields results that are made publicly available are not legally subsidies. The test of a subsidy in this case is whether the result of a government-funded research project in biotechnology is published and made available. Loans are deemed subsidized to the extent that the borrower obtains a better interest rate for the loan than that which would otherwise be available to him for a loan of similar size and terms. As for government equity ownership, the U.S. position is that government ownership implies a subsidy only when it is inconsistent with commercial considerations. If the government buys shares either directly from the company or the stock market, a subsidy arises to the extent the government pays more than the market price. Given the favorable market for most biotechnology stocks, even for those issues that have shown no operating profits to date, it seems unlikely that government investment in biotechnology companies such as Celltech would be classified as a subsidy under U.S. practice.

Price regulation and government procurement

Price regulation is central in importance to the world market in pharmaceuticals and may be an important means of discouraging foreign suppliers to enter particular domestic pharmaceutical markets. Thus, price regulation, particularly of new drugs, will be important to the marketing and profitability of biotechnology pharmaceuticals. The basis for price regulation is the local or national social insurance scheme, which pays for all or part of the beneficiaries' drug cost. Although the basic motivation for price regulation is health care cost containment, price regulation can be used to reward manufacturers for local production, local R&D, and other desired behavior. Thus, in countries where drug costs are paid or reimbursed by the government, in a real sense the government creates the market. Furthermore, inclusion on the government list of approved drugs, at a profitable price, is essential to market access for foreign drugs.

GATT Article III requires that products of GATT signatories be given treatment equal to that given local products with regard to price regulation, internal taxes, and other regulations. If there is a

*Serious prejudice relates to effects of subsidies in third-country markets but is not defined in the Code or GATT.

**Countervailing duties are imposed by governments to offset subsidies found to benefit imports into countries where the subsidized imports cause or threaten material injury to a domestic manufacturer producing a like product.

clear factual case of discrimination, enforcement of this requirement is straightforward.

In the United States, Federal and State funds pay for only 8 percent of out-of-hospital drug costs. The Maximum Allowable Cost program instituted in 1979 by the Department of Health, Education, and Welfare sets price ceilings on drugs paid for by federally financed health-care programs such as Medicaid. In addition, a growing number of States are instituting open or closed formulary systems (recommended or mandatory drug lists) for prescriptions paid for by State funds (22).

In the Federal Republic of Germany, all drugs are dispensed through the pharmacies or hospitals, which are reimbursed by the insurance plan to which the patient belongs. An official price list is set by the pharmaceutical manufacturers association, and the Government regulates the wholesalers' and pharmacies' markup.

In the United Kingdom, the Government pays for approximately 90 percent of drugs consumed (22). Dispensing of drugs is through the pharmacies, which are reimbursed on the basis of ingredient cost, profit, professional fee, and container allowance. The Department of Health allows a larger profit margin for companies that manufacture or perform R&D locally, a provision which may be inconsistent with GATT Article III and the Treaty of Rome (21).

In France, drugs are distributed primarily through pharmacies, and patients are reimbursed by the social insurance system at a set percentage (40, 70, or 100 percent) of the official list price. The Government sets not only the retail price for each drug on the official price list, but also the markups in the distribution chain. One report states that health care cost containment concerns have led to drug prices too low to finance R&D by the local pharmaceutical industry (21).

In Switzerland, dispensing of drugs is through pharmacies and doctors. Price regulation is the responsibility of the Federal Social Insurance Office which maintains two lists of drugs for reimbursement: 1) generic drugs, for which reimbursement is required; and 2) the "SL List," a list of specialty drugs for which reimbursement is not required but usually happens anyway. For imported drugs, sales prices abroad are carefully monitored; the Federal Social Insurance Office will allow a 25-percent margin over the selling price in the country of origin (excluding tax) (22).

The Japanese drug distribution system is unique. Almost all drugs in Japan are dispensed by physicians whose drug lists and markup are regulated by the national health insurance system. The doctor buys drugs from the wholesaler at a price that varies depending on the size of order, size of clinic, and other commercial factors. The doctor then resells the drug to his patients at the regulated price. The difference is the doctor's profit, which averages between 20 to 50 percent of the regulated price (12). Japan's price regulation system is used to encourage R&D. The recently revised (April 1983) method of drug price reimbursement allows a larger profit margin depending on desirability and efficacy of the drug; this, in combination with the more generous official prices set for new drugs, may be used to reward R&D and favor new drug (including biotechnology drug) development (14).

Government procurement

Under GATT, governments may buy products as they wish for their own consumption and target their procurement to favor local suppliers. However, the GATT Procurement Code, negotiated in the Tokyo Round reciprocally, opens bidding opportunities on certain procurement and provides fair procurement procedures. For biotechnology products, government procurement would have substantial impact only where consumption by the government is large relative to the total market or has a significant demonstration effect. It is unlikely that government procurement will play a role in biotechnology development comparable to the role of the U.S. Defense Department or the Japanese Government in the semiconductor industry. While governments do buy pharmaceuticals, many drug companies avoid bidding on government tenders for commercial reasons, and in developed countries, procurement markets are not significant relative to total pharmaceutical demand.

Customs classification

Customs classification might be a problem only for those biotechnology products for which classification is an open issue—i.e., either those products that are genuinely new or existing products assigned to a different classification due to their biotechnologically based production. Over the next several years, most biotechnology products with the exception of some vaccines will probably be replacement products for existing products. If the trading partners of the United States reclassify biotechnology products and raise tariffs, such strategic protectionism could raise new barriers around foreign markets. Since only a few products developed through biotechnology are traded at present, it is not clear whether the competitor countries will reclassify the biotechnology products under different (higher tariff) categories. There is, however, no reason to believe that they will be reassigned.

Trade laws

Trade laws may offer a means to improve the competitive position of U.S. firms using biotechnology. This section reviews the array of trade law actions relevant to U.S. firms using biotechnology and assesses whether biotechnology raises particular issues as to the adequacy of present trade laws.

While trade in biotechnology products is in its infancy, some factors will influence the likely interaction between biotechnology and trade law. First, to the extent that trade in a product is wholly under a licensing agreement or is an intracompany transfer of a patented substance (or organism), there are likely to be few problems with import competition. Second, there is no reason to believe that biotechnology products will trade differently or be classified differently from other products; human insulin will have the same distribution channels as animal-derived insulin, for instance. Third, since the efficacy of any type of import relief is tied to the pace of product obsolescence, which differs by industry, the import relief concerns of other industries such as the semiconductor industry will be of limited importance to industries using biotechnology.

Section 337 of the Tariff Act of 1930

The U.S. import trade statute most immediately relevant to firms using biotechnology is section 337 of the Tariff Act of 1930 (19 U.S.C. §1337), which provides for relief against unfair competition in import trade, including imports found to be infringing intellectual property rights, where such practices injure an efficiently operated industry in the United States, prevent the establishment of such an industry, or restrain trade. If the U.S. International Trade Commission (ITC) finds a violation, it may issue either an exclusion order prohibiting the import of the goods in question or a cease and desist order to proscribe specific conduct by parties over which ITC has jurisdiction. Investigations under section 337 are conducted by ITC. The President may disapprove such a determination for policy reasons within 60 days.

There are several points to note about section 337. If an import is found to violate section 337, it can be completely excluded from importation; ITC need not get jurisdiction over the foreign manufacturer. Second, section 337 investigations are faster (18 months maximum) and generally less expensive than other types of litigation (e.g., patent or trade secret infringement litigation). Third, where there is multiple-source infringement, ITC can issue a general exclusion order, excluding all infringing products made by any firm. Section 337a (19 U.S.C. §1337a) provides that section 337 can be used to enforce process patents; ITC in past process patent cases has been willing to issue broad exclusion orders, particularly where infringing and noninfringing goods are physically indistinguishable.

Section 337's greatest relevance for biotechnology is that at present, section 337 is the most effective means of enforcing process patents against foreign producers. It could, for instance, be used to enforce the Cohen-Boyer process patent against imports from firms that have not taken a license from Stanford (although Stanford would run the risk that its patent might be found invalid by ITC). Furthermore, a firm need not have patented the intellectual property in question; section 337 applies as well to misappropriations of trade secrets. A firm that has elected to take the trade secret route instead of patenting its research results could use section 337 against goods incorporating stolen trade secrets.

A section 337 investigation concerning allegations of patent infringement and trade secret misappropriations with respect to "certain limited charge cell culture microcarriers" is now in progress.*

Section 301 of the Trade Act of 1974

The other important trade remedy for firms using biotechnology is section 301 of the Trade Act of 1974 (19 U.S.C. 2411 ff). Under section 301, firms can petition the U.S. Government to enforce U.S. rights under trade agreements or to negotiate to eliminate foreign government actions that unreasonably limit market access abroad. Section 301 also provides authority for the President to retaliate against any foreign government action that is "unjustifiable, unreasonable, or discriminatory" and burdens or restricts U.S. commerce (14).

*Certain Limited Charge Cell Culture Microcarriers, Investigation No. 337-TA-129, instituted Aug. 17, 1982, concerning allegations of: misappropriation of trade secrets; refusal to sell sieved beads; false and deceptive advertising; false and disparaging comments about complainants; direct, contributory, and induced patent infringement; and unauthorized manufacture abroad in violation of process claims of a U.S. patent. Complainants are Flow General Inc. and Massachusetts Institute of Technology; respondents are AB Fortia, Pharmacia AB, Pharmacia Fine Chemical of Sweden, and Pharmacia Inc. of New Jersey.

An investigation under section 301 of the Trade Act of 1974 is conducted by the U.S. Trade Representative and is normally initiated in response to a petition by any interested person.* The Trade Representative, with the advice of other U.S. Government agencies, recommends what action should be taken by the President. Firms using biotechology can use section 301 to gain U.S. Government action against foreign government actions that restrict market access or violate GATT, the Standards Code, or bilateral or multilateral agreements. For such problems, section 301 is often the best or only formal remedy. However, section 301 would not apply if there were no foreign government involvement (e.g., dumping or illegal private cartels). Also, even without formal section 301 action, the assistance of the U.S. Government is available for resolving market access problems abroad.

Countervailing and antidumping duty laws

Countervailing duties (19 U.S.C. 1671 ff) are imposed to offset subsidies found to benefit imports into the United States where the subsidized imports cause or threaten material injury to U.S. industry producing a like product. Similarly, antidumping duties (19 U.S.C. 1673 ff) are imposed to offset injurious dumping of foreign merchandise in the United States.** The U.S. Department of Commerce makes preliminary and final findings concerning subsidization or dumping, and ITC makes preliminary and final findings concerning material injury to the U.S. industry. Biotechnology products are unlikely to raise novel issues for these laws.

*Interested persons include any person representing a significant economic interest affected by the complained policy or actions.

**"Dumping" exists when goods are sold for export below their cost of production or more cheaply than for the home market.

Findings

Export control laws restrict outward technology transfer for national security, economic, or foreign policy reasons. Of the six countries studied, the United States is the only country that controls the export of medicines for foreign policy reasons. The United States also has imposed more far reaching controls on the export of micro-organisms than have the European nations and Japan, which appear to limit their concern to biological warfare agents. With these broader commodity controls come commensurately broader controls over the export of technical data. These controls may have a slightly adverse effect on the competitiveness of U.S. companies commercializing biotechnology because they could cause delays that result in sales being lost to foreign competitors.

All of the countries studied, except the United States, have compulsory licensing provisions of general applicability for patents. The United States has special compulsory licensing provisions in some statutes, notably the Plant Variety Protection Act. In addition, compulsory licensing has been imposed in patent misuse cases. It should have little effect on U.S. competitiveness in biotechnology.

Exchange controls may delay or limit the remittance of royalties. Investment controls may obstruct inward foreign investment or licensing and technical assistance activities. The United States, the United Kingdom, the Federal Republic of Germany, and Switzerland do not have significant controls. France formerly required prior review of investments and licensing agreements but now requires only notification after the fact. However, non-EEC residents who plan to invest in France must submit a declaration of their proposed activity to the Ministry of Economics and Finance, which may order the suspension of the proposed action. Japan has had a prior notification system since 1980. However, both the French and Japanese systems give the Government the ability to object or order alteration of the transaction. This system may increase the leverage of French and Japanese prospective licensees of biotechnology transfers. It might also provide protection for domestic firms against foreign competition in the local market.

For biotechnology products such as pharmaceuticals, tariffs are relatively insignificant as a barrier to trade. The significant trade barriers are nontariff trade barriers, such as standards and certification systems, subsidies, and the use of price regulation to discriminate against imports.

Multilateral trade agreements such as GATT provide rules aimed at eliminating nontariff barriers to trade. Similarly, the Standards Code prohibits its parties from discriminating against imports in their standards and certification systems. The Subsidies Code prohibits certain forms of subsidies. All of the competitor countries belong to all three of these agreements. U.S. rights under these agreements can be enforced through dispute settlement proceedings before an impartial panel of arbitrators.

Biotechnology products may face significant nontariff barriers to trade because of the desirability of the technology and because of health and safety regulation likely to surround the product. For instance, certification of safety requirements may be difficult to gain, especially for imported biotechnology products. Additionally, price regulation in important overseas markets such as France and Japan may on occasion significantly impair return on R&D investment for biotechnology pharmaceuticals.

The U.S. trade remedy of greatest interest to U.S. firms engaging in biotechnology is section 337 of the Tariff Act of 1930, which provides a remedy against imports that create unfair competition, including those that infringe intellectual property rights. A firm using biotechnology producing a product in the United States can use section 337 to gain exclusion of infringing imports, even in the case of those made by a process patented only in the United States or where the firm has chosen the trade secret route rather than patenting. Section 337 proceedings are administrative (before the U.S. International Trade Commission) and can be much speedier than other types of litigation.

The other significant trade remedy for U.S. firms using biotechnology is section 301 of the Trade Act of 1974. This statute provides a window for U.S. parties to get the Government to negotiate to enforce U.S. rights abroad. Antidumping and countervailing duty laws may be of significance in the future as well.

Since trade in biotechnology products has barely begun, it is too soon to assess definitively whether the present trade laws are adequate to address the trade problems of this industry. However, since there are no trade issues peculiar to biotechnology and biotechnology products are likely to trade similarly to other products, biotechnology is not likely to raise new issues for trade law.

Issue

ISSUE: How could Congress respond to the international transfer of biotechnology?

Biotechnical knowledge is being rapidly transferred both domestically and internationally, but there is no empirical evidence showing the amount and net direction of the transfer. Much of the new knowledge is being generated in the United States, primarily in research universities and NBFs, and because of the openness of the university scientific establishment and the many joint R&D ventures between NBFs and larger manufacturing companies, particularly foreign ones, this knowledge is being disseminated worldwide. At the same time, however, high-quality research in molecular biology, immunology, and bioprocess engineering is done in many foreign countries, and the published results are available to U.S. scientists. The technique for making hybridomas, for example, was developed in the United Kingdom. Furthermore, patents granted in the United States and abroad to foreign inventors and companies make technology available to all. Finally, R&D joint ventures between NBFs and large companies presumably have resulted in the transfer of some technology to NBFs in the United States, although this is not certain because of the proprietary nature of these agreements. Despite the lack of empirical evidence showing the amount and net direction of biotechnology transfer, most observers would agree that currently the net flow of biotechnology transfer is outward from the United States. However, the net flow outward may change as foreign companies enter the U.S. market (via subsidies or foreign manufacturing operations) bringing with them foreign technology. The long-term effect of what appears to be an outward flow of technology on the international competitiveness of U.S. companies applying biotechnology is unknown.

Although certain laws affect the international technology transfer and will therefore affect the transfer of biotechnology, biotechnology raises few, if any, unique issues in this context. Similarly, since there are no trade issues peculiar to biotechnology, and biotechnology products are likely to trade similarly to the products they replace, biotechnology is not likely to raise new issues for trade law. The laws most relevant to biotechnology now are the export control laws, which could have a modest effect on U.S. competitiveness.

As this study went to press, the debate over replacement of the Export Administration Act of 1979 was still in progress. Although the delay and commercial uncertainty created by current U.S. export controls may adversely affect the development of biotechnology in the United States, these problems are general ones that are now under consideration in Congress, and as such, are beyond the scope of this report. The reader is referred to the OTA report *Technology and East-West Trade: An Update* (19) for a full discussion of the export control issue.

Chapter 19 references

1. Baker, R., and Bohlig, R., "The Control of Exports: A Comparison of the Laws of the United States, Canada, Japan, and the Federal Republic of Germany," *International Lawyer* 1:163:172-173, 1966.
2. Baxter, J. W., *World Patent Law and Practice,* 2:117 (New York: Matthew Bender & Co., 1983).
3. Bent, S., "Export Controls and U.S. Biotechnology," *Biotechnology Law Report* 2:6-14, 1983.
4. *Biotechnology Newswatch,* "White House Gets Study on Biocompetitiveness: Will It Surface or Sink," June 20, 1983, p. 1.
5. Business International Corp., *Investing, Licensing, and Trading Conditions Abroad* (France), November 1981.
6. Cooper, I., "U.S. Export Controls on Biotechnology," *Recombinant DNA Technical Bulletin* 6:12, 1983.
7. Defense Science Board Task Force on Export of U.S. Technology, *An Analysis of Export Control of*

U.S. Technology—A DOD Perspective (Bucy Report) (Washington, D.C.: U.S. Government Printing Office, 1976).

8. Executive Office of the President, Office of Science and Technology Policy, Interagency Working Group on Biotechnology, *Report of a Working Group on Competitive and Transfer Aspects of Biotechnology,* Washington, D.C., May 27, 1983.
9. France, Décret 70-441, *Journal Officiel de la République Française* (Official Journal of the Republic of France), May 26, 1970, p. 4992.
10. France, Arrete of Aug. 9, 1973, "Arrêté of Decree 68-1021," *Journal Officiel de la République Française* (Official Journal of the Republic of France), Aug. 9, 1973, as reported in "Controle Des Relations Financiéres Avec L'Etranger," *Bulletin Official Des Douanes,* Vol. 81-1061/G.11, June 5, 1981, p. 5.
11. Ikle, F., *East-West Trade and Technology Transfer,* statement in hearings before the Subcommittee on International Finance and Monetary Policy of the House Committee on Banking, Housing, and Urban Affairs, 1982 (Washington, D.C.: U.S. Government Printing Office, 1982).
12. Kyoikusha Publishing Co. (ed.), *Management Comparison of the Drug Industry* (Tokyo: Kyoikusha Publishing Co., 1980).
13. National Science Board, *Science Indicators 1980* (Washington, D.C.: U.S. Government Printing Office, 1981).
14. Porges, A., "Trade and Investment Laws and U.S. Competitiveness in Biotechnology," contract report prepared for the Office of Technology Assessment, U.S. Congress, June 1983.
15. Schwaab, R., Jeffery, D., and Conlin, D., "U.S. and Foreign Intellectual Property Law Relating to Biological Inventions," contract report prepared for the Office of Technology Assessment, U.S. Congress, February 1983.
16. Switzerland, "Surveillance Des Exportations: Ordinance No. 1," *Recueil Officiel* (Official Record), June 18, 1951, p. 531.
17. *United States* v. *Edler Industries,* 579 F.2d 516 (9th Cir. 1978).
18. U.S. Congress, Office of Technology Assessment, *Technology and East West Trade,* OTA-ISC-101, Washington, D.C., November 1979.
19. U.S. Congress, Office of Technology Assessment, *Technology and East-West Trade: An Update,* OTA-ISC-209, Washington, D.C., May 1983.
20. U.S. Department of Agriculture, U.S. Department of Commerce, U.S. Department of State, and Office of the U.S. Trade Representative, *Report to the United States Congress on the Agreement on Technical Barriers to Trade—"Standards Code,"* January 1980 to December 1982.
21. Vaquin, M., "Biotechnology in France," contract report prepared for the Office of Technology Assessment, U.S. Congress, January 1983.
22. *World Drug Market Manual,* 1982-83, IMS World Publications, 1983.

Chapter 20

Targeting Policies in Biotechnology

Contents

Figure

Chapter 20

Targeting Policies in Biotechnology

Introduction

During the past few years, some governments in countries other than the United States have designated the commercial development of biotechnology as essential to their nations' continued economic well-being. Unlike the U.S. Government, which has relied on a policy of funding basic research in the life sciences and encouraging research and development (R&D) in all industries with tax credits,* these governments have instituted targeting policies in biotechnology designed to promote the commercial development of biotechnology. In the context of this report, a targeting policy for biotechnology is defined as any policy that singles out the indigenous development of biotechnology for special attention from the central government. Foreign targeting policies in biotechnology may have the potential both to enhance the international competitiveness of foreign firms and to weaken that of U.S. firms.

This chapter examines the targeting policies in biotechnology of Japan, the Federal Republic of Germany, the United Kingdom, and France.* The targeting policies of most foreign governments are directed toward both "old" and "new" biotechnology. This chapter focuses on the aspects of these policies applicable to new biotechnology, as defined at the outset of this report. Although it does not address the issue of whether the U.S. Government should adopt a targeting policy for biotechnology, it does identify which targeting mechanisms could most readily be adopted in the United States if the U.S. Government chose to target biotechnology.

*See *Chapter 12: Financing and Tax Incentives for Firms* and *Chapter 13: Government Funding of Basic and Applied Research.*

*Switzerland is not considered in this chapter, because the Swiss Federal Government has no central policy for the industrial development of biotechnology.

Timing and coordination of policies

The biotechnology targeting policies of Japan and the Federal Republic of Germany have evolved out of at least a decade of interest in the commercialization of life-science-related technologies; these policies have more recently emphasized the incorporation of the new recombinant DNA (rDNA) and hybridoma/monoclonal antibody (MAb) technologies, as well as advances in bioprocess engineering. The biotechnology targeting policies of the United Kingdom and France, in contrast, have developed since about 1980, largely in response to the recent developments that have occurred in the field of molecular biology. The extent and degree of coordination of targeting policies differ among countries.

Japan

As early as April 1971, the Council for Science and Technology, Japan's highest science and technology policymaking body, including government, business, and academic leaders, stressed the importance of promoting life science on a nationwide basis because of its commercial potential (16). Since then, three governmental departments in Japan—the Science and Technology Agency

(STA), the Ministry of International Trade and Industry (MITI), and the Ministry of Agriculture, Forestry, and Fisheries (MAFF)—have specifically targeted the development of biotechnology.

STA responded in 1973 by establishing the Office for Life Science Promotion to plan and coordinate STA's R&D programs in life sciences. Until MITI's entry into major biotechnology programing in 1980, STA's R&D programs in fields related to biotechnology were the largest and the best funded in Japan. Even today, STA's programs are comparable in scale to those of MITI (25).

STA, in addition to being responsible for carrying out its own R&D program in the fields related to biotechnology, is responsible for interministerial coordination. It should be pointed out, however, that STA's influence on the formulation and implementation of Japanese biotechnology policy is not as pervasive as it might appear on paper. Interministerial rivalries and competition are common in Japan, and as described below, MAFF and MITI, each with substantially larger in-house staffs and laboratories than STA, have independently formulated their own biotechnology targeting policies. Nevertheless, STA's foresight with respect to the development of biotechnology has accorded the agency a more authoritative position for biotechnology than for other high-technology fields.*

MITI did not enter the biotechnology area until 1981. In that year, MITI reorganized itself to deal comprehensively with the challenges of new developments in technology and established its "System for Promotion of Research on Next-Generation Industrial Technologies," an overall plan to promote "next-generation" industrial technologies (25). Three "next-generation" projects in biotechnology were established within MITI's Basic Industries Division, and an Office of Biotechnology Promotion was established within this division to provide policy oversight for MITI's biotechnology effort and to serve as liaison between MITI's Biotechnology Long-Term Vision Advisory Group and possible MITI efforts to obtain from the Japanese Diet special legislation governing the promotion of biotechnology in Japan (25).*

MAFF has more recently established the Committee on Biological Resources Development and Utilization, which compiled a report recommending actions MAFF could take to promote biotechnology's development (21).

In addition to STA, MITI, and MAFF, three other Japanese Government agencies are funding R&D in biotechnology: the Ministry of Health and Welfare, the Ministry of Education, and the Environment Agency (26).

Federal Republic of Germany

The West German Government's interest in the development of old biotechnology, like that of the Japanese Government, is more than 10 years old. In 1968, the old Federal Ministry for Scientific Research explicitly recognized the potential commercial importance of old biotechnology by including it in a program to promote new technologies (15). In 1972, the newly reorganized Ministry for Research and Technology (BMFT, Bundesministerium für Forschung und Technologie), along with the Ministry of Education, commissioned a report on old biotechnology from the German Society for Chemical Engineering (DECHEMA, Deutsche Gesellschaft für Chemisches Apparatewesen) (7). The DECHEMA study, completed in 1974, laid the groundwork for a comprehensive Federal policy for the development of old biotechnology (15). In 1980, in light of increasing evidence suggesting potential commercial applications of advances in both scientific and engineering aspects of biotechnology, BMFT presented its *Leistungsplan: Biotechnologie,* a performance plan for biotechnology (5). This plan identified and targeted for support specific areas in which West German industry could commercially exploit both old and new biotechnology (15).

BMFT makes policy and coordinates German governmental activity for all biotechnology. BMFT funds basic and generic applied research in biotechnology through a number of public and non-

*STA was involved from the beginning with its own program and had the central role in the setting of rDNA regulations. The agency has a policy of reviewing on a case-by-case basis scaled-up production of genetically manipulated micro-organisms beyond 20 liters and has been reluctant to relinquish this authority (4).

*Several factors, including visible American concern with Japanese Government aid to high-technology industries, have made the passage of such programs unlikely (25).

profit research centers (15). Its most important function, however, is to oversee the development efforts of various industries in biotechnology, and it aids such efforts with a strong funding program (15).

United Kingdom

The formulation of official Government interest in the commercialization of biotechnology in the United Kingdom dates from March 1980, with the publication of the Spinks' report (1). This report identified major weaknesses in the country's biotechnology commercialization efforts and suggested ways of correcting them. The document elicited almost immediate Government action on its recommendations and sparked a spirited dialog among the various sectors with an interest in developing and incorporating the latest advances in this set of technologies into British industries.*

The Department of Industry is the United Kingdom's lead department for biotechnology. Other Government departments involved in health, energy, the environment, agriculture, and food, however, contribute to the advancement of biotechnology within their respective sectors, primarily by funding basic research (8). In April 1982, the Department of Industry established the Interdepartmental Committee on Biotechnology to strengthen the existing coordinating arrangements by focusing the Government's effort on the commercial development of biotechnology. This committee coordinates the activities of other related bodies, such as the Research Councils, the British Technology Group (BTG), and the Public Health Laboratory Service, and serves as a point of contact for those outside Government.

*_Biotechnology and Education: Report of a Working Group_, The Royal Society, 1981; _Biotechnology_, Cmnd 8177 (London: H. M. Stationery Office, March 1981); _The Strategy for Biotechnology in Britain_, BCCB Seminar, London, October 1981, series of unpublished papers, widely circulated at the time; _Biotechnology: Interim Report on the Protection of the Research Base in Biotechnology_, Sixth Report from the Education, Science and Arts Committee, Session 1981-82, House of Commons Paper 289 (London: H. M. Stationery Office, July 29, 1982).

France

Official interest in the commercialization of biotechnology in France was marked by the appearance of the Pelissolo report (23) in December 1980. Since the election of the socialists in 1980, the French Government has resolved to push the development of several new technologies in French industries and has accorded a privileged position to biotechnology within this scheme.

In July 1982, the old Ministry of Research and Technology in France was reorganized into a new, more powerful Ministry of Research and Industry (Ministère de la Recherche et de l'Industrie) based on the model of Japan's MITI (29). Furthermore, a wide-ranging research law adopted by the French National Assembly in July 1982 stipulated a real increase in the civilian R&D budget of 17.8 percent per year for 5 years, economic conditions permitting, and set up seven technological "programmes," on which the majority of all civilian research funds are now to be focused (30).

Biotechnology was one of the seven "programmes," and a Biotechnology Mission (Mission des Biotechnologies), established in August 1981, produced a planning document for biotechnology in France in July 1982. This document, the "Programme Mobilisateur: l'Essor des Biotechnologies," called for the restructuring of biotechnology policymaking into three separate coordinating bodies: 1) a national committee, presided over by the Minister of Research and Industry; 2) an interministerial coordinating committee; and 3) a program team to work in daily liaison with other Government organizations most closely involved in distributing research funds (18).

Since the publication of the "Programme Mobilisateur," the Ministry of Research and Industry has undergone a further restructuring. The new name of this ministry, Ministry of Industry and Research (Ministère de l'Industrie et de la Recherche), further reflects the efforts of French policymakers to focus on the commercialization of research results, including those in biotechnology (9).

Industrialists' role in policy formulation

Formulating a policy with the assistance of the parties whose activities it is intended to affect usually makes its implementation far more effective. Foreign nations competing with the United States in the commercialization of biotechnology have various mechanisms which incorporate industrialists into the formulation of a government targeting policy.

Japan

In Japan, technological strategy is usually formed by a "bottom-up" process, and the formulation of the strategy for biotechnology was no exception. After the announcement of the Cohen-Boyer patent for the basic rDNA process in 1980, five major Japanese chemical companies organized a joint study group called the Biotechnology Forum. The Biotechnology Forum was instrumental in lobbying for the establishment of MITI's three major "next-generation" biotechnology R&D projects: rDNA technology, bioreactors, and mass cell culture (25).* Furthermore, discussions with industrialists helped narrow MITI's focus. A planned "next-generation" R&D project in cell fusion was dropped, because the chemical companies working with the Basic Industries Division of MITI were already rather advanced in this area and because MAFF and the Ministry of Health and Welfare were developing their own programs in the field (25).

Federal Republic of Germany

The biotechnology policy of the Federal Republic of Germany was formulated with industry consultation. As noted above, a report on old biotechnology from DECHEMA, the private sector research association of the German chemical industry (7), laid the groundwork for a comprehensive Federal policy. Much of BMFT's funding goes to nonprofit research centers such as the Society for Biotechnology Research (GBF, Gesellschaft für Biotechnologische Forschung) that conduct generic applied research useful to industry (13). The research institutes of these organizations have boards of directors with strong industrial representation, so their research strategy is thus usually formed by a "bottom-up" process.*

United Kingdom

The Department of Industry launched in November 1982 a new 3-year, $30 million program of support for biotechnology in industry (2). To promote and monitor its funding initiatives, the Laboratory of the Government Chemist, part of the Department of Industry, set up a Biotechnology Unit. The unit is headed by one official from the Laboratory of the Government Chemist and three full-time biotechnologists on loan from industry. The purpose of this group is to provide industrial biotechnology expertise previously unavailable in the Department of Industry (12). The establishment of the Biotechnology Unit in 1982 marks the first time the British Government has incorporated the industrial sector on a regular basis into the policymaking process for biotechnology. Previously, the direction of the United Kingdom's informal involvement in biotechnology was determined largely by Government officials and scientists acting through already existing committees, with only occasional input from the private sector.

France

The presentation of the "Programme Mobilisateur" in July 1982 followed an intensive period of analysis and discussion between French Government officials, research scientists, and industrialists. A product of the plan was a National Biotechnology Committee, presided over by the

*In fact, following the award of the Cohen-Boyer patent, the Committee on Life Sciences of the Japan Federation of Economic Organizations met in alarm to discuss a Japanese response. Included at this meeting were representatives of 30 major Japanese companies with an interest in biotechnology. The Cohen-Boyer patent was seen as a matter of concern because, according to their company sources, the patent would affect almost any product application of rDNA technology. Ironically, it was suggested that the United States was designating biotechnology as a strategic national industry and was weaving about it a network of protective patents (27).

*OTA's report *U.S. Industrial Competitiveness: A Comparison of Steel, Electronics, and Automobiles* (28) presents a general description of structural integration of business into West Germany's policymaking apparatus, pp. 196-200.

Minister of Research and Industry, with 30 to 40 members from the Government, academia, and industry responsible for providing general guidance in implementing the Government's policy. In the past, the industrial policy of France has been more autocratic than that of West Germany or Japan (31). For biotechnology, enthusiastic French Government officials advocated generalized support of R&D projects regardless of the prospects for successful exploitation, to the dismay of industrialists who doubted the viability of some of the projects designated to receive Government support (29).

Policy goals

An examination of the goals of foreign biotechnology policies indicates that the domestic development of biotechnology, rather than the advancement of knowledge per se, is their foremost objective.

Japan

Japanese Government programs for biotechnology R&D are concerned specifically with the development of Japanese industry.

MITI's interest in biotechnology has been almost exclusively related to a more general program of structural adjustment for Japan's extremely depressed basic chemicals industry (24,25). MITI's three "next-generation" biotechnology R&D projects are part of a 10-year program that is specifically designed to develop and diffuse biotechnology among Japanese companies. According to a recent MITI policy statement, it is not feasible to rely on the private sector for biotechnology-related research that involves huge economic risks, so "the Government itself must take the initiative in such R&D, while at the same time offering assistance to private corporations in various forms to expedite this R&D" (19).

STA also is directly concerned with providing the technological underpinning for industrial advancement in Japan. The essential distinction between the STA and the MITI biotechnology projects is that the former concentrate on medical applications and longer term development of advanced bioreactors, whereas the latter are mainly concerned with fine chemicals, biological routes to production, fertilizers, and enzyme technology (25).

Federal Republic of Germany

According to a September 1979 BMFT statement, a primary goal of Germany's Federal biotechnology policy is "to establish the preconditions for industrial innovation in this key area of technology" (15). Another goal is "to strengthen the performance and competitive capacity of the German economy in long-range growth-oriented areas, in the process, correcting weaknesses revealed through international comparison and preventing distortions in Germany's competitive position" (15).

United Kingdom

While the British Government recognizes the potential of biotechnology, it is fairly guarded about the objectives of its biotechnology policy. The Minister of Industry has stated that "many developments are only now beginning to emerge from the research phase, and the direction of development for commercial exploitation remains uncertain. In addition, new biotechnological techniques and processes may well emerge over the next 20 years with benefits as yet unforeseen" (8). Clearly, however, the British Government intends to assist the country's industries in realizing the commercial potential of biotechnological developments as such developments appear (8).

France

The French Government's "Programme Mobilisateur" plans to remedy the present deficiencies in qualified personnel and spending levels for

R&D in biotechnology in French industry and the lack of public sector applied research in 5 years. According to the document, French companies should account for 10 percent of the world market in the "bioindustries" (not defined) in 1990, compared with an estimated 7.5 percent now (18).

Policy implementation

Examples of the mechanisms used to implement biotechnology targeting policies in Japan and other countries illustrate the variety of forms which biotechnology targeting policies can take. Several examples are cited below. For more information on government funding, see *Chapter 12: Financing and Tax Incentives for Firms* and *Chapter 13: Government Funding of Basic and Applied Research.*

Japan

The activities of STA's Office for Life Science Promotion are shown in figure 32. As shown in the figure, the Office is funding two goal-oriented research projects in biotechnology. These projects are to be carried out in 10 years by research groups whose members are affiliated with Japanese universities and research institutes (26). One of the projects, the project on the development of bioreactors, aims to develop what the Japanese call "second generation" bioreactors and includes computer control, biochemistry, and systems design. STA has encouraged an interdisciplinary approach to the project by inviting a variety of Japanese companies skilled in various aspects of biotechnology to participate. This approach has been very productive (24). As shown in figure 32, the Office for Life Science Promotion is providing sup-

Figure 32.—Activities of STA's Office for Life Science Promotion

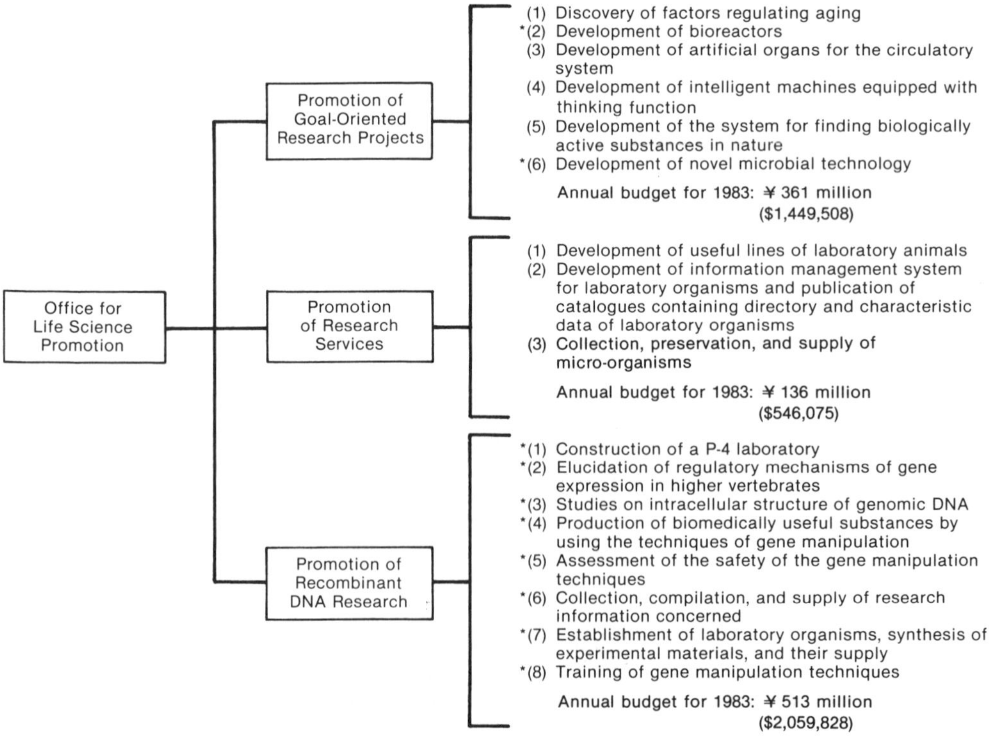

* = Biotechnology activities.

SOURCE: Office of Technology Assessment, adapted from *Science and Technology in Japan,* April/June 1983.

port for rDNA research. This support includes funding for the construciton of facilities. In 1982, construction was begun on a P-4 (highest physical containment level) facility in which experiments in genetic manipulation can be performed in Tsukuba Science City (26). In 1980, as one of the Office for Life Science Promotion's projects for the promotion of research services, the Japan Collection of Micro-Organisms was constructed to collect, preserve, and supply micro-organisms (26).

STA is implementing its policy in part through the general New Technology Development Fund. This fund has already commenced funding a number of biotechnology-related projects. A $4 million grant to the pharmaceutical company Green Cross in March 1980, for example, launched Green Cross into the international arena of competition in pharmaceuticals by enabling it to conduct research on rDNA methods for the production of alpha interferon (25).

MITI's three next-generation biotechnology projects, which are targeted to establish and diffuse scale-up techniques among companies, are even more illustrative of Japanese Government cooperation with industry. MITI has invited 14 companies to participate in the projects on a long-term (10-year) basis* and will provide allocations over 10 years of $43 million each to both the rDNA and bioreactor projects and $17 million to $22 million for the mass cell culture project (10). Although some 10 percent of the R&D work (by expenditure) for MITI's biotechnology projects is being conducted in the national laboratories** of the Agency for Industrial Science and Technology, the bulk of the work (90 percent) is conducted in industry laboratories. To facilitate coordination by the Office of Biotechnology Promotion and the Next-Generation Research Coordination Bureau of MITI's Agency for Industrial Science and Technology, the 14 companies receiving grants under the next-generation biotechnology projects have been organized into the Biotechnology Development Research Association. This association has its own central office through which the various companies communicate with MITI, but otherwise there are no intercompany institutions (e.g., there are no common laboratories being maintained by the companies). MITI subsidies to these companies cover 100 percent of all direct expenses (salaries and laboratory expenses) for biotechnology R&D, but no overhead is allowed and any capital equipment purchased is nominally the property of the Japanese Government. Furthermore, all patents resulting from the work belong to the Japanese Government, which, MITI has assured other companies, both domestic and foreign, will be freely available (14).

MAFF also is actively promoting cooperative research with private industry at its laboratories and is currently funding work with both Nippon Shokuhin Kako and Oriental Yeast at the National Food Research Institute and with Kao Soap at the National Institutes of Agricultural Sciences. Further joint research is planned in the areas of plant breeding and species improvement with private seed companies. Achievements from the research are used jointly by Government and industry, but those companies that participate in the research projects receive exclusive licensing rights to the patents resulting from these projects for 3 years (22).

Federal Republic of Germany

BMFT implements its biotechnology targeting policy in the Federal Republic of Germany through three categories of support. One category is funding for already existing schemes for industrial development. Another category is funding for third-party organizations to which BMFT contributes as part of more generalized funding programs for all areas of public research. GBF is the foremost example of such an organization. Originally founded to conduct generic bioprocessing research to meet the needs of industries (17), GBF employs 365 people and has a budget (1982) of $13 million (DM31 million), of which 89 percent came from BMFT (13). GBF's current activities include general development of bioprocess technology, scale-up of laboratory processes, screening

*The bioreactor project has been divided into two subprojects with Mitsubishi Chemicals as the overall leader. Sumitomo Chemicals is the leader of the rDNA project, and Kyowa Hakko is the leader of the mass cell culture project (25).

**These include the Fermentation Research Institute, National Chemical Laboratory for Industry, Research Institute for Polymers and Textiles, Government Industrial Research Institute, and Institute of Physical and Chemical Research (25).

of micro-organisms and plant and animal cell cultures, support of other research groups in biotechnology, participation in joint biotechnology projects with industry, and advanced interdisciplinary training for scientists, engineers, and technicians.

A third category of support is funding for biotechnology programs specifically designated by BMFT. For these programs, BMFT has funded a wide spectrum of projects with about $35 million (DM55 million) in 1982 (15): food requirements, biological pesticides, plant and animal cell culture techniques, biomass, metal refining, bioprocesses for commodity chemicals, bioreactors, and principles of biotechnological procedures (5).

The list of BMFT's grant recipients for these biotechnology programs includes every major German chemical and pharmaceutical company (5). BMFT's support for research on the development of interferon is particularly noteworthy. Between 1975 and 1977, BMFT gave Merck, Ltd., $300,000 (DM0.6 million) for the study of interferon induction. Rentschler, Inc., has been supported since 1976 with about $9 million (DM18.54 million) for its R&D effort on fibroblast interferon (6).

United Kingdom

In the United Kingdom, the Department of Industry has launched a new, 3-year $30 million "Biotechnology in Industry" program. The British Government also funds BTG, which encourages cooperative projects between industry and public sector laboratories. Government laboratories, such as the Centre for Applied Microbiology Research, carry out both applied research of potential interest to industry and specific industrial contracts.

In 1981, the British Government, through BTG and in association with four private investors, established Celltech, Ltd., to develop and market products made by some of the new technologies. In an arrangement similar to that of Immunotech in France, Celltech has a total initial capital of $20 million and the right of first refusal* on all work done in the Medical Research Council (20). In 1983, BTG, Advent Eurofund (a venture capital group), and Ultramar (a petroleum and financial group) established the firm Agricultural Genetics with a total initial capitalization of $28 million. This firm has the right of first refusal on all work done in the Agricultural Research Council (3).

France

The French Government is supporting R&D in various governmental agencies, including the National Institute of Health and Medical Research (INSERM, Institut National de la Santé et de la Recherche Medicale), the National Center for Scientific Research (Centre National de la Recherche Scientifique), and the Institut Pasteur. Government funding in applied areas is intended to benefit the pharmaceutical, food, and agricultural industries.

In 1982, the French Government supplemented its applied research program by creating a company, Immunotech, to facilitate the commercialization of biotechnology and transfer the results of immunology research, a traditional French strength, to French industry. Immunotech does applied research on bioprocessing and hybridoma technology for the production of immunoassay and immunopurification systems. The Ministry of Research and Industry contributed $3.2 million to its formation. Immunotech has the right of first refusal on all work financed by INSERM.

*This is the right to choose whether or not to produce and market any good or service without having to bid competitively with other firms.

Findings

The governments of four leading industrialized competitors of the United States—Japan, the Federal Republic of Germany, the United Kingdom, and France—have instituted programs to target the development of certain areas of biotechnology. The targeting policies are intended to reduce economic risk and lessen corporate duplication in biotechnology R&D.

The governments of these four countries took an interest in biotechnology at different times. The governments of both the Federal Republic of Germany and Japan identified the life sciences in the early 1970's as an area worthy of special government and private sector assistance. Those of France and the United Kingdom, on the other hand, realized the industrial importance of biotechnology only recently, primarily as a result of the recent advances in molecular biology.

The centralization of government activities varies among countries. In France and the Federal Republic of Germany, the direction of all activities, from basic research to industrial development, is centralized in a single ministry: the Ministry of Industry and Research in France and BMFT in Germany. In the United Kingdom, the Department of Industry is responsible for articulating and executing the Government's policy to commercialize biotechnology, but it must work with other departments that are concerned with the development of science in specific fields. In Japan, at least three Government departments have major biotechnology policies of their own.

These four foreign countries have various processes by which industrialists are brought into the formulation of their commercial biotechnology policies. Japan, France, and West Germany have a long history of involving industrialists. The United Kingdom, on the other hand, has only been officially involving industrialists in the formulation of its biotechnology policy for a short period.

The mix of policy measures to encourage industrial innovation in biotechnology assumes a variety of forms within each country. In Japan and the Federal Republic of Germany, the governments carry out their policies partly in the form of joint R&D projects with industry. These projects concentrate the resources of the government and private companies to meet specific objectives set by the government. In some cases, the companies have exclusive rights to the resulting patents; in other cases, the patents are made available to all interested parties. The British and French Governments, in addition to providing support for specific projects, have adopted a different sort of approach: the organization and support of small firms, such as Celltech in the United Kingdom and Immunotech in France, to commercialize the results of government-funded basic and generic applied research.

At this early stage, any evaluation of the foreign targeting programs' probability for success is preliminary. History has shown that even the best thought-out targeting policies do not guarantee competitive success. Whether the targeting policies of Japan, the Federal Republic of Germany, the United Kingdom, or France are superior to the U.S. Government policy of funding basic research in the life sciences and encouraging R&D in all industries with tax credits remains to be seen. The United States currently leads the world in the commercialization of biotechnology. Although targeting policies may not be of great importance when compared with other competitive factors, they could tip the balance of equivalent competitive situations in the future.

Issue

ISSUE: How could the U.S. Government target biotechnology?

It is beyond the scope of this report to evaluate whether the commercialization of biotechnology is of sufficient importance to the U.S. economy as a whole to warrant targeting efforts by the U.S. Federal Government. If such efforts are undertaken, however, several targeting mechanisms might be considered.

The mechanisms for targeting biotechnology in France, the United Kingdom, the Federal Republic of Germany, and Japan range from highly coordinated to loosely organized, but all reflect some combination of the following:

- *Firm-specific assistance.* Firm-specific assistance involves choosing a single company or group of companies for assistance from the government in jointly agreed upon areas of high-risk R&D. The companies chosen sometimes perform the subsidized research in consortia.
- *Industrywide assistance.* Industrywide assistance involves providing government assistance to all companies that perform R&D in a particular area (or funding R&D in a national laboratory open to all interested industry participants). Low-interest loans or tax credits for R&D and procurement of new products are methods commonly used.
- *An interagency coordinating committee.* An interagency oversight committee without the authority to set goals or grant subsidies facilitates coordination of the policies and actions of government agencies and periodically recommends action through the appropriate agencies to address problems hindering the development of biotechnology.

The U.S. Government would probably have to avoid actions in the category of firm-specific assistance. If the U.S. Government were to select a few companies for subsidies, demands for equal assistance would probably arise from the companies that did not receive subsidies.

For U.S. Government policies in the category of industrywide assistance, there are historical precedents. The types of U.S. Government support that were provided for the U.S. semiconductor industry in its early years are described in *Appendix C: A Comparison of the U.S. Semiconductor Industry and Biotechnology.* As it did in the case of the U.S. semiconductor industry, the U.S. Government could provide or guarantee low-interest loans for high-risk R&D in biotechnology. It could also guarantee Government procurement of certain products to eliminate some market size uncertainties. A commitment by the Federal Government to purchase certain drugs developed by biotechnology could spur R&D that otherwise might not be undertaken.

The third mechanism, an interagency coordinating committee, would probably raise the fewest objections in the United States but would also be the least substantial. The defunct Interagency Working Group on Biotechnology of the White House Office of Science and Technology Policy temporarily served this function and presented its recommendations to the Office of Science and Technology Policy in June 1983 (11).

Earlier chapters of this report have outlined options that could improve U.S. competitiveness in biotechnology. The adoption of the most acceptable of these options in a coordinated fashion would be one way in which the U.S. Government could target biotechnology.

Chapter 20 references

1. Advisory Council for Applied Research and Development, Advisory Board for the Research Councils, and the Royal Society, *Biotechnology: Report of a Joint Working Party* (Spinks' Report) (London: H. M. Stationery Office, March 1980).
2. Beardsley, T., "More Help," *Nature* 300:394, 1982.
3. Beardsley, T., "U.K. Company's Delayed Budding," *Nature* 304:269, 1983.
4. *Biotechnology Newswatch,* "120 Companies Form Trade Organization," Mar. 7, 1983, p. 1.
5. Bundesministerium für Forschung und Technologie (German Ministry for Research and Technology), *Leistungsplan: Biotechnologie* (Performance Plan: Biotechnology), Bonn, January 1980.
6. Bundesministerium für Forschung und Technologie (German Ministry for Research and Technology), *Interferon,* Bonn, 1981.
7. Deutsche Gesellschaft für Chemisches Apparatewesen (German Society for Chemical Engineering), *Biotechnologie* (Biotechnology), report prepared under contract to the Federal Ministry for Education and Science and the Federal Ministry for Research and Technology, Frankfurt, 1974.
8. Department of Education and Science (U.K.), "Bio-

technology: Interim Report," Government response to the House of Commons Select Committee on Education, Science, and the Arts interim report on the protection of the research base in biotechnology, London, July 29, 1982.
9. Dickson, D., "Research Chief Quits French Cabinet," *Science* 220:177, 1983.
10. Elkington, J., "The Climate for Biotechnology: National Strategies Contrasted," *Bio/Technology* 1(6):490-492, 1983.
11. Executive Office of the President, Office of Science and Technology Policy, *Report of a Working Group on Competitive and Transfer Aspects of Biotechnology,* Washington, D.C., May 27, 1983.
12. Fishlock, D., "Powerhouse for British Genes," *Financial Times,* London, Apr. 26, 1983.
13. Gesellschaft für Biotechnologische Forschung (Society for Biotechnology Research), *Programmbudget* (Program Budget), Braunschweig, 1982.
14. *Japan Economic Journal,* Mar. 23, 1982.
15. Jasanoff, S., "Public and Private Sector Activities in Biotechnology: The Federal Republic of Germany," contract report prepared for the Office of Technology Assessment, U.S. Congress, January 1983.
16. Kagaku Gijutsu Kaigi (Council for Science and Technology), *1970 Nendai Ni Okeru Sogo-Teki Kagaku Gijutsu Seisaku No Kihon Ni Tsuite* (The Basics of a Comprehensive Science and Technology Policy for the 1970's), Tokyo, 1971.
17. Maddox, J., "Research Labs Galore," *Nature* 297:279, 1982.
18. Ministère de la Recherche et de l'Industrie (Ministry of Research and Industry), Mission des Biotechnologies (Biotechnology Mission), "Programme Mobilisateur: l'Essor des Biotechnologies" (Mobilizing Program: The Expansion of Biotechnology), Paris, July 1982.
19. Ministry of International Trade and Industry, Japan External Trade Organization, "Research on Biotechnology in Japan," *The Japan Industrial and Technological Bulletin,* November 1982.
20. National Enterprise Board, Information Office, London, press release issued on behalf of Celltech, Ltd., Nov. 11, 1980.
21. *Nihon Keizai Shimbun* (Japan Economic Journal), "Promotion of Biotechnology Development," Sept. 2, 1982, translated by the Foreign Broadcast Information Service, in *Japan Report* (Joint Publications Research Service, Arlington, Va.).
22. *Nikkei Biotechnology,* "Ministry of Agriculture, Forestry, and Fisheries," Feb. 14, 1983, translated by the Foreign Broadcast Information Service, in *Japan Report* (Joint Publications Research Service, Arlington, Va.).
23. Pelissolo, J. C., *La biotechnologie, demain?* (Biotechnology, tomorrow?) (Paris: La Documentation Française, December 1980).
24. Rogers, M. D., "The Role of the Japanese Government in Biotechnology Research and Development," paper presented to conference of Biotechnology Group of the British Coordinating Committee for Biotechnology, London, June 1982.
25. Saxonhouse, G., "Biotechnology in Japan," contract report prepared for the Office of Technology Assessment, U.S. Congress, June 1983.
26. *Science and Technology in Japan,* "Research and Development of Life Sciences in Japan," April/June 1983, pp. 8-29.
27. Tawara, S., *Idenshi Sangyo Kakumei* (Genetic Industry Revolution) (Tokyo: Bungei Shunju, 1982).
28. U.S. Congress, Office of Technology Assessment, *U.S. Industrial Competitiveness: A Comparison of Steel, Electronics, and Automobiles,* OTA-ISC-135, Washington, D.C., July 1981.
29. Vaquin, M., "Biotechnology in France," contract report prepared for the Office of Technology Assessment, U.S. Congress, February 1983.
30. Walgate, R., "*La Loi* at Last," *Nature* 298:110, 1982.
31. Zysman, J., *Political Strategies for Industrial Order: State, Market, and Industry in France* (Berkeley, Calif.: University of California Press, 1977).

Chapter 21

Public Perception

Contents

Chapter 21
Public Perception

Introduction

Public perception of genetic research and technology is a factor that could influence the rate of commercialization of biotechnology. This chapter considers the factors that may affect public *perception* of genetic research and technology. As it does not consider the many ways by which the public might *express* its perceptions, it does not describe various methods that have been or could be used for public *participation* in decisionmaking processes, nor does it consider the arguments advanced for each.

Most of the discussion in this chapter is centered on the United States. One of the final sections considers the relative influence of public perception on the commercialization of biotechnology in the United States and foreign countries. For issues and policy options, readers are referred to OTA's April 1981 report *Impacts of Applied Genetics: Micro-Organisms, Plants, and Animals* (29).

The discussion in this chapter goes beyond biotechnology as defined in the rest of this report, and, for that reason, uses the broader terms "genetic research" and "genetic technology." These broader terms include directed manipulation of genes in human beings. Biotechnology, as defined in this report, does not include directed change of genes in human beings and is limited to industrial applications of new genetic technologies to produce useful substances, to improve the characteristics of economically significant plants and animals, and to act on the environment in useful ways. Because the public does not always make a clear distinction between industrial applications of novel genetic technologies and the manipulation of genes in humans, biotechnology can elicit public concerns that are based on incomplete knowledge and sometimes erroneous assumptions. Regardless of the *accuracy* of public perceptions about biotechnology, however, these perceptions could influence the rate of commercialization.

Public perception in the United States

The discussion that follows begins by considering the U.S. policymaker vis-a-vis the public on issues related to science and technology. It then describes various factors that influence public perception of biotechnology in the United States. It also reviews some arguments frequently raised in debates over genetic research and technology, considers difficulties in assessing risks and benefits of genetic research and technology, discusses the influence of the media on public perception of biotechnology, and provides some survey data.

The public and the policymaker

In a democratic society, where decisions are made by elected representatives, the public plays a vital role in the acceptance of new technology and the directions in which it will be applied (2). That public beliefs can significantly influence U.S. policymakers with respect to biotechnology is illustrated by the changing attitudes of policymakers in Massachusetts. In 1976, Boston Mayor Alfred Vellucci argued strongly for major controls on research and development (R&D) using recombinant DNA (rDNA) technology in Boston and Cambridge. As a result, the Cambridge Experimental Review Board was established to determine whether additional protection for citizens was needed beyond that provided by the National Institutes of Health (NIH) Guidelines for Research Involving Recombinant DNA Molecules (NIH Guidelines).* Mayor Vellucci's position may be con-

*The NIH Guidelines for Research Involving Recombinant DNA Molecules are discussed along with the rDNA research guidelines of other countries in *Chapter 15: Health, Safety, and Environmental Regulation* and *Appendix F: Recombinant DNA Research Guidelines, Environmental Laws, and Regulation of Worker Health and Safety.*

trasted with that taken by then Massachusetts Governor King when he addressed Harvard University's symposium on "New Partnerships in Biotechnology" in 1982. Governor King pledged his assistance to the establishment of commercial biotechnology firms in the State. The different positions taken by Mayor Vellucci and Governor King reflect, in part, the changes in public concern over the risks posed by rDNA technology.

Although the level of U.S. public concern about R&D involving rDNA appears lower now than it was in the late 1970's, it is not nonexistent. As of June 1982, two States and nine municipalities had passed laws and resolutions relating to control of rDNA R&D. The two States are New York and Maryland. With the exception of Princeton, N.J., the municipalities are located in Massachusetts (Amherst, Boston, Cambridge, Newton, Somerville, and Waltham) and California (Berkeley and Emeryville). It is interesting to note that all local municipalities involved in formulating laws or resolutions are the sites of, or located near, major centers of corporate and university research activity in rDNA. Although most of this legislative activity took place in the late 1970's, several municipalities in Massachusetts either amended or originated ordinances or laws in 1981. At a minimum, the laws extend the NIH Guidelines from institutions receiving NIH funds to all public and private institutions conducting rDNA research. Some of them also establish additional occupational and environmental safety requirements (15).

In light of the developments noted above, U.S. policymakers probably can expect to be increasingly involved in biotechnology issues. One issue in biotechnology is the amount of consideration that should be given to the unanticipated consequences of deliberately releasing into the environment products of rDNA technology (e.g., modified plants or microbes with improved capability for mineral leaching or pollution control). But this is just the opening wedge to a wider range of societal concerns that are emerging as new knowledge leads to new capabilities. The potential capabilities of genetic research and technology include human gene therapy, gene surgery, and estimation of differential susceptibility to disease based on differences in genetic traits.

An accident or perceived negative consequence involving genetic research or technology could stir up public fears and have a sizable impact on biotechnology's further development. This observation is true especially in the United States, where public involvement in the debates surrounding rDNA technology in its early years was very strong compared with public involvement in other Western democracies.

Factors influencing public perception of genetic research and technology

The Organisation for Economic Co-Operation and Development identified the following characteristics of science and technology issues that distinguish these issues from other public controversies (18):

- rapidity of change;
- the raising of *new* issues;
- scale, complexity, and interdependence among technologies;
- irreversibility of effects;
- strong public sensibilities about real or imagined threats to human health; and
- challenging of deeply held social values.

OTA's April 1981 report *Impacts of Applied Genetics: Micro-Organisms, Plants, and Animals* (29) noted that these factors were especially applicable to advances in genetics and that they helped to explain the public controversy over the safety of rDNA technology. The same factors remain applicable to advances in genetics today. Some are discussed below, along with other factors that may elicit positive, negative, or mixed public reactions to developments in genetic research and technology.

THE TECHNOLOGY IS PERCEIVED TO ENDANGER BASIC HUMAN NEEDS

Some new developments in science and technology are far more threatening to the societies in which they arise than are other developments. In an attempt to understand and predict which emerging technologies will be most threatening, and hence be most likely to raise issues for policymakers, E. W. Lawless makes the reasonable assumption that public concern with a new technology will vary in direct proportion to the degree that the technology is perceived to affect basic human needs (16). The greater the importance of an individual or societal need, and the greater the impact of the new technology on that need, the greater will be public concern.

At the top of the list of important *individual* needs developed by Lawless are the functions controlled by the nervous system, and particularly by the brain. Genetic technology has the potential to alter the functioning of the human brain, affecting attitudes, emotions, learning, and memory. Besides the concerns associated with the technology's potential to alter these characteristics per se, genetic technology may arouse deeper concerns that relate to an individual's sense of self. Aspects of self derive from each person's most basic characteristics—tendencies to elation or depression, ambition or sloth, and extroversion or introversion, to name a few. If these characteristics can be modified, what happens to an individual's unique, inviolate self?

The most fundamental *societal* need identified by Lawless is sexual activity, reproduction, and family organization. He notes (16):

> . . . any events or practices which portend a threat to man's reproduction or care of children cause immediate and serious alarm. Technologically related cases involving materials which are mutagenic (cause genetic damage) or teratogenic (cause congenital deformities) receive wide coverage by the news media and attention by the public—the announcement that LSD may cause chromosome breakage apparently caused much more concern to its users than other stated hazards, and the thalidomide case is almost classic.

The application of genetic technology to the production of useful industrial substances is not always clearly distinguished from the genetic manipulation—or "genetic engineering"—of higher organisms. Following Lawless, if biotechnology is associated with the capability to alter human reproductive cells, and hence future human generations, it is likely to be perceived as threatening.

TERMINOLOGY

As has been pointed out by various authors (20,21), some of the terminology of applied genetics has negative overtones. The phrase "genetic engineering," for example, may raise Frankenstein-like subconscious fears when associated with human application. "Cloning" of genes, a basic technique of rDNA technology, can be confused in the minds of those who are not expert in the field with the cloning of individual human beings. Because language is widely understood to influence perception, the problem of terminology is not a minor one. Terms that are widely used, however, even though inaccurate, misleading, or imprecise, are not easily changed.

PERCEPTION OF BENEFITS FROM BIOTECHNOLOGY

Biotechnology appears to offer potentially major positive contributions to diverse aspects of life. Economic benefits (e.g., cheaper chemicals and drugs), health benefits (e.g., cures for cancer, schistosomiasis, and herpes; improved diagnostic tools), agricultural benefits (e.g., saline-tolerant or pest-resistant plants, a vaccine for foot-and-mouth disease), and even decreased dependence on foreign oil (e.g., substitution of biomass for petroleum feedstocks, production of fuel alternatives) are envisioned.* To the extent that these benefits are perceived by the public, their perceptions of biotechnology are likely to be positive.

NIH GUIDELINES FOR RECOMBINANT DNA RESEARCH

Biological scientists were instrumental in bringing about the NIH Guidelines for Research Involving Recombinant DNA Molecules that established safety procedures for rDNA research conducted with NIH funds. The NIH Guidelines apply only to work supported by NIH funds, but other U.S. Government agencies have adopted them voluntarily. As far as is known, private industry observes them as well.

On the one hand, the history of the NIH Guidelines should produce a positive perception of responsible action with regard to genetic research and technology by the scientists concerned and the Federal Government. On the other hand, NIH is in a position of potentially conflicting interests. It serves both as a quasi-regulator of genetic research through the NIH Guidelines and as a promoter of genetic research through its sizable

*For a review of the state of the art in achieving these benefits, the reader is referred to chs. 5 through 10 of this report.

funding of genetic research. The degree to which the public perceives a potential conflict of interest and its influence on public perception of biotechnology are unknown.

THE IMAGE OF THE SCIENTIST

Some members of the public appear to be dismayed by the fact that some scientific researchers have turned into entrepreneurs. The question of the appropriateness of private gain from research supported by public funds was aired as part of joint hearings in 1981 and again in 1982 by the Subcommittee on Investigations and Oversight and the Subcommittee on Science, Research, and Technology of the U.S. House of Representatives (26,27).

There is no reason that scientists should not share in financial rewards that accompany application of the results of their research, but the *deliberate* pursuit of profits makes a scientist also a businessman. It can be argued that a major reason for supporting research with public funds is that such research leads to commercial products that benefit society and also generates more public funds through taxes levied on new businesses. However, the fact that some scientists have become millionaires through corporations they have helped to establish has disturbed some people. Simple envy is not the sole reason for unease; more important may be the public image of the scientist. Although U.S. cultural tradition has supported, and even encouraged, the entry of engineers and inventors into the business world (e.g., Edison), it has not done the same for individuals with established careers in pure science.*

COMMENT

A fundamental reason that rDNA technology may be "so inflammatory" is that it elicits a *mixture* of concerns from many categories (9). These concerns range from perceived positive benefits to fears associated with research on human subjects. The point for the policymaker is that, because of the wide range of concerns, genetic research and technology is a volatile area, one where the smallest incident may raise heated public emotions.

*For a discussion of university/industry relationships in biotechnology, see *Chapter 17: University/Industry Relationships.*

Arguments raised in debates over genetic research and technology

Five broad categories of arguments that are frequently raised in debates over genetic research and technology are briefly summarized below. It should be noted that the discussion that follows is in the simplest possible terms. The purpose is to indicate some topics of controversy rather than to describe the considerable subtlety of some of the positions that have been taken.

FREEDOM OF INQUIRY

Some people argue that scientists should be free to pursue any inquiry they choose, and hence that genetic research should not be restricted in any way. Others disagree and feel that at least some forms of research are subject to restraint. H. Jonas takes the latter position and argues that unqualified free inquiry ceases as a preeminent right when science moves from *contemplation* to *action* (12). As soon as science involves action (e.g., conducting experiments with real apparatus and real subjects) rather than just thought, it is subject to legal and moral restraints, as all *actions* are.

RISK OF CATASTROPHIC CONSEQUENCES

Some people argue that genetic research should be banned unless the risk of catastrophic consequences can be shown to be zero. At the other extreme, some people argue that any level of risk is acceptable. Although either of these extreme positions may be taken by individuals, neither is likely to be taken by society. What constitutes an acceptable level of risk of catastrophic consequences, however, is a major societal issue, in part because of the difficulty of assessing both risks and benefits. The fundamental disagreement on both this and the preceding topic is where the line is to be drawn between two extreme positions that can be taken. The position of the line is a societal decision that is never permanent and that varies across cultures and over time.

THE TECHNOLOGICAL IMPERATIVE

Some people argue that what is technologically possible will eventually be done, regardless of moral and ethical guidelines. Others disagree. As S. P. Stich points out, successful animal breeding has been carried out for centuries, yet controlled breeding is not done in humans even though it has been known for a long time that it could be (24). Thus, people have differing views on whether society is capable of deciding when genetic manipulation of traits is and is not permissible.

"WE SHOULD NOT PLAY GOD"

Some opponents of genetic research argue that humans should not "play God" by manipulating the genes of other organisms or themselves. Despite its use of the term "playing God," this argument is based on areligious as well as on religious grounds. Both types of arguments are briefly considered below.

To opponents of genetic research who argue on religious grounds that humans should not manipulate genes, proponents respond that humans have manipulated the genes of other organisms for thousands of years. Long before the laws of genetics were known, humans were successful in changing the characteristics of plants and animals by selectively breeding them for desired characteristics. In addition to altering the genes of other organisms, humans also have altered their own gene pool. Throughout history, because some persons are more desirable than others as mates, some genes have tended to increase in the gene pool while others have tended to decrease. More recently, medical advances have permitted persons with genetic diseases, such as hemophilia and phenylketonuria, to live and reproduce (17).

But, opponents argue, the genetic changes that have been brought about so far have been limited and *did not involve crossing fundamental species barriers.* So far, this argument is correct in that species are *defined* by the fact that fertile hybrids between them do not occur in nature. However, some opponents of research involving genetic manipulation further argue that the forces of evolution have led to separation of the species and that *breaking down the separation will be deleterious or separation would not have occurred in the first place.* The accuracy of this argument is not known.

As noted above, arguments for a prohibition against genetic research are sometimes based on religious grounds. Fundamentalist and religious objections have played a major role in U.S. debates over genetic research and technology in the past and are likely to continue to do so in the future. Recognizing the importance of religious views in such debates, the President's Commission for the Study of Ethical Problems in Medicine and Biomedical and Behavioral Research (hereafter referred to as the President's Commission) asked the General Secretaries of the National Council of Churches, the Synagogue Council of America, and the United States Catholic Conference to "elaborate on any uniquely theological considerations underlying their concern about gene splicing in humans" (21). The scholars concluded (21):

> . . . contemporary developments in molecular biology raise issues of responsibility rather than being matters to be prohibited because they usurp powers that human beings should not possess. The Biblical religions teach that humans beings are, in some sense, co-creators with the Supreme Creator.

Furthermore, Pope John Paul II, who has been critical of genetic manipulation, "recently told a convocation on biological experimentation of the Pontifical Academy of Science of his approval and support for gene splicing when its aim is to 'ameliorate the conditions of those who are affected by chromosomic diseases, because this offers hope for the great number of people affected by these maladies' " (21).

It should be noted, however, that the religious community's position is in a state of flux. As illustration, a resolution was issued on June 8, 1983, that urged the U.S. Congress to ban genetic changes affecting human reproductive cells. The resolution was signed by 64 religious leaders representing several faiths. The actual positions of the signatories of the resolution are difficult to decipher, because some church officials who signed the resolution appear to be in favor of genetic changes that would repair the effects of genetic diseases. Some forms of genetic defect, such as Tay Sachs disease, may be best eliminated through changes that affect the reproductive cells. Such changes would be banned by the resolution (3,11,14,19).

GENETIC DIVERSITY

Another area of controversy is the potential effect directed genetic manipulation may have on genetic diversity, i.e., the total number of different kinds of genes available to a population. All members of a given species can mate with any other member of that species, so the total number of genes available to the species population is the sum of all the different kinds of genes in all members of the population. Nevertheless, certain combinations of genes may be perceived as particularly desirable. In agriculture, for example, most farmers in a given location often plant the same strain of a particular crop that they perceive as especially desirable; then all members of that crop in a given location are genetically identical. When a new pest threatens the crop, much of the crop will be lost, because the genetic similarity of the plants results in a similar susceptibility to disease. The corn blight of 1970 is a case in point (10).

Opponents of directed genetic manipulation fear that it may result in increased genetic uniformity with a consequent loss of a species' resistance to future threats. Whether such fears are justified depends, of course, on how the organisms resulting from genetic manipulation are used.

COMMENT

Genetic technology, particularly when direct applications to humans are considered, raises strong public concerns. The degree to which public concerns about direct human applications of genetic research and technology are likely to influence the commercial development of biotechnology as defined at the outset of this report is unclear. Some influence is likely, however, because of a failure on the part of the public to make a clear distinction between human and nonhuman applications of genetic technology, a problem that is exacerbated by multiple uses of terms such as cloning.

Difficulties in weighing the risks, costs, and benefits of genetic research and technology

The central question raised by genetic research and technology is how risks, costs, and benefits are to be weighed. This is a question surrounded by problems.

One problem is that of establishing the probabilities that various risks and benefits will occur. Some probabilities can be estimated more accurately than others because of differences in the assumptions that must be made and in the availability of data that are useful in making estimates. Estimating the probability that an organism will escape from a laboratory, for example, involves different assumptions than estimating the probability that an organism released to the environment (e.g., a genetically modified plant or a microbe designed to control oil spills) will adversely affect that environment.

Then, there is the problem of measuring benefits, risks, and costs. First, it is necessary to decide whether the measure should be in economic terms (i.e., dollars) or human terms (e.g., lives saved or lost, illnesses prevented, or some measure of quality of life). If a measure can be selected, then there is the problem of applying it. Furthermore, if different measures are appropriate for costs, benefits, and risks, how should they be compared? Although methods have been developed to deal with these questions, including cost-benefit and cost-effectiveness analysis, they are always fraught with assumptions that become particularly acute with a new technology.*

Finally, like most new technologies, some applications of the new genetic technologies will have consequences that cannot be envisioned. These

---

*For a discussion of some of the limitations of techniques such as cost-benefit analysis and cost-effectiveness analysis, see OTA's 1980 report *The Implications of Cost-Effectiveness Analysis of Medical Technology* (28).

consequences may be high in benefit or high in cost, but some are certain to alter significantly any calculations that are made today.

In sum, assessment of benefits, risks, and costs, except where empirical data are available, is a subjective rather than an objective process, as is the assigning of relative value to various benefits, risks, and costs. Unfortunately, the most interesting and significant contributions of genetic research are those for which there are no empirical data. While risk assessment analysis was helpful several years ago when concern focused on the safety of laboratory research with rDNA, it may be of little use in considering many issues that may emerge as the technology matures, such as whether to release genetically modified organisms to the environment.

What, then, can be done? In a thoughtful analysis of gene splicing as applied to humans, the President's Commission recommends that an oversight group be established (21):

> . . . through which the issues generated by genetic engineering can continue to receive appropriate attention. These issues are not matters for a single day, deserving of only occasional attention. They will be of concern to the people of this country—and of the entire globe—for the foreseeable future; indeed, the results of research and development in gene splicing will be one of the major determinants of the shape of that future. Thus, it is important that this field, with its profound social and ethical consequences, retain a place at the very center of "the conversation of mankind."

The President's Commission suggests several objectives to guide the oversight group. Education, it states, should be a primary responsibility—education of the public about science and education of the scientific community about the social and ethical implications of emerging capabilities in genetic technology.

That Congress may perceive that the recommendation of the President's Commission for an oversight body reflects a broader public interest is suggested by the introduction of H.R. 2788 to the 98th Congress (Apr. 27, 1983) by Representative Albert Gore. H.R. 2788 would establish the President's Commission on the Human Applications of Genetic Engineering. The proposed Commission would review developments in "genetic engineering" that have implications for human application and examine the medical, legal, ethical, and social issues that might accompany such application. As of this writing, H.R. 2788 has been incorporated into the Health Research Education Act of 1983, H.R. 2350.

Influence of the media on public perception of genetic research and technology

The media bring knowledge of new discoveries and applications of genetic research to the attention of the public and thereby play a role in public perception of biotechnology. The role of the media extends beyond simple reportage of facts, however, because television, radio, and print media have time or space limitations that result in selective coverage. In selecting items for coverage, the media impose value judgments on the relative worth of possible news items. The media also determine how the items they consider newsworthy will be covered and thus vary the amount of coverage and the tone of coverage. Thus, it is helpful to explore the role of the media in public perception of biotechnology further.

June Goodfield, in an essay entitled "Reflections on Science and the Media" (8), traces the shifting relationship between scientists and the media in American society and the reasons for present day dissatisfaction between these two groups. Goodfield's orientation is to the public, which, she believes, both professions serve. The media and scientists, Goodfield observes, share a common aim in their respective spheres, namely, "the public expression of truth." Different pressures, however, constrain achievement of this ideal for each profession. Constraints on the print media include the need to create interest, the basic structure of newspaper reports, and the constant need for newness. The problems are exaggerated for radio and television. Scientists, on the other hand, are constrained by the nature of their work and their methodology. No scientist likes to "go public" before being sure that his or her findings are reproducible. The tendency among scientists, therefore, is toward caution. There is also, for a variety of reasons, an aversion among scientists to popularization. Thus, the different forces acting on each profession tend to polarize scientists and the media rather than bring them together.

In considering the relationships among scientists, the public, and the media, Goodfield is particularly concerned with three aspects: 1) the obligation of science to inform, 2) the duty of the public to become informed, and 3) the appropriate role of the journalist relative to science and the public. The journalist, she believes, not only must help the public distinguish what is factual from what is speculative but also must help people judge between scientists who differ.

Some of Goodfield's observations are echoed by William Stockton, former Director of *Science Times* of the *New York Times*. At a recent New York Academy of Sciences meeting, Stockton cited an increasing number of science publications, such as *Science 80* and the *Science Times,* as indicators that scientific journalism is moving into an era of scientific interpretation (25).

The possible roles for the media vis-a-vis genetic research and technology include:

- reporting the facts;
- separating facts from speculation;
- presenting issues;
- indicating which individuals or groups have a stake in each side of an issue and why;
- promoting, or downplaying, specific aspects of genetic technology; and
- educating the public in genetic science and technology, both their methods and their content.

Although many media people would probably claim that their role is limited to reporting the facts and separating these from speculation, their role is clearly larger. The media promote or downplay a technology, if only by virtue of the fact that some news items are selected for print or featured in a radio or television spot while others are rejected. Furthermore, the media's promotional role is sometimes far more active than simple selection.

Surveys of public perception

Given all the above, it is reasonable to ask for actual data on public perception of biotechnology, or at least of the broader area of genetic research and technology. Unfortunately, such data are extremely limited.

Two early surveys of the U.S. population were conducted in the 1970's with the following results (6):

- In 1977, the National Assessment of Educational Progress surveyed the attitudes of adults 26 to 35 years in age toward rDNA technology. About two-thirds of the respondents opposed its use on any life form.
- In 1979, the National Science Board conducted a survey of 1,635 adults. Sixty-five percent of the respondents believed that studies relating to creating new life forms should not be conducted.

In the 1980's, Cambridge Reports, Inc., included five questions on "genetic engineering" in its survey for the first quarter report of 1982 (5) and one question on behalf of the American Chemical Society in its survey for the first-quarter report of 1983 (1). The responses to the five questions in the 1982 survey showed (5):

- About half the people surveyed either hadn't heard the phrase "genetic engineering" or wouldn't guess what it meant.
- Of those who had heard of private corporations "getting into the field of 'genetic engineering' or biotechnology" (roughly 40 percent), and who were willing to take a position as to whether this was good or bad, positive sentiments (15 percent) outweighed negative (8 percent) by almost two to one.
- Of those expressing an opinion about "genetic engineering," 25 percent believed it would bring major benefits to society; 11 percent believed it would endanger public health and safety; 44 percent didn't know; and 20 percent believed it would bring both benefits and dangers.

- Respondents with higher income levels and/or higher levels of education were more likely to expect major benefits from "genetic engineering" than those with lower incomes and/or less education.
- Of respondents able to choose between government regulation and self-regulation, 28 percent favored the former and 16 percent the latter. Combination of both government regulation and self-regulation and "don't know" made up the balance.

The single question in the 1983 survey by Cambridge Reports, Inc., asked what respondents thought of when the term "DNA" was mentioned. Sixty-three percent didn't know; 27 percent responded with relevant but incomplete answers; 2 percent gave an accurate definition; and 2 percent said it was "poison" (1).

In 1981 and 1982, Yankelovich, Skelly, and White surveyed the general public with regard to "genetic engineering" (13). Their survey population is a nationwide stratified random sample of 2,500 persons aged 16 and over. Results are considered predictive of the U.S. population as a whole at a confidence level of 98 percent. The results showed the following:

- The percentage of the general public believing that the benefits of "genetic engineering" outweigh the risks increased from 31 percent in 1981 to 39 percent in 1982.
- Seventy percent of the public had heard of "genetic engineering" in 1982.
- Sixty-two percent of the public were very or somewhat concerned about "genetic engineering" in 1982.
- In 1982, those who had heard of "genetic engineering" were asked how it would be applied (by responding to a list of possible application areas). Health was selected most frequently (61 percent), followed closely by test tube babies (58 percent), and farming (57 percent). Responses to other application areas were: food processing (33 percent), forestry (31 percent), waste management (30 percent), chemical research (28 percent), pollution control (20 percent), and energy (19 percent).

Yankelovich, Skelly, and White believe that, although the intensity of public concern with "genetic engineering" is low at present, there is a significant latent level of public concern that could surface if adverse consequences associated with applied genetics were reported (13).

The survey data just cited suggest several things:

- A relatively small fraction of the American public is fully informed about genetics in general and, undoubtedly, about biotechnology in particular.
- The more informed public is more likely to view applied genetics favorably than unfavorably.
- There are real concerns about applied genetics.

Implications of public perception for competitiveness in biotechnology

As a factor influencing competitiveness in biotechnology, the importance of public perception varies greatly both across and within countries. Considering first democratic v. nondemocratic countries, public perception as a factor influencing competitiveness will be more important in the democracies than in those countries without such forms of government, simply because of the greater public input permitted by democratic, representative forms of government and the independence of the media.

Among democratic nations, variability in the importance of public perception as a factor influencing the commercialization of biotechnology is a function of many cultural characteristics. Of these characteristics, the traditions of the media, the degree to which the public participates in decisionmaking on scientific and technological issues, and the level of public education in science and technology are particularly important.

Of the six countries examined in this assessment—the United States, Japan, the Federal Republic of Germany, the United Kingdom, Switzerland, and France—public perception appears to have the greatest importance in the United States. The basis for this statement is that public debate over the establishment of rDNA R&D laboratories in the late 1970's was much greater in the United States than in the other countries. The behavior of the public and the media in the United States and other countries in the years since has changed little, and thus, public involvement as a factor in competitiveness currently remains of greatest importance in the United States.

Public perception will be a factor in determining competitiveness of the United States in the commercialization of biotechnology primarily in the event that genetic research or technology results in actual or perceived adverse consequences. In the case of an accident or perceived negative consequence, several factors would operate to make public perception of genetic research and technology of particular importance in the United States compared to other countries: the role of the media, traditions regarding public participation in scientific and technological issues, and the public's level of education in such issues. In this context, "level of education" requires further elaboration.

A technologically literate public can discriminate between different uses of genetic research and technology; this is important because different uses are associated with different issues. Some uses do not raise any new issues; others do. Thus, use of rDNA technology to produce drugs and biologics that replace similar products produced by chemical synthesis or extraction is simply an alternate means of production and in itself raises no ethical issues (17). An ethical issue for the pharmaceutical industry may be allocation of resources to produce drugs using biotechnology with markets that are potentially large and profitable v. drugs for treating rare diseases or diseases endemic to the Third World, where profits are more limited. Ethical issues are also raised if rDNA technology permits the manufacture of drugs that influence learning, memory, and personality traits, for decisions will be needed on whether such substances should be produced and perhaps on how their distribution should be handled and controlled.

Use of normal DNA to treat the body cells of patients with genetic diseases such as sickle cell anemia is another area where rDNA raises no new ethical issues beyond those associated with other treatment of sick persons. As geneticist A. G. Motulsky points out, this therapy is (17):

> . . . conceptually no different from any therapy in medicine that attempts to improve the health of a sick patient. The only difference is that DNA, rather than other biologicals, drugs, or surgery, is used as the therapeutic modality.

An application of genetic research and technology that does involve new ethical issues is use of genetic markers for diagnosis of susceptibility to disease. This application raises questions pertaining to private v. societal goals and confidentiality. Similarly, any genetic manipulation that alters the reproductive cells is "a qualitative departure from previous therapies since this would affect future generations" (17).

Rational consideration of issues raised by genetic research and technology is often confounded by failure to discriminate between different types of applications. The problem is compounded, because, as pointed out in *Chapter 14: Personnel Availability and Training,* scientific education in the United States is falling behind that of many industrialized nations. These factors could act to the disadvantage of the United States in the worldwide commercial development of biotechnology should an accident or other adverse consequence occur.

Findings

Public perception of the risks and benefits of biotechnology is of greater importance in countries with representative, democratic forms of government than it is in countries with other forms of government, simply because of the greater attention paid to public opinion in the democracies, and the independence of the media. As a factor influencing competitiveness, public perception is probably of greater importance in the United States than it is in Japan, the Federal Republic of Germany, the United Kingdom, Switzerland, or France.

A number of factors influence the relative importance of public perception as a factor influencing competitiveness. In all countries, the importance of public perception will be greatly increased in the event of an accident or perceived negative consequence of biotechnology. In such a case, the level of scientific and technological literacy in the various competitor countries becomes important, as judgments must be made concerning complex issues. Unfortunately, at least in the United States, survey data show that only a small fraction of the public is fully informed concerning genetics in general and therefore, undoubtedly, about biotechnology in particular. Survey data also suggest that there are real concerns in the public mind concerning applied genetics.

Given the lack of public knowledge, it is particularly important that the media play a responsible role with respect to biotechnology. The role of the media extends beyond mere reporting of the facts. How far the media should go beyond such reportage deserves consideration.

Issues and policy options

OTA's first assessment in the field of genetics, *Impacts of Applied Genetics: Micro-Organisms, Plants, and Animals* (29), was published in April 1981 and contained a chapter titled "Genetics and Society." The issues that arise from the material presented in the preceding pages are similar to the ones developed in the chapter on genetics and society. Since the issues in this report and OTA's earlier report are similar, the reader is referred to that earlier report for issues, options, and arguments relevant to them.

Chapter 21 references

1. Bigger, F., American Chemical Society, Washington, D.C., personal communication, June 27, 1983.
2. Blank, R. H., *The Political Implications of Human Genetic Technology* (Boulder, Colo.: Westview Press, 1981).
3. Briggs, K. A., "Clerics Urge U.S. Curb on Gene Engineering," *New York Times,* June 9, 1983, p. 1.
4. Bronowski, J., *Science and Human Values* (New York: Harper & Row, 1956).
5. Cambridge Reports, Inc., *The Cambridge Report (30)—First Quarter, 1982,* Cambridge, Mass., 1982.
6. Edwards, C., "Consumers Must Be Prepared for Biotech Products," *Biotechnology* 1(2):137, April 1983.
7. Goodfield, J., *Genetic Engineering and the Manipulation of Life* (New York: Harper & Row, 1977).
8. Goodfield, J., "Reflections on Science and the Media," AAAS Publication No. 81-5 (Washington, D.C.: American Association for the Advancement of Science, 1981).
9. Graham, L. R., "Concerns About Science and Attempts To Regulate Inquiry," *Contemporary Issues in Bioethics,* 2d ed., T. L. Beauchamps and L. Walters (eds.) (Belmont, Calif.: Wadsworth, Inc., 1982).
10. Harlan, J. R., "Our Vanishing Genetic Resources," *Science* 188:618-621, 1975.
11. Hilts, P. J., "Clergymen Ask Ban on Efforts To Alter Genes," *Washington Post,* June 8, 1983, p. A-1.

12. Jonas, H., "Freedom of Scientific Inquiry and the Public Interest," *Contemporary Issues in Bioethics,* 2d ed., T. L. Beauchamps and L. Walters (eds.) (Belmont, Calif.: Wadsworth, Inc., 1982).
13. Kaagan, L., "Assessment of Current Public Views of Biotechnology and Their Implications," presented at the Seminar on Social and Ethical Issues, Industrial Biotechnology Association, Denver, Colo., June 21-22, 1983.
14. Kalette, D., and Fink, D., "Hopes To End Birth Defects, Cure Cancer," *USA Today,* June 10, 1983.
15. Krimsky, S., Baeck, A., and Bolduc, J., *Municipal and State Recombinant DNA Laws,* prepared for The Boston Neighborhood Network, June 1982.
16. Lawless, E. W., *Technology and Social Shock* (Brunswick, N.J.: Rutgers University Press, 1977).
17. Motulsky, A. G., "Impact of Genetic Manipulation on Society and Medicine," *Science* 219:135-140, 1983.
18. Nichols, G. K., "Technology on Trial: Public Perception in Decisionmaking Related to Science and Technology" (Paris: Organisation for Economic Co-Operation and Development, 1979).
19. Norman, D., "Clerics Urge Ban on Altering Germline Cells," *Science* 220:1360-1361, 1983.
20. Padwa, D., "Let's Clean Up Our Language," *Genetic Engineering News* 2(1):4, 1982.
21. President's Commission for the Study of Ethical Problems in Medicine and Biomedical and Behavioral Research, *Splicing Life: A Report on the Social and Ethical Issues of Genetic Engineering With Human Beings* (Washington, D.C.: U.S. Government Printing Office, 1982).
22. *Science News,* "What is DNA?," 123:366, 1983.
23. Sinsheimer, R., "Troubled Dawn for Genetic Engineering," *Contemporary Issues in Bioethics,* 2d ed., T. L. Beauchamps and L. Walters (eds.) (Belmont, Calif.: Wadsworth, Inc., 1982).
24. Stich, S. P., "The Recombinant DNA Debate," *Contemporary Issues in Bioethics,* 2nd ed., T. L. Beauchamps and L. Walters (eds.) (Belmont, Calif.: Wadsworth, Inc., 1982).
25. Stockton, W., "Making Editorial Decisions for *Science Times," Science and Public Policy II, Ann. N.Y. Acad. Sci.* 387:23-27, 1982.
26. U.S. Congress, House Committee on Science and Technology, *Commercialization of Academic Biomedical Research,* hearings before the Subcommittee on Investigations and Oversight and the Subcommittee on Science, Research and Technology, June 8-9, 1981 (Washington, D.C.: U.S. Government Printing Office, 1982).
27. U.S. Congress, House Committee on Science and Technology, *University-Industry Cooperation in Biotechnology,* hearings before the Subcommittee on Investigations and Oversight and the Subcommittee on Science, Research and Technology, June 16-17, 1982 (Washington, D.C.: U.S. Government Printing Office, 1982).
28. U.S. Congress, Office of Technology Assessment, *The Implications of Cost-Effectiveness Analysis of Medical Technology,* OTA-H-126, Washington, D.C., August 1980.
29. U.S. Congress, Office of Technology Assessment, *Impacts of Applied Genetics: Micro-Organisms, Plants, and Animals,* OTA-HR-132, Washington, D.C., April 1981.

Appendixes

Appendix A

Definitions of Biotechnology

The following is a list of definitions of biotechnology used by the governments and organizations of various countries in assessments of the developing field within their jurisdictions. Most of these definitions encompass both old and new biotechnology.*

Australia

[Biotechnology is] "the devising, optimising, and scaling-up of biochemical and cellular processes for the industrial production of useful compounds and related applications. This definition envisages biotechnology as embracing all aspects of processes of which the central and most characteristic feature is the involvement of biological catalysts" (2).

"In its broadest sense, biotechnology encompasses industrial processes based on biological systems involving naturally occurring micro-organisms, micro-organisms that have been modified by genetic engineering, or isolated cells of plants or animals, and the genetic manipulation of cells to produce new strains of plants or animals" (4).

Canada

[Biotechnology is] "the application of biological organisms, systems, or processes to manufacturing or service industries" (9).

[Biotechnology is] "the utilization of a biological process, be it via microbial, plant or animal cells, or their constituents, to provide goods and services" (11).

European Federation of Biotechnology

[Biotechnology is] "the integrated use of biochemistry, microbiology, and engineering sciences in order to achieve technological (industrial) application of the capabilities of micro-organisms, cultured tissue cells, and parts thereof" (3).

Federal Republic of Germany

"Biotechnology deals with the introduction of biological methods within the framework of technical processes and industrial production. It involves the application of microbiology and biochemistry together with technical chemistry and process engineering" (5).

*The distinction between old and new biotechnology as used in this report is noted in *Chapter 1: Executive Summary.*

France

"Biotechnology consists of the industrial exploitation of the potential of micro-organisms, animal and plant cells, and subcellular fractions derived from them" (6).

International Unions of Pure and Applied Chemistry (1981)

[Biotechnology is] "the application of biochemistry, biology, microbiology, and chemical engineering to industrial processes and products (including here the products in health care, energy, and agriculture) and on the environment" (3).

Japan

[Biotechnology is] "a technology using biological phenomena for copying and manufacturing various kinds of useful substances" (7).

The Netherlands

[Biotechnology is] "the science of the production processes based on the action of micro-organisms and their active components, and of production processes involving the use of cells and tissues from higher organisms. Medical technology, agriculture, and traditional crop breeding are not generally regarded as biotechnology" (10).

Organisation for Economic Co-Operation and Development

Biotechnology consists of "the application of scientific and engineering principles to the processing of materials by biological agents to provide goods and services" (3).

Switzerland

The Swiss Government uses the same definition the European Federation of Biotechnology uses (8). (See definition above.)

United Kingdom

[Biotechnology is] "the application of biological organisms, systems or processes to manufacturing and service industries" (1).

Appendix A references

1. Advisory Council for Applied Research and Development, Advisory Board for the Research Councils, The Royal Society, *Biotechnology: Report of a Joint Working Party* (London: H.M. Stationery Office, March 1980).
2. *Biotechnology for Australia,* Report to the Executive of the Commonwealth Scientific and Industrial Research Organisation by a Special Committee of Review, Sydney, Australia, June 1981.
3. Bull, A. T., Holt, G., and Lilly, M. D., *Biotechnology: International Trends and Perspectives* (Paris: Organisation for Economic Co-Operation and Development, 1982).
4. Commonwealth Scientific and Industrial Research Organisation, *Biotechnology Research and Development,* Sydney, Australia, 1981.
5. Deutsche Gesellschaft für Chemisches Apparatewesen (German Society for Chemical Engineering), "Biotechnology: A Study of Research and Development," contract report prepared for the Federal Ministry for Education and Science and the Federal Ministry for Research and Technology, Frankfurt, 1974.
6. Delegation Generale a la Recherche Scientifique et Technique (General Delegation of Scientific and Technical Research) (1980), cited in M. Vaquin, "Biotechnology in France," contract report prepared for the Office of Technology Assessment, U.S. Congress, December 1982.
7. Japan External Trade Organization, "Research on Biotechnology in Japan," *Japan Industrial and Technological Bulletin* (Tokyo: Ministry of International Trade and Industry, 1982).
8. Jasanoff, S., "Public and Private Sector Activities in Biotechnology: Switzerland," contract report prepared for the Office of Technology Assessment, U.S. Congress, January 1983.
9. Miller, J., Fish, K., Basuk, J., et al., *Biotechnology in Canada: Promises and Concerns* (Ottawa: Science Council of Canada, 1981).
10. Netherlands Study Centre for Technology Trends, *Biotechnology: A Dutch Perspective* (Delft, The Netherlands: Delft University Press, 1981).
11. Task Force on Biotechnology, *Biotechnology: A Development Plan for Canada* (Ottawa: Minister of State for Science and Technology, 1981).

Appendix B
Country Summaries

OTA identified five foreign countries as the major potential competitors of the United States with respect to the commercialization of biotechnology: Japan, the Federal Republic of Germany, the United Kingdom, Switzerland, and France. This appendix summarizes information about those countries presented elsewhere in this report. It also describes the activities in biotechnology of Sweden, the Netherlands, Australia, Israel, Canada, the U.S.S.R., and Brazil.

Japan

INTRODUCTION

The commercialization of biotechnology in Japan is accelerating over a broad range of industries, many of which have extensive experience in bioprocessing. Leading Japan's drive to commercialize biotechnology are large established Japanese companies such as Takeda Pharmaceutical, Shionogi Pharmaceutical, Mitsubishi Chemical, Sumitomo Chemical, Toray Industries, Suntory, and Ajinomoto. The general chemical and petrochemical firms especially are leaning strongly to biotechnology, and some of them are making rapid advances in research and development (R&D) through their efforts to make biotechnology a key technology for the future.

The Japanese Government, which fell behind in starting to form a national support structure, has embarked on building a foundation for R&D and is demonstrating ambitious movement by forming Government and private collaborative projects with the motto "catch up, get ahead" (8). As biotechnology product markets begin to develop, Japan's expertise in the art of bioprocessing will provide Japanese companies with significant competitive strengths.

INDUSTRY

All of the large private sector Japanese companies using biotechnology have come from established industries. In this respect, Japan differs from the United States, where more than 100 new biotechnology firms (NBFs)* have been started specifically to exploit biotechnology.

Japanese companies did not start investing in new biotechnology until after 1980, when publicity spread about its potential applications to the pharmaceutical industry. Since then, led by the promise of interferon and monoclonal antibodies (MAbs) in cancer treatment and the potential of producing unlimited quantities of each through biotechnology, more than 150 Japanese companies have rapidly reorganized their R&D systems, equipped research institutes, and recruited new staff to evaluate the applications of biotechnology. The breakdown by funding sector of Japan's total expenditures for recombinant DNA (rDNA) related R&D for fiscal year 1981 is illustrated in figure B-1.

Japanese pharmaceutical companies, whose penetration of international markets heretofore has been low, show promise of becoming increasingly competitive with the United States in world pharmaceutical markets. The Japanese pharmaceutical market is currently second only to the U.S. market in size. In addition to the pharmaceutical companies, Japanese companies from the food, chemical, textile, and pulp and paper industries have also begun to further exploit their accumulated experience in bioprocessing by diversifying into newly developing pharmaceutical product

Figure B-1.—Breakdown of Japan's Expenditures for Recombinant DNA Technology R&D, Fiscal Year 1981

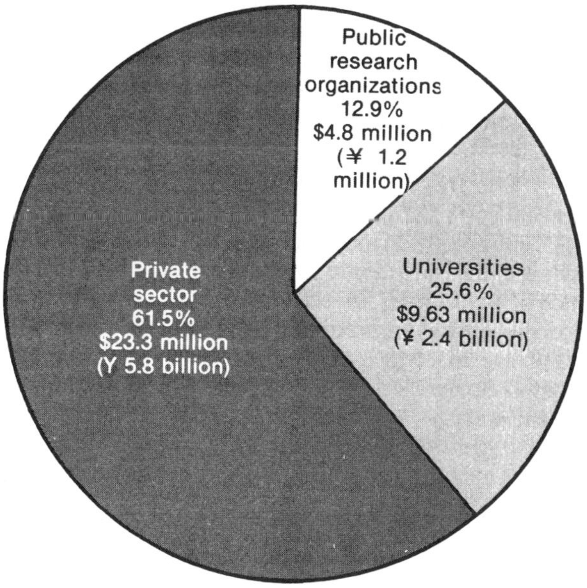

Total rDNA expenditure = $38.1 million (¥ 9.5 billion)

SOURCE: Office of Technology Assessment, based on data from *Science and Technology in Japan*, April/June 1983.

*NBFs, as defined in *Chapter 4: Firms Commercializing Biotechnology*, are firms that have formed specifically to capitalize on developments in biotechnology.

markets.* The field of specialty chemicals will be another highly competitive area of Japanese involvement. Japan is already the dominant international force in amino acid production, and two of the largest amino acid producers, Ajinomoto and Kyowa Hakko Kogyo, have production plants in the United States. Japanese companies' current emphasis on research in specialty chemicals such as enzymes and amino and organic acids reflects efforts to pull the Japanese petrochemical industry out of its present decline in international markets. The urgency of this task is greater in Japan than in the United States, because Japanese petroleum-based industries such as chemicals and textiles are solely dependent on imported petroleum feedstocks. Although some specialty chemicals have traditionally been made by bioprocesses, opportunities for using bioprocesses to make specialty chemicals previously made from petroleum-derived feedstocks have arisen with biotechnology. Producing specialty chemicals using biotechnology offers Japanese companies in these industries an opportunity to reduce their dependence on petroleum and at the same time switch from the production of high-volume, low value-added products to products with higher profit margins.

GOVERNMENT TARGETING POLICIES AND FUNDING OF BASIC AND APPLIED RESEARCH

Within the Japanese Government, a consensus regarding the importance of biotechnology to the future health of the Japanese economy has been achieved. Three Government departments in Japan—the Science and Technology Agency (STA), the Ministry of International Trade and Industry (MITI), and the Ministry of Agriculture, Forestry, and Fisheries (MAFF)—have specifically targeted the development of biotechnology.

STA was the first to demonstrate an interest. As early as April 1971, STA's advisory group, the Science and Technology Council, composed of government, business, and academic leaders, stressed the importance of promoting life science on a nationwide basis because of its commercial potential (4), and STA responded in 1973 by establishing its Office for Life Science Promotion. This office, which is Japan's highest science and technology policymaking body, also manages and coordinates R&D projects in biotechnology. Until the early 1980's, STA's basic, generic applied research,* and applied programs in biotechnology were the largest and best funded Government programs in Japan, and even today STA's programs are comparable in scale to those of MITI (see below). The agency is currently funding corporate generic applied research projects to develop DNA synthesis techniques, bioreactors, immobilized enzyme processes, screening techniques for new micro-organisms, and new medicines.

MITI did not enter the biotechnology field until 1981. That year, MITI established its "System for Promotion of Research on Next-Generation Industrial Technologies," an overall plan to promote "next-generation" industrial technologies, including biotechnology (11). To focus MITI's overall biotechnology effort and to oversee its three next-generation biotechnology projects, an Office of Biotechnology Promotion was established within MITI's Basic Industries Division.

MITI's three next-generation projects in biotechnology— bioreactors, rDNA technology, and mass cell culture—are a part of a 10-year program that is specifically designed to develop and diffuse new biotechnology among Japanese companies.** MITI has invited 14 companies to participate in the projects and will provide allocations over 10 years of $43 million each to the rDNA and bioreactor projects and $17 million to $22 million for the mass cell culture project (2). Some 10 percent of the R&D work (by expenditure) for MITI's biotechnology projects is conducted in the national laboratories of MITI's Agency for Industrial Science and Technology. Ninety percent of the work is conducted in industry laboratories.

To facilitate coordination, the 14 companies that MITI has invited to participate in the biotechnology projects have been organized into the Biotechnology Development Research Association. This association has its own central office through which the various companies communicate with MITI, but otherwise maintains no intercompany institutions or laboratories. MITI subsidies to the companies cover 100 percent of all direct expenses (salaries and laboratory expenses) for biotechnology R&D, but no overhead is allowed, and any capital equipment purchased is nominally the property of the Japanese Government. Furthermore, all patents resulting from the work belong to the Japanese Government. MITI has assured both domestic and foreign companies access to the patents (11).

*The first Japanese companies to enter the field of rDNA-produced pharmaceuticals, Green Cross, Hayashibara, and Suntory, were led by pioneering entrepreneurial managers. For example, the Hayashibara venture into producing interferon with hamsters was possible only because the owner owns or controls 12 companies (hotels, gas stations, and candy manufacturing) and does about $150 million (¥ 37.4 billion) worth of business a year (14). Suntory's (a whiskey company) diversification into rDNA-produced pharmaceuticals is a similar situation.

*Basic, generic applied, and applied research are defined in *Chapter 13: Government Funding of Basic and Applied Research*.

**The Biotechnology Forum, a group of five major Japanese chemical companies that had organized independently after the announcement of the Cohen-Boyer rDNA process patent, was instrumental in lobbying for the establishment of the biotechnology projects.

The third Japanese Government agency that is taking an active role in biotechnology, MAFF, recently established the Committee on Biological Resources Development and Utilization, which compiled a report recommending actions MAFF could take to promote biotechnology development (7). Currently, MAFF is actively promoting cooperative biotechnology research with private industry at its laboratories and is funding work both with Nippon Shokuhin Kako and Oriental Yeast at the National Food Research Institute and with Kao Soap at the National Institutes of Agricultural Sciences. It is also planning cooperative research with Japanese seed companies in the areas of plant breeding and species improvement. Although achievements from the cooperative research are used jointly by Government and industry, these companies that participate in the research projects receive exclusive licensing rights to the patents resulting from these projects for a 3-year period (9). MAFF funding for biotechnology R&D is comparable to that of MITI and STA (11).

In addition to STA, MITI, and MAFF, three other Japanese Government agencies are funding basic and generic applied research in biotechnology: the Ministry of Health and Welfare, the Ministry of Education, and the Environmental Protection Agency. Total Japanese Government funding for biotechnology R&D in 1983 is $67 million (11). Although the level of Japanese funding may be slightly lower than Government funding in both the Federal Republic of Germany and the United Kingdom and is dwarfed by that of the United States, a far greater proportion of Japanese than U.S. funding goes to applied research.

The importance of the Japanese Government's investment in applied research relevant to biotechnology, however, should not be overstated. Of greater importance than the Government's investment in research per se is the Japanese Government's success in encouraging industry's involvement in and long-term commitment to biotechnology. The strength of Japan's biotechnology policy lies in its emphasis on the sensible development of mutually agreed on research strategies, horizontal organization and coordination within the private sector, and timely funding of the necessary high technologies (known in Japan as the "seed corn" policy).

FINANCING AND TAX INCENTIVES FOR FIRMS

Private sector financing in Japanese biotechnology is still mostly indirect and mediated through the Japanese banking system. At present, most Japanese firms using biotechnology are very thinly capitalized. The ratio of debt to equity is still far higher in Japan than it is in the United States. As far as can be determined, however, the financing of R&D efforts is not a major problem for the large companies in Japanese biotechnology. The Japanese companies involved in biotechnology R&D have either their own internal sources of funds or close relations with the banks (11).

Certain weaknesses in Japan's financial system have been especially evident in biotechnology. Despite many changes in recent years, capital remains heavily concentrated in the Japanese banking system, and stock markets play a relatively small role in allocating capital. Only 111 Japanese companies currently have their securities traded over the counter, and total venture capital investments amount to no more than $84 million (1).* Mostly because of the lack of venture capital and the cultural factors inhibiting risk-taking entrepreneurialism, Japan does not have a large class of startup companies that specialize in biotechnology R&D such as that found in the United States.

Japan's private sector has recently taken some initiative in developing a source of "venture capital" by pooling corporate resources. The Japan Associated Finance Corp. (JAFCO) is a private venture capital fund that was organized by Nomura Securities Co. One French, three Hong Kong, and 10 Japanese firms are involved in JAFCO, which plans to offer financial help to new businesses until they qualify for listing as a joint stock company. When the firm reaches this stage of maturity, its income gains will be distributed among the partners of the fund according to the ratio of the capital contribution to the fund (3). These new sources of venture capital may or may not succeed in increasing the supply of venture capital in Japan. In any case, the amount of venture capital these sources currently provide is very small when compared to the amount available in the United States.

The Japanese Government is interested in changing the country's financial system. In 1982, MITI set up a new Office of Venture Enterprise Promotion in parallel with the creation of the Office of Biotechnology Promotion (6). In fiscal year 1981, a Government-related organization called the Center for Promoting R&D Type Corporations guaranteed approximately $3.7 million (¥ 750 million) in loans (a total of 24 loans), and beginning in 1982, this center began making its own loans as well as guaranteeing other lender's loans. In an equally significant development, MITI and the Ministry of Finance (MOF) have recently begun discussing an "automated over-the-counter share transaction system" to make it easier for enterprising small and medium-sized firms that lack business experience to raise funds in the finance market. Currently, MOF's evaluation standards are so strict from the standpoint of protecting investors that venture businesses find

*Institutions such as Japan Godo Finance, Sogo Finance, and Universal Finance Corp. are viewed as nascent venture capital companies.

it difficult to have their shares sold when they want to go public.

In the past, Government-funded banks like the Japan Development Bank (JDB) have played a key role in providing large amounts of low interest loans to heavy industries. Certain funds within the JDB loan portfolio are targeted for "technology promotion," and loans from the fund are made at interest rates between 7.5 and 8.4 percent. Currently, however, these funds are not being channelled into biotechnology (11).

Japan's corporate tax code exhibits a uniformity across industrial sectors that is not evident in the United States. Furthermore, corporate taxes are generally lower in Japan than they are in the United States (13). A number of Japanese tax code provisions are aimed at benefiting R&D activity and technological innovation across the board.

One Japanese tax break of particular relevance to the development of biotechnology is the special depreciation schedule used for companies that are members of a MITI-approved National Research Association (e.g., the Biotechnology Development Research Association). Such companies can take an immediate 100-percent depreciation deduction on all fixed assets used in connection with their research association activities. Because of the decentralized character of most National Research Association R&D—90 percent of it is performed separately in corporate laboratories—the tax writeoffs directly encourage R&D activity within corporate laboratories.

Figure B-2.—Breakdown of Japanese Personnel Engaged in Recombinant DNA Technology R&D by Type of Research Organization, Fiscal Year 1981

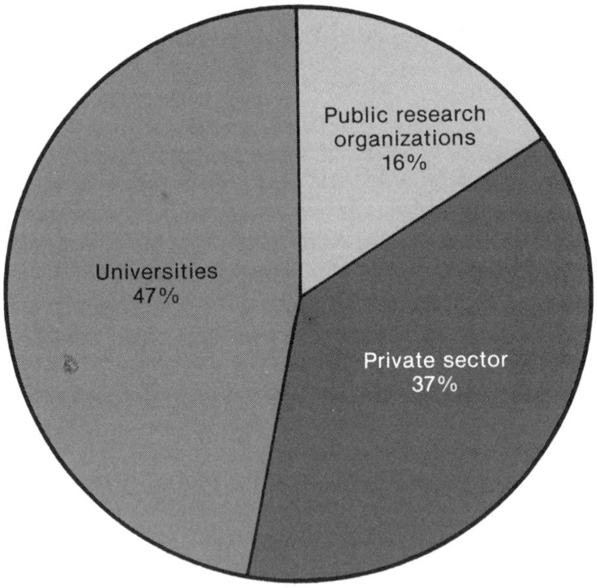

Total 1,353 researchers

SOURCE: Office of Technology Assessment, based on data from *Science and Technology in Japan*, April/June 1983.

PERSONNEL AVAILABILITY AND TRAINING

Since World War II, the training of industrial microbiologists and bioprocess engineers has been encouraged by both Government and industry funding in Japan, and as a result, a steady supply of these personnel has been maintained. In fact, Japan is considered the world leader in this area. On the other hand, largely because of its weak basic biological science research base, Japan is experiencing a shortage of molecular biologists and immunologists. Some Japanese companies have addressed this problem by sending some of their personnel to the United States for training in molecular biology. Other companies have had success in repatriating Japanese workers already trained overseas. Figure B-2 gives a breakdown of Japanese personnel engaged in rDNA R&D by type of research organization.

Retraining of corporate workers in biotechnology is being pursued actively in Japan. In Japan, more than in any other industrialized country, worker training is the responsibility of the corporation. Japan's ability to adjust rapidly to weaknesses in its labor force, based primarily on the Japanese corporations' funding of worker retraining, is truly extraordinary. In 1981, for example, no more than 10 private Japanese companies had more than 10 researchers working on rDNA technology; a year later, surveys revealed that 52 out of the 60 leading companies surveyed had obtained 10 or more research workers in that area (11).

UNIVERSITY/INDUSTRY RELATIONSHIPS AND DOMESTIC TECHNOLOGY TRANSFER

In applied research areas such as bioprocessing and microbiology, Japanese university/industry relations and the transfer of information from universities to industry are generally very good. In basic research areas, however, the transfer of information from universities to industry is impeded by the fact that almost all university rDNA and hybridoma research in Japan takes place in "basic" science departments, and these departments pride themselves on independence from industrial influence. The Japanese Government has launched new programs designed to cross the barriers between university basic science departments and industry, but their future success is questionable (11).

The movement of knowledge across industrial sectors in Japan is facilitated by the unique "keiretsu" structure (a group of companies with historical ties, which usually consists of a company from each industrial sector and a bank or trading company which plays a dominant role by virtue of its contact with other companies within the group). The transfer of information among companies within sectors, however, is inhibited by extreme secrecy and a lack of mobility of personnel from one company to another. MITI's "next generation" projects in biotechnology are designed in part to compensate for this problem and to diffuse knowledge among companies using biotechnology. In part because they suspect they would have to sacrifice proprietary positions in some commercially important research areas, however, some Japanese companies have not joined the MITI projects in the areas in which they have comparative advantages (11). For example, Kyowa Hakko, a leader in work on rDNA, is not participating in the "next generation" project in this area.

OTHER FACTORS

Historically, Japan's guidelines for rDNA research have been among the most restrictive in the world. Although the guidelines have recently been relaxed somewhat, they are still quite restrictive. Japanese companies have mounted intensive lobbying efforts to get the guidelines changed. Although companies have had extreme difficulty in obtaining approval to do work with more than 20 liters of culture, this situation is expected to change soon.

Although estimates are difficult to obtain, the cost of gaining approval for new pharmaceuticals is believed to be lower in Japan than the United States. In Japan, the cost of obtaining approval for a new drug is about $12 million to $20 million (¥ 3 billion to ¥ 5 billion), compared to about $87 million in the United States. The time required for drug development and approval is similar (about 10 years) in both the United States and Japan (5).

The basic law governing worker health and safety in Japan is the Industrial Safety and Health Law. This law imposes on employers the obligation of preventing health impairment caused by substances and conditions found in the workplace. Substantial criminal penalties and fines are imposed for violations. At the present time, no regulations are addressed specifically to biotechnology. Furthermore, specific measures governing environmental effects of biotechnology applications have not been prepared by the Japanese Government.

Because the United States is considered a world leader in the commercial applications of biotechnology, Japanese companies have been actively importing technology from the United States and other countries through R&D joint ventures and licensing agreements. NBFs in the United States in need of financial support widely accept research contracts from Japanese companies, often because U.S. partners cannot be found.

An issue brought up in recent U.S.-Japan trade negotiations was U.S. access to the technologies developed by the MITI-sponsored National Research Associations. MITI has promised to abandon its past policy and disclose the patents obtained in National Research Associations to foreign firms. MITI is also promising membership in National Research Associations to U.S. companies that have Japanese subsidiaries or substantial technological expertise.

Japan is engaged in international efforts to secure sources of biomass* in the event that biomass becomes the favored route to meeting energy needs. In cooperation with developing countries (mostly Asian), Japan is organizing biomass centers. This foresight may operate to Japan's advantage in the future.

Nontariff trade barriers in Japan, especially in the area of pharmaceuticals, may hinder U.S. companies' penetration of Japanese markets. The Japanese Ministry of Health and Welfare has not yet begun to accept clinical test data from the United States, although as of April 1983, Japan did begin accepting foreign test data on animals. Foreign stability test data and data on specifications and test methods will be accepted from October 1983 onward (10).

Unlike the United States, Japan has constraints inhibiting foreign acquisition of domestic companies. Foreign acquisitions in Japan require the unanimous approval of the Japanese company's board of directors and also the approval of MOF. Recently, however, the regulation surrounding the establishment of foreign subsidiaries in Japan has noticeably eased; large numbers of European pharmaceutical companies have established wholly owned subsidiaries in Japan during the past year. The case of foreign acquisition of domestic companies in the United States is an important issue to consider, because Japanese companies very often acquire foreign companies to gain access to their technology, markets, and distribution networks.

CONCLUSIONS

Because of its present competitive strength in biologically produced specialty chemicals, Japan can be expected to be a major competitor in future specialty chemical markets defined by biotechnology. The fu-

*Biomass, discussed further in *Chapter 9: Commodity Chemicals and Energy Production*, is all organic matter that grows by the photosynthetic conversion of solar energy.

ture competitive position of Japanese companies in future pharmaceutical markets is more difficult to assess. Japanese companies traditionally have not had a significant presence in world pharmaceutical markets, but Government promotion of the pharmaceutical industry, rising investments in pharmaceutical R&D (including related biotechnology applications), and increased competition in the domestic pharmaceutical market all portend a greater role for Japanese companies in future international markets.*

Federal Republic of Germany

INTRODUCTION

A powerful private sector, a well-developed administrative infrastructure, an extensive research base, a generous funding program, and an adequate supply of personnel all contribute to the potential of the Federal Republic of Germany to compete with the United States and other industrialized countries in biotechnology. The overall West German effort does have certain deficiencies (e.g., an inflexible research grants system), however, and the ability to correct them will be a factor that influences the country's competitive position.

The ability to correct these deficiencies, however, will not by itself guarantee competitive success. Politics, for example, and its most powerful ally, public perception, could influence the course of biotechnology development more immediately in the Federal Republic of Germany than in any other country. The West German environmentalists, embodied in the political party of the Greens, have yet to focus their attention on risks specifically associated with biotechnology, but the leading German companies using biotechnology have already aroused public protest as major chemical polluters. The Greens, now incorporated in the Federal parliamentary process, represent a potential threat, especially in the event of a mishap, to the progress of biotechnology in the Federal Republic of Germany (24).

INDUSTRY

The Federal Republic of Germany's competitive position in biotechnology will be determined by the ability of large, established West German companies to develop and market biotechnologically produced goods and services. Responsibility for most of the development of the country's industrial capabilities in biotechnology to date rests largely with chemical companies such as Hoechst, Bayer, and BASF, three of the four largest in the world, and with the slightly smaller pharmaceutical companies such as Boehringer and Schering. Small and medium-sized West German companies have played no significant role in biotechnology innovation, despite the West German Government's efforts to encourage this through the provision, for example, of startup funding for high-risk undertakings (24).

To speed the transition to new biotechnological techniques and processes, the large West German companies that are developing biotechnology have sought outside expertise. Hoechst, for example, signed a 10-year, $70 million contract with Massachusetts General Hospital to support work in molecular biology (18). Hoechst, criticized in Germany for a breach of faith with national science and in the United States for the appropriation of U.S. technology, apparently entered into this agreement with the objectives of getting a "window on the technology" and gaining access to a large, state-of-the-art laboratory in which to train its scientists (18).

GOVERNMENT TARGETING POLICIES

A government policy for the commercialization of biotechnology rates as one of the Federal Republic of Germany's strengths. According to a 1979 statement by the Federal Ministry for Research and Technology (BMFT, Bundesministerium für Forschung und Technologie), the German Government has an obligation to establish the preconditions for industrial innovation in key areas of technology in order to strengthen the competitive performance and competitive capacity of the German economy in long-range growth areas, and in the process, correct weaknesses revealed through international comparisons (24).

The present biotechnology targeting policy has evolved from the West German Government's historical interest in the life sciences. In 1972, BMFT commissioned a report on old biotechnology from the German Society of Chemical Engineering (Deutsche Gesellschaft für Chemisches Apparatewesen) (19), and in 1979, BMFT presented its first official policy specifically for biotechnology (16). This "performance plan" (Leistungsplan) outlined biotechnology research programs with specific objectives, such as the development of unconventional feed and foodstuffs, bioinsecticides, and pharmaceuticals from plant cell cultures. BMFT's more recent statements continue to promote the development of specific product areas (e.g., pharmaceuticals, plant agriculture) and particular proc-

*For example, in 1981, Japanese companies ranked first in terms of the largest number of major new drugs introduced into world markets. In 1982, not only did Japanese companies account for over 16 percent of all U.S. patents issued for pharmaceutical and medicinal products, but 38 percent of all U.S. medicinal patents granted to foreign firms went to Japanese originators. See *Chapter 4: Firms Commercializing Biotechnology* for a more detailed description of Japanese pharmaceutical activity.

esses (e.g., cell culture), but they also focus attention on the importance of basic research and the need for greater interdisciplinary cooperation between biologists, chemists, medical experts, and engineers, disciplinary areas which are important to the development of biotechnology (24).

BMFT implements its policy primarily through a strong and varied funding program. Types of BMFT support fall into three broad categories: 1) funds specifically set aside for the development of biotechnology, 2) grants that fall into already existing schemes for industrial development work, and 3) funds distributed by third-party organizations to which BMFT contributes as part of more generalized funding programs for all areas of public research. For its own biotechnology program alone, BMFT in 1982 spent $29 million (DM70 million), up $5 million (DM12 million) from 1981. In 1981, BMFT also contributed to the German Research Society (DFG, Deutsche Forschungsgemeinschaft) (25) and to the Max Planck Society (Max Planck Gesellschaft) (15). It is impossible to calculate the exact proportion of these other funds dedicated to biotechnology research, but a reasonable estimate might range from $20 million to $40 million (DM50 million to DM100 million). Since data are unavailable to support this estimate, a total BMFT biotechnology funding figure of $50 million to $70 million (DM120 million to DM170 million) for 1982 should be regarded with caution.

GOVERNMENT FUNDING OF BASIC AND APPLIED RESEARCH

The Federal Republic of Germany maintains an extensive public research base. Both basic and generic applied research are generally good. Three different types of nonindustry laboratories conduct basic research in biotechnology: 1) laboratories belonging to the universities, 2) laboratories dependent on BMFT for operating expenses and on DFG for project support, and 3) laboratories supported by the Max Planck Society (which, in turn receives support from BMFT).

The operating costs of the universities are supported by the individual States (Länder). Highly publicized deficiencies in German university research have resulted from budget cuts and university reform laws. With the current shortage of funds, grant allocations go to tenured professors (27) and to replace used equipment, not to the young researchers (29). University reform laws have created excessive administrative duties for university professors, making it difficult for them to dedicate sufficient time to their research (20). Despite such problems, however, universities such as those at Heidelberg, Munich, and Cologne continue to conduct research fundamental to the development of biotechnology (21).

Although laboratories supported jointly by BMFT and DFG, such as the Cancer Research Center at Heidelberg, carry out important biotechnology-related work, laboratories funded by the Max Planck Society are responsible for the bulk of the basic research advances in biotechnology. The Max Planck Institute for Plant Breeding Research in Cologne boasts some of the best plant genetics teams in the world (24). Other leading Max Planck institutes working in basic research related to biotechnology include those in biochemistry at Martinsried, biology and virus research in Tubingen, genetics in Berlin, and cell biology in Ladenburg (21).

Some of the Max Planck institutes conduct generic applied biotechnology research, but the center for such research is the Society for Biotechnological Research (GBF, Gesellschaft für Biotechnologische Forschung). GBF is a Government-supported though private institution that was originally founded to conduct generic bioprocessing research to meet the needs of industries (26). GBF employs 365 people (249 permanent and 116 temporary), and its 1982 budget was $13 million (DM31.6 million), of which 89 percent came from BMFT, 9 percent from the Länder, and 2 percent from its own earnings (Gesellschaft für Biotechnologische Forschung, 1982). GBF's current activities include the general development of bioprocess technology, the scale-up of laboratory processes, the screening of micro-organisms and plant and animal cell cultures, the support of other research groups in biotechnology, the participation in joint biotechnology projects with industry, and the advanced interdisciplinary training for scientists, engineers, and technicians. GBF suffers from the usual rigidity of a large German research organization—funds, once allocated, cannot be shifted from one area of research to another. Nevertheless this well-equipped and well-staffed Government-supported applied research facility in West Germany is one of Europe's best

FINANCING AND TAX INCENTIVES FOR FIRMS

There is no parallel in the Federal Republic of Germany to the U.S. venture capital industry. The powerful and rather rigid banking structure in the Federal Republic of Germany virtually inhibits the formation of venture capital, though there is apparently little demand for it (24). Commercial banks provide most of the funds used for industrial expansion, and it is common for such banks to have equity participation in companies in which they invest. The commercial banking sector is dominated by three banks, and the linkages between the banking and corporate structures are so close that the Monopoly Commission in 1976 concluded that the banks effectively utilize management functions to the detriment of competition (24).

In 1975, a consortium of 28 banks recognized that the German banking system was not conducive to funding high-risk innovative, startup firms and formed a venture capital concern called the Risk Financing Society (WFG, Deutsche Wagnisfinanzierungs-Gesellschaft) (17). The principal objective of this organization was to aid small and medium-sized firms in commercializing their products. So far, however, this concern has not shown much interest in biotechnology companies, a major reason being that since 1980 it has been looking for innovations that could achieve success within 24 months. If this continues to be the criterion for a firm to receive funds from WFG, it would be surprising if many biotechnology startup firms were established in the Federal Republic of Germany with WFG funds.

Tax incentives are a less important source of financing for private sector innovation in the Federal Republic of Germany than direct Government subsidies. This country maintains the highest nominal corporate tax rate of the six countries analyzed in this report (56 percent on retained earnings and 36 percent on distributed earnings). Measures such as an investment grant provision allowing a company to recover up to 20 percent of the cost of R&D capital expenditures contribute to lower the effective tax rate, although the United Kingdom, Switzerland, and Japan still have the lowest effective tax rates of the competitor countries.

PERSONNEL AVAILABILITY AND TRAINING

The Federal Republic of Germany has sufficient personnel to compete with the United States and other competitor countries in biotechnology. Molecular biologists with expertise in rDNA and hybridoma research are in short supply, but the training of such specialists is now a high priority (24). Like Japan, the Federal Republic of Germany maintained a steady supply of both industrial and government funding for applied microbiology and bioprocess engineering after World War II. Thus, the supply of personnel in these areas appears to be adequate.

The Max Planck Society's senate and the present Minister of Research and Technology have indicated that there is a significant drain of German researchers from the Federal Republic of Germany to the United States (21,28). The "brain drain" of scientists from West Germany, however, appears to be less serious than that from the United Kingdom (see below).

UNIVERSITY/INDUSTRY RELATIONSHIPS AND DOMESTIC TECHNOLOGY TRANSFER

The Federal and State Governments and the private sector in the Federal Republic of Germany use several mechanisms to accomplish the transfer of technology developed in public research laboratories into domestic industries. The Max Planck Institute for Plant Breeding Research and GBF both have several contract arrangements with private companies. On a much larger scale, the pharmaceutical company Schering joined with the State of Berlin and its two universities to establish a biotechnology research institute (Biotechnikum). Though the institute will undertake primarily basic research in rDNA technology, it will also support industrial microbiology research and the production of hormones and amino acids (22). Bayer, BASF, and Hoechst have also established cooperative research programs with West German universities and other research institutes.

OTHER FACTORS

In general, the West German regulatory environment is comparable to that in the United States and poses no additional barriers to the commercial development of biotechnology for either domestic or foreign firms. Guidelines for rDNA research, food and drug testing regulations, intellectual property law, and international trade laws in West Germany are approximately equivalent to those in the rest of the competitor countries.

CONCLUSIONS

The Federal Republic of Germany could become one of the principal competitors of the United States in the commercialization of biotechnology. West Germany's extensive research base would be one of the most well-balanced in the world, were it not for the funding and administrative problems in the universities and the resulting effects on the quality of research. Another problem is that the Government bureaucracy for implementing biotechnology policy is somewhat inflexible. Once funding has been granted for specific projects, money cannot be shifted to other potentially more promising studies. One of the Federal Republic of Germany's strengths, however, is the country's private sector. The size and international market penetration of established German chemical and pharmaceutical companies suggests that these companies are likely to be competitive in the commercial use of biotechnology.

United Kingdom

INTRODUCTION

In many respects, the United Kingdom has the capabilities to compete in biotechnology on an equal basis with the United States, Japan, and the Federal

Republic of Germany. Government initiatives, national science and technology resources, both human and material, and efforts by a few individual companies to commercialize biotechnology place the United Kingdom on a par with these other competitor countries. A relative lack of experience in joint government, industry, and public research cooperation compared to the United States and, with some exceptions, a generally risk-averse private sector, however, could become obstacles to the smooth development of biotechnology in the United Kingdom.

INDUSTRY

A number of NBFs have been started to commercialize biotechnology in the United Kingdom. These include Celltech, Agricultural Genetics, Plant Sciences, Imperial Biotechnology, IQ (Bio), and other companies that were founded specifically to exploit results of basic research in biotechnology-related disciplines. Although the United Kingdom has more NBFs than do other European countries or Japan, the importance of NBFs to the commercialization of biotechnology in the United Kingdom does not generally rival that of their U.S. counterparts. The 1983 marketing by Celltech of MAbs to detect and isolate interferon (34) and of two blood-typing kits using MAbs (47), however, demonstrates a certain dynamism within the United Kingdom's NBF sector.

The large established U.K. companies such as ICI, Burroughs-Wellcome, G. D. Searle, Unilever, Glaxo, and others will play the major role in determining the United Kingdom's competitiveness in the commercialization of biotechnology. These companies, like established companies in the other competitor countries, are better equipped than the NBFs to absorb the high costs of large-scale production, health and safety testing, and marketing, in fields such as pharmaceuticals, food, or agriculture. Although they appear to be investing large sums in biotechnology R&D (44), it remains to be seen whether established companies in the United Kingdom can generate the same level of innovation from in-house research and arrangements with universities as the NBFs in the United States.

GOVERNMENT TARGETING POLICIES

Until recently, many analysts in the United Kingdom believed that biotechnology products would reach markets only after 10 to 20 years (36) and that the British Government should maintain its traditional functions with respect to developing technologies, i.e., limit itself to supporting basic R&D, training qualified personnel, and creating a propitious climate for industry to capitalize on discoveries made in public research facilities (35).

In 1980, a Government committee published a report that identified weaknesses in the development of biotechnology and recommended that the Government take specific corrective actions to assist the transfer of the results of public sector research to industry and to expand existing programs supporting training, research, and innovation (30). The British Government has responded to this report, commonly known as the Spinks' Report, by increasing funds both for the British Technology Group (BTG) for investment in innovative private sector projects in biotechnology and for the Research Councils and Government departments for the support of basic life science research.

In 1981, the British Government, through BTG and in association with four private investors, established Celltech, Ltd., to develop and market products made by some of the new technologies. In 1982, the Department of Industry launched a new 3-year, $30 million program of support for biotechnology in industry (31). The Government has also encouraged the creation of university centers of expertise in biotechnology to bring together experts in different disciplines within a single field and has established a Biotechnology Directorate at the Science and Engineering Research Council (SERC) to coordinate biotechnology R&D in all public sector research laboratories.

GOVERNMENT FUNDING OF BASIC AND APPLIED RESEARCH

The United Kingdom has a strong and well-established basic research base. The Research Councils and the universities possess considerable depth in basic research fields such as immunology and plant genetics. Although the economic recession has forced cuts in both university and Research Council grants (46), the Government has attempted to protect the basic science research budget and to redirect resources within this budget to priority areas such as biotechnology. Research Council funds for biotechnology have actually increased. University funds have been reduced in some areas, but the Government has encouraged universities to protect basic research, and the University Grants Committee has been funding the establishment of new posts at many different universities (37).

Generic applied research in biotechnology has been receiving strong support in the United Kingdom. The British Government sponsors generic applied research at a number of locations, including the Centre for Applied Microbiology Research in Porton Down (bioprocess engineering); Warren Spring Laboratory in Harwell (downstream processing); and the Biotechnology Institute and Studies Centre Trust (enzymes). These and other programs all contribute to make development a strength of the Government's support for biotechnology.

Definitional problems make it difficult to arrive at a figure for overall Government expenditures for biotechnology R&D. Though the British Government uses the Spinks' Report definition,* research institutes tend to classify work in scientific terms such as rDNA technology, hybridoma technology, and others. A conservative estimate of biotechnology funding for all phases of R&D would fall between $56 million and $60 million for 1982 (46), though the Government expects to increase this level substantially during 1983. The 1982 figure roughly equals spending in Japan, the Federal Republic of Germany, and France.

FINANCING AND TAX INCENTIVES FOR FIRMS

Views on whether there is a shortage of funds available for biotechnology firms in the United Kingdom vary depending on the source of information. Financial institutions say funds are not in short supply; rather, the shortage is in well-presented ideas with commercial value that are capable of earning the relatively high rates of return desired by investors with risk capital. Entrepreneurs say that there is a shortage of funds, because institutions demand more evidence than they can supply to prove that their products are capable of earning high profits.

Funds for the industrial development of biotechnology, especially for NBFs, are available from both public and private sources. The major public source of venture capital is BTG (see above). Private venture capital groups with either investments or plans to invest include Biotechnology Investments, Prutec, Advent Eurofund, Cogent, Technical Development Capital, and others. Of these, Biotechnology Investments, a branch of N. M. Rothschild Asset Management, is the largest, with an initial capital pool of $55 million (39). Most of the fund's investments to date have been in U.S. NBFs and in primarily foreign quoted companies (39), although the company recently purchased equity in Celltech (33) and is now considering more project proposals from British firms than from U.S. companies (43). Other sources of capital for NBFs include banks and other financial institutions, whose project loans are guaranteed by the Government, and the Unlisted Securities Market, for companies with profits of less than $1 million.

Tax law in the United Kingdom tends to favor established companies with programs in or plans to implement biotechnology R&D rather than NBFs. Most of the Government's tax incentives apply to companies earning taxable income (i.e., the large established companies) and are used primarily to encourage additional expenditures on R&D or on plants and equipment required for research or scale-up. The tax code allows the largest and most rapid depreciation allowance of capital expenditures for scientific research of all the competitor countries (100 percent in the first year of use). This provision contributes to making the effective corporate tax rate in the United Kingdom among the lowest of the countries analyzed by this report.

Few of the tax incentives in the United Kingdom, on the other hand, encourage the formation of capital, a necessary precondition for starting an NBF. Both the taxation of long-term capital gains (30 percent) and of income resulting from the sale of technology (in the form of patent sales or licensing royalties) are the most unfavorable of the competitor countries. The British Government recently introduced new measures designed to encourage the private sector to make equity investments in startup firms by offering tax relief at the top marginal rate to investors in new (up to 5 years old) qualifying trades, but the effect of this policy remains to be seen.

PERSONNEL AVAILABILITY AND TRAINING

Like the United States, the United Kingdom boasts both qualified personnel and excellent training and education programs for personnel in the basic life sciences. Personnel supported by the Medical Research Council are internationally prominent in the development of rDNA and hybridoma technologies.* Also like the United States, the United Kingdom is experiencing personnel shortages in areas related to scale-up. The shortage in the United Kingdom in part results from the fact that very limited opportunities in British universities have led some scientists to leave their posts in academia for positions in foreign biotechnology companies. Approximately 70 Ph. D.s have left the United Kingdom in the past few years. Slightly less than two-thirds of these scientists have come to the United States, though some of them may not be working exclusively in biotechnology. About 30 of the 70 have joined commercial enterprises (13 now work at Biogen S.A. in Switzerland). This "brain-drain" also affects another class of professionals, i.e., individuals skilled in applying the new technologies such as bioprocess and chemical engineers and masters-level microbiologists. Analysts estimate that a total of between 100 and 1,500 experts in some aspect of biotechnology have left the United Kingdom over the past several years (45).

The effects of this outflow on the overall British effort are difficult to determine; no one really knows

*This and other definitions of biotechnology are presented in *Appendix A: Definitions of Biotechnology*.

*British researchers Georges Köhler and Cesar Milstein at the Medical Research Council were the first to develop hybridomas.

whether the United Kingdom may be losing visionaries as well as scientists or whether 100 people represent a significant portion of the available specialized personnel in the United Kingdom (41). In an effort to correct a situation which often obliges some younger researchers and engineers to emigrate, the British Government has recently launched a program to make room for "new blood" in the life sciences in the universities. The creation of these new positions will raise the number of lecturers and create new openings for postdoctoral research and postgraduate courses. In addition, SERC maintains a list of British biotechnologists outside the United Kingdom and may be taking measures to encourage them to return (45).

UNIVERSITY/INDUSTRY RELATIONSHIPS AND DOMESTIC TECHNOLOGY TRANSFER

The universities in the United Kingdom have had very few ties with industry in biotechnology. As a result, the transfer of technology from public research to the industrial sector in the United Kingdom has not always been effectively accomplished. In 1975, for example, the Government failed to patent Kohler and Milstein's technique for making hybridomas, the specialized cells which produce MAbs, and the Americans were the first to recognize the commercial potential of MAbs (40).

With the growth of biotechnology and of public support for these technologies, however, the British Government has taken steps to encourage the process of domestic technology transfer. BTG, which encourages cooperative projects between industry and public sector research and serves as a public source of venture capital, has committed $21 million in support for biotechnology projects so far, with $6.5 million annual increases expected for the next few years (44). In addition, the Department of Industry launched in 1982 a new, 3-year, $30 million "Biotechnology in Industry" program, independent of BTG's activities. Directed by the Laboratory of the Government Chemist, this program sets aside funds for consultancies and project feasibility studies, supports demonstration plant construction, and sponsors joint industry-research centers (31). SERC has initiated several collaborative research programs and promoted, for example, the Leicester Biocentre. The British Government's establishment of NBFs such as Celltech and Agricultural Genetics Co. in association with private investors and BTG's loss of the rights of first refusal* on inventions in public research (32) may help stimulate direct relationships between researchers and industrialists.

*This is the right to choose whether or not to produce and market any good or service, without having to bid competitively with other firms.

OTHER FACTORS

The regulatory environment in the United Kingdom poses little threat to the development of biotechnology in that country. Approval for the marketing of a new drug in the United Kingdom, for example, occurs twice as quickly as in the United States (46).*

The public body that has been responsible for setting and enforcing the United Kingdom's guidelines for rDNA research is the Genetic Manipulation Advisory Group (GMAG). GMAG's status was recently reviewed by the Health and Safety Executive, and the subsequent report recommended the relocation of the group from the Department of Education and Science to the Department of Health and Social Security (42). GMAG, now called the Health and Safety Commission Advisory Committee on Genetic Manipulation, has been moved to the Department of Health and Social Security and will advise the Health and Safety Commission and Executive on general questions, giving advice, when requested, to Government departments. This change in status of the old GMAG reflects a belief by the Government that those responsible for agriculture, environment, and industry need the committee's advice now more than those in charge of education and science (44). Only in exceptional instances will the Advisory Committee on Genetic Manipulation actually review project proposals. The burden of this task has been passed on to Government officials (42).

British patent law in general conforms to European standards. The lack of case law specific to biotechnology inventions, however, precludes an assessment of whether certain patents that are issued in the United States would receive the same treatment in the United Kingdom. Antitrust laws are approximately equivalent to U.S. statutes.

CONCLUSIONS

The United Kingdom could be a major competitor of the United States in specific product markets in biotechnology. The country's strong basic and generic applied research base, the British Government's strong interest in direct measures to stimulate the commercial development of biotechnology, the excellent university system, and the relatively positive regulatory environment all contribute to allow domestic industries a competitive foothold in biotechnology. The future of commercial biotechnology will be decided in part by the speed, content, and scale both of political decisionmaking with respect to biotechnology and of industrial commitment to developing the technologies.

Although the number of NBFs has grown in the United Kingdom because of an increasingly positive

*For further discussion, see (38).

public attitude toward high technology in general, the development of high-technology fields in the United Kingdom may lack some of the dynamism of similar enterprises in the United States. The causes of what appears to be a lack of entrepreneurialism fall outside the scope of conventional modes of analysis and may be due in part to cultural factors which defy measurement.

The ability of all interested parties to adopt recent Government measures to encourage technology transfer from public institutions to industry and to solve other problems will, to a large extent, determine whether the country can challenge the United States, Japan, and West Germany in this new set of technologies. The United Kingdom's affinity with the United States and longstanding commercial ties to the Pacific Basin could very well be assets.

Switzerland

INTRODUCTION

Switzerland reveals an impressive national commercial potential in the area of biotechnology. It has a good university system and several renowned research institutions. A strong financial sector and a technology-based, export-oriented economy also contribute to Switzerland's potential competitiveness in biotechnology. Swiss companies produce 10 percent of the world's pharmaceuticals (53), and, by reinvesting large proportions of sales revenues in R&D, they achieve high rates of innovation essential to competitive success.

Switzerland is organized as a federation of 26 relatively autonomous regions (cantons), and a liberal economic tradition constrains the Federal Government's role in industrial policymaking. Consequently, the Swiss Federal Government has not developed a central policy for biotechnology. A number of steps have been taken to promote innovation through Government loans to highly focused, small-scale projects, but these have not been focused on biotechnology (53). In fact, in 1982, a proposal to establish a national research program specifically for biotechnology under the auspices of the Swiss National Science Foundation (Schweizerischer Nationalfund zur Forderung der Wissenschaftlichen Forschung) was voted down by this organization.

INDUSTRY

Private sector biotechnology R&D in Switzerland is concentrated among three large pharmaceutical companies (Ciba-Geigy, Hoffmann-La Roche, and Sandoz), an NBF (Biogen S.A.*), and, to a lesser extent, several companies involved with bioprocess engineering and biomass conversion for producing chemicals and for energy production (Bioengineering AG, Chemap AG (now owned by Alfa Laval), Petrotec Holding Co. AG, and Batelle Geneva Research Center (U.S. owned).

All of the three large Swiss pharmaceutical companies spend a substantial portion of their R&D expenditures abroad. Ciba-Geigy has made the greatest in-house commitment to biotechnology R&D by improving current production lines such as antibiotics with genetic manipulation. Ciba-Geigy's commitment to the development of biotechnology can be seen in its new $19.5 million biotechnology research center employing 150 people and in its extensive program of support for research in local universities and its own institute laboratories (53). Ciba-Geigy, spent about 8 percent of its 1981 total sales of $1.8 billion (SFr 3.8 billion), on overall R&D. Of this amount, almost 60 percent was spent within Switzerland, while expenditures in U.S. facilities comprised 23 percent and those in the rest of Europe and Asia accounted for 20 percent of the total outlays (49).

In comparison with Ciba-Geigy, Hoffmann-La Roche and Sandoz look more toward the United States for developing biotechnology expertise through contracts and R&D subsidiaries. Hoffmann-La Roche, in conducting biotechnology R&D in its research institutes throughout the world (especially New Jersey) and forming partnerships with NBFs in the United States, spent $59 million on biotechnology R&D in 1981 (50). Approximately one-third of Hoffmann-La Roche's biotechnology R&D budget goes to rDNA experiments (48). Similarly, Sandoz pursues biotechnology through a half-million dollar contract with the Wistar Institute (Philadelphia), a contract with NPI (Salt Lake City), a $5 million investment in the Genetics Institute (Boston), and the purchase of Zoecon (Palo Alto), in addition to research conducted in its Austrian institutes. Though only $5 million of the $226 million Sandoz R&D budget has been spent on biotechnology since 1977, biotechnology will account for an increasing share in the future (48). For example, a biotechnology research institute recently established by Sandoz at University College, London, a center of neurobiology and neuro-

*Biogen, S.A., a Swiss company, is one of the four principal operating subsidiaries of Biogen N.V., which is the parent company of the Biogen group and is registered in the Netherlands Antilles. Biogen N.V. is about 80 percent U.S. owned. The other three subsidiaries include: Biogen Research Corp. (a Massachusetts corporation) which conducts R&D under contract with Biogen N.V. and Biogen B.V. (a Dutch corporation) and Biogen, Inc. (a Delaware corporation) both of which perform marketing and licensing operations. Biogen's principal executive offices are located in Geneva, Switzerland. Biogen N.V. is largely U.S. owned.

chemistry, will receive $7.6 million over the next 3 years.

While the established pharmaceutical companies are beginning to explore new applications of biotechnology in the area of pharmaceuticals, the NBF Biogen S.A. is applying biotechnology to several industrial sectors with a diverse R&D program. Biogen was established in 1978, largely at the initiative of venture capitalists from the United States, with funds from International Nickel Co. Biogen currently has three other principal shareholders: Monsanto (U.S.), Schering-Plough (U.S.), and Grand Metropolitan Limited (U.K.). Biogen S.A. has yet to sell any products made from biotechnology, but it was the first firm to obtain expression of hepatitis B surface antigens, leukocyte interferon, and the viral antigen of foot-and-mouth disease from rDNA technology. The diverse background of its scientific board suggests a flexible R&D policy with widespread applications of biotechnology to mining and metals refining, pharmaceuticals, chemicals, energy, agriculture, and food and beverage production (54). In 1982, through $20.5 million generated from contract research (primarily with Schering [F.R.G.], Shionogi [Japan], and Fujisawa [Japan]), Biogen S.A. supported an $18.4 million R&D program (48).

GOVERNMENT FUNDING OF BASIC AND APPLIED RESEARCH

Though the Swiss Federal Government has no specific biotechnology policy, its funding for biotechnology-related research is increasing (48). The Swiss National Science Foundation serves as a clearinghouse for Federal funds for the support of basic research related to biotechnology at specific universities and other institutions. Much of the fundamental research in the life sciences, however, is carried out in the largely canton-supported universities (52). Out of Switzerland's total biological and biomedical research budget of about $73 million (SF150 million), about 4 percent or $980,000 (SF2 million) goes to biotechnology.

The major Government source of applied research funds is the Commission for the Encouragement of Scientific Research (Kommission zur Forderung der Wissenschaftlichen Forschung). This commission provides grants for applied research projects of proven interest to industry, normally contributing 50 percent of the costs. The Department of Biotechnology (Institut für Biotechnologie) at the Swiss Federal Institute of Technology at Zurich (ETH-Zurich, Eidgenossische Technische Hochschule) receives strong support from the commission. ETH-Zurich, with an additional complex at Honggerberg, conducts research in the areas of basic biological research, bioprocess engineering, and water and sludge treatment. In addition to funding these activities, the Commission for the Encouragement of Scientific Research itself plays an active role in identifying potential industrial partners and interesting them in particular research projects (53). Given the proprietary nature of much of the work, funding figures are unavailable (52).

TAX INCENTIVES FOR FIRMS

Beause of low corporate tax rates, Switzerland provides a favorable environment for established companies in biotechnology. Though corporations conducting business in Switzerland are subject to both Federal and cantonal taxes, the Swiss effective corporate tax rate is the lowest in Europe (51).

PERSONNEL AVAILABILITY AND TRAINING

The access to distinctive universities and the high standard of living in Switzerland, attract highly qualified personnel from around the world to participate in Swiss biotechnology. Although the availability of personnel may not be important for the large pharmaceutical companies, which conduct a large proportion of their R&D in other countries, it is crucial to the Swiss advancement of biotechnology in other sectors. The attraction of talent from other industrialized countries may help the competitive efforts of Swiss companies in biotechnology in the future.

OTHER FACTORS

Swiss antitrust laws preventing monopolies present no serious problems for R&D joint ventures. In Government-industry joint projects, Swiss law assigns patents to industry, though holders of inventions whose R&D was supported by a Federal grant must repay the Federal contribution from license fees generated by the patent.

Health and safety laws in Switzerland do not generally impose barriers to biotechnology development. Although Switzerland is following a previous, and more restrictive version of the U.S. guidelines for rDNA research, there are no requirements covering large-scale work. The licensing of pharmaceuticals is more streamlined in Switzerland than in other countries. There is no requirement for Government approval before initiation of clinical trials, and the drug approval process generally takes from 6 to 10 months.*

*The Swiss pharmaceutical industry exports roughly 90 percent of its products. Thus, the drug and other product regulations of importing countries cause more concern to these companies than Switzerland's relatively relaxed regulatory framework (53).

CONCLUSIONS

The factors cited above and a growing commitment to biotechnology by the private sector suggest that biotechnology is advancing in Swiss industries. Both the Federal Government and most companies have been slow to initiate R&D programs in biotechnology, although the Swiss pharmaceutical industry and especially four companies have boosted their activities in these fields. For several reasons, Switzerland has only recently begun to dedicate its collective efforts to biotechnology (53):

- financial experts and bankers have lacked the technical expertise to evaluate high risk technologies;
- manufacturers have been averse to incorporating biotechnology into some Swiss industries because of the high financial risks and uncertainies caused by public and professional concern about the safety of rDNA research;
- Swiss industrial scientists have trailed Swiss and non-Swiss academic scientists in recognizing the widespread potential of biotechnology; and
- Swiss industries are highly oriented toward chemical synthesis and thus have underestimated the commercial implications of new biological processes.

In conclusion, the majority of Swiss biotechnological expertise rests in the large pharmaceutical companies and in Biogen S.A. and a few other small firms. The large companies generally conduct their R&D in foreign subsidiaries or in the form of proprietary research at in-house facilities and make no concerted effort to support domestic basic research outside industry (48). Thus, technology transfer between large Swiss firms and the universities is limited. Nevertheless, given the quality of Swiss educational institutions teaching the knowledge needed for the development of biotechnology, the attraction of foreign talent to Switzerland, and a new Government focus toward biotechnology development, the industrial use of biotechnology by Swiss companies is likely to become more widespread in the near future.

France

INTRODUCTION

France is currently in a less favorable position to compete with the United States than Japan and the other European countries analyzed in this report. The country's research system and industries generally lack a critical mass of qualified personnel in many disciplines important to the development of biotechnology. In addition, attempts by the socialist government to increase R&D expenditures have met with frustration because of an adverse macroeconomic situation in France during the last 2 years. However, the existence of isolated centers of excellence in scientific disciplines such as immunology, molecular biology, and bioprocessing, and of a few companies with bioprocessing expertise and a strong commitment to developing biotechnology, such as Elf Aquitaine and Rhone Poulenc, may help France to compete with other industrialized companies in selected product markets.

INDUSTRY

Three large French companies have R&D programs in biotechnology—Elf Aquitaine (67-percent Government owned), Rhône Poulenc (100-percent Government owned), and Roussel Uclaf (40-percent Government owned and a Hoechst subsidiary). Of these three, Elf Aquitaine has committed the most effort and money to biotechnology. It owns Sanofi, a pharmaceutical company that has the right of first refusal on all development research at Institut Pasteur Production (the scale-up branch of the Institut Pasteur), and has established Elf Bioindustries and Elf Bioresearch to develop biotechnology in the foodstuff and agricultural sectors. Medium-sized French companies, especially in the foodstuff sector, spend very little in overall R&D (about 0.1 to 0.2 percent of revenues) and have hesitated to devote their energies to biotechnology (62). Furthermore, France has only a few NBFs (e.g., Genetica, Transgene, Hybridolab, and Immunotech), and most of them are subsidiaries of large companies or commercializing arms of research institutes. Thus, the ability of large companies to commercialize biotechnology products will determine France's competitiveness in certain product markets.

GOVERNMENT TARGETING POLICIES

Official interest in the commercialization of biotechnology in France emerged only recently, with the appearance of the Pelissolo report in December 1980 (59). Since the election of the socialists, the French Government has resolved to push the development of several new technologies in its national industries and has accorded a privileged position to biotechnology within this scheme.

The French socialist government has established the most highly coordinated policy for the development of biotechnology of any of the six major competitor countries identified in this assessment. This policy rests on two cornerstones:

- a general research law (Loi de Programmation et d'Orientation) adopted by the French National Assembly in the first week of July 1982, and

- a program specifically for biotechnology ("Programme Mobilisateur: L'Essor des Biotechnologies") presented toward the end of the same month (58).

The general research law sets two objectives: 1) to stimulate French effort in new technologies by "guaranteeing" real increases in the overall civilian R&D budget of 17.8 percent per year, economic conditions permitting, and setting up seven technological "programmes," including one for biotechnology, on which a major portion of research funds are now to be directed; and 2) to open up French science to industry and education by encouraging scientists in research institutes to work in collaboration with private sector colleagues and to teach in universities (65). The Programme Mobilisateur, presented in July 1982 by the Biotechnology Mission of the newly organized Ministry of Research and Industry (now the Ministry of Industry and Research), outlines in detail the steps the Government should take to strengthen French biotechnology. This document calls for intervention from Paris through a myriad of organizations and committees in all aspects of research, education, technology transfer, and industrial development.

Both the research law and the Programme Mobilisateur demonstrate the French Government's determination to promote the necessary multidisciplinary approach to the various technologies and to establish vertical chains (filières) that incorporate all the relevant expertise in basic research, generic applied research, and large-scale production necessary to bring a product to market (60). The effectiveness of the French policy, however, will depend in part on the extent of voluntary cooperation among the various Government groups implementing the policy and the sectors the plans affect (i.e., public research centers, universities, and private industry).

GOVERNMENT FUNDING OF BASIC AND APPLIED RESEARCH

Most basic research in France is conducted in public research centers ("grands organismes"), similar in principle to the British Research Councils, or in a few university laboratories associated with these centers.* One of the three major "grands organismes," the National Center for Scientific Research (CNRS, Centre National de le Recherche Scientifique), conducts basic research related to biotechnology in three different divisions, and some of the projects CNRS sponsors overlap with similar work both at the center itself and at other centers and universities (62).

*For a more detailed description of the research infrastructure in France, see R. Walgate, "Great Schools, Great Contradictions" (63) and "CNRS—The Core of Research (64).

Little public sector generic applied research takes place in France. There are no national applied research laboratories, and with the exception of isolated programs at the universities at Compiègne (enzymology and bioprocess engineering) and Toulouse (biotechnology), the Government of France supports almost no generic applied research of benefit to its domestic industries.

Until recently, Government funding of both public and industrial R&D counted as a French strength. Although it should be noted that definitions of biotechnology differ from one organization to the next, funding estimates vary according to referred sources, and many research projects receiving biotechnology money have nothing to do with biotechnology (62), the French Government probably spent between $35 million and $60 million on biotechnology R&D in 1982.* Notwithstanding the Government's strong initial effort to fund biotechnology, increases planned for 1983 were effectively reduced. The National Assembly reduced the scheduled 17.8-percent real increase in the 1983 civil research budget to about 10 percent (66), and the reduction for researchers in biotechnology related fields was even greater. CNRS saw its original 1983 budget cut by 12.5 percent, and the Programme Mobilisateur research has lost a quarter of its allocation. These austerity measures allow research funding to keep pace with inflation, but little more. In spite of the reductions, the overall research budget still represents a 7.5-percent real increase over 1980 levels, and the Ministry of Industry and Research continues to support its policy of increasing allocations for science (56).

FINANCING AND TAX INCENTIVES FOR FIRMS

A new law enacted in February 1983 created a legal structure allowing the formation and investment of venture capital (67), but the venture capital market in France is poorly developed. Banks are the major source of financing in France, and have always hesitated to take major equity positions in industry. The financing that French banks provide, however, is designed for long-term projects, thus eliminating the problem, encountered by companies in the United States, of finding sources for second- and third-round financing.

With the exception of one provision, tax law in France generally conforms to European and American standards. A generous depreciation allowance in the tax code permits a company in France to write off 50

*This estimate is based on a 3-year (1983-85) projected total of $175 million, with a guaranteed (by law) 17.8-percent annual increase in the civil research budget, plus increased support for industry through existing schemes.

percent of its expenditures on R&D capital assets during the first year following the acquisition of these assets.

PERSONNEL AVAILABILITY AND TRAINING

France has a serious shortage of qualified personnel that could well undermine the country's basic and applied science base and prevent France and its industries from competing successfully in the world biotechnology marketplace. Specialists in the fields of general and industrial microbiology, rDNA and hybridoma technologies, enzymology, plant and animal cell culture, and bioprocess engineering are few (55). Although some French research centers boast internationally recognized teams, such as the enzymology and bioprocess technology teams at the technical University of Compiègne or the immunology groups at the Institut Pasteur (62), these are isolated clusters of expertise and will have difficulty matching the total output of the large and balanced national research bases of other competitor countries.

The scarcity of personnel in France cuts across several sectors of R&D in these technologies and applies equally to different categories of personnel, from scientists and bioprocess engineers with advanced degrees to skilled laboratory and production technicians. In order to correct this situation, the French Government has given special attention to the education and training of qualified personnel. The research law passed in July called for the active involvement in the educational process of public sector researchers outside universities (65), and the Programme Mobilisateur presents educational guidelines for all stages of schooling from secondary to postdoctoral levels, placing special emphasis on an interdisciplinary approach within the universities (58). The education of a specialist in rDNA technology, nonetheless, takes many years, as does the implementation of such training programs. As a short-term solution to its present lack of personnel, therefore, France imports foreign experts (58).

UNIVERSITY/INDUSTRY RELATIONSHIPS AND DOMESTIC TECHNOLOGY TRANSFER

Universities in France have had very few ties with industry in biotechnology. Large firms in France actively seek out developments in basic research, either by locating plants near research centers or through an office that monitors current developments in biotechnology research in France and other countries.

The French Government encourages domestic technology transfer through the National Agency for the Evaluation of Research (ANVAR, L'Agence National de la Valorisation de la Recherche). ANVAR, which has no right of first refusal on the results of research in public laboratories, acts as a catalyst for the direct interaction between these institutes and private firms (e.g., through publications on the status of innovation with applications in different industrial sectors).*

OTHER FACTORS

The French legal and regulatory environment, with one exception, poses no real barriers to the commercial development of biotechnology. France maintains the most rigid investment control laws in Europe (61). These regulations allow the French Government to prevent strategic companies from being acquired by foreign concerns and may well hinder foreign firms' ability to penetrate French markets.

Health and safety regulations, as well as patent and antitrust laws in France, however, are approximately equivalent to those in other European countries.

CONCLUSIONS

At present, France lags somewhat behind the United States, Japan, the Federal Republic of Germany, the United Kingdom, and Switzerland in the commercial development of biotechnology. If the country can solve its personnel problems, however, French industries could well gain a competitive footing in selected product markets. The Government's well-coordinated formal policy and adequate but precarious funding program represent a strong commitment to the development of biotechnology that needs to be completed with the necessary qualified personnel. Although the French private sector until rather recently has hesitated to develop its biotechnology capabilities, large companies do have the money and the means of uncovering the latest technological developments. Therefore, the ability of both the public and private sector to recruit and train scientists and technicians and the maintenance of sufficient Government allocations for R&D in the face of adverse macroeconomic conditions may ultimately determine the competitiveness of French biotechnology in the international marketplace.

Sweden

Sweden is a technologically progressive country, but adverse public opinion toward rDNA technology has resulted in the imposition of Government restrictions on the use of rDNA in research and industry. Furthermore, a lack of trained personnel in basic sciences has restrained the commercialization of biotechnology.

---

*For a general review of ANVAR's functions and activities, see "Commentary on the National Agency for the Evaluation of Research," *Le Monde* (57).

Swedish public opinion and Government policies may be changing to encourage biotechnology in Sweden. If this proves to be the case, Sweden may market products in areas such as the following:

- *Support sector.* Swedish scientific instrumentation, filtration, and industrial separation systems are used around the world and are important in the commercialization of biotechnology.
- *Bioprocess engineering.* A large portion of Sweden's combined public and private sector R&D efforts is devoted to heterogeneous bioprocessing systems, stabilization of immobilized cell systems, membrane technology, and downstream purification and regeneration (76).
- *Pharmaceutical industry.* Swedish pharmaceutical companies maintain aggressive export policies and are active in innovation. The five largest Swedish companies have a gross annual income of about $1 billion, with 70 percent derived from exports (76). It is not known to what extent Swedish pharmaceutical companies will use biotechnology, given Sweden's shortage of trained personnel in rDNA technology and other areas. In the near term, most Swedish companies will probably rely on licensing arrangements with NBFs in the United States to gain access to biotechnology (76).

Among the Swedish companies that appear to have the potential to use biotechnology for producing goods and services are Pharmacia AB, KabiGen/KabiVitrum, and Alfa-Laval.

Pharmacia AB concentrates on pharmaceuticals, separation products, diagnostics, and cosmetic products, and derives 90 percent of its revenues from exports; the U.S. subsidiary, Pharmacia, accounts for 25 percent of these sales. With demonstrated abilities to serve specialty markets, this company is a leader in separation science and is working to establish rDNA capabilities.

KabiGen/KabiVitrum, operated by the Swedish Government, is currently the world's largest supplier of pituitary-derived human growth hormone (hGH). In order to protect its hGH market from foreign competition, Kabi has entered into a licensing arrangement with Genentech (U.S.) to market rDNA-produced hGH outside of the United States. KabiGen is also moving to establish its own rDNA capabilities, intending to pursue projects on human insulin, methanol production, bacterial metal enrichment from ores, interferon, and anticoagulant pharmaceuticals (71,72). Furthermore, Kabi is involved with the development of support equipment, including a polynucleotide synthesizer (69).

Alfa-Laval has large-scale fermentation capabilities and is currently working to establish rDNA capabilities through its subsidiary AC Biotechnics, in which it shares ownership with Cardo Co. Biotechnics has a budget of $8 million to $10 million for an unspecified length of time to produce specialty chemicals and ethanol using rDNA technology.

Other Swedish companies interested in biotechnology include Sorigona AB, which produces chemicals and foods; Astra, which is working in collaboration with U.S. researchers to develop long-acting anesthetics (74); and approximately a dozen additional firms.

Funding for high technology in Sweden is available from several Government sources. Since each department of the Swedish Government establishes its own R&D budget, however, overall R&D funding estimates are difficult to obtain. Some degree of R&D coordination is maintained by the National Swedish Industrial Board (Statens Industri Verk), which is responsible for promoting technological development, organizing training, and orchestrating Government actions, and the National Swedish Board for Technical Development (STU, Styrelsen för Teknisk Utveckling). STU, which is the main source of Government funds for biotechnology, granted an estimated $4 million for biotechnology in 1982, and Swedish industry probably spent an additional $15 million (72).

The manner in which STU distributes R&D funds reflects a Swedish Government policy of directly promoting strategic industries. STU works through joint Government/private ventures with foundations established by Swedish and foreign companies interested in a particular field of development. STU provides half the R&D funding as provisional grants and the foundation provides the other half. If the venture is successful, the funding is treated as an interest-free loan; otherwise, it is considered a grant. Research grants/loans are limited to $100,000, and those for product development to $600,000. In 1973, 20 Swedish, 2 Danish, 2 Finnish, and 1 Norwegian company established a specific foundation to promote biotechnology called the Biotechnology Research Foundation (SBF, Stiftelsen Bioteknisk Forskning) (72). SBF, in conjunction with STU, is currently conducting research on heterogeneous bioprocessing systems, immobilized cell systems, membrane biotechnology, and regeneration of coenzymes (76).

Private industry R&D in Sweden is encouraged by corporate tax incentives, which include a 10-percent deduction for R&D and a 20-percent deduction for any increase in R&D from the previous year.

Sweden's Central Investment Bank and commercial banks provide risk capital in promising technological areas. Information about the banks' views toward new biotechnology is not available, but in 1982, $300 million for all R&D loans in Sweden were tendered. Capital for risk ventures from other sources is limited in Sweden, and the larger Swedish companies, such as Fortia, rely primarily on internal funds and Biotechnology Research Foundation loans (73).

The Swedish Government has encouraged high-technology, export-directed growth for many years and has promoted relations among the Government, industry, and the universities. Seven Swedish universities have liaison officers with industry whose salaries are paid by STU. A 6-year, $7 million agreement has been established between the University of Uppsala, the University of Agriculture, the Swedish Veterinary Institute, and Fortia AB, that is intended to develop expertise in rDNA technology and to create the ". . . most intensive programme of biotechnology in the world" (68).

Although extensive interaction between the sectors is encouraged and funded, Swedish efforts to commercialize biotechnology suffer most from a shortage of certain types of trained personnel. Estimates of the number of Swedes working in biotechnology vary from 30 to 40 people (72) to as many as 200 workers at Uppsala alone (68), but shortages of personnel in key areas such as rDNA technology hamper wider commercial applications (75).

Personnel training for biotechnology has been largely inhibited by negative Swedish public attitudes toward rDNA experimentation. As a result of the restrictive rDNA guidelines, which required the Swedish National Recombinant DNA Advisory Committee's permission to conduct any rDNA research, there was little need for trained personnel, and Sweden's private sector relied on foreign companies for developing products requiring rDNA processes (70). In a joint project between KabiVitrum and Genentech (U.S.) to develop and produce hGH, for example, the first actual cloning of the hGH gene was performed in the United States by Genentech. Since the relaxation of the guidelines, however, the need for qualified engineers and scientists has increased, and some Swedish universities have instituted training programs in biotechnology.

The Swedish Government's identification of biotechnology as an industrially strategic area, as exemplified by the establishment of joint programs with SBF and other promotional activities for research, indicates that Swedish views may be changing. With Sweden's demonstrated ability to successfully exploit new technologies, Swedish companies may prove to be competitive in the future in the support and bioprocess sectors, as well as in pharmaceutical markets.

Netherlands *

The Dutch Innovation Programme on Biotechnology, started in May 1981, is aimed at filling the gap between basic research and applied development work in Dutch universities. Funds supplied by the Government of the Netherlands will be used to develop research in areas where current national effort is insufficient. The program will be coordinated by the Dutch Programme Committee on Biotechnology. The program will last until the end of the 1980's, after which the existing research budgets of universities and institutes will furnish Dutch industry with the needed basic research.

The Programme Committee on Biotechnology (Programma Commissie Biotechnologie) requested $11.2 million (NLG30 million) to be spent on basic biotechnology research from 1983 until 1988. This amount is in addition to the $11.2 million to $15 million (NLG30 million to NLG40 million) which the Government spends yearly on research projects in the fields of molecular and classical genetics, microbiology, cell biology, biochemistry, enzymology, and bioprocess and bioreactor engineering.

In addition to the aforementioned sums, $2.6 million (NLG7 million) will be used by university/institute and industry groups in the Netherlands for multidisciplinary biotechnological research projects. According to the Programme Committee, these projects should be in the following areas:

- host vector systems for industrial and agricultural applications,
- somatic cell hybridization,
- second generation of biotechnological reactors and processes, and
- downstream processing.

Established Dutch companies that are setting up inhouse R&D efforts in biotechnology include the following:

- Gist-Brocades N.V.
- Akzo-Pharma N.V.
- Unilever H.V.
- N.V. DSM
- Heineken N.V.
- Dupher N.V.

---

*This summary is based on a personal communication with Dr. Ir. R. R. Van der Meer, Secretary-Coordinator, Programme Committee on Biotechnology, Gravenhague, April 1983 (78).

Gist-Brocades N.V., one of the two companies in the world that supply more than 60 percent of the world's enzymes for industrial use, is devoting almost all of its $20.6 million (NLG55 million) budget for R&D to biotechnology. Intervet International, a subsidiary of Akzo-Pharma, was the first company to market vaccines produced through rDNA technology. Intervet's vaccines, introduced in March 1982, prevent scours (infectious diarrhea) in calves and piglets (77).

The Programme Committee on Biotechnology forecasts no personnel shortages. In fact, there is an excess of biochemical and microbiology students for the available Dutch jobs in industry. There are no tax policies aimed at encouraging biotechnology in Dutch industries. The Dutch have eased their regulatory guidelines for working with rDNA technologies to conform to U.S. guidelines.

Australia *

The Australian Government supports a highly respected basic research system, especially in plant breeding and molecular biology, but it regards the development of biotechnological applications, including scale-up development and bioprocess engineering, as the responsibility of the private sector. Owing to a historic dearth of capital for high-risk ventures and a lack of trained personnel in applied technology, commercial biotechnology in Australia is not well developed. Australia's problems are exacerbated by the emigration of some of its top scientists to other countries where attractive jobs exist, although there is some indication that this situation might change in the future. The Australian Government is taking steps to implement incentives to help retain scientists and encourage venture capital formation to help foster promising applications of biotechnology.

Australian efforts are not expected to have an immediate impact on the markets discussed in this report. Nevertheless there is a strong possibility that, by using biotechnology to help solve local problems, Australia will find new markets for biotechnology products. Areas of biotechnology application in Australia being pursued include the following:

- plant improvement programs to develop agricultural species that are adapted for higher yields in Australian conditions;
- animal health products, particularly veterinary and nutritional products that improve the marketability of Australia's animals and animal products (especially wool) for export;*
- microbiological mineral recovery to reduce extraction and separation costs for certain minerals that Australia exports in great quantities;
- biomass conversion to ethanol and chemicals, based on Australia's large resources of grain crops and sugar cane residues.

Other applications of biotechnology in Australia include animal breed improvements through embryonic gene transfer, MAb-based diagnostic reagents for a number of human diseases, and, on a small scale, interferon and other rDNA projects to develop pharmaceutical products.

Government funding for biotechnology in Australia is administered through several Government agencies, including the Australian Science and Technology Council, which emphasizes expanded manufacturing and agricultural production with biotechnology, and the Commonwealth Scientific and Industrial Research Organization (CSIRO), the main research agency in Australia, which provided $4.6 million ($A4.5 million) for biotechnology research in 1981. Other sources are the National Health and Medical Research Council, which distributed $19.0 million ($A18.7 million) in research funds in 1980/81 (some of which benefited Australian biotechnology); the Energy Research, Development, and Demonstration Program which distributed $3.9 million ($A3.8 million) in 1980/81, partly for biotechnology project development; the Department of Health, which gave $1.88 million ($A1.85 million) to the Commonwealth Serum Laboratories to conduct research on interferon from 1980/81 to 1983/84; and the Australian Research Grants Scheme, which awarded $18.3 million ($A18 million) in 1982 to individual research scientists, some of whom use biotechnology in their work. In addition, financial assistance for general industry R&D projects is provided under the Australian Industrial Research and Development Incentives Scheme which in 1980/81, distributed $9.8 million ($A9.7 million) in commencement grants and $36.6 million (A$36.1 million) in project grants.

Other Australian incentives include tax policies that give minor benefits to firms undertaking R&D activities. Buildings used solely for scientific research purposes are depreciable over a 3-year period, compared to general industry's 40-year depreciation schedule. New equipment used for scientific research is also depreciable over a 3-year period, as opposed to a 5-year period for general industry equipment.

*This summary is based on information presented in "Biotechnology Research and Development, the Application of Recombinant DNA Techniques in Research and Opportunities for Biotechnology in Australia" (79) and "Genetic Engineering—Commercial Opportunities in Australia" (80).

*To date, most rDNA efforts have centered on cloning the genes that encode sheep wool keratin and other wool constituents in an effort to improve wool quality and lessen treatment costs of wool.

In addition to basic research funding and tax incentives to businesses, liaisons between Australian universities and industry are encouraged. In some cases, academic researchers have financial equity in biotechnology firms. In other cases, the relationship is through contracts with the universities. One example is an agreement under which Agrigenetics Research Association, Ltd. provided $2 million for biotechnology research at Australian National University.* Although Australia has the infrastructure to support healthy biotechnology development, lack of capital for high-technology firms retards growth. The Government and Australian banks make loans available to small businesses at low interest rates, but these loans are not generally available to high-risk enterprises such as NBFs. High-risk ventures are hampered by a smaller capital base in Australia than in the six major competitor countries. With increased Government interest in commercial biotechnology, more capital may become available. This increase in capital might in turn encourage increased efforts by existing NBFs to find applications for new biotechnology, as well as the formation of more NBFs. It should be noted, however, that Australia has some of the most restrictive drug licensing laws in the world, and these regulations may impede Australian applications of biotechnology to the pharmaceutical industry.

Biotechnology companies in Australia include the following:

- Biotechnology Australia Pty., Ltd. (a subsidiary of CRA Ltd.). Projects include animal feed additives and health care products, specialty chemicals, biomass conversion, and mineral extraction schemes.
- Austgen Pty., Ltd. (includes Biojet International [Australia] Pty. Ltd.). Projects include nutritional additives and waste treatment systems. Much of Biojet International's R&D is oriented towards products that can be exported.
- Australian Genetic Engineering Pty., Ltd. Projects focus on MAbs for diagnosis (a $5 million per year market for MAbs for diagnosis currently exists in Australia; a $15 million market is expected by 1986).
- Bioclone Australia Pty., Ltd. This firm markets MAbs made by the Garvan Institute and CSIRO on a worldwide basis. Its best known product is an antiprolactin MAb. Eleven additional MAb products have been or will soon be marketed.
- Australian Monoclonal Development Pty., Ltd. This company supplies MAbs primarily for research purposes.
- Fielder Gillespie, Ltd. This milling company funds MAb and biomass conversion projects.

In conclusion, Australia has the potential to develop and commercialize several applications of biotechnology successfully. A good Australian research base exists, but increased infusions of capital are necessary for new commercial startups if the potentials of biotechnology in Australia are to be realized. Australian Government policies have targeted the development of biotechnology, but the effect of the policies remains to be seen. Some Australian products, such as MAb diagnostic products, may prove to be competitive in world markets, but overall, major competition in the pharmaceutical and specialty chemical industries is unlikely.

Israel

For several reasons, Israel may be unique among developed nations in fostering a strong basic and applied research capability in biotechnology without having a large industrial infrastructure to exploit the successes of research endeavors. Israeli scientists train in U.S. institutions prominent in biotechnology and have become well-versed in molecular biology and immunology. Except for small brewery plants and one bioprocess plant (Gadot, which manufactures about 7,000 tons of citric acid per year), however, Israel does not have companies using old biotechnological techniques. Furthermore, Israel's tax and financial structures do not encourage financial risk-taking or the formation of new firms. Therefore, there are few industrial positions available for scientists trained in biotechnology.

As a result of the lack of depth in industrial expertise in Israel, Israeli universities, through their University-Connected Research and Development Organizations (UCRDOs),* turn to foreign companies that have the expertise to evaluate Israeli research and the resources needed to commercialize the results of this research. The number of joint ventures between Israeli UCRDOs and foreign firms is fairly large.

Noteworthy basic research in biotechnology is taking place at several Israeli universities and institutes, among them Hebrew University, Technion Institute at the Israel Institute of Technology, the Center for Biotechnology at Tel-Aviv University, and the Weizmann Institute of Science.

At Hebrew University, 12 departments in the medical school are conducting biotechnology-related research projects, ranging from cellular biology to can-

*The goal of this research is to incorporate the nitrogen-fixing genes of bacteria into plants adapted to Australian conditions.

*UCRDOs are set up by Israeli universities to promote commercialization and applied research. These organizations may enter into joint ventures or own equity in spinoff firms.

cer research. The agriculture department has initiated several projects and has received over $410,000 (more than DM1 million) from the Minerva Fund in the Federal Republic of Germany for cooperative projects on improving plant tissue culture techniques, rDNA and protoplast fusion in plant breeding, nitrogen fixation and control of soil-borne plant pathogens by microorganisms, and new uses of algae (84). Hebrew University's UCRDO Yissum signed a $5 million agreement on nitrogen-fixation research with Biotechnology General (Israel), Ltd. (82), an Israeli NBF, and another $3 million agreement has been signed with International Genetic Sciences Partnership (U.S.) (85).

Technion Institute, at the Israel Institute of Technology, is doing research on biotechnology instrumentation and on blood and blood plasma substitutes (82). Tel-Aviv University, Center for Biotechnology, conducts research on MAbs, enzyme systems of anaerobic bacteria, and immobilized enzymes (82).

The Weizmann Institute of Science is Israel's main center for rDNA research and is especially noted for its work with interferons. Additionally, research is proceeding with MAbs, antiviral vaccines, synthetic antigens, and new genetic forms of wheat, within seven departments.

Applied research using new biotechnology began in Israel in 1978. As of 1981, 17 universities, institutes, and venture firms in Israel had been identified as performing or funding applied research in biotechnology. Of the 17, perhaps 10 use the new technologies in their work (87). The four universities and institutes cited above, in addition to conducting basic research, also do applied work.

Israeli companies noted for their applied R&D include Biotechnology General, Interpharm, Inter-Yeda, Kibbutz Beit Ha'Emek. Biotechnology General develops research findings from the Weizmann Institute and Hebrew University. Its main emphasis is on foot-and-mouth disease vaccine, bovine growth hormone, biological disease control agents, and nitrogen fixation (82).

Interpharm, a subsidiary of Applied Research Systems (ARES), a Dutch multinational firm based in Geneva, sold over 1.15 million shares of common stock on the United States over-the-counter market in 1981. At the time of offering, Interpharm had a contract to supply ARES with hGH. Further, Interpharm may soon market human fibroblast interferon for labial and genital herpes, depending on results of clinical trials, produced by its R&D susidiary Inter-Yeda (83). Other projects with commercial possibilities include an immunoassay separation technology, extraction technologies for follicle-stimulating hormone and luteinizing hormone, and research on hybridomas (81).

Inter-Yeda, a joint venture firm owned 60 percent by Interpharm and 40 percent by Yeda, will concentrate on four areas: production of interferon using rDNA techniques, identification and isolation of interferon-associated proteins, artificial production of interferon, and MAb research (82). Inter-Yeda is shipping human fibroblast interferon to the Serono Corporation in the United States (81).

Kibbutz Beit Ha'Emek hired researchers in order to use advanced tissue culture techniques to "develop plant varieties resistant to herbicides, diseases and other environmental hazards" (86). The kibbutz claims a $1 million income from "tissue-culture-derived products," of which 65 percent are exported, mainly to the Netherlands and the Federal Republic of Germany.

At present, there is no central planning of any R&D by the Israeli Government, and thus the Government has no national targeting policy for biotechnology. Each Ministry within the Israeli Government determines and funds the R&D it deems necessary. The major source of Government funds for biotechnology R&D is the Office of the Chief Scientist in the Ministry of Industry and Trade. The Israeli Ministry of Industry and Trade plans to invest $25 million in biotechnology R&D over the next 5 years (85).

Canada

Canada's economy relies greatly on its natural resources such as agriculture, livestock, mining, and forestry. In the past 3 years, Canada's Federal and Provincial governments, as well as a few Canadian companies, have worked to incorporate biotechnology specifically as it relates to the development and exploitation of the country's natural resources. A focus on improving domestic capabilities in the necessary technologies and avoiding dependence on imported products and processes, however, represents an attempt by both the public and private sectors in Canada to compete in selected world markets. Whether Canada becomes internationally competitive in areas of biotechnology such as agricultural plant strain development, mineral leaching, or lignocellulose conversion, for example, will depend to a large extent on the rapidity with which it can exploit national expertise before other countries with extensive R&D programs in these fields.

Interest in the commercial development of biotechnology has evolved slowly in Canada. In June 1980, the Canadian Federal Government commissioned a Task Force on Biotechnology to evaluate the opportunities available to Canada in this area. This task force, in its report to the Minister of State for Science and Technology, identified specific weaknesses in

Canada's research base, Federal Government programs, regulations, and industry, and made specific recommendations to help correct these deficiencies (96). The Canadian Federal Government took more than 2 years to act on these recommendations. In early May 1983, it announced two separate yet complementary initiatives to help promote biotechnology in Canada.

First, as part of a broader plan to support the development of emerging technologies in general, the Ministry of State for Science and Technology designated biotechnology as one of the priority technologies targeted for development (94). The plan to support emerging technologies consisted of five basic components. The first was identification of strategic areas of development most important to Canada. Adopting the recommendations of the Task Force on Biotechnology, the Federal Government will concentrate efforts on research in nitrogen fixation, plant strain development, cellulose utilization, mineral leaching and metal recovery, and animal and human health care products. The second component was creation of research networks. Individual Federal departments will establish and promote networks of research projects in biotechnology and researchers in areas relevant to their mandates. The third component of the plan was establishment of a cost-sharing program. Under the program, and with $7.7 million per year, the Federal Government will match funds invested by industry in universities or Provincial research organizations. The funds could be used for purposes such as specific biotechnology research projects, the replacement of equipment, and the establishment of research chairs. The fourth component of the plan was strengthening of overall Federal research capacity in biological sciences ($3.1 million). The funds will be used to establish and promote networks, to promote interactions between Federal departments and universities and industry, and to strengthen existing programs within Government research organizations. Finally, the fifth component of the plan was the creation of advisory and coordinating committees. A National Biotechnology Advisory Committee, chaired by a member from the private sector with 25 representatives from industry, academia, and Federal Government departments, will monitor the course of the biotechnology policy and advise the Minister of State for Science and Technology on the program's progress. An Interdepartmental Committee on Biotechnology, which functions at the Deputy Ministry level, will control the allocation of funds to departments participating in the Federal plan and will deal with a wide range of issues such as patenting and regulation in biotechnology (92).

Parallel to and coordinated with the five-pronged program outlined above, the Ministry of State for Science and Technology has charged the National Research Council (NRC)* with responsibility for the promotion of centers of expertise in biotechnology. Under this program, NRC will undertake three separate projects:

- construction in Montreal of a $61 million biotechnology institute which will probably conduct generic applied research on bioprocessing and enzyme technology (95);
- refurbishment and reorientation of the Prairie Regional Laboratory in Saskatoon (95); and
- strengthening of the NRC Biological Sciences Division in Ottawa.

In addition to the Canadian Federal Government, many Canadian Provinces have begun to promote the development of biotechnology. Quebec, Ontario, Saskatchewan, Alberta, and British Columbia have shown an increasing interest in the commercial opportunities offered by biotechnology. Quebec, for example, has developed an explicit policy which gives high priority to biotechnology. Sasketchewan is also in the process of developing such a policy. Quebec and Ontario have invested in commercial ventures in biotechnology (Bio-Endo and Allelix, respectively).

Several problems may limit the commercial development of biotechnology in Canada. First, there is a generalized shortage of personnel trained in the relevant technologies (only 200 to 300 Ph. D. s), and many of those who do graduate with degrees, for example, in molecular biology or biochemical engineering are lured to the United States to work in the private sector (97). Furthermore, very few private firms have directed their efforts to developing an expertise in new biotechnology; most rely instead on more traditional techniques in research, development, and production.**Canada also has very little experience in joint university, industry, and Government cooperation (93), though current Federal initiatives are addressing this problem.

---

*NRC is an independent Crown Corp. with considerable influence on Federal science and technology policy. Though not a Government Department, the Council is funded primarily by the Federal Government. Because of the scientific expertise NRC possesses, it will administer $120 million for the technology support program (of which biotechnology forms a part). NRC currently employs a total of over 600 persons (including support staff) in biotechnology alone when their program is in full operation.

**Allelix Corp. appears to be one of the few companies devoted entirely to developing new biotechnology. Started by the Provincial Government of Ontario, the Canadian Development Corporation, and John Labatt Ltd. with a total initial capitalization of $105 million (89), this company is currently concentrating on the development of new plant strains, using both cell-fusion and rDNA techniques (88). For further information on private sector activities in biotechnology in Canada, see "Biotechnology Research and Development in Canada" (90).

The current Canadian patent law requires compulsory licensing of all human therapeutic drugs developed by one company to other general generic pharmaceutical companies (92). As a result of the implementation of this law in 1969, all multinational pharmaceutical companies in Canada closed their research operations (91). There is no equivalent in Canada to the U.S. Plant Variety Protection Act, even though certain mechanisms do exist in Canada to protect the ownership of new plant strains (88).

Canadian tax law favors the development of biotechnology. One provision allows for a 100-percent, first year deduction on all current and capital expenditures for R&D. Additionally, corporations in Canada may deduct a further 50 percent for incremental R&D expenditures (calculated from a moving 3-year average). R&D expenditures are also eligible for a 10 percent investment tax credit (small businesses and investments in some provinces receive a higher percentage rate) up to a limit of $12,200 ($C15,000). R&D limited partnerships are also permitted in Canada (94).

U.S.S.R.

It is extremely difficult to obtain information on development plans for biotechnology in the U.S.S.R. Although it is known that biotechnology R&D is carried out in the Soviet Union, information about the extent of these activities is unavailable to the general public. The following summary formed part of the report on competitive and technology transfer aspects of biotechnology by a working group for the White House Office of Science and Technology Policy (98):

> The Soviet Union is actively supporting biotechnology R&D and has established an Interagency Technical Council to organize and stimulate its progress across a broad spectrum of disciplines. There is no information regarding the budget for biotechnology R&D. However, the rate of growth of the Soviet research establishment mirrors that which occurred in the United States 3 to 5 years ago. Their stated interests are directed toward domestic concerns such as the development of medical/pharmaceutical preparations and agriculutural applications. Soviet establishment of U.S. patents covering an amino acid producing organism and the enabling technology suggests an interest in international commercial competition as well.
>
> Although Soviet research is often hampered by difficulties in obtaining equipment and reagents, the Soviet system offers one major advantage over the free enterprise system of the United States; i.e., R&D is supported from inception through production and distribution. The financing gap between completion of basic research which has potential for application, and actual development, the costs of which in the United States must be borne by industry, receives full support of the Government in the Soviet Union. The advantages of this system are:
>
> - risks are taken by the Government;
> - costs of development are borne by the Government;
> - the Government's financial base can support an extended period of development; and
> - the Government can support long-term price control to facilitate international market entry.

It is too early to project potential Soviet success in the international biotechnology market. Much depends on successful completion of research programs now underway, and, most importantly, continued support by the Soviet Government.

*Brazil**

Brazil is the only developing country that has a formal government policy for biotechnology. This policy was developed because relations among the university, industry, and government sectors in Brazil tend to be adversarial, inhibiting communication among the sectors. Brazilian industry tends not to fund risky projects, concentrating its efforts instead on already existing products and processes. Historically, Brazilian industry has relied on the purchase of foreign technology and on joint ventures with foreign companies. Brazil's universities have little contact with either industry or the Government and conduct little multidisciplinary research. These historical relationships suggest that the government (both Federal and State) will have to play a strong role in Brazil to develop the R&D infrastructure necessary to develop biotechnology and to aid the commercialization of biotechnological applications.

In general, the major weaknesses for biotechnology development in Brazil are as follows:

- Brazil's human resource base trained in advanced biotechnology techniques is limited. In 1982, six qualified and experienced researchers in the field of rDNA and MAbs were identified.
- Brazil's national industrial sector is fairly underdeveloped and has little in-house R&D capability and little inclination to pursue high-risk, new product operation.
- There is uncertainty about the interpretation of Brazilian patent statutes with respect to biotechnological products and processes.
- Import and bureaucratic delays make it difficult for both public and private laboratories to obtain the necessary R&D equipment and supplies not available on the Brazilian market.

*This summary is based on "An Analysis of Current and Projected Biotechnological Activity in Brazil," a contract report prepared for the Office of Technology Assessment, U.S. Congress, by Robert Goodrich, July 21, 1982.

- Adequate analyses of market needs and opportunities are lacking, leading to inadequate orientation of research activities.

Three major Federal agencies in Brazil are involved in the funding of biotechnology: 1) the National Research Council, now known as the Council for the Development of Science and Technology (CNPq, Conselho Nacional de Desenvolvimento Científico e Tecnológico); 2) the National Funding Agency for Studies and Projects (FINEP, Financiadora de Projetos); and 3) the Secretariat of Industrial Technology of the Ministry of Industry and Commerce. CNPq will devote about 5 percent of its annual budget to biotechnology or $1.12 million (BCr200 million) during 1982-83. FINEP will spend approximately $1.5 million to $2 million (BCr270 million to BCr360 million) during 2 years to aid the commercializing of R&D in biotechnolgy by supporting economic analyses, commercialization ventures, and marketing studies. The Secretariat of Industrial Technology of the Ministry of Industry and Commerce is responsible for the National Alcohol Program and is already funding extensive R&D in bioprocesses and enzymology.

The Brazilian Federal Government plans to fund the development of two biotechnology research centers. The first, in Sao Paulo, will be an educational facility for multidisciplinary training. Its research program will focus on bioprocesses and enzyme research. The second center, the Biotechnology Center in Porto Alegre, will receive $0.97 million (BCr175 million) in funding and will concentrate on microbiology and applied genetics with little or no concern for product development. It will have an initial staff of four Ph. D. and four M.S. researchers and trainees.

Appendix B references

JAPAN REFERENCES

1. *Business Week,* "Smoothing the Way for Venture Capital—Again," Oct. 11, 1982.
2. Elkington, J., "The Climate for Biotechnology: National Strategies Contrasted," *Bio/Technology* 1(6):490-492, 1983.
3. *Japan Economic Journal,* "Japan Associated Finance Forms U.S. Type Venture Capital Funds," Mar. 23, 1982.
4. Kagaku Gijutsu Kaigi (Council for Science and Technology), *1970 Nendai Ni Okeru Sogo-Teki Kagaku Gijutsu Seisaku No Kihon Ni Tsuite* (The Basics of a Comprehensive Science and Technology Policy for the 1970's), Tokyo, 1971.
5. Kojima, N., Chairman, Japanese Pharmaceutical Manufacturers Association, personal communication, to L. Santucci, Pharmaceutical Manufacturers Association, 1983.
6. *Nihon Keizai Shimbun* (Japan Economic Journal), "Dainiji Bencha Boomu Torai" (The Second Venture Capital Boom is Coming), Sept. 9, 1982.
7. *Nihon Keizai Shimbun* (Japan Economic Journal), "Promotion of Biotechnology Development," Sept. 2, 1982, translated by the Foreign Broadcast Information Service in *Japan Report* (Joint Publications Research Service, Arlington, Va.).
8. *Nihon Kogyo Shimbun* (Japan Industrial Daily), "Four Applications of Biotechnology," Feb. 10, 1982, translated by the Foreign Broadcast Information Service in *Japan Report* (Joint Publications Research Service, Arlington, Va.).
9. *Nikkei Biotechnology,* "Ministry of Agriculture, Forestry, and Fisheries," Feb. 14, 1983, translated by the Foreign Broadcast Information Service, in Japan Report (Joint Publications Research Service, Arlington, Va.).
10. Porges, A., "Trade and Investment Laws and U.S. Competitiveness in Biotechnology," contract report prepared for the Office of Technology Assessment, U.S. Congress, April 1983.
11. Saxonhouse, G., "Biotechnology in Japan," contract report prepared for the Office of Technology Assessment, U.S. Congress, June 1983.
12. *Science and Technology in Japan,* "Research and Development of Life Sciences in Japan," April 1983, p. 8.
13. U.S. Congress, Congressional Research Service, *Tax Rates in Major Industrial Countries: A Brief Comparison,* No. 80-224E, Washington, D.C., December 1980.
14. Yamamura, K., University of Washington, personal communication, July 1982.

FEDERAL REPUBLIC OF GERMANY REFERENCES

15. Boldt, D., Counsellor, Office of Science and Technology, West German Embassy to the United States, personal communication (from 1982 Yearbook of the Max Planck Society, in German), May 1983.
16. Bundesministerium für Forschung und Technologie (Federal Ministry for Research and Technology), *Leistungsplan: Biotechnologie* (Performance Plan: Biotechnology), Bonn, 1979.
17. *Business Week,* "High Tech Tries an End Run Around Germany's Banks," July 18, 1983, p. 74.
18. Culliton, B., "The Hoechst Department at Mass General," *Science* 316:1201, 1982.
19. Deutsche Gesellschaft für Chemisches Apparatewesen (German Society of Chemical Engineering), *Biotechnologie* (Biotechnology), a contract report to the Federal Ministry for Education and Science and the Federal Ministry for Research and Technology, Frankfurt, 1974.
20. Dunnett, J., "Time To Count Costs," *Nature* 301:367, 1983.
21. Dunnett, J. S., "Brain Drain Threatens Progress," *Nature* 302:644, 1983.
22. *Europa Chemie,* "Genetic Engineering Institute Established in Berlin," Dusseldorf, October 1982, translated by the Foreign Broadcast Information Service in *West Europe Report: Science and Technology* (Joint Publications Research Service, Arlington, Va.).
23. Gesellschaft für Biotechnologische Forschung (Society for

Biotechnological Research), *Programmbudget* (Program Budget), Braunschweig, 1982.
24. Jasanoff, S., "Public and Private Sector Activities in Biotechnology: The Federal Republic of Germany," contract report prepared for the Office of Technology Assessment, U.S. Congress, January 1983.
25. Maddox, J., "Where To Get Research Grants," *Nature* 297:265, 1982.
26. Maddox, J., "Research Labs Galore," *Nature* 297:279, 1982.
27. *Nature,* "Crunch in West Germany," 298:108, 1983.
28. Strack, G., "Bonn Entices Bio-Scientists," *Frankfurter Rundschau,* Frankfurt/Main, February 1983, translated by the Foreign Broadcast Information Service in *West Europe Report: Science and Technology* (Joint Publications Research Service, Arlington, Va.).
29. Tooze, T., "Few Changes for West Germany," *Nature* 299:569, 1982.

UNITED KINGDOM REFERENCES

30. Advisory Council for Applied Research and Development, Advisory Board for the Research Councils, and the Royal Society, *Biotechnology: Report of a Joint Working Party* (The Spinks' Report) (London: H.M. Stationery Office, March 1980).
31. Beardsley, T., "More Help," *Nature* 300:394, 1982.
32. Beardsley, T., "Cans and Can'ts for Government," *Nature* 305:172, 1983.
33. *Business Week,* "A Biotech Venture Lures Some Big Bettors," Aug. 15, 1983, pp. 15-16.
34. Cookson, C., "Biotechnology Begins To Flower," *The Times,* June 9, 1983.
35. Department of Education and Science (U.K.), "Government Response to the Select Committee on Biotechnology Report of July 29, 1982," Feb. 2, 1983.
36. Department of Industry (U.K.), "Minutes of Evidence to Education, Science and Arts Committee on Biotechnology," 289:vi:119:338 (London: H.M. Stationery Office, May 24, 1982).
37. Dickson, D., "British Universities in Turmoil," *Science* 217:813, 1982.
38. Diggle, G. E., and Griffen, J. P., "Licensing Times in Granting Market Authorizations for Medicines—A Comparison Between the U.K. and USA," *Pharmacy International,* July 1982.
39. Fishlock, D., "Biotechnology Fund Still Seeks British Investment," *Financial Times,* Jan. 11, 1983.
40. Fishlock, D., "How To Make a Nobel Prize Pay," *Financial Times,* May 13, 1983.
41. Haydon-Baillie, W., L.H. Fermentation, London, personal communication, June 1983.
42. *Nature,* "Watchdog To Bark Less Often," 302:467, 1983.
43. *New Scientist,* "Biotech Fund Finds Hope in Britain," 99:748, Sept. 15, 1983.
44. Norton, M., First Secretary Science, British Embassy to the United States, personal communication, May 1983.
45. Pearson, R., Institute of Manpower Studies, University of Sussex, personal communication, June 1983.
46. Vaquin, M., "Biotechnology in Great Britain," contract report prepared for the Office of Technology Assessment, U.S. Congress, December 1982.
47. Yanchinsky, S., "British Leader in Biotechnology Faces Cash Crisis," *New Scientist* 97:704, Mar. 17, 1983.

SWITZERLAND REFERENCES

48. *Biotechnology Newswatch,* "Biotechnology in Switzerland—Parts I and II," 3(8):6-7 and 3(9):6-7, 1983.
49. Ciba-Geigy, Ltd., *Annual Report,* Basel, 1981.
50. E. F. Hutton, "Biotechnology Overview," Jul. 7, 1983, p. 6.
51. Ernst and Ernst, *International Business Series,* a series of pamphlets discussing the tax laws of various countries, New York, 1975.
52. Executive Office of the President, Office of Science and Technology Policy, "Report of a Working Group on Competitive and Transfer Aspects of Biotechnology," Washington, D.C., May 27, 1983.
53. Jasanoff, S., "Public and Private Sector Activities in Biotechnology: Switzerland," contract report prepared for the Office of Technology Assessment, U.S. Congress, January 1983.
54. Sharp, P., Biogen, Inc., Cambridge, Mass., personal communication, May 1983.

FRANCE REFERENCES

55. *AFP Sciences,* "Scientific Research Policy and Organization," July 22 1982, pp. 4-5.
56. Dickson, D., "The Party is Over for French Science," *Science* 220:935, 1983.
57. *Le Monde* (French), "Commentary on the National Agency for the Evaluation of Research," Jan. 2, 1983, p. 6.
58. Ministère de la Recherche et de l'Industrie (Ministry of Research and Industry), Mission des Biotechnologies (Biotechnology Mission), "Programme Mobilisateur: l'Essor des Biotechnologies" (Mobilizing Program: The Expansion of Biotechnology), Paris, July 1982.
59. Pelissolo, J., *La biotechnologie, demain?* (Paris: La Documentation Française, 1981).
60. Pinon, J. C., Scientific Attaché, Embassy of France to the United States, personal communication, May 1983.
61. Porges, A., "Trade and Investment Laws and U.S. Competitiveness in Biotechnology," contract report prepared for the Office of Technology Assessment, U.S. Congress, April 1983.
62. Vaquin, M., "Biotechnology in France," contract report prepared for the Office of Technology Assessment, U.S. Congress, December 1982.
63. Walgate, R., "Great Schools, Great Contradictions," *Nature* 296:287-288, 1982.
64. Walgate, R., "CNRS—The Core of Research," *Nature* 296:292-293, 1982.
65. Walgate, R., *"La Loi* at Last," *Nature* 298:110, 1982.
66. Walgate, R., "Let the Good Times Stop," *Nature* 298:110, 1982.
67. Zotov, A., and Wilson, T., "French Biotechnology in the New Regime," *Bio/Technology* 1:3, 1983.

SWEDEN REFERENCES

68. Dampier, W., "The Biotech Boom Town," *Sweden Now,* pp. 23-25, May 1982.
69. Erland, R., "World Interest in New Version of Swedish DNA Machine," *NY Teknik,* Stockholm, Dec. 3, 1981, translated by the Foreign Broadcast Information Service in *West Europe Report: Science and Technology* (Joint Publications Research Service, Arlington, Va.).
70. Fairlamb, D., "Scandanavian Setback," *Dun's Business Month* 120(6):86-88, 1982.
71. Haijlen-Rylander, M. L., mailgram from U.S. Embassy, Stockholm, U.S. Department of State, July 18, 1980.
72. Henriksson, K., "Swedish Industries' Interest in Biotechnology Increasing," *Svenska Dagbladet,* May 18, 1982, translated by the Foreign Broadcast Information Service in *West Europe Report: Science and Technology* (Joint Publications Research Service, Arlington, Va.).
73. Malmgren, Inc., and Developing World Industry and Technology, Inc., "Technology and Trade Policy, Issues and Agenda for Action," prepared for the Department of Labor, Bureau of International Labor, and the Office of the United States Trade Representative, Washington, D.C., October 1981.
74. *Marketplace Sweden,* "Swedish Pharmaceuticals: Cresting on Product Development," 14(4):11-12, 1982.
75. Royal Swedish Academy of Engineering Sciences, "Industry and Biotechnology, Developments Internationally and in Sweden," Report –214, Stockholm, December 1981.
76. Skagius, K., "Can SBF Offer Danish Companies New Policies New Possibilities?," speech, Nov. 25, 1982.

NETHERLANDS REFERENCES

77. Schuuring, C., "New Era Vaccine," *Nature* 296:792, 1982.
78. Van Der Meer, Ir. R. R., Secretary—Coordinator, Programme Committee on Biotechnology, Gravenhague, personal communication, April 1983.

AUSTRALIA REFERENCES

79. Commonwealth Scientific and Industrial Research Organization, *Biotechnology Research and Development, The Application of Recombinant DNA Techniques in Research and Opportunities for Biotechnology in Australia,* Canberra, November 1981.
80. Department of Science and Technology (Australia), Bank of New South Wales, Australian Institute of Management, NSW Division, "Genetic Engineering—Commercial Opportunities in Australia," proceedings of a symposium held in Sydney, Nov. 18-20, 1981 (Canberra: Australian Government Publishing Service, 1982).

ISRAEL REFERENCES

81. *Bioengineering News,* "Yet Another Offering: Interpharm," June 1981, p 6.
82. *Biotechnology Newswatch,* "Biotechnology Research and Development in Israel," Oct. 5, 1981, p. 7.
83. *Biotechnology Newswatch,* "Interferon From Interpharm Enters Anti-Herpes Trials," Mar. 1, 1982, p. 1.
84. *Biotechnology Newswatch,* "Israel, West Germany Join Forces in Plant Biobreeding," May 16, 1983.
85. Executive Office of the President, Office of Science and Technology Policy, "Report of a Working Group on Competitive and Transfer Aspects of Biotechnology," Washington, D.C., May 27, 1983.
86. Meyers, N., "Farming Today," *Nature* 297:354, 1982.
87. Zilinskas, R., "Biotechnology in Israel," a report prepared for the Office of Technology Assessment, U.S. Congress, 1983.

CANADA REFERENCES

88. Bates, A., President, Allelix Corp., Toronto, personal communication, July 1983.
89. *Biotechnology Newswatch,* "$105 Million Starts 'Allelix,' Canadian R&D Firm," Jan. 4, 1982, p. 1.
90. *Biotechnology Newswatch,* "Biotechnology Research and Development in Canada," Jan. 4, 1982, pp. 4-5.
91. *Chemical Week,* "Canada's Biotech Plan Gets Chilly Reception," 132(21):17, 1983.
92. Farina, C., Director University Science and Technology Branch, Ministry of State for Science and Technology, Ottawa, personal communication, July 1983.
93. Miller, J., "Biotechnology Studies in Canada," notes for a presentation to the Canada-Europe Conference on Technological Planning, Ottawa, June 1-3, 1983.
94. Ministry of State, Science and Technology, "Towards 1990: Technology Development for Canada," Government brochure describing the new technology policy (Ottawa: Minister of Supply and Services Canada, 1983).
95. National Research Council Canada, "Institute for Plant Biotechnology in Saskatoon," communique', Ottawa, May 12, 1983.
96. Task Force on Biotechnology, *Biotechnology: A Development Plan for Canada,* a report to the Minister of State for Science and Technology (Ottawa: Minister of Supply and Services Canada, 1981).
97. Zeman, Z., Institute for Research on Public Policy, Montreal, personal communication, July 1983.

U.S.S.R. REFERENCES

98. Executive Office of the President, Office of Science and Technology Policy, "Report of a Working Group on Competitive and Transfer Aspects of Biotechnology," Washington, D.C., May 27, 1983.

BRAZIL REFERENCES

99. Goodrich, R., "An Analysis of Biotechnological Activity in Brazil," contract report prepared for the Office of Technology Assessment, U.S. Congress, July 21, 1982.

Appendix C

A Comparison of the U.S. Semiconductor Industry and Biotechnology*

Introduction

A parallel is sometimes drawn between the early development of the U.S. semiconductor industry and biotechnology. There are similarities. Semiconductors and biotechnology each showed promise for major advances. Whereas semiconductors immediately showed promise for major advances in electronics, biotechnology shows promise for major advances in many industries, from agriculture to oil recovery. Furthermore, developments in semiconductors and in biotechnology have both been characterized by the pioneering efforts of small startup companies, which have played a major role in technological innovation. Another reason for drawing a parallel between the U.S. semiconductor industry and firms using biotechnology is probably the hope that the development of biotechnology will be accompanied by the same kind of intense competition, continuing innovation, wide commercial diffusion, and spectacular financial returns that characterized the U.S. semiconductor industry.

As will be seen in this appendix, the early history of the U.S. semiconductor industry and the history of biotechnology to date are in fact characterized more by differences than by similarities. Nevertheless, studying the history of the U.S. semiconductor industry may aid the healthy development of biotechnology in the United States. Some of the actions that fostered the development of the U.S. semiconductor industry could be applied to the further development of biotechnology, thereby increasing its similarity to the semiconductor industry. The clear success of the U.S. semiconductor industry suggests that such actions deserve consideration for their applicability to biotechnology, although biotechnology is not an industry, but a set of technologies that can potentially be used by many industries.

The purpose of this appendix is to clarify the similarities and differences between the early history of the U.S. semiconductor industry and the development of biotechnology, to identify factors contributing to the successful development of the semiconductor industry, and to consider the relevance of these factors to the further development of biotechnology.

*The primary source for this comparison was a contract report prepared for OTA by Michael Borrus and James Millstein (2).

Semiconductor devices: terminology and evolution

Semiconductors are materials such as silicon and germanium with electrical conductivities intermediate between good conductors, such as copper, and insulators, such as glass. By appropriate manipulations, these materials can be made into *semiconductor devices* that have special properties. Such devices include diodes and transistors.

One of the most important properties of a *transistor* is its ability to amplify an electrical current flowing through it. A transistor is a compact, reliable replacement for the vacuum tube, which was the foundation of the early electronics industry. While transistors substantially improved the reliability and performance of electronic devices such as computers, they were simply components in electrical circuits connected by wires to other components.

Integrated circuits were the next major advance in semiconductor technology. Integrated circuits are "chips" or single components that perform functions that had previously required groups of components wired together.

The next step in semiconductor technology involved increasing the density of circuit elements on each chip. The integrated circuit era began in the early 1960's. By the end of the decade, medium-scale integration (MSI) had been achieved (10 to 100 digital logic gates on one chip). Large-scale integration (LSI) (100 to 1,000 gates) was achieved in the mid-1970's, and the industry is now working on very large-scale integration (VLSI) (circuit complexity exceeding 1,000 gates) (9).

Advances in semiconductor technology have resulted in extraordinary gains in reliability and performance, with simultaneous reductions in component size and cost. In the 1950's, for example, the cost of computer memory capacity was about $1 per bit, but by 1981, a bit could be purchased for only $0.0001 (9).

The U.S. *semiconductor industry* is comprised of the companies that manufacture semiconductor devices such as transistors and integrated circuits. Two types of firms can be differentiated: 1) firms that develop and manufacture semiconductor devices for sale to other firms that use them to manufacture computers and other end products; and 2) firms that develop and manufacture semiconductor devices for in-house use in the manufacture of final products. Both types of firms have been important to the development of the industry.

The following material describes the early development of the U.S. semiconductor industry and compares it to the short history of biotechnology. For the semiconductor industry, the period covered is from 1947 (the invention of the transistor) to the early 1960's. For biotechnology, which began in the mid-1970's, the period covered is from the mid-1970's to the present. In part because of the different time periods in which the semiconductor industry and biotechnology initially developed, an immediate difference between the two can be identified. The early development of the U.S. semiconductor industry occurred primarily in the context of the U.S. domestic market, whereas biotechnology is evolving in a world marketplace. International competition, which is an important factor in the development of biotechnology, is a far more important factor in the semiconductor industry now than it was in the early history of the industry. Both differences and similarities between the development of the U.S. semiconductor industry and biotechnology are indicated in the material that follows.

Development of the U.S. semiconductor industry

Two major influences in the development of the U.S. semiconductor industry were Bell Telephone Laboratories (Bell Labs) and the U.S. Government. These two influences are intimately related, because the Federal Government played a major role in shaping Bell Labs' contribution to the preeminence of the United States in high-technology electronic products including semiconductors, lasers, and computers. These industries have been built, in large measure, on the results of research undertaken at Bell Labs.

The role of Bell Labs in the development of the U.S. semiconductor industry is briefly described below. The multifaceted role of the Federal Government is discussed in the section that follows.

THE ROLE OF BELL TELEPHONE LABORATORIES

As part of the American Telephone & Telegraph Co. (AT&T), Bell Telephone Laboratories does fundamental and applied research in many areas to benefit its parent company. Bell Labs also serves a broader constituency. During World War II, for example, Bell Labs undertook about 2,000 research and development (R&D) projects for the U.S. Army, U.S. Navy, and National Defense Research Council (11). Federal funding of research at Bell Labs and AT&T's manufacturing arm Western Electric from 1949 to 1959 amounted to about $609 million—or about 48 percent of all AT&T research (17). The quality of research at Bell Labs and the level of funding available from corporate and Government sources attracted the most competent electronics scientists and engineers to work there.

In the late 1930's, the electronics industry depended on the vacuum tube for amplification of electric currents. The advantages of a smaller, more reliable device that would generate less heat were obvious, however, and because of military and aerospace needs, there was strong motivation to invent an alternative. Also clear was the potential importance of the transistor to commercial communications and computer applications. It is not surprising, given Bell Labs' commanding position in fundamental and applied electronics research, that the first new device that could compete with the vacuum tube in the marketplace, i.e., the transistor, was invented in 1947 at Bell Labs. This invention gave Bell Labs a lead in what would ultimately become the semiconductor industry.

Semiconductor R&D by Bell Labs was supported with corporate funds from AT&T. Between 1946 and 1964, Bell Labs' annual expenditures on semiconductor R&D rose from less than $1 million to about $22 million. In 1959, the funding of semiconductor R&D at Bell Labs represented about 30 percent of all privately funded semiconductor R&D in the United States (14).

The fact that Bell Labs was part of AT&T also contributed to Bell Labs' leadership in the semiconductor industry (2). The research done at Bell Labs was linked to real-world problems through AT&T's manufacturing arm, Western Electric. Western Electric involved Bell Labs in the solution of engineering problems associated with conversion from vacuum tube to semiconductor technology in communications systems. Western Electric also involved Bell Labs in research to improve *production* of semiconductor devices. In addition to conducting research that led to new devices, therefore, Bell Labs did research that led to process innovations. It was these process innovations that dramatically decreased the cost of semiconductor devices (2).

Federal and corporate investment in Bell Labs produced significant return. Between 1947 (invention of the transistor) and 1959 (invention of the integrated circuit at Texas Instruments and Fairchild), Bell Labs obtained 339, or more than 25 percent, of the patents related to the development of semiconductors. During this period, Bell Labs also was responsible for a disproportionate share of the most important product and process innovations (14).

In summary, market pull for an alternative to the vacuum tube favored the development of the semiconductor industry. The key invention, the transistor, arose from fundamental R&D in an industrial laboratory. That laboratory was an arm of a major corporation that also would be a significant user of the new technology.

The history of biotechnology is quite different from the early history of the U.S. semiconductor industry. Biotechnology arose from basic research in universities—research supported by Federal funds for basic biomedical research. Probably most significant were Federal funds for research associated with the "war on cancer." Because of the "war on cancer," a great deal of research was done on tumors and tumor viruses. One of the simplest viruses, SV40, causes tumors in hamsters and mice. Researchers went to great effort to locate the genes in SV40 that enabled it to cause tumors. A need to improve on tedious genetic selection procedures for mapping genes led to the identification and use of restriction enzymes that cut DNA in specific locations, and thus enabled physical mapping of genes. Restriction enzymes also produce the "sticky ends" that are fundamental to recombinant DNA (rDNA) experiments. Physical mapping of an entire genome (an organism's complete set of genes) using restriction enzymes was first accomplished with SV40. And it was a proposed rDNA experiment using SV40 that gave rise to the Asilomar meeting that eventually led to the National Institutes of Health (NIH) Guidelines for Research Involving Recombinant DNA Molecules.*

Other researchers concentrated on myelomas (neoplastic growth of certain white blood cells). Thus, cancer research probably also contributed to the discovery of hybridomas** and the monoclonal antibodies they make possible.

In summary, cancer research played a significant role in the history of biotechnology and is another example of how fundamental research may produce unexpected results. In the development of biotechnology, "science push," rather than the "market pull" that gave impetus to the U.S. semiconductor industry, was particularly important.

THE ROLE OF THE U.S. GOVERNMENT

The actions of the U.S. Government that influenced the development of the U.S. semiconductor industry were many and diverse. Undoubtedly, not all the effects of the Federal Government's actions were intended or anticipated. With the benefit of hindsight, however, it is apparent that these actions helped to produce a dynamic, healthy U.S. semiconductor industry. Similar actions by the Federal Government could encourage the development of companies in other high-technology fields such as biotechnology.

Federal Funding of Semiconductor Research and Development To Encourage Competition.—In the late 1940's, the U.S. Department of Defense (DOD) wanted to miniaturize and increase the reliability of electronic devices so that a new generation of defensive weapons could be developed. Defensive missile systems, in particular, required these advances. To ensure achievement of its objectives, DOD distributed R&D funds to many research houses, including Bell Labs. The provision of funding to many research houses encouraged the competitive development of semiconductor technology throughout the U.S. electronics industry. It also had the effect of leveraging private funding of semiconductor R&D (2).

The same forces driving military interests—miniaturization and reliability—also applied to the U.S. aerospace program. In addition to DOD, therefore, the National Aeronautics and Space Administration (NASA) also became a major source of funding for semiconductor R&D.

It is important to note that the early development of semiconductor technology was dominated by the interests of the U.S. military and NASA (2). Civilian applications followed. This early predominance of military interests driving the development of semiconductors contrasts with the development of biotechnology, for although there are military applications of biotechnology, civilian commercial interests have driven its development.

Federal Funding of Demonstration Projects, Production, and Consumption of Semiconductor Devices.—Demonstration projects using semiconductor technology were financed by the Federal Government. The U.S. Air Force, for example, funded a demonstration in which a small digital computer using integrated circuits was built by Texas Instruments (1). Demonstration projects such as this convincingly demonstrated to both military and civilian users the feasibility of using integrated circuits in electronic systems (2).

In addition to funding demonstration projects, the Federal Government funded the development of semiconductor production capability and provided a market for semiconductor products under industrial preparedness contracts in 1952-53 and 1956-57. In 1952-53, $11 million of DOD funds were used to build pilot transistor production lines at five sites operated by Western Electric, General Electric (GE), Raytheon, RCA, and Sylvania (10). In 1956, DOD provided major assistance to production technology with $40 million

*These U.S. guidelines for rDNA research are discussed in *Chapter 15: Health, Safety, and Environmental Regulation*.

**Hybridomas are made by fusing an antibody-producing spleen cell with a myeloma cell.

in transistor production contracts to 12 firms. Because early production was often faulty and about 90 percent of the devices produced could not meet Federal specifications, the 12 firms had to build production facilities potentially capable of manufacturing 10 to 12 times the number of devices the Government wanted, thus assuring the Government of the number of devices it needed (19). As processes improved, more and more usable devices came off each assembly line, and the search for new commercial markets was stimulated by the need to absorb increases in production capacity.

The actions of the Federal Government just outlined helped to demonstrate the value of semiconductor technology to users other than the Federal Government, greatly reduced the risk of developing and producing semiconductor devices, and helped to develop industry capacity to produce semiconductor devices at levels that would meet the needs of new users as well as those of the Federal Government.

The Federal Government could support in biotechnology, just as it did in the semiconductor industry, the development of process and production technology. These are the very areas in biotechnology where needs for funds and for innovation are high. It is also in process and production capability and capacity that the United States is least competitive with Japan, its major competitor in biotechnology (2). One area of biotechnology that might be stimulated by a bioprocess production and demonstration project is the production of commodity chemicals. Large-scale bioprocess facilities, and hence large financial investments, will be necessary for U.S. firms using biotechnology to successfully enter the commodity chemical market. Cetus Corp. made an attempt to enter this market with its fructose-alkene oxide process using Standard Oil of California (SOCal) as financial backer. The attempt was frustrated when SOCal decided to terminate its backing (2). Federal funds could help new biotechnology firms (NBFs)* enter commercial markets requiring large-scale production. Alternately, rather than funding specific projects at particular firms, the Federal Government could support R&D in generic technology underlying bioprocessing. Regardless of the particular form of support, the Government should ensure that new knowledge of bioprocess technology gained with the assistance of Federal funding is made available to other potential users.

Federal Government support of field and clinical trials necessary for approval of some products of biotechnology by the U.S. Department of Agriculture (USDA) and the Food and Drug Administration (FDA) would be somewhat analogous to the federally funded semiconductor demonstration projects. Such trials are very expensive and beyond the financial resources of many small firms.

The 1956 Consent Decree.—In the development of the semiconductor industry, the Federal Government provided more than dollars, useful as these were to fund R&D, build production lines, demonstrate their products, and provide a first market. Substantial Federal dollar investments were accompanied by less direct policy decisions that helped shape a highly competitive U.S. semiconductor industry. The 1956 consent decree is a case in point.

In 1949, the U.S. Department of Justice initiated an antitrust suit against AT&T. Resolved in 1956, the consent decree (20) required AT&T's manufacturing arm Western Electric to license existing Western Electric patents to U.S. firms without royalty and to establish reasonable rates for licenses under future patents. AT&T was permitted to retain its vertically integrated structure but was prohibited from entering new product markets; in other words, AT&T was restricted to its existing markets of basic common carrier communications and Government defense and aerospace. Thus, AT&T was prohibited from using the results of research at Bell Labs to enter additional commercial markets that semiconductor technology promised to advance, such as commercial electronic computers.

Given the consent decree, one option for AT&T would have been to redirect Bell Labs' research so that it would not benefit fields AT&T could not enter. However, semiconductor R&D directed to enhancing AT&T's major interests in the telecommunications, military, and aerospace markets was not separable from R&D applicable to areas such as commercial computers from which AT&T was prohibited. In addition, Bell Labs had a history of open communication regarding its research. As a result, AT&T conformed not only to the letter but also to the spirit of the 1956 decree. The effect was to transform Bell Labs, for a time, into a sort of national laboratory for semiconductor R&D.

Continuing its open practices begun prior to the consent decree, Bell Labs actively contributed to the diffusion of the technology that it helped develop. Symposia to educate Government users and small and large firm licensees were begun in 1951, and a liberal license policy was begun in 1952. Also important, Bell Labs and Western Electric personnel moved freely to new employment in firms exploiting the results of Bell Labs R&D without fear of suit for theft of trade secrets (18). Such movements transferred know-how developed at Bell Labs and Western Electric to other firms.

*NBFs, as defined in *Chapter 4: Firms Commercializing Biotechnology*, are new, generally small firms that have been formed specifically to capitalize on new biotechnology.

Liberal licensing, the educational activities of AT&T, and personnel mobility encouraged by Federal antitrust activity resulted in wide diffusion of semiconductor technology. Diffusion was facilitated by the fact that data acquired under DOD R&D contracts were subject to unlimited use by the Government, including their supply to other contractors working in related areas. Various DOD offices and agencies, and DOD-funded centers at universities, served as information centers for research findings. The U.S. Department of Commerce (6), National Science Foundation, National Bureau of Standards (4), and NASA (13) served as clearinghouses for semiconductor information and transferred knowledge derived from military contracts to civilian users. Government agencies held symposia and colloquia to inform industrial contractors of the results of federally funded research and of future military and space requirements. The result was an acceleration in the pace, and hence the competitiveness, of the U.S. semiconductor industry, in civilian as well as military markets. In 1961, the Army Signal Corps estimated that defense R&D had made possible many civilian applicatons of semiconductor technology in a period perhaps 75 percent shorter than that which would have occurred without Government support (17).

In biotechnology, there is no institutional equivalent to Bell Labs, which served as a national resource for semiconductor research, development, education, and personnel. Furthermore, the scope and magnitude of Federal actions facilitating diffusion of knowledge and know-how in the area of semiconductors have no parallel in biotechnology at present. Finally, the diffusion of technology by personnel mobility that occurred in the semiconductor industry because of the commanding position of Bell Labs, which was restrained by the 1956 consent decree, is unlikely to occur to the same degree in biotechnology, where knowledge is spread among many competing firms.

Federal Loan and Tax Policies.—In the 1950's and 1960's, the U.S. Government also encouraged the development of the U.S. semiconductor industry through Federal loan guarantees and tax policies. Although not developed specifically for the semiconductor industry, these general policies made funds available for operations, plant investment, and new equipment.

The Defense Procurement Act of 1950 established the V-loan program and was a major source of Federal loan guarantees to defense contractors from 1950 to 1958. This act provided Federal loan guarantees that obligated the Federal agency guaranteeing the loan to purchase a stated percentage of the loan if the borrower defaulted. Thus, the Federal agency shared any potential losses up to the amount of the guaranteed percentage (16). Such guaranteed loans accomplished several things:

- They encouraged private investors by decreasing their risk of loss.
- Because they were granted at lower than prevailing interest rates, they decreased the cost of capital.
- They served as a system of revolving credit. Guarantees were not tied to particular loans but instead were guarantees against loss of a particular level of debt. As periodic repayments reduced outstanding debt, therefore, additional loans could be taken out as long as repayments kept debt within the face amount of the authorization. Thus, authorizations of only $2.9 billion allowed loans totaling about $11.6 billion to be made to defense contractors.
- They returned a net profit to the Federal Government of about $24.5 million (15). This profit resulted because the Federal guaranteeing agent was entitled to a portion of the interest paid on the loan.

Most of the funding leveraged by the V-loan program was used for working capital rather than facilities. Other Government financial aids produced additional working capital. Progress payments, advance payments, and direct loans were made to companies involved in defense production (16).

A particularly important financial instrument encouraging investment in defense production capability was a program permitting accelerated depreciation. In the 1950's, the Office of Defense Mobilization awarded Certificates of Necessity that provided a 5-year writeoff (compared to the usual 20- to 25-year amortization schedule) of the percentage of the cost of certified production facilities that could be attributed to major defense production needs. From November 1950 through April 1957, 21,925 Certificates of Necessity permitted the accelerated writeoff of almost $23 billion on facilities costing $39.2 billion (15). Although these figures include more than semiconductor firms and data do not permit isolation of their share, semiconductor firms definitely received Certificates of Necessity and their writeoff was surely substantial (5).

The growth of the U.S. semiconductor industry was further spurred in 1962 by two changes in general U.S. tax policy (2). One change was that the Revenue Act of 1962 permitted all manufacturing industries an investment tax credit of up to 7 percent of qualified investment in machinery and equipment. This investment tax credit stimulated investment in semiconductor production capacity just when integrated circuit procurement began to expand. The second change was adoption by the U.S. Treasury Department in 1962 of new regulations that shortened depreciation guidelines by 15 to 20 percent.

Clearly, Federal tax and loan policies can stimulate substantially the growth of emerging industries. Consideration might be given to whether current tax and loan policies are stimulating development of biotechnology adequately or whether additional Government financial instruments are needed.

Defense Laboratory Research.—During the 1950's and early 1960's, each branch of DOD developed intramural programs for semiconductor R&D. Although these defense facilities produced relatively few significant semiconductor discoveries (with some major exceptions) (21), they nonetheless played a major role in the development of the semiconductor industry. In addition to serving as centers for information and technical liaison, these laboratories tested theories and ideas considered too speculative by private industry. Those that turned out to be practical were then developed by industry (7). Furthermore, personnel movements from defense establishments to private industry served to transfer knowledge, sometimes at critical points in the development of the U.S. semiconductor industry (23). Especially important, defense laboratory researchers provided the Federal Government with an independent view of the state-of-the-art of semiconductor technology and the capacity to verify, assist, and at times lead industrial efforts.

In terms of level of expertise and dynamic interaction between Federal agencies and industry, the closest analogs in biotechnology are NIH and FDA. Because it issues the U.S. guidelines for rDNA research, however, NIH is a quasi-regulator of biotechnology. This role puts NIH in a conflict of interest position vis-a-vis both its substantial funding of basic research in biotechnology and any additional role it might assume in commercialization. NIH, which has been forced to be aware of developments in the commercialization of biotechnology by the guidelines, however, nevertheless has a major potential role in biotechnology transfer. The degree to which and how best to involve NIH in commercial development of biotechnology deserve consideration.

FDA has developed expertise in biotechnology because of its regulatory function. Its major contribution to the development of biotechnology to date has been in providing a favorable regulatory climate for new products. However, the present regulatory climate is highly subject to administration views on industry regulation. Whether U.S. regulatory agencies should be better insulated from the effects of changes in administrations so that biotechnology evolves in a relatively stable environment deserves thought. In any case, an increased role for FDA in fostering the development of biotechnology is probably prohibited by conflict of interest with its significant regulatory responsibilties.

Other relevant U.S. Government agencies, such as DOD, the Environmental Protection Agency (EPA), the National Bureau of Standards, the National Science Foundation, the Occupational Safety and Health Administration (OSHA), and USDA have so far been less involved in the development of biotechnology than either NIH or FDA.

In sum, the substantial role that DOD and NASA played in encouraging the early development of the U.S. semiconductor industry is a role that is not being played by the U.S. Government in the commercial development of biotechnology.

THE ROLE OF UNIVERSITIES

During World War II, the successful funding of defense developments at universities gave rise to a conscious national policy of U.S. Government funding of university basic research. Although Federal funds for joint research at universities and industrial laboratories in solid-state physics and materials helped provide the basis for the U.S. semiconductor industry (22), the key discovery leading to the transistor was made in an *industrial* laboratory.

In the early 1950's, university electrical engineering departments lagged behind industry in the area of semiconductors by a considerable margin.* Federal funds were provided to universities to help reduce this gap and build the university expertise and training capacity that would be needed to support the expansion of the U.S. semiconductor industry.

These Government expenditures were fruitful. By roughly 1960, the major research universities in the United States had highly trained electronics personnel, creative basic research programs, and faculty members who served as expert consultants to industry.

Furthermore, the U.S. semiconductor industry became concentrated geographically around the major university recipients of Federal dollars, in particuiar, in Boston and San Francisco. The geographic proximity of semiconductor firms and these universities fostered productive interchange and insured the continued buildup of university expertise.

Increasingly cooperative ties between U.S. universities and the semiconductor industry resulted in the part-time employment by the industry of significant numbers of students. Many university faculty mem-

*Massachusetts Institute of Technology's Lincoln Laboratories is an exception.

bers served as directors of semiconductor corporations, and some even held positions such as board chairman and part-time company president (2). Some faculty members became millionaires through equity participation in the companies with which they were associated (2). In comparison with the protests that have been raised in reaction to similar arrangements in biotechnology, public protests against these arrangements were small.

In sum, in the early history of the U.S. semiconductor industry, few innovations emerged from federally funded university research. The universities used Federal dollars to bring their expertise up to a level commensurate with industry's and to become geographic foci for the development of the new semiconductor industry. In the case of biotechnology, by contrast, innovations have emerged directly from university research. New semiconductor firms tended to locate near major university research institutions. This colocation occurred fairly gradually as Federal dollars flowed to universities and helped build their expertise. In the case of biotechnology, the colocation of new firms and universities occurred immediately, because the universities were the site of biotechnology expertise (2).

The lack of public and congressional concern over equity ownership of semiconductor companies by university professors is in stark contrast to the reaction to similar arrangements in biotechnology. Some of the factors that may account for the differences include the following:

- The locations from which biotechnology and semiconductor technology emerged and the source of their expertise, coupled with patterns of Federal spending, are different. Semiconductor R&D was dominated by industry, especially in its early years, and Federal funds went to industry for the development of the technology. Federal funds to the universities were used very differently from Federal funds to industry, namely, to build the scientific infrastructure necessary to support the new industry. Thus, the roles played by universities and industry and the use of Federal funds in the two sectors were more distinct in the early years of the semiconductor industry than they have been in biotechnology.
- Many recent advances in research in biotechnology immediately suggest commercial products. Although there are many problems to be solved between, for example, cloning the gene for human insulin and market success, the potential marketability of the product of the research is obvious immediately. In addition, the DNA organism that makes insulin, is, in a sense, itself the product. A transistor, on the other hand, is of no value unless it is used with other electronic components to make an end product such as a missile guidance system. Thus, in biotechnology, the contributions of the universities and industry are less distinct than they were in the semiconductor industry.
- The semiconductor industry had obvious contributions to make to aerospace and defense. Defense and aerospace are seen as national objectives and national commitment to them tends to be stronger and more focused than commitment to other sectors of the economy, where biotechnology is making its first contributions. Actions that would be protested otherwise may be tolerated when they relate to meeting defense and aerospace needs.

Structure of the U.S. semiconductor industry

Industries develop unique structures in response to their own characteristics and the effects of external forces acting upon them. The forces that have been described in this appendix shaped the U.S. semiconductor industry so that its particular structure evolved from a myriad of possible structures, much as biological systems evolve in response to pressures of selection. The structure that emerged in the semiconductor industry consisted of three types of companies:

- small, new entrepreneurial firms that developed and manufactured semiconductor devices, the so-called "merchant" firms;
- generally larger, established companies that obtained most or all of their semiconductor devices from the merchant firms and incorporated them into electrical systems; and
- one very large, vertically integrated company, AT&T, that manufactured semiconductor devices for use in its telecommunications systems but was constrained by antitrust policy from dominating other markets.*

The role of AT&T, along with its affiliates Bell Laboratories and Western Electric, has already been discussed. The rest of this section describes the relationships between the other two groups of firms.

The emergence of new entrepreneurial firms in the U.S. semiconductor industry was facilitated by U.S. Government policies and actions, such as the 1956 consent decree and military and aerospace demands. Information on semiconductor technology was widely available, and personnel mobility was not effectively discouraged. AT&T's liberal licensing policy, a U.S.

*Later in the history of the semiconductor industry, a second vary large, vertically integrated firm, IBM, was added to this group. IBM manufactured semiconductor devices for its own use in the computer industry.

Government market for new products, and the fact that transistors could be *substituted* for vacuum tubes meant that an entrepreneur could start a new semiconductor firm and move immediately to market with a few million dollars of capital, a license from AT&T, and a DOD or NASA contract.

Larger U.S. companies were helped in establishing their position in the semiconductor industry by the patterns of DOD development and procurement established during World War II that favored large corporations. "Even as late as 1959 the old-line vacuum tube companies were awarded 78 percent of the federal R&D funds devoted to improving the performance and reducing the cost of the transistor although they accounted for only 37 percent of the product market" (3). In contributing to the development of transistors and integrated circuits, the large defense electronics companies were speeding the obsolescence of a technology in which they had a very large investment, vacuum tubes. The large companies were forced into this position, however, by the presence of small entrepreneurial firms that managed to obtain DOD funds by their more flexible and rapid response to DOD's demands for miniaturization and reliability. The small, new firms undoubtedly contributed to the speed of entry of the large companies into semiconductor technology.

Small entrepreneurial firms did contribute to innovation in semiconductors, but preeminence in that role went to Bell Laboratories. In the development of the U.S. semiconductor industry the major contributions of small firms were to diffuse semiconductor technology and to stimulate competition. Diffusion of semiconductor technology occurred because the small firms exploited new markets. It was they who most "quickly and successfully (took) new technology from the laboratory and adapted it for large-scale production" (14). The small firms also stimulated competition. In effect, the small firms, as independent sources of advanced semiconductor technology, introduced an element of dynamic uncertainty into the U.S. semiconductor industry. And because Federal policies helped them to *produce and market* their products, the small firms stimulated semiconductor R&D among all companies in the industry, large and small.

Biotechnology, as it now stands, presents a very different picture. Small NBFs in the United States, in order to spread risk and raise capital, have had to turn to complex cooperative arrangements with large domestic and foreign companies.* On the surface, the arrangements between NBFs and established companies may appear analogous to the relationship between the small new semiconductor firms and the Federal Government. An essential difference, however, is that small new semiconductor firms and the Federal Government did not compete for markets; NBFs *would* like to compete with established companies.

In the absence of support from the Government for producing and marketing its products or processes, an NBF is likely to turn to a large established company that has expertise in scale-up technology and regulatory clearance procedures. The established company is likely to have gained this expertise by developing a product similar to the one the NBF wants to bring to market. If the new product threatens an existing product of the established company, the established company's marketing of the new product is likely to be less than optimal. This is not to say that the established company will refuse to undertake the clinical trials, marketing, and distribution of the new product developed by the NBF. Indeed, the motivation of the established company is just the opposite. By obtaining a license for the NBF's new product, the established company ensures that another large competitor does not obtain the biotechnology product that threatens its own market. Furthermore, the established company can control the market environment of the new product. By entering into an agreement with an NBF, the established company also gains access to the new technology.

The arrangements between Eli Lilly and the NBF Genentech with respect to the new biotechnology product Humulin® are illustrative.* Eli Lilly has licensed this rDNA-produced human insulin product from Genentech. Humulin® is a competitor of insulin obtained from animals, and Lilly currently holds about 85 percent of the U.S. insulin market. Thus, the pace of market development in Humulin® can be controlled by the very company whose monopoly position Humulin® sales otherwise might challenge. A consequence of arrangements of this kind could be to slow market development and to reduce the flow of royalties to NBFs. Yet royalties may be necessary to NBFs' survival and certainly are anticipated by the new firms to assist them in expansion. Arrangements like that between Eli Lilly and Genentech in biotechnology go against the lessons to be learned from the evolution of the U.S. semiconductor industry. Both the pace of technological development and the growth of small, innovative semiconductor firms such as Texas Instruments might have been quite different had Texas Instruments found it necessary to license GE or RCA to get its transistor products on the market.

*These arrangements are discussed in *Chapter 4: Firms Commercializing Biotechnology*.

*These arrangements are discussed in *Chapter 5: Pharmaceuticals*.

Like the semiconductor industry in its early stages, biotechnology currently is restricted by its need for process technology. The history of process development in the evolution of the U.S. semiconductor industry is of relevance to biotechnology. As has been shown, large electronic defense contractors such as GE were assisted in developing production lines for semiconductor devices by large infusions of DOD dollars. But the history of the U.S. semiconductor industry demonstrates that small firms are not automatically foreclosed from process advances. Thus, the early growth of Fairchild Semiconductor, for example, was tied largely to its development of the planar process, which dramatically increased the firm's production yield and helped compensate it for its lack of production experience.

In the case of biotechnology, firms that exploit possibilities in both new product development and process innovation clearly will have more growth opportunities than those that restrict themselves to one or the other. In biotechnology, as in semiconductors, process know-how is probably transferable across a range of potential products and markets. Thus, if NBFs can surmount the financial hurdles to commercial production, the pace of technological advance and market development likely will be accelerated significantly, and the competitiveness of U.S. firms using biotechnology probably will be increased.

Other differences

Two other differences between the early history of the U.S. semiconductor industry and biotechnology are noteworthy. The first difference is the range of economic sectors each technology was perceived potentially to affect. For semiconductors, military, aerospace, communications, and computer applications were foreseen. All these draw primarily on the disciplines of electronics and engineering. The applications of biotechnology are perceived to be broader—pharmaceuticals, plant and animal agriculture, chemicals, pollution control, energy production, mining, oil recovery, and biosensors/biochips are areas where applications are being pursued. Not only is the array of sectors expected to be affected by biotechnology broader, the technical disciplines required for effective application of biotechnology are more numerous. Developing an effective infrastructure to support the commercialization of biotechnology, therefore, may be more complex than was developing an infrastructure to support the semiconductor industry.

The second difference is the prominent role of Federal regulation in biotechnology. NIH, through the rDNA research guidelines, is in a quasi-regulatory position with regard to both R&D and scale-up to commercial production. And for specific products of biotechnology, FDA, which regulates food ingredients and human drugs and biologics, and USDA, which regulates animal biologics, are particularly important. EPA and OHSA also may have significant regulatory authority, although their exact authority is somewhat unclear.* U.S. Government regulation in research, development, and marketing of many products of biotechnology, for which there is no parallel in the semiconductor industry, makes effective commercialization of the products of biotechnology relatively more complex.

Conclusions

Certain differences between the early history of the U.S. semiconductor industry and biotechnology are particularly important from a policy perspective:

- The U.S. semiconductor industry arose from a fundamental invention (the transistor) made at a major industrial laboratory, AT&T's Bell Telephone Laboratories, in 1947, and most of the subsequent product and process innovations in the period from 1947 to the early 1960's also were made by industry. Biotechnology arose from fundamental biomedical research in universities, and its early subject matter experts were primarily unversity professors.
- The need for development of the U.S. semiconductor industry to meet military and aerospace needs was clear. The tie between biotechnology and national objectives is less clear. The U.S. Government's role in support of basic biomedical research has been, and remains, clear, but its role in the commercialization of biotechnology is far less defined.
- At Bell Labs, early commercial exploitation of semiconductor discoveries was strictly limited to one industrial sector, communications (despite the much wider applicability of semiconductor technology). In effect, Bell Labs became, for a time, something like a national laboratory for the semiconductor industry. There is no equivalent in biotechnology.
- Many new semiconductor firms in the United States were formed to market a definite product, and, because of the availability of Federal contracts, relatively little capital was required to enter the market. Most NBFs were started as R&D houses, with the objective of determining how to make a product. With certain exceptions (e.g., in vitro monoclonal antibody diagnostic products),

*This issue is discussed in *Chapter 15: Health, Safety, and Environmental Regulation.*

the capital required to produce a biotechnology product and bring it to market will be greater than that needed by early semiconductor firms. For NBFs attempting to commercialize a new drug or biological for human use, capitalization requirements may be $50 million to $100 million.*

- The early U.S. semiconductor industry was characterized by multifaceted Federal encouragement of commercialization through a variety of policies ranging from antitrust to Federal loan and tax policies. There is no parallel to this in biotechnology.
- Biotechnology differs from the U.S. semiconductor industry in that the Federal Government is not providing substantial funds for process engineering and development of pilot and production facilities. Nor is the Federal Government serving as a "creative first market" for the products of biotechnology as it did for the semiconductor industry.
- Biotechnology also differs from semiconductor technology in the wider array of economic sectors it is perceived potentially to affect and in the larger role of the Federal Government in regulating many products of biotechnology.

Thus, NBFs currently face a very different, and much more complex, market environment than did the new entrants in the semiconductor industry. The industrial sectors in which biotechnology appears to be making its first contributions are human and animal health care, and the pharmaceutical sector has special characteristics. The market for a particular pharmaceutical product is often dominated by one or a few major corporations, as, for example, the U.S. insulin market is dominated by Eli Lilly.** The product of the dominant corporation is supported by extensive advertising in medical journals, by a complex distribution system involving detail men who provide product samples and are recognized by the physicians they serve, and by the reluctance of physicians to switch to a product with less familiar properties. The established company is also skilled in the clinical testing procedures necessary to obtain market approval. An NBF with a competing product, but without production capacity, experience in regulatory compliance, and an established marketing and distribution system within the medical community, has little choice but to license the new product to an established company that already produces a similar product. Such licensing, however, will tend to reduce the competitive stimulation to the industry that the NBF might otherwise provide.

The Federal Government was clear about its role in the development of the U.S. semiconductor industry. DOD and NASA funded the industry to produce products needed in military and aerospace applications. The Federal Government has funded basic biomedical research in university settings, but as yet it has no explicit role in the commercialization of biotechnology. Unlike semiconductor technology, biotechnology has sprung primarily from academia. As biotechnology moves to the market, universities of necessity have played a role in commercializing the fruits of public funding of research, because they were the sole source of basic knowledge. Moreover, the role of the universities has been further complicated in biotechnology by the close association between basic and applied research in this area. The traditionally distinct roles of the university as source of research and training and of industry as source of commercialization, which were clear with respect to semiconductors, are blurred for biotechnology.*

In the early history of the U.S. semiconductor industry, the Federal Government and industry were partners, with industry providing know-how and the Federal Government supplying public funds for R&D, demonstration projects, production, and consumption of semiconductor devices. Direct returns to the Federal Government, in the form of advances in defense and aerospace electronics, were obvious. In the case of biotechnology, however, not the Federal Government but the public health organizations and universities that were the sources from which biotechnology arose have been industry's partner in commercialization. As a result, an impression is left that the public is ceding the biotechnology research infrastructure and discoveries brought about by public moneys to private industry without corresponding return. The problem has been exacerbated because biotechnology emerged so quickly from the academic setting. Basic biomedical research nourished by Federal dollars is applicable suddenly to the development of commercial products.

Consideration of the differences between the early history of the U.S. semiconductor industry and biotechnology suggests several areas of need for biotechnology:

- One need is for the Federal Government clearly to distinguish basic research from commercialization and to define its different roles with regard to each.

*For discussion of the financial needs of firms using biotechnology, see *Chapter 12: Financing and Tax Incentives for Firms*.

**A profile of Lilly's share in U.S. and foreign insulin markets is presented in *Chapter 5: Pharmaceuticals*.

*University/industry relationships in biotechnology are explored at greater length in *Chapter 17: University/Industry Relationships*.

- A second need, suggested by the successful history of the U.S. semiconductor industry, is for the Federal Government to facilitate the development of NBFs so that they can compete effectively in the marketplace. As in the semiconductor industry, small firm competition would stimulate innovation by all companies, large and small.
- Related to the above is the need to develop effective mechanisms for the diffusion of knowledge developed in biotechnology.

The last is very important and is really the central issue with respect to ensuring a return to the public for the financial investment that the public has made in biotechnology.

Appendix C references

1. Asher, N., and Strom, L., *The Role of the Department of Defense in the Development of Integrated Circuits* (Arlington, Va.: Institute for Defense Analysis, 1977).
2. Borrus, M., and Millstein, J., "Technological Innovation and Industrial Growth: A Comparative Assessment of Biotechnology and Semiconductors," contract report prepared for the Office of Technology Assessment, U.S. Congress, March 1983.
3. Braun, E., and MacDonald, S., *Revolution in Miniature* (Cambridge, England: Cambridge University Press, 1975).
4. Diamond, H., "Research and Development Programs of the National Bureau of Standards," unpublished paper, Dec. 1, 1947.
5. *Electronic News*, "9 Firms Get Tax Writeoff Certificates," July 15, 1957, p. 6.
6. *Electronic News*, "2 Services-Backed Research Studies Available by OTS," Jan. 27, 1958, p. 14.
7. *Electronic News*, "Navy Electronics Lab Turns Equipment Ideas Into Reality," Mar. 28, 1960, p. 17.
8. Ginzberg, E., Kuhn, J. W., Schnee, J., et al., *Economic Impact of Large Public Programs: The NASA Experience* (Salt Lake City, Utah: Olympus Press 1976).
9. Levin, R. C., "The Semiconductor Industry," *Government and Technical Progress: A Cross Industry Analysis*, R. R. Nelson, ed. (New York: Pergamon Press, 1982).
10. Linville, J. G., and Hogan, L. C., "Intellectual and Economic Fuel for the Electronic Revolution," *Science* 195:1107, 1977.
11. Lustig, L. K. (ed.), *Impact* (Short Hills, N.J.: Bell Telephone Laboratories, 1981).
12. Organisation for Economic Co-Operation and Development, *Gaps in Technology, Electronic Components*, Paris, 1968.
13. Organisation for Economic Co-Operation and Development, *United States Industrial Policy*, Paris, 1970.
14. Tilton, J., *International Diffusion of Technology: The Case of Semiconductors* (Washington, D.C.: Brookings Institution, 1971).
15. U.S. Congress, Joint Committee on Defense Production, *Hearings on the Defense Production Act of 1950, Program Report No. 39* (Washington, D.C.: U.S. Government Printing Office, May 28, 1957).
16. U.S. Congress, Senate Committee on Banking and Currency, *Review of Voluntary Agreements Programs Under the Defense Production Act (Guaranteed Loans)* (Washington, D.C.: U.S. Government Printing Office, 1958).
17. U.S. Congress, Senate Committee on the Judiciary, Subcommittee on Patents, Trademarks, and Copyrights, *Patent Practices of the Department of Defense*, (Washington, D.C.: U.S. Government Printing Office, 1961).
18. U.S. Department of Commerce, *A Report on the U.S. Semiconductor Industry* (Washington, D.C.: U.S. Government Printing Office, 1977).
19. U.S. National Commission on Technology, Automation, and Economic Progress, "Technology and the American Economy," in *The Employment Impact of Technological Change*, vol. 2 (Washington, D.C.: U.S. Government Printing Office, 1966).
20. *U.S. vs. Western Electric Co., 1956 Trade Case* (CCH 68,246 (D.N.J., 1956).
21. Utterback, J., and Murray, A., *The Influence of Defense Procurement and Sponsorship of Research and Development on the Development of the Civilian Electronics Industry* (Boston: Center for Policy Alternatives, MIT, 1977)
22. Weiner, C., "How the Transistor Emerged," *I.E.E.E. Spectrum*, January 1973, pp. 24-33.
23. Wolff, M. F., "The Genesis of the Integrated Circuit," *I.E.E.E. Spectrum*, August 1976, pp. 44-53.

Appendix D

Firms in the United States Commercializing Biotechnology

Table 4 in *Chapter 4: Firms Commercializing Biotechnology* listed firms in the United States commercializing biotechnology and their product markets. Their names and addresses are provided below. In order for a company to be listed in table 4, the existence of the company and the fact that the company is pursuing the development of biotechnology as defined by OTA had to be confirmed by at least two sources (e.g., company directories, individuals, trade journals). The existence and commercial application areas of many of the companies listed also were confirmed through the survey of firms' personnel needs conducted by the National Academy of Sciences and OTA.*

The number of companies listed in table 4 is a very conservative estimate of the number of companies commercializing biotechnology in the United States. More than five established companies thought to be applying novel bioprocessing technology (e.g., G. B. Fermentation Industries, Inc.) are missing from the list, because sufficient information to confirm their activities could not be obtained. Like the biotechnology research of established companies, the existence of new biotechnology firms (NBFs) is often difficult to confirm. More than 10 new companies, not included here, are thought by OTA to be operating but with very little public visibility. Some established companies and NBFs regard the application of their biotechnology research to be proprietary, and others will not even publicly confirm whether or not they are involved in biotechnology. Approximately 10 companies are not listed for this reason. Various other companies are not listed, because their existence and involvement in biotechnology were not confirmed by at least two sources.

Most support firms are not included in the table, because they are not applying biotechnology to the production of their products. Those support companies that, in addition to supplying support products (e.g., restriction enzymes and oligonucleotides), are applying biotechnology to the development and production of such products as vaccines and monoclonal antibodies are included.

*All but 33 of the firms listed were sent the OTA/NAS survey questionnaire, which is reproduced in *Appendix E: OTA/NAS Survey of Personnel Needs of Firms in the United States.*

Abbott Laboratories
14th St. & Sheridan Rd.
North Chicago, Ill. 60064

Actagen
Rm. 802
99 Park Ave.
New York, N.Y. 10016

Advanced Biotechnology Associates, Inc.
177 Post St., Suite 700
San Francisco, Calif. 94108

Advanced Genetic Sciences, Inc.
42 Maher Ave.
Greenwich, Conn. 06830

Advanced Genetics Research Institute
2220 Livingston St.
Berkeley, Calif. 94606

Advanced Mineral Technologies, Inc.
P.O. Box 1339
Socorro, N. Mex. 87801

Agrigenetics Corp.
3375 Mitchell Lane
Boulder, Colo. 80301

Allied Chemical Corp.
Columbia Rd. & Park Ave.
P.O. Box 4000R
Morristown, N.J. 07960

Alpha Therapeutic Corp.
5555 Valley Blvd.
Los Angeles, Calif. 90032

Ambico, Inc.
P.O. Box M, Route 2
Dallas Center, Iowa 50063

American Cyanamid Co.
One Cyanamid Plaza
Wayne, N.J. 07470

American Diagnostics Corp.
1600 Monrovia Ave.
Newport Beach, Calif. 92663

American Qualex
14620 Firestone Blvd.
La Mirada, Calif. 90638

Amgen
1892 Oak Terrace Lane
Newbury Park, Calif. 91320

Angenics
100 Inman St.
Cambridge, Mass. 02139

Animal Vaccine Research Corp.
3333 Torrey Pines Ct., Suite 120
La Jolla, Calif. 92037

Antibodies, Inc.
P.O. Box 442
Davis, Calif. 95617

Applied DNA Systems, Inc.
4415 Fifth Ave.
Pittsburgh, Pa. 15213

Applied Genetics, Inc.
5 Jules Lane
New Brunswick, N.J. 08901

ARCO Plant Cell Research Institute
6560 Trinity Ct.
Dublin, Calif. 94568

Atlantic Antibodies
10 Nonesuch Rd.
P.O. Box 60
Scarborough, Maine 04074

Axonics
1500 Salado Dr., Suite 202
Mountain View, Calif. 94043

Baxter-Travenol Laboratories, Inc.
One Baxter Parkway
Deerfield, Ill. 60015

Becton Dickinson & Co.
Corporate Research Center
P.O. Box 12016
Research Triangle Park, N.C. 27709

Bethesda Research Laboratories, Inc.
P.O. Box 577
Grovemont Circle
Gaithersburg, Md. 20760

Biocell Technology Corp.
220 East 23rd St.
New York, N.Y. 10010

Biochem Technology, Inc.
66 Great Valley Parkway
Great Valley Corporate Center
Malvern, Pa. 19355

Bio-con, Inc.
3601 Gibson St.
P.O. Box 5277
Bakersfield, Calif. 93388

Biogen, Inc.
241 Binney St.
Cambridge, Mass. 02142

BioGenex Laboratories
6529 Sierra Lane
Dublin, Calif. 94566

Biological Energy Corp.
P.O. Box 766
2650 Eisenhower Ave.
Valley Forge, Pa. 19482

Bio Response, Inc.
550 Ridgefield Rd.
Wilton, Conn. 06987

Biotech Research Laboratories, Inc.
1600 East Gude Dr.
Rockville, Md. 20850

Biotechnica International, Inc.
85 Bolton St.
Cambridge, Mass. 02140

Bio-Technology General Corp.
280 Park Ave.
New York, N.Y. 10017

Brain Research
46 East 91st St.
New York, N.Y. 10028

Bristol-Myers Co.
Industrial Division
P.O. Box 657
Syracuse, N.Y. 13201

BTC Diagnostics, Inc.
61 Moulton St.
Cambridge, Mass. 02138

Calgene
1910 Fifth St.
Davis, Calif. 95616

California Biotechnology, Inc.
2450 Bayshore Frontage Rd.
Mountain View, Calif. 94303

Cambridge Bioscience Corp.
495 Old Connecticut Path
Framingham, Mass. 01701

Campbell Institute for Research and Technology
Campbell Soup Co.
Campbell Rd.
Camden, N.J. 08101

Celanese Research Co.
86 Morris Ave.
Summit, N.J. 07901

Cellorgan International, Inc.
300 Park Ave.
New York, N.Y. 10010

Celtek, Inc.
102 West Eufala
Norman, Okla. 73069

Centaur Genetics Corp.
120 South LaSalle St., Suite 825
Chicago, Ill. 60603

Centocor
3508 Market St.
Philadelphia, Pa. 19104

Cetus Corp.
600 Bancroft Way
Berkeley, Calif. 94710

Cetus Immune Corp.
3400 West Bayshore Rd.
Palo Alto, Calif. 94303

Cetus Madison Corp.
2208 Parkview Rd.
Middleton, Wis. 53562

Chiron Corp.
4560 Horton St., Suite 0214
Emeryville, Calif. 94608

Ciba-Geigy
444 Saw Mill River Rd.
Ardsley, N.Y. 10502

Clonal Research
1598 Monrovia Ave.
Newport Beach, Calif. 92630

Codon
430 Valley Dr.
Brisbane, Calif. 94005

Collaborative Genetics, Inc.
128 Spring St.
Lexington, Mass. 01273

Collagen, Inc.
2455 Faber Pl.
Palo Alto, Calif. 94303

Cooper Diagnostics, Inc.
1230 Wilson Dr.
West Chester, Pa. 19380

Cooper-Lipotech, Inc.
1030 Curtis St.
Menlo Park, Calif. 94025

Corning Glass Works
Corning Biotechnology Department
Baron Steuben Plaza
Corning, N.Y. 14830

Crop Genetics International
7170 Standard Dr.
Dorsay, Md. 21076

Cutter Laboratories, Inc.
2200 Powell St.
P.O. Box 8817
Emeryville, Calif. 94662

Cytogen Corp.
201 College Rd., East
Princeton Forrestal Center
Princeton, N.J. 08540

Cytox Corp.
954 Marcon Blvd.
Allentown, Pa. 18103

Dairyland Foods Corp.
620 Progress Ave.
Waukesha, Wis. 53187

Damon Biotech, Inc.
115 Fourth Ave.
Needham Heights, Mass. 02194

Dart & Kraft, Inc.
2211 Sanders Rd.
Northbrook, Ill. 60062

Davy McKee Corp.
10 South Riversider Plaza
Chicago, Ill. 60606

DeKalb Pfizer Genetics
Sycamore Rd.
DeKalb, Ill. 60115

Diagnon Corp.
225 Main St.
Westport, Conn. 06880

Diagnostic Technology, Inc.
240 Vanderbilt Motor Parkway
Hauppauge, N.Y. 11788

Diamond Laboratories
2538 S.E. 43rd St.
Des Moines, Iowa 50316

Diamond Shamrock Corp.
T. R. Evans Research Center
P.O. Box 348
Painesville, Ohio 44077

DNA Plant Technology
2611 Branch Pike
Cinnaminson, N.J. 08077

DNAX Corp.
1454 Page Mill Rd.
Palo Alto, Calif. 94304

Dow Chemical Co.
2030 Dow Center
Midland, Mich. 48640

Ean-tech, Inc.
699-A Cerramonte Blvd.
Dale City, Calif. 94015

Eastman Kodak Co.
343 State St.
Rochester, N.Y. 14650

Ecogen, Inc.
c/o Johnston Associates, Inc.
1101 State Rd., Bldg. O
Princeton, N.J. 08540

E. I. du Pont de Nemours & Co.
Central Research and Development
Department
1007 Market St.
Wilmington, Del. 19898

Electro Nucleonics Laboratories, Inc
12050 Tech Rd.
Silver Spring, Md. 20904

Eli Lilly & Co.
Lilly Research Laboratories
307 East McCarty St.
Indianapolis, Ind. 46285

EnBio, Inc.
Union Ave. #408A
Fairfield, Calif. 94533

Endorphin, Inc.
1000 Seneca St.
Seattle, Wash. 98111

Engenics, Inc.
2 Palo Alto Sq., Suite 500
Palo Alto, Calif. 94304

Enzo Biochem, Inc.
325 Hudson St.
New York, N.Y. 10013

Enzyme Bio-systems, Ltd.
Box 8000
Englewood Cliffs, N.J. 07632

Enzyme Center, Inc.
33 Harrison Ave.
Boston, Mass. 02111

Enzyme Technology Corp.
783 U.S. 250 East, Route 2
Ashland, Ohio 44805

Ethyl Corp.
P.O. Box 341
Baton Rouge, La. 70821

Exxon Research & Engineering Co.
180 Park Ave.
Florham Park, N.J. 07932

Fermentec Corp.
301 Saratoga Ave.
Los Gatos, Calif. 95030

FMC Corp.
2000 Market St.
Philadelphia, Pa. 19103

Frito-Lay, Inc.
Frito-Lay Tower
Exchange Park
P.O. Box 35034
Dallas, Tex. 75235

Fungal Genetics, Inc.
14721 Cottonwood Pl.
Bothell, Wash. 98011

Genencor
Baron Steuben Pl.
Corning, N.Y. 14870

Genentech, Inc.
460 Point San Bruno Blvd.
South San Francisco, Calif. 94080

General Electric Co.
Research and Development
Laboratories
One River Rd.
Schenectady, N.Y. 12345

General Foods Corp.
555 South Broadway
Tarrytown, N.Y. 10591

General Genetics
15400 West 44th Ave.
Golden, Colo. 80403

General Molecular Applications
1834 Elmwood Ave.
Columbus, Ohio 43212

Genetic Diagnostics Corp.
160 Community Dr.
Great Neck, N.Y. 11021

Genetic Replication Technologies, Inc.
1533 Monrovia Ave.
Newport Beach, Calif. 92663

Genetic Systems Corp.
3005 First Ave.
Seattle, Wash. 98121

Genetics Institute
225 Longwood Ave.
Brookline, Mass. 02115

Genetics International, Inc.
50 Milk St., 15th Floor
Boston, Mass. 02109

Genex Corp.
6110 Executive Blvd.
Rockville, Md. 20852

Gentronix Laboratories, Inc.
15825 Shady Grove Rd.
Rockville, Md. 20850

Genzyme
1 Bishop St.
Norwalk, Conn. 06851

W. R. Grace & Co.
Research Division
7379 Route 32
Columbia, Md. 21044

Hana Biologics, Inc.
626 Bancroft Way
Berkeley, Calif. 94710

Hem Research
12220 Wilkins Ave.
Rockville, Md. 20852

Hoffmann-La Roche, Inc.
340 Kingsland St.
Nutley, N.J. 07110

Hybridoma Sciences, Inc.
4761 Hugh Howell Rd., Suite D
Tucker, Ga. 30084

Hybritech, Inc.
11085 Torreyana Rd.
San Diego, Calif. 92121

Hytech Biomedical, Inc.
1440 Fourth St.
Berkeley, Calif. 94710

IBM Corp.
Thomas J. Watson Research Center
Yorktown Heights, N.Y. 10598

IGI Biotechnology, Inc.
9110 Red Branch Rd.
Columbia, Md. 21045

Immulok, Inc.
1019 Mark Ave.
Carpinteria, Calif. 93013

Immunetech, Inc.
8950 Villa La Jolla Dr., Suite 2132
La Jolla, Calif. 92037

Immunex Corp.
51 University Bldg., Suite 600
Seattle, Wash. 98101

Immuno Modulators Laboratories, Inc.
10511 Corporate Dr.
Stafford, Tex. 77477

Immunogen
c/o T. A. Associates
111 Devonshire St.
Boston, Mass. 02109

Immunotech Corp.
11 Blackstone St.
Cambridge, Mass. 02139

Imreg, Inc.
P.O. Box 56643
New Orleans, La. 70156

Indiana BioLab
Palmyra, Ind. 47164

Integrated Genetics, Inc.
51 New York Ave.
Framingham, Mass. 01701

Interferon Sciences, Inc.
783 Jersey Ave.
New Brunswick, N.J. 08901

International Genetic Engineering, Inc. (INGENE)
1701 Colorado Ave.
Santa Monica, Calif. 90404

International Genetic Sciences Partnership
155-25 Styler Rd.
Jamaica, N.Y. 11433

International Minerals & Chemical Corp.
Biochemical Division
1401 South Third St.
Terre Haute, Ind. 47808

International Plant Research Institute
853 Industrial Rd.
San Carlos, Calif. 94070

Kallestad Laboratories, Inc.
Austin National Bank Tower, Suite 2000
Austin, Tex. 78701

Kennecott Copper Corp.
One Stanford Forum
Stanford, Conn. 06904

Lederle Laboratories
One Cyanamid Plaza
Wayne, N.J. 07470

The Liposome Co., Inc.
1 Research Way
Princeton Forrestal Center
Princeton, N.J. 08540

Liposome Technology, Inc.
1030 Curtis St.
Menlo Park, Calif. 94025

Litton Bionetics
5516 Nicholson Lane
Kensington, Md. 20895

3M Co.
3M Center
St. Paul, Minn. 55144

Mallinckrodt, Inc.
675 McDonald Blvd.
P.O. Box 5840
St. Louis, Mo. 63134

Martin Marietta
1450 South Rolling Rd.
Baltimore, Md. 21227

Meloy Laboratories, Inc.
6715 Electronic Dr.
Springfield, Va. 22151

Merck & Co., Inc.
Merck Sharp and Dohme Research Laboratories
P.O. Box 2000
Rahway, N.J. 07065

Microlife Genetics
P.O. Box 2399
1817 57th St.
Sarasota, Fla. 33578

Miles Laboratories, Inc.
1127 Myrtle St.
Elkhart, Ind. 46515

Miller Brewing Co.
3939 West Highland Blvd.
Milwaukee, Wis. 53201

Molecular Biosystems, Inc.
1118-A Roselle St.
San Diego, Calif. 92121

Molecular Diagnostics
400 Morgan Lane
West Haven, Conn. 06516

Molecular Genetics, Inc.
10320 Bren Rd., East
Minnetonka, Minn. 55343

Monoclonal Antibodies, Inc.
2319 Charleston Rd.
Mountain View, Calif. 94043

Monsanto Co.
500 N. Linbergh
St. Louis, Mo. 63167

Multivac, Inc.
P.O. Box 575
Seal Beach, Calif. 90740

Nabisco, Inc.
River Rd. and De Forest Ave.
East Hanover, N.J. 07936

National Distillers & Chemical Co.
99 Park Ave.
New York, N.Y. 10016

Neogen Corp.
Nisbet Bldg., Suite 22
1407 S. Harrison Rd.
East Lansing, Mich. 48824

New England Biolabs
32 Tozer Rd.
Beverly, Mass. 01915

New England Monoclonal Resources
267 Plain St.
Providence, R.I. 02905

New England Nuclear Corp.
85 Wells Ave.
Newton Center, Mass. 02159

Norden Laboratories
601 West Cornhusker Highway
Lincoln, Nebr. 68521

Novo Laboratories, Inc.
59 Danbury Rd.
Wilton, Conn. 06897

NPI
417 Wakara Way
University Research Park
Salt Lake City, Utah 84108

Nuclear & Genetic Technology, Inc.
172 Brook Ave.
Deer Park, N.Y. 11729

Ocean Genetics
1990 N. California Blvd., Suite 830
Walnut Creek, Calif. 94596

Oncogen
3005 First Ave.
Seattle, Wash. 98121

Oncogene Science Inc.
Nassau Hospital
Professional Bldg., Suite 330
222 Station Plaza North
Mineola, N.Y. 11501

Organon, Inc.
375 Mt. Pleasant Ave.
West Orange, N.J. 07052

Ortho Pharmaceutical Corp.
Route 202
Raritan, N.J. 08869

Petrogen, Inc.
2452 East Oakton St.
Arlington Heights, Ill. 60005

Pfizer, Inc.
25 East 42nd St.
New York, N.Y. 10017

Phillips Petroleum Co.
Research Center
Bartlesville, Okla. 74004

Phytogen
101 Waverly Dr.
Pasadena, Calif. 91105

Phyto-tech Lab
21822 South Vermont Ave.
Torrance, Calif. 90502

Pioneer Hybrid International Corp.
1206 Mulberry St.
Des Moines, Iowa 50308

Plant Genetics, Inc.
1930 Fifth St., Suite A
Davis, Calif. 95616

Polybac Corp.
1251 S. Cedar Crest Blvd.
Allentown, Pa. 18103

PPG Industries
One Gateway Center
Pittsburgh, Pa. 15222

Purification Engineering, Inc.
9505 Berger Rd.
Columbia, Md. 21046

Quidel Home
11077 North Torrey Pines
La Jolla, Calif. 92037

Replicon
P.O. Box 27053
South San Francisco, Calif. 94127

Repligen Corp.
101 Binney St.
Cambridge, Mass. 02142

Ribi Immunochem Research, Inc.
P.O. Box 1409
Hamilton, Mont. 59840

Rohm & Haas Co.
Independence Mall
West Philadelphia, Pa. 19105

Salk Institute Biotechnology/
Industrial Associates, Inc.
3333 Torrey Pines Ct., Suite 140
La Jolla, Calif. 92037

Sandoz, Inc.
Route No. 10
East Hanover, N.J. 07936

Schering-Plough Corp.
2000 Galloping Hill Rd.
Kenilworth, N.J. 07033

SDS Biotech Corp.
7528 Auburn Rd.
Painesville, Ohio 44077

G. D. Searle & Co.
Box 1045
Skokie, Ill. 60076

Serono Laboratories, Inc.
280 Pond St.
Randolph, Mass. 02368

SmithKline Beckman
One Franklin Plaza
P.O. Box 7929
Philadelphia, Pa. 19101

E. R. Squibb & Sons, Inc.
P.O. Box 4000
Princeton, N.J. 08540

A. E. Staley Manufacturing Co.
2200 Eldorado St.
Decatur, Ill. 62525

Standard Oil Co. of California
225 Bush St.
San Francisco, Calif. 94104

Standard Oil Co. of Indiana
Amoco Research Center
P.O. Box 400
Naperville, Ill. 60566

Standard Oil Co. of Ohio
1424 Midland Bldg.
Cleveland, Ohio 44115

Stauffer Chemical Co.
Nyala Farm Rd.
Westport, Conn. 06881

Summa Medical Corp.
4272 Balloon Park Rd., N.E.
Albuquerque, N. Mex. 87109

Sungene Technologies Corp.
3330 Hillview Ave.
Palo Alto, Calif. 94304

Sybron Biochemical
Birmingham Rd.
Birmingham, N.J. 08011

Synbiotex Corp.
348-B Rancho Dr.
San Marcos, Calif. 92069

Syncor International
12847 Arroyo St.
Sylmar, Calif. 91342

Synergen
1885 Thirty Third St.
Boulder, Colo. 80301

Syngene Products & Research, Inc.
225 Commerce Dr.
P.O. Box 2211
Fort Collins, Colo. 80524

Syntex Research
c/o Syntex Corp.
3401 Hillview Ave.
Palo Alto, Calif. 94304

Syntro Corp.
11095 Torreyana
San Diego, Calif. 92121

Syva Co.
900 Arastradero Rd.
Palo Alto, Calif. 94303

Techniclone International Corp.
3301 South Harbor Blvd., Suite 104
Santa Ana, Calif. 92704

Unigene Laboratories, Inc.
110 Little Falls Rd.
Fairfield, N.J. 07006

Universal Foods Corp.
433 East Michigan St.
Milwaukee, Wis. 53202

University Genetics Co.
537 Newtown Ave.
Norwalk, Conn. 06852

U.O.P., Inc.
10 UOP Plaza
Des Plaines, Ill. 60016

The Upjohn Co.
7000 Portage Rd.
Kalamazoo, Mich. 49001

Viral Genetics
10 Cutter Mill Rd., Rm. 403
Great Neck, N.Y. 11021

Wellcome Research Laboratories
3030 Cornwallis Rd.
Research Triangle Park, N.C. 27709

Worne Biotechnology, Inc.
Medford Medical Bldg.
Stokes Rd., Box 458
Medford, N.J. 08055

Xenogen, Inc.
557 Wormwood Rd.
Mansfield, Conn. 06250

Xoma Corp.
3516 Sacramento St.
San Francisco, Calif. 94118

Zoecon
975 California Ave.
P.O. Box 10975
Palo Alto, Calif. 94304

Zymed Laboratories
P.O. Box 1856
Burlingame, Calif. 94010

Zymos Corp.
2121 North 35th St.
Seattle, Wash. 98103

Appendix E

OTA/NAS Survey of Personnel Needs of Firms in the United States

As noted in *Chapter 14: Personnel Availability and Training,* OTA and the National Academy of Sciences' (NAS) Committee on National Needs for Biomedical and Behavioral Research Personnel cosponsored a survey of the personnel needs of U.S. firms using biotechnology. The purpose of the OTA/NAS survey was twofold. First, OTA was interested in identifying the companies that were using new biotechnology as defined at the outset of this report. Second, OTA and NAS were interested in the number of employees engaged in industrial biotechnology, how that number would grow, and where shortages of personnel, if any, are occurring. The cover letter and survey questionnaire reproduced in this appendix were sent to 286 U.S. companies. Of the 133 firms that responded, 18 indicated that they were not engaged in biotechnology activities, and 20 others were determined not to be engaged in biotechnology from their answers to the questionnaire. The remaining 95 indicated that they were engaged in biotechnology activities. The responses of these 95 firms, which are tabulated on the survey questionnaire reproduced in this appendix, are the basis of the characteristics described for the respondents. The distribution of size of firms was not significantly different between respondents and nonrespondents. Because the survey response rate was low, however, only general trends in the data have been used in the discussion of personnel needs in chapter 14.

Cornell University
Ithaca, New York 14853

March 4, 1983

Dear :

The Congressional Office of Technology Assessment (OTA) and the National Academy of Science's (NAS) Committee on National Needs for Biomedical and Behavioral Research Personnel have a mutual interest in determining the nation's need for research personnel. I am chairman of the NAS Committee's Panel on Basic Biomedical Sciences. We are particularly concerned that there be an adequate number of people trained in areas of the new biotechnology.

I am writing to ask your assistance in collecting some information on this issue. You could help us greatly in our efforts to get a profile of current employment opportunities and a sense of future demand in biotechnology and related industries by responding to the three questions on the attached page. To be useful in our report to the Congress, we need your answers before March 14, 1983. The tabulated data from the questionnaire will be published. Only OTA and the NAS panel will have access to the individual responses.

If you have additional comments or suggestions that you think would assist us, please include them with your response. A self-addressed envelope is enclosed. Also, if you have any questions concerning the questionnaire, don't hesitate to call me at (607) 256-3374.

With thanks for your help.

Yours sincerely,

Robert Barker, Ph.D.
Director, Division of Biological Sciences
Cornell University

RB:db
Enclosures

COMPANY NAME AND ADDRESS:

____

____

____

____

PERSON COMPLETING THIS FORM:
Name:
____
Phone Number:
____

For the purpose of this questionnaire, Biotechnology is defined as the application of novel biological strategies (rDNA, cell-fusion, mobilized cells or enzymes) for biochemical processing.

1. What year did your company begin research or development in activities related to the new biotechnology? ________

2. Please check all areas of biotechnology application in which your company is involved:

a) ____ fine chemicals
b) ____ bulk chemicals
c) ____ pharmaceuticals
d) ____ animal agriculture
e) ____ biomass conversion
f) ____ human diagnostics
g) ____ plant agriculture
h) ____ mineral leaching and mining
i) ____ pollution control
j) ____ enhanced oil recovery
k) ____ other; specify ____________

PLEASE INCLUDE ANY COMMENTS ON REVERSE

3.

SPECIALTIES	(1) Check if you are experiencing personnel shortages in any of these specialties.			(2) No. scientists currently on staff (list each employee only once).			(3) No. you intend to retain during next 18 months.			(4) Expected No. of scientists to be hired in next 18 months.			(5) For vacant positions, do you expect to:		
	Ph.D.	MS	BS	Ph.D.	MS	BS	Ph.D.	MS	BS	Ph.D.	MS	BS	Hire from Industry	Hire from Academia	Retrain current Staff
a) Recombinant DNA/molecular genetics	10	7	4	293	126	167				143	65	92	22	43	4
b) Hybridomas/monoclonal antibodies	6	5	6	89	50	108				38	29	79	16	30	9
c) animal reproduction/embryotransplantation	1	0	0	1	1	1		THE SAME AS (2)		0	0	0	1	1	1
d) classical genetics	1	1	1	30	8	12				6	5	3	4	4	1
e) gene synthesis	10	5	3	26	5	14				18	7	17	9	18	3
f) enzymology/immobilized systems	6	3	3	97	36	86				23	17	19	10	13	4
g) industrial microbiology	7	2	2	58	34	43				34	17	17	16	18	3
h) bioprocess engineering	12	4	3	71	47	68				44	31	25	21	15	2
j) analytical biochemistry	2	3	3	37	20	28				21	6	5	8	16	4
k) biochemistry, general	3	2	5	117	41	83				30	19	44	14	25	5
l) Cell culture	5	5	7	42	47	98				17	24	25	14	22	5
m) Cell fusion	3	2	2	11	4	11				6	3	8	3	5	2
n) Cell biology/physiology	4	1	3	18	7	18				10	5	10	3	9	1
o) plant molecular biology	8	0	1	40	18	18				31	12	20	7	17	2
p) plant biology/physiology	3	1	0	30	7	20				8	5	3	1	8	3
q) pharmacology	1	0	0	9	7	14				16	7	11	2	3	0
r) toxicology	1	0	0	13	2	16				2	2	1	3	1	0
s) microbiology, general	3	0	1	82	33	84				22	14	20	14	15	1
t) physiology	0	1	2	8	3	15				1	1	5	4	1	0
u) Other biotechnology specialties (specify) ________	4	3	4	51	23	44				8	7	9	8	11	2

Appendix F

Recombinant DNA Research Guidelines, Environmental Laws, and Regulation of Worker Health and Safety

Chapter 15: Health, Safety, and Environmental Regulation discussed the regulatory policies of the United States, the Federal Republic of Germany, the United Kingdom, France, Switzerland, and Japan as they pertain to biotechnology. This appendix elaborates on the material presented in that chapter.

Recombinant DNA research guidelines

UNITED STATES

The National Institutes of Health "Guidelines for Research Involving Recombinant DNA Molecules" (NIH Guidelines) apply to all research involving recombinant DNA (rDNA) in the United States or its territories conducted at or sponsored by any institution receiving support for rDNA research from NIH (28). All Federal agencies require their own scientists to comply with the guidelines, and Federal agencies other than NIH funding rDNA research also require their grantees to comply. Compliance is enforced by the authority of the agency to suspend, terminate, or place restrictions upon its financing of the offending project or all rDNA projects at the institution receiving support.

Although the NIH Guidelines are not legally binding on private companies (unless the company receives Federal funds), the private sector has espoused voluntary compliance. Some States and localities have required industry to comply by law.

Administrative Framework.—The NIH Guidelines create an administrative framework for oversight that specifies the responsibilities of scientists, their institutions, and the Federal Government. The primary responsibility for ensuring compliance lies with the institutions and scientists doing the research. The institution must establish an Institutional Biosafety Committee (IBC) meeting certain requirements, appoint a biological safety officer if certain experiments are done, ensure appropriate training, and implement health surveillance, if appropriate. The principal investigator has the initial responsibility for determining and implementing containment and other safeguards and for training and supervising the staff.

The IBC oversees all rDNA work at the institution for compliance with the NIH Guidelines. The IBC must consist of at least five members who collectively have the expertise to assess the safety of rDNA experiments. Two members must be otherwise unaffiliated with the institution and must represent the community's interest with respect to health and the environment. Institutions are encouraged to open IBC meetings to the public, and minutes of IBC meetings and certain other documents must be made available to the public on request. The institution must register the IBC with NIH by providing information about its members.

At the Federal level, the responsible parties are the Director of NIH, the NIH Recombinant DNA Advisory Committee (RAC), the NIH Office of Recombinant DNA Activities, and the Federal Interagency Advisory Committee on Recombinant DNA Research (Interagency Advisory Committee). The Director of NIH is the final decisionmaker under the guidelines. For major actions, he or she must seek the advice of the RAC and must provide the public and other Federal agencies at least 30 days to comment on proposed actions. Every action taken by the Director of NIH must present "no significant risk to health or the environment." RAC is a diverse group of experts that meets three or four times a year to advise the Director of NIH on the major technical and policy issues.* The NIH Office of Recombinant DNA Activities performs NIH's administrative functions under the guidelines. Additional oversight is provided by the Interagency Advisory Committee. This committee, which is composed of representatives of approximately 20 agencies, coordinates all Federal rDNA activities, and its members are nonvoting members of RAC.

Substantive Requirements.—The NIH Guidelines classify all experiments into four categories: 1) exempt, 2) those requiring RAC review and NIH approval before initiation, 3) those requiring IBC approval before initiation, and 4) those requiring IBC notification at the time of initiation. The first cate-

*In accordance with its charter, RAC is composed of not more than 25 members. At least eight must specialize in molecular biology or related fields; at least six must be experts in other scientific disciplines; and at least six must be authorities on law, public policy, the environment, public or occupational health, or related fields. As of June 30, 1983, RAC was composed of 10 molecular biologists, 6 experts from other scientific disciplines, and 9 persons in the third category (6). An industry trade association has requested that an industry representative be appointed to the RAC as a nonvoting member.

gory—exempt—covers an estimated 80 percent to 90 percent of all rDNA experiments. Examples include work with *E. coli* K-12, *S. cerevisiae*, and asporogenic *B. subtilis* host-vector systems.

NIH approval is required for experiments involving formation of rDNA containing genes for the synthesis of certain toxins lethal to vertebrates, deliberate release of recombinant organisms into the environment, and transfer of drug resistance to certain micro-organisms under certain conditions.

IBC approval is required for experiments involving certain pathogenic organisms, whole organisms or plants, or more than 10 liters of culture (except for certain exempt experiments). The last category—experiments requiring IBC notification—is a catch-all category. Containment levels are specified for each category except the one requiring NIH approval, where containment is set on a case-by-case basis.

Application to Industry.—In the absence of legal authority over industry's work with rDNA, NIH has taken several steps to encourage voluntary compliance and provide a modest degree of Federal oversight. Part VI of NIH Guidelines, added in January 1980, sets up a mechanism for voluntary compliance. It creates a parallel system of project review and IBC registration, modified to protect proprietary information.* In addition, RAC established a subgroup in May 1979 to deal with large-scale work. "Physical Containment Recommendations for Large-Scale Uses of Organisms Containing Recombinant DNA Molecules" (Large-Scale Recommendations) (27) developed by that subgroup, RAC, and NIH specify physical containment requirements, suggest the appointment of a biological safety officer, and suggest the establishment of a worker health surveillance program for work done at higher containment levels. (They were added to the NIH Guidelines as Appendix K in June 1983.)

According to industry spokespeople, the NIH Guidelines are accepted and followed by the private sector.** Compliance with the Large-Scale Recommendations also appears to be widespread, but there have been few, if any, definitive statements by industrial spokespeople on this point. Regarding present Large-Scale Recommendations, one industry group stated that its experience has indicated that "the present [recommendations] are reasonable and workable, although they are quite stringent for work at the P1-LS level. The design requirements in the Recommendations make sense to us and are consistent with other regulations relating to the manufacture of products for use with human subjects" (4). The group went on to state that it also saw difficulties arising from the recommendation that the primary containment system not be opened until all micro-organisms are inactivated because that could compromise that product in some cases (4).

Impact on Biotechnology.—The impact of the NIH Guidelines on biotechnology appears to be minimal. As essentially voluntary codes of practices that are fairly consistent with previously established good laboratory and manufacturing practices, they add little in the way of additional restrictions. Moreover, an estimated 80 to 90 percent of the experiments are exempt. On the basis of past history and what experts continue to learn about risks, the NIH Guidelines are likely to be further liberalized and may even disappear. In fact, whatever burdens they impose are probably offset by the gains in public confidence and the likelihood that they have headed off more restrictive mandatory controls.

EUROPEAN ECONOMIC COMMUNITY COUNTRIES: FEDERAL REPUBLIC OF GERMANY, UNITED KINGDOM, AND FRANCE

European Economic Community.—The European Economic Community (EEC) has considered at length the problems and prospects for rDNA research and the need for common Communitywide action to regulate and promote its development (13), but only a nonbinding recommendation has been made by the Council of the European Communities to member states on the question of guidelines applicable to rDNA research. The nonbinding EEC Guidelines were adopted in June 1982 (2). By that time, most of the individual member states with any significant amount of rDNA research had already adopted their own national guidelines. The EEC Guidelines impose no stricter requirements on rDNA research than those of the individual member states. They principally provide that any laboratory wishing to conduct rDNA research notify the competent national or regional authority in the member state and that the member states adopt a common definition of work involving rDNA (secs. 1-3).

More particularly, the EEC Guidelines suggest that notification of any rDNA research be given before work is commenced, except for research of very low-risk potential.* The notification should include infor-

*Proprietary information is protected in several ways. First, there is a presubmission review of data as to availability under the Freedom of Information Act. Second, NIH must consult with institutions applying for exemptions or approvals about the content of any public notice to be issued, if the application contains proprietary information. Finally, applications involving proprietary information are considered by RAC in nonpublic sessions.

**Although there is no means for NIH to monitor compliance with the NIH Guidelines or Large-Scale Recommendations, there is no evidence suggesting noncompliance.

*The EEC Guidelines do not define the term "very low risk potential," but indicate that this be determined by the competent national authorities. The United Kingdom, France, and the Federal Republic of Germany have adopted somewhat different methodologies in their guidelines for defining risk potential.

mation about the experimental protocol, the protective measures to be taken, and the general education and training of the staff working on the experiment or monitoring it. Such notification is thought desirable because it creates records that will be helpful in what the Commission of the European Communities believes to be the highly unlikely event of an accident or other misfortune involving rDNA (2). The authority receiving the notification must also, under the recommendation, protect the confidentiality of the information submitted (2). The EEC Guidelines do not call for specific approval of rDNA research of any type. As is discussed below, certain member states do require specific approval.

The EEC Guidelines do not address many issues which national guidelines, including those of the United States, have attempted to cover. The EEC Guidelines do not discuss the question of whether private laboratories should be subject to regulation, leaving this decision to the discretion of national authorities. Neither do they address how large-scale rDNA research should be regulated.

The fact that the EEC issued its rDNA guidelines despite the existence of more comprehensive guidelines in the member states reflects both the continuing concern over the safety of rDNA research and the difficulty in obtaining agreement on such matters. It is clear that the EEC has not yet determined its proper role in the regulation of rDNA research. Although discussions concerning rDNA as well as biotechnology generally are continuing within the EEC, it is likely to be some time before any agreement is reached concerning the respective roles of the EEC and the member states.

Federal Republic of Germany.—The Federal Republic of Germany has issued guidelines for rDNA research (3) that borrow heavily from the NIH Guidelines of the United States. The West German guidelines are theoretically broader than the NIH Guidelines because the German guidelines nominally apply to all research activities involving DNA. The only enforcement mechanism, however, is control over research funding from the German Federal Government.

The West German guidelines, like the NIH Guidelines, provide that the physical and biological containment measures required for particular experiments be determined according to the risk of the experiment. Risk is evaluated largely in terms of the source of the DNA. The German guidelines also prohibit certain specified experiments in the host organism *E. coli* K12 and other *E. coli* strains discussed in the NIH Guidelines (and the corresponding bacteriophages and plasmids of these strains), thereby requiring that the higher biological containment measures be used, regardless of the source of the DNA.* The guidelines also specify the appropriate containment methods required for various rDNA experiments. Physical and biological containment measures are divided into four and two levels (L1 to L4 and B1 to B2), respectively.

The German guidelines for rDNA research are administered by the Central Commission for Biological Safety (Zentrale Kommission für die Biologische Sicherheit),** a biological safety officer or committee at each laboratory, and a project leader for each experiment.*** The guidelines specify that the Central Commission must be notified of all rDNA experiments except those at the lowest physical containment level. For research at the next level, the Central Commission must authorize one of its scientist members to supervise the work and to keep the Commission informed. Experiments using mid-level containment measures require the prior approval of two members of the Commission. Prior approval must be sought from the Central Commission for all experiments using vertebrate cells as the host and for experiments using DNA from pathogenic organisms. In the case of the latter, the Central Commission must find that the expected benefits clearly outweigh the conceivable hazards. On request, the Central Commission will also authorize the use of new host-vector systems not enumerated in the German guidelines. The Central Commission also gives advice on research and safety measures.

United Kingdom.—The U.K. guidelines for rDNA research (26)† are similar to the NIH Guidelines in broad conceptual terms but differ with respect to

*These specially restricted experiments are: 1) the production of recombinant DNA for the biosynthesis of powerful bacterial exotoxins such as botulinus toxin, tetanus toxic, diphtheria toxin, and snake toxin; 2) the use of genomes of extremely pathogenic viruses such as Lassa, small pox, and hepatitis B; and 3) the transmission of genes which confer resistance to an antibiotic between micro-organisms that do not naturally exchange genes when the resistance gene has not previously been known in the receptor cell.

**West Germany's Central Commission for Biological Safety, the only Government body, has 12 members, 4 rDNA experts, 4 experts from related field of biology, and 4 "outstanding individuals" from unions, industry, or research-promoting organizations, all appointed by the Federal Minister for Research and Technology.

***The officer or at least one member of the committee must have the appropriate license, if the research work involves pathogenic or toxin-producing organisms. The project leader must possess adequate experience in microbiology and, for certain higher containment level work, knowledge about pathogens. The project leader is responsible specifically for planning and conducting the research, health monitoring of laboratory workers, informing the Central Commission and the biological safety officer or committee of the research and the planned safety measures, implementing Commission instructions, making regular reports to the Commission, maintaining a record of safety instruction, and training laboratory personnel.

†The term used in the United Kingdom to describe rDNA research is "genetic manipulation." Genetic manipulation is defined in the Genetic Manipulation Regulations as: the formation of new combinations of heritable material by the insertion of nucleic acid molecules produced by whatever means outside the cell, into any virus, bacterial plasmid, or other vector system so as to allow their incorporation into a host organism in which they do not naturally occur but in which they are capable of continued propagation.

scope, risk assessment, and enforcement. Like the NIH Guidelines, the U.K. guidelines have been gradually relaxed. Nevertheless, the guidelines in the United Kingdom are still regarded as more restrictive than those of the United States.

The guidelines for rDNA research in the United Kingdom are promulgated and administered by the Genetic Manipulation Advisory Group (GMAG) under the Health and Safety at Work Act of 1974 (24).* The guidelines apply to all research in the United Kingdom, not just that funded by the Government. Enforcement is the responsibility of the Health and Safety Executive (HSE), which is comparable to the U.S. Occupational Safety and Health Administration. HSE has taken no enforcement action to date.

As do the guidelines in all of the competitor countries, the GMAG guidelines establish four progressively more restrictive physical containment levels based on the perceived hazards of the research.** Facilities for the highest two levels must be examined by HSE inspectors before any rDNA research can be conducted to ensure that the GMAG requirements are met.

The GMAG guidelines also adopt the two-level biological containment approach of most of the other countries*** which is based on the degree of disability of the host-vector system being used. However, GMAG has also developed special rules for rDNA research involving experimental animals† and for work that involves the introduction of foreign nucleic acid into higher plants or into any plant pest.*

GMAG asssess the degree of potential hazard in a way somewhat different from the other countries, including the United States. GMAG considers three factors: access, expression, and damage.* As a general matter, the British classification system appears to require less stringent containment measures for some types of research than would be required in other countries. For example, the damage factors associated with interferon and insulin are quite low and work with these products would be classified as less risky in the United Kingdom than in some other countries (22).

The administrative framework for implementing the GMAG guidelines relies on institutional and governmental oversight. GMAG and HSE must be given advance notice of work involving rDNA except for certain self-cloning experiments.** Most work at the lowest two physical containment levels can go forward after notice. Although no express provision prohibits work at containment levels three and four before GMAG issues its advice, such premature work might violate the Health and Safety at Work Act, which carries criminal penalties. In addition, each institution conducting rDNA research is required to have certain personnel responsible for the research*** review, to forward notifications of proposed rDNA research to GMAG, and to suggest other health and safety actions that the institution might take.

Industrial or large scale applications of rDNA—that is, research involving the growth of self-propagating products of genetic material in volumes of 10 liters or more—are subject to special rules. GMAG reviews proposals to conduct such large-scale research on a case-by-case basis and visits each site, commenting on the safety measures proposed. GMAG expects that this review will involve "integration" of questions about physical and biologic containment. Whether this means that review of large-scale work will be stricter or more relaxed is unclear. GMAG has stated, however, that vaccine and antibiotic production can be done safely using ordinary chemical engineering measures—measures probably more relaxed than the containment-level measures required for small-scale research (20).

GMAG has recognized the potential commercial and industrial importance of genetic manipulation by establishing special confidentiality requirements for work that raises questions about commercial proper-

*The Government department with responsibility for GMAG policy is the Department of Education and Science, although this department has little expertise in such areas, particularly in comparison to the Department of Health and Social Security, which has a very limited role, via the Medical Research Council, in the oversight of genetic manipulation safety (25). GMAG's status was recently reviewed by the Health and Safety Executive, and the subsequent report recommended the relocation of the group to the Department of Health and Social Security (12). GMAG, now called the Health and Safety Commission Advisory Committee on Genetic Manipulation, has been moved to the Department of Health and Social Security.

**Certain DNA research is considered so safe as to not require containment. Laboratories conducting this research must instead follow simply the Guidelines for Microbiological Safety.

***France has four levels of biological containment.

†These require isolation of the animal, safe disposal of refuse and waste, and stricter rules for research in category III and IV laboratories (23).

*Plant pest is defined as "any living organism, other than a vertebrate animal, or any pathogen which is injurious to any plant, and includes any culture of such organism or pathogen." The work requires a special license from the Agricultural Ministries. The license will be issued only if the research is conducted according to the containment recommendations of GMAG, which include special rules for the handling of plants and preventing the dissemination of pollen and seed. The special plant rules do not cover experiments involving the introduction of plant nucleic acid into bacteria or other microorganisms (except plant pests), which are covered by the existing GMAG guidelines (21). It should be noted that the United Kingdom has adopted specific restrictions on the importation of such pests.

*"Access" is the possibility that escaped organisms will enter the human body and eventually reach susceptible cells. "Expression" is the possibility that a foreign gene incorporated into the gene sequence of an organism will be able to carry on or "express" its normal function, such as secretion of a toxin that the organism formerly did not secrete. "Damage" is the chance that a new gene sequence will cause physiological damage in the body to which it gains access once it is expressed (15,18,19,22).

**These include experiments using *E. coli* K12, *B. subtilis*, and *S. cerevisae* (17).

***These include a Biological Safety Officer familiar with the safety procedures for rDNA work and a Safety Committee to consider the containment and other safety measures proposed for genetic manipulation.

ty or patents. While confidentiality arrangements may vary from case to case, GMAG generally treats as confidential any material so labeled. Members of GMAG who have commercial interests in DNA work are prohibited from seeing such material or taking part in the discussion about it (17,20).

France.—The French guidelines for rDNA research (5) largely follow those of the United States. The guidelines were promulgated and are administered by the National Control Commission (Commission de Controle), which reports to the General Delegation of Scientific and General Research (Delegation Generale de la Recherche Scientifique et Technique). The French guidelines apply only to Government-funded research and require that scientists conducting such research notify the Control Commission of the planned research and in some cases obtain approval of the research. Local safety committees monitor ongoing research. The principal sanctions for failure to comply with the French guidelines are loss of Government funding or denial of approval to conduct research.

As in the United States, rDNA experiments in France must be conducted using certain physical and biological containment measures. The degree of containment depends on the risk of the work. Risk is assessed using a method very similar to that used in the United States. Research with DNA from oncogenic or highly pathogenic viruses is reviewed on a case-by-case basis but generally must be conducted according to the most stringent containment measures unless the oncogenic or highly pathogenic genes are eliminated before cloning.

In certain respects, the physical and biological containment requirements in the French guidelines differ from those in the United States. Although the French guidelines use four levels of physical containment as in the United States, they appear to be more flexible than the U.S. guidelines with respect to upgrading containment. In some cases, the French guidelines permit a laboratory's containment level to be upgraded without requiring construction of a new facility. Use of an approved safety cabinet will give the laboratory the next higher rating. If a safety cabinet is used to render a P3 laboratory equivalent to a P4 laboratory (the laboratory with the highest degree of containment), however, the National Control Commission must certify the facility. This upgrading system should expand the ranges of research that a French laboratory can do, as well as make research at higher containment levels less expensive. With respect to biological containment, the French guidelines use four levels, unlike the U.S. guidelines, which use two levels. Biological containment is based on the safety of the host-vector system. In effect, the French approach to biological containment appears quite similar to that of the United States, with the four levels of containment in France being finer gradations of the two levels used in the United States.

France allows biological agents containing rDNA to be imported and exported freely, although the French guidelines specify that certain measures must be met to safely transport and import rDNA materials. Large-scale research—i.e., experiments involving volumes of 10 liters or more—is not covered by the French guidelines for rDNA research, but Government oversight exists on a case-by-case basis.

SWITZERLAND

The Swiss have basically adopted the U.S. guidelines as their national rDNA research guidelines. Although the Swiss generally have amended their guidelines whenever the NIH Guidelines are amended, they are currently using a version based on the NIH Guidelines in effect in April 1982 (14).

There are other basic differences. The Swiss Government has no direct role in regulation of rDNA research; Swiss scientists instead have established a system of complete self-regulation. The Commission for Experimental Genetics (Kommission für Experimentelle Genetik) created by the Swiss Academy of Medical Sciences, is responsible for monitoring rDNA research. The guidelines that this commission has promulgated apply to all research involving rDNA in Switzerland, not only that funded by the Government. Moreover, the Swiss guidelines do not require special approval for work using cell culture volumes in excess of 10 liters.

The adminstrative structure for oversight in Switzerland is quite similar to that in the United States. The Commission for Experimental Genetics must approve certain experiments in advance, such as those involving the deliberate release into the environment of any organism containing rDNA. For two other classes of experiments, scientists must notify the commission but need not obtain approval. A final class of experiments are exempt from the guidelines. Principal investigators, safety officers, and institutional safety committees also bear oversight responsibility.

JAPAN

Japan's guidelines for rDNA research (11) are promulgated by the Ministry of Education (on recommendation by the Science Council). The guidelines apply only to publicly funded research, but private industry has followed them on a voluntary basis.

Each research institution is required under these guidelines to have laboratory supervisors, a safety committee, and a safety officer. The head of each research institution is also charged with specific duties in supervising the rDNA work. The laboratory supervisor must submit plans of experiments and changes in plans to the head of the research institution for his or her approval. The head of the institution then consults with the safety committee—a body consisting of "members representing the relevant fields, and having high standards of both professional and technical knowledge and judgment"—to determine whether the plans comply with the guidelines, what training will be necessary, and other issues relevant to the safety of the research. The safety officer's role is to monitor the safety of ongoing work and to make appropriate reports to the safety committee.

The Japanese Government monitors rDNA research through two bodies: 1) the Council for Science and Technology, which advises the Prime Minister and which oversees work by private institutions; and 2) the Science Council, which advises the Ministry of Education and which supervises Government-funded university research. The Science Council and the Ministry of Education formerly had to approve university rDNA research; now it is only necessary that the university safety committee and the university president approve the experiment (7,9). Ministry authorization is still required, however, for experiments involving specified "especially dangerous" organisms and the release of such organisms into the environment.*

Certain experiments are effectively prohibited in Japan, because the Japanese guidelines for rDNA research specify no safety or containment rules for them. Effectively prohibited experiments include large-scale research (more than 20 liters of cell culture) and experiments in which recombinant organisms infect individual animals and plants, in which the source of the DNA is other than specified cells or host-vector systems. Such experiments can be done once containment standards are set, but setting such standards depends under the guidelines on confirmation of the safety of these experiments, which has not been completed for most types of this research. Large-scale research is possible if special permission is granted by the Ministry of Education; few companies have sought it successfully. Japanese companies using biotechnology are now lobbying heavily for relaxation of restrictions on large-scale research.

For permissible experiments, the Japanese rDNA research guidelines require physical and biological containment based on the perceived risk of each experiment. Under the guidelines, risk is assessed principally according to a phylogenic scale** but also according to the biological characteristics of the source of the DNA,* the purified or unpurified nature of the DNA,** the size of the clone number,*** and the scale of the cultivation.† Required physical containment measures resemble those under the NIH Guidelines and are categorized in a similar P_1 to P_4 scale. Similarly, the Japanese guidelines provide for two levels of biological containment.

Historically, the Japanese guidelines have been among the most restrictive in the world. Although Japan's guidelines have recently been relaxed considerably to bring them more into line with the guidelines in other countries, they are still the most restrictive of the ones surveyed in this appendix. Japanese companies applying biotechnology consider themselves handicapped in competition against their foreign counterparts for two principal reasons. First, hosts are limited in Japan, with a few exceptions, to *E. coli* and *B. subtilis*; other micro-organisms such as the actinomycetes, which is effective in producing antibiotics, therefore cannot be used. Second, work in Japan is limited to volumes of 20 liters or less, and successful commercial development requires larger fermenters (8). Japanese companies using biotechnology have mounted an intensive lobbying campaign to eliminate the 20-liter rule (10).

Environmental laws and regulations

UNITED STATES

The United States has no laws specifically directed toward biotechnology, but, as discussed in *Chapter 15: Health, Safety, and Environmental Regulation,* the Toxic Substance Control Act (15 U.S.C. §§2601-2629) and the Federal Insecticide, Fungicide, and Rodenticide Act (47 U.S.C. §136(a)-(y)) will play a major role in preventing any adverse environmental impacts from biotechnology products. In addition, there are several statutes dealing with pollution that would apply because they generally define pollutants or wastes so as to cover biological materials. They are:

- The Federal Water Pollution Control Act, as amended by the Clean Water Act of 1977 (33

*"Especially dangerous" experiments include the transplant of manipulated genes with toxicity into animal and plant cells. University presidents may still approve work with disease pathogens, including influenza and hepatitis viruses (7).

**DNA donor organisms closer phylogenically to humans are considered riskier.

*The relevant biological characteristics of the DNA are pathogenicity, toxin-producing ability, carcinogenicity, parasitic quality, drug resistance, likelihood of becoming an allergen, masked infective factors such as nucleic acids related to C-type virus, vulnerability to contamination by viruses, bacteria, or other parasites, ability to produce substances such hormones or metabolic intermediates affecting the metabolism of human beings, and possibility of causing ecological disturbances.

**Purified DNA, proved to carry only nonhazardous genes, is deemed safer than unpurified DNA.

***The fewer the number of clones, the safer the experiment is, on the reasoning that a lower number will reduce the probability that harmful genes will appear.

†Smaller-scale experiments are considered safer than large-scale ones.

U.S.C. §§1251-1376, *as amended by* Public Law No. 95-217, 91 Stat. 1566 (1977)).
- The Marine Protection, Research and Sanctuaries Act of 1972 (33 U.S.C. §§1401, 1402, 1411-1421, 1441-1445).
- The Clean Air Act (42 U.S.C. §§7401-7508, 7521-7574, 7601-7626).
- The Solid Waste Disposal Act, as amended by the Resource Conservation and Recovery Act of 1976 (42 U.S.C. §§6901-6987, *as amended by* Public Law No. 94-580, 90 Stat. 2795 (1976)).

Under the Federal Water Pollution Control Act, as amended, the Environmental Protection Agency has promulgated regulations on wastewater from the manufacture of pharmaceuticals by fermentation (4 C.F.R. Part 439 (1982)).

EUROPEAN ECONOMIC COMMUNITY COUNTRIES: FEDERAL REPUBLIC OF GERMANY, UNITED KINGDOM, AND FRANCE

European Economic Community.—Although the EEC has issued no directives or taken any other action specifically to regulate the environmental effects of biotechnology, several general directives concerning waste disposal and water pollution will be applicable to biotechnological products (30,31,33). The EEC's environmental regulations are general and flexible, giving maximum discretion and authority to the bureaucrats that implement them.

Companies using biotechnology will encounter environmental regulation in manufacturing biotechnological products in the EEC member states and in exporting products to those states. Under the premarketing notification requirement imposed by the Sixth Amendment to the EEC's dangerous substances directive (32),* a firm must test a new chemical before marketing, must provide the proper authorities in the member states where the product is to be marketed with the results of the "base test" (minimum testing requirements), and must conduct such further tests as those authorities may deem necessary before approval may be granted. Since many biotechnology products will likely qualify as "new chemicals," the Sixth Amendment's requirements would apply. Of course, a firm seeking to build a plant to manufacture biotechnology products in a member state would be required not only to secure "new chemical" approval, but also to comply with the more comprehensive system of environmental regulation in the member state.

Federal Republic of Germany.—The Federal Republic of Germany is a federal state, and under its Constitution, the 11 Länder (States) share power with the Federal Government. In controlling pollution, poisonous substances, and waste, the Federal Government and the Länder have concurrent jurisdiction, but the Länder may pass laws in these areas only if the Federal Government has not done so. In environmental protection, land use, and water law, the Federal Government may enact broad "framework" legislation, but the Länder must implement the general Federal laws by enacting detailed legislation adapted to the conditions of each State.

The Ministry of the Interior (Bundesministerium Des Innern) coordinates the environmental policies of the West German Federal Government, including environmental planning, waste and water management, and control of air pollution. The Federal Environmental Agency (Umweltbundesamt), which is more concerned with environmental protection, furthers Federal environmental policies by developing planning programs and performing research. Coordination of Federal environmental programs also is conducted by a Cabinet Committee for Environmental Questions (Arbeitsgemeinschaft für Umweltfragen E.V.).

The only environmental regulations directed specifically at biotechnology are contained in the Federal Republic of Germany's guidelines for rDNA research (39). The German guidelines impose requirements on disposal of waste from rDNA experiments, requirements that depend on the containment level of the work involved. In no case may biological agents containing rDNA be released into the environment. Experimental plants and animals containing rDNA must be kept under conditions of isolation. All rDNA material may be removed from the laboratory only in airtight packaging and must eventually be destroyed, usually by incineration. All wastes containing microorganisms or nucleic acids must be sterilized or denatured. Waste water from experiments at the L3 or L4 level must be decontaminated.

Apart from the rDNA research guidelines, it appears that the Federal Republic of Germany's legislation and implementing regulations do not specifically regulate environmental impacts from biotechnological products and processes. Instead, companies using biotechnology would appear to be subject, like other firms in West Germany, to a series of general environmental protection laws and regulations.

The most general of these laws is the Chemicals Act (40), which is designed to protect humans and the environment from all types of dangerous substances. This law set up compulsory testing of substances and compulsory classification, labeling, and packaging of dangerous substances and materials. It implements in the Federal Republic of Germany the Sixth Amendment to the EEC's environmental protection directive.

*The first directive in the field of dangerous substances was Council Directive of June 27, 1967 (29).

Other relevant statutes in the Federal Republic of Germany are the following: the Law for the Prevention of Harmful Effects on the Environment Caused by Air Pollution, Noise, Vibration, and Similar Phenomena (Federal Emission Control Law) (37), the Law on Disposal of Wastes (36). Act on Regulation of Matters Relating to Water (Federal Water Act) (35), and the Waste Water Charges Act (Waste Water Law) (38).

A Committee of the German Society for Chemical Engineering (Deutsche Gesellschaft für chemisches Apparatewesen E.V.) completed a study of the risks specifically associated with biotechnology and of the relevant statutory and regulatory provisions that could be used to control those risks (34). The study concluded that adequate legal authority exists in the Federal Republic of Germany for regulating the kinds of hazards most likely to arise in connection with biotechnology.

United Kingdom.—Responsibility for protection of the environment in the United Kingdom lies primarily with the Department of the Environment. In addition, a Royal Commission on Environmental Pollution was established in 1970 to advise the government on environmental issues. As in the United States, much environmental regulation in the United Kingdom is the responsibility of local governments.

Although the United Kingdom has an extensive statutory environmental protection scheme, there is no legislation or regulation specifically concerned with environmental impacts of biotechnological products and processes. Companies using biotechnology, therefore, would be subject to the general environmental protection laws and regulations.

The Control of Pollution Act of 1974 (53), provides in chapter 40 for licensing of sites for the disposal of "controlled waste," defined as household, industrial, and commercial waste, both on land and in water. The penalties for unlicensed disposal are fines and imprisonment. The law is to be phased in between July 1983 and July 1986. Waste products of biotechnological processes would appear to be covered by this law.

France.—The principal environmental protection agency in France is the Ministry of the Environment (Ministére de l'Environnement). Environmental protection legislation applies broadly to activities that degrade the environment in a variety of ways. The touchstone of most regulation is not the nature of a particular activity, but whether it produces environmentally adverse effects. To the extent that biotechnological products and processes produce such effects, they would be subject to these laws.

The most general environmental statute is the Law on Installations Classified for Purposes of Environmental Protection (44). This law covers all types of risk to humans and the environment resulting from the activities of various types of facilities, including but not limited to industrial and commercial establishments. These facilities are subject to requirements specific to the type of danger or inconvenience involved. This determination rests largely in the hands of local authorities, who have a continuing right of access to the regulated facilities. Failure to comply with the law may result in administrative and criminal penalties. No rules specifically aimed at biotechnology facilities have yet been adopted under the authority of this law.

The Chemicals Control Law of France (45), which predates the Sixth Amendment to the EEC's dangerous substances directive, would apply to chemical compounds produced by biotechnology. This law aims to protect human beings and the environment against risks arising from both naturally occurring and industrially produced chemicals. Any producer or importer seeking to import or manufacture commercially a chemical which has never been placed on the French market before must notify the relevant authority, provide certain information, and submit to whatever conditions may be imposed.*

Two other statutes would be particularly relevant to biotechnology. They are the Law on Waste Disposal and Recovery of Materials (43) and the Act on the Administration and Classification of Waters and the Control of Water Pollution (42).

SWITZERLAND

Although the Swiss rDNA research guidelines prohibit the release of biological agents containing rDNA into the environment, they do not mention effects on the environment from other forms of waste which may result from applications of biotechnology. These would presumably be regulated in Switzerland under Article 24 *septies* (seventh) of the Federal Constitution, which gives the Federal Government far-reaching powers to pass environmental laws.

Legislation under this article has been sparse, however, and there are apparently no confederal rules in Switzerland on air pollution, noise abatement, or waste disposal. Only in the area of water pollution has legislation been enacted. The Water Protection Act of October 8, 1971 (51), seeks to ensure the quality of the nation's water by means of sweeping protective measures which cover all natural, artificial, ground, and surface waters.

In addition, Article 6 of the Federal Act on Work in Industry, Trade, and Commerce (52) requires employers to protect the area surrounding their business enterprise from harm or discomfort by taking all measures shown necessary by experience and found to be technically feasible and appropriate.

*Decree No. 79-35 describes the technical dossier to be provided when providing notice concerning a new chemical substance (41).

JAPAN

Specific measures governing environmental effects of biotechnology applications have not been prepared by the Japanese Government. The regulations applicable to biotechnology are those applicable to all industry. The agencies with responsibility for environmental protection in Japan include the Environmental Protection Agency, the Ministry of International Trade and Industry (MITI), the Ministry of Health and Welfare, and the Ministry of Agriculture, Forestry, and Fishery. The Environmental Protection Agency has jurisdiction over basic policy, general coordination of governmental pollution control activity, budgetary policy, and research and investigation.

The Basic Law for Environmental Pollution Control (46) establishes fundamental national principles and policies and establishes the basic regulatory framework for environmental protection in Japan. The law applies to air, water, soil, and other pollution. It empowers the Central Government to promulgate and enforce environmental quality standards necessary to protect the public health and conserve natural resources. This and other environmental laws are supplemented by and implemented through Cabinet orders issued by the Prime Minister, and through ministerial orders and Environmental Protection Agency notifications. Administrative guidance is used to regulate pollution from specific industrial plants and industries. Local governments have responsibility with the Central Government in monitoring pollution and for regulation, and they may set more stringent standards than those set by the Central Government.

Japan's Basic Law for Environmental Pollution Control is supplemented by laws aimed at specific types of pollution. These include the Air Pollution Control Law (47), the Water Pollution Control Law (49) and the Waste Management Law (48).

Finally, the Chemical Substances Control Law (50) requires manufacturers to notify the Japanese Government and to test all new chemical substances to be produced in quantities exceeding 100 kilograms. Chemicals are tested for their biodegradability and bioaccumulation. Manufacturers and importers of chemical substances must notify MITI of their intent to use or market a new chemical. Japan's Environmental Protection Agency monitors the effect of chemicals in the air and water, and the Ministry of Health and Welfare administers laws on chemical products.

Regulation of worker health and safety

UNITED STATES

The Occupational Safety and Health Administration (OSHA), which is part of the U.S. Department of Labor, is the agency primarily responsible for worker safety and health. OSHA's authority derives from the Occupational Safety and Health Act of 1970 (29 U.S.C. §§651-678) which creates a broad mechanism for protecting workers from workplace hazards. Section 5(a)(1) of the act requires U.S. employers to furnish their employees with a workplace "free from recognized hazards that are causing or are likely to cause death or serious physical harm." Section 5(a)(2) requires employers to comply with safety and health standards set by the U.S. Secretary of Labor. Under a recent U.S. Supreme Court decision (62), the Secretary of Labor can promulgate permanent standards for toxic substances or harmful physical agents only after a finding that the standard is "reasonably necessary to remedy a significant risk of material health impairment." Section 6(c) of the act permits the Secretary of Labor to promulgate emergency temporary standards after a finding that employees are "exposed to grave danger." The statute also creates the National Institute for Occupational Safety and Health to gather data, assess risks, and recommend safety and health standards to OSHA. Other sections grant OSHA authority to require record keeping and medical surveillance and to enforce the act and its regulations through civil and criminal penalties.

Given the language quoted above regarding risk and hazard, the applicability of the Occupational Safety and Health Act to biotechnology would be limited when risk is conjectural. However, the act would be applicable to large-scale processes using known human toxins, pathogens, or their DNA. It also would be applicable to physical hazards presented by the fermentation process, such as temperature, pressure, and toxic solvents. OSHA has not promulgated health and safety standards for bioprocesses and has made no statements on how it might apply the act to biotechnology.

OSHA arguably has authority to require a medical surveillance program, although this is not clear cut. Section 8(c)(1) of the Occupational Safety and Health Act requires employers to "make, keep and preserve" such records as the U.S. Secretary of Labor prescribes by regulation as "necessary or appropriate for the enforcement of this act or for developing information regarding the causes and prevention of occupational accidents and illness." Further, section 8(c)(2) of the act authorizes the Secretary of Labor to require employers to "maintain accurate records of, and to make periodic reports on, work-related deaths, injuries and illnesses other than minor injuries . . . ." Since the purpose of a surveillance program would be to develop information on any occupational disease related to biotechnology, section 8(c)(1) of the Occupational Safety and Health Act would seem to apply. In addition, the information developed in such a program would also be the kind of information necessary for compliance with regulations promulgated under section 8(c)(2). Employers, on the other hand, might argue that both sections require an initial showing that biotechnology causes occupational disease.

EUROPEAN ECONOMIC COMMUNITY COUNTRIES: FEDERAL REPUBLIC OF GERMANY, UNITED KINGDOM, AND FRANCE

European Economic Community.—Although its powers in the area of worker health and safety regulation are limited and indirect, the EEC has attempted to ensure at least minimal protection for most industrial workers. In 1980, the EEC adopted a directive that required each member state to adopt a variety of measures to protect workers' health and safety (54).* The directive covers work that does or may involve a "chemical, physical or biological agent . . . likely to be harmful to health." The directive is quite general; the specific content and substance is left to the discretion of the member states.

The directive does not refer explicitly to rDNA work or other applications of biotechnology. Thus, the question of how worker health and safety laws will affect the biotechnology industry is left to the discretion of each member state.

Federal Republic of Germany.—The rDNA research guidelines of the Federal Republic of Germany (57) provide specifically for the health-monitoring and training of laboratory workers. Each worker at an rDNA laboratory that is above the lowest containment level must have a pre-employment examination by an authorized doctor. If the results of this examination reveal a susceptibility to hazards which may be involved in the contemplated research, the worker may not be employed. Appropriate immunizations are required for work with pathogenic micro-organisms. Blood serum from the worker must be taken at the first examination and at the end of employment and stored until at least 2 years after the end of participation in the research. All workers must receive instruction before the research begins and annually thereafter in the methods to be used, the conceivable hazards of the experiment, and the protective measures to be applied.

The Federal Republic of Germany's general worker health and safety regulations would also apply to commercial uses of biotechnology. At the Federal level, substantive workplace health and safety requirements are stated in the Act Respecting Plant Physicians, Safety Engineers, and Occupational Safety Specialists (55),* in the Ordinance Respecting Workplaces (56),** and in rules that are issued by the Dangerous Industrial Substances Committee (Ausschus für Gefährliche Arbeitsstoffe) of the Federal Ministry of Labor and Social Affairs (Bundesministerium für Arbeit und Sozialordnung) concerning the marketing and handling of dangerous substances (70).

Within this Federal framework, a significant regulatory role is played in the Federal Republic of Germany by accident insurance funds. These funds are authorized by statute to issue regulations setting standards for workplace health and safety (58). When approved by the Federal Minister of Labor and Social Affairs, the regulations become binding on covered employers. The funds, which are organized by indus-

*The required measures include the following:
1. limitations on the use of chemical, physical or biological agents in the workplace;
2. limitations on the number of workers exposed or likely to be exposed to such agents;
3. engineering controls;
4. establishment of exposure limit values for such agents and methods of assessing their level;
5. safe working procedures and methods;
6. collective protection measures;
7. individual protection measures, where exposure cannot reasonably be avoided by other means;
8. hygiene measures;
9. information for workers on potential risks associated with the exposures to such agents, technical preventive measures workers should take, and precautions to be taken by the employer and the workers;
10. use of warning and safety signs;
11. surveillance of workers' health;
12. maintenance of current records of exposure levels, workers exposed, and medical records;
13. emergency procedures; and
14. if necessary, general or limited bans on an agent from which protection cannot be adequately ensured.

*The Act Respecting Plant Physicians, Safety Engineers, and Occupational Safety Specialists requires each employer to appoint a plant physician and an occupational safety specialist. The appointed physician must conduct medical examinations of employees, advise the employer concerning health and safety precautions (including technical equipment and personal protective devices), supervise workplace safety, investigate and report to the employer on the causes of work-related illnesses, and instruct employees concerning the dangers to which they are exposed in the course of their work and the measures available to avert such dangers.

**Section 3(1)1 of the Ordinance Respecting Workplaces imposes a general obligation on employers to operate workplaces in accordance with both the law and the "generally recognized rules of safety engineering, occupational medicine and hygiene and any other scientifically established findings in the labor field." Its specific requirements, however, relate to physical design and construction.

try, are authorized not only to promulgate the applicable standards, but also to enforce them through inspections and fines. Because all employers must carry accident insurance, the funds have a large role in occupational safety and health.

United Kingdom.—Guidelines promulgated by GMAG contain specific requirements regarding the health and safety of laboratory workers who are involved in rDNA research (67,68,69) (see discussion of "Recombinant DNA Research Guidelines" above). Each laboratory must appoint a supervisory medical officer with experience in public health, infectious diseases, or occupational medicine, and conduct health reviews of all workers before they start work involving genetic manipulation. The reviews are designed to check workers for particular susceptibilities and to assist in determining whether any laboratory-contracted illnesses have developed. If a worker's medical history indicates that the worker's participation in genetic manipulation may be particularly hazardous, appropriate steps may be required to prevent his or her exposure to genetic manipulation work. The institution must also investigate any unexplained illiness, and if a laboratory-contracted infection is suspected, the institution must inform both the worker and the worker's physician as well as GMAG and other authorities.

Companies using biotechnology in the United Kingdom must also fulfill the obligations imposed on virtually all employers and manufacturers by the Health and Safety at Work Act of 1974 (66). In general, an employer must ensure so far as reasonably practicable that employees are not exposed to health and safety risks and to inform them of the risks that are created. Employees also have certain obligations under the act.

Health and safety regulations in the United Kingdom, under the Health and Safety at Work Act, are promulgated by the Secretary of State, on the advice of the Health and Safety Commission. The Health and Safety Commission also supervises efforts to improve worker health and safety, makes necessary investigations, and may approve codes of practice for particular industries.

There is no code of practice for biotechnology other than the GMAG guidelines for rDNA research. If a broader code were developed, it would be only advisory; violation of a code is not per se a violation of the Health and Safety Work Act but is only evidence tending to show a violation of the act. HSE (and local authorities) enforce the act through appointed inspectors, who may issue "notices" prohibiting certain activities as too risky or requiring remedial actions. Violators of the act are subject to civil and criminal penalties.

France.—The guidelines for rDNA research in France contain no provisions dealing expressly with the health or the health-monitoring of laboratory workers. The guidelines do require, however, that scientists and technicians be familiar with the physical and biological containment measures involved in rDNA research and be prepared to take emergency action in the event of an accident.

The formulation and implementation of general policy on the prevention of occupational hazards in France is the responsibility of the Central Council for the Prevention of Occupational Hazards (Conseil Central pour la Prevention des Risques Professionalle). So far, the council has not specifically addressed worker health problems arising from biotechnology.

Specific employee health and safety regulations are promulgated and enforced in France by the Minister of Labor, who is in charge of conditions in industrial and commercial establishments, and by the Minister of Agriculture, who is granted the same authority over agricultural facilities.

An occupational safety and health committee must be set up in any industrial establishment normally employing 50 or more workers (59,60). The committee advises management on safety procedures and periodically inspects the establishment to ensure that the safety laws and regulations are being applied. It also is supposed to take immediate action to avert imminent danger at the facility and to conduct an inquiry into the causes of any accident or serious occupational disease.

The manufacture of chemical substances potentially harmful to workers is also regulated by statute (61). Prior to the marketing of any substance or preparation that may involve a danger to workers, the manufacturer, importer, or seller must file with a Government-approved laboratory the information necessary to assess the risks of the manufacturing process. If the chemical substance has already been placed on the market, its manufacture, sale, transfer, or use may be restricted or prohibited in the interests of occupational health and safety.

SWITZERLAND

By following the U.S. guidelines for rDNA research, Switzerland applies to rDNA work the worker health and safety rules set out therein. Thus, each research institution in Switzerland must ensure that laboratory workers receive appropriate training, determine whether a health surveillance program is appropriate, and report to the Commission for Experimental Genetics any work-related accidents or illness. The responsibility for assessing the training provided to personnel and the adopting emergency plans for accidental spills and personnel contamination rests with the institution's biohazards committee. The biological safety officer must report work-related accidents or illnesses and assist in developing emergency plans. The group leader is obligated to train and supervise his or her staff.

Worker health and safety not specifically related to rDNA research is regulated in Switzerland on the cantonal rather than the confederal level. In one canton, Geneva, an advisory committee has been established to serve as a channel of communication between public authorities and business and to develop proposals on worker health and safety. The committee meets four times a year (63). The other cantons do not have such committees.

JAPAN

The basic law governing worker health and safety in Japan is the Industrial Safety and Health Law of 1972 (64).* This law imposes on employers health and safety obligations which are comprehensive in scope but very general in actual language. Among these obligations is the duty to take necessary measures to prevent health impairment caused by substances, agents, and conditions found in the workplace. The law vests broad discretion in the Japanese Ministry of Labor to determine when regulation is appropriate and what kinds of precautions an employer must take. Employers who manufacture, import, or use "chemical substances" may be subject to special requirements. Medical examinations must be conducted on all employees, but employers may also be required to provide special tests for employees engaged in harmful work. At the present time, no regulations have been addressed specifically to biotechnology.

The Industrial Safety and Health Law includes a stringent enforcement mechanism. Substantial criminal penalties and fines are imposed for violations. For the most serious violations, offending employers may also be ordered to close or alter their operations.

Appendix F references**

RECOMBINANT DNA RESEARCH GUIDELINES REFERENCES

1. Commission of the European Communities, "Draft Council Recommendation Concerning the Registration of Recombinant DNA (deoxyribonucleic acid) Work," *Official Journal of the European Communities* 23(C214):7-8, Aug. 21, 1980.
2. Council of the European Communities, "Council Recommendation of 30 June 1982, No. 82/472/EEC, Concerning the Registration of Work Involving Deoxyribonucleic Acid (DNA)," *Official Journal of the European Communities* 25 (L213):15-16, July 21, 1982.
3. Federal Republic of Germany, Bundesministerium für Forschung und Technologie (Federal Ministry for Science and Technology), "Richtlinien zum Schutz vor Gefahren durch in-vitro neukombinierte Nukleinsäuren" (Guidelines for the Safe Handling of Recombinant Nucleic Acid (DNA)), appearing in 4. u"berarbeitete Fassung, Bonn, Aug. 7, 1981.
4. Fildes, R. A., Chairman, Industrial Biotechnology Association Committee on Industrial Applications and Scale-Up, remarks in *Recombinant DNA Technical Bulletin* (NIH) 5:196-197, December 1982.
5. France, "Les lois Gouvernant la Récherche du DNA Recombinaisant" (Guidelines on Recombinant DNA Research), adopted Dec. 13, 1977.
6. Gartland, W., Director, Office of Recombinant DNA Activities, National Institutes of Health, Bethesda, Md., personal communication, Aug. 22, 1983.
7. *Genetic Engineering News,* "Regulatory Update: Universities To Follow in Easing DNA Rules," September/October 1982, p. 23.
8. *Japan Economic Journal,* "Biotechnology Development Role Mounts With New Entrants," Mar. 16, 1982, p. 16.
9. *Japan Economic Journal,* "Easing Recombinant DNA Rules to Western Levels is Recommended," Aug. 3, 1982, p. 9.
10. *Japan Economic Journal,* "Biotechnology is Entering Second Stage of Its Boom," Sept. 28, 1982, p. 9.
11. Japan, "Kumikae DNA jikken shishin" (Guidelines for Recombinant DNA Experiments) (Cabinet Prime Minister, Aug. 27, 1979, as amended on Apr. 7, Nov, 7, 1980, and Apr, 1, 1981, Tokyo, Japan).
12. *Nature,* "Watchdog To Bark Less Often," 302:467, 1983.
13. Rörsch, A., *Genetic Manipulations in Applied Biology—A Study of the Necessity, Content, and Management Principles of a Possible Community Action,* contract no. 346-77-7 ECI, NL, submitted to Directorate General "Scientific and Technical Information and Information Management Management," Commission of the European Communities, Brussels, Belgium, 1979.
14. Switzerland, "Les lois Gouvernant la Reserche du DNA Recombinaisant" (Guidelines on Recombinant DNA Research).
15. United Kingdom, Genetic Manipulation Advisory Group, "Prospective New Basis for the Categorisation of Genetic Manipulation in the United Kingdom," *Nature* 276:104-108, 1978.
16. United Kingdom, Genetic Manipulation Advisory Group, GMAG Note No. 1, "Local Safety Committees in Genetic Manipulation Laboratories," London.
17. United Kingdom, Genetic Manipulation Advisory Group, GMAG Revised Note No. 7, "Information and Advice on the Completion of Proposal Forms for Centres (Part A) and Projects (Part B)," London, May 1979.

*This discussion is based on a 1980 publication of the Japanese Government entitled *Labour Laws of Japan* (65), and does not reflect changes that may have been subsequently enacted.

**Note: S.Ct. = Supreme Court Reporter.

18. United Kingdom, Genetic Manipulation Advisory Group, GMAG Note No. 8 (March 1979), Note No. 8-Supplement 1 (December 1979), Note No. 8-Supplement 2 (December 1980), "Self-Cloning Experiments," London, 1979 and 1980.
19. United Kingdom, Genetic Manipulation Advisory Group, GMAG Note No. 11, "The GMAG Categorisation Scheme," London, January 1980.
20. United Kingdom, Genetic Manipulation Advisory Group, GMAG Note No. 12, "Large-Scale Uses of the Products of Genetic Manipulation—Work Involving GMAG Volumes of 10 Litres or More," London, November 1979.
21. United Kingdom, Genetic Manipulation Advisory Group, GMAG Note No. 13, "Genetic Manipulation of Plants and Plant Pests," London, January 1980.
22. United Kingdom, Genetic Manipulation Advisory Group, GMAG Note No. 14, "Revised Guidelines for the Categorisation of Recombinant DNA Experiments," London, 1980.
23. United Kingdom, Genetic Manipulation Advisory Group, GMAG Note No. 15, "Code of Practice for Genetic Manipulation Containment Facilities," London, June 1981.
24. United Kingdom, Health and Safety at Work Act of 1974, *Statutory Instruments,* Chapter 37.
25. United Kingdom, House of Parliament, House of Commons Select Committee on Science and Technology, *Recombinant DNA Research—Interim Report,* vii (London: H.M. Stationery Office, Apr. 3, 1979).
26. United Kingdom, Health and Safety (Genetic Manipulation) Regulations, Statutory Instrument 1978 No. 752 (London: H. M. Stationery Office, 1978).
27. U.S. Department of Health, Education and Welfare, National Institutes of Health, "Physical Containment Recommendations for Large-Scale Uses of Organisms Containing Recombinant DNA Molecules," *Federal Register* 45:24968, Apr. 11, 1980.
28. U.S. Department of Health and Human Services, National Institutes of Health, "Guidelines for Research Involving Recombinant DNA Molecules," *Federal Register* 48:24556-24581, June 1, 1983.

ENVIRONMENTAL LAWS AND REGULATIONS REFERENCES

29. Council of the European Communities, Council Directive of 27 June 1967, No. 67/548/EEC, on the Approximation of Laws, Regulations and Administrative Provisions Relating to the Classification, Packaging and Labelling of Dangerous Substances, *Official Journal of the European Communities* 10(196):234, Aug. 16, 1967.
30. Council of the European Communities, Council Directive of 4 May 1976, No. 76/464/EEC, on Pollution Caused by Certain Dangerous Substances Discharged into the Aquatic Environment of the Community, *Official Journal of the European Communities* 19(L129):23, May 18, 1976, reprinted in 2 Int'l. *Env't. Rep. Ref. File* (Bureau of National Affairs, Washington, D.C.) 151:2101.
31. Council of the European Communities, Council Directive of 20 March 1978, No. 78/319/EEC, on Toxic and Dangerous Wastes, *Official Journal of the European Communities* 21(L84):43, Mar. 31, 1978, reprinted in 2 *Int'l. Env't. Rep. Ref. File* (Bureau of National Affairs, Washington, D.C.) 151:1201.
32. Council of the European Communities, Council Directive of 18 September 1979, No. 79/831/EEC, Amending for the Sixth Time Directive 67/548/EEC on the Approximation of the Laws, Regulations and Administrative Provisions Relating to the Classification, Packaging and Labelling of Dangerous Substances, *Official Journal of the European Communities* 22(L259):10, Oct. 15, 1979.
33. Council of the European Communities, Council Directive of 17 December 1979, No. 80/684/EEC, on the Protection of Groundwater Against Pollution Caused by Certain Dangerous Substances, *Official Journal of the European Communities* 22(L20):43-47, Jan. 26, 1980, reprinted in 2 *Int'l. Env't. Rep. Ref. File* (Bureau of National Affairs, Washington, D.C.) 151:2101.
34. Deutsche Gesellschaft für chemisches Apparatewesen, *Arbeitmethoden für die Biotechnologie* (Work Procedures for Biotechnology), Frankfurt, 1982.
35. Federal Republic of Germany, Act on Regulation of Matters Pertaining to Water, Law of July 27, 1957, 1957 Bundesgesetzblatt [BGB1] I:1110, as amended by Fourth Act for Amendment of Federal Water Act, Apr. 26, 1976, BGB1 at 1109, reprinted in 3 *Int'l. Env't. Rep. Ref. File* (Bureau of National Affairs, Washington, D.C.) 241:2201.
36. Federal Republic of Germany, Abfallbeseitigungsgesetz (Law on Disposal of Wastes), Law of June 7, 1972, 1972 Bundesgesetzblatt [BGB1] at 873:1601, summarized in 3 *Int'l. Env't. Rep. Ref. File* (Bureau of National Affairs, Washington, D.C.) 241:0105.
37. Federal Republic of Germany, Law for the Prevention of Harmful Effects on the Environment Caused by Air Pollution, Noise, Vibration, and Similar Phenomena, Law of March 15, 1974, 1974 Bundesgesetzblatt [BGB1], 721, 1193, as amended by Law Amending the Law To Introduce the Penal Code, August 15, 1974, 1974 BGB1 at 1942, reprinted in 3 *Int'l. Env't. Rep. Ref. File* (Bureau of National Affairs, Washington, D.C.) 241:1001.
38. Federal Republic of Germany, Waste Water Charges Act, Law of Sept. 13, 1976, 1976 Bundesgesetzblatt [BGB1] I:2721, reprinted in 3 *Int'l. Env't. Rep. Ref. File* (Bureau of National Affairs, Washington, D.C.) 241:2301.
39. Federal Republic of Germany, Bundesministerium für Forschung und Technologie (Federal Ministry for Science and Technology), "Richtlinien zum Schutz vor Gefahren durch in-vitro neukombinierte Nukleinsäuren" (Guidelines for the Safe Handling of Recombinant Nucleic Acid (DNA)), 4. überarbeitete Fassung, Bonn, Aug. 7, 1981.
40. Federal Republic of Germany, Chemikaliengesetz (Chemicals Act), Law of Sept. 16, 1980, 1980 Bundesgesetzblatt [BGB1] (*Fed. L. Gaz.* I), I:1718, translation reprinted in 3 *Int'l. Env't. Rep. Ref. File* (Bureau of National Affairs, Washington, D.C.) 241:4101.
41. France, Decree No. 79-35 *Journal Officiel de la République Française* (Official Journal), Jan. 17, 1979.
42. France, La loi no. 64-1245 du 16 decembre 1964 relative au regime et répartitions des eaux et à la lutte contre la pollution (Law No. 64-1245 of Dec. 16, 1964 on the

Administration and Classification of Waters and the Control of Water Pollution), *Journal Officiel de la République Française* (Official Journal), Dec. 16, 1964, pp. 11,258 et seq.
43. France, Law No. 75-633 on Waste Disposal and Recovery of Materials, *Journal Officiel de la République Française, (Official Journal) July 15, 1975, excerpts reprinted in 3 Int'l. Env't. Rep. Ref. File* (Bureau of National Affairs, Washington, D.C.) 231:7001.
44. France, La loi no. 76-663 du 19 juillet 1976 relative aux installations classes pour la protection de l'environment (Law No. 76-663 of July 19, 1976, on Installations Classified for Purposes of Environmental Protection), *Journal Officiel de la République Française* (Official Journal), pp. 4320 et seq., July 19, 1976, reprinted in 3 *Int'l. Env't. Rep. Ref. File* (Bureau of National Affairs, Washington, D.C.) 231:2001.
45. France, Law No. 77-771 on the Control of Chemicals, *Journal Officiel de la République Française* (Official Journal), July 12, 1977, excerpts reprinted in 3 Int'l. Env't. Rep. Ref. File (Bureau of National Affairs, Washington, D.C.) 231:5001.
46. Japan, Basic Law for Environmental Pollution Control, Law No. 132 of 1967, as amended, reprinted in 2 *Int'l. Env't. Rep. Ref. File* (Bureau of National Affairs, Washington, D.C.) 91:0501.
47. Japan, Air Pollution Control Law, Law No. 97 of 1968, as amended, reprinted in 2 *Int'l. Env't. Rep. Ref. File* (Bureau of National Affairs, Washington, D.C.) 91:0901.
48. Japan, Waste Management Law, Law No. 137 of 1970, as amended, reprinted in 2 *Int'l. Env't. Rep. Ref. File* (Bureau of National Affairs, Washington, D.C.) 91:2401.
49. Japan, Water Pollution Control Law, Law No. 138 of 1970, as amended, reprinted in 2 *Int'l. Env't. Rep. Ref. File* (Bureau of National Affairs, Washington, D.C.) 91:1401.
50. Japan, Chemical Substances Control Law, Law No. 117 of 1973, reprinted in 2 *Int'l. Env't. Rep. Ref. File* (Bureau of National Affairs, Washington, D.C.) 91:6401.
51. Switzerland, loi Federal du 8 Octobre 1971 sur la Protection des Eaux Contre la Pollution, RS 814.20 (Water Protection Act of Oct. 8, 1971).
52. Switzerland, loi Federal du 3 Mars 1974 sur le Travaille d'en l'Industrie, l'Artisant, et les Commerce, RS 822.11 (Federal Act on Work in Industry, Trade, and Commerce for March 3, 1974.
53. United Kingdom, Control of Pollution Act of 1974, reprinted in 3 *Int'l. Env't. Rep. Ref. File* (Bureau of National Affairs, Washington, D.C.) 291:7301.

REGULATION OF WORKER HEALTH AND SAFETY REFERENCES

54. Council of the European Communities, "Council Directive of 27 November 1980, No. 80/1107/EEC, on the Protection of Workers From the Risks Related to Exposure to Chemical, Physical, and Biological Agents at Work," *Official Journal of the European Communities* 23(L327):8-13, Dec. 3, 1980.
55. Federal Republic of Germany, Act of 12 December 1973 Respecting Plant Physicians, Safety Engineers, and Occupational Safety Specialists, as amended, Bundesgestzblatt [BGB1] 105:1885, Dec. 15, 1973, translation reprinted in *Legislative Series,* 1973-Ger. F.R.2 (International Labor Office, Geneva), January-February 1975.
56. Federal Republic of Germany, Ordinance of 25 March 1975 Respecting Workplaces, Bundesgesetzblatt [BGB1] 32:729, Mar. 25, 1975, translation reprinted in *Legislative Series,* 1975-Ger. F.R. 2 (International Labor Office, Geneva), September-October 1976.
57. Federal Republic of Germany, Bundesministerium für Forschung und Technologie (Federal Ministry for Science and Technology), "Richtlinien zum Schutz vor Gefahren durch in-vitro neukombinierte Nukleinsäuren" (Guidelines for the Safe Handing of Recombinant Nucleic Acid (DNA)), in 4. überabeitete Fassung, Bonn, Aug. 7, 1981.
58. Federal Republic of Germany, Federal Insurance Code (as amended), translation reprinted in *Legislative Series,* 1980-Ger. F.R. 1 (International Labor Office, Geneva) §708 (as last amended on 18 August 1980), Sozialgestzbuch-Verwaltungsverfahren BGBLI:1467, Aug. 18, 1980.
59. France, Act No. 73-1195 of 27 December 1973 Respecting the Improvement of Working Conditions, *Journal Officiel de la République Française* (Official Journal) 304:14-146, Dec. 30, 1973, translation reprinted in *Legislative Series,* 1973-Fr. (International Labor Office, Geneva) 3:1-7, May-June 1974.
60. France, Decree No. 74-274 of 1 April 1974, *Journal Officiel de la République Française* (Official Journal), Apr. 5, 1974.
61. French Labour Code, Article L. 231-7, translation reprinted in *Legislative Series* 1981-Fr. 1 (International Labor Office, Geneva) 1:78, 90, January-February 1982.
62. *Industrial Union Dept., AFL-CIO* v. *American Petroleum Inst.,* 100 S. Ct. 2844 (1980).
63. International Labour Organisation, "Cantonal Advisory Committee on Occupational Health and Safety Set Up," *Soc. & Lab. Bull.* 4:368, December 1976.
64. Japan, Industrial Safety and Health Laws, Law No. 57 of June 8, 1972, as amended by Law No. 28 of May 1, 1975 (Working Environment Measurement Law) and Law No. 76 of July 1, 1977, translation available in Japan, Ministry of Labor, *Labour Laws of Japan,* Tokyo, 1980.
65. Japan, Ministry of Labor, *Labour Laws of Japan,* Tokyo, 1980.
66. United Kingdom, Health and Safety at Work etc. Act of 1974, *Statutory Instruments,* ch. 37.
67. United Kingdom, Genetic Manipulation Advisory Group, GMAG Note No. 6, "Health Monitoring," London, revised January 1980.
68. United Kingdom, Genetic Manipulation Advisory Group, GMAG Note No. 14, "Revised Guidelines for the Categorisation of Recombinant DNA Experiments," London, 1980.
69. United Kingdom, Genetic Manipulation Advisory Group, GMAG Note No. 14, Supplement 1, "Guidelines for the Categorisation of Recombinant DNA Experiments Involving Eukaryotic Viruses," London, January 1981.
70. Versen, P., "Employer-Worker Cooperation in the Prevention of Employment Injuries," *Int'l. Labour Rev.* 118:75, January-February 1979.

Appendix G
Intellectual Property Law

Chapter 16: Intellectual Property Law discusses three areas of intellectual property law that are particularly relevant to the commercialization of biotechnology: patents, trade secrets, and plant breeders' rights. That chapter focuses initially on the United States and then discusses the laws of the other countries by comparing them to the U.S. laws. This appendix elaborates on the intellectual property laws of the five countries likely to be the major competitors of the United States in the commercialization of biotechnology—Japan, the Federal Republic of Germany, the United Kingdom, Switzerland, and France—and is the basis for the comparisons in chapter 16. The first section examines the laws of the four European countries, and the second section considers the intellectual property law of Japan.

Intellectual property laws of the Federal Republic of Germany, the United Kingdom, Switzerland, and France

The Federal Republic of Germany, the United Kingdom, Switzerland, and France, have created an intellectual property law similar to that of the United States. Important differences exist, however, especially on a country-by-country basis. Patent laws, laws of trade secrets, and plant breeders' rights in these countries are reviewed in the sections that follow.

PATENT LAW

Eleven European countries, including the Federal Republic of Germany, the United Kingdom, Switzerland, and France, have agreed to a treaty, the European Patent Convention (EPC), that creates a European patent system (8). These countries also have patent systems created by national laws.

European Patent Convention.—The EPC entered into force on October 7, 1977, and as of January 1, 1983, the treaty had been ratified by Belgium, the Federal Republic of Germany, France, the United Kingdom, Luxembourg, the Netherlands, Switzerland, Sweden, Italy, Austria, and Liechtenstein. The EPC establishes a legal system for granting European patents through a single supranational European Patent Office and a uniform procedural system with respect to patent applications. The single European patent application, if granted, becomes a bundle of individual European patents, one for each of the countries designated by the applicant.

The EPC system and the resulting patents exist in parallel with the patent systems of the member countries. The ultimate goal is for each of the member countries to adopt in its national law the same substantive law of patents set forth in the EPC; in the beginning, however, and perhaps always to a certain extent, differences in substantive law will exist between countries. Enforcement of European patents is handled by the same national authorities that are responsible for handling enforcement of national patents in the EPC member countries (EPC art. 64(3)).

Patentable Subject Matter. Under the EPC, patents can be granted for any invention susceptible of industrial application* that is new and involves an inventive step (EPC art. 52(1)). This broad definition is narrowed by specific exclusions. Discoveries, scientific theories, and naturally occurring products, for example, are not considered patentable inventions. Methods of treating humans or animals and related diagnostic methods are similarly excluded from patentability, although products so used are not. Finally, plant or animal varieties and essentially biological processes for the production of plants or animals are not patentable; however, their exclusion does not apply to microbiological processes or the products of such processes (EPC art. 53(a) and (b)). The question of whether a process is "essentially biological" depends on the extent to which there is technological intervention by humans in the process. Under the Guidelines for Examination of the European Patent Office, if such intervention plays a significant part in determining or controlling the result it is desired to achieve, the process would not be excluded (G.E. pt. C(IV)(3.4)).

Under EPC articles 52(1) and 53(b), as interpreted by the European Patent Office, microbiological inventions of the following kind would be patentable: 1) micro-organisms (including viruses and cell lines), 2) processes for making them, 3) processes using them, 4) products obtained from microbiological processes, and 5) DNA and RNA molecules or subcellular units (e.g., plasmids) (G.E. pt. C(IV)(3.5-3.6)). The European Patent Office also stated that the term "micro-organism" covers plasmids.

One major area that will require further clarification, however, is whether naturally occurring micro-organisms, subcellular units, or DNA and RNA molecules are patentable. Under the EPC, there appears to be no absolute bar, it will simply be a question of

*The term industrial application includes agricultural applications (EPC art. 57). This is actually the standard for utility under the EPC.

the degree to which such subject matter is naturally available and of the effort required to identify and/or isolate it (G.E. pt. C(IV)(2.1)).

Novelty. Under the EPC, an invention is new if it is not part of the state-of-the-art on the effective filing date of the patent application (EPC art. 54(1)). The EPC provides that the state-of-the-art comprises everything made available to the public by means of a written or oral description, by use, or in any other way, before the date of filing of the European patent application (EPC art. 54(2)).* There are no restrictions as regards the geographical location where, or the language or manner in which, the relevant information is made available to the public.

This is known as an "absolute novelty standard" because certain public disclosures even by the inventor himself/herself before the filing can result in loss of patent rights. The absolute novelty standard is a major distinction of European patent law from that of the United States.

Standard of Invention. The EPC defines inventive step as follows (EPC art. 56):

> An invention shall be considered as involving an inventive step if, having regard to the state of the art, it is not obvious to a person skilled in the art.

This definition parallels the definition of nonobviousness under section 103 of the U.S. patent law (35 U.S.C. 103), except that §103 refers to a person of *ordinary* skill in the art and also to the differences between the invention and the prior art.

The European Patent Office's Guidelines for Examination indicate that the test of obviousness to be applied by the European patent examiners is consistent with the objective test under section 103 (G.E. pt. C(IV)(9.9)). In particular, the European Patent Office apparently will consider such factors as unexpected advantages, evidence of immediate commercial success, and evidence of long felt need (18,30).

Disclosure Requirements. The basic disclosure requirement under the EPC is as follows (EPC art. 83):

> The European patent application must disclose the invention in a manner sufficiently clear and complete for it to be carried out by a person skilled in the art.

This enablement requirement has as its essential element the concept of the reproducibility or repeatability; i.e., the making of the invention must not be dependent on chance. For micro-organisms, enablement generally is satisfied by depositing a culture of the micro-organism in a depository to which the public has access and referencing the depository and file number in the patent application. However, a deposit need not be made if the micro-organism is already publicly available or can be described so as to be reproducible.

Deposit Requirements. If a deposit is required, it must be made with a recognized depository not later than the date the application is filed. The European Patent Office publishes a list of recognized depositories, and, since it adheres to the Budapest Treaty, the European Patent Office also recognizes deposits made pursuant to the treaty. Cultures must be maintained for at least 30 years.

Since all European patent applications are published approximately 18 months after their filing date (unless previously withdrawn) (EPC art. 93(1)), the deposited micro-organism can become publicly available before the patent has been issued. The EPC sets up certain safeguards on access to prevent abuse.*

Claims. Claims in an EPC application must define the subject matter for which protection is sought, be clear and concise, and be supported by the description (EPC art. 84; G.E. A(III)(4)-(6)).

Enforcement. Under the EPC, European patents are granted for a term of 20 years. Enforcement is handled by the national courts of the EPC member countries. The question of infringement is considered under national law principles, but taking account of treaty requirements regarding claim interpretation. European patents may be revoked by a national court on certain specified grounds (EPC arts. 138(1) and 139(2)).

Patent Laws of the Federal Republic of Germany, the United Kingdom, Switzerland, and France.—As described below, the patent laws in the Federal Republic of Germany, the United Kingdom, Switzerland, and France vary with respect to certain provisions regarding patentable subject matter, novelty, disclosure requirements, or enforcement.

Patentable Subject Matter. The provisions defining patentable subject matter in the patent law of the Federal Republic of Germany are virtually identical with the corresponding provisions of the EPC. Regarding biological inventions, the Federal Republic of Germany has been a pioneer in recognizing the patentability of micro-organisms per se. After deciding in 1969 that patents could be obtained for inventions in the field of biology (22), the German Federal Supreme Court specifically held in 1975 that micro-organisms per se constituted patentable subject matter (2). Therefore, in line with EPC law, the same categories of biologi-

*Implicit in the concept of the state-of-the-art is the concept that the public disclosure must be enabling.

*The safeguards are as follows: 1) the recipient may not pass the sample on to anyone else unless or until the application is abandoned or all European patents have expired; 2) the recipient can only use the micro-organism for experimental purposes until the application is abandoned or a patent issues; 3) the patentee can elect to permit samples to be given only to certain neutral experts (EPC Rule 28(3)-(4)).

cal inventions are patentable in principle according to West German law.*

In its Patent Act of 1977, the United Kingdom adopted the EPC definition of patentable subject matter. The British Patent Office has taken the position that all of the five general categories of biological subject matter listed above are patentable (27).

Section 1a of the Swiss Patent Law corresponds to EPC Article 53(b), stating that "micro-biological methods and products obtained thereby shall be patentable." There is no specific provision in the law which states that "discoveries" are not patentable subject matter, although prior case law recognizes such an exclusion (5).

Nevertheless, it appears that Swiss practice varies considerably from that under the EPC. According to the Swiss Patent Office, micro-organisms per se are not patentable, including human-made ones. The Patent Office has apparently not yet taken a position on the patentability of DNA and RNA molecules or subcellar units (7).

As to the remaining categories of subject matter involving micro-organisms, the Swiss law provides for patent protection in the same manner as the EPC. Furthermore, since microbiological processes are explicitly patentable, some protection is obtainable for micro-organisms per se under Swiss law, because section 8 of the Swiss Patent Act provides that the protection of a patent claiming a process shall extend also to the immediate products of the process.

The substantive law regarding patentable subject matter in France corresponds to the EPC, specifically in all respects which are relevant to micro-organisms. However, article 7 of the French patent law (1978) excludes patents on plant varieties to the extent to which such varieties are protectable under French plant protection legislation.

Utility. All of the EPC countries have adopted the EPC requirement for utility—that the invention be useful in industry (including agriculture) (24). However, Swiss law restricts the concept of industrial use by excluding private use and use for research (15).

Novelty. The Federal Republic of Germany, United Kingdom, and France have adopted the EPC absolute novelty standard in their latest national patent laws (24). Switzerland also has adopted the absolute novelty standard with one technical exception relating to prior filed Swiss or EPC applications (Swiss Patent Law, art. 7a).

Disclosure and Deposit Requirements. The statutory provision of West German law governing disclosure requirements (West German patent law sec. 35(2), 1981) is identical to article 83 of the EPC, i.e., enablement of a person skilled in the art. However, there are certain differences in practice regarding biological inventions. By court decision, a new micro-organism cannot be patented unless the application discloses a *reproducible* method of producing it. Thus, a deposit without an enabling written description is inadequate to support a claim to the micro-organism itself (3,26). This is in marked contrast to the law of the other countries. On the other hand, a deposit alone is sufficient to support a claim to a method of using a new micro-organism (32). A required deposit must be made no later than the filing date (or the priority date) (32). Although the applicant must furnish samples of the deposit to third parties after publication of the application, the applicant can require that the samples not be removed from the Federal Republic of Germany and not be passed on to others.

The British Patents Act, in section 14(3), has the same enablement standard as the EPC. In the case of an invention involving a micro-organism, the application as filed must contain the relevant information on the characteristics of the micro-organisms, to the extent known to the applicant. The required deposit must be made no later than the filing date or the priority date (British Patent Office Rule 17(1) (1978)). Samples will be publicly available when the application is published 18 months after the priority date. Those who request samples must undertake not to pass them on to others and to use them only for experimentation until the patent is granted or the application is abandoned (British Patent Office Rule 17(2) (1978)).

The Swiss Patent Act, in section 50 (1978), contains the same enablement standard as the EPC. The Patent Ordinance, section 26(6) (1977), also requires that the description explain how the invention may be used industrially. In the case where the micro-organism is not publicly available or cannot be described in an enabling manner, a deposit in a recognized depository is required. The application must identify the depository, the deposit number, and the date of the deposit (Swiss Guidelines for Examination, Z-14.3 and 14.4, May 12, 1980). In the case of a micro-organism that is available to the public, identification of a known source need not be disclosed in the application as originally filed. Such information (e.g., reference to a deposit that was publicly available on the application filing date) can be added to the application after the filing date (Swiss Guidelines for Examination, Z-13.2, May 12, 1980). Since Swiss applications are not published before the patent is granted, culture samples are not required to be furnished until the patent is granted. Then samples are released only to identified

*The German Federal Patent Court has also upheld a patent on a micro-organism obtained as a pure culture from an unpurified, naturally occurring state through a selective culture process (16).

parties, who undertake not to pass them on (Swiss Patent Ordinance, sec. 27(6)).

The French patent law, in article 14bis (1978), sets forth the same standard of enablement as the EPC. Publicly available micro-organisms need not be deposited. Required deposits must be made in a Government-authorized depository no later than the priority date. A regulation under the statute (Decree No. 79-822 on Sept. 19, 1979, amended by Decree No. 81-865 issued on Sept. 11, 1981) contains provisions regarding the content of a French patent application relating to a micro-organism that are consistent with EPC Rule 28. Thus, the application must contain (French patent law, art. 10):

- the available information as to the characteristics of the micro-organism, and
- an identification of the depository and deposit number.

Access to the deposit, which is granted at the time of publication, can be limited to recognized experts until the patent is granted or the application is abandoned (French patent law, art. 31).

Claim Practice. Claims acceptable under EPC practice should be acceptable in the four countries. Switzerland, however, will not accept claims to a micro-organism per se.

*Enforcement.** Subject to specific requirements contained in the EPC regarding claim interpretation, European patents as well as national patents are interpreted and considered with respect to the questions of both infringement and validity in accordance with national law in the EPC member countries.

In the Federal Republic of Germany, an infringer is broadly defined as any person who makes use of a patented invention. Protection for a patented process extends to the product directly obtained by that process. Provisional rights for reasonable compensation are given for applications which have been published but not yet granted.

Infringement was defined for the first time in the new British law, and a separate Patent Court was established for the purpose of trying patent infringement cases. Infringement includes the acts of making, using, importing, disposing of, or offering to dispose of an infringing product. Similar provisions are provided with regard to a process and with regard to a product obtained by a patented process. Provisional rights are given for published applications, and full recovery for damages from the date of publication may be obtained after grant. The 1977 act also provides that the scope of the patent may extend beyond the literal meaning of the words of the claims.

*The discussion in this section is based substantially on ch. III in Schwaab and Thurman (24).

Swiss law defines infringement to include any unlawful utilization of the patent invention, including imitation. Patent protection for a process also extends to products which are directly made from the process. The patent rights begin at publication, but suit for damages may be initiated only after grant. Criminal sanctions may also be imposed as well as confiscation and destruction of the infringing goods.

Infringement in France is defined broadly to include the acts of manufacture, offer, commercial disposal, use, or import of the patented product. However, for actions other than manufacturing or importing, there is no liability unless the acts were committed with knowledge of infringement. Process patents extend coverage to products obtained directly by the process. Provisional rights for published applications are limited to reasonable compensation. Suit may be brought before grant but will probably be suspended until after grant.

In countries with national laws providing for provisional protection after preliminary publications—namely, the Federal Republic of Germany, the United Kingdom, and France—there should be no difference in treatment between published national applications and published European applications. In Britain and France, damages may be recovered for published national or European applications. Moreover, in France, damages are recoverable from the time of notification to the infringer of the patent application contents. Only reasonable compensation may be obtained in West Germany.

The EPC also provides for provisional protection after publication of a European patent application. Generally, the right is limited to recovery of damages after the patent issues.

In Switzerland, on the other hand, provisional protection is not provided. But, in ratifying the EPC, Switzerland has provided a provisional remedy for European patent applications.

Remedies for infringement include injunctions and monetary damages. In addition, as a general rule, the loser pays most or all of the costs of litigation of the winning party. Finally, in most cases, the infringing goods will be destroyed or handed over to the patentee.

Criminal sanctions exist in the national patent laws of the Federal Republic of Germany and Switzerland, but they are not of much practical importance.

LAW OF TRADE SECRETS

National laws that protect trade secrets, confidential information, and know-how (hereinafter sometimes referred to collectively as "proprietary information") are designed to prevent the misappropriation of a competitor's technical and commercial information. These laws coexist with the patent laws of the various countries and are a necessary adjunct to those laws in order to provide basic protection in many areas where the patent laws do not reach.

There are no treaties, such as the EPC for patents, dealing with the international protection of proprietary information. Thus, when a question involving trade secrets comes before the European Court of Justice, it will be decided generally in accordance with the national laws of the member states, much like U.S. Federal Courts are governed by State law in trade secret cases.

Federal Republic of Germany.—The West German law dealing with trade secrets has at least two components, "industrial secrets" and "commercial secrets." Although no distinction is made in enforcement of rights as to one type or the other, the fact that both are protected makes it clear that not only technical secrets are protected, but also secret commercial or business information.

With respect to the elements for establishing protectable industrial and commercial secrets, the German Supreme Court has stated on several occasions that such a secret may be any fact that is: 1) connected with a business, 2) known only to a small number of persons, 3) for which its possessor has a justifiable interest in keeping secret, and 4) for which its possessor has manifested an express or recognizable intent to keep secret (33).

The West German law is more liberal than the U.S. law as to the degree of public knowledge required to destroy a trade secret. In the Federal Republic of Germany, if information is discernible only with a great deal of work and expense, it is still protectable as an industrial or commercial secret. Thus, for example, even the purchase of a machine does not destroy the secret nature of its contents if the purchaser must dismantle, tear apart, and put in substantial time and effort to uncover its secrets (33). Further, knowledge by a small group of persons, particularly if they are not competitors, will not destroy the secret nature of an industrial or commercial secret.

As in the United States and the United Kingdom, neither novelty nor technical advance need be established in order for information to be classified as an industrial or commercial secret in West Germany.

One element of a trade secret is whether the information gives its possessor an advantage in competition which would be lost if it were disclosed to competitors. But at least one commentator has suggested that the industrial or commercial secret need not be actually industrially or commercially utilized at the time of its loss (4). Thus, it would appear that research data that would or potentially could give the holder a competitive advantage would satisfy the requirements for an industrial secret.

Substantial civil and criminal liabilities for violation of trade secret rights are written in statutory law. The most pertinent provisions are in the German Unfair Competition Law of 1909 (UWG, Gesetzgegen den unlauteren Wettbewerb). An employee who wrongfully communicates an industrial or commercial secret may be imprisoned for up to 3 years and fined. If the employee uses the secret abroad, or knows it is to be used abroad, the prison sentence is increased to up to 5 years. Civil penalties and a civil right of action for damages or an injunction are also available (6,20,33).

United Kingdom.—The British courts, much like their American counterparts, have refrained in most instances from adopting a hard and fast definition of the term "trade secret." One definition is as follows (31):

1. It consists of information;
2. The information must be secret either in an absolute or a relative sense;
3. The possessor must demonstrate that he has acted with an intention to treat the information as a secret;
4. The secret information must be capable of industrial or commercial application; and
5. The possessor must have an interest in the information worthy of legal protection, bearing in mind English principles of equity. This will generally be an economic interest.

The English (as well as the other Europeans) are rather parochial in their approach to the question of whether something is secret. They are concerned most with public knowledge in their own country. For example, knowledge by other people outside of the United Kingdom would not be as great a threat as knowledge of a few people inside of the United Kingdom (31).

One possible problem for biotechnology in Great Britain is the requirement that information must have some industrial or commercial use in order to qualify as a trade secret. Thus, research data or abstract ideas not capable of being used commercially in the near future may not be a trade secret (31). Such information may be protectable, however, as "confidential information" (23). While English legal scholars have debated the degree of secrecy necessary for information to be protected as confidential, it is clear that the degree necessary to protect such information pursuant to a confidentiality agreement is less than that required to establish a trade secret. The British "confidential information" approach might well be the way to avoid the problem raised by some U.S. cases which have indicated that technical information will not be protected if it is not developed to the stage of practical application (9).

Enforcement of trade secret law in the United Kingdom is by way of civil actions for damages. Unlike other major industrialized countries, the United Kingdom has no specific statute making misappropriation of trade secrets a crime, and there has been no significant prosecution under more general theft or conspiracy statutes.

Switzerland.—Swiss law recognizes "industrial secrets" and "commercial secrets."* The elements of protectable industrial and commercial secrets are quite similar to those under West German law. Knowledge by a small number of people, or public availability, but only after substantial expense or effort, does not defeat the secrecy of the information (19,20). There must be an intention to maintain the secrecy of the information and an intent in maintaining its secret for the purpose of enhancing economic or competitive position (19). One additional element to the Swiss law, however, is that the secret must have a relationship to a particular business enterprise. Secrets held by professors, scientists, factory workers, and others not engaged in business do not qualify as industrial and commercial secrets, unless, of course, they own or participate in a business and the secret is possessed by the enterprise rather than themselves as individuals (19).

Switzerland's Unfair Competition Law of 1943 specifically prohibits the misappropriation of industrial or commercial secrets, and contains sections establishing both civil and criminal liability. One who is injured by an act of unfair competition may obtain injunctive relief and damages (19).

Switzerland has a wide variety of criminal statutes prohibiting misappropriation of industrial and commercial secrets and various other types of industrial espionage. The Unfair Competition Law provides that those guilty of the same acts of unfair competition discussed above shall be punished by a fine or imprisonment, on complaint of the aggrieved party (19).

Thus, Switzerland has a formidable array of civil and criminal liabilities to discourage industrial espionage and misappropriation of propriety information.

*The Swiss Supreme Court has defined "industrial secret" as (BGE 64 II 66) (19):

> All facts related to a manufacturing process or method and neither in the public domain nor generally available, in the secrecy of which the holder has a justified interest and which he actually wishes to be maintained secret, can be the subject matter of an idustrial secret.

and "commercial secret" as (BGE 74 IV 103) (19):

> The term "commercial trade secret" encompasses basically all facts of economic life in the maintenance of secrecy of which an interest worthy of protection exists.

France.—French law, like West German law, rather than following the single concept of "trade secret" found in the U.S. and English law, segregates the secrets into "manufacturing secrets" (secrèt de fabrique) and "commercial secrets" (secrèt de commerce) (10). A commercial secret is treated by the commentators similarly to a manufacturing secret, although there is no direct reference to commercial secret in the French Code (10). For information to be a manufacturing secret, it must be: 1) relatively secret, 2) of industrial application, 3) of commercial or market value, 4) a secret of the factory; and 5) the misappropriator must know it is a secret (10).

The difficulty for researchers is the requirement of industrial application. The majority view seems to be that to be a manufacturing secret, the secret information must either be suitable for immediate industrial application or have already been used industrially. For example, a process not yet applied industrially, but used only in research and experimentation cannot be a manufacturing secret. Mere unapplied, theoretical ideas of a technical or scientific nature do not qualify (10).

Misappropriation of manufacturing secrets by an employee is a criminal violation under article 418 of the French Penal Code, if the employee has the requisite criminal intent for doing the act for his or her own benefit (10). Disclosure to aliens or non-French residents is punishable by significantly higher fines and much longer prison terms.

PLANT BREEDERS' RIGHTS

The important provisions of the plant breeders' rights laws of the Federal Republic of Germany, the United Kingdom, Switzerland, and France are as follows.

Federal Republic of Germany.—Article 2(3) of the Federal Republic of Germany's Law on the Protection of Plant Varieties (text of May 20, 1968) covers both sexually and asexually reproduced varieties. The variety must be new, sufficiently homogeneous, and stable. Novelty exists when the variety is clearly distinguishable by at least one important morphological or physiological characteristic from any other variety, the existence of which is a matter of common knowledge at the time for which protection is applied. Common knowledge is defined in terms of absolute novelty in Germany, with commercialization of the variety in Germany prior to filing the application constituting a statutory bar (art. 2(3)). Homogeneous means plants of the variety must be identical in all their essential characteristics (art. 5). Stability is demonstrated when plants of the variety retain their essential characteristics true to the definition of the variety after each successive reproduction or reproductive cycle (art. 6).

Article 36 provides that as a part of the examination procedure, the variety must be grown, either by the Federal Office of Plant Varieties or a delegated outside service. The holder of the protection right also is required to submit to the Federal Office of Plant Varieties, upon request, material for establishing the continued existence of that variety. If the holder is unable to do so, the protection right ceases (arts. 16 and 20).

The duration of protection or grant is for 20 years, except for certain varieties for which it is 25 years (art. 18). The law provides for criminal penalties comprising fine or imprisonment of a term of up to 1 year (arts. 48 and 49). The holder of the protection right may claim remuneration from any person who has propagated material without authorization in the interval between the publication of the application and the grant of title of protection (art. 47(4)).

United Kingdom.—The Plant Varieties and Seed Act of 1964 covering United Kingdom is the basis for adherence to the UPOV 1961 Convention, with ratification being effective September 17, 1965.* The act covers both sexually and asexually reproduced plant materials.

The new variety must be distinct, uniform, and stable. To meet the first requirement, it must be clearly distinguishable by one or more important morphological, physiological, or other characteristics from any other variety whose existence is a matter of common knowledge at the time of the application (pt. II, 1(1)). The variety must be sufficiently uniform or homogeneous (pt. II(4)). The variety must be stable in its essential characteristics—i.e., it must remain true to its description after repeated reproduction or propagation (pt. II(5)).

There is an absolute novelty requirement, that is, the variety may not have been offered for sale or sold in the United Kingdom prior to the filing of the application. Where such sales or offers for sales are made outside the United Kingdom, a grace period of 4 years is provided prior to the filing of the application (pt. II, (2)(1) and (2)).

The scope of protection afforded by the rights include the exclusive right to produce or propagate the variety for the purpose of selling the variety or parts or products of the variety (pt. II, 3(1) and (2)). The term of protection ranges from 15 to 25 years, depending on the type of plant.

A growing trial is required during the examination period, thus requiring the submission of plant material. Further, every holder of plant breeders' rights must ensure that, throughout the period for which the rights are exercisable, he or she is in a position to provide reproductive material that is capable of producing the variety, and the holder must provide such information and facilities as the plant variety rights office may request for the purpose of fulfilling the maintenance requirements. If plant material cannot be so provided, the protection rights shall be terminated (pt. I(6)).

The law provides for a Plant Variety Rights Tribunal having jurisdiction over cases brought under the act, with the tribunal being authorized to sit in any designated place in Great Britain to hear any proceedings.

Switzerland.—Switzerland ratified the 1978 UPOV Text on June 17, 1981. Under Swiss law, sexually and asexually reproduced varieties are covered. Protected varieties must be novel, stable, and sufficiently homogeneous. The variety is considered novel unless, at the time the application is filed, the variety has already been offered for sale or marketed in Switzerland or for more than 4 years outside of Switzerland. A "variety" refers to any cultivar, clone, line, stock, or hybrid and is considered new if it is clearly distinguished by one or more important features from any other variety whose existence is generally known at the time the application is filed.

Variety protection precludes another, without the consent of the holder, from producing propagation material of the protected variety with a view to marketing it, offering it for sale, or selling it in the course of business. Propagation material includes seeds, fruits, or vegetative material. Protection is for a term of 20 years following issue, but it can be extended in certain cases.

The applicant is required to deposit propagation material for purposes of conducting examination for verifying the stated characteristics of the plant. The title of protection can be annulled when the title holder cannot supply a propagation material capable of producing the new variety with its morphological and physiological characteristics as defined when the right was granted.

Action for variety infringement is brought in the canton of the defendant's place of residence in Switzerland. Intentional infringement can be punished by imprisonment for up to 1 year or by a fine.

France.—Although France was an early ratifier of the 1961 UPOV Convention Text, and a signatory to the 1978 Text, it has not yet ratified the latter. France continues to operate under the Law on the Protection of New Plant Varieties, Law No. 70-489 of June 11, 1970.

Both sexually and asexually produced plant materials of all species are covered, including bacteria, although

*For further information about UPOV, see *Chapter 16: Intellectual Property Law.*

the schemes are limited to specified varieties. For a variety to be "new," it must be distinct from similar known varieties, by reason of one characteristic that is important, specific, and subject to little fluctuation, or more than one characteristic where the combination thereof is such as to give it the quality of a new plant variety (ch. I, sec. 1). Further, the variety must not have been exploited in France, or appear in specified publications, before the filing of the application in France; if so, a valid certificate cannot be issued. The variety must be homogeneous in all of its characteristics, and must remain stable—i.e., it must remain identical with its original definition at the end of each propagating cycle (ch. 1, sec. 1). An application for each new variety fulfilling the above requirements must be given a denomination and a sample to be left in a collection (ch. II, sec. 2).

The plant variety certificate confers on the certificate owner the exclusive right to produce, import into France, sell, or offer for sale all or part of the plant (ch. II, sec. 3). The certificate is valid for 20 years from the date of issue, although this period shall be extended to 25 years if the constitution of the elements for production of the species requires a considerable time.

The breeder must at all times keep a vegetative collection of the plant variety (ch. I, sec. 9). If the owner is unable to furnish the administration at any time with the elements of reproduction or vegetative propagation so that the specified characteristics of the variety can be ascertained, the rights of the owner will be forfeited (ch. IV, sec. 22).

Chapter IV, section 23 relates to infringement, which is broadly defined. It provides that any violation of the rights of the owner of a new plant variety certificate shall constitute an infringement for which the offender shall be liable.

Intellectual property law of Japan

Having discussed the patent law, trade secret law, and plant breeders' rights in the European competitor countries, we turn now to Japan.

PATENT LAW

Patentable Subject Matter.—The Japanese Patent Act contains the following broad definition of patentable subject matter (art. 29(1), 1976):

> Any person who has made an invention which can be utilized in industry may obtain a patent . . . .

Until 1979, the Japanese Patent Office took the position that micro-organisms were unpatentable because they are not industrially applicable. After reversing that position, the Japanese Patent Office issued a set of Working Standards for micro-organism inventions in November 1979, and in August of 1980, it issued a Classification of Inventions Relating to Genetic Engineering (14).* According to these guidelines, recombination of the genes of higher animals is not permitted, so that inventions in that area are thought to not be patentable (14).

In the intervening years, the greatest obstacle to securing patent protection for microbiological inventions in Japan was the rDNA research safety guidelines published by the Science and Technology Council and the Ministry of Education. These guidelines originally permitted only *E. coli* bacteria to be genetically modified. In January 1980, yeast strains were also included. Since then, other micro-organisms have been included.** Any rDNA inventions that were not directed to subject matter approved by the safety guidelines were considered to fall into the category of inventions "likely to injure the public health" and thus were precluded from patenting under article 32(2) of the patent law (13).

Subject to the above considerations, therefore, the following five basic categories of biotechnological inventions appear to be patentable: 1) micro-organisms, 2) processes for producing micro-organisms, 3) processes using micro-organisms, 4) products obtained from microbiological processes, and, 5) DNA and RNA molecules or subcellular units.

Novelty.—Under article 29 of the Japanese Patent Act (1978), an invention is novel if it is not worked or publicly known in Japan, or it is not described in a publication anywhere prior to the application filing date (or priority date). A 6-month grace period is provided in article 30 (1978) for: 1) experimentation, publication, and papers presented before scientific organizations by the applicant, 2) unauthorized disclosure by a third party, and 3) displays at authorized exhibits.

Utility.—The standard of utility is one of industrial applicability, similar to the EPC. Processes in the field of medicine, diagnosis, therapy, and pharmacology in which the human body is an indispensible element are excluded from patentability by the Japanese Manual for Examination and by court decision, as not being usable in industry (11).

Standard of Invention.—Under article 29(2) of the Japanese Patent Act, a claimed invention is not patentable, even if novel, if it "could easily have been made, at the filing of the application, by a person with ordinary skill in the art to which the invention pertains." This standard is similar to the concept of obviousness under U.S. law, except that U.S. law focuses

*The guidelines also mention vectors, DNA molecules, and enzymes.

**See *Chapter 15: Health, Safety, and Environmental Regulation* for details.

on the difference between the claimed invention and the prior art.

Disclosure Requirements.—Disclosure requirements for inventions under article 37 of the Japanese Patent Act (1976) require that an application be accompanied by a specification setting forth a detailed explanation of the invention including the purpose, construction, and effect of the invention to the extent that any person having an ordinary knowledge in the technical field to which the invention belongs may easily make it. This is basically an enablement standard.

Deposit Requirements.—A micro-organism must be deposited *except* in the case where:

- it cannot be preserved in a depository for technical reasons or cannot be controlled under safe conditions; or
- it is readily available to those skilled in the art (e.g., a commercially available micro-organism or one constituting a stock culture listed in a catalog published by a reliable depository) (35).

The situation is unclear in the case of micro-organisms for which an enabling disclosure is presented in the patent application (35).

Japan is bound by the Budapest Treaty, and therefore, it must accept deposits made thereunder, without requiring deposit in Japan. For those deposits not made under the Budapest Treaty, the minimum required maintenance period for the culture deposit is the lifetime of the Japanese patent (28).

Generally, no sample of a deposited culture will be furnished to third parties (without consent of the depositor) until the patent application is accepted and published for opposition. After publication, access is granted on the condition that the party will not furnish the sample to others (28).

Claims Practice.—There are no formal limitations on the basic type, style, or category of claims (1).

Enforcement.—Infringement is defined in article 101 and 3(2) of the Japanese Patent Act (1978) to include the acts of making, using, selling, and importing the patented article and/or patented process, including importing an article produced by a patented process. There is a presumption that a claimed process for producing a *novel* product has been used to produce the product whenever found in Japan (art. 104, 1978).

It is the predominant view that claims in a Japanese patent define the outer boundary of the invention and that only in rare instances is it possible to establish infringement for anything outside of the literal language of the claims, i.e., there is no traditional doctrine of equivalents (29).

Article 65(3) of the Japanese Patent Act provides that after the first publication of a Japanese application, the applicant has a right to reasonable compensation. After acceptance and grant, the patentee has the right to injunctive relief as well as monetary damages and, in theory, criminal sanctions (29).

LAW OF TRADE SECRETS

There are no specific statutes assigning liability for misappropriation of trade secrets;* thus, one must rely on general principles of Japanese civil law (see 17). That is, an injured party may sue under general tort law principles.** Employees, however, are viewed as having an implied contractual obligation not to misappropriate or improperly disclose trade secrets of their employer.

The Japanese Penal Code does not contain a provision specifically punishing misappropriation of trade secrets, manufacturing secrets, or commercial secrets. Criminal liability can only attach through the general sections of the penal code dealing with larceny, embezzlement, receiving stolen property, fraud, etc. Misappropriation of trade secrets has been successfully criminally prosecuted under such general statutes in Japan (see 12).

Trade secret protection in Japan for any type of technology is seen as very unsatisfactory. Liability for misappropriation has been the exception rather than the rule. In fact, one commentator has described Japan as the world's leading country for industrial espionage (34).

PLANT BREEDERS' RIGHTS

A Seed and Seedling Law in Japan, enacted July 10, 1978, conforms to the provisions of the UPOV Treaty, which Japan has signed (21). The details of the Japanese legislation are similar in essential respects to the EPC countries discussed previously.

Appendix G references***

1. Aoki, A., et al., *Japanese Patent and Trademark Law*, (Lausanne: Seminar Services, 1976), distributed by Bureau of National Affairs, Washington, D.C.
2. *Baekerhefe* (Baker's Yeast), *International Review of Industrial Property and Copyright Law* 6:207 (1975).
3. *Bakterienkonzentrat, Gewerblicher Rechtsschutz und Urheberrecht,* 263, 1981.
4. Baumbach-Hefernahl, A., *Wettbewerbs und Warenzeicher,* 9th ed. (Munich: C. H. Beck'she Verlag, 1964).

*While the term "trade secret" is sometimes used in Japanese law, one is more likely to find the terms "industrial secret" and "commercial secret" utilized, in a manner similar to that of German law (34).

**The general tort principle is set out in art. 709 of the Japanese Civil Code as follows: "A person who, willfully or negligently, has injured the right of another is bound to compensate him for the damage which has arisen therefrom."

***Note: R.P.C. = Reports of Patent, Design, and Trade Mark Cases (Great Britain).

5. Braendli, P., "Das neu schweizerische Patentrecht," *Gewerblicher Rechtsschutz und Urheberrecht, International Part,* 1979, p. 1.
6. Cohn, E. J. (ed.), *Manual on German Law* §7.182, §7.194 (London: Ocean Publications, Inc., 1971).
7. Comte, J. L., Deputy Director, Swiss Patent Office, personal communication to R. Schwaab, OTA contractor, Feb. 16, 1982.
8. Convention on the Grant of European Patents, 2d. ed. (Munich: European Patent Office, 1981).
9. Cornish, W. R., "Protection of Confidential Information in English Law," *International Review of Industrial Property and Copyright Law* 6:43-46, 1975.
10. Harle, Y. R., "Trade Secrets and Know-How in French Law," *Trade Secrets and Know-How Throughout the World,* vol. 3, A. N. Wise (ed.) (New York: Clark Boardman Co., Ltd., 1981).
11. Hayashi, I., "A Japanese Prospective on Patenting Microorganisms: Prospects and Considerations," *APLA Q.J.* 7:306-308, 1979.
12. *Japan* v. *Telenchef,* 7 Kakyu Keishu 1319, 419 Hanrei Jiho 14 (Tokyo District Court 1965).
13. *Japan Patents and Trademarks,* No. 25 (Tokyo: Suzuye Institute of I.P.P.R., 1980).
14. *Japan Patents and Trademarks,* No. 27 (Tokyo: Suzuye Institute of I.P.P.R., 1980).
15. Katzarov, K., *Manual on Industrial Property* (Geneva: Prof. K. Katzarov S.A., 1981).
16. *Lactobacillis bavaricus, Gewerblicher Rechtsschutz und Urheberrecht,* 586, 1978.
17. Lyon, J., and Doi, T., "The Protection of Unpatented Know How and Trade Secrets in the United States and Japan," *Patent and Know How Licensing in Japan and the United States,* T. Doi, et al. (ed.) (Seattle: University of Washington Press, 1977).
18. Pagenberg, J., "The European Patent System—More Objective Standards Needed," *International Review of Industrial Property and Copyright Law* 9:1-20, 121-152 (1978).
19. Pestalozzi, P., "Trade Secrets and Know-How in Swiss Law," *Trade Secrets and Know-How Throughout the World,* vol. 4, A. N. Wise (ed.) (New York: Clark Boardman Co., Ltd., 1981).
20. *Pinners World Unfair Competition Law,* §74 (Alpen and Rijn: Sijthoff & Noordhoff, 1978).
21. *Plant Variety Protection*—Number 28, Newsletter, UPOV, April 1982.
22. Rote Taube (Red Dove), *International Review of Industrial Property and Copyright Law, 135,,* 1970.
23. *Saltman Engineering Co., Ltd.* v. *Campbell Engineering Co., Ltd.,* 65 R.P.C. 203, 1948.
24. Schwaab, R., and Thurman, R., *International Law: EPC & PCT Practice and Strategy* (Washington D.C.: Patent Resources Institute, 1980).
25. Schwaab, R., Jeffery, D., and Conlin, D., "U.S. and Foreign Intellectual Property Law Relating to Biological Inventions," contract report prepared for the Office of Technology Assessment, U.S. Congress, February 1983.
26. *7-chor-6-dimethyltetracyclin, Gewerblicher Rechtsschutz und Urheberrecht,* 162, 1978.
27. Sharrock, J., Principal Examiner, British Patent Office, personal communication to R. Schwaab, OTA contractor, Mar. 2, 1982.
28. Suzuye, T., Japanese patent attorney, Tokyo, personal communication to R. Schwaab, OTA contractor, Feb. 25, 1982.
29. Tanabe, T., and Wegner, H., *Japanese Patent Law* (Tokyo: AIPPI Japan, 1979).
30. Ullrich, H., "Standards of Patentability for European Inventions," *International Review of Industrial Property and Copyright Law Studies* 1:35 (Weinheim/Bergstr.: Verlag Chemie, 1977).
31. Vitoria, M., "Trade Secrets, Know-How and Confidential Information in English Law," in *Trade Secrets and Know-How Throughout the World,* vol. 2, A. N. Wise (ed.) (New York: Clark Boardman Co., Ltd., 1981).
32. Vossius, V., "Patent Protection for Biological Inventions—Review of Recent Case Law in EEC Countries," *European Industrial Property Reporter* 1979:278-279.
33. Wise, A. N., "Trade Secrets and Know-How in German Law," *Trade Secrets and Know-How Throughout the World,* vol. 3, A. N. Wise (ed.) (New York: Clark Boardman Co., Ltd., 1981).
34. Wise, A. N., "Trade Secrets and Know-How in Japanese Law," *Trade Secrets and Know-How Throughout the World,* vol. 3, A. N. Wise (ed.) (New York: Clark Boardman Co., Ltd., 1981).
35. *Yuasa and Hara Patent News,* "Procedures for Deposit of Microorganisms in Japan," 6:26-30, spring 1979.

Appendix H

Selected Aspects of U.S. University/ Industry Relationships in Biotechnology

University/industry relationships in biotechnology were the focus of the discussion in *Chapter 17: University/Industry Relationships.* Material on selected university/industry agreements is presented below. Also described are guidelines for university/industry research adopted by the National Association of Land Grant Colleges and the 1982 Pajaro Dunes Conference, selected statements on patent rights and commingling of research funds, and university policies on patents, consulting, and sponsored research in the United States.

Selected university/industry agreements in biotechnology

Selected university/industry arrangements in biotechnology are discussed below. The arrangements were selected for discussion because they represent different approaches to university/industry relationships, because they are relatively large agreements, and because some of them have raised issues central to university/industry agreements. The agreement between the Whitehead Institute and the Massachusetts Institute of Technology (MIT) is not strictly a university/industry agreement, but has been included because it raises issues in university/industry relationships and because it is a product of industrial interest in biotechnology research.

RESEARCH PARTNERSHIPS

Monsanto/Washington University.—Washington University and Monsanto (U.S.) have a 5-year renewable contract totaling $23.5 million. Under the contract, individual research projects in biotechnology will be carried out by cooperative arrangements involving Washington University faculty and Monsanto scientists. About 30 percent of the research will be fundamental research (terminology of the agreement) and 70 percent will be special research directly applicable to human disease. The contract between Washington University and Monsanto establishes an 8-person advisory committee made up of 4 members from each institution. This committee will solicit research proposals from the faculty of Washington University and from researchers at Monsanto, review and approve the proposals on the basis of individual merit, distribute appropriate funding, and act as a liaison between the university and Monsanto.

Monsanto's participation in the program will begin with a $3 million grant the first year (1982) and rise annually to accommodate expansion in the number and scope of research projects involved. Washington University faculty members will be at liberty to publish results of any research done under the Monsanto funding. Monsanto will exercise the right of prior review of draft materials, because they may contain potentially patentable technical developments. If they contain patentable developments, Monsanto can request a short delay of submission for publication or other public disclosure in order to begin the patent process. Monsanto will pay for and carry out the entire patent application process. If Monsanto does not elect to license a patent, the university is free to license the patent to others.

Washington University will retain patent rights, while Monsanto will have exclusive licensing rights. Royalties will go to Washington University for support of its education and research programs—not to individual researchers. The portion of royalty normally going to the individual will instead be channeled to his/her laboratory to support more research.

During the third year of the 5-year agreement, the entire program will be reviewed by a panel of distinguished scientists who are independent of both Monsanto and Washington University.

The schedule for funding in millions of dollars is as follows (11,13):

The schedule for funding in millions of dollars is as follows (11,13):

Contract year	*Exploratory projects*	*Specialty projects*	*Contract year total budget*
1982-83	$1.5	$1.5	$3.0
1983-84	1.6	2.2	3.8
1984-85	1.7	3.0	4.7
1985-86	1.8	3.8	5.6
1986-87	1.9	4.5	6.4
Total	$8.5	$15.0	$23.5

The process by which the agreement between Washington University and Monsanto came about had some major strengths. First, individuals from Monsanto and Washington University met continually over a period of 2½ years to discuss the project. Second, members of the university faculty and administrative staff and representatives of the company held a 3-day retreat to discuss the interactions and what form they should take. Furthermore, the Washington University/Monsanto agreement is unlike other agreements in that it is intended to be a cooperative research agreement with industrial and university scientists working together on research projects. In other agreements, the explicit purpose has been to allow industry to gain a window on the technology and educate its personnel.

Hoechst/Massachusetts General Hospital.—A $70 million agreement between Massachusetts General Hospital, a teaching hospital associated with Harvard University, and the West German company Hoechst will create a department of molecular biology at Mass General and will provide support for the department for 10 years. The department of molecular biology will be headed by Dr. Howard Goodman and will eventually have a staff of about 100 persons. Hoechst will fund basic research in eukaryotic cell gene regulation, somatic cell genetics, microbial genetics, virology, immunology, and plant molecular biology.

The department faculty will be regular members of the staff of Mass General and will be nominated for membership on the faculty of the Harvard Medical School. Faculty duties will primarily consist of research for the department of molecular biology. Faculty may "also devote a reasonable amount of time to faculty duties other than research and to consulting for non-profit-making entities so long as such activities do not interfere materially with their research activities under the agreement."

Hoescht has the right to send up to 4 individuals to work and be trained in the department at any one time and to send up to 40 individuals over the life of the contract. The individuals that Hoechst sends, however, must have qualifications acceptable to the department.

The contract between Hoechst and Mass General states that the scientists in the department of molecular biology are free to collaborate with others but that "research collaborations funded in part by the Company and in part by others shall take into account the interest of the Company in obtaining exclusive, worldwide licenses." If Hoechst cannot obtain an exclusive license from such collaboration, it must be assured of a nonexclusive license.

All faculty in the department have the right to publish but must submit early drafts of all manuscripts from Hoechst-sponsored research not less than 30 days prior to the submission of the manuscript for publication.

Mass General will hold any patents that may arise out of the Hoechst-sponsored research. The hospital will grant Hoechst an exclusive worldwide license for the life of the patent. Hoechst will pay the hospital royalties at rates that give "due consideration" to the fact that Hoechst paid for the research (2,10).

The exclusive funding may preclude department scientists from seeking grants from the U.S. National Institutes of Health (NIH), thereby taking them out of the peer review process. The department will report to a scientific advisory committee of two members from Mass General, two from Hoechst, and two from elsewhere. The committee's review, however, may not be the equivalent of the critical peer review of proposals at NIH. The department will be physically separate from Mass General, and all equipment and physical plant will be paid for by Hoechst. Department scientists will generally be free to collaborate with others but will have to obtain written permission from Hoescht. Dr. Goodman hopes to collaborate with Dr. Philip Leder who has a 5-year $6 million research agreement with DuPont. Whether Hoechst will grant this request will probably depend on the nature of the collaboration.

Whitehead Institute/Massachusetts Institute of Technology.—Whitehead Institute, a biomedical research institute administratively separate from MIT, has been provided for by Edwin C. Whitehead, the President of Technicon Corporation. Whitehead has bequeathed about $20 million to build the structure, $5 million annually to operate it through the year 2003, and a gift of $100 million upon his death. Whitehead has also given $7.5 million to MIT plus support moneys estimated to be worth about $1 million a year for faculty, graduate students, and research assistants in MIT's biology department.

The Whitehead Institute is headed by a 14-member board of directors that includes 3 MIT directors, 3 of the Whitehead children, and David Baltimore, the director of Whitehead Institute who is serving a renewable 5-year term. Whitehead Institute faculty will have joint appointments with MIT but will be paid entirely from Whitehead Institute funds. Faculty appointments will be proposed by Whitehead Institute according to the research needs of the institute and in consultation with the appropriate MIT department. Appointees will follow the rules and regulations of MIT with regard to teaching, consulting, tenure, benefits, salaries, etc. It is expected that 10 to 15 appointments will be made during the first 7 or 8 years. Graduate students will also be supported.

Whitehead Institute will retain the patent rights on any inventions arising from the research. After deduction for expenses, the royalty will be shared according to the following formula: one-half to the inventor and one-half shared by Whitehead Institute and MIT. The term of the agreement is 10 years, with a 5-year renewal and 2 years written notice necessary for termination. If the agreement should be terminated, faculty will be given the choice of joining the MIT or Whitehead Institute faculty.

Prior to the signing of the agreement, the agreement was extensively discussed by MIT faculty and administrators. Some were concerned that an imbalance in the MIT biology department might result from the addition of 15 new faculty members in molecular biology; other important specialties would have less representation. Since the members of the faculty of Whitehead Institute, though approved by MIT, would be chosen for their research contributions to Whitehead Institute rather than to MIT's educational or research needs, there was also concern over the possibility that the loyalty of the Whitehead Institute faculty would be divided. Other concerns centered on conflict of interests. Some faculty thought that the findings of Whitehead Institute could turn up in the investment portfolio of Whitehead Associates, Edwin H. Whitehead's venture capital firm. Furthermore, since David Baltimore has equity in the Collaborative Research Company, and several other proposed faculty of Whitehead Institute consult for the company, there were concerns that the link between Collaborative and Whitehead Associates might be too close. After extensive discussions, the MIT faculty decided that the positive aspects of the arrangement outweighed these concerns and voted overwhelmingly to approve the agreement. MIT's Board of Trustees would not have approved the arrangement without faculty support. Furthermore, a special committee will be appointed to monitor the arrangement so that any misunderstandings can be avoided (3).

PRIVATE CORPORATIONS

Engenics/Center for Biotechnology Research and Stanford University.—The for-profit company Engenics was established in September 1981, along with the nonprofit Center for Biotechnology Research (CBR). The purpose of CBR is to support basic and applied biotechnology research at universities, disseminate the results of such research to the public, and facilitate the conversion of knowledge into products and processes. The purpose of Engenics is to carry out research and process development and to establish new businesses. Although the two organizations are separate, they have the same six corporate sponsors and will work in close cooperation.

CBR is receiving $2.5 million from its six corporate sponsors over a period of 4 years. The six sponsors of CBR are Elf Technologies, Inc. (a U.S. venture capital subsidiary of Elf Aquitaine), General Foods, Koppers Co. Inc., Bendix Corp., Mead Corp., and McLaren Power and Paper Co. (a subsidiary of Noranda Mines). CBR holds about 30-percent of the equity in Engenics, equity that was issued to Engenics in exchange for options to licenses under university patents. The same six corporations that sponsored CBR paid $7.5 million for a total of about 30 percent of the equity in Engenics. The remaining 30 percent of the equity in Engenics is shared by the line officers and the consultants Channing Robertson (Chairman of the Chemical Engineering Department at Stanford), Abdul Matin (Professor at Stanford's Medical School), and Harvey W. Blanch (Professor of Chemical Engineering at the University of California, Berkeley).

CBR can use all capital appreciation or dividends from the equity in Engenics only for the support of university research. Any patents resulting from CBR-sponsored research will be held by the university at which the work was performed, with CBR, Engenics, and the six corporate sponsors receiving royalty-bearing licenses. Negotiations at the time of the patent will determine the terms of the license. Investigators performing CBR-sponsored research will retain the right to publish their findings.

CBR is currently funding three university research contracts. One is a 4-year $970,000 contract with Drs. Robertson and Matin, both of Stanford, as principal coinvestigators. The second contract is a 5-year $783,000 contract with Dr. Blanch, of the University of California, Berkeley, as the principal investigator. This contract is funded by both CBR and Engenics, because University of California policy stipulates that licensing agreements cannot be made with nonprofit organizations. The third contract is with Anthony Sinsky at MIT. No data on the amount of this contract are available. Dr. Sinsky is on the Scientific Advisory Board of Engenics (68).

Neogen/Michigan State University Foundation and Michigan State University.—Neogen was founded in July of 1981 by the Michigan State University (MSU) Foundation, an independent nonprofit fundraiser and disburser of donations and royalty income to MSU. Neogen, which was organized to seek venture capital for limited partnerships to develop and market innovations arising out of research at MSU, was formed for several reasons: MSU wished to retain faculty members who are getting lucrative offers from other small companies; MSU would like to allow faculty to develop their entrepreneurial talents and remain at the university; and a company such as Neogen can help diversify Michigan's economy. The company was organized with full knowledge of the board of trustees, the administration, and the faculty of MSU.

Neogen limited partnership purchases are being managed by an investment firm in Detroit. The MSU Foundation, which purchased $100,000 of stock when the company was founded, will soon purchase another $130,000 of stock, and Doan Resources is buying $250,000 in stock.

One project (a parasite diagnosis project) is ready to begin (funded at $455,000) and two projects are awaiting funding. Neogen will be able to buy title to any resulting patents from MSU for the parasite diagnosis project. The money will be paid through the MSU Foundation to Neogen.

Patents will usually be applied for by MSU. The patents will be assigned by MSU to the MSU Foundation for subsequent sale to Neogen in exchange for stock. Inventors will receive a 15-percent royalty or can exchange this for a 1-to 2-percent stock option in Neogen.

Because Neogen is tied to the MSU Foundation, MSU receives moneys from successful commercialization of products or processes and the individuals are rewarded commensurate with their efforts. The basic research takes place on the campus of MSU, but commercialization will be moved off-campus to a nearby research park in order to avoid conflicts of interest.

The MSU faculty and administration were aware of and/or participated in the founding of the company, and there is a scientific advisory board that reviews the projects, thus preserving the principle of peer view.

Guidelines for industrial sponsorship of university research in biotechnology

NATIONAL ASSOCIATION OF STATE UNIVERSITY AND LAND GRANT COLLEGES, DIVISION OF AGRICULTURE

A document titled "Genetic Engineering Policy for the State Agricultural Experiment Stations" was adopted by the Experiment Station Committee on Policy (ESCOP) in November 1981 at a meeting held in conjunction with the fall meeting of the National Association of State University and Land Grant Colleges (NASULGC). ESCOP, headed by Dean Clarke of Texas, was brought together after Clarke and several other members observed that several State Agricultural Experiment Stations (SAES) were being simultaneously approached by industry to do genetic research. Since there were no policy guidelines for the new field of biotechnology, SAES often found themselves in a weak position during contract negotiations. Thus, ESCOP was formed to draw up guidelines.

Because the field of "genetic engineering" is changing rapidly, the November 1981 ESCOP policy document is regarded by ESCOP as an interim document subject to annual revision, if necessary. In addition, Clarke is collecting copies of legal documents from SAES institutions and will develop an aggregate summary of appropriate components of general agreements to be made available to all members of NASULGC's Division of Agriculture. Work is now underway to draw up guidelines for NASULGC's Division of Agriculture. The committee that is drawing up these guidelines is headed by Dean F. A. Wood of the University of Florida.

The ESCOP document of November 1981 is summarized below because it addresses issues that are common to most industry/university relationships in biotechnology. As noted in that document, the SAES have five general concerns (5):

1. As publicly supported institutions, the SAES will need to assure that industrial relationships generate an end result in the interest of the general public. This end result should reward the industrial investor but avoid placing such an investor in an unwarranted position of financial advantage through privileged use of information or technology partly derived from research using public funds; nor should a curtailment of new information to the public occur.
2. The SAES are greatly concerned about the curtailment of communication on early research results and about the constraints on sharing of germplasm emerging due to concerns . . . for protecting potentially patentable research results. . . .
3. There is general concern in the academic community about the drain of scientific manpower from the universities to industry. . . .
4. There is concern that individual scientists may place themselves in the positions of compromise or conflict of interest as they establish personal relationships with industry as contractors, consultants or institutional officers.
5. There is concern on the part of both scientists and the SAES that through industrial sponsorship of research, there may be introduced an undesirable level of direction of effort by industry.

The guidelines set forth in the ESCOP document are subsumed under the three major issue areas outlined below (5):

A. Institutional relationships

1) Maintain SAES management control of research:

Consensus: SAES should retain the ability to manage research programs, and control the direction of new investigations, regardless of the source of support, including situations in which one or several firms may sponsor research at several institutions.

2) Strong basic research and graduate education capability:

Consensus: SAES should maintain and expand the basic research capability in genetic engineering and related areas within the domain of publicly supported institutions.

3) Faculty-industry relationships:

Consensus: Scientists should maintain close communication with institutional administrators in development of relationships and commitments with the commercial sector. Institutional guidelines should be developed which assist the scientists in avoiding institutional or personal conflicts of interest.

B. Technical relationships

4) Publication and communication:

Consensus: The ability to publish and exchange information is essential and must be secured in agreements. In some instances, publications or information exchange may need to be temporarily delayed to allow time for an institution or sponsor to assure adequate patent protection. The final decision to defer or modify a publication should reside with the public institution.

5) Trade secrets and confidential information:

Consensus: Protection of "trade secrets" or "confidential information" for more than a very limited period should be avoided by public institutions. Advance review by a private sponsor, to avoid premature release of information, may be advisable but should not become a mechanism to "shelve" useful information or unpatentable technology.

6) Patent rights and premature disclosure:

Consensus: SAES should retain the right to participate in the decisions related to the disposition of intellectual and real property and patent rights resulting from research. Retained ownership of patents by the SAES is preferred. In any agreement, the SAES should retain the right to use discoveries and inventions from SAES research to extend and enhance public research and education. The need of private sponsors to obtain a return on investment must be recognized, and agreements may provide for special licenses for patents originating from sponsored research.

7) Biosafety of recombinant DNA:

Consensus: SAES must retain responsibility for review and decisions in the release or distribution of laboratory research products, although some research may be supported by outside sponsors.

C. Fiscal and management relationships

8) Grants and income earnings:

Consensus: Extending knowledge and developing new technology while serving the public interest should be the prime motivations in agreements between SAES and the private sector. Royalty income from discoveries originating under such agreements should be recognized as a secondary consideration.

9) Licensing responsibilities and performance expectations:

Consensus: SAES should assure that "due diligence" clauses are included in contracts to assure that new technology is not shelved and the public interest is served while private investment in commercialization is respected. Assignments of rights or licensing of patents for commercial use should be considered separately from contractual definition of research to be conducted. Initial or developmental processes and pervasive technology ultimately leading to improved biological materials generally should not be assigned for sole use by a sponsoring firm.

10) Tax code implications:

Consensus: When sponsored research is motivated by certain interpretations of Tax Code Section 1235, exclusive licensing or co-ownership of patent rights is a preferred alternative for the institution, since the institution maintains a vested interest and some ownership of patent rights involving the scientist, the institution, and the firm may require unique documentation. Careful attention to these rights and relinquishments is suggested.

PAJARO DUNES CONFERENCE, MARCH 1982

The March 1982 Pajaro Dunes Conference on university/industry relationships in biotechnology, which was financed by the Henry J. Kaiser Family Foundation, was organized principally by Donald Kennedy, the President of Stanford, and included the Presidents of Harvard, Derek Bok; California Institute of Technology, Paul Gray; and the University of California, David Saxon. Also invited were an administrator and two faculty from each university. Leading industrialists were also invited, among them representatives from Beckman Instruments, Inc.; Syntex Corp.; Cetus Corp.; Cabot Corp.; Applied Biosystems, Inc.; Damon Corp.; Gillette Corp.; Eli Lilly and Co.; E. I. du Pont de Nemours; and Genentech Corp. A statement drafting guidelines and principles emerged from the conference, although Kennedy and others stressed its role as a draft of the process of policy formation rather than a statement of policy.

The premise of the Pajaro Dunes Conference was that collaboration between universities and industry will benefit all parties, including the general public, if the university's ideals are not distorted. The general consensus was as follows (9):

> . . . research agreements and other arrangements with industry (must) be so constructed as not to promote a secrecy that will harm the progress of science; impair the educational experience of students and postdoctoral fellows; diminish the role of the university as a credible and impartial resource; interfere with the choice by faculty members of the scientific questions they pursue, or divert the energies of faculty members and the resources of the university from primary educational and research missions.

The consensus of the Pajaro Dunes Conference with respect to specific issues is discussed further below.

Disclosure of Research Agreements.—On this topic, the following views were expressed (9):

> In order to satisfy the faculty and general public that the role of the university is being maintained, contracts should be made public. This could involve publication of relevant provisions of research contracts with industry or, alternatively, examination by a facul-

ty committee or some other competent body of all research contracts to assure that terms are consistent with university values.*

Patents and Licenses.—The consensus on patents and licenses was as follows (9):

> The traditions of open research and prompt transmission of research results should be maintained. However, it is appropriate for the institution to file for patent coverage; actions which might require brief delays in publication or other public disclosure. Receipt of proprietary information may occasionally be desirable to facilitate research. These situations must be handled on a case-by-case basis so as not to violate the educational process or the traditions of openness.

There was a disagreement on the issue of whether exclusive rights should be given, although the document does appear to favor the granting of exclusive licenses (9):

> Some people fear that allowing a single firm the sole right to develop a patent will necessarily remove competition, slow the development of the patent, or even prevent development altogther. This fear is exaggerated. . . . Thus, universities should be able to negotiate exclusive licenses provided that exclusivity seems important to allow prompt, vigorous development of the patent to occur.

In license negotiations, the consensus was that the university should insist on a requirement of due diligence on the part of the licensee in developing and using the patent.

The situation is more difficult when a sponsor requests the right to exclusive licenses on all discoveries made as a result of the research funded by the company (9):

> Some of us believe that such exclusive rights are an appropriate quid pro quo for the funds provided for research. Others believe that the university should be willing to agree to provide instead nonexclusive royalty-free licenses to the sponsor, but should not give up its right to examine the appropriateness of exclusivity for each invention on a case-by-case basis.

Conflict of Interest.—Discussion focused on two aspects of the problem. The first was the propriety of a university's taking an equity position in a company in which one of its faculty is a major stockholder or officer. Most were against such investments (9):

> It is not advisable for universities to make such investments unless . . . there are sufficient safeguards to avoid adverse effects on the morale of the institution . . . .

The second and really complex issue, conflict of interests, was avoided by participants entirely. Issues related to university/industry relationships are not new, and the Pajaro Dunes Conference participants were all experienced with and knowledgeable about these relationships. Rather than producing some definite guidelines regarding the structuring of such relationships, however, Pajaro Dunes Conference participants provided only general principles underlying general university policies.

*Harvard has elected to keep its contracts confidential and Stanford is following an informal policy of full disclosure (1).

Selected statements on patent rights and commingling of research funds

Since one of the purposes of the 1980 U.S. patent law (Public Law 96-517) is to foster cooperative research arrangements among government, universities, and industry, one question that immediately arises is how the establishment of patent rights is affected by potential commingling of funds. Circular A-124 issued by the U.S. Office of Management and Budget (OMB) sets out some guidance on this (4):

> Notwithstanding the right of research organizations to accept supplemental funding from other sources for the purpose of expediting or more comprehensively accomplishing the research objectives of the government sponsored project, it is clear that the Act would remain applicable to any invention "conceived or first actually reduced to practice in performance" of the project. Separate accounting for the two funds used to support the project in this case is not a determining factor.
>
> To the extent that a non-government sponsor establishes a project which, although closely related, falls outside the planned and committed activities of a government funded project and does not diminish or distract from the performance of such activities, inventions made in performance of the non-government sponsored project would not be subject to the conditions of the Act. An example of such related but separate projects would be a government sponsored project having research objectives to expand scientific understanding in a field with a closely related industry sponsored project having as its objectives the application of such new knowledge to develop usable new technology. The time relationship in conducting the two projects and the use of new fundamental knowledge from one in the performance of the other are not important determinants, since most inventions rest on a knowledge base built up by numerous independent research efforts extending over many years. Should such an invention be claimed by the performing organization to be the product of non-government sponsored research and be challenged by the sponsoring agency as being reportable to the government as a "subject invention," the challenge is appealable . . . .
>
> An invention which is made outside of the research activities of a government funded project but which in its making otherwise benefits from such project without adding to its cost is not viewed as a "subject invention," since it cannot be shown to have been "con-

ceived or first actually reduced to practice" in performance of the project. An obvious example of this is a situation where an instrument purchased with government funds is later used, without interference with or cost to the government funded project, in making an invention all expenses of which involve only non-government funds.

Members of the Advisory Committee to the Director of NIH asked Mr. Dietrich of OMB for some guidance on problems posed by commingled funds. Dietrich noted that application of OMB and the Department of Health and Human Services cost-accounting and auditing principles can resolve some of the issues. He stated that one good way to distinguish between commingled funds is to determine whether a project was supported through direct costs (in which case the patent regulations would likely apply) or by indirect costs (in which case the regulations would likely not apply). He then provided an assessment of some specific cases (12).

- In a situation where privately supported work is done in a building previously constructed with Government funds, the Government obtains no patent rights in inventions developed through those private funds.
- Similarly, in a situation where privately supported work is done using equipment previously purchased with Government funds, the Government obtains no patent rights in inventions developed through those private funds; however, it does if the equipment is currently operated under Government support.
- If a single individual spends one-half time on a project supported with Government funds and one-half time on a privately supported project, the Government obtains patent rights only if the privately supported project is directly dependent on ideas or materials generated in the publicly supported project.
- Similarly, if a scientist spends 10 years on a publicly supported project and then 10 years on a privately supported project, the Government obtains no patent rights to the invention developed under private support unless it is clear the idea was conceived with public funds.
- In the case of a team working on a single project with both public and private support, the Government would obtain patent rights.
- For inventions resulting from normal intellectual intercourse in which two individuals, one privately and one publicly supported, exchange information, the Government would obtain no patent rights unless there is intent to commit fraud (e.g., the scientist on public funds provides information to the scientist in the private sector to increase the marketability of an invention and then shares in the profits).

Selected university policies

UNIVERSITY PATENT POLICIES

To analyze the patent policies of universities in the United States, OTA reviewed documents on the patent policies of the following 32 universities:

1. Alabama/Birmingham, University of
2. Arizona, University of
3. Boston University
4. California Insitute of Technology
5. California, University of
6. Case Western Reserve University
7. Colorado, University of
8. Connecticut, University of
9. Cornell University
10. Georgia, University of
11. Indiana University
12. Iowa, University of
13. Johns Hopkins University
14. Maryland, University of
15. Massachusetts Institute of Technology
16. Miami, University of
17. Michigan, University of
18. Minnesota, University of
19. Northwestern University
20. Ohio State University
21. Pennsylvania, University of
22. Purdue University
23. Rochester, University of
24. Rockefeller University
25. Rutgers University
26. Southern California, University of
27. Stanford University
28. Vanderbilt University
29. Virginia, University of
30. Washington University
31. Washington, University of
32. Wisconsin, U. of

In general, the patent policies of the 32 universities OTA sampled define the obligations and rights of the university and the university researchers who produce inventions that have commercial potential. They also recognize the rights of outside sponsors. Typically, university patent policy documents state that the relationships defined between the university and inventor are subject to the obligations that the inventor has made in return for outside support from either private or government sources. In some cases, an industrial sponsor may have retained the right to the invention (because most universities grant only nonexclusive licenses if they own the patent, subject to a short exclusive licensing period to help commercialize the invention) and also may have defined how royalties are to be shared. Thus, for example, the Stanford patent policy document notes:

> In practice, the great majority of inventions arise from externally funded research covered by agreements containing patent provisions. Some agreements permit the University to retain title and grant license rights to the sponsor; some provide for the reverse or defer allocation of rights.

The crucial issue, therefore, seems *not* to be the patent agreements between universities and their researchers (i.e., what is covered in the documents OTA reviewed), but the terms of contracts from external sponsors to individual researchers.

Most university patent policies cover anybody working with university facilities, although individual universities vary in the degree of specific identification of personnel types. Most of them also cover students, although MIT excepts students from the provision and Johns Hopkins invites students to "take advantage of the mechanisms set forth herein." University employees who produce inventions on their own time and without substantial use of university resources own their inventions, but all 32 universities invite them to use the university's commercialization mechanism.

All 32 universities require researchers to report inventions with potential commercialization promptly so that the university can assess their potential and file for a patent. Some universities (e.g., University of Pennsylvania) also require delay in publication of the findings to allow for filing of a patent. Since publications prior to patenting can make an invention nonpatentable, the practice of requiring a delay in publication is probably common even at universities whose documents do not explicitly mention it.

University administrative mechanisms have been set up to evaluate inventions, to settle disputes, and to attempt commercialization. Many universities use the services of commercialization firms such as Research Corporation of New York and Battelle Development Laboratories. Other universities have their own commercialization ventures (e.g., the Wisconsin Alumni Research Fund at the University of Wisconsin).

The sharing of royalties varies with each university. Almost all the universities use the U.S. Government's stipulation that no more than 15 percent of gross royalty income is to go to the inventor, but they usually set this as the minimum share (i.e., many give the inventor a bigger share if the stipulations of outside sources do not apply). Private universities have a greater propensity than public universities to give ownership of the invention to the inventor, while the university is given a license. This may not be a substantive difference, as the other provisions in university policies (commercialization, royalty sharing, etc.) do not seem to be related to whether the inventor rather than the university owns the invention. On the question of ownership, universities having the right to take ownership have the *option* to do so. Conversely, the inventor can petition to have the invention assigned to him/her if the university does not diligently pursue its commercial applications.

Royalties, after deduction for expenses and the inventor's share, may be assigned to a number of university activities. Some universities place the remaining royalty income in their general operating funds; often, however, royalties are assigned to "research" or to "research and training" either through stipulation or through a separate fund set up for that purpose (e.g., Cal Tech's California Institute Research Foundation). Some universities also allocate a share to the inventor's department, division, and/or area of activitiy (e.g., the University of Colorado allocates a 25-percent share each to the discoverer, to an account for support of the discoverer's research, to the discoverer's department or primary administrative unit, and to the university).

The crucial issue is the commercialization stipulations that are attached to funds provided by outside sponsors, public and private. The patent policies discussed here are subject to these external conditions, and, as the Stanford document states, external sponsorship of university research is more the rule than the exception.

UNIVERSITY POLICIES ON CONSULTING

The policies on consulting of five major U.S. universities (Harvard, MIT, Johns Hopkins, Stanford, and the University of California) are summarized in table H-1 below.

UNIVERSITY POLICIES ON SPONSORED RESEARCH

The policies on sponsored research of three major U.S. universities (Harvard, MIT, and Johns Hopkins) are summarized in table H-2 below.

Table H-1.—Summary of Selected University Policies on Consulting

Harvard University	Harvard Medical School	Massachusetts Institute of Technology	Johns Hopkins University	Johns Hopkins Medical School	Stanford	University of California (all campuses)
Conflict of interest:						
• Time for outside involvement regulated • Primary commitment to the university required • Disclosure of potential conflict required	• Time for outside involvement regulated • Primary commitment to the university required • Disclosure of potential conflict required	• Outside activities may not conflict with their obligations to the institute • For all those in decisionmaking roles required annual acknowledgement in writing of the policy • Required disclosure of all outside activities, including financial interests, to institute officers • Requirement: To seek advice of department head if a potential conflict exists	• No formal policy	• Time for outside involvement regulated • Primary commitment to the university required • Financial gain regulated	• Overriding professional allegiance to the university • Disclosure of potential conflict situations urged • Prewritten clause to be inserted into all agreements stating that university conditions of employment prevail before all other agreements	• Primary responsibilities to university stressed • Outlines specific examples of conflict-of-interest situations
Time regulation:						
• 20%	• 2x salary	• 1 day/week • No dollar amount	• 20%	• 20%	• 13 days per academic quarter (13-week quarter)	• No limit on consulting days unless time conflicts with primary responsibility to the university
Disclosure:						
• Not required, unless potential conflict exists	• Required annually—reported to the dean's office	• Faculty are required to keep their department heads continuously informed on all outside activities	• Not required	• Monthly reporting	• Disclosure of names of companies you request of dean, provost, etc.	• California Political Reform Act of 1982, requires disclosure of faculty member financial interest in industrial sponsor of his/her research • Annual reports of consulting activities to be supplied to heads of units

Table H-1.—Summary of Selected University Policies on Consulting (Continued)

Harvard University	Harvard Medical School	Massachusetts Institute of Technology	Johns Hopkins University	Johns Hopkins Medical School	Stanford	University of California (all campuses)
Policy enforcement:						
• Essentially self-enforced	• Essentially self-enforced	• Department heads are required to register once yearly faculty members outside commitments in terms of: — number of days spent — nature of the relationship — any significant financial interest the faculty member may have in the company	• Self-enforced	• By department director and dean	• Essentially self-enforced	• By department dean, variable enforcement among campuses and departments
• Minimally by department chairman	• By dean					
	• By department					

SOURCE: Management Analysis Center, Inc., "Study of University/Industry Relationships in Biotechnology," contract report prepared for the Office of Technology Assessment, U.S. Congress, January 1983; and P.R. Lee, W. Levinson, L.H. Butler, et al., "Industrial-Academic Relationships in Biotechnology at Stanford University, University of California, Berkeley, and University of California, San Francisco," contract report prepared for the Office of Technology Assessment, U.S. Congress, July 1982.

Table H-2.—Summary of Selected University Policies on Sponsored Research

Harvard University (includes Medical School and Mass. General Hospital)	Massachusetts Institute of Technology	Johns Hopkins University (includes Medical School)
Patent rights:		
• Retained by the university	• Retained by the university	• Retained by the university
License:		
• Generally nonexclusive encouraged	• Generally nonexclusive encouraged	• Generally nonexclusive encouraged
Publication rights:		
• Guaranteed	• Guaranteed	• Guaranteed
• Sponsor preview	• Sponsor preview deferrals up to 30 days	• Sponsor preview deferrals up to 120 days
Confidentiality:		
• No confidentiality of results	• No confidentiality of results	• No confidentiality of results
Choice of research topics:		
• Selected by researcher	• Selected by researcher	• Selected by researcher
• Reviewed by department chairman	• Reviewed by department head	• Reviewed by committees (by Biosafety Committee at the Medical School)
Policy enforcement:		
• Review by the department chairman. Approval by the Committee on Patents and Copyright required	• A three-stage approval process is utilized. The stages are: — review by department head — review by the Office of Sponsored Programs — review by dean or provost	• Review by the dean and Office for Sponsored Research (Office of Research Administration at the Medical School)
• Required disclosure to dean of faculty of all personal and remunerative commitments to potential industrial sponsor		
Policy development:		
• Currently underway at all faculties	• Centrally developed policies already in existence	• Being developed by divisions under the direction of central administration
• Decentralized development, moving toward greater centralization		

SOURCE: Management Analysis Center, Inc., "Study of University/Industry Relationships in Biotechnology," contract report prepared for the Office of Technology Assessment, U.S. Congress, January 1983.

Appendix H references

1. Broad, W., "Pajaro Dunes: The Search for Consensus," *Science* 216:155-157, Apr. 9, 1982.
2. Culliton, B. J., "The Hoechst Department at Mass General," *Science* 216:1200-1203.
3. Culliton, B. J., "New Bology Foundation Off to a Good Start," *Science* 220:803, 1983.
4. Executive Office of the President, Office of Management and Budget, OMB 82-5 Public Affairs Cover Memo for OMB Circular A-124 (Patents—Small Firms and Nonprofit Organizations), Washington, D.C., released Feb. 12, 1982.
5. Experiment Station Committee on Policy, "Genetic Engineering Policy for the State Agricultural Stations," November 1981.
6. Lee, P. R., Levinson, W., Butler, L. H., et al., "Industrial-Academic Relationships in Biotechnology at Stanford University, University of California, Berkeley, and University of California, San Francisco," contract report prepared for the Office of Technology Assessment, U.S. Congress, July 1982.
7. Management Analysis Center, Inc., "Study of University/Industry Relationships in Biotechnology," contract report prepared for the Office of Technology Assessment, U.S. Congress, January 1983.
8. Muto, F. T., Cooley, Godward, Castro, Huddleson, and Tatum, San Francisco, Calif., personal communication, 1982.
9. Pajaro Dunes Conference, press release, Stanford University, March 1982.
10. Socolar, M. J., Comptroller General of the United States, letter to Congressman Albert Gore, Oct. 16, 1981.
11. U.S. Congress, House Committee on Science and Technology, *University/Industry Cooperation in Biotechnology,* hearings before the Subcommittee on Investigations and Oversight and the Subcommittee on Science, Research and Technology, House Committee on Science and Technology, U.S. Congress, June 16-17, 1982 (Washington, D.C.: U.S. Government Printing Office, 1982).
12. U.S. Department of Health and Human Services, National Institutes of Health, Advisory Committee to the Director, "NIH Cooperative Research Relationships With Industry," Bethesda, Md., October 1981.
13. Washington University, St. Louis, Mo., news release, distributed at *University/Industry Cooperation in Biotechnology,* hearings before the Subcommittee on Investigations and Oversight and the Subcommittee on Science, Research, and Technology, House Committee on Science and Technology, U.S. Congress, Washington, D.C., June 16-17, 1982.

Appendix I

List of Acronyms and Glossary of Terms

List of acronyms

AAU —American Association of Universities
AcNPV —*Autographa californica* nuclear polyhedrosis virus
ACS —American Cancer Society
AHF —antihemophilic factor
AIDS —acquired immune deficiency syndrome
ANDA —Abbreviated New Drug Application
ANVAR —L'Agence Nationale de la Valorisation de la Recherche; National Agency for the Funding of Research (France)
ARES —Applied Research Systems (Netherlands)
ARS —Agricultural Research Service (U.S.)
AT&T —American Telephone & Telegraph Co. (U.S.)
BCr —Brazilian cruzeiros
BGA —Bundesgesundheitsamt: Federal Health Office (F.R.G.)
BISCT —Biotechnology Institute and Studies Centre Trust (U.K.)
BMFT —Bundesministerium für Forschung und Technologie; Federal Ministry of Science and Technology (F.R.G.)
BRL —Bethesda Research Laboratories (U.S.)
BTG —British Technology Group (U.K. Department of Industry)
CAMR —Center for Applied Microbiology and Research (U.K.)
CCL —Commodity Control List (U.S.)
CDC —Centers for Disease Control (U.S.)
C.F.R. —Code of Federal Regulations (U.S.)
CNPq —Conselho Nacional de Desenvolvimento Cientifico e Tecnologico; National Research Council, now known as the Council for Development of Science and Technology (Brazil)
CNRS —Centre National de la Recherche Scientifique; National Center for Scientific Research (France)
CoCom —Coordinating Committee for Multi-Lateral Export Controls
CODIS —Comité d'Orientation des Industries Stratégiques; Committee for the Organization of Strategic Industries (France)
COGENE —Committee on Genetic Experimentation (international)
CRGO —Competitive Research Grants Organization (U.S.)
CSIRO —Commonwealth Science and Research Organisation (Australia)
CSRS —Cooperative State Research Service (U.S.)
DARPA —Defense Advanced Research Projects Agency (U.S. Department of Defense)
DECHEMA —Deutsche Gesellschaft für Chemisches Apparatewesen; German Society for Chemical Engineering (F.R.G.)
DESAT —Defense Business Advanced Technology program (U.S. Department of Defense)
DFG —Deutsche Forschungsgemeinschaft; German Research Society (F.R.G.)
DHHS —Department of Health and Human Services (U.S.)
DM —Deutsche mark
DNA —deoxyribonucleic acid
DOD —Department of Defense (U.S.)
DOE —Department of Energy (U.S.)
DSM —Deutsche Sammlung von Mikroorganismen; German Collection of Micro-Organisms (F.R.G.)
EAA —Export Administration Act of 1979 (U.S.)
ECUT —Energy Conversion and Utilization Technologies program (U.S.)
EEC —European Economic Community
EMBL —European Molecular Biology Laboratory
EPA —Environmental Protection Agency (U.S.)
EPC —European Patent Convention
EPO —European Patent Office (supranational)
ETH —Eidgenössiche Technische Hochschule; Federal Institute of Technology (Switzerland)
FDA —Food and Drug Administration (U.S.)
FFDCA —Federal Food, Drug and Cosmetic Act (U.S.)
FIFRA —Federal Insecticide, Fungicide, and Rodenticide Act U.S.)
FINEP —Financiadora de Projetos National Funding Agency for Studies and Projects (Brazil)
FMD —foot-and-mouth disease
FRI —Fermentation Research Institute (Japan)
F.R.G. —Federal Republic of Germany
FTC —Federal Trade Commission (U.S.)
GAO —General Accounting Office (U.S.)
GBF —Gesellschaft für Biotechnologische Forschung; Society for Biotechnological Research (F.R.G.)
GE —General Electric Corp. (U.S.)
G.E. —Guidelines for Examination
GENBANK —Genetic Sequence Data Bank (U.S.)
GG —gamma globulin

GH —growth hormone
GMAG —Genetic Manipulation Advisory Group (U.K.)
GRAS —generally recognized as safe by qualified experts
GWB —Gesetz gegen Wettbewerbsbeschrankungen; Act Against Restraints of Competition (F.R.G.)
HBsAg —hepatitis B surface antigen
hCG —human chorionic gonadotropin
HFCS —high fructose corn syrup
hGH —human growth hormone
hI —human insulin
HPLC —high-performance liquid chromatography
H.R. —House of Representatives (U.S. Congress)
HSA —human serum albumin
HSE —Health and Safety Executive (U.K.)
HSV —herpes simplex virus
HSV2 —herpes simplex virus type 2
IBC —Institutional Biosafety Committee
IBM —International Business Machines Corp. (U.S.)
ICI —Imperial Chemical Industries (U.K.)
Ifn —interferon
IMC —International Minerals & Chemicals Corp. (U.K.)
IND —Notice of Claimed Investigational Exemption for a New Drug
Ingene —International Genetic Engineering, Inc. (U.S.)
INSERM —Institut National de la Santé et de la Recherche Médicale; National Institute of Health and Medical Research (France)
IOCM —Interkantonale Kontrollstelle für Heilmittel: Intercantonal Office for the Control of Medicaments (Switzerland)
IRS —Internal Revenue Service (U.S.)
ITC —International Trade Commission (U.S.)
JAFCO —Japan Associated Finance Corporation
JDB —Japan Development Bank
JETRO —Japan External Trade Organization
JFTC —Japanese Fair Trade Commission (Japan)
LSI —Large-Scale Integration
MAbs —monoclonal antibodies
MAFF —Ministry of Agriculture, Forestry, and Fisheries (Japan)
MCC —Microelectronics Computer Corp. (U.S.)
MCTL —Militarily Critical Technologies List (U.S. Department of Defense)
MEOR —microbial enhanced oil recovery
MGH —Massachusetts General Hospital
MGI —Molecular Genetics, Inc. (U.S.)
MIT —Massachusetts Institute of Technology
MITI —Ministry of International Trade and Industry (Japan)
MOF —Ministry of Finance (Japan)
MRC —Medical Research Council (U.K.)
mRNA —messenger RNA
MS —multiple sclerosis
MSG —monosodium glutamate
MSH —melanoycte-stimulating hormone
MSI —Medium-Scale Integration
NAS —National Academy of Sciences (U.S.)
NASA —National Aeronautics and Space Administration (U.S.)
NBFs —new biotechnology firms
NCDRH —National Center for Devices and Radiologic Health (U.S.)
NCI —National Cancer Institute (U.S.)
NDA —New Drug Application
NIBSC —National Institute of Biological Standards and Controls (U.K.)
NIH —National Institutes of Health (U.S.)
NIOSH —National Institute for Occupational Safety and Health (U.S.)
NLG —Netherlands guilder
NRC —National Research Council (Canada)
NSF —National Science Foundation (U.S.)
NYU —New York University
OECD —Organisation for Economic Co-Operation and Development
OMB —Office of Management and Budget (U.S.)
OSHA —Occupational Safety and Health Administration (U.S.)
OSRD —Office of Scientific Research and Development (U.S.)
OSTP —Office of Science and Technology Policy (Executive Office of the President, U.S.)
OTA —Office of Technology Assessment (U.S.)
PAL —phenylalanine ammonia lyase
PEPCase —phosphoenol pyruvate carboxylase
PHB —polyhydroxybutyrate
PMA —Pharmaceutical Manufacturers Association (U.S.)
PTO —Patent and Trademark Office (U.S.)
PVPA —Plant Variety Protection Act of 1970 (U.S.)
RAC —Recombinant DNA Advisory Committee (U.S.)
R&D —research and development
rDNA —recombinant DNA
Ri —root-inducing
RuBPCase —ribulose bisphosphate carboxylase
SAES —State Agricultural Experiment Stations (U.S.)
SBA —Small Business Administration (U.S.)

SBF	—Stiftelsen Bioteknisk Forksning; Biotechnology Research Foundation (Sweden)
SBIC	—Small Business Investment Corporation
SBIR	—Small Business Innovation Research
SCP	—single-cell protein
SERC	—Science and Economic Research Council (U.K.)
SFr	—Swiss francs
SNDA	—Supplemental New Drug Application
SOCal	—Standard Oil of California
SRBCs	—sheep red blood cells
STA	—Science and Technology Agency (Japan)
STU	—Stryelsen for Teknisk Utveckling; National Swedish Board for Technical Development
TDC	—Technical Development Corporation (U.K.)
T-DNA	—transferred-DNA
THMs	—trihalomethanes
Ti	—tumor-inducing
tPA	—tissue plasminogen activator
TRP	—tangible research property
TSCA	—Toxic Substance Control Act (U.S.)
UCLA	—University of California, Los Angeles
UCRDO	—University Connected Research and Development Organization (Israel)
UCSD	—University of California, San Diego
UCSF	—University of California, San Francisco
U.K.	—United Kingdom
UPOV	—International Convention for the Protection of New Varieties and Plants
U.S.C.	—United States Code
USDA	—U.S. Department of Agriculture
USM	—Unlisted Securities Market (U.K.)
UWG	—Gesetz gegen den unlauteren Wettbewerb; Unfair Competition Law of 1909 (F.R.G.)
VLSI	—Very-Large Scale Integration
VOCs	—volatile organic compounds
VST Act	—Virus, Serum, Toxin Act of 1913 (U.S.)
WARF	—Wisconsin Alumni Research Fund
WFG	—Deutsche Wagnisfinanzierungs-Gesellschaft; Risk Financing Society (F.R.G.)
WHO	—World Health Organization

Glossary of terms

Accession: In biotechnology, the addition of germplasm deposits to existing germplasm storage banks.

Acclimatization: The biological process whereby an organism adapts to a new environment. Describes process of developing micro-organisms that degrade toxic wastes in the environment.

Active immunity: Disease resistance in a person or animal due to antibody production after exposure to a microbial antigen following disease, inapparent infection, or inoculation. Active immunity is usually long-lasting. (Compare *passive immunity*.)

Adsorption: The taking up of molecules of gases, dissolved substances, or liquids by the surfaces of solids or liquids with which they are in contact.

Aerobic: Living or acting only in the presence of oxygen.

Affinity chromatography: The use of compounds, such as antibodies, bound to an immobile matrix to "capture" other compounds as a highly specific means of separation and purification.

Amino acids: The building blocks of proteins. There are 20 common amino acids.

Amino acid sequence: The linear order of amino acids in a protein.

Anaerobic: Living or acting in the absence of oxygen.

Antibiotic: A specific type of chemical substance that is administered to fight infections, usually bacterial infections, in humans or animals. Many antibiotics are produced by using micro-organisms; others are produced synthetically.

Antibody: A protein (immunoglobulin) produced by humans or higher animals in response to exposure to a specific antigen and characterized by specific reactivity with its complementary antigen. (See also *monoclonal antibodies*.)

Antidumping laws: Laws that prevent a country from exporting goods to another country and selling those goods below cost or more cheaply than in the home market. Antidumping duties may be imposed by a country to offset damages sustained from dumping. In the United States, the antidumping law most relevant to biotechnology is Section 337 of the Tariff Act of 1930 (19 U.S.C. 1337).

Antigen: A substance, usually a protein or carbohydrate which, when introduced in the body of a human or higher animal, stimulates the production of an antibody that will react specifically with it.

Antihemophilic factor (AHF): The fraction of whole blood that contains blood clotting agents. AHF is used to treat hemophilia, a set of hereditary disorders that prevent blood clotting.

Antimicrobial agent: See *antibiotic*.

Antiserum: Blood serum containing antibodies from animals that have been inoculated with an antigen. When administered to other animals or humans, antiserum produces passive immunity.

Applied research: Research to gain knowledge or understanding necessary for determining the means by which a recognized and specific need may be met (National Science Foundation definition). (See also *generic applied research*.)

Aromatic compound: A compound containing a benzene ring. Many specialty and commodity chemicals are aromatic compounds.

Ascites: Liquid accumulations in the peritoneal cavity. Used as a method for producing monoclonal antibodies.

Assay: A technique that measures a biological response.

Attenuated vaccine: Whole, pathogenic organisms that are treated with chemical, radioactive, or other means to render them incapable of producing infection. Attenuated vaccines are injected into the body, which then produces protective antibodies against the pathogen to protect against disease.

Autotrophic: Capable of self-nourishment (opposed to heterotrophic).

Bacillus subtilis (B. subtilis): An aerobic bacterium used as a host in rDNA experiments.

Bacteria: Any of a large group of microscopic organisms having round, rodlike, spiral, or filamentous unicellular or noncellular bodies that are often aggregated into colonies, are enclosed by a cell wall or membrane, and lack fully differentiated nuclei. Bacteria may exist as free-living organisms in soil, water, organic matter, or as parasites in the live bodies of plants and animals.

Bacteriophage (or phage)/bacterial virus: A virus that multiplies in bacteria. Bacteriophage lambda is commonly used as a vector in rDNA experiments.

Basic research: Research to gain fuller knowledge or understanding of the fundamental aspects of phenomena and of observable facts without specific applications toward processes or products in mind (National Science Foundation definition).

Batch processing: A method of bioprocessing in which a bioreactor is loaded with raw materials and micro-organisms, and the process is run to completion, at which time products are removed. (Compare *continuous processing.*)

Betaendorphin: A neuro-active polypeptide with analgesic properties similar to opiate compounds such as morphine.

Biocatalyst: An enzyme that plays a fundamental role in living organisms or industrially by activating or accelerating a process.

Biochemical: Characterized by, produced by, or involving chemical reactions in living organisms; a product produced by chemical reactions in living organisms.

Biochip: An electronic device that uses biological molecules as the framework for molecules that act as semiconductors and functions as an integrated circuit.

Bioconversion: A chemical conversion using a biocatalyst.

Biodegradation: The breakdown of substances by micro-organisms, mainly aerobic bacteria.

Biological oxygen demand (BOD): The oxygen used in meeting the metabolic needs of aerobic organisms in water containing organic compounds.

Biological response modifer: Generic term for hormones, neuroactive compounds, and immunoactive compounds that act at the cellular level; many are possible targets for production with biotechnology.

Biological warfare agents: Biological products or processes that are determined to be useful in military applications and whose export is restricted for national security reasons.

Biologics: Vaccines, therapeutic serums, toxoids, antitoxins, and analogous biological products used to induce immunity to infectious diseases or harmful substances of biological origin.

Biomass: All organic matter that grows by the photosynthetic conversion of solar energy.

Biooxidation: Oxidation (the loss of electrons) catalyzed by a biocatalyst.

Biopolymers: Naturally occurring macromolecules that include proteins, nucleic acids, and polysaccharides.

Bioprocess: Any process that uses complete living cells or their components (e.g., enzymes, chloroplasts) to effect desired physical or chemical changes.

Bioreactor: Vessel in which a bioprocess takes place.

Biosensor: An electronic device that uses biological molecules to detect specific compounds.

Biosurfactant: A compound produced by living organisms that helps solubilize compounds such as organic molecules (e.g., oil and tar) by reducing surface tension between the compound and liquid.

Biosynthesis: Production, by synthesis or degradation, of a chemical compound by a living organism.

Biotechnology: Commercial techniques that use living organisms, or substances from those organisms, to make or modify a product, and including techniques used for the improvement of the characteristics of economically important plants and animals and for the development of micro-organisms to act on the environment. In this report, biotechnology is used to mean "new" biotechnology, which only includes the use of *novel* biological techniques—specifically, recombinant DNA techniques, cell fusion techniques, especially for the production of monoclonal antibodies, and new bioprocesses for commercial production.

Callus: An undifferentiated cluster of plant cells that is a first step in regeneration of plants from tissue culture.

Capacitor: A device that consists of two conductors insulated from each other by a dielectric. A capacitor stores electrical energy, blocks the flow of direct current, and introduces alternating current into a circuit.

Carboxylation: The addition of an organic acid group (COOH) to a molecule.

Catalysis: A modification, especially an increase, in the rate of a chemical reaction induced by a material (e.g., enzyme) that is chemically unchanged at the end of the reaction.

Catalyst: A substance that induces catalysis; an agent that enables a chemical reaction to proceed under milder conditions (e.g., at a lower temperature) than otherwise possible. Biological catalysts are enzymes; some nonbiological catalysts include metallic complexes.

Cell: The smallest structural unit of living matter capable of functioning independently; a microscopic mass of protoplasm surrounded by a semipermeable membrane, usually including one or more nuclei and various nonliving products, capable alone, or interacting with other cells, of performing all the fundamental functions of life.

Cell culture: The in vitro growth of cells isolated from multicellular organisms. These cells are usually of one type.

Cell differentiation: The process whereby descendants of a common parental cell achieve and maintain specialization of structure and function.

Cell fusion: Formation of a single hybrid cell with nuclei and cytoplasm from different cells.

Cell line: Cells that acquire the ability to multiply indefinitely in vitro.

Cellulase: The enzyme that digests cellulose to sugars.

Cellulose: A polymer of six-carbon sugars found in all plant matter; the most abundant biological compound on earth.

Centrifuge: A machine for whirling fluids rapidly to separate substances of different densities by centrifugal force; also, to whirl in a centrifuge.

***Chakrabarty* decision:** *Diamond* v. *Chakrabarty*, U.S. Department of Commerce, PTA, sec. 2105, 1980; landmark case in which U.S. Supreme Court majority held that the inventor of a new microorganism, whose invention otherwise met the legal requirements for obtaining a patent, could not be denied a patent solely because the invention was alive.

Chemostat selection: Screening process used to identify micro-organisms with desired properties, such as micro-organisms that degrade toxic chemicals. (See also *acclimatization*.)

Chloroplasts: Cellular organelles where photosynthesis occurs.

Chromatography: A process of separating gases, liquids, or solids in a mixture or solution by adsorption as the mixture or solution flows over the absorbent medium, often in a column. The substances are separated because of their differing chemical interaction with the absorbent medium.

Chromosomes: The rodlike structures of a cell's nucleus that store and transmit genetic information; the physical structure that contain genes. Chromosomes are composed mostly of DNA and protein and contain most of the cell's DNA. Each species has a characteristic number of chromosomes.

Clinical trial: One of the final stages in the collection of data for drug approval where the drug is tested in humans.

Clone: A group of genetically identical cells or organisms produced asexually from a common ancestor.

Cloning: The amplification of segments of DNA, usually genes.

Coding sequence: The region of a gene (DNA) that encodes the amino acid sequence of a protein.

Cofactors: Additional molecules needed for enzymatic function.

Colibacillosis: A bacterial disease that causes diarrhea, dehydration, and death in calves and piglets.

Commodity chemicals: Chemicals produced in large volumes that sell for less than $1 per pound (50¢ per kg). (Compare *specialty chemicals*.)

Commodity controls list (CCL): Large roster of items that have been identified under the Export Administration Act by the U.S. Department of Commerce to require a "validated license" before they can be exported to certain countries.

Complementary DNA (cDNA): DNA that is complementary to messenger RNA; used for cloning or as a probe in DNA hybridization studies.

Compulsory licensing: Laws that require the licensing of patents, presumably to ensure early application of a technology and to diffuse control over a technology.

Continuous processing: Method of bioprocessing in which raw materials are supplied and products are removed continuously, at volumetrically equal rates. (Compare *batch processing*.)

Corn wet milling: The processing of corn, including hydrolysis of starch, to yield products used for food and chemicals.

Cosmid: A DNA cloning vector consisting of plasmid and phage sequences.

Countervailing duties: Duties charged to importers when their product is determined to cause or threaten material injury to domestic industries producing similar products.

Corporate venture capital: Capital provided by

major corporations exclusively for high-risk investments.

Culture deposits: See *accession.*

Culture medium: Any nutrient system for the artificial cultivation of bacteria or other cells; usually a complex mixture of organic and inorganic materials.

Cytoplasm: The "liquid" portion of a cell outside and surrounding the nucleus.

Cytotoxic: Damaging to cells.

Debt financing: The use of outside or borrowed capital to finance business activities.

Deoxyribonucleic acid (DNA): A linear polymer, made up of deoxyribonucleotide repeating units, that is the carrier of genetic information; present in chromosomes and chromosomal material of cell organelles such as mitochondria and chloroplasts, and also present in some viruses. The genetic material found in all living organisms. Every inherited characteristic has its origin somewhere in the code of each individual's DNA.

Deposit requirements: Patent requirements for inventors to turn over at the time of patent application a sample of the invention which is maintained throughout the life of the patent.

Diagnostic products: Products that recognize molecules associated with disease or other biologic conditions and are used to diagnose these conditions.

Dicots (dicotyledons): Plants with two first embryonic leaves and nonparallel veined mature leaves. Examples are soybean and most flowering plants.

Disclosure requirements: A patent requirement for adequate public disclosure of an invention that enables other people to build and use the invention without "undue" experimentation.

DNA: Deoxyribonucleic acid.

DNA base pair: A pair of DNA nucleotide bases. Nucleotide bases pair across the double helix in a very specific way. adenine can only pair with thymine; cytosine can only pair with guanine.

DNA probe: A sequence of DNA that is used to detect the presence of a particular nucleotide sequence.

DNA sequence: The order of nucleotide bases in the DNA helix; the DNA sequence is essential to the storage of genetic information.

DNA synthesis: The synthesis of DNA in the laboratory by the sequential addition of nucleotide bases.

Downstream processing: After bioconversion, the purification and separation of the product.

Drug: Any chemical compound that may be administered to humans or animals as an aid in the treatment of disease.

Elution: The removal of adsorbed material from an adsorbent, such as the removal of a product from an enzyme bound on a column.

Emulsification: The process of making lipids soluble in water.

Enablement requirement: A patent requirement for adequate public disclosure of an invention, enabling others in the relevant field of technology to build and use the invention.

Endorphins: Opiate-like, naturally occurring peptides with a variety of analgesic effects throughout the endocrine and nervous systems.

Enkephalins: Small, opiate-like peptides with analgesic effects in the brain.

Enzyme: Any of a group of catalytic proteins that are produced by living cells and that mediate and promote the chemical processes of life without themselves being altered or destroyed.

Equity capital: Capital proceeds arising from the sale of company stock.

Equity investment: An investment made in a company in exchange for a part ownership in that company.

Escherichia coli (E. coli): A species of bacteria that inhabits the intestinal tract of most vertebrates. Some strains are pathogenic to humans and animals. Many nonpathogenic strains are used experimentally as hosts for rDNA.

Eukaryote: A cell or organism with membrane-bound, structurally discrete nuclei and well-developed cell organelles. Eukaryotes include all organisms except viruses, bacteria, and blue-green algae. (Compare *prokaryote.*)

Export controls: Laws that restrict technology transfer and trade for reasons of national security, foreign policy, or economic policy.

Fatty acids: Organic acids with long carbon chains. Fatty acids are abundant in cell membranes and are widely used as industrial emulsifiers.

Feedstocks: Raw materials used for the production of chemicals.

Fermentation: An anaerobic bioprocess. Fermentation is used in various industrial processes for the manufacture of products such as alcohols, acids, and cheese by the action of yeasts, molds, and bacteria.

Fibrinolytic agents: Blood-borne compounds that activate fibrin in order to dissolve blood clots.

Flocculating agent: A reagent added to a dispersion of solids in a liquid to bring together the fine particles into larger masses.

Food additive (or food ingredient): A substance that becomes a component of food or affects the characteristics of food and, as such, is regulated by the U.S. Food and Drug Administration.

Foot-and-mouth disease: A highly contagious virus disease of cattle, pigs, sheep, and goats that is characterized by fever, salivation, and formation of vesicles in the mouth, pharynx and on the feet and is transmissable to humans.

Fractionation (of blood): Separation of blood by centrifugation, resulting in components sold as plasma, serum albumin, antihemophilic factor, and other products.

Free-living organism: An organism that does not depend on other organisms for survival.

Fungus: Any of a major group of saprophytic and parasitic plants that lack chlorophyll, including molds, rusts, mildews, smuts, and mushrooms.

Gamma globulin (GG): A protein component of blood that contains antibodies and confers passive immunity.

Gene: The basic unit of heredity; an ordered sequence of nucleotide bases, comprising a segment of DNA. A gene contains the sequence of DNA that encodes one polypeptide chain (via RNA).

Gene amplification: In biotechnology, an increase in gene number for a certain protein so that the protein is produced at elevated levels.

Gene expression: The mechanism whereby the genetic directions in any particular cell are decoded and processed into the final functioning product, usually a protein. See also *transcription* and *translation*.

Generic applied research: Research along the continuum between the two poles of basic and applied. This research may be characterized as follows: 1) it is not committed to open-ended expansion of knowledge as university basic research typically is but is less specific (more widely applicable or "generic") than the typical industrial product or process development effort; 2) it has more well-defined objectives than basic research but is long term relative to product and process development; and 3) it is high risk, in the sense that the stated objectives may fail and the resources committed may be lost for practical purposes.

Gene transfer: The use of genetic or physical manipulation to introduce foreign genes into host cells to achieve desired characteristics in progeny.

Genome: The genetic endowment of an organism or individual.

Genus: A taxonomic category that includes groups of closely related species.

Germ cell: The male and female reproductive cells; egg and sperm.

Germplasm: The total genetic variability available to a species.

Glycoproteins: Proteins with attached sugar groups.

Glucose: A 6-carbon sugar molecule used as a basic energy source by the cells of most organisms.

Glycosylation: The attachment of sugar groups to a molecule, such as a protein.

Government procurement: The acquisition by a government of goods or services. Government procurement may stimulate development of technology.

Growth hormone (GH): A group of peptides involved in regulating growth in higher animals.

Helminth: Parasitic worm.

Herbicide: An agent (e.g., a chemical) used to destroy or inhibit plant growth; specifically, a selective weed killer that is not injurious to crop plants.

High performance liquid chromatography (HPLC): A recently developed type of chromatography that is potentially important in downstream processing.

Hormone: A chemical messenger found in the circulation of higher organisms that transmits regulatory messages to cells.

Host: A cell whose metabolism is used for growth and reproduction of a virus, plasmid, or other form of foreign DNA.

Host-vector system: Compatible combinations of host (e.g., bacterium) and vector (e.g., plasmid) that allow stable introduction of foreign DNA into cells.

Human chorionic gonadotropin (HCG): A hormone produced by human placenta, indicating pregnancy; widespread target of MAb developers to diagnose pregnancy at an early stage.

Human insulin (hI): Hormone that stimulates cell growth via glucose uptake by cells. Insulin deficiency leads to diabetes.

Human serum albumin (HSA): Abundant protein in human blood; as a product, used in highest quantities in medicine, primarily in burn, trauma, and shock patients.

Hybrid: The offspring genetically dissimilar parents (e.g., a new variety of plant or animal that results from cross-breeding two different existing varieties, a cell derived from two different cultured cell lines that have fused).

Hybridization: The act or process of producing hybrids.

Hybridoma: Product of fusion between myeloma cell (which divides continuously in culture and is "immortal") and lymphocyte (antibody-producing cell); the resulting cell grows in culture and produces monoclonal antibodies.

Hybridoma technology: See *monoclonal antibody technology*.

Hydrolysis: Chemical reaction involving addition of water to break bonds.

Hydroxylation: Chemical reaction involving addition of hydroxyl (-OH) group to chemical compound.

Immobilized enzyme or cell techniques: Techniques used for the fixation of enzymes or cells onto solid supports. Immobilized cells and enzymes are used in continuous bioprocessing.

Immune response: The reaction of an organism to invasion by a foreign substance. Immune responses are often complex, and may involve the production of antibodies from special cells (lymphocytes), as well as the removal of the foreign substance by other cells.

Immunoassay: The use of antibodies to identify and quantify substances. The binding of antibodies to antigen, the substance being measured, is often followed by tracers such as radioisotopes.

Immunogenic: Capable of causing an immune response. (See also *antigen*.)

Immunotoxin: A molecule attached to an antibody capable of killing cells that display the antigen to which the antibody binds.

Interferons (Ifns): A class of glycoproteins (proteins with sugar groups attached at specific locations) important in immune function and thought to inhibit viral infections.

In vitro: Literally, in glass; pertaining to a biological reaction taking place in an artificial apparatus; sometimes used to include the growth of cells from multicellular organisms under cell culture conditions. In vitro diagnostic products are products used to diagnose disease outside of the body after a sample has been taken from the body.

In vivo: Literally, in life; pertaining to a biological reaction taking place in a living cell or organism. In vivo products are products used within the body.

Joint venture: Form of association of separate business entities which falls short of a formal merger but unites certain agreed on resources of each entity for a limited purpose; in practice most joint ventures are partnerships.

Leaching: The removal of a soluble compound such as an ore from a solid mixture by washing or percolating.

Lignin: A major component of wood.

Lignocellulose: The composition of woody biomass, including lignin and cellulose.

Lignolytic: Pertaining to the breakdown of lignin.

Linker: A small fragment of synthetic DNA that has a restriction site useful for gene cloning, which is used for joining DNA strands together.

Lipids: A large, varied class of water-insoluble organic molecules; includes steroids, fatty acids, prostaglandins, terpenes, and waxes.

Liposome transfer: The process of enclosing biological compounds inside a lipid membrane and allowing the complex to be taken up by a cell.

Lymphocytes: Specialized white blood cells involved in the immune response; B lymphocytes produce antibodies.

Lymphokines: Proteins that mediate interactions among lymphocytes and are vital to proper immune function.

Medical device: An instrument or apparatus (including an in vitro reagent such as MAbs) intended for use in the diagnosis or treatment of a disease or other condition and which does not achieve its intended purpose through chemical action within or on the body.

Messenger RNA (mRNA): RNA that serves as the template for protein synthesis; it carries the transcribed genetic code from the DNA to the protein synthesizing complex to direct protein synthesis.

Metabolism: The physical and chemical processes by which foodstuffs are synthesized into complex elements, complex substances are transformed into simple ones, and energy is made available for use by an organism.

Metabolite: A product of metabolism.

Metallothioneins: Proteins, found in higher organisms, that have a high affinity for heavy metals.

Methanogens: Bacteria that produce methane as a metabolic product.

Micro-organisms: Microscopic living entities; micro-organisms can be viruses, prokaryotes (e.g., bacteria), or eukaryotes (e.g., fungi).

Microencapsulation: The process of surrounding cells with a permeable membrane.

Mixed culture: Culture containing two or more types of micro-organisms.

Molecule: A group of atoms held together by chemical forces; the smallest unit of matter which can exist by itself and retain its chemical identity.

Monoclonal antibodies (MAbs): Homogeneous antibodies derived from a single clone of cells; MAbs recognize only one chemical structure. MAbs are useful in a variety of industrial and medical capacities since they are easily produced in large quantities and have remarkable specificity.

Monoclonal antibody technology: The use of hybridomas that produce monoclonal antibodies for a variety of purposes. Hybridomas are maintained in cell culture or, on a larger scale, as tumors (ascites) in mice.

Monocots (monocotyledons): Plants with single first embryonic leaves, parallel-veined leaves, and simple stems and roots. Examples are cereal grains such as corn, wheat, rye, barley, and rice.

Multigenic: A trait specified by several genes.

Mutagenesis: The induction of mutation in the genetic material of an organism; researchers may use physical or chemical means to cause mutations that improve the production of capabilities of organsims.

Mutagen: An agent that causes mutation.

Mutant: An organism with one or more DNA mutations, making its genetic function or structure different from that of a corresponding wild-type organism.

Mutation: A permanent change in a DNA sequence.

Myeloma: Antibody-producing tumor cells.

Myeloma cell line: Myeloma cells established in culture.

Neurotransmitters: Small molecules found at nerve junctions that transmit signals across those junctions.

New biotechnology firm (NBF): A company formed after 1976 whose sole function is research, development, and production using biotechnological means.

NIH Guidelines: Guidelines established by U.S. National Institutes of Health to regulate the safety of NIH-funded research involving recombinant DNA.

Nitrate: A compound characterized by a NO_3-group. Sodium nitrate and potassium nitrate are used as fertilizers.

Nitrogen fixation: The conversion of atmospheric nitrogen gas to a chemically combined form, ammonia (NH_3) which is essential to growth. Only a limited number of micro-organisms can fix nitrogen.

Nodule: The anatomical part of a plant root in which nitrogen-fixing bacteria are maintained in a symbiotic relationship with the plant.

Nodulins: Proteins, possibly enzymes, present in nodules; function unknown.

Nontariff trade barrier: A government regulation, other than a tariff (see below), that directly alters the volume or composition of international trade. Examples include quotas (restrictions on the quantity of goods imported), orderly marketing agreements (by which exporters agree to restrict the volume of goods exported), exchange controls (which constrain the value of foreign exchange spent rather than the number of units purchased), government preferences in purchases, and standards and certification systems.

Nucleic acids: Macromolecules composed of sequences of nucleotide bases. There are two kinds of nucleic acids: DNA, which contains the sugar deoxyribose, and RNA, which contains the sugar ribose.

Nucleotide base: A structural unit of nucleic acid. The bases present in DNA are adenine, cytosine, guanine, and thymine. In RNA, uracil substitutes for thymine.

Nucleus: A relatively large spherical body inside a cell that contains the chromosomes.

Oligonucleotides: Short segments of DNA or RNA.

Organelle: A specialized part of a cell that conducts certain functions. Examples are nuclei, chloroplasts, and mitochrondria, which contain most of the genetic material, conduct photosynthesis, and provide energy, respectively.

Organic compounds: Molecules that contain carbon.

Organic micropollutant: Low molecular weight organic compounds considered hazardous to humans or the environment.

Passive immunity: Disease resistance in a person or animal due to the injection of antibodies from another person or animal. Passive immunity is usually short-lasting. (Compare *active immunity.*)

Patent: A limited property right granted to inventors by government allowing the inventor of a new invention the right to exclude all others from making, using, or selling the invention unless specifically approved by the inventor, for a specified time period in return for full disclosure by the inventor about the invention.

Pathogen: A disease-producing agent, usually restricted to a living agent such as a bacterium or virus.

Peptide: A linear polymer of amino acids. A polymer of numerous amino acids is called a *polypeptide.* Polypeptides may be grouped by function, such as "neuroactive" polypeptides.

pH: A measure of the acidity or basicity of a solution on a scale of 0 (acidic) to 14 (basic). For example, lemon juice has a pH of 2.2 (acidic), water has a pH of 7.0 (neutral), and a solution of baking soda has a pH of 8.5 (basic).

Pharmceuticals: Products intended for use in humans, as well as in vitro applications to humans, including drugs, vaccines, diagnostics, and biological response modifiers.

Photorespiration: Reaction in plants that competes with the photosynthetic process. Instead of fixing CO_2, RuBPCase can utilize oxygen, which results in a net loss of fixed CO_2.

Photosynthesis: The reaction carried out by plants where carbon dioxide from the atmosphere is fixed into sugars in in the presence of sunlight; the transformation of solar energy into biological energy.

Plant Patent Act of 1930 (35 U.S.C. §5161-164): Confers exclusive license on developer of new and distinct asexually produced varieties other than tuber-propagated plants for 17 years.

Plant Variety Protection Act of 1970 (7 U.S.C. §2321): Provides patent-like protection to new plants reproduced sexually.

Plasma: The liquid (noncellular) fraction of blood. In vertebrates, it contains many important proteins (e.g., fibrinogen, responsible for clotting).

Plasmid: An extrachromosomal, self-replicating, circular segment of DNA; plasmids (and some viruses) are used as "vectors" for cloning DNA in bacterial "host" cells.

Polymer: A linear or branched molecule of repeating subunits.

Polypeptide: A long peptide, which consists of amino acids.

Polysaccharide: A polymer of sugars.

Prior art: Publicly known technology; patent requirements include the demonstration of the novelty of an invention, as distinguished from prior art.

Probe: See *DNA probe.*

Proinsulin: A precursor protein of insulin.

Prokaryote: A cell or organism lacking membrane-bound, structurally discreet nuclei and organelles. Prokaryotes include bacteria and the blue-green algae. (Compare *eukaryote.*)

Promoter: A DNA sequence in front of a gene that controls the initiation of "transcription" (see below).

Prophylaxis: Prevention of disease.

Protease: Protein digesting enzyme.

Protein: A polypeptide consisting of amino acids. In their biologically active states, proteins function as catalysts in metabolism and, to some extent, as structural elements of cells and tissues.

Protoplast fusion: The joining of two cells in the laboratory to achieve desired results, such as increased viability of antibiotic-producing cells.

Protozoa: Diverse phylum of eukaryotic microorganisms; structure varies from simple single cells to colonial forms; nutrition may be phagotropic or autotrophic; some protozoa are pathogenic.

Pyrogenicity: The tendency for some bacterial cells or parts of cells to cause inflammatory reactions in the body, which may detract from their usefulness as pharmaceutical products.

Public offering: The Securities and Exchange Commission approved sale of company stock to the public.

R&D limited partnership: A risk capital source and tax sheltered mechanism for funding the R&D of new products. It raises the potential rate of return to investors without adding extra cost to the corporation.

Reagent: A substance that takes part in a chemical reaction.

Recombinant DNA (rDNA): The hybrid DNA produced by joining pieces of DNA from different organisms together in vitro.

Recombinant DNA technology: The use of recombinant DNA for a specific purpose, such as the formation of a product or the study of a gene.

Recombination: Formation of a new association of genes or DNA sequences from different parental origins.

Regeneration: The laboratory process of growing a whole plant from a single cell or small clump of cells.

Regulatory sequence: A DNA sequence involved in regulating the expression of a gene.

Replication: The synthesis of new DNA from existing DNA and the formation of new cells by cell division.

Resistance gene: Gene that provides resistance to an environmental stress such as an antibiotic or other chemical compound.

Resistor: A device designed to limit electron flow in an electric circuit by a definite amount, resulting in a limited current or a voltage drop.

Restriction enzymes: Bacterial enzymes that cut DNA at specific DNA sequences.

Ri-plasmid: Plasmid from *Agrobacterium rhizogenes* used as plant vector.

RNA: Ribonucleic acid. (See also *messenger RNA.*)

RuBPCase (ribulose bisphosphate carboxylase): An enzyme that catalyzes the critical step of the photosynthetic CO_2 cycle.

Saccharification: The degradation of polysaccharides to sugars.

Scale-up: The transition of a process from an experimental scale to an industrial scale.

Selection: A laboratory process by which cells or organisms are chosen for specific characteristics.

Semiconductor: A material such as silicon or germanium with electrical conductivities intermediate between good conductors such as copper wire and insulators such as glass.

Semiconductor device: An electronic device that uses a semiconductor to limit or direct the flow of electrons. Examples are transistors, diodes, and integrated circuits.

Semiconductor industry: Companies that manufacture semiconductor devices. As used in this report, the description of the semiconductor industry is that deriving from the period between 1947 (discovery of the transistor) to the early 1960's.

Single cell protein: Cells, or protein extracts, of micro-organisms grown in large quantities for use as human or animal protein supplements.

Slimes: Aggregations of microbial cells that pose environmental and industrial problems; may be amenable to biologic control.

Sludge: Precipitated solid matter produced by water and sewage treatment or industrial problems; may be amenable to biologic control.

Small Business Investment Corporations (SBICs): Private companies licensed by the Small Business Association (SBA) and owned by stockholders who have made investments in exchange for equity. SBICs are required by SBA to invest or loan money exclusively to U.S. small businesses.

Somaclonal variation: Genetic variation produced from the culture of plant cells from a pure breeding strain; the source of the variation is not known.

Specialty chemicals: Chemicals, usually produced in small volumes, that sell for more than $1 per pound (50¢ per kg). (Compare *commodity chemicals.*)

Species: A taxonomic subdivision of a genus. A group of closely related, morphologically similar individuals which actually or potentially interbreed.

Spectrometer: An instrument used for analyzing the structure of compounds on the basis of their light-absorbing properties.

Starch: A polymer of glucose molecules used by some organisms as a means of energy storage; starch is broken down by enzymes (amylases) to yield glucose, which can be used as a feedstock for chemical or energy production.

Startup financing: Financing usually supplied by venture capitalist to fund the early R&D, production, sale of a new company's products.

Steroid: A group of organic compounds, some of which act as hormones to stimulate cell growth in higher animals and humans.

Storage protein genes: Genes coding for the major proteins found in plant seeds.

Strain: A group of organisms of the same species having distinctive characteristics but not usually considered a separate breed or variety. A genetically homogenous population of organisms at a subspecies level that can be differentiated by a biochemical, pathogenic, or other taxonomic feature.

Subsidy: A government intervention in the form of either grants, loans, or tax preferences that are directed to a particular domestic industry.

Substrate: A substance acted upon, for example, by an enzyme.

Subunit vaccine: A vaccine that contains only portions of a surface molecule of a pathogen. Subunit vaccines can be prepared by using rDNA technology to produce all or part of the surface protein molecule or by artificial (chemical) synthesis of short peptides.

Symbiont: An organism living in symbiosis, usually the smaller member of a symbiotic pair of dissimilar size.

Symbiosis: The living together of two dissimilar organisms in mutually beneficial relationships.

Tariff: Charges levied on importers of a particular good by a government in return for granting access to the government's domestic markets, which may occur at the expense of domestic industry; sometimes high tariffs are used to discourage importation and protect domestic industry.

T-DNA: Transfer DNA; that part of Ri or Ti plasmids that is transferred to the plant chromosome.

Technology transfer: The movement of technical information and/or materials, used for producing a product or process, from one sector to another; most often refers to flow of information between public and private sectors or between countries.

Therapeutics: Pharmaceutical products used in the treatment of disease.

Thermophilic: Heat loving. Usually refers to micro-organisms that are capable of surviving at elevated temperatures; this capability may make them more compatible with industrial biotechnology schemes.

Thrombolytic enzymes: Enzymes such as streptokinase and urokinase that initiate the dissolution of blood clots.

Thrombosis: Blockage of blood vessels.

Ti plasmid: Plasmid from *Agrobacterium tumefaciens* used as a plant vector.

Totipotency: The capacity of a higher organism cell to differentiate into an entire organism. A totipotent cell contains all the genetic information necessary for complete development.

Toxicity: The ability of a substance to produce a harmful effect on an organism by physical contact, ingestion, or inhalation.

Toxin: A substance, produced in some cases by disease-causing micro-organisms, which is toxic to other living organisms.

Toxoid: Detoxified toxin, but with antigenic properties intact.

Trade secret: An invention used continuously by its holder in his or her business to maintain a competitive edge over other competitors who do not know or use it. Trade secrets are often used instead of patents to protect production information.

Transcription: The synthesis of messenger RNA on

a DNA template; the resulting RNA sequence is complementary to the DNA sequence. This is the first step in gene expression. (See also *translation.*)

Transformation: The introduction of new genetic information into a cell using naked DNA.

Transistor: An active component of an electrical circuit consisting of semiconductor material to which at least three electrical contacts are made so that it acts as an amplifier, detector, or switch.

Translation: The process in which the genetic code contained in the nucleotide base sequence of messenger RNA directs the synthesis of a specific order of amino acids to produce a protein. This is the second step in gene expression. (See also *transcription.*)

Transposable element: Segment of DNA which moves from one location to another among or within chromosomes in possibly a predetermined fashion, causing genetic change; may be useful as a vector for manipulating DNA.

Trihalomethanes (THMs): Organic micropollutants and potential carcinogens, consisting of three halide elements attached to a single carbon atom; their destruction during water purification may be done biologically.

Turbid: Thick or opaque with matter in suspension.

Vaccine: A suspension of attenuated or killed bacteria or viruses, or portions thereof, injected to produce active immunity. (See also *subunit vaccine.*)

Vector: DNA molecule used to introduce foreign DNA into host cells. Vectors include plasmids, bacteriophages (virus), and other forms of DNA. A vector must be capable of replicating autonomously and must have cloning sites for the introduction of foreign DNA.

Venture capital (venture capital funds): Money that is invested in companies with which a high level of risk is associated.

Virus: Any of a large group of submicroscopic agents infecting plants, animals, and bacteria and unable to reproduce outside the tissues of the host. A fully formed virus consists of nucleic acid (DNA or RNA) surrounded by a protein or protein and lipid coat.

Viscosity: A measure of a liquid's resistance to flow.

Volatile organic compounds (VOCs): Group of toxic compounds found in ground water and that pose environmental hazards; their destruction during water purification may be done biologically.

Wild-type: The most frequently encountered phenotype in natural breeding populations.

Yeast: A fungus of the family Saccharomycetacea that is used especially in the making of alcoholic liquors and as leavening in baking. Yeast are also commonly used in bioprocesses.

Appendix J

Currency Conversion Factors

The following is a list of conversion factors for currencies from the countries studied in the report. All figures are averages from calendar year 1982 and were provided by the International Monetary Fund.

1 dollar = 249.05 Japanese yen (¥)

1 dollar = 179.51 Brazilian cruzeiros (BCr)

1 dollar = 24.267 Israeli shekels (IS)

1 dollar = 6.5724 French francs (F)

1 dollar = 6.2826 Swedish kroner (Skr)

1 dollar = 2.67021 Netherlands guilder (NLG)

1 dollar = 2.4266 German marks (DM)

1 dollar = 2.0303 Swiss francs (SwF)

1 dollar = 1.23370 Canadian dollars ($C)

1 dollar = 0.98586 Australian dollars ($A)

1 dollar = 0.5713 British Pounds (£)

Appendix K

Other Contractors, Contributors, and Acknowledgments

Other contractors

Robert Barker
Cornell University
Ithaca, N.Y.

Keith Gibson
Medical Research Council
London, England

Loren Graham
Washington, D.C.

Kozo Yamamura
University of Washington
Seattle, Wash.

Contributors

James E. Bailey
California Institute of Technology

Harvey Blanch
University of California

Hugh Bollinger
NPI

Charles Cooney
Massachusetts Institute of Technology

Morris Danzig
CPC International

Wensley Haydon-Baillie
Porton/LH International

Ed Lipinsky
Battelle Columbus Laboratories

Hunt McCauley
Big Timber, Mont.

Roger Morris
University of Minnesota

Patrick O'Connor
Genetic Sciences International

Robert Rabson
U.S. Department of Energy

Mark Radcliffe
Brobeck, Phleger & Harrison

John W. Schlicher
Townsend and Townsend

John Weil
(formerly of) Bendix Corp.

Acknowledgments

Lynne Alexander
Office of Technology Assessment

William Amon, Jr.
Cetus Corp.

Mary Anderson
Office of Technology Assessment

Roger Andewelt
U.S. Department of Justice

Paul Baumann
University of California

William R. Beisel
U.S. Army Medical Research Institute
for Infectious Diseases

Frederick Betz
U.S. Environmental Protection Agency

Judy Blakemore
Cetus Corp.

Detlef Boldt
Embassy of the Federal Republic of Germany

Lawrence J. Brady
U.S. Department of Commerce

Corale Brierley
Advanced Mineral Technologies, Inc.

Mark F. Cantley
Commission of the European Communities

Ronald Cape
Cetus Corp.

Peter Carlson
Crop Genetics International N.V.

Earl Cilly
Stanford University

Mary Clutter
U.S. National Science Foundation

Stanley Cohen
Stanford University

Eileen Collins
U.S. National Science Foundation

Brian Cunningham
Genentech, Inc.

Linda Curran
Battelle Columbus Laboratories

Michael Curran
Micro Industries

Eugene P. Damm, Jr.
International Business Machines Corp.

Lee Davenport
Greenwich, Conn.

William Davies
(formerly at) Armos Corp.

John Dearden
Johns Hopkins University

Arnold Demain
Massachusetts Institute of Technology

Roger G. Ditzel
University of California

Donald K. Dougall
University of Tennessee

Sidney Draggan
U.S. National Science Foudnation

George H. Dummer
Massachusetts Institute of Technology

Diana Dutton
Stanford University

James J. Eberhardt
U.S. Department of Energy

Gerry J. Elman
Elman, Shaiman and Moskowitz

Frederic Eustis
Biogen, Inc.

L. Peter Farkas
Craig and Burns

Norman Fast
Capital Publishing Corp.

Alan Fechter
U.S. National Science Foundation

Andrew J. Ferrara
Collaborative Research, Inc.

Philip Filner
ARCO Plant Cell Research Institute

William Finnerty
University of Georgia

James E. Flinn
Battelle Columbus Laboratories

Gene Fox
ARCO Plant Cell Research Institute

Orrie M. Friedman
Collaborative Research, Inc.

James Glavin
Genetic Systems Corp.

Norm Goldfarb
Calgene, Inc.

June Goodfield
East Sussex, England

Michael Gough
Office of Technology Assessment

Tom Graham
Kilpatrick & Cody

James D. Grant
CPC International

Harold P. Green
George Washington University

Lorance Greenlee
Agrigenetics Corp.

Joe Griffin
Wald, Harkrader, and Ross

Charles Groginski
Vega Biochemicals

Ben Hall
University of Washington School of Medicine

Albert P. Halluin
Cetus Corp.

Gerald Hickman
Collaborative Research, Inc.

Richard Hill
U.S. Environmental Protection Agency

Steven Hochhauser
Redwood City, Calif.

Anne Hollander
U.S. Environmental Protection Agency

John Holt
Eli Lilly & Co.

Jack Houck
Endorphin, Inc.

John Jennings
Pharmaceutical Manufacturers Association

Elmima Johnson
U.S. National Science Foundation

David Kabikoff
Hybritech, Inc.

Paul Kesselring
Swiss Federal Institute of Reactor Research

George H. Kidd
(formerly of) International Plant Research Institute

Wayne Kindsvogel
Zymos Corp.

Jay Kingham
Pharmaceutical Manufacturers Association

Jay Kraemer
Fried, Frank, Harris, Shriver, & Kampelman

Sheldon Krimsky
Tufts University

Lawrence P. Kunstadt
Novatech Resource Corp.

Stephen E. Lawton
Pierson, Ball, and Dowd

Jeff Lepon
Kornmeier, McCarthy, Lepon & Harris

Morris Levin
U.S. Environmental Protection Agency

Warren Levinson
Massachusetts General Hospital

Herman Lewis
U.S. National Science Foundation

Malcolm D. Lilly
University College, London

Charles R. Lipsey
Finnegan, Henderson, Farabow, Garrett and Dunner

Max Lyon
Genetic Systems Corp.

Hugh Macki
Cruachem, Inc.

Bruce F. Mackler
Meyer, Faller, and Weisman

William E. Marshall
General Foods Corp.

James McAlear
Gentronix International, Inc.

James McCullough
U.S. Congressional Research Service

Donald E. McElmury
CPC International

Thomas O. McGarity
University of Texas at Austin

Bruce Merchant
U.S. Food and Drug Administration

Larry Miike
Office of Technology Assessment

Teri Miles
Office of Technology Assessment

Elizabeth Milewski
U.S. National Institutes of Health

Henry Miller
U.S. Food and Drug Administration

Charles Muscoplat
Molecular Genetics, Inc.

Kay Noel
Cetus Corp.

John R. Norell
Phillips Petroleum Co.

Michael G. Norton
British Embassy

Thomas C. O'Brien
(formerly of) U.S. National Bureau of Standards

David V. O'Donnell
Commonwealth of Virginia

Sandra Panem
Brookings Institution

Richard Pearson
University of Sussex

Joseph Perpich
(formerly of) Genex Corp.

George Pierce
Battelle Columbus Laboratories

Jean-Claude Pinon
Embassy of France

Harvey S. Price
Industrial Biotechnology Association

Frank Pugliese
U.S. Food and Drug Administration

William H. Rastetter
Genentech, Inc.

Neils Reimers
Stanford University

David Rosenbloom
Boston, Mass.

Roy Rothwell
University of Sussex

Robert Ryan
U.S. Office of Naval Research

Louis G. Santucci
Pharmaceutical Manufacturers Association

Ken Sargeant
Commission of the European Communities

Holly Schauer
U.S. Department of Agriculture

Stanley Schlosser
U.S. Patent and Trademark Office

Charles D. Scott
Oak Ridge National Laboratory

David Seligman
Hoffmann-La Roche, Inc.

Al Shapero
Ohio State University

Stuart Shemtob
U.S. Department of Justice

Gerald D. Shockman
Temple University

Alexander W. Sierck
Beveridge and Diamond

David Silberman
Agrigenetics Corp.

Allen Singer
National Academy of Sciences

Ira Skurnick
U.S. Defense Advanced Research Projects Agency

Hedy Sladovich
(formerly of) E. F. Hutton

Ed Soule
Weyerhauser Co.

Gary Steele
Genentech, Inc.

R. W. Stieber
Merck & Co., Inc.

Eliezer Tal
Inter-American Development Bank

Masami Tanaka
Ministry of International Trade and Industry (Japan)

Fred Tarpley
Georgia Institute of Technology

Ron Taylor
Hybritech, Inc.

Albert H. Teich
American Association for the Advancement of Science

Elizabeth Terry
U.S. Central Intelligence Agency

Eleanor Thomas
U.S. National Science Foundation

Susan Thompson
U.S. Food and Drug Administration

Sue Tolin
U.S. Department of Agriculture

Ezra Vogel
Harvard University

Don Wallace
Georgetown University Law Center

William J. Walsh, III
U.S. Department of State

Amy Walton
U.S. National Science Foundation

Daniel Wang
Massachusetts Institute of Technology

Robert Warren
U.S. Naval Air System Command

Rick Weber
U.S. National Institute of Mental Health

William A. Weigand
(formerly of) U.S. National Science Foundation

Barry Weiner
Enzo Biochem

Bruno O. Weinschel
Weinschel Engineering

Philip J. Whitcome
Amgen

Bill Williams
ARCO Plant Cell Research Institute

John Wilson
(formerly of) Bendix Research

Phyllis Windle
Office of Technology Assessment

Zachary Wochok
Monsanto Co.

James Womack
Massachusetts Institute of Technology

Jim Young
Zoecon

Oskar Zaborsky
(formerly of) U.S. National Science Foundation

Michael E. Zacharia
U.S. Department of Commerce

Index

Index

Office of Technology Assessment

The Office of Technology Assessment (OTA) was created in 1972 as an analytical arm of Congress. OTA's basic function is to help legislative policymakers anticipate and plan for the consequences of technological changes and to examine the many ways, expected and unexpected, in which technology affects people's lives. The assessment of technology calls for exploration of the physical, biological, economic, social, and political impacts that can result from applications of scientific knowledge. OTA provides Congress with independent and timely information about the potential effects—both beneficial and harmful—of technological applications.

Requests for studies are made by chairmen of standing committees of the House of Representatives or Senate; by the Technology Assessment Board, the governing body of OTA; or by the Director of OTA in consultation with the Board.

The Technology Assessment Board is composed of six members of the House, six members of the Senate, and the OTA Director, who is a nonvoting member.

OTA has studies under way in nine program areas: energy and materials; industry, technology, and employment; international security and commerce; biological applications; food and renewable resources; health; communication and information technologies; oceans and environment; and science, transportation, and innovation.